Urbanization

Urbanization

Second Edition

An Introduction to Urban Geography

PAUL L. KNOX

University Distinguished Professor and Dean,
College of Architecture & Urban Studies,
Virginia Tech.

LINDA McCARTHY

Associate Professor
Department of Geography
University of Wisconsin–Milwaukee

PEARSON
Prentice
Hall

Upper Saddle River, New Jersey 07458

Library of Congress Cataloging in Publication Data

Knox, Paul L.
    Urbanization : an introduction to urban geography / Paul L. Knox and
Linda McCarthy. 2nd ed.
        p. cm.
    Includes bibliographical references and index.
    ISBN 0-13-142450-5 (alk. paper)
    1. Urban geography. 2. Urbanization. I. McCarthy, Linda. II. Title.
GF125.K56 2005
307.76 dc22   2004024420

Editor-in Chief: *John Challice*
Executive Editor: *Dan Kaveney*
Executive Managing Editor: *Kathleen Schiaparelli*
Assistant Managing Editor: *Beth Sweeten*
Production Editor: *nSight/Erin Connaughton*
Director of Creative Services: *Paul Belfanti*
Art Director: *Jayne Conte*
Cover Designer: *Bruce Kenselaar*
AV Editor: *Abby Bass*
Marketing Manager: *Robin Farrar*
Assistant Manufacturing Manager: *Michael Bell*
Manufacturing Buyer: *Lynda Castillo*
Cover Images: (left): Ron Sanford/CORBIS; (center): Bill Ross/CORBIS; (right): Paul Knox

© 2005, 1994 Pearson Education, Inc.
Pearson Prentice Hall
Pearson Education, Inc
Upper Saddle River, NJ 07458

Pearson Prentice Hall™ is a trademark of Pearson Education, Inc. Microsoft and PowerPoint are registered trademarks of Microsoft
Corporation.

Printed in the United States of America

10 9 8 7 6 5 4 3 2 1

ISBN 0-13-142450-5

Pearson Education Ltd., *London*
Pearson Education Australia Pty. Ltd., *Sydney*
Pearson Education Singapore, Pte. Ltd.
Pearson Education North Asia Ltd., *Hong Kong*
Pearson Education Canada, Inc., *Toronto*
Pearson Educación de Mexico, S.A. de C.V.
Pearson Education—Japan, *Tokyo*
Pearson Education Malaysia, Pte. Ltd.

CONTENTS

PREFACE

Towns and cities are in constant flux. They are hives of industry and crucibles of social, cultural, and political change, where there is always something happening. At times, circumstances accelerate the restlessness of urban change, with the result that the function, form, and appearance of cities are transformed. Such was the case a hundred years ago, when a combination of economic, social, and technological changes were turning cities in Europe and America inside out and upside down, forging, in the process, the physical, economic, and political framework for the evolution of the "modern" city.

We are currently living through another phase of transformation, this time involving global processes of economic, cultural, and political change. Within the cities of the developed world, the classic mosaic of central city neighborhoods has become blurred as cleavages of income, race, and family status have been fragmented by new lifestyle and cultural preferences. The long-standing distinction between central cities and suburbs has become less and less straightforward as economic reorganization has brought about a selective *re*centralization of commercial and residential land uses in tandem with a selective *de*centralization of commerce and industry. Outlying centers big enough to be called "edge cities" and "boomburbs" have appeared, as if from nowhere.

Meanwhile, cities in less developed countries have grown at unprecedented rates, with distinctive processes of urbanization creating new patterns of land use and posing new sets of problems. A pressing problem today for many less developed countries is a process of overurbanization in which cities are growing more rapidly than the jobs and housing that they can sustain. There has been a "quartering" of cities into spatially partitioned, compartmentalized residential enclaves. Luxury homes and apartment complexes correspond with an often dynamic formal sector of the economy that offers well-paid jobs and opportunities; these contrast sharply with the slums and squatter settlements of people working in the informal sector—in jobs not regulated by the state—who are disadvantaged by a lack of formal education and training and the often rigid divisions of labor shaped by gender, race, and ethnicity.

Urban geography allows us to address these trends, to relate them to our own individual lives and concerns, and to speculate on how they play a role in other fields of study such as economics, history, sociology, and planning. The study of urban geography can help us better understand the marketplace and appreciate the interdependencies involved in local, national, and international economic development. It can provide us with an appreciation of history and the relationships among art, economics, and society. It can illuminate the interplay of science and technology with economic and social change, reveal important dimensions of race and gender, raise important issues of social inequality, and point to important lessons for governance and policy. Most of all, of course, it can help us to understand, analyze, and interpret the landscapes, economies, and communities of towns and cities around the world.

In this book we attempt to capture the changes in the nature and outcomes of urbanization processes as well as the development of new ways of thinking about urban geography. A dynamic approach to the study of urban geography is the most distinctive feature of the book: unraveling the interlocking processes of urbanization to present a vivid and meaningful explanation of constantly changing urban geographies. An important advantage of such an approach is that it provides a framework that is able to capture recent changes while addressing much of the "traditional" subject matter of urban geography. The dynamic approach also allows for the integration of theory with fact. In this book, key concepts and theories are presented in relation to prior events and ideas. In this way, we can appreciate the logic of particular theories and their relevance to particular circumstances. In writing this book, we have aimed at providing a coherent and comprehensive introduction to urban geography that offers a historical and process-oriented approach with a North American focus that also provides a global context and comparative international perspectives.

The text of this second edition has been completely revised and updated with a large number of new illustrations, Follow Up exercises, Key Sources and Suggested Reading, helpful websites, and a glossary. The focus on North American cities has been augmented with material on cities in other developed countries (in Europe, Australia, and Japan, as well as Russia). New material has been added on urban environmental issues such as brownfields and urban sustainable development, and on the interdependence between globalization and urbanization (including such topics as terrorism and cities and the future of cities in an increasingly interconnected world).

A new chapter (Chapter 2) has been added on the origins and growth of cities from Mesopotamia, through Greek, Roman, and Medieval cities to the cities of the Industrial Revolution, together with three new chapters on urbanization in less developed countries: Chapter 7 deals with the legacy of colonial urbanization and contemporary urbanization trends in less developed countries; Chapter 8 deals with urban form and land use in less developed countries; and Chapter 9 deals with urban problems (poverty, inadequate housing, lack of urban services, transportation problems, and environmental degradation) and responses (by governments, private agencies, non-profits, and communities) in less developed countries.

We are grateful to many individuals for their help in forming and testing our ideas. Our gratitude is wide and deep, and we take this opportunity to acknowledge the contributions of Brian Berry, University of Texas at Dallas; Martin Cadwallader, University of Wisconsin, Madison; Bill Clark, University of California, Los Angeles; Ron Johnston, University of Bristol; Peter Taylor, Loughborough University; and Helga Leitner and Roger Miller at the University of Minnesota. We have also been fortunate in being able to call on the talents and energies of Ceylan Oner, Michael Peragine, and Joel Schneider in searching for material and checking data, and Erin Taylor Connaughton and Joe Gustaitis at nSight, Inc. in preparing the book for publication. Dan Kaveney at Prentice Hall provided a constant source of advice, enthusiasm, encouragement, and support.

PAUL L. KNOX
LINDA MCCARTHY

Urbanization

1

URBANIZATION AND URBAN GEOGRAPHY

This book introduces urban geography by focusing on the processes and outcomes of urbanization. Cities are products of many forces. They are engines of economic development and centers of cultural innovation, social transformation, and political change. At the same time, cities vary in everything from employment opportunities to patterns of land use, racial composition, and social behavior. Given the considerable range of issues that urban geography encompasses, we need to try to develop a consistency in our approach. Understanding theories about cities and the way they change—rather than simply listing their various attributes—will help to ensure that we maintain a consistency and will give us greater insight into the way cities work. Although we will sometimes need to look closely at specific events, our goal is to focus on understanding how to read the economic, social, and political "blueprints" that give shape and character to various kinds of cities. By "generalizing" in this way, we will have a more immediate and richer understanding of each new aspect of urbanization that we encounter.

CHAPTER PREVIEW

The fundamental task of the student of urban geography is to make sense of the ways that towns and cities have changed and are changing, with particular reference to the differences both between urban places and within them. As a beginning student, you will understandably find yourself asking some fundamental questions. What exactly is urban geography, and how does it relate to other aspects of geography and to other social science subjects? What is involved in "urbanization," and what outcomes and consequences of urbanization are important in studying urban geography? This first chapter answers those questions and introduces some of the organizing principles and concepts that recur throughout the book.

First, we address the question of urban geography as a subject for academic study, emphasizing the importance of concepts of space, territoriality, distance, and place. We then review the various approaches that have been taken to the study of cities from a geographical perspective. We will see that they are all encompassed by the overall framework for study that is adopted in this book: a framework that allows us to deal with various aspects of urban geography in terms of their relationship to urbanization as a process.

The chapter's second half addresses this question of urbanization as a process. Here we see how urban geographies and urban change are the result (and sometimes the cause) of long-term changes in economic development and in the interrelated dynamics of technological, demographic, political, social, cultural, and environmental change. Because these changes are being framed increasingly at the global scale, our framework for studying urban geography also views urbanization from a global perspective. Finally, we look ahead to the contents and organization of the rest of the book.

THE STUDY OF URBAN GEOGRAPHY

Like other aspects of human geography, urban geography is concerned with "local variability within a general context."[1] This means that it is concerned with an understanding of both the *distinctiveness* of individual places (in this case, towns and cities) and the *regularities* within and between cities in terms of the spatial relationships between people and their environment. We should immediately note that environment here includes not only the natural physical environment but also the built environment (everything from homes, factories, offices, and schools to roads and bridges), the economic environment (economic institutions, the structure and organization of economic life, and so on), and the social environment (including norms of behavior, social attitudes, and cultural and political values that shape interpersonal relations).

For urban geographers some of the most important questions therefore include the following: What attributes make cities and neighborhoods distinctive? How did these distinctive identities evolve? Are there significant regularities in the spatial arrangement of towns and cities across a country or region of the world? Are there significant regularities in the spatial organization of land use within cities and in the patterning of neighborhood populations by social status, household type, or race? The urban geographer will also need to know about the causes of any regularities that do exist. How, for example, do people choose where to live and what are the constraints on their choices? How do people's area of residence affect their behavior? What groups, if any, can manipulate the spatial organization of towns and cities? And who profits from such manipulation?

In posing these questions, urban geographers have learned that the answers are ultimately to be found in the wider context of economic, social, and political life. Cities must be viewed as part of the economies and societies that maintain them. The study of cities cannot be undertaken in isolation from the study of history, economic development, sociocultural change, or the increasing interdependence of places within the world economy. A proper understanding of cities requires a cross-disciplinary approach. The traditional focus of geography—the interrelationships between people and their physical and social environments—requires geographers to draw on the work of researchers in those related disciplines. Accordingly, the chapters you are about to read will be peppered with references to the work of economists, sociologists, political scientists, historians, and city planners.

SPACE, TERRITORIALITY, DISTANCE, AND PLACE

Urban geography is a coherent and distinctive framework of study through the central themes of space, territoriality, distance, and place. For the geographer, *space* is not simply a medium in which economic, social, political, and historical processes are expressed. It is also a factor that influences patterns of urban development

FIGURE 1.1   New suburban development, in Las Vegas, Nevada. An urban geographer looking at this neighborhood would want to discover what it has in common with similar neighborhoods in other cities and what makes it distinctive from them.

and the nature of the relationships between different social groups within cities. From this perspective, cities are simultaneously the products and the shapers of economic, social, and political change. Partitioning space through the establishment of legal boundaries is also important because it affects the dynamics of cities in several ways. For example, the placement of municipal boundaries affects a city's capacity to raise revenue, while electoral boundaries affect the outcome of local elections and politics.

Territoriality is important for the geographer because it is often the basis for individual and group behavior that creates distinctive spatial settings. These spatial settings in turn mold the attitudes and behavior of their inhabitants.

Distance is important for several reasons. It affects the behavior of both producers and consumers of all goods and services. It influences patterns of social interaction and the shape and extent of social networks. Variations in people's physical accessibility to opportunities and to amenities such as jobs, schools, stores, parks, and hospitals are important in determining the local quality of life.

Finally, *place* is important because of the geographer's traditional and fundamental concern with areal differentiation and the distinctiveness of regions and localities. The distinctiveness of particular metropolitan regions, cities, districts, and neighborhoods is central to the analytical heart of urban geography: mapping variability, identifying regularities in spatial patterns, and establishing the linkages that constitute functional regions and subareas. In addition, the sense of place associated with certain cities and localities is important because it can influence people's decisions: where to live, where to locate an office or a factory, whether to hire someone from a particular place, or whether to walk alone through a certain part of town, for example.

APPROACHES TO URBAN GEOGRAPHY

Urban geography has evolved to encompass several approaches to its subject matter. A more general intellectual evolution has, in part, caused this evolution. For example, a widespread "quantitative revolution" has occurred both in urban geography and in the social sciences as a whole since the early 1960s. Two developments spurred this revolution. Large quantities of reliable socioeconomic data about cities and city

FIGURE 1.2   A wall mural, "The Great Wall of L.A.," in Los Angeles, depicts Japanese interment during World War II. Community art such as this provides a remarkably clear expression of the sense of place and territoriality that are fundamental to the spatial organization of people in cities.

neighborhoods became available from sources such as the censuses of population and housing in many countries. At the same time, tools to analyze and shape this information were becoming widely available in the wake of the dawning computer revolution. Modern analytical and modeling techniques have made a decisive contribution to the social sciences. They have allowed the urban geographer to see farther, with more clarity, and have provided the means by which to judge theories about urbanization.

Like other fields, urban geography has also been influenced by changing social values. As each society's comprehension of urban problems has grown, attitudes toward research in urban geography have become more flexible. Much of the research undertaken today in urban geography has relevance far beyond the ivory towers of academia and involves a more active engagement with the needs and concerns of local communities, private companies, and government. Urban geographers are likely to be consulted on issues that range, for example, from optimal political redistricting (redrawing the boundaries of local voting districts) to the evaluation of government policies aimed at enhancing local economic development.

Finally, changes in cities themselves and in the nature of urbanization have also contributed to the evolution of approaches to urban geography. As we have become aware of changes in cities and have looked more closely at those changes, new topics for study have emerged. A vivid example is the interest that geographers took in urban poverty after the riots that broke out in many American cities during the latter part of the 1960s. An equally striking, more recent example is the interest by geographers in the link between rapid urbanization amid poverty and the urban character of the HIV/AIDS problem in sub-Saharan Africa.

The details of the evolution of urban geography as an academic subject are beyond the scope of this book. It is important, however, to establish a few central points about that evolution. During the 1950s the approach to studying cities was highly descriptive. Some of the work in urban geography (or "settlement geography," as it was more commonly known) saw towns and cities as adaptations to natural physical circumstances.[2] Attributes of settlements were interpreted as responses to local sites, regional resources, and the opportunities and constraints surrounding them. Thus, for example, the growth of Pittsburgh as a steel town can be interpreted spatially, from this perspective, in terms of the availability of local sources of coal, iron ore, limestone, and water, along with proximity to large markets for iron and steel products.

Another body of work within this *spatial description approach* that dominated the early development of the subject was focused on the "morphology" of towns and cities—their physical form, their plan, and their various townscapes and "functional areas" (that is, districts with a distinctive mixture of interrelated land uses). Keeping with the example of Pittsburgh, these studies would likely have emphasized the influence of the city's hilly topography on the layout of its streets and neighborhoods, and perhaps shown how urban growth and land use were "fixed" by the limited amounts of flat land (along the Monongahela River) that was suitable for large ironworks.

After the 1950s scientific principles increasingly influenced attitudes toward knowledge, and both settlement and morphology studies fell out of favor. In their place there emerged a *spatial analysis approach* based on the philosophy and methodology of *positivism* that had been developed in the natural sciences. This philosophy was founded on the principle of verification of facts and relationships through accepted scientific methods. The rise of positivism affected all of geography and most of the social sciences, and the "quantitative revolution" that began in the 1960s reinforced it. Its practitioners came to redefine urban geography as the *science* of urban spatial organization and spatial relationships, and, as a science, it focused on the construction of testable models and hypotheses. One example of the many possible positivistic approaches to aspects of Pittsburgh's urban geography would be an attempt to quantify the relationship between neighborhood social status and proximity to steel mills. The hypothesis might be that neighborhood social status tends to increase steadily with distance from a steel mill. If this were demonstrated with verifiable evidence, the next step might be to begin to develop a theory by replicating the study in other industrial cities.

This more abstract approach has contributed a great deal to our understanding of cities, and it continues to be a mainstay of urban studies. But sometimes its abstractions and overdependence on statistical data can seem flat and lifeless in the face of the urban realities to which they are applied. The abstractions tend to leave unanswered many of the important questions concerning underlying *processes* and *meanings*.[3] Thus, although it might be useful to be able to use census tract data to establish and quantify a tendency for neighborhood social status to increase with distance from an industrial plant, we are left without any sense of how the relationship came about: Who made what decisions, and why, in the process of establishing the

relationship? How might any exceptions to the overall relationship be explained?

In response to such questions, a *behavioral approach* emerged by the 1970s. This approach focuses on the study of individual people's activities and decision making in urban environments. Although the behavioral approach continued to use a positivist methodology, explanatory concepts and analytical techniques were also derived from social psychology and some key ideas came from social philosophy, with its insights into human needs and impulses.

The behavioral approach's relative neglect of the importance of cultural context for understanding people's actions and the meanings attached to those actions led to the emergence of the *humanistic approach*. The behavioral approach's positivist methodology was replaced by methods, such as ethnography and participant observation, that attempted to answer questions that capture people's subjective experiences. For Pittsburgh, what kind of meaning does the presence of a steel mill carry for different groups within the city? How do people's feelings about steelworkers' neighborhoods affect their decisions about where to live in Pittsburgh? Like the behavioral approach, though, the humanistic approach has been criticized for not paying enough attention to the constraints on people's decision making and behavior.

As a result, another approach, generally referred to as the *structuralist approach*, gained momentum within urban geography in the late 1970s and 1980s. This approach is cast, in contrast to the behavioral and humanistic approaches, at the scale of macroeconomic, macrosocial, and macropolitical changes. It focuses on the implications of such changes for urbanization and on the opportunities and constraints they present for the behavior and decision making of different groups and classes of people. At its broadest level, this approach draws on a combination of macroeconomic theory, social theory, and the theories and concepts of political science, and includes the *political economy approach*. For example, a political economy approach to Pittsburgh's urban geography would certainly want to relate the patterns of growth and decline of both steel mills and their associated blue-collar neighborhoods to the broader structure of, among other things, access to capital for investment in industry, the availability of skilled labor, the framework of government policies affecting industrial and residential development, and deindustrialization associated with corporate restructuring within the global economy.

The structuralist analyses of the structure of social inequality paved the way for explicitly incorporating the experience of women into urban geography. The *feminist approach* deals with the inequalities between men and women and the way in which unequal gender relations are reflected in the spatial structure of cities.

For Pittsburgh, the feminist approach might be interested in examining changing gender roles as the urban labor force has restructured following the closing of the steel mills, such as the trend for some women in two-earner households to be overrepresented in certain fast-growing part-time occupations.

The *structure-agency approach* was an attempt to unite the structuralist approach's concern with macrolevel social, economic, and political structures with the humanistic approach's emphasis on human agency. Structuration theory[4] sees society's social structures as created and recreated by the social practices of human agents, whose actions are themselves constrained by these social structures. As a result, it is impossible to predict the exact outcome of the interactions between social structure and human agency. Despite the elegance of this theorization, empirical investigation has proved difficult because it is not easy to analyze the continuous and complex interrelationships between structure and agency. A study of Pittsburgh might involve questions about how the **gentrification** of some central city neighborhoods is the result of an intricate set of interactions between human agents, including landowners, mortgage lenders, planners, and realtors, and so on; institutions such as the city government; and social structures involving planning regulations such as **land use zoning** and building codes.

Finally, the influence of architecture and poststructuralist literary theory (the French Deconstructionist School) led to the emergence of the *poststructuralist approaches*, including the *postmodern approach*. The postmodern approach strongly opposes the idea that any general theories can explain the world. Instead, it accepts the shifting and unstable nature of the world and concentrates on questions of who defines meaning, how this meaning is defined, and to what end. It is concerned with understanding the power of symbolism, images, and representation as expressed in language, communication, and the urban landscape. Again, using the example of Pittsburgh, a postmodern approach would likely examine the city government's attempts to "reinvent" the city within the global economy as it restructures away from steel and toward jobs in the high-tech and services industries. This approach would draw attention to the city government's use of language and communication to create certain images of reality that, in turn, influence the views and decisions of potential investors and residents.

For the most part, these approaches can be regarded as complementary. Although it is neither possible nor desirable to merge them all into some kind of all-encompassing model or theory of urbanization, it is possible to gain insights from each. To that extent, all these approaches are represented in this book. This does not mean, however, that the reader will be faced with sudden or unexpected shifts. The material in this book is organized to emphasize urban geography as the outcome of urbanization as a process. So, for example, the macrolevel structures addressed by the political economy approaches will be used at certain times as framing perspectives, with the other approaches being brought in to illuminate particular outcomes or dynamics.

BOX 1.1   CENSUS DEFINITIONS

Just what is meant by an "urban" settlement varies a good deal from one country to another. For the Bureau of the Census in the United States, the term urban applies to the territory, population, and housing units located within urbanized areas and urban clusters.[5]

- An *urbanized area* (UA) is a densely settled area (whether or not the territory is legally incorporated as a city) with at least 50,000 people and a density of at least 1,000 people per square mile at the urban core and at least 500 people per square mile in the surrounding territory.

- An *urban cluster* (UC) consists of an urban core with a population density of at least 1,000 people per square mile and at least 500 people per square mile in the surrounding territory that together encompass a population of at least 2,500 people, but fewer than 50,000 people.

Since such definitions of *urbanized* depend on administrative boundaries, they do not capture the concentrations of population that exist where a number of contiguous jurisdictions form one continuous metropolitan sprawl. The U.S. Bureau of the Census began to use standardized definitions of metropolitan areas in the 1950 census under the designation of standard metropolitan area (SMA). The term was changed to standard metropolitan statistical area (SMSA) by the 1960 census and to metropolitan statistical area (MSA) in 1983.

The expansion of large neighboring metropolitan areas led to the term metropolitan area (MA) being adopted in 1990. In the 2000 U.S. census an MA referred to a large population nucleus (a central city or twin cities), together with adjacent communities that had a high degree of economic and social integration with the nucleus (as measured by commuting patterns). The term MA referred collectively to three

(Continued)

BOX 1.1   CENSUS DEFINITIONS   *(Continued)*

large statistical areas: (1) an MSA, which was an MA that was not closely associated with another MA; an MSA consisted of one or more counties (or, in New England, cities and towns); (2) two or more primary statistical areas (PMSAs) could be defined within an area that qualified as an MA if the MA had a population of 1 million or more and local governments favored that designation based on local perceptions of relative independence within the larger metropolitan complex; and (3) the MA could then be designated a consolidated metropolitan statistical area (CMSA) when PMSAs were established within it.

In June 2000 the term *core based statistical area* (CBSA) became effective (or, in New England, NECTA (New England City and Town Area)).[6] CBSA replaced the MSAs, PMSAs, and CMSAs. The new CBSA designation uses 2000 census data as a reference point. Each CBSA contains at least one urban area of 10,000 or more people. Counties (or, in New England, cities and towns) form the geographic "building blocks" for CBSAs. A county or counties in which at least 50

percent of the population lives in urban areas of 10,000 or more or that contain at least 5,000 people residing within a single urban area of 10,000 or more is identified as a "central county" or "central counties." Additional "outlying counties" are included in a CBSA if they meet specified requirements of commuting to or from the central county or counties.

The term CBSA refers collectively to micropolitan and metropolitan statistical areas. The general concept of a micropolitan or metropolitan statistical area is that of a core area containing a substantial population nucleus, together with adjacent communities having a high degree of social and economic integration with that core. Micropolitan and metropolitan statistical areas comprise one or more entire counties.

- *Micropolitan statistical areas*, or *Micros*, have at least one urban cluster of at least 10,000 but fewer than 50,000 people. As of March 2004, there were 554 Micros in the continental United States.[7]

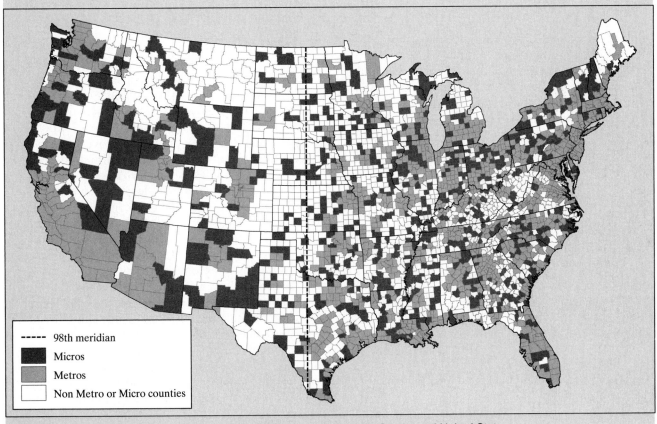

Legend:
- - - - - 98th meridian
- Micros
- Metros
- Non Metro or Micro counties

FIGURE 1.3   Metropolitan and Micropolitan Statistical Areas in the Continental United States.

(Continued)

Box 1.1 Census Definitions (Continued)

- *Metropolitan statistical areas*, or *Metros*, have at least one urbanized area of 50,000 or more inhabitants. As of March 2004, there were 276 Metros in the continental United States.[8] If specified criteria are met, Metros containing a single core with a population of 2.5 million or more (specifically, Boston, Chicago, Dallas, Los Angeles, Miami, New York, Philadelphia, San Francisco, Seattle, and Washington, D.C.) may be subdivided to form smaller groupings of counties referred to as *metropolitan divisions*.

These designations, then, are based on the size of the urban center rather than on the total population. Micros can be populous regions without huge centers; Metros have large centers but may not necessarily be surrounded by large populations. Since the introduction of the new definitions in 2000, much more of the land area of the United States falls into urban areas and a new milestone has been reached: Over half (53 percent) of all the land area of the 48 contiguous states falls within either Metros or Micros, meaning that for the first time rural areas comprise a minority share of this land area.[9]

The largest city in each Metro or Micro is designated a "principal city." Additional cities qualify if specified requirements are met concerning population size and employment. The title of each Metro or Micro consists of the names of up to three of its principal cities and the name of each state into which the Metro or Micro extends. The titles of "metropolitan divisions" also typically are based on principal city names but in certain cases consist of county names.

If specified criteria are met, adjacent Metros and Micros, in various combinations, may become the components of a new set of areas called combined statistical areas (CSAs). For instance, a CSA may comprise multiple Metros and Micros, two or more Metros, a Metro and a Micro, or two or more Micros. The areas that combine retain their own designations as Metros or Micros within the larger CSA.

Within urban areas, detailed census information is available by *census tracts, block groups*, and *census blocks*:

- *Census tracts.* These geographical subareas have boundaries that were drawn with the objective of delineating small populations that are relatively uniform in terms of their demographic and socioeconomic characteristics. They vary a good deal in territorial extent and population size, although tracts within metropolitan areas contain between 1,000 and 8,000 people. As metropolitan America has grown, so has the number of census tracts recognized by the Census Bureau. The boundaries of many longer-established census tracts have remained the same over several decades, making it possible to use census data to analyze neighborhood change in certain areas. Elsewhere, however, modifications to tract boundaries and the addition of new tracts make intercensal comparison difficult.

- *Block groups.* Every census tract is divided into as many as 9 block groups, each of which contains an average of 10 census blocks. A block group consists of all blocks whose numbers begin with the same digit in a census tract. Block groups generally contain between 300 and 3,000 people. Block groups are important because they are the smallest geographical area for which detailed data are tabulated.

- *Census blocks.* Census blocks generally correspond to the physical configuration of city blocks and are bounded by streets or other prominent physical features. They are the smallest statistical unit for which census data are available, although the range of data is limited to basic population and housing characteristics. The census blocks were completely renumbered using 4-digit numbers in the 2000 census.

In preparing the 1990 census, the U.S. Bureau of the Census developed an electronic database called the Topologically Integrated Geographic Encoding and Referencing (TIGER) system. TIGER files contain address ranges, latitude and longitude coordinates, and the location of roads, railways, rivers, and other physical features. They can be linked with small-area census data to form the basis of sophisticated Geographic Information System (GIS) applications that offer a great deal of potential for research in urban geography. GIS—organized collections of computer hardware, software, and geographic data that are designed to capture, store, update, manipulate, and map geographically referenced information—has grown rapidly to become an important method of urban geographic analysis. GIS technology allows an enormous range of urban problems to be analyzed. For instance, it can be used to identify the most efficient evacuation routes from all or part of a city in the event of a terrorist attack, to monitor the spread of infectious diseases within and between cities, to analyze the impact of proposed changes in the boundaries of legislative districts, to identify potential customers in the vicinity of a new business, and to provide a basis for urban and regional planning.

URBANIZATION: PROCESSES AND OUTCOMES

Figure 1.4 provides a useful outline of urbanization as a process. It is clear from the diagram that urbanization involves much more than the mere increase in the number of people living and working in towns and cities. It is driven by a series of interrelated processes of change—economic, demographic, political, cultural, technological, environmental, and social. It is also modified by local factors such as topography and natural resources. It is true that the overall result has been a tendency for more and more people to live and work in ever-larger towns and cities (although, as we will see, this is not a necessary condition of urbanization). At the same time, urbanization results in some important changes in the character and dynamics of the **urban system** (the complete set of towns and cities within a region or country[10]), and within towns and cities, it causes changes in patterns of land use, in **social ecology** (the social and demographic composition of neighborhoods), in the *built environment*, and in the nature of **urbanism** (the forms of social interaction and ways of life that develop in urban settings). Certain groups of people might view some of these outcomes as problems. Government policies, legal changes, city planning, and urban management might eventually address those problems, often resulting in changes (sometimes unanticipated) that in turn affect the dynamics that drive the overall urbanization process.

As suggested by Figure 1.4, the relationships between these processes are complex. Urbanization is not only influenced by the direct effects of these dynamics,

but it also experiences "feedback" effects. Meanwhile, almost every aspect of urban change is itself to some extent interdependent with some of the others. This complexity can be confusing! We will eliminate much potential confusion with concepts and examples introduced in this chapter. Bear in mind, though, that Figure 1.4 presents a very broad framework that covers the entire subject matter of urban geography. The further you get into the book, therefore, the more meaningful Figure 1.4 will become. The economic, demographic, political, cultural, social, technological, and environmental processes and relationships implied in the diagram will be elaborated, and urban outcomes will be detailed. It is recommended, therefore, that you refer to Figure 1.4 from time to time, putting each new set of material into overall perspective.

ECONOMIC CHANGE

At the heart of the dynamics that drive and shape urbanization are economic changes. The sequence and rhythm of economic change will be a recurring theme as we trace and retrace the imprint of urbanization. It is the evolution of capitalism itself that has structured this imprint. Figure 1.5 summarizes the main features of this evolution. In the United States, there have already been two broad phases in the nature of capitalism, and we are now entering a third phase.[11] The earliest phase, lasting from the late eighteenth century until the end of the nineteenth, was a phase of

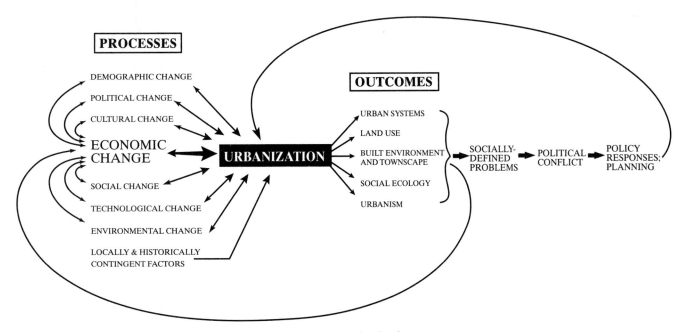

FIGURE 1.4   A framework for the study of urban geography: urbanization as a process.

Box 1.2 Globalization and Cities

The urbanization processes that produce the urban outcomes shown in Figure 1.4 operate at different spatial scales. The rapidly increasing interdependence of the **world-system** means that the economic and social well-being of cities everywhere depends increasingly on complex interactions that are framed at a global scale. Globalization has had profound effects on cities and systems of cities because of the close interaction between global and local forces—a process that has been called *glocalization*[12] or the *global-local nexus*. The process and its outcomes involve **uneven development** both within and among cities. Globalization has led, for example, to the emergence of so-called **world cities**—command centers such as New York, London, and Tokyo that are key players in the new concentrated financial system.

Although globalization is a complicated and controversial topic, we can identify a number of interrelated dimensions—economic, cultural, and political—associated with its processes and urban outcomes. Economic globalization reflects the fact that, although urban, regional, and national circumstances remain very important, what happens in any given city is broadly determined by its role in systems of production, trade, and consumption that have become global in scope. The term globalization is usually associated with the growing importance of transnational corporations operating across a number of countries. The activities of these companies, in the spheres of both production and marketing, are increasingly integrated at a global scale in a new international division of labor. Products are made in multiple locations from components manufactured in different places in order to take advantage of the full range of geographical variations in costs. With a global assembly system, labor-intensive work can be done where labor is cheap, raw materials can be processed near their source of supply, and final assembly can be done close to major urban markets.

Cultural globalization is associated with the development of a broader global culture. This is a controversial idea, but it essentially involves the widespread diffusion of Western values of materialism. Globalization can be seen in the popularity of Hollywood films or the spread of the hamburger. Indeed, Ritzer[13] coined the term "McDonaldization" to denote the ways in which processes of mass consumption are eroding cultural differences around the world. It is argued that globalization involves the homogenization of culture—the development of cultural interrelatedness throughout the world. New telecommunications systems—especially within urban areas—that allow rapid transmission of information and images have facilitated this process. But there has been resistance to these global forces through the assertion of local cultural identities, including various popular movements seeking greater autonomy at the level of regions and cities.

Political globalization has been associated with the reduced power of national governments to shape their own destinies. In large measure this has been bound up with the globalization of financial markets so that money can now flow rapidly across national boundaries. In addition, because transnational corporations can switch investment from one country to another, public bodies, including city governments, have little choice but to adopt incentive strategies in order to compete in attracting and keeping internationally mobile companies and their investment.

competitive capitalism, the heyday of free enterprise and laissez-faire economic development, with markets characterized by competition between small family businesses and with few constraints or controls imposed by governments or public authorities. In the earlier years of this phase, the dynamism of the entire system rested on the profitability of agriculture and, increasingly, manufacture and "machinofacture" (industrial production that was based less on handicraft and direct labor power than on mechanization, automation, and intensively used skilled labor).

After the Civil War, the extensive railroad system and a strengthened federal system helped to create an economy that was truly national urban scale. The most successful family businesses grew bigger and began to take over their competitors. Business became more organized as corporations set out to serve regional or national consumer markets rather than local ones. Labor markets became more organized as wage norms spread, and government began to be more organized as the need for regulation in public affairs became increasingly apparent.

By the turn of the twentieth century, these trends had reached the point where the nature of capitalist enterprise had changed significantly. It could now be characterized as **organized capitalism**—a label that came to be increasingly appropriate with the evolution of the economy over the next 75 years or so. In the early decades of the twentieth century, the system's dynamism (that is, the basis of profitability) shifted away from industrial manufacture and machinofacture as a new labor process took hold. This process was **Fordism**, named after Henry Ford, the automobile manufacturer who was a pioneer of the principle of mass production, based on assembly line techniques and "scientific" management (known as **Taylorism**), together with mass consumption, based on higher wages and sophisticated advertising techniques.

The success of Fordism was associated with the development of a rather tense but nevertheless workable relationship between business interests and the labor unions, whose new strength was in itself another important element of "organization." Meanwhile, the role of government had also expanded—partly to regulate the unwanted side effects of free-enterprise capitalism and partly to mediate the relationship between organized business and organized labor. After the Great Depression of 1929–1934, government's role expanded dramatically to include responsibility for

full employment, the management of the national economy, and the organization of various dimensions of social well-being.

After World War II, another important transformation in the nature of capitalism became evident. The developed countries like the United States and Canada experienced a shift away from industrial production and toward services, particularly sophisticated business and financial services, as the basis for profitability. It is denoted in Figure 1.5 as an evolution from **industrial capitalism** to **advanced capitalism**. This shift began to transform occupational structures, sparking **deindustrialization**—a decline in manufacturing jobs but *not* in manufacturing production. Meanwhile, the increasing **globalization** of the economy (which allowed huge **transnational corporations** to outmaneuver the national scope of both governments and labor unions by moving routine production and assembly operations to lower-cost, less developed parts of the world as part of a

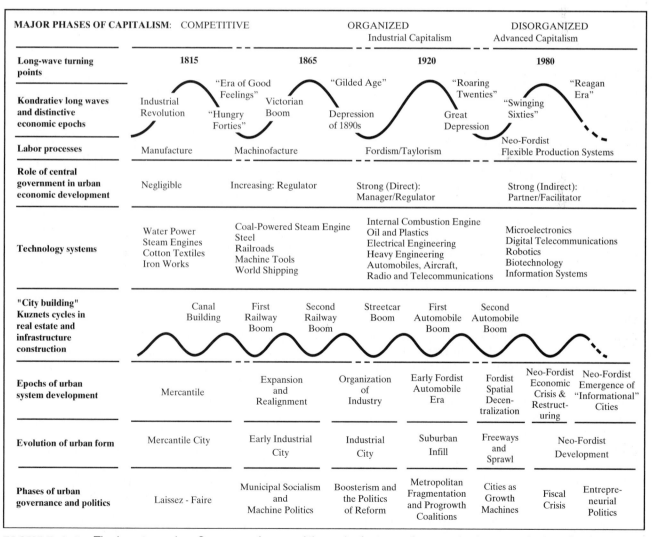

FIGURE 1.5 The long-term view. Summary of some of the major features of economic change and urban development in the United States.

new international division of labor) contributed to a destabilization of the relationship between business, labor, and government in the developed countries.

Meanwhile, Fordism began to be a victim of its own success, with mass markets for many products becoming saturated. As it became increasingly difficult to extract profits from mass production and mass consumption, many enterprises sought profitability through serving specialized market niches. Instead of standardization in production, in **neo-Fordism** specialization required variability and, above all, **flexible production systems**. The net result has been labeled **disorganized capitalism**, not so much because of the lack of organization or purpose in business, government, or labor but because of the contrast with the orderly interdependence of all three during the organized capitalism phase.[14]

The significance of this evolution for urbanization and urban geography is fundamental. Each new phase of capitalism saw changes in what was produced, how it was produced, and where it was produced. These changes called for new kinds of cities, while existing cities had to be modified. At the same time, of course, cities themselves played important roles in the transformation of capitalist enterprise. As centers of innovation, cities and towns have traditionally functioned as engines of economic growth that provide opportunities for new forms of livelihood and improved prosperity. But despite producing the bulk of national wealth, cities are also locations of exploitation and unemployment. Within a global economy in particular, the costs and benefits of globalization are unevenly distributed in cities. For that reason, city governments have increasingly been called upon to play a key role in mediating some of the adverse impacts of globalization. Through good governance and effective partnerships with communities and businesses, it might be possible to help mitigate urban inequalities and poverty. The challenge for cities in the global economy of the twenty-first century is to function not only as engines of economic growth but also as agents of change for greater social justice and urban sustainability.[15]

BOX 1.3   LONG-WAVE ECONOMIC FLUCTUATIONS AND URBAN CHANGE

Parallel with the overall evolution of capitalism, the character and dynamics of urbanization have been shaped by—and have contributed to—*economic long waves* (cycles of expansion and contraction in the rate of economic development), another important feature of long-term economic change. Economic long waves have been an important aspect of the capitalist economy since its earliest phase. At the most fundamental level, they can be regarded as the result of market imbalance: the consequence of overproduction or underproduction as producers have sought to match output to demand under constantly changing conditions. Long waves are very complex and have been a controversial topic in macroeconomic theory. It is important, however, to identify the main features of long-wave changes insofar as they relate to the broader sweep of urbanization in Europe and the United States.[16]

The best place to begin is by observing that the economy has been carried along by a succession of technology systems that have been fundamental to the changing conditions that producers have had to confront. These are indicated in Figure 1.5 in terms of the clusters of energy sources, transportation technologies, and key industries that characterized each system:

- Early mechanization based on waterpower and steam engines; the development of cotton

textiles, pottery, and iron working; and the development of river systems, canals, and turnpike roads for the assembly of raw materials and the distribution of finished products.

- The development of coal-powered steam engines, steel products, railroads, world shipping, and machine tools.

- The development of the internal combustion engine, oil and plastics, electrical and heavy engineering, automobiles, aircraft, radio, and telecommunications.

- The exploitation of nuclear power, the development of limited access highways, durable consumer goods industries, aerospace industries, electronics, and petrochemicals.

- The most recent (and still incomplete) technology system, which is based on microelectronics, digital telecommunications, robotics, biotechnology, fine chemicals, and information systems.

These technology systems gave shape and direction not only to the evolving national economy but also to the pace and character of urbanization. What is especially intriguing about the timing of these technology systems is that they are closely related to fluctuations in the overall rate of change of prices. These fluctuations in the rate of price changes are known as

(Continued)

Box 1.3   Long-Wave Economic Fluctuations and Urban Change   (*Continued*)

Kondratiev cycles after the Russian economist who first identified them. There have been four complete Kondratiev cycles, and we are now in the fifth. Each one has been marked by a progressive acceleration in the rate of price increases for about 20 years, followed by a rapid inflationary spiral. After the peak, prices collapse, eventually reaching a trough some 50 to 55 years after the start of the cycle. This periodicity, seemingly linked in a cause-and-effect manner with technology systems, has given the political economy of capitalism a remarkable

rhythm that has reverberated widely to influence social and institutional structures, politics, demographics, and even culture. The point, of course, is that cities have been in the middle of these reverberations, echoing changes while simultaneously transforming themselves.

The Kondratiev rhythm is not the only important economic long wave in the urban context. There are other rhythms, of which the most significant is the **Kuznets cycle**. This is a cycle of regular changes in the rate of economic growth, as

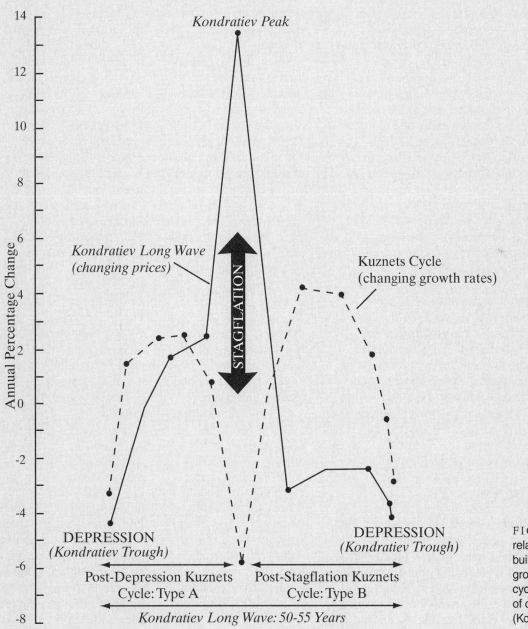

FIGURE 1.6 The relationship of city-building economic growth cycles (Kuznets cycles) to long waves of changing prices (Kondratiev cycles).

(*Continued*)

BOX 1.3   LONG-WAVE ECONOMIC FLUCTUATIONS AND URBAN CHANGE   *(Continued)*

measured by indicators such as per capita gross national product. Kuznets, a Ukrainian-born economist, identified 25-year cycles of economic growth, each characterized by an acceleration phase of 11–15 years, followed by a deceleration phase of similar duration. These cycles of economic growth are particularly important to students of urbanization because it is during bursts of economic growth that new elements of urban geography tend to be inscribed most vigorously into the urban landscape.

Kuznets cycles are business cycles that reflect the inner laws of free-market dynamics, in which expansion (production and supply increasing to meet increased consumer demand) is followed inevitably by overshoot (when overly optimistic producers and suppliers overestimate the rate of increase in demand) and then collapse (as overcapacity and excess inventory lead to falling profits, reduced asset values, pessimism, and reduced levels of investment). These cycles have affected many aspects of economic development, including the rhythm of investment in transport infrastructure, in city building, and in immigration. As suggested by Figure 1.5, these cycles have been synchronized, the whole trajectory of U.S. economic development following the same 25-year oscillation.

Brian Berry, one of the founders of modern urban geography, has pointed to a further aspect of synchronicity with important implications for economic and urban development: a striking relationship between Kondratiev cycles and Kuznets cycles.[17] Figure 1.6 summarizes this relationship, showing how two infrastructure-investment and city-building Kuznets cycles are embedded within each Kondratiev cycle. As the economy moves out of recession and price collapse, there is a "post depression" Kuznets cycle that begins with Type A growth—growth that coincides with a price recovery. Prices continue to accelerate, however, even after the overshoot and collapse of Type A economic growth. This produces a combination of economic slowdown or stagnation combined with price inflations: **stagflation**. Stagflation represents a crisis point in economic development that is usually marked by banking crises. There has always followed a second, Type B, Kuznets cycle, during which a cycle of economic expansion, overshoot, and collapse takes place under conditions of price deflation.

As we look back at the history of U.S. economic development, we can identify the rhythm of these synchronized long waves in distinctive epochs of economic, political, social, and urban development. The "Roaring Twenties" (1920s), "Swinging Sixties" (1960s), and the "Reagan Era" (1980s), for example, were all episodes of Type B economic growth during price deflation, marked by significant bursts of city building and urban redevelopment. Similarly, the "Hungry Forties" (1840s), the "Great Depression" (1929–1934) and the recession of the early 1990s can all be seen as episodes of Type B economic decline during price deflation, marked by stagnation in city building, acute social distress, and pressure for economic and social reform. Such episodes are sometimes called **overaccumulation crises** because three surpluses exist simultaneously: surplus labor (the unemployed and underemployed), surplus productive capacity (idle factories and machinery), and surplus capital (because owners of profits, interest, and dividends cannot find enough reasonably safe, profitable investment opportunities in productive enterprises). Such epochs make convenient markers that help in understanding urbanization as a process. History, however, should not be forced into neat periodizations—there are always leading and lagging aspects of urban change that fixed chronologies cannot capture.

DEMOGRAPHIC CHANGE

One of the most important subsets of interdependence suggested in Figure 1.4 is that between demographic change and urbanization. Cities are, in a fundamental way, the product of their people. Put another way, the size, composition, and rate of change of urban populations significantly shape the character of urbanization. Yet the condition of cities themselves can in turn influence those characteristics. Crowded and degraded slums, for example, can lead to higher death rates; cities with good amenities tend to attract particularly large numbers of migrants; and border towns and big cities with international ports and airports tend to attract a disproportionate share of immigrants. Meanwhile, urban economic well-being often mediates the relationships between demographic change and urbanization. Thus, for example, both birth rates and migration rates depend a great deal on people's perceptions and expectations of economic opportunities.

POLITICAL CHANGE

The broad ideological swings and shifts that occur from time to time are an important aspect of the influence of political change on urbanization. One well-known example is the reform movement that emerged in the United States in the 1870s and 1880s in response to a variety of social problems. As we will see in Chapter 16, the reform movement had an important and lasting influence on urban affairs. A very different and more recent example is that of the political shift at the national and international level before the beginning of the "War on Terrorism": the end of the Cold War, which had a marked effect on the economies of some Sunbelt cities that had been heavily dependent on defense-related industries (see p. 86).

In this last example we see again the mediating role of economic change. Indeed, politics has become intimately related to economic development. Economic issues almost always appear in local elections, while both the need for local services and the ability to pay for them are functions of local economic prosperity. As suggested by the direction of the arrows in Figure 1.4, urbanization also directly affects political change in some ways. One example is the way that coalitions of urban voters shaped the basis of modern party politics at the national level during the 1930s and 1940s in the United States (see p. 463). Another is the electoral significance of suburban voters who have more recently formed an important foundation of support for the Republican Party. Urbanization also affects political change indirectly through people's perceptions of the problems associated with various dimensions of urban change, because their perceptions inform and frame many of the issues that are contested in the political arena.

FIGURE 1.7 The economic and demographic profile of San Diego, California, has been influenced by its military bases, climate, scenery, recreational opportunities, and proximity to the Mexican border.

CULTURAL CHANGE

We can find parallel examples of the interdependence of urbanization and cultural change. The broad cultural shift toward "postmodernity" (see p. 325) that began in the 1970s and 1980s, for example, brought, among other things, a renewed interest in the past that has found expression in urban form through historic preservation and the recycling of architectural styles. Meanwhile, urbanization has contributed to cultural dynamics through the youth subcultures that have flourished in certain urban settings. Other aspects of the interdependence between urbanization and culture involve still further processes of change. The materialism of mainstream American culture, for example, has affected urbanization through patterns of residential development, but it has depended on processes of economic change that have resulted in modifications to the distribution of income (see p. 83). It is not only (or always) economic change that is important in mediating such interrelationships. As we will see (p. 100), demographic change is also important, the "Baby Boom" generation having been the innovators and "carriers" of successive aspects of cultural change from the counterculture of the 1960s to the yuppie materialism of the 1980s and the more restrained "inconspicuous" consumption of the 1990s.

TECHNOLOGICAL CHANGE

There are many examples of the interdependence of urbanization and technological change, although it is often difficult to disentangle cause and effect. Many technological changes, although not strictly causing—or being caused by—changes in urbanization, have been important preconditions for change: the streetcars that facilitated the first widespread suburbanization in the United States, for example (p. 126). In addition, we have already seen the importance of successive **technology systems** in relation to the rhythms of economic development that have been imprinted, layer by layer, on American cities. Meanwhile, the broader web of interdependence implied by Figure 1.4 is well illustrated by the impact of new birth control technology (the contraceptive pill) in the mid-1960s. Not only did this technology help put an end to the baby boom, it also helped change attitudes toward sex, marriage, and female participation in the labor force—thus simultaneously affecting several dimensions of urban life.

ENVIRONMENTAL CHANGE

The complexity of the interactions between urbanization and environmental change creates problems of local through global proportions. The area of the Earth's surface needed to absorb the waste products of a large city is likely to exceed that city's boundaries—the ecological footprint—although this can depend on the fuels used for heating, energy generation, and manufacturing; the amount of motorized traffic; the technologies used for disposing of solid and liquid wastes; and local climatic conditions. Most large cities cannot assimilate their waste products, and by burning them they contribute to pollution.[18] The role of cities in producing greenhouse gas emissions from increased automobile use and coal-fired power plants has global climate change implications. At the local scale, cities involve changes in land use and land cover that can produce a diversity of environmental problems. In the United States, Europe, and Russia, for example, countless **brownfields**—abandoned or underused traditional manufacturing facilities with actual or potential contamination—are a legacy of a weakly regulated early industrialization process that now complicates redevelopment efforts in many central cities.

SOCIAL CHANGE

Still following Figure 1.4, we move now to some brief illustrations of the interdependence between urbanization and social change. Here we can cite the changes that have occurred over the past 30 years in terms of people's behavior toward racial minorities—changes that have carried over to affect educational achievement, occupational composition, and, ultimately, urban residential patterns. Black suburbanization, for example, is largely attributable to such changes. Urbanization can also induce social change. The physical and socioeconomic attributes of urban settings, for example, foster certain behavioral changes, such as the social isolation and withdrawal that seems to be generated among the "lonely crowd" of inner-city districts (see p. 381). The most important changes in American society, for example, have been changes in social status that have been driven by occupational changes resulting from the structural transformation of the economy: the growth of the middle-income workers that accompanied the emergence of industrial capitalism and their subsequent decline with the shift toward advanced capitalism.

THE PLAN OF THE BOOK

A rapid sequence of examples such as these can do no more than illustrate the scope and diversity of the interrelationships that surround the central dynamic between urbanization and economic and other change. Urbanization is clearly a multidimensional phenomenon, driven by multiple, interdependent processes. In subsequent chapters we will see how these processes produce both regularity and distinctiveness in the geography of cities. *Concepts and theories will be presented in relation to the overall context of urban change.* Such an approach allows us to integrate theory with fact, to appreciate the logic of particular theories, and to understand their relevance to particular circumstances.

We begin in Chapter 2 with an overview of the growth of cities from their earliest origins some 5500 years ago, which allows us to discuss some of the concepts and theories related to processes and outcomes of urban-system change over the long historical term. We continue in Chapter 3 with a survey of the foundations of the American urban system, which lets us consider some of the concepts and theories related to more recent processes and outcomes of urban-system change. In Chapter 4 we examine recent changes in the urban system in more detail, focusing on the consequence of economic restructuring for patterns of urban growth and decline. Chapters 5 and 6 provide a parallel review of the changes that have occurred in the overall form and spatial organization of individual metropolitan areas. Here we trace the evolution of urban form from the beginnings of land use specialization, through the emergence of distinctive sectors and zones that came with industrialization, to the metropolitan sprawl of the automobile era and contemporary **neo-Fordist** development.

In the next three chapters we turn our attention to urbanization in the less developed countries. In Chapter 7 we consider some of the historic and contemporary changes in social, cultural, economic, political, technological, and environmental processes—including colonialism, industrialization, **rural-to-urban migration**, and **overurbanization**—affecting cities in the less developed countries. Chapter 8 discusses the outcomes of these processes by examining the various patterns of urban form and land use in Latin American, African, Islamic, South Asian, Southeast Asian, and East Asian cities. The discussion in Chapters 7 and 8 leads conveniently to a review of some of the major problems of urbanization in the less developed countries (Chapter 9): poverty, inadequate housing, lack of urban services, transportation problems, and environmental degradation.

At this point we get much closer to the detail of urban form in Chapter 10, which shows how the architecture and design of cities can be "read" in relation to the broader sweep of urbanization processes. We will see how the city's built environment is like a "text" that mirrors the successive phases of economic development, technological systems, social and cultural change, and so on, that have been inscribed into the overall metropolitan forms described in Chapters 5 and 6. Chapter 11 provides an understanding of urban development as a market-oriented process of investment and production, focusing on the behavior of "city makers" such as speculators, developers, builders, investors, and financial managers. Chapter 12 deals with the overall patterns of residential differentiation in American cities, including a discussion of the reasons for residential segregation and an examination of how changes in urbanization have affected patterns of urban social segregation. A closer look at the dynamics of neighborhood change is provided by Chapter 13. Here we focus on the nature and operation of housing markets, on households' behavior in selecting homes, and on the rhythm of neighborhood life cycles.

The geographer's traditional concerns with space and **territoriality** are pursued in Chapter 14, where we see how our daily lives shape—and are shaped by—the larger structures of urban social life and how the nature of urban spaces conditions various aspects of social organization and disorganization. This discussion leads conveniently to a review of some of the major problems of urbanization in the developed countries (Chapter 15): slums, ghettos, and poverty areas; criminal violence; homelessness; and infrastructure and environmental problems.

Managing these problems is, of course, a task of urban governance, and in Chapter 16 we trace the evolution of urban governance and the changing emphases of urban politics in relation to the sequence and patterns of urbanization established in earlier chapters. Chapter 17 focuses on the evolution of urban planning (a particular type of policy making) as ways of coping with the outcomes of urbanization and of modifying or managing some of the processes that contribute to urban change. Finally, Chapter 18 looks to the future, presenting some visions of urban development in the coming decades.

FOLLOW UP

At the end of each chapter there will be some ideas for following up on the topics and ideas covered in that chapter, followed by a list of key sources and suggested reading, and some related websites. Since this is an introductory chapter, we need not suggest further reading here. It would be a good idea, though, to follow up on the material in this chapter and to begin to prepare for the subject matter of the rest of the book.

1. First, be sure that you understand the following key terms:

   competitive capitalism (early phase of industrial capitalism)

   deindustrialization

   disorganized capitalism (advanced capitalism), neo-Fordism, and flexible production systems

   globalization

   Kondratiev cycles

   Kuznets cycles

   new international division of labor

   organized capitalism (later phase of industrial capitalism) and Fordism

   overaccumulation crisis

   social ecology

   stagflation

   technology systems

   territoriality

   uneven development

   urban system

   urbanism

2. Consider the statement (p. 3) that "cities are simultaneously the products and the shapers of economic, social, and political change" and review the discussion of such changes (pp. 9–16). Can you identify examples that illustrate how cities are both the products and the shapers of change? Can you think of any additional examples?

3. Begin a *portfolio* of your own work that builds on your reading. A portfolio can be compiled in many ways; yours should reflect your own reactions to the material in the book and your own urban interests, experiences of cities, and impressions of city life.

   It is a good idea to compile the portfolio in such a way that you can easily add new material. Ideally, the text should be written with word-processing software so you can easily insert or change material. Modern word-processing programs let you include images, sound, Internet links, and other multimedia in a fairly seamless

manner. A loose-leaf format will give you a lot of flexibility. Another possibility would be to build your portfolio using some sort of Web publishing tool, with the ultimate goal of displaying your work on a website. You could also consider creating your portfolio in Microsoft PowerPoint or some other presentation software.

The content of the portfolio should consist of three elements (although they do not necessarily have to be kept separate):

a) A summary of the most important aspects of the material in each chapter, together with a record of the questions that the material raises for you. Note how the material is related to the overall framework provided by Figure 1.4 and how it is related to topics and ideas covered elsewhere in the book. Note any issues that you think need further clarification and record people's ideas and opinions that you believe are particularly provocative or challenging.

b) Library, Web, and other material that illustrates or explains the issues you have identified in item (a) above. This material might include notes, diagrams, maps, photos, or tables taken from the key sources and suggested readings at the end of each chapter that are available from your library or from other sources, such as newspaper or magazine articles (older editions might only be available in the library, not online), reports from government or research institution websites, and so on.

c) Additional material that reflects your own interests and reactions to the themes and ideas that you have encountered. This material might take the form of prose commentary, short essays, poetry (your own or others'), quotations or short extracts, drawings, photographs, brief video animations, short sound recordings, or data in the form of maps, charts, or graphs.

Subsequent follow-up sections will contain specific suggestions on what you can do to make your portfolio an interesting project.

4. Get in the mood—watch a movie! There are many movies with "urban" themes that you can rent or borrow. A great nonfiction movie is *Koyaanisqatsi* (1983; Pacific Arts Video Records, directed by Godfrey Reggio, music by Philip Glass), which features a stream of images that begins with primeval natural landscapes and steadily moves on to ever more complex and frenetic scenes from urban life. The title is a Hopi word meaning "life out of balance."

RELATED WEBSITES

Association of American Geographers: *http://www.aag.org/*
   The AAG's website for the subject and discipline of geography has many useful hyperlinks, including ones to the websites of specialty groups with a focus on issues related to cities, especially the Urban Geography Specialty Group.

Royal Geographical Society with the Institute of British Geographers: *http://www.rgs.org/*
   The RGS/IBG website for the subject and discipline of geography offers many useful hyperlinks to websites of interest to urban geographers.

U.S. Census Bureau Home Page: *http://www.census.gov/*
   The 2000 census was the first one in the United States for which the Internet was the primary means of disseminating information (formatted tables, maps, data sets for downloading, printing, viewing, and manipulating). American FactFinder is the Census Bureau's main online data retrieval tool for census data (*http://factfinder.census.gov/*). The Topologically Integrated Geographic Encoding and Referencing (TIGER) system and small-area census data offer a great deal of potential for research in urban geography (*http://tiger.census.gov/cgi-bin/mapbrowse/*).

The Globalization Website: *http://www.emory.edu/SOC/globalization/*
   This Emory University website offers a wealth of information on globalization and includes a glossary of terms related to that topic.

2

THE ORIGINS
AND GROWTH
OF CITIES

A fascinating aspect of urban geography involves trying to understand the processes that led to the development and growth of individual cities and systems of cities. Since urban areas are the result of a long evolution, we need to develop a historical perspective. We can see history's powerful legacy in the surviving fragments that document the sequence of events that helped produce today's towns and cities. As Ildefons Cerdà, a nineteenth-century Spanish civil engineer and town planner, put it: "Our cities are like historical monuments to which every generation, every century, every civilization has contributed a stone." Since the evolution of the first cities about 5,500 years ago, changes in social, cultural, economic, political, technological, and environmental processes—including long-distance trade, overseas colonization, and industrialization—have helped fuel urban growth and change. These changes are visible in the internal structure and land use within cities and in the development of regional, continental, and, later, global urban systems, as well as world cities.

CHAPTER PREVIEW

This chapter follows the evolution of cities from their earliest origins about 5,500 years ago through the Industrial Revolution that began in the English Midlands in the mid-1700s. During this long span of urban development and redevelopment, changes in social, cultural, economic, political, technological, and environmental processes—including merchant capitalism, overseas colonization, and industrialization—helped drive urban growth and change. The impacts of these changes can be seen in the internal structure and patterns of land use within cities, and in the development of regional, continental, and later global urban systems, as well as world cities.

The term **urban system** refers to the complete set of urban settlements of different sizes that exists within a given territory. Territorial limits set the bounds of an urban system, though only a global scale can be justified as defining a true system in the sense that it captures all the functional relationships among cities. Conventionally, however, urban systems are analyzed and discussed at the scale of regions or countries.

With cities and urban life being such relatively recent features in the long span of human existence, we need to consider first the environmental, demographic, and other preconditions needed before cities could even begin to emerge. We then review the various theories of urban origins that together offer the reasons for why cities actually originated. The earliest towns and cities developed independently in regions of the world where the transition to agricultural food production had taken place. Five

regions provide the earliest evidence for urbanization and urban civilization: Mesopotamia, Egypt, the Indus Valley, northern China, and Mesoamerica. A look at the internal structure of these cities—street patterns, religious precincts, different neighborhoods, and so on—reveals a great deal about their evolution and the political, economic, and social changes that went on in them.

Urbanization spread out from the five regions of urban origin so that by about 1000 A.D. successive generations of city-based empires—including those of Greece, Rome, and Byzantium—had emerged in Southwest Asia, China, and parts of Europe. But urban expansion was a precarious and uneven process. For example, although urbanization continued in other parts of the world, the **Dark Ages** that followed the collapse of the Roman Empire in Western Europe was a time of stagnation and decline in economic and city life. Not until the eleventh century did the regional specializations and long-distance trading patterns emerge that provided the foundations for a new phase of urbanization based on merchant capitalism. Colonization and the expansion of trade around the world eventually allowed Europeans to shape the world's economies and urban societies. The Industrial Revolution later generated new kinds of cities—and many of them. Together, European colonization and the Industrial Revolution created unprecedented concentrations of people in cities that were connected in networks and hierarchies of interdependence around the world.

THE DEFINITION OF A CITY

Although most people recognize a city when they see one, no single definition can apply to all cities across space and throughout time. Wheatley's definition of **urbanism** captures the remarkable social and political changes surrounding the emergence of cities that resulted in a

> particular set of functionally integrated institutions which were first devised some five thousand years ago to mediate the transformation of relatively egalitarian, ascriptive, kin-structured groups into socially stratified, politically organized, territorially based societies.[1]

Sjoberg's definition highlights important physical and economic attributes that define a city:

> It is a community of substantial size and population density that shelters a variety of nonagricultural specialists, including a literate elite.[2]

V. Gordon Childe[3] attempted to characterize the distinctive features of cities with a list of distinguishing features of urban civilization (see the Box entitled "A Summary of Childe's Characteristics of Urban Civilization"). Childe used the term *urban civilization* because civilization and cities historically have gone hand in hand—the Latin word *civitas* (cities) is the word from which *civilization* is derived. From the beginning, cities have been crucibles of innovation that have produced some of the most incredible breakthroughs in human achievement. Ancient Sumer (in southern Iraq) is called the Cradle of Civilization because of the countless inventions in the earliest cities there. The legacies of these early urban innovations continued throughout subsequent civilizations and are seen in contemporary cities—in the use of writing, mathematics, the wheel, and recording time in multiples of 60.

BOX 2.1 A SUMMARY OF CHILDE'S CHARACTERISTICS OF URBAN CIVILIZATION

Harold Carter summarized V. Gordon Childe's list of distinguishing features of urban civilization as follows:[4]

- *Size.* Settlements were significantly larger in population size than anything that had existed previously.
- *Structure of the population.* Occupational specialization—with the transition from the old agricultural order—meant that the employment of full-time administrators and craftspersons was possible. Consequently, residence rather than kinship became the qualification for citizenship. Inevitably, the rule of the priest-kings, who guaranteed peace and order, involved social stratification.
- *Public capital.* The emergence of public capital allowed monumental public buildings to be erected and full-time artists to be supported.
- *Records and the exact sciences.* The need to keep records promoted the beginnings of a written script and mathematics, both of which became intimately bound up with urban civilization.
- *Trade.* By no means an urban innovation, establishing and maintaining a network of trade routes have become hallmarks of urbanization.

PRECONDITIONS FOR URBANIZATION

The preconditions necessary for cities to emerge came about with the transition from mobile food collection—hunting, gathering, and fishing—to sedentary food production based on agriculture. The increased volume and reliability of an agricultural food supply allowed higher numbers and densities of people to live permanently in one place. This population increase promoted the proliferation of agricultural villages.

The villages had to be located in regions where the environmental conditions—climate, water supply, topography, natural resources, and soil conditions—were favorable for agriculture. Early breakthroughs in technology and in farming practices—innovations in river and water management, crop and animal strains, and food transportation and storage techniques—were needed to support improvements in food production. Increasingly complex social organizational structures were required to handle the growing population and exchange of agricultural and other products among the village communities.

THEORIES OF URBAN ORIGINS

Although it would have been difficult for cities to emerge in the absence of certain basic environmental, demographic, social, and other preconditions, there are various explanations for why cities originated. Although no one theory provides a full account, each allows insights into the role of different factors in promoting early urbanization.

AGRICULTURAL SURPLUS

Archaeologists, including V. Gordon Childe[5] and Sir Leonard Woolley,[6] argued for the importance of an agricultural surplus. Once early farmers could each produce more food than was needed to feed their own families, they could support a growing sedentary population. The need to administer the agricultural surplus called for the more centralized structures of social organization found in cities. New, stratified social structures and institutions were needed to assign rights over resources, exact tributes and impose taxes, deal with the ownership of property, and administer the formal exchange of goods. Elite groups stimulated urban development because they used their wealth to build palaces, arenas, and monuments as displays of their power and status. This building construction demanded a greater degree of occupational specialization in nonagricultural activities, such as crafts, engineering, and administration, that could be organized effectively only in an urban setting.

This interpretation has been criticized as simplistic—an agricultural surplus alone might not have been enough to trigger all the societal and other changes necessary to produce cities. Some experts disagree on the cause-and-effect relationship and believe that fundamental changes in social organization would have been required before an agricultural surplus could be produced.

HYDROLOGICAL FACTORS

Karl August Wittfogel[7] pointed out that many early cities emerged in areas of agriculture that often depended on irrigation and the control of the regular spring floods. He contended that elaborate irrigation projects demanded new **divisions of labor**, cooperation on a large scale, and the intensification of cultivation. These demands would stimulate urban development by promoting occupational specialization, a centralized social organization, and population growth based on the production of an agricultural surplus.

Again, this interpretation has been criticized by those who believe that major changes in social organization would have been necessary, if not before, then at least in tandem with, the development of major irrigation projects. Other experts have questioned whether a complex social organizational structure was even necessary in order to undertake large-scale irrigation. Still others have pointed out that not all early cities, including some in Mesoamerica, depended on massive irrigation schemes.

POPULATION PRESSURES

Ester Boserup[8] believed that increasing population densities and/or a growing scarcity of wild food sources from hunting and gathering—that previously had provided adequate levels of subsistence from a relatively low workload—brought on the transition to agricultural food production and urban life. Again, the relationship is unclear in terms of whether food production and urban life caused—or were the result of—increased population densities. In certain cases demographic growth pressures may have disturbed the balance between population and resources and forced some people to move to areas with more marginal environmental conditions for agriculture. This scenario could have promoted early breakthroughs in agricultural technologies and practices or the establishment of nonagricultural activities, such as trade, defense, or religion, that would have supported the establishment of further urban settlements.

TRADING REQUIREMENTS

Some experts, observing how countless urban centers had evolved around marketplaces, have interpreted the emergence of cities primarily as a function of long-distance trade.[9] Participation in large-scale trading networks would call for a system to administer the formal exchange of goods that in turn would promote the development of centralized structures of social organization. Increasing occupational specialization and economic competitiveness would encourage more urban development. What remains unclear is the extent to which trade was the cause or the consequence of urban development.

DEFENSE NEEDS

Some theorists, including Max Weber,[10] have contended that cities originated because of the need for people to gather together for protection inside the safety of military defenses. Wittfogel[11] pointed out that a comprehensive system of defense was needed to protect valuable irrigation systems from attack. But despite widespread evidence of walls and other fortifications, not all early cities had defenses. As Wheatley acknowledged, although not necessarily being the initial reason for the evolution of cities, "warfare may often have made a significant contribution to the *intensification* of urban development by inducing a concentration of settlement for purposes of defense and by stimulating craft specialization."[12]

RELIGIOUS CAUSES

The presence of temples and other religious structures reflects the importance of religion in the lives of the people living in the earliest cities. Sjoberg[13] suggested that the control of altar offerings by the religious elite conferred economic and political power that allowed this group to influence the social changes that helped initiate urban development. Wheatley[14] maintained that a pervasive institutional structure like religion would have been needed to reinforce the changes in social organization that were associated with the economic, technological, and military transformations involved in early urban growth.

A MORE COMPREHENSIVE EXPLANATION

More recently, the consensus has been that our understanding of the origin of cities should be based on a combination of these separate yet interrelated explanatory factors. As Wheatley put it:

It is doubtful if a single autonomous, causative factor will ever be identified in the nexus of social, economic, and political transformations which resulted in the emergence of urban forms.[15]

Understanding both the complexity of the many changing processes and the interactions among them are more important than identifying the cause-and-effect relationship for any one explanatory factor. The desire for this kind of comprehensive understanding reflects a growing conceptualization that the origin of cities represented a gradual transformation involving incremental change over time rather than an abrupt urban revolution.

URBAN ORIGINS

The earliest towns and cities developed independently in regions of the world where the transition to agricultural food production had taken place. Five regions provide the earliest evidence for urbanization and urban civilization (Figure 2.1). Over time, the regions of urban origin produced successive generations of urbanized world-empires.

MESOPOTAMIA

Mesopotamia (the land between the Tigris and Euphrates rivers), in the area of modern Iraq, provides the earliest evidence for urbanization—from about 3500 B.C. This was the eastern part of the so-called **Fertile Crescent** (Figure 2.1). The significant growth in size of some of the agricultural villages located on the rich alluvial soils of the river floodplains in the region formed the basis for the large, relatively autonomous, and often-rival **city-states** of the Sumerian Empire from about 3000 B.C. They included Ur, in southern Iraq, the capital from about 2300 to 2180 B.C., as well as Eridu, Uruk, and Erbil (ancient Arbela). These fortified city-states contained tens of thousands of inhabitants, social stratification, with religious, political, and military classes, innovative technologies, including massive irrigation projects, and extensive trade connections. By 1885 B.C., the

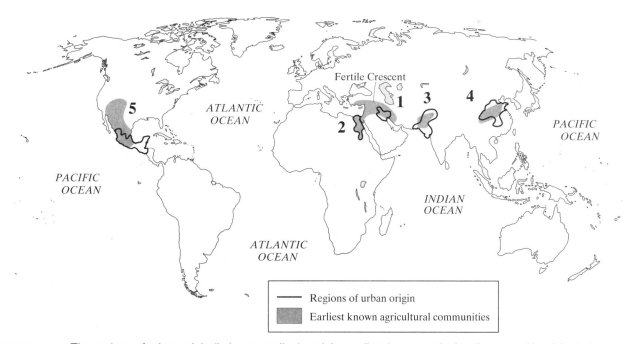

FIGURE 2.1   The regions of urban origin (in heavy outline) and the earliest known agricultural communities (shaded areas): 1) Mesopotamia, 2) Egypt, 3) Indus Valley, 4) Northern China, and 5) Mesoamerica.

Sumerian city-states had been taken over by the Babylonians and then the Neo-Babylonians, who governed the region from their capital city, Babylon.

EGYPT

The Fertile Crescent stretched in an arc as far west as Egypt, which became a unified state from about 3100 B.C. (Figure 2.1). Large irrigation projects controlled the Nile's waters for agricultural and other uses. Despite the importance of early urbanization in the Nile Valley, only limited archaeological evidence of the first towns and cities has survived.[16] Internal peace in Egypt meant that there was no need to occupy the same site continuously to justify massive investments in a city's defensive fortifications. Given this potential for urban mobility, the lifespan of the largest city, the capital, was relatively short. Each pharaoh was free to locate a new capital at any site he selected for his tomb. After his death the city was usually abandoned to the priests. The main surviving structures are the stone tombs and temples that were the primary focus of construction. Few of the other buildings—public, commercial, and residential—survive, having been constructed of more perishable materials, such as sun-dried brick and timber. Without a tradition of long-term building and rebuilding, Egyptian cities did not generate the rich evidence of urban development and redevelopment characteristic of the tells of Mesopotamia (see Figure 2.2). Nevertheless, between 2000 and 1400 B.C., urbanization continued with the founding of capital cities like Thebes, Akhetaten (Tell el-Amarna), and Tanis.

THE INDUS VALLEY

As in Mesopotamia, but later, by about 2500 B.C., the Indus Valley in modern Pakistan contained relatively large urban communities that were supported on the fertile alluvial soils and extensive irrigation systems of the river plains. This region had a single ruler and two capital cities: Harappa in the north, after which the civilization was named, and Mohenjo-Daro in the south (Figure 2.1). A network of trade extended as far as the Sumerian Empire in Mesopotamia. Much about Harappan civilization, including its origin, is unknown, partly because the Indus Valley script has not been deciphered.

NORTHERN CHINA

The Shang dynasty developed in the fertile plains of the Huang He (or Huang Ho, Yellow River) by about

FIGURE 2.2   Erbil (ancient Arbela) in northeast Iraq is located atop a *tell*, a mound, visible as a hill rising high above the surrounding plain—representing the remains of generations of sun-dried mud-brick buildings—that reflects millennia of human occupation in one place involving new structures being built upon the collapsed rubble of demolished ones. The 100-foot high Erbil tell is believed to represent perhaps 6,000 years of continuous occupation.

1800 B.C. (Figure 2.1). As in the Fertile Crescent and Indus Valley, Shang cities, such as Chengchow (Zhengzhou) and An'yang, the capital city by 1384 B.C., were supported by irrigated agriculture. There is evidence of social stratification and occupational specialization, including a hereditary leader as well as a warrior elite with absolute control over the agricultural peasants.

MESOAMERICA

The earliest urban settlements in Mesoamerica date to only about 500 B.C. (Figure 2.1). The Zapotec civilization was based on small-scale irrigated farming and was centered on Monte Albán in Mexico. This city was surrounded by a wall and contained pyramids and temples. The later city of Teotihuacán, near modern Mexico City, was larger and at its height—between about 300 and 700 A.D.—contained about 200,000 people.

Mayan cities such as Tikal and Uaxactun date from about 100 B.C. The Mayan cities were located in lowland areas of modern Mexico, including the Yucatan peninsula, as well as in Guatemala and Belize. Some reached populations as large as 50,000. These settlements were the centers of small states that united occasionally into loose confederations.[17] A supreme ruler and an urban elite, including religious and military groups, administered the settlements. Some of the breakthroughs in technology and farming practices in the other regions of urban origin are not found in Mesoamerica. The agriculture here was based on maize (corn) cultivation that did not require metal agricultural implements, domesticated animals to pull plows, or extensive irrigation systems. Mayan civilization reached its peak between about 300 and 900 A.D.; by the time of the Spanish conquest in the sixteenth century, it had been in decline for several centuries because of droughts, warfare, and population pressures.

INTERNAL STRUCTURE OF THE EARLIEST CITIES

A common approach to examining the internal structure of cities is to identify whether the layout of an urban area was largely unplanned or planned. This categorization distinguishes between cities that evolved in an unplanned—**organic growth**—process and those that were laid out in a predetermined way based on some planned approach, such as a **gridiron** street pattern. The planned layout of streets and transportation arteries may indicate the presence of significant central control from an early stage. But an unplanned street layout does not mean the absence of a central authority. In Mesopotamia, for example, the street patterns were not planned but reflected the organic street layouts that survived as the early agricultural villages grew into towns; the archaeological evidence of massive walls and irrigation systems, however, indicates central planning of defense and water management (see Box 2.2 entitled "Internal Structure of the Earliest Cities").

A city's internal structure is never static, being the product of development and redevelopment over time. Cities that were founded with a strong hand of planning can contain later sections of organic growth, as in some cities of Roman origin, such as London, where later unplanned urban growth during the medieval period obliterated the gridiron street pattern of the former Roman core. Similarly, cities that evolved with an organic growth pattern can have later planned sections, as in Vienna where the planned redevelopment of the area of the former wall in the late nineteenth century contrasts with the organic growth of the earlier medieval street pattern.

BOX 2.2   INTERNAL STRUCTURE OF THE EARLIEST CITIES

MESOPOTAMIA

Sir Leonard Woolley's excavations at Ur in the 1920s and early 1930s revealed the organic growth pattern characteristic of the Mesopotamian city-states (Figure 2.3). Woolley described Ur in about 1700 B.C.:

- *The walled city* contained about 35,000 people. The mud-brick wall was about 25 feet high and at least 77 feet thick. It was oval shaped and about three-quarters of a mile long by half a mile wide. The Euphrates River ran along the wall in the west, a navigable canal along the north and east, and harbors were in the north and west. Near the center were the palaces and residences of the royal officials. Social differentiation was less orderly in other sections of the city. Woolley's description of an excavated middle-income neighborhood illustrates how income and climate were reflected in the size and design of the housing, while the unplanned nature of the streets provided a measure of privacy and defense:

  > The unpaved streets are narrow and winding, sometimes mere blind alleys leading to houses hidden away in the middle of a great block of haphazard buildings; large houses and small are jumbled together, a few of them flat-roofed tenements one storey high, most of them of two storeys . . . The basic plan was of a house built round a central courtyard . . . that gave light and air to the house.[18]

- *A religious area* measuring about 270 by 190 yards, surrounded by a huge mud-brick wall, was located in the northwest of the walled city. A 68-foot high terraced tower (ziggurat) containing the shrine of Nannar, the Moon god, the owner of the city-state, would have been visible for miles. This religious and administrative core—reserved for the priests and royal

(Continued)

Box 2.2   Internal Structure of the Earliest Cities   (*Continued*)

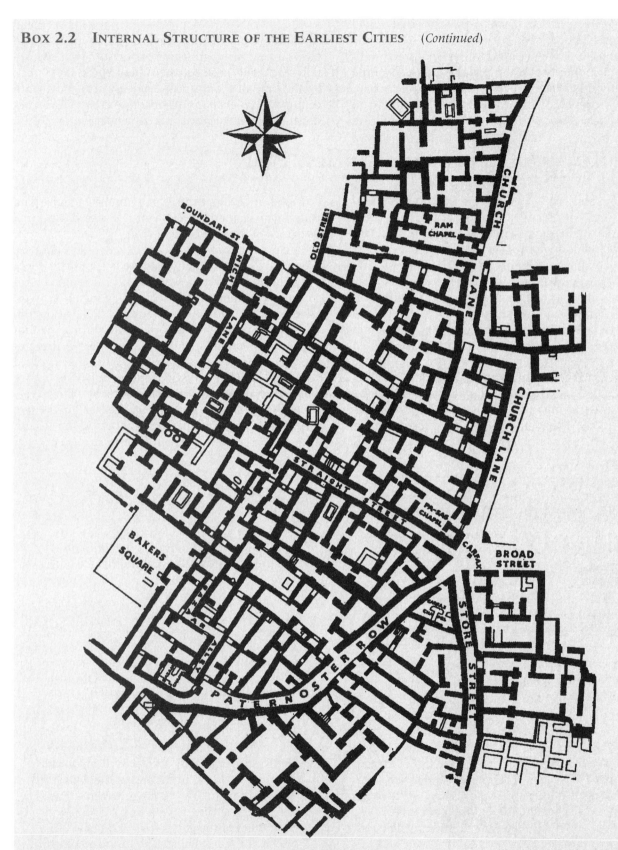

FIGURE 2.3   A residential section of Ur to the southeast of the religious area in about 1700 B.C. Note how the narrow and winding streets and the irregular size and shape of the lots reflect an unplanned—organic growth—process of urban development in Mesopotamia.

(Continued)

Box 2.2 Internal Structure of the Earliest Cities (*Continued*)

household—had a great courtyard surrounded by temples, a court of law, tax offices, and storage buildings for religious offerings.

- *The outer city* or *suburbs* comprised the remainder of the city-state. The houses and farms there contained an estimated 200,000 people.

EGYPT

Towns, and even capital cities, in Egypt remained small because most people were agricultural workers who lived on the land. The cities contained the markets and the residents who worked in retail, craft, government, and religious activities. With civil war and armed invasions rare, urban settlements in Egypt had little need for defensive walls or fortifications. The capital was the home of the pharaoh and the royal court. Akhetaten (modern Tell el-Amarna), built about 1400 B.C., is representative of capital cities in Egypt. This city ran for about 5 miles along the east bank of the Nile and was about a half mile to a mile wide. Social stratification is reflected in the city's layout:

- A walled temple and palace at the center, with other temples, government offices, military barracks, and storage buildings nearby.
- Areas of slum housing throughout the city.
- Northern and southern suburbs.
- A workmen's village based on a gridiron plan to the east of the city.

The roughly rectangular city blocks were laid out when the city was founded, with infilling left to the residents. The wealthiest residents took the best lots that fronted onto the two main streets that ran parallel to the river. The typical middle-income house was built in the center of a walled enclosure and had a porch and central living room.

THE INDUS VALLEY

Although not the first to use a gridiron plan, the Harappan cities likely were the first planned towns because they are believed to be the first system of cities that used the same town planning approach (Figure 2.4). Despite being located hundreds of miles apart, cities like Harappa and Mohenjo-Daro shared similarities in their basic internal structure. Each covered at least a square mile in area and contained about 35,000 people. An imposing walled citadel built on a mud-brick platform was located to the west of each city. In contrast to the religious area of Mesopotamian cities, this citadel did not contain a dominant religious building like a ziggurat. The citadel contained some structures that might have had a ritual use (such as the bath at Mohenjo-Daro), as well as buildings used for administrative purposes (including offices and grain storage buildings). The citadel's western location would have allowed urban residents to see the rooftop civic and ceremonial gatherings silhouetted dramatically against the backdrop of the setting sun.[19] The main east-west streets in the gridiron plan led to the citadel. The houses varied from small one-story one-roomed buildings to larger two-story houses with central courtyards. The workmen's quarters contained rows of identical two-roomed houses.

NORTHERN CHINA

The archaeological evidence is relatively limited, but excavations at the capital city of Anyang have revealed thick walls of beaten earth surrounding the city. At the center was a walled palace. The nearby houses of the wealthy were wooden and constructed on raised platforms of beaten earth. Poorer residents lived in pit-dwellings. The city's layout was probably planned because all the buildings that were excavated were oriented toward the north.

MESOAMERICA

The most imposing buildings in Mayan cities were the religious and other ceremonial structures, such as the tall pyramids that were planned around broad plazas or courtyards. The temples and the palaces and residences of the ruler and the military and religious elite were laid out at the center of the city. Nearby were the homes of the wealthy citizens. Further out were low-density areas of simple thatched wooden huts that housed the majority of the people who worked in agriculture or crafts.

(*Continued*)

BOX 2.2 INTERNAL STRUCTURE OF THE EARLIEST CITIES   *(Continued)*

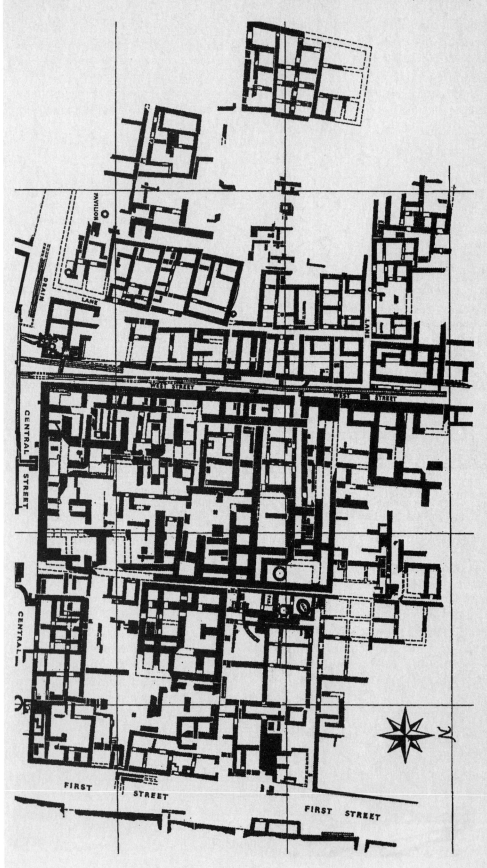

FIGURE 2.4   Part of the town plan of Mohenjo-Daro. Note how the use of a gridiron plan in Mohenjo-Daro (as in Harappa and other Indus Valley cities) reflects a planned approach in which fairly wide and straight main streets intercept cross streets at right angles to form large city blocks that contained a number of houses.

URBAN EXPANSION FROM THE REGIONS OF URBAN ORIGIN

The spread of urbanization outward from the regions of urban origin involved **uneven development**—over time and within and between different parts of the world (Figure 2.5).[20] There was a progression of growth, expansion, and succession of early urban empires that was gradual and incremental. The spread of urbanization was a precarious process in which many civilizations lapsed into ruralism before being revived or recolonized.

The Persians helped spread cities from Mesopotamia to Central Asia. To the north, the Assyrians established a system of cities that extended west from their capital city, Assur, to Syria and Asia Minor (the Asian part of Turkey, also called Anatolia). By about 1700 B.C., other groups, including the Hittites, had displaced the Assyrians and established their own cities. Further west, by 1600 B.C., Mycenaeans had founded urban settlements, including the legendary cities of Troy in Asia Minor and Mycenae on mainland Greece.

The small "Canaanite" city-states that grew up in what is now Israel and Syria by 2000 B.C., such as Tyre, Beirut, Jericho, Gaza, and Damascus, were taken over by the Egyptians and Hittites. After 1200 B.C., with the collapse of Egyptian and Hittite control, the Israelites founded small urban centers that grew to become large cities such as Jerusalem. To the west, the Phoenicians helped spread urbanization by sea as far as Spain.

Further east in what became modern India and Pakistan, it took a thousand years for urban life—dislodged by the Aryan invaders in about 1750 B.C.—to reestablish itself. Hindu cities were founded, and from the end of the fourth century B.C., the Maurya Empire built cities across India and laid the foundations for urban life throughout southeast Asia. The Arab invasions of the eighth century A.D. began the period of Muslim rule in India; cities like Lahore were established, and the city of Delhi became an important administrative and cultural center.

In China the Chou dynasty succeeded the Shang dynasty in 1122 B.C. and over the next nine centuries spread urbanization from the Huang Ho region to the east and south of the Yangtze River. Between the third century B.C. and the third century A.D., the Ch'in and later Han dynasties helped spread urbanization throughout East Asia, including along the Silk Road (see the Box entitled "The Silk Road: Long-Distance Trade and Urban Expansion"). The next major period of urbanization in China came at the time of the Mongol invasions with the establishment of cities as part of a new Mongol empire. Urbanization came later to Korea and Japan through Chinese influence.

In Japan urbanization began with the founding of Osaka in 400 A.D., followed by a succession of royal capitals in the seventh and eighth centuries. This culminated by the ninth century with Kyoto, whose status as capital for almost a millennium led to its growth to an unprecedented size for cities of the time.

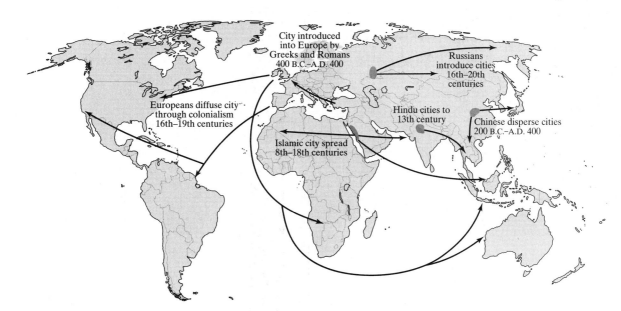

FIGURE 2.5   Urban expansion in conjunction with the expansion of selected empires.

Following a period of decline, there was a resurgence of urbanization in the late fourteenth century with the establishment of "castle towns," including some, like Edo (Tokyo), that would become huge urban centers.

In Mesoamerica, the Maya and neighboring groups, including the Zapotecs and, later, Aztecs to the north and west, as well as the Incas further south, continued to build cities. As we will see later in this chapter (and in Chapter 7), aggressive European colonization, beginning with the Spanish conquest in the sixteenth century, brought drastic changes to these urban civilizations.

Urban growth and expansion outward from the regions of urban origin were fueled by critical innovations, especially in technology and economic organization,

allied with changes in social organization. Demographic changes—including setbacks from epidemics and wars—were a factor. Adequate numbers of workers were needed to maintain the social and economic infrastructures that supported urbanization. This proved critical in the Indus Valley in about 1750 B.C., where population decline allowed the Aryan invaders to bring the Harappan urban civilization to an abrupt end.

Changes in the balance between a population and its resource base could help fuel urban growth or promote decline. Ultimately, self-propelled urban growth is limited by the size of a society's resource base. The maintenance of irrigation systems, for example, combined with the necessity for increasing productivity

---

BOX 2.3   THE SILK ROAD: LONG-DISTANCE TRADE AND URBAN EXPANSION

The Silk Road evolved out of trade routes that developed initially along the Fertile Crescent in Mesopotamia in the west and later in the area of the Ch'in dynasty in the east. The Silk Road is an excellent example of how long-distance trade fueled urban expansion and produced an extensive system of cities along a route that eventually connected China with Europe between about 500 B.C. and 1500 A.D. (Figure 2.6). The overland portion started in the east from the ancient Chinese capital of Changan (modern Xi'an), a bustling center of trade and industry. This collection and distribution node served the surrounding region, from which goods were assembled and some processed before being exported westward along the Silk Road. Leaving Changan, the route followed the Great Wall of China until it forked around the Takla Makan Desert before crossing the Pamir Mountains and continuing on through Afghanistan and Iran to reach the Mediterranean, where the goods continued on to Europe by ship.

The caravans could be enormous, with as many as 1,000 camels carrying up to 500 pounds of goods each. In addition to silk, which initially appeared so extraordinary to Europeans, green and white jade, blue lapis lazuli, ceramics, gunpowder, iron, spices, fruit, and flowers went west toward Europe, while gold, silver, amber, ivory, cotton, and wool headed east. Perhaps as important as trade was the exchange of ideas about new technologies, scientific skills, lan-

guage, art, and religion—not least, writing, printing, and papermaking—along these routes.

The need for safe overnight campsites for the caravans and their valuable cargos led to the growth of cities that offered security and trade opportunities along the route. Caravan cities that grew to considerable size, like Samarkand, now in eastern Uzbekistan, and Kashgar (Kashi) at the foot of the Pamir Mountains in extreme western China, were heavily fortified. Regular markets were held in these cities, especially near the city gates serving the Silk Road.

Flourishing trade along the route also generated rich profits that helped support entire empires and urban civilizations. By about 100 B.C., the Roman Empire in the west, the Han dynasty in the east, and the Parthian Empire in Persia (modern Iran) in between were all benefiting from the commercial activity that crisscrossed the Silk Road.

With Europe's emergence from the Dark Ages, steady population growth and limited amounts of usable land helped trigger the transition from feudalism to merchant capitalism. With time, the Europeans' growing technological superiority on the seas allowed them to increasingly dominate world trade. European naval discoveries opened up new trade opportunities, including a sea route to India around Africa that bypassed southwest Asia and ultimately contributed to the demise of ancient trade networks like the Silk Road.

(Continued)

BOX 2.3   THE SILK ROAD: LONG-DISTANCE TRADE AND URBAN EXPANSION   *(Continued)*

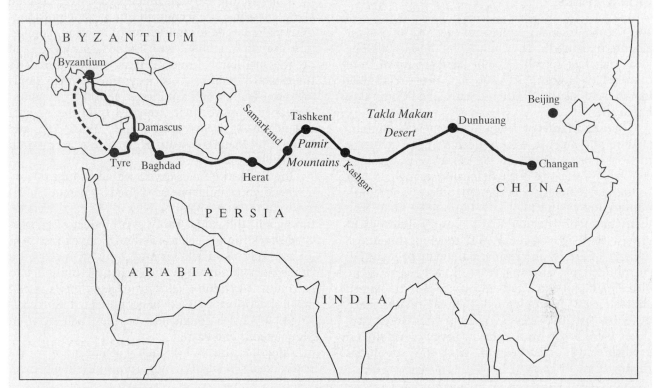

FIGURE 2.6   The Silk Road, an ancient trade network of caravan routes that extended across Central Asia from as early as 500 B.C. until about 1500 A.D., is an excellent example of how long-distance trade fueled urban expansion and produced a extensive system of cities that connected the four major regions containing the great empires at the time: Europe, Southwest Asia, India, and China.

The political and military situation in countries like Iraq, Iran, and Afghanistan would make any attempt at crossing Central Asia along the 5,000 miles of the Silk Road an incredibly difficult, if not impossible, journey today. Besides, most of the Silk Road is gone. Only some surviving remnants attest to the importance of this ancient trade network of caravan routes that were studded with towns and cities like strings of pearls extending across the inhospitable deserts and mountain ranges of Central Asia.

to sustain a growing population, could put incredible pressures on the agricultural workforce. After a while, investments could be neglected, armies reduced in size, and an empire's strength and cohesion fatally undermined. This kind of sequence may have been behind the eventual collapse of the Sumerian Empire and may have contributed to the abandonment of many Mayan cities hundreds of years before the Spanish conquest.

One response to a limited resource base—territorial expansion and colonization—often reinforced and extended the urbanization process. The need for increasing numbers of control centers and improved transportation networks to support colonization and long-distance trade tended to produce a hierarchical urban system. In this way the expansion of the Greek and Roman empires laid the foundations for an urban system in Western Europe.

THE ROOTS OF EUROPEAN URBAN EXPANSION

GREEK CITIES

The Greeks originally came into the Aegean Sea region from the north. They built upon the city-building ideas that had spread into the Mediterranean from the Fertile Crescent. By 800 B.C. the Greeks had founded cities such as Athens, Sparta, and Corinth, to name a few.

The importance of religion, commerce, administration, and defense were reflected in the layout of the Greek city-states. At the center was the high city—the "acropolis"—the defensive stronghold that contained temples, government offices, and storehouses. Below the high city ("sub-urbs" in Latin), were the "agora" for the markets and political gatherings, more government and religious buildings, military quarters, and residential neighborhoods, all surrounded by a defensive wall. Athens and the older mainland cities had an organic growth pattern characteristic of unplanned Mesopotamian urban development. The street systems of later Greek city-states were based on a gridiron pattern, usually on a north-south axis, regardless of site conditions (Figure 2.7).

Many of the earliest Greek cities were located along coastlines, reflecting the importance of long-distance sea trade for this urban civilization. Population growth combined with limited cultivable land on the mainland drove overseas colonization and the establishment of a Greek system of cities. Bands of colonists and their families established new independent city-states that stretched from the Aegean Sea to the Black Sea, around the Adriatic Sea, and along the Mediterranean as far west as modern Spain (Figure 2.8).

The Greek city-states developed new forms of government whose influence is reflected in subsequent democratic and participatory modes of urban governance throughout the world. Partly as a product of enlightened Greek culture, political authority came to reside in an assembly of—albeit male—citizens who elected a city leadership. Although Greek civic life continued to be conducted within a religious context, the laws and political decisions were no longer presented as unchallengeable divine commands, as they had been in Mesopotamia and Egypt.[21]

After the Greeks lost their independence to the Macedonians in 338 B.C., overseas colonization became

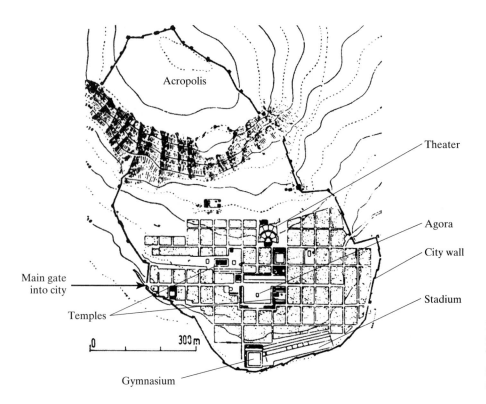

FIGURE 2.7   General plan of a typical Greek city-state—Priene (in Turkey). Note how the use of the gridiron plan reflects a planned approach.

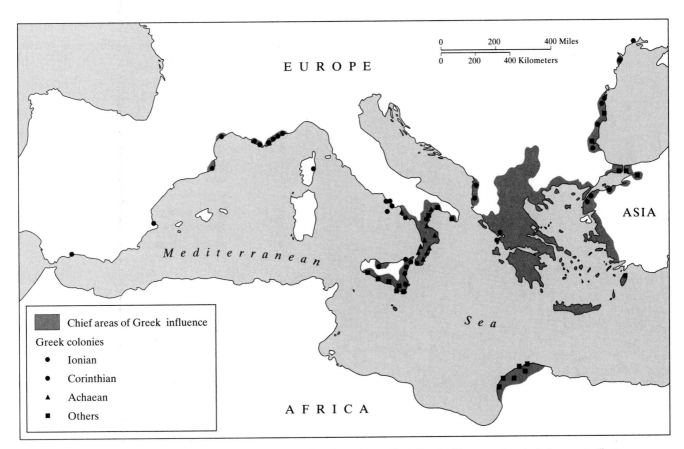

FIGURE 2.8   Greek city-states in the Mediterranean. Note how the earliest Greek cities were located along coastlines, reflecting the importance of long-distance sea trade for this urban civilization.

more centrally organized and based on a "mass production of cities" that extended eastward toward Central Asia:

> Alexander the Great and his Successors founded on geopolitically important sites and built up as strategic strongholds, a whole network of towns and cities, which evolved into centres spreading Greek culture and civilization, over the greatest part of the then known world. It was one of the greatest 'colonial' town planning and building actions which was ever accomplished in history.[22]

But Greek cities remained quite small by today's standards. Although Athens probably reached a population of about 150,000, most "large" cities ranged from 10,000 to 15,000 inhabitants, while most urban settlements had only a few thousand people.

ROMAN CITIES

The expanding Roman Empire displaced Greek civilization during the second and first centuries B.C. By the second century A.D., the Romans had established towns across southern Europe and laid the foundation for the Western European urban system (Figure 2.9).

Roman cities were similar in certain respects to some of their Greek predecessors (Figure 2.10). They were based on the grid system, contained a central "forum" for markets and political gatherings, were encircled by a defensive wall, were deliberately established in newly colonized territories, were part of an extensive system of long-distance trade, and remained fairly small. Although the population of Rome likely reached the one million mark by 100 A.D., large Roman towns usually contained only about 15,000 to 30,000 inhabitants, while most places had no more than 2,000 to 5,000 people.

An important difference between Greek and Roman cities was that Roman cities were not independent. They functioned within a well-organized empire centered on Rome and were designed along hierarchical lines, reflecting the Roman rigid class system. Another difference was the greater concentration of Roman cities in inland locations, reflecting their predominant function as control centers. Many modern European cities can trace their origins to the Roman period, including London, Brussels, Paris, Cologne, Vienna, Sofia, and Belgrade.

The Romans achieved impressive feats of civil engineering. The most important towns were directly

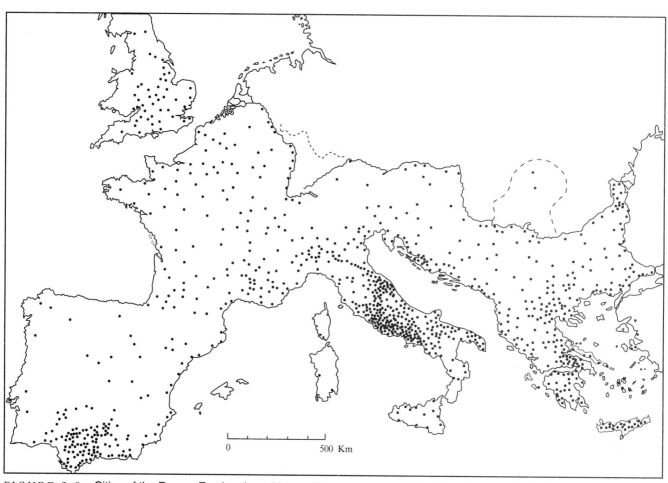

FIGURE 2.9 Cities of the Roman Empire about 200 A.D. The Romans established a well-integrated urban system and transportation network that laid the foundation for the Western European urban system.

connected to one another and to Rome by a magnificent system of roads that facilitated strategic military and trading communications. The underground sewer and surface water supply infrastructures in Roman cities contributed to impressive health improvements that set the standard for later cities. In Rome public latrines served the majority of ordinary people. The system of elevated aqueducts and freshwater reservoirs that brought water for drinking and bathing into the city ran for over 300 miles and carried 60 million cubic feet of water per day.[23]

The Romans used cities as a mechanism to impose and maintain their authority and legal system throughout their vast empire. They understood that any attempts to hold newly acquired territories by military force could invite guerrilla warfare that would distract the army from its task of maintaining and extending the empire's borders; it could also hurt the development of commerce. Native tribes, therefore, had to be brought into the empire on advantageous terms—by equating Romanization with urbanization. Tribal centers were redeveloped as Roman towns of varying political status. Other towns were also established for

economic and political reasons and populated by ex-soldiers and emigrant settlers from Rome and other older towns.

Morris described three main classes of imperial towns: *coloniae*—either newly founded settlements or native towns, allied to Rome with full Roman status and privileges; *municipia*—usually important tribal centers, taken over with formal chartered status but only partial Roman citizenship for their inhabitants; and *civitates*—market and administrative centers for tribal districts which were retained in a Romanized form.[24]

At the time of the fall of Rome in the fifth century A.D., the Romans had established a well-integrated urban system and transportation network that stretched from England in the northwest to as far as Babylon in the east. By as early as the second century A.D., however, the empire's population had already begun to decline, causing labor shortages, abandoned fields, and depopulated towns and facilitating the incursions of "barbarian" settlers and tribes from the Germanic lands of east-central Europe that helped topple the empire.

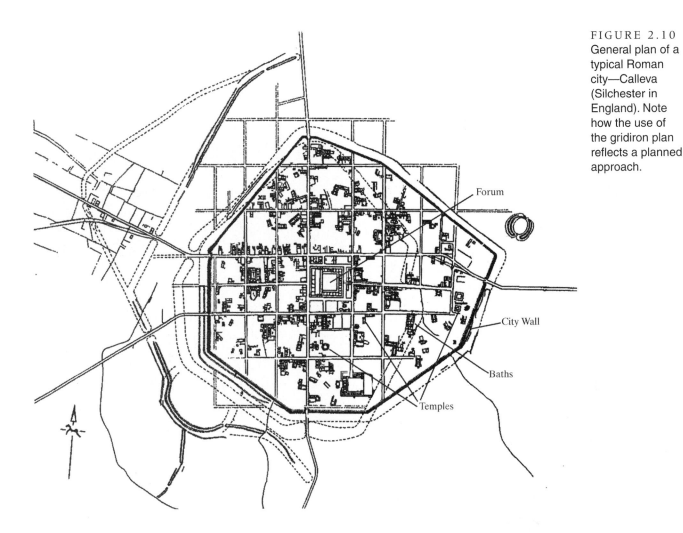

FIGURE 2.10
General plan of a typical Roman city—Calleva (Silchester in England). Note how the use of the gridiron plan reflects a planned approach.

Forum

City Wall

Baths

Temples

DARK AGES

The Dark Ages in Western Europe was a period of stagnation and decline in city life after the collapse of the Roman Empire until about 1000 A.D. Of course, urban life continued to flourish in other parts of the world, including the city building associated with the "explosion of Islam"[25] beginning in the latter half of the 7th century A.D. In the following centuries, existing towns like Mecca, Medina, Baghdad, and Damascus experienced a dramatic rebirth, while new cities were founded, including Teheran in Iran, Basra, Mosul, and Karbala in Iraq, and Cairo and Tangiers in North Africa, and further south in sub-Saharan Africa, cities in the west, such as Timbuctoo (Timbuktu) in modern Mali and Kano in Nigeria, as well as cities in the east, such as Mombasa in Kenya.

Urban life flourished in parts of Europe where long-distance trade continued, as in those cities under Muslim influence, including Cordova, Granada, and Seville in Spain, or in cities under Byzantine control, most notably Byzantium (Constantinople, later Istanbul) in modern Turkey. By the end of the fourth century, as Rome was falling into decline, Constantine had moved the capital of the Roman Empire to Constantinople. As the capital of the Byzantine Empire between about 360–650 A.D., with its strategic location for trade between Europe and Asia, the city grew to become the largest in the world at the time, with about half a million people.

In the rest of Europe the Germanic invaders and, later, Viking raiders from the north took advantage of the vacuum created by the Roman Empire's collapse. This politically unstable situation made long-distance trade virtually impossible, cutting off the lifeblood of cities, and creating isolated, crumbling, and depopulated urban centers. Most of the urban places that survived were ecclesiastical or university centers, defensive strongholds, or administrative hubs.

- *Ecclesiastical or university centers:* Some Roman towns continued to be occupied because of the church practice of designating certain urban centers as bishoprics—the seats of bishops—and building a cathedral within the old Roman town's walls. Other towns survived because of their importance as educational and subsequently university centers. Examples include

St. Andrews in Scotland; Canterbury, Cambridge, and Coventry in England; Rheims and Chartres in France; Liège in Belgium; Bremen in Germany; Trondheim in Norway; and Lund in Sweden (Figure 2.11).

- *Defensive strongholds:* The constant threat of attack spurred the construction of castles and other fortifications in cities in general and in some parts of Eastern Europe that had previously been underrepresented by urban development. Examples include the hilltop towns of central Italy such as Foligno, San Gimignano, and Urbino (Figure 2.12).

- *Administrative hubs:* Administrative centers for the upper tiers of the feudal hierarchy included Cologne, Mainz, and Magdeburg in Germany; Winchester in England; and Toulouse in France (Figure 2.13).

Feudalism curtailed the development of European cities because its highly structured and self-contained nature favored the self-sufficient country manor as the basic building block of settlement. Feudalism was a rigid, mostly rural form of economic and social organization based on the communal chiefdoms of the Germanic tribes who had invaded the disintegrating Roman Empire. Each feudal estate was more or less self-sufficient in the provision of food, and each

FIGURE 2.11   St. Andrews, Scotland, an important ecclesiastic center. The cathedral was built in the twelfth century, the castle (an episcopal residence), about 1200. The university was founded in 1410.

FIGURE 2.12   San Gimignano, Italy, on a classic hilltop defense site, was originally settled by the Etruscans in the third century B.C. The town was named for Gimignano, the bishop of Modena and future saint, who saved the town from barbarian pillaging in the tenth century. It grew prosperous because of the nearby Francigena Way, a busy trade and pilgrim route, and became a city-republic in 1199. Rival families built the towers during the eleventh and thirteenth centuries as symbols of their wealth and status—the higher and more imposing the better. Fifteen of the original 72 medieval towers survive.

FIGURE 2.13   Cologne, Germany, a medieval administrative hub. When this woodcut was made in the late 1400s, Cologne had a population of less than 25,000 but was already an important administrative, commercial, and manufacturing center, with an important cathedral and a university that was already more than 100 years old.

kingdom or principality was more or less self-sufficient in the provision of raw materials needed to craft simple products. Yet in spite of this unlikely beginning, an elaborate urban system evolved whose largest centers eventually grew into what would become the **nodal centers** of a global world-system.

URBAN REVIVAL IN EUROPE
DURING THE MEDIEVAL PERIOD

Beginning in the eleventh century, the feudal system weakened and began to collapse in the face of successive demographic, economic, and political crises. The fundamental cause of these crises was steady population growth in conjunction with only modest technological improvements and limited amounts of cultivable land. To bolster their incomes and raise armies against one another, the feudal nobility began to levy increasingly higher taxes. As a result, the peasants, most of whom were serfs (descended from slaves and not free) or tenants (whose freedom to move, marry, leave property to their heirs, buy goods, or sell their labor was closely circumscribed by public law), were forced to sell more of their produce for cash on the market. This fostered a more extensive money economy and the beginnings of a pattern of trade in basic agricultural produce and craft manufactures. Some long-distance trade even began in luxury goods, such as spices, furs, silks, fruit, and wine. This trade caused towns to begin to grow in size and vitality.

Medieval towns can be classified into five categories on the basis of their origin.[26]

- *Towns of Roman origin* survived during the Dark Ages or were reestablished after being deserted. Examples include London, Chichester, and York in England and Regensburg in Germany (Figure 2.14).
- *Burgs* were fortified military bases that evolved into towns as they acquired commercial functions. Examples include Oxford, Nottingham, and Wallingford in England and Magdeburg in Germany (Figure 2.15).
- *Towns that evolved from village settlements through organic growth.* Examples in England include Wycombe in Buckinghamshire and Wickham in Hampshire (Figure 2.16).
- *Bastides* were planned new towns in France, England, and Wales. The initial motive for establishing these towns was usually strategic, but their sponsors also saw them as an investment that could yield income. Bastide towns were typically laid out around a castle and protected by heavily fortified walls. The towns provided essential services for their military garrisons and helped stabilize the surrounding countryside, but they also provided a source of income for their sponsors through market tolls, rents, and court fines. In this regard, bastide towns were part of a deliberate and fairly widespread policy of town plantation that coincided with an unprecedented boom in urban growth in the twelfth and thirteenth

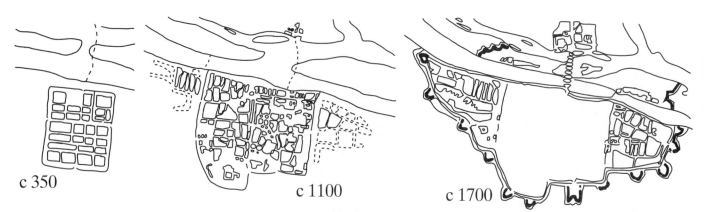

c 350   c 1100   c 1700

FIGURE 2.14   Regensburg, Germany, a medieval town of Roman origin on the southern bank of the Danube at three different stages in the development of the city. The original Roman gridiron street pattern is clearly evident in 350 A.D. Although not totally deserted during the Dark Ages, the layout of the city in 1100 A.D. shows little sign of the original gridiron. By 1700, the city's fortifications are a strong determinant of urban form.

centuries as increased trade across northern Europe marked the transition from feudalism to merchant capitalism. As planned towns, most were laid out with a gridiron street plan. The main inducement to attract settlers was the grant of a house plot within the town and farming land nearby. Examples include Ludlow and Kingston upon Hull in England, Caernarvon in Wales, and Aigues-Mortes and Carcassone in southwestern France (Figure 2.17).

- *Planted towns* included other planned new towns throughout Europe, with or without a predetermined layout. Most were founded on a roadside or riverside location for commercial purposes to take advantage of the general reestablishment of long-distance trade. Examples include Offenburg and Freiburg im Breisgau in Germany and Berne in Switzerland (Figure 2.18).

All medieval towns shared common features of internal structure. At the center was an open square for markets, surrounded in larger cities by the cathedral or church, the town hall, the guildhalls, the palaces, and the houses of prominent citizens. Close to the center were streets or districts that specialized

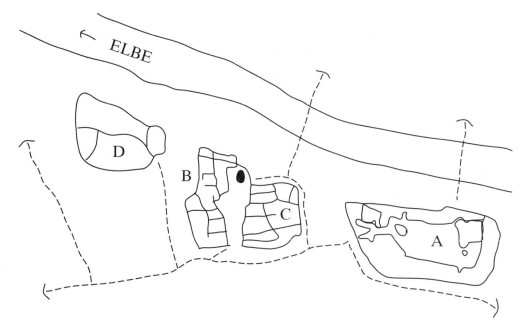

ELBE

FIGURE 2.15
Magdeburg, Germany. The medieval administrative hub with adjacent burg were located on high ground on the west bank of the Elbe. This plan shows: A, the castle site of 805, with the monastery of St. Moritz that was added in 937; B, the cathedral of 968; C, the marketplace; and D, the burg that was located further downstream.

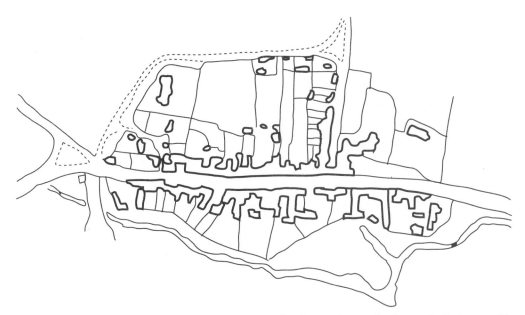

FIGURE 2.16 Wycombe in Buckinghamshire, England, a medieval town that evolved organically from a village settlement. The linear plan reflects how the original village grew up along a roadway. A characteristic of this urban form is the long back garden that was often approached from the rear by an access lane. With time, the back lanes were often upgraded into roads that provided access to new houses during the early stages in the growth of the village. In smaller market towns today, many of the back gardens of the original cottages have yet to be developed. In the towns that prospered and grew, the gardens were usually built over by the advent of the Industrial Revolution.

in particular functions, such as banking or the production and sale of items like furniture or metalwork. The organic growth towns could have streets and alleys that were unplanned and quite narrow. The city defenses became probably the most important determinant of urban form. Urban development

had to take place in stages, each of which was normally preceded by the construction of a new wall.[27]

With urban growth concentrated inside the city wall, population density was high, and the constrained space caused different socioeconomic groups to be stratified *vertically* within the same building. For

FIGURE 2.17 Aigues-Mortes, France, a medieval bastide. This aerial view shows the gridiron street pattern inside the rectilinear walls of the town and the surrounding fields and vineyards.

FIGURE 2.18   Berne, Switzerland, a medieval planted town. This aerial view shows the gridiron street pattern of the original town that formed the nucleus of the contemporary city and how the layout reflects the influence on urban form of its location on a riverbend in the Aare.

example, a merchant's or craftsperson's shop would be on the main floor, and because there were no elevators, the family's living quarters would be above on the next floor, the apprentice quarters would be above, with the servants up the stairs in the attic (Figure 2.19). In order to maximize street frontage, residential and commercial buildings were often aligned with their narrow or gable end fronting onto the main streets.

The emerging regional specializations and trading patterns provided the foundations for a new phase of urbanization based on **merchant capitalism**. The key actors in this system were the merchants who supplied the capital required to reestablish a vibrant system of long-distance trade—hence the label *merchant capitalism*.

> The merchants made the towns. They needed walls and wall builders, warehouses and guards, artisans to manufacture their trade goods, caskmakers, cart builders, smiths, shipwrights and sailors, soldiers and muleteers. They needed farmers and herdsmen outside the walls to feed them; and bakers, brewers and butchers within. They bought the privilege of self-government, substituting a money economy for one based on land . . . . Towns recruited manpower by offering freedom to any serf who would live within their walls for a year and a day.[28]

Beginning with commercial networks established by the merchants of Venice, Pisa, Genoa, and Florence (in northern Italy) and the trading partners of the

FIGURE 2.19   A 1928 view of cobblestone streets in the Alstadt, the medieval section of Hannover, Germany. Note how the "vertical" social stratification within the buildings is reflected in the size and ornateness of the windows.

Hanseatic League, a federation of city-states around the North Sea and Baltic coasts (see the Box entitled "Hanseatic League Cities"), a trading system of immense complexity soon spanned Europe from Bergen to Athens and from Lisbon to Vienna. Long-distance trade became firmly reestablished, but it was based more on bulky staples such as grain, wine, salt, wool, cloth, and metals rather than on the luxury goods of the pioneer merchants. The increased volume of trade fostered a great deal of urban development as the merchants began to settle at locations to take advantage of the major trade routes that crisscrossed Europe and as local economies everywhere came to focus on market exchange.

At the close of the medieval period in the late thirteenth century, Europe had about 3,000 cities containing some 4.2 million people, representing between about 15 to 20 percent of the total population. Most of these urban centers were small—with fewer than 2,000 people. Paris was the dominant European city, with a population of about 275,000. Besides Constantinople and Cordoba, only Milan, Genoa, Venice, Florence, and Bruges had more than 50,000. This then was the Europe that stood poised to extend its grasp to a global scale.

URBAN EXPANSION AND CONSOLIDATION DURING THE RENAISSANCE AND BAROQUE PERIODS

Between the fourteenth and eighteenth centuries, fundamental changes transformed not just the cities and urban systems of Europe but also the entire world economy. The Protestant Reformation and the scientific revolution of the Renaissance (from the mid-fourteenth to the mid-seventeenth centuries) stimulated economic and social reorganization. While the Church and religious doctrine had been dominant during medieval times, the glorification of human reason and achievement was foremost during the Renaissance. The scale and sophistication of merchant capitalism increased. Aggressive overseas colonization allowed Europeans to shape the world's economies and societies.

Spanish and Portuguese colonists were the first to connect the world's **peripheral regions** into the European urban system. Beginning in 1520, it took just 60 years for these colonial powers to establish the basis of a Latin American urban system. The Spanish colonists founded their cities in the western parts of Latin America by rebuilding at the sites of conquered indigenous centers like Oaxaca and Mexico City in Mexico, Cuzco in Peru, and Quito in Ecuador or in regions of dense indigenous population, as in Puebla and Guadalajara in Mexico and Lima in Peru. These **colonial cities** were established primarily as administrative and military centers from which the Spanish Crown could occupy and exploit the New World. In contrast, the cities of the Portuguese colonists further east, such as São Paulo and Rio de Janeiro, were more commercial in nature. Although also motivated by exploitation, the Portuguese colonists located their colonial towns in the best places commercially for collecting and exporting the products of their mines and plantations.

A major aspect of this urbanization and expansion of trade was the establishment of **gateway cities** around the world to act as links between one country or region and others. These **control centers** commanded entrance to, and exit from, their particular country or region (for North America, see the Box entitled "Vance's Mercantile Model" in Chapter 3). The Europeans founded or enlarged thousands of towns in other parts of the world as they extended their trading networks and established their colonies. The great majority were ports protected by fortifications and European naval power. Beginning as colonial trading posts and administrative centers, some grew rapidly as gateways for colonial expansion into the continental interiors. European settlers came in through these cities, and produce from the continental interiors went out. Rio de Janeiro grew on the basis of gold mining; São Paulo on coffee; Buenos Aires on mutton, wool, and cereals; Kolkata (Calcutta) on jute, cotton, and textiles; Accra (Ghana) on cocoa; and so on.

These products in turn fueled urban growth in Europe. In terms of location, the exploitation of the New World gave a decisive advantage to the port cities along the North Sea and the Atlantic Coast. By 1700 London had grown to 500,000, while Lisbon and Amsterdam had each reached about 175,000. The cities of continental and Mediterranean Europe grew at a more modest rate. Between 1400 and 1700 Venice grew only from 110,000 to 140,000; Milan's population did not grow at all.

More integrated national urban systems evolved along with the centralization of political power and the formation of national states that characterized the Renaissance period. Best exemplified by the capital cities of Paris and Madrid, their central location nationally facilitated the process of political consolidation, which, in turn, gave both cities a further impetus for growth as they acquired important administrative functions. Regional capitals and seats of county government emerged to fill out the evolving national urban systems.

The overall appearance and internal structure of cities in Europe changed during the Renaissance with the introduction of new forms of art, architecture, and urban planning. Especially in the capital cities, the flourishing of artistic and architectural expression

BOX 2.4 HANSEATIC LEAGUE CITIES

With its roots dating back to the twelfth century, the Hanseatic League was a trading association of independent German city-states around the North Sea and Baltic coasts. A trading arrangement initially between Hamburg and Lübeck provided a cooperative model for the merchants in other German cities. It involved an alliance between these two northern German towns that were located on either side of the Danish peninsula, Hamburg in the west and Lübeck in the east. Lübeck had access to the Baltic herring spawning grounds. Fish, which could be eaten on Fridays when meat could not, made up a high percentage of the Christian European diet at the time. Without refrigeration or canning, shipping this perishable product needed salt for salting, which was easily accessible to Hamburg from the nearby salt mines at Kiel. The merchants in Lübeck and Hamburg opened a trade route along the canal that was constructed between them and named for the source of the salt, Kiel.

Although not approaching the level and extent of economic and political cooperation among the countries of today's European Union, the Hanseatic League became the first great cooperative effort that united cities across a region of Europe into an economic association. These city-states entered into commercial agreements to promote trade

through special privileges for members and to protect themselves against pirates and robber barons. Over time, more and more cities joined in pursuit of the trade security and increased opportunities that membership provided. Conversely, the League engaged in negotiation, bribes, blockades, embargoes, and even war against port cities that were hostile to the organization and wanted to break its monopoly.

At its height, as many as 200 towns participated in this association that extended from Amsterdam to Reval (Tallinn) in Estonia and from Stockholm on Sweden's Baltic Sea coast to important inland port cities along rivers, such as Lübeck, Hamburg, Köln (Cologne), and Magdeburg. Their foreign trading outposts, or counting houses, the forerunners of today's stock exchanges, extended the influence of this powerful trading association to Bergen on Norway's North Sea coast, Visby on the island of Gotland, London, Bruges in Flanders (Belgium), and Novgorod in Russia (Figure 2.20).

Merchant capitalism and long-distance trade fueled urban expansion and the development of this subcontinental system of trading cities, some of which, like London, later grew to global dominance. The trading association controlled the shipping of fish, salt, grain, timber, amber, fur, flax,

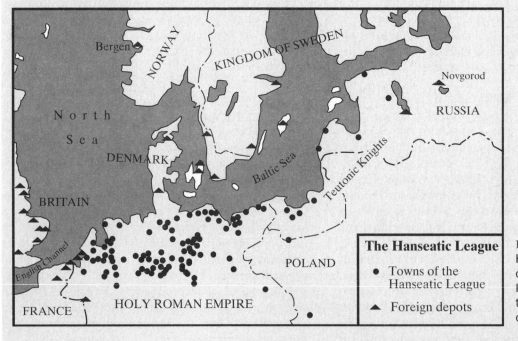

FIGURE 2.20 Hanseatic League cities. The towns of the Hanseatic League and their foreign trading depots.

The Hanseatic League
- Towns of the Hanseatic League
- ▲ Foreign depots

(Continued)

BOX 2.4 HANSEATIC LEAGUE CITIES (*Continued*)

and honey from Russia and the Baltic coast to the west, and cloth and manufactured goods from Flanders and England to the east. The Hanseatic League had its own financial and legal systems, as well as strong traditions of civic and individual rights.

By the early sixteenth century, the Hanseatic League had begun to disintegrate—a casualty of factors that weakened its power, such as rivalry and internal struggles among League members, the new trade opportunities opened up by the discoveries of Columbus and Vasco da Gama, declining Baltic fish stocks, the social and political insecurities of the Reformation, the growing strength of trade competitors like the Dutch and later the English, and the encroachment into its trade routes by the Ottoman Empire. The difficulties of long-distance trade during the Thirty Years War (1618–48) further weakened the League and led to its demise.

Interestingly, although the Hanseatic League met for the last time in 1669, it was never officially dissolved. It lives on in the names of German cities like 'Hansestadt Lübeck' and 'Freie und Hansestadt Hamburg.' Even today, Hamburg and Bremen are individual city-states within Germany. The League's legacy can been seen in the widespread use of German merchant and administrative terminology and in standardized sea travel and trade regulations, a first for the League. The League's strong architectural legacy is also visible in the "stepped gable" design of the buildings in the Hanseatic League towns (Figure 2.21).

FIGURE 2.21 The facades of Hanseatic League buildings with their characteristic stepped gables. Note how, in order to maximize street frontage, residential and commercial buildings were aligned with their narrow or gable end fronting onto the main streets.

brought about greater use of sculpture in public places, other forms of urban beautification, such as fountains, and the embellishment of monumental buildings that reached a peak with the Baroque period that began in the sixteenth century. Merchant capitalism generated great wealth for the nobility of the various monarchies and countries, and their burgeoning spending power was used to build opulent palaces in many cities. City walls were built with more complex and costly designs during the Renaissance, such as star shaped, that allowed the use of greater firepower against an attacking army (Figure 2.22).

FIGURE 2.22   Sabbioneta, Italy, was built by Vespasiano Gonzaga in the mid-sixteenth century as an ideal town, with a central piazza, a ducal palace, churches, garden palace, theater, and residences all encompassed within a star-shaped plan, bounded by thick walls bearing the Gonzaga family crest.

URBANIZATION AND THE INDUSTRIAL REVOLUTION

Large-scale manufacturing began in the English Midlands in the mid-1700s. The Industrial Revolution was a powerful impetus for urban growth because it brought fundamental changes in how and where goods were produced. Previously, individual rural workers (e.g., spinners and weavers) in their own cottages had carried out the different stages of the production process by hand. Now, all the stages of production were mechanized and combined under one roof—in factory buildings. Initially, access to a water source that could produce power for the machinery dictated the factories' locations. In previously rural areas, urban settlements grew up around the new factories as the early industrialists provided housing to attract workers whose long working hours forced them to live nearby.

Industrialization and cities grew hand in hand (Figure 2.23). The Industrial Revolution generated new kinds of cities—and many of them. Industrial economies needed what cities had to offer: the physical infrastructure of factories, warehouses, stores, and offices; the transportation networks; the large labor pools; and the consumer markets. In turn, industrialization changed the appearance, internal structure, and functioning of cities. Whole districts of factory buildings grew up with their grimy smokestacks, deafening machinery, and general hustle-bustle of industrial activity (see the Box entitled "Manchester: Shock City of European Industrialization"). With industrial inputs and products transported mainly by trains, new tracks, stations, and rail traffic began to play a significant role in these cities, as did the new

public transportation systems of trolleys and subways. The industrial period also heralded the development of the **central business district (CBD)** with its office buildings and corporate headquarters for the new companies. Large tracts of worker housing were built.

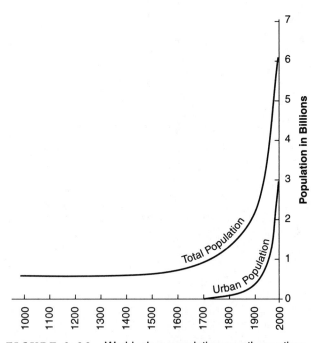

FIGURE 2.23   World urban population growth over time. Note the tremendous explosion in the number of people living in towns and cities that was triggered by the Industrial Revolution.

BOX 2.5   MANCHESTER: SHOCK CITY OF EUROPEAN INDUSTRIALIZATION

Manchester was the **shock city** of European industrialization in the nineteenth century. It grew from a small town of 15,000 in 1750 to a city of 70,000 in 1801, a metropolis of 500,000 in 1861, and a world city of 2.3 million by 1911. A shock city is seen at the time as the embodiment of surprising and disturbing changes in economic, social, and cultural life. Manchester was an archetypal form of an entirely new kind of city—the *industrial city*—whose fundamental reason for existence was not its military, political, ecclesiastical, or trading functions, as in earlier generations of cities. Instead, Manchester existed to assemble raw materials and to fabricate, assemble, and distribute manufactured goods. The city had to cope with record rates of growth and associated unprecedented economic, social, and political problems (Figure 2.24). Manchester was also a **world city**, in which a disproportionate share of the world's most important business—economic, political, and cultural—is conducted. At the top of a global urban system, such cities experience growth largely as a result of their role as key nodes in the world economy.

Friedrich Engels surveyed the dreadful living conditions of the working poor in the shock city of Manchester in 1844:

> One walks along a very rough path on the river bank, in between clothes-posts and washing lines to reach a chaotic group of little, one-storied, one-roomed cabins. Most of them have earth floors, and working, living and sleeping all take place in one room. In such a hole, barely six feet long and five feet wide, I saw two beds—and what beds and bedding!—which filled the room, except for the fireplace and the doorstep. Several of these huts, as far as I could see, were completely empty, although the door was open and the inhabitants were leaning against the door posts. In front of the doors filth and garbage abounded. I could not see the pavement, but from time to time, I felt it was there because my feet scraped it. This whole collection of cattle sheds for human beings was surrounded on two sides by houses and a factory and on a third side by a river . . . All along the Irk [river] slums of this type abound.[29]

Johann Kohl, who traveled through the English Midlands in the 1840s, captured how the profound changes associated with industrialization played out in the working lives of the countless factory workers in Manchester:

> It was a cold, damp, foggy morning in December, that I took my leave of Manchester. I rose earlier than usual; it was just at the hour when, from all quarters of the busy town, the manufacturing labourers crowded the streets as they hurried to their work. I opened the window and looked out. The numberless lamps burning in the streets, sent a dull, sickly, melancholy light through the thick yellow

FIGURE 2.24
Manchester, the shock city of European industrialization.

(Continued)

BOX 2.5 MANCHESTER: SHOCK CITY OF EUROPEAN INDUSTRIALIZATION   *(Continued)*

mist. At a distance I saw huge factories, which, at first wrapt in total darkness, were brilliantly illuminated from top to bottom in a few minutes, when the hour of work began. As neither cart nor van yet traversed the streets, and there was little other noise abroad, the clapping of wooden shoes upon the crowded pavement, resounded strangely in the empty streets. In long rows on every side, and in every direction, hurried forward thousands of men, women, and children. They spoke not a word, but huddling up their frozen hands in their cotton clothes, they hastened on, clap, clap, along the

pavement, to their dreary and monotonous occupation. Gradually the crowd grew thinner and thinner, and the clapping died away. When hundreds of clocks struck out the hour of six, the streets were again silent and deserted, and the giant factories had swallowed the busy population. All at once, almost in a moment, arose on every side a low, rushing, and surging sound, like the sighing of wind among trees. It was the chorus raised by hundreds of thousands of wheels and shuttles, large and small, and by the panting and rushing from hundreds of thousands of steam-engines.[30]

This row housing, typical of English cities, was often cramped and poorly constructed.

Rural development and urban growth were intimately connected in the industrialized **core regions** of Europe and North America (Figure 2.25). Agricultural productivity benefited from the mechanization and the innovative techniques that had been developed in cities. The higher productivity released rural dwellers to work in the growing manufacturing sector in the towns and cities, at the same time providing the additional food to support a growing urban population. This process, reinforced by the agricultural tools, machinery, fertilizer, and other products made in the cities, allowed even greater increases in agricultural productivity. This kind of urbanization is a special case of **cumulative causation**, where particular places enjoy a spiral buildup of advantages due to the development of **external economies**, **agglomeration economies**, and **localization economies**.

As industrialization spread across Europe, the pace of urbanization increased (Figure 2.26). The higher wages and greater opportunities in the cities attracted a massive influx of rural workers. Declining death rates associated with the **Demographic Transition** in Europe contributed to the rapid urban population growth. This growth in population in turn provided a huge increase in the labor supply during the nineteenth century, further boosting the rate of urbanization not only in Europe but also in the United States, Canada, Australia, New Zealand, and South Africa, as emigration spread industrialization and urbanization to the frontiers of the world-system.

With time, however, the breakthrough technology of coal-fired steam power freed industry to locate in existing population centers or close to resources, like coal mines. Industrialists no longer had to attract

workers to their factories. There was also an overabundance of workers due to fewer farm workers being needed because of increased agricultural productivity and the large number of small landholders left landless by the consolidation of smaller farms into larger, more efficient ones. Improvements in technology caused overproduction, while international competition increased as industrialization spread from England to the European mainland and beyond. Prices dropped, producers cut costs, and wages fell. Long hours for little pay made it a struggle to pay for housing, which led to overcrowding. Living conditions in the slums were appalling (see the Box entitled "Residential Segregation in Mid-Nineteenth Century Glasgow, Scotland"). Public sanitation and water systems were poor or nonexistent. Outbreaks of

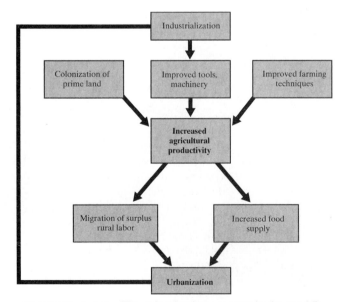

FIGURE 2.25   The urbanization process in the world's core regions.

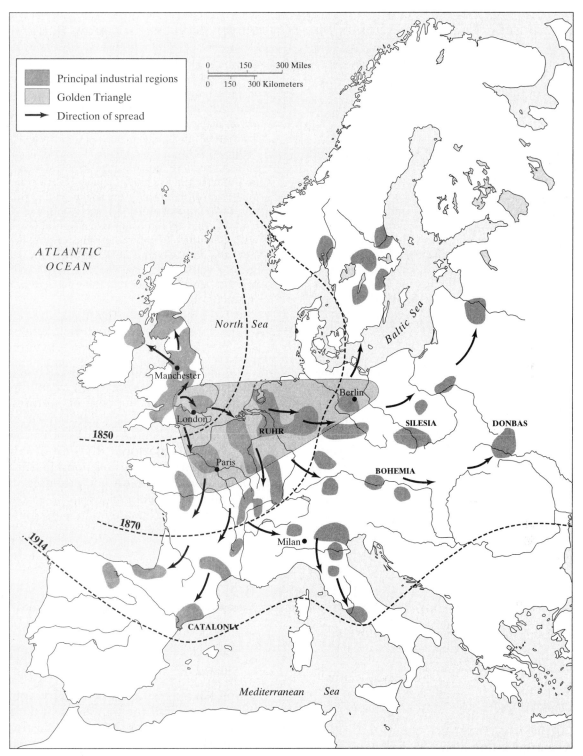

FIGURE 2.26   The spread of industrialization and industrial cities in Europe. European industrialization began with the emergence of small industrial regions in several parts of Britain, where early industrialization drew on local mineral resources, water power, and early industrial technologies. As new rounds of industrial and transportation technologies emerged, industrialization spread to other regions with favorable locational attributes (access to raw materials and energy sources, good communications, and large labor markets). The "Golden Triangle" is Europe's economic core region that centers on the area between London, Paris, and Berlin, and includes the early industrial regions of southeastern England, northeastern France, and the Ruhr district of Germany.

BOX 2.6   RESIDENTIAL SEGREGATION IN MID-NINETEENTH CENTURY GLASGOW, SCOTLAND

As the Industrial Revolution progressed, so too did residential segregation. Michael Pacione provides a stark description of living conditions in the tenement slum areas of central Glasgow by the middle of the nineteenth century:

> The westward migration of the wealthy from the old town not only introduced socio-spatial polarisation into what had been in the eighteenth century a heterogeneous urban structure, but freed land and housing for other uses. Those parts of the old city abandoned by the elite were colonised by a working-class population that was burgeoning in response to the new industries' demand for labour. Residences previously occupied by a single wealthy family were 'made down' to accommodate large numbers of the poor in grossly crowded conditions . . . . In 1861 two-thirds of

Glasgow's population of 394,864 lived in houses with only one or two rooms, and of these 60 per cent shared a room with at least four others. Overcrowding was intensified by property speculators building on any available open space to produce 'backjams' and 'backlands' tenements, which were either added to existing buildings or erected in the erstwhile gardens of the formerly wealthy burgher residences. By the mid-nineteenth century the old city was a congeries of poor-quality housing. One shelter visited [in 1858] in a dark ravine of a close housed a husband, wife and two children and comprised a 'sort of hole in the wall' measuring six shoe lengths in breadth, between eight and nine in length from the bed to the fireplace, and of a height which made it difficult to stand upright. Densities of 1,000 persons per acre were commonplace.[31]

water-borne diseases such as cholera and typhoid were common.

By the nineteenth century, urbanization had become an important dimension of the **world-system** in its own right. In 1800 less than 5 percent of the world's 980 million people lived in towns and cities. By 1850, 16 percent of the world's population was urban, and there were more than 900 cities of 100,000 or more around the world. The Industrial Revolution and European colonization had created unprecedented concentrations of people in cities that

were connected in networks and hierarchies of interdependence. As was the case for merchant capitalism, the industrialization of the **core economies** of Europe and North America depended on the exploitation of peripheral regions. As we will see in Chapters 7, 8, and 9, the **international division of labor** that inevitably resulted from this relationship had a significant impact on the patterns and processes of urbanization in the peripheral regions of the world. In the next chapter we turn our attention to the foundations of the American urban system.

FOLLOW UP

1. Be sure that you understand the following key terms:

  central business district (CBD)

  city-state

  colonial cities

  Dark Ages

  Demographic Transition

  Fertile Crescent

  gateway cities

  gridiron street pattern

  Hanseatic League

  Industrial Revolution

  merchant capitalism

  Mesopotamia

  organic growth

  shock city

  Silk Road

2. If you have access to a video library, you should arrange to watch the first video in the six-part *Legacy* series that was written and hosted by historian Michael Wood and produced by Maryland Public Television and Central Independent Television, UK (1991). This video, *Iraq: The Cradle of Civilization*, provides an excellent account of the beginning of urban civilization in southern Iraq more than 5,000 years ago. It illustrates some of the important changes associated with the shift to urban life such as occupational specialization, organized religion, bureaucratic government, and international trade.

3. "A common approach to examining the internal structure of cities is to identify whether the layout of an urban area was largely unplanned or planned" (p. 27). Look in a book, atlas, or library for a map of an ancient or contemporary city from a region of the world that particularly interests you. While being aware that all cities are subject to change over time, study the layout of the streets, transportation routes, and other features of your city's internal layout to try to determine if the city was largely planned or unplanned, or whether over time different parts appeared to have been planned while others were unplanned. Think about what kinds of political, economic, social, technological, and environmental processes may have influenced the layout of your city. Then find a book or surf the World Wide Web to find out more about the history of urban planning for your city that can shed light on what you have concluded from your analysis of the map.

4. Chicago was the shock city of North American industrialization. Do some research in the library and on the Web to find out why. How important was this city's location? Then find some information about the tremendous changes that were associated with Chicago's industrial growth—in population growth, modes of transportation, and the kinds of raw materials that were brought in for fabrication, assembly, and distribution as part of the manufacturing process.

5. Work on your *portfolio*. Look for supplementary materials that can help you flesh out what you have learned in this chapter. Research some examples of Mesopotamian, Greek, Roman, and medieval towns to find out how and why they were established and whether they have survived and why or why not.

Think about how you can make the aspects of urbanization described in this chapter meaningful to you personally. Find some contemporary accounts of life in cities at different points during the extensive span of time covered in this chapter. Think about how very different life was for different urban residents in earlier centuries.

KEY SOURCES AND SUGGESTED READING

Brunn, S. D., J. F. Williams, and D. J. Zeigler, eds. 2003. *Cities of the World: World Regional Urban Development*, 3rd ed. Lanham, Md.: Rowman & Littlefield.

Brunt, B. 1997. Urbanisation and Urban Development. pp. 263–89 in *Western Europe: A Social and Economic Geography*, ed. D. Gillmor, 2nd ed. Dublin: Gill & Macmillan.

Carter, H. 1983. *An Introduction to Urban Historical Geography*. London: Edward Arnold.

Chant, C., and D. Goodman, eds. 1999. *Pre-industrial Cities & Technology*. London: Routledge.

Childe, V. G. 1950. The Urban Revolution. *Town Planning Review* 21: 3–17.

Dennis, R. 1984. *English Industrial Cities of the Nineteenth Century: A Social Geography*. Cambridge, UK: Cambridge University Press.

Hohenberg, P. M., and L. Hollen Lees. 1985. *The Making of Urban Europe, 1000–1950*. Cambridge, Mass.: Harvard University Press.

Lawton, R. 1972. An Age of Great Cities. *Town Planning Review* 43: 199–224.

Morris, A. E. J. 1994. *History of Urban Form: Before the Industrial Revolution*, 3rd ed. London: Pearson Education.

Pounds, N. J. G. 1990. *An Historical Geography of Europe*. Cambridge, UK: Cambridge University Press.

Sjoberg, G. 1960. *The Preindustrial City: Past and Present*, New York: Free Press.

Sjoberg, G. 1965. The Origin and Evolution of Cities. *Scientific American* (September): 55–63

Verhulst, A. 1999. *The Rise of Cities in North-West Europe*. Cambridge, UK: Cambridge University Press.

RELATED WEBSITES

National Geographic: *http://www.nationalgeographic.com/*
The National Geographic website offers an abundance of information and photos on world history and culture related to cities, as well as a searchable encyclopedia and online atlas called MapMachine.

Exploring Ancient World Cultures: An Introduction to Ancient World Cultures on the World-Wide Web: *http://eawc.evansville.edu/*
This University of Evansville (Illinois) website contains a wealth of resources on the classical world and its cities.

Ancient World History: *http://members.aol.com/TeacherNet/Ancient.html*
This website contains numerous hyperlinks to a variety of topics related to cities, including one covering Ancient China to Modern Times.

Internet Ancient History Sourcebook: *http://www.fordham.edu/halsall/ancient/asbook.html*
This searchable online bibliography (maintained by Fordham University, New York) offers a variety of resources on ancient history and cities, including material on Mesopotamia and China.

3

THE FOUNDATIONS
OF THE AMERICAN
URBAN SYSTEM

In the United States urbanization has evolved in step with the economy. American towns and cities have played a central role in the economy ever since the seventeenth century, when the first outposts of European settlement were established on North American soil. They have been crucial in a rapid transition from a dependent, preindustrial economy to an advanced form of capitalism where cities function as nodes in a *global* economic system. During each major phase of this evolution, new resources, technologies, and business organizations were developed and cities changed to accommodate the new economic order. In some cities these changes occurred sequentially and the outcomes were superimposed one on the other. In others the new economic systems were not always profitable or appropriate. As a result, the impact of each phase of economic development was felt in different ways and to different degrees in different cities. Nevertheless, it is possible to identify several distinctive epochs in the evolution of the urban system, each roughly coincident with the rhythm of the economic long waves outlined in Chapter 1.

CHAPTER PREVIEW

This chapter follows the evolution of the U.S. urban system through five distinctive epochs that together established the foundations for the contemporary **urban system**. As we will see, this set of urban settlements is both the *product of*, and a continuing *framework for*, processes of economic, technological, demographic, political, and social change. Each successive epoch of urbanization brought new patterns of settlement, new kinds of towns and cities, and new patterns of trade and migration between towns and cities. Each change also brought new challenges in terms of understanding the underlying processes of change. As we follow the development of the urban system, therefore, we will also follow the emergence of key ideas, concepts, and theories about urbanization.

We begin with the earliest epoch of urban development, the frontier urbanization around which the U.S. economy was organized until independent nationhood. The second distinctive epoch (1790–1840) was a period of merchant trading, or mercantilism, during which there emerged a more extensive system of **central places**, or local marketing and service centers. The third epoch (1840–1875) was characterized by an expansion and realignment of the urban system in response to early industrialization, the mechanization of agriculture, and immigration. With these changes in mind, we will discuss some of the key principles of urban growth, the attributes of urban systems, and the principles that underpin the spatial pattern of central places.

With the fourth distinctive epoch, industrialization (1875–1920), we will see the effects of principles of industrial location on the development and adaptation of the urban system. We will also see how and why urban-industrial development is inherently uneven and unstable—part of a constantly changing landscape of investment, disinvestment, and reinvestment.

The fifth and final epoch corresponds with the emergence of Fordism and mass-produced automobiles, trucks, and aircraft (1920–1945), which significantly changed the spatial organization of the urban system. This was also a period that saw some important changes in corporate organization and a severe economic depression. We will see how these developments affected the fundamental organization of the U.S. political economy and, therefore, the very foundations of urban development.

FRONTIER URBANIZATION

Although there were many small urban settlements in North America before the sixteenth century (Figure 3.1), the first large towns and cities in the United States were those that were established as outposts of European economies. Spanish colonialism was the first to leave its imprint. Guided by the Laws of the Indies (dating from 1583), Spanish settlers in Florida and the American Southwest planted towns that were laid out with a rectangular grid of streets surrounding a large central plaza. The earliest of these planned communities was La Villa Real de Santa Fe (Santa Fe, New Mexico), founded in 1610 as the administrative center for New Spain's northernmost frontier. Over the next 150 years or more, Spanish settlers founded a series of "pueblos" (centers of commerce and administration), "missions" (centers for religious conversion), and "presidios" (military outposts), all of which, as they grew, acquired a mixture of commercial, administrative, religious, and military functions. Among the settlements founded by Spanish colonists were Los Angeles, Saint Augustine, San Antonio, Santa Barbara, San Diego, and San Francisco. Not long after the founding of Santa Fe, Dutch settlers sailed into the Hudson estuary and established a fur trading post they called New Amsterdam. The French were less interested in colonization than in opportunities for trade, but as they foraged through the Great Lakes and along the Mississippi River system they established trading posts that slowly grew into small towns. These included Québec, Montréal, Detroit, St. Louis, and New Orleans.

FIGURE 3.1   Tyuonyl Anasazi Pueblo in Frijoles Canyon, Bandelier National Monument, New Mexico. Tree-ring analyses date construction of this settlement of community houses and central plaza within a stone oval-shaped enclosure to between A.D. 1383 and 1466.

It was English colonization, however, that established the most vigorous roots of the American urban system. In Virginia, the Jamestown colony, founded in 1607, initiated the cultivation of tobacco for export to Europe and established the first representative government on the continent. Nearby Williamsburg, established in 1663 as a stockaded refuge from Native Americans, became the capital of the Virginia colony in 1699 after fire destroyed much of Jamestown. Although profitable, tobacco cultivation was labor intensive. In 1619 the first Africans were brought to Virginia to work in the tobacco plantations. By 1807, when Britain banned its subjects from participating in the slave trade, 600,000 to 650,000 Africans had been forcibly transported to North America.

In New England, Boston, founded in 1630 as a "City upon a Hill" in a celebration of spirituality, soon became a prosperous center of trade and commerce. Newport, on Rhode Island, was founded by a group of religious dissenters in 1639 as a haven for the persecuted, but, with the best natural harbor in southern New England, it too developed into a trading port. In 1664 the British took New Amsterdam from the Dutch, renamed it New York, established it as the capital of the New York colony, and set about developing the best natural harbor on the entire Atlantic seaboard into a major port. Charles Town (later Charleston) was established as the capital of the Carolinas in 1680, and William Penn established Philadelphia as the capital of Pennsylvania in 1682 (Figure 3.2).

The embryonic urban system operated as a string of **gateway cities**: control points for (1) the assembly of staple commodities for export, (2) the distribution of imported manufactured goods, and (3) the civil administration of the new territories. For some time each gateway port operated quite independently, each having more linkages with European cities than with one another. As colonization extended, a hierarchy of settlements began to develop. Places with better resources and better accessibility became larger and acquired a broader range of services. Meanwhile, the need to rationalize transatlantic shipping schedules prompted the development of a network of coastal shipping between the largest ports. As a result, a few cities—Boston (Figure 3.3), Charleston, Newport, New York, and Philadelphia—were able to establish themselves as major **entrepôts**: intermediary centers of trade and transshipment. As these entrepôts grew in stature, they came to dominate larger and larger **hinterlands** (market areas) in which smaller settlements emerged as local market towns. In time these market towns became *inland gateways* that acted as bulking points and provided an array of services for the frontier agriculturalists. By the time of the Revolution in 1775, the most notable of these were Hartford, Middletown, and Norwich (Connecticut), Albany (New York), Lancaster (Pennsylvania), and Richmond (Virginia). As the 13 colonies merged into an American union, the largest city was New York, with about 25,000 inhabitants. Philadelphia was almost as large, with about 24,000; Boston was next biggest, with about 16,000; and Baltimore, Charleston, and Newport all had between 10,000 and 12,000 inhabitants. The inland gateway cities were all relatively small, none exceeding 10,000 people.

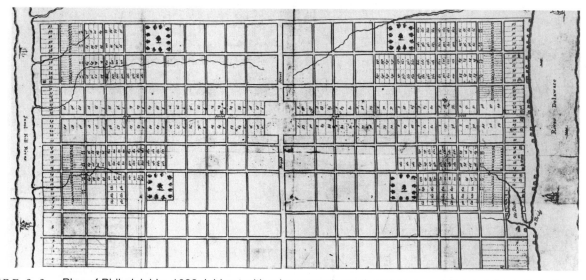

FIGURE 3.2   Plan of Philadelphia, 1683, laid out with primary north-south and east-west streets intersecting at a central town square containing public structures such as government buildings and churches, with grid blocks delineated by secondary streets and four minor squares. The preplanned layout facilitated settlement by allowing colonists to select a parcel for their new home before leaving Europe.

FIGURE 3.3   Boston in 1770. Boston was important during the early development of the American urban system because of its role as an entrepôt, an intermediary center of trade and trans-shipment between colonial America and cities in northwestern Europe.

THE MERCANTILE EPOCH (1790–1840)

Although only 1 in 20 of the infant country's population lived in towns or cities, city-based newspapers and city-based lawyers and merchants greatly influenced the Constitution that was framed in Philadelphia in May 1787. The result was a Constitution that favored city-based manufacture and trade—*and therefore stimulated city growth*—by preventing the potential fragmentation of the economy that would have come from allowing states to set their own tariffs, coin their own money, or issue their own bills of credit. The political independence gained by the colonies as a result of the Constitution stimulated the development of the urban system in several other ways:

- It became practical—and necessary—for economic links to be forged between towns and cities whose main linkages under the colonial system had been with European ports.
- It meant that a much greater proportion of investment was financed by American capital, with the result that fewer profits "leaked" back to the European urban system.
- It required a proliferation of government functions, from county courthouses and town halls to state capitals and, of course, the development of the District of Columbia, chosen in 1790 as the site for a permanent seat of federal government.
- Westward expansion (following the Northwest Ordinances of 1785 and 1787, the Louisiana Purchase of 1803, and the addition of Texas and the territories acquired from Mexico in 1846) required

frontier towns (like Santa Fe: Figure 3.4) that subsequently matured into local service centers, or central places.

The most striking growth occurred in gateway cities located at strategic points along rivers that linked the new western territories with the larger cities of the Atlantic seaboard. Roads and turnpikes provided a relatively slow and expensive way to move freight in comparison to river and sea. New Orleans (Figure 3.5), St. Louis, and other river ports grew quickly because of their strategic location on major waterways. East

FIGURE 3.4   An engraving of Santa Fe, taken from the *Report of Lt. J. W. Albert of His Examination of New Mexico in the Years 1846–1847.*

FIGURE 3.5   The most striking urban growth during the Mercantile Epoch took place in river ports. New Orleans grew very rapidly because of its situation as a gateway to the entire Mississippi system. This view is from about 1862.

Coast merchants, however, were not content to let the lucrative trade from western and southern hinterlands slip from their grasp. Their response was to tap the waterways of the Great Lakes and the Hudson, Ohio, and Mississippi river systems with networks of canals. As a result, two east-west corridors of trade merged. The first stretched from New York up the Hudson and the Erie Canal (opened in 1825) to the eastern Great Lakes, where Buffalo, Cleveland, Detroit, Chicago, and Milwaukee emerged as important wholesaling centers. The second stretched across the mountains from Philadelphia and Baltimore to Pittsburgh and the Ohio Valley, where Cincinnati and Louisville became important inland gateways.

As the urban system expanded and trade between cities increased, particular cities and regions were able to specialize according to their **comparative advantage** (i.e., the economic activity which in local conditions could be undertaken most efficiently, compared to other possible activities). Cincinnati, for example (Figure 3.6), began to specialize in hog processing, thus earning the nickname of "Porkopolis." Manufacturing began to contribute to the growth of the leading eastern ports, while some towns in the more heavily populated and intensively developed northeast—Albany, Lowell, Newark, Poughkeepsie, Providence, Springfield, and Wilmington—became the cradle of the Industrial Revolution in the United States.

Immigration provided an important source for peopling both the frontier and the growing ports and inland

FIGURE 3.6   Cincinnati, shown here in 1848, an important river port that had already acquired an industrial specialization—hog processing—by the mid-nineteenth century.

gateways. By the American Revolution (1775–1783), the colonial population had reached about 2.5 million people, of which more than 500,000 were African-American slaves, 250,000 were Scots-Irish, and 200,000 were German. Meanwhile, an increasingly important source of urban growth was improved agricultural productivity. Urbanization was fueled by advances in farm productivity that (1) provided the extra food to support the increased numbers of townspeople and (2) released farmers and agricultural workers who then moved to towns and cities to swell the numbers of producers and consumers. Improved agricultural productivity came about partly through the colonization of prime land, partly through the application of improved tools, machinery, and techniques, and partly through forcing out surplus labor in order to maximize the gains from these applications. Labor displaced in this way ends up consuming food rather than producing it, but this disparity is more than compensated for by the

increases in agricultural productivity and by the greater capacity of urban labor forces to produce agricultural tools, machinery, fertilizer, and so on that will make for still further increases in agricultural productivity. This self-reinforcing sequence of events is sometimes referred to as the *urbanization process* (Figure 3.7).

By 1840 the American urban system had become independent and was on the way to becoming economically integrated. New York had grown to 391,114 and had increased its lead over Philadelphia (93,665) by a significant margin. Meanwhile, Baltimore and New Orleans had topped 100,000 and Boston had grown to more than 93,000. A few regional centers such as Cincinnati and Albany had populations between 25,000 and 50,000, while larger central places such as Louisville and Richmond stood at about 20,000 (roughly the same size as Washington, D.C.). Most towns had populations of 15,000 or less; Chicago had fewer than 5,000 inhabitants.

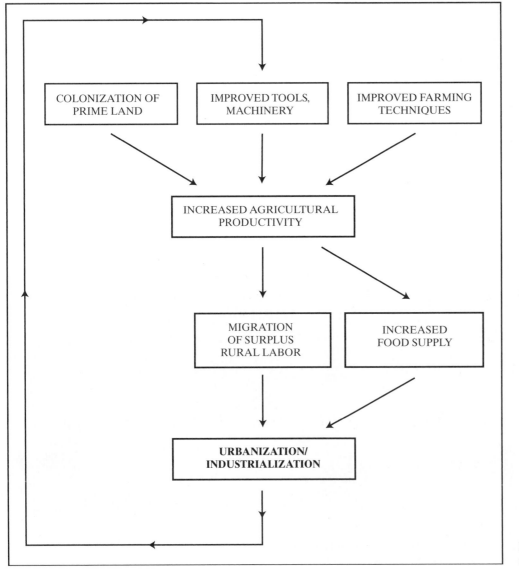

FIGURE 3.7 The urbanization process.

FIGURE 3.8 St. Louis, the most important inland gateway port of the Mercantile Epoch, was already past its prime by the time this photograph was taken in 1903, when railroad hubs like Chicago were handling the bulk of all passenger and cargo traffic.

Box 3.1 Vance's Mercantile Model

One model of the development of the North American urban system is the *mercantile model* proposed by urban geographer James Vance, Jr. According to Vance, external influences and long-distance trade have been particularly important in forging the geometry of the urban hierarchy through five distinctive stages (Figure 3.9).

1. *Exploration.* The search for knowledge and economic opportunities from the Old World, where an urban system had emerged that was based on localized market functions.

2. *Harvesting of natural resources.* This stage involves the establishment of colonists' settlements in order to exploit natural resources such as the codfish of the Grand Banks off Newfoundland and the timber and beaver pelts from New England.

3. *Emergence of farm-based staple production.* The colonization of land for agricultural production and permanent settlement is associated with the development of an embryonic urban system geared to the export of staple products such as grain, salted meat, indigo, tobacco, and cotton and the import of manufactured goods and luxury items from Europe. Gateway cities along the coast ("points of attachment" in Vance's terminology) become the focus of the emergent system.

4. *Establishment of interior depot centers.* Continuing demand for staple exports and increasing colonization combine to draw settlement further into the interior, requiring the development of long-distance routes and the emergence of towns serving as "depots of staple collection" at strategic locations along these routes. This stage corresponds to the "frontier urbanization" described above, with entrepôts and inland gateway cities functioning primarily as wholesale collection centers, the "spearheads of the frontier."

5. *Economic maturity and central place infilling.* This stage depends on the emergence of a domestic market large enough and affluent enough to sustain the growth of domestic manufacturing industry. With their established populations and their superior accessibility, the gateway ports, entrepôts, and inland gateways (the "depots of staple collection") attract much of this manufacturing activity. Meanwhile, agricultural settlement in the regions between established long-distance routes begins to support subregional "central place systems" (systems of market towns: see below). But the established ports, entrepôts, and inland gateways, with much larger markets and better-developed distribution networks, are the places from which higher-order goods and services are offered. Long-distance trade, then, rather than local markets, "fixes" the spatial pattern of the cities that come to function as the leading central places of the maturing urban system.

(Continued)

BOX 3.1   VANCE'S MERCANTILE MODEL   (*Continued*)

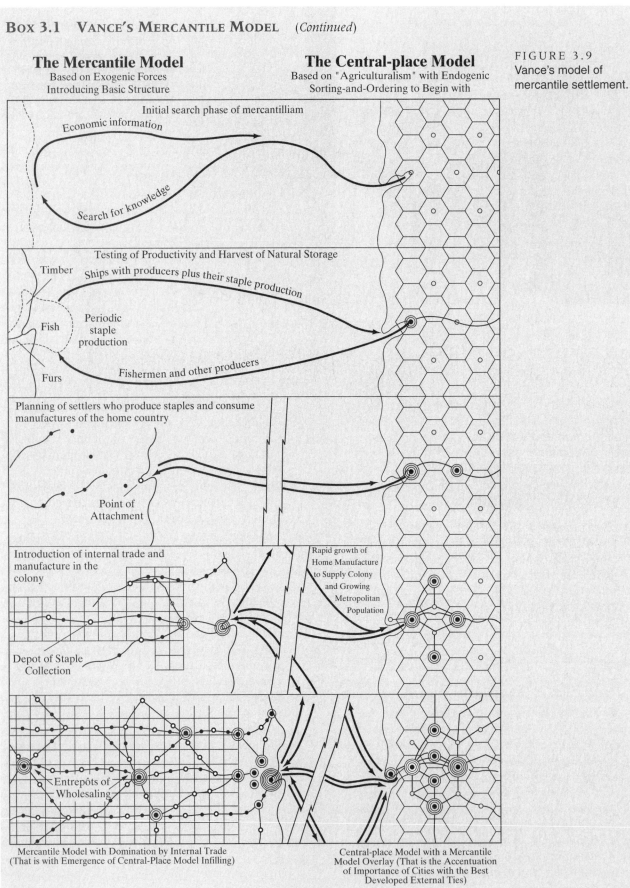

FIGURE 3.9
Vance's model of
mercantile settlement.

The Mercantile Model
Based on Exogenic Forces
Introducing Basic Structure

The Central-place Model
Based on "Agriculturalism" with Endogenic
Sorting-and-Ordering to Begin with

Initial search phase of mercantilliam

Economic information

Search for knowledge

Testing of Productivity and Harvest of Natural Storage

Timber

Ships with producers plus their staple production

Fish

Periodic
staple
production

Furs

Fishermen and other producers

Planning of settlers who produce staples and consume
manufactures of the home country

Point of
Attachment

Introduction of internal trade and
manufacture in the
colony

Rapid growth of
Home Manufacture
to Supply Colony
and Growing
Metropolitan
Population

Depot of Staple
Collection

Entrepôts of
Wholesaling

Mercantile Model with Domination by Internal Trade
(That is with Emergence of Central-Place Model Infilling)

Central-place Model with a Mercantile
Model Overlay (That is the Accentuation
of Importance of Cities with the Best
Developed External Ties)

EARLY INDUSTRIAL EXPANSION AND REALIGNMENT (1840–1875)

The transition from a trading economy to a mature agricultural and embryonic industrial economy took place during the 1840s for several reasons. One reason was the arrival of industrial technology and methods of industrial and commercial organization from the hearth of the Industrial Revolution in northwestern Europe. Another was the improvement in agricultural productivity and the ability to colonize formerly marginal land that came about as a result of mechanization. With less labor needed on farms, **rural-to-urban migration** fueled the urbanization process. At the same time, higher levels of agricultural production were able to feed the constantly growing numbers of immigrants who were also fueling urban growth and providing the labor needed in new factories (Figure 3.10).

Beginning in 1840, a great influx of immigrants began to arrive in the United States from Europe (Figure 3.11). Severe famines in Ireland in the 1830s and 1840s, combined with a British "laissez faire" policy and the rationalization of agriculture from small farms to larger estates, fueled mass migration. By 1850 in continental Europe, the economic and social changes associated with rapid industrialization and the mechanization of agriculture had prompted an even greater exodus of people from the German states, France, and Belgium, extending to Scandinavia by the late 1870s and sparking waves of Danish, Norwegian, and Swedish immigrants, and spreading to southern and eastern Europe—involving Italians and Jews, as well as Slavs from the Austro-Hungarian Empire—during the 1880s and 1890s (Figure 3.11).

Because many of the new industrial technologies had specific locational requirements (proximity to large quantities of raw materials, for example, or accessibility to a variety of different suppliers or markets), this phase of urbanization brought the emergence of some new towns and cities and the rapid growth of

FIGURE 3.10 New arrivals. Crowding the deck on arrival in New York harbor, these immigrants would have been processed on Ellis Island before going on to find jobs in America's fast-growing cities.

what previously had been very small settlements. There were four categories of such newcomers:

1. *Power sites* that attracted industries that consumed significant quantities of energy. Before the widespread use of coal-fired steam technology and, later, electricity, the power of falling or running water was particularly important. In the United States, this factor led to the appearance of a series

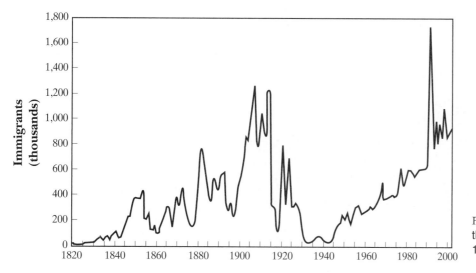

FIGURE 3.11 Immigration to the United States, by fiscal year, 1820–2002.

Box 3.2 Immigrant Housing Conditions and the American City, 1840–1875[1]

By the second half of the nineteenth century, many of the immigrants who had flooded into larger American cities were living in horrific housing conditions. The term *slum* conveyed the interrelationships among the environmental limitations, spatial isolation, and social problems of the largely immigrant populations of the central parts of large American cities (Figure 3.12). A survey of housing in Manhattan in 1867 by the Metropolitan Board of Health inspected over 18,000 structures that housed more than three families. Many of these buildings were tenements—the former homes of wealthy single families that had been partitioned into cheap multifamily units for recent immigrants. More than half of the 18,000 structures surveyed were considered to be in bad condition, and in one-third of these tenements neglect was found to be the main reason.

Although anxieties at the time about urban society publicized the coincidence of destitution, depravity, and dreadful living conditions, some observers recognized that slum development reflected the underlying economics of property investment that set rent levels for adequate housing well above the means of the poor. In 1857 it was reported in New York Assembly Document No. 205, p. 11, that

> in its beginning, the tenant house became a real blessing to that class of industrious poor whose small earnings limited their expenses and whose employment in workshops, stores or about the wharves and thoroughfares, rendered a near residence of much importance. At this period rents were

moderate, and a mechanic with family could hire two or more comfortable even commodious apartments, in a house once occupied by wealthy people, for less than half he is now obliged to pay for narrow and unhealthy quarters.

By the second half of the nineteenth century, however, overcrowded rooms, subdivided buildings, congested lots, and inadequate facilities were all common strategies used to keep rents low without reducing the return on a property. The reports on the condition of housing for the poor at the time regularly mentioned how buildings were sublet to an agent for a fixed return. The agent would then obtain the maximum yield from a building by neglecting repairs and services and demanding the highest possible rent. For their part, this arrangement allowed the absentee owners to shirk direct responsibility for their property or tenants.

In 1870 the newly formed Massachusetts Bureau of Statistics of Labor—reflecting similar observations made about the tenements in New York—concluded in its first annual report that:

> low-paid laborers are not earning sufficient wages to justify their living in [improved] tenement houses . . . and, as there are no intermediate houses between these and the lowest class of houses . . . the laborers of this class are practically compelled to crowd into the miserable refuges in which they now congregate. In fact, so far as we have been able to ascertain, there are no places within the settled portions of the city of Boston where the low-paid toiler can find a house of decency and comfort.

FIGURE 3.12   Back alley in New York City in the late 1800s. By the second half of the nineteenth century, many of the immigrants who had flooded into larger American cities were living in horrific housing conditions. Many buildings were tenements—the former homes of wealthy single families that had been partitioned into cheap multifamily units for recent immigrants.

of industrial towns along the Fall Line in New England and along the eastern margins of the Appalachians. Examples include Allentown and Harrisburg in Pennsylvania.

2. *Mining towns* that sprang up to provide the coal and ores needed by the industrial economy. The most important of these in this epoch were Appalachian coalfield towns such as Norton, Virginia.

3. *Transportation centers* that emerged at strategic locations within the "new" economic space made accessible by the canal and railroad network (Roanoke, Virginia, is a classic example of a railroad town).

4. *Heavy manufacturing towns* whose dependence on large volumes of raw materials tied them to the source areas of those inputs. The classic and most important cases involve steelmaking and the associated heavy engineering. Pittsburgh, which was to become *the* steeltown, was, of course, already an important river port and wholesaling center; it just happened to be near coalfields and deposits of iron ore.

The development of steam-powered riverboats and the growth of the railroad system were central to the evolution of the new industrial economy and its new urban system. Ports and lakeside cities such as Chicago, Cincinnati, Memphis, and Nashville prospered because they were able to operate as interfaces between established trading routes and the budding railroad network. Within a few years of the initiation of work on the first railroad (the Baltimore and Ohio, in 1828), rail networks began to be developed around all the established ports on the inland waterways that reached into the agricultural lands of the interior plains. Initially, therefore, the railroads were complementary to the waterways as long-haul carriers of general freight.

In 1869 the railroad network reached the Pacific, when, at Promontory, Utah, the Union Pacific railroad (built west from Omaha) met the Central Pacific railroad (built east from Sacramento). By 1875 intense competition between railroad companies had begun to open up the western prairies as far as Minneapolis–St. Paul and Kansas City. The "iron horse" epoch had extended the urban system to a continental scale. The significance for economic development (and therefore for urbanization) was enormous. The railroad allowed a loose-knit collection of regional economies to develop into a national economy within which American enterprise could fully exploit the commercial advantages and **economies of scale** of a huge market and an apparently unlimited resource base.

Meanwhile, the development of the railroads realigned the spatial organization of the urban system (Figures 3.13, 3.14). The westward penetration of railroad lines allowed large quantities of corn and wheat to be moved directly eastward, rather than being shipped by water via St. Louis and New Orleans. As a result, these cities, along with smaller, intermediate ports, experienced slowed rates of growth and came to rely increasingly on regional trade and service functions. In contrast, cities along the two major east–west wholesaling alignments (New York–Buffalo–Detroit–Chicago–Milwaukee and Philadelphia–Pittsburgh–Cincinnati–Louisville) benefited both from the extra trade and from reduced freight prices that resulted from fierce competition between the railroads and waterborne transportation. Their growth helped lay the foundations of what was to become the Manufacturing Belt.

This chronology points us to another important generalization about the development of urban systems: *Most early industrial growth, and therefore most urban growth during the early phases of industrialization, occurred in the largest existing towns and cities.* The chief exceptions were the textile towns that were tied to waterpower sites in New England and along the Fall Line at the eastern margin of the Appalachians.

By 1875 the urban system had expanded to the point where more than 15 cities, together with their adjacent counties, each accounted for more than 100,000 people. New York, with 1.3 million people, stood at the top of the urban hierarchy, followed by five cities with populations in the range of 350,000 to

FIGURE 3.13 The development of the railroad system contributed to the expansion and realignment of the urban system, while local railroad stations became important nuclei for commercial development. This photograph shows the station at Valdosta, Georgia.

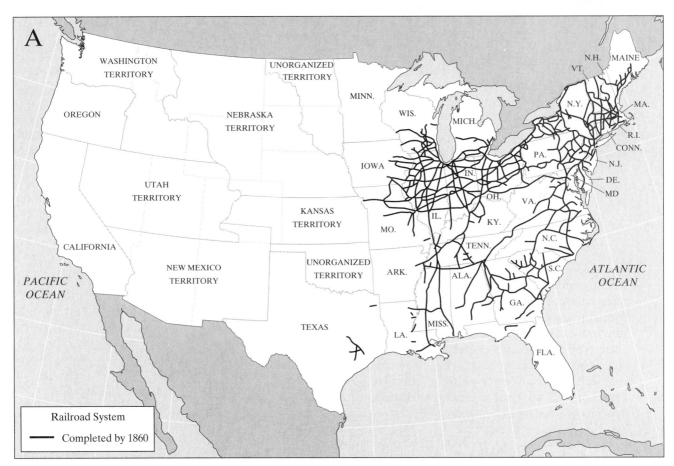

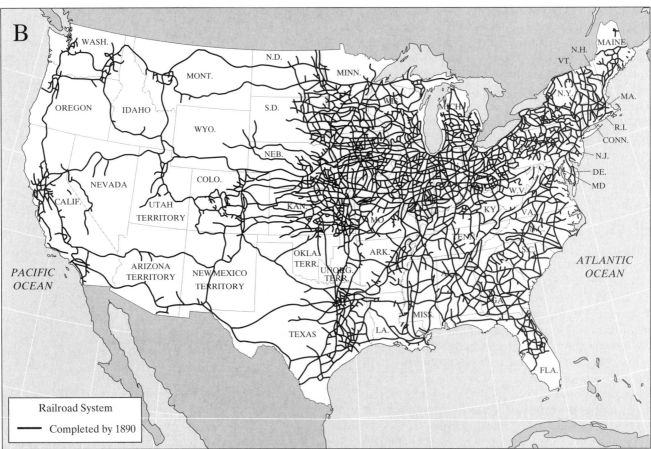

FIGURE 3.14   (A) The railroad system in the United States in 1860. This map shows over 30,000 miles of track, much of it in the form of short lines. (B) The railroad system in the United States in 1890. The railroads had expanded rapidly in the 1880s, when more than 70,000 miles of track were added. By 1890, the total network was 163,597 miles.

450,000: Baltimore, Chicago, Philadelphia, Pittsburgh, and St. Louis. Cities with populations of between 100,000 and 150,000 included Athens (Georgia), Boston, Buffalo, Cincinnati, Cleveland, Detroit, Manchester (New Hampshire), New Orleans, Providence, Rochester, and Syracuse.

SOME PRINCIPLES OF URBAN GROWTH

The reasons for the concentration of growth in the largest of the existing towns and cities stem from the principle of **initial advantage**. This principle was based on the fact that existing cities had:

- The craft industries and wholesaling and transportation activities that provided the basis for industrial enterprises as their owners used their capital and accumulated profits to invest in factories and machinery
- Traditions and skills of entrepreneurship, investment, and lending that lent themselves easily to industrial development
- The largest pools of labor
- The largest and most affluent markets, which made attractive outlets for manufactured products

These observations are related to some fundamental principles inherent to the process of urban-industrial development. The principle of initial advantage is itself a special case of the operation of **external economies**: cost advantages that accrue to individual companies because of their locational setting. External economies can be derived from the availability of a pool of labor with appropriate skills, from accessibility to good specialized business services, or from the quality of the physical infrastructure of roads, harbors, and utilities—any advantage, in fact, that stems from the collective rather than exclusive use of any of the elements necessary to profitable activity.

Urban settings, with large pools of labor with a wide range of skills, intensively developed infrastructural components, specialized educational and training facilities, research institutions, and concentrations of business services, provide companies in many industries with opportunities for external economies. In this context, external economies are often referred to as **agglomeration economies** or **urbanization economies**. Where external economies are limited to companies involved in a particular industry, they are known as **localization economies**. Examples include the sharing of a pool of labor with special skills or experience (e.g., shipwrights, precision instrument makers), supporting specialized technical schools, joining together to develop a research institute or a marketing organization, and drawing on specialized subcontractors, maintenance firms, suppliers, distribution agents, and legal counsel. These localization economies go a long

way toward explaining the tendency for cities to maintain economic specializations: Pittsburgh's attractiveness for the iron and steel industry, Akron's attractiveness to manufacturers of rubber products, and Dayton's attractiveness to manufacturers of fabricated metal and machinery, for example.

INTERPRETING AND ANALYZING THE URBAN HIERARCHY AND THE CENTRAL PLACE SYSTEM

What was emerging was an urban system that had much in common, in terms of the size distribution and the spatial patterning of towns and cities, with urban systems in many other countries. What is of special importance to social scientists is that there emerged certain regularities that cried out for explanation. Among the regularities to appear in the mercantile and early industrial expansion epochs were consistent relationships between (1) cities' rank (by population size) in the urban system and their actual population—the so-called rank-size rule, and (2) the size and spacing of towns and cities—the basis of central place theory.

The Rank-Size Rule

The relationship between cities' size and their rank within an urban system is expressed in a simple formula, known as the **rank-size rule**, set out by G. K. Zipf in 1949:[2]

$$P_i = P_1 \div R_i$$

where P_i is the population of city i, R_i is the rank of city i by population size, and P_1 is the population of the largest city in the urban system. Thus, if the largest city in a particular system has a population of 1 million, the fifth-largest city should have a population of 200,000, the hundredth-ranked city should have a population of 10,000, and so on. Plotting this relationship on a graph with a logarithmic scale for population sizes would produce a straight line, reflecting a *log-linear* relationship. The rank-size distribution of the U.S. urban system plotted in this fashion for several points in time in Figure 3.15 shows clearly that the system as a whole has conformed fairly consistently to the rank-size rule. The slope moves to the right on the graph, reflecting increased populations brought about by urbanization in cities at every level in the urban hierarchy. This conformity to the rule, however, does not necessarily mean that individual cities have maintained their relative position along each curve. The growth of some cities (e.g., San Diego) has sent them from the lower end of the hierarchy to the very top, while other cities (e.g., Savannah) have declined, at least in relative terms, so that they have fallen down the hierarchy. *It is the overall relationship between cities' size and their rank within the urban system that stays fairly constant.*

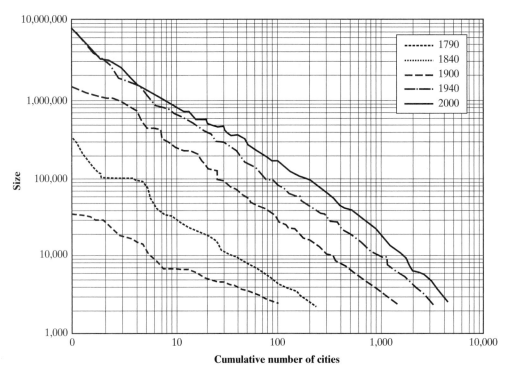

In some urban systems (though not the U.S. urban system), the disproportionate size of the largest city in the system distorts the log-linear character of the rank-size distribution. In the United Kingdom, the city of London is more than seven times the size of Birmingham, the second-largest city, while in Ireland, Dublin is more than five times the size of the second-largest city, Cork. Such cases led to the "law" of the **primate city**: "A country's leading city is always disproportionately large and exceptionally expressive of national capacity and feeling."[3] But for every case of **primacy** there is an urban system where the largest city approximates to the rank-size rule. The search for explanations of primacy has dwelt upon levels of economic development and the size of a country's "ecumene" (the extent of its effectively settled area), but neither factor withstands empirical testing. More likely, primacy is a good example of the *contingency* noted in Figure 1.4, with each case arising from a particular combination of circumstances.

Innovation and Urban Growth

In pursuing the causes of the rank-size characteristics of urban systems, scholars have noted that city growth rates are related to city size. Specifically, it has been observed that the *variability of urban growth rates decreases with urban size*: Larger cities, in other words, have more consistent and predictable rates of growth. The explanation, supported by empirical tests, is based on the dynamics of *innovation diffusion* within the urban system. These dynamics work as follows. Over the medium to long term, urban population growth is a function of the number of innovations originating within a city's economy or successfully adopted from outside sources. Larger cities, with larger economies, are likely to produce more innovations, thus attracting migrants to the city and helping to ensure that residents do not leave for other cities. Therefore, the sheer size of large cities ensures a strong and steady stream of innovations, which in turn contributes to a steady source of population growth. Smaller cities must depend more on adopting innovations that diffuse from other, usually larger, centers.

The probability of innovations successfully diffusing to a particular city is a function of

- city size—the "hierarchy effect," and
- distance from the hearth of the innovation—the "neighborhood effect"

It is also recognized that there is a small random effect that, for example, accounts for smaller, remote places sometimes producing important innovations or adopting innovations at an early stage in the process. Because smaller cities tend to be spread throughout an urban system at different distances from innovation hearths in large cities, there is much greater variability among them as to the strength of the neighborhood effects that help to determine their rate of innovation adoptions. The result is that population change is more variable among smaller cities than among larger ones.

Central Place Theory

The significance of trade and marketing in the economy of the early and mid-nineteenth century also introduced certain regularities in the spacing and spatial patterning of towns and cities within the urban system. In part, this regularity was a result of transportation technology: Settlements evolved to match the distances that could be covered by river, canal, or turnpike in a day's travel. In addition, the role of towns and cities as local service centers—central places—brought a new logic that helped to shape the spacing of towns and cities by size. Smaller settlements, offering a limited range of services for their residents and the residents of nearby areas, were quite numerous and were located at relatively short, fairly even distances from one another. Large cities were fewer and farther between but offered a much greater variety of services, catering to residents from many surrounding areas. In between were intermediate-sized central places serving intermediate-sized markets with middling packages of service functions.

It was not until much later that such patterns came to be explained in terms of underlying principles and idealized outcomes. The seminal work was that of Walter Christaller, a German geographer who had observed striking regularities in the size and spacing of settlements in southern Germany.[4] For the sake of clarity and simplicity of argument, Christaller first postulated a uniform topographic landscape, uninterrupted by rivers, roads, or canals: Accessibility is assumed to be a direct function of distance, in any direction. From this starting point, the foundations of central place theory depend on elementary principles concerning the "range" and "threshold" of goods and services.

The *range* of a particular product or service is the maximum distance that consumers will generally travel to obtain that product or service. "High-order" goods and services such as specialized equipment, professional sports, and specialized medical care (relatively costly and generally required infrequently) might have a range of a hundred miles or more, whereas "low-order" goods and services such as bakeries, barber shops, and saloons (perishable or required in relatively large amounts at frequent intervals) will have a much shorter range. It follows that the maximum area to be served by a particular service from a particular central place will be a circular market area, or hinterland, whose radius is equal to the range of the service.

The *threshold* of a particular product or service is the minimum market size required to keep the product or service on offer. It can be thought of in terms of the lower or minimum range of a product or service: the minimum distance necessary to circumscribe a hinterland with enough potential consumers to make the enterprise profitable. High-order services such as hospitals have thresholds in the tens of thousands; low-order services such as grocery stores have thresholds of just 200–300 people.

From these principles we can hypothesize an urban system in which the smallest settlements, all offering

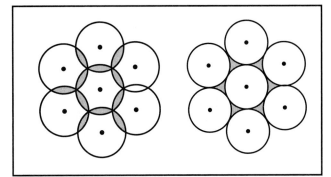

FIGURE 3.16   Central places (dots) and their market areas, or hinterlands (circles). Circular hinterlands like this were a logical assumption for theorists but were problematic, since they meant that some areas would be underserved (right) or that there would be an overlap of service provision (left).

similar packages of low-order goods and services—such as a general store, a post office, and a saloon—are arrayed across the landscape in a regular lattice, as in Figure 3.16. It is immediately apparent, however, that the circular hinterlands derived from these basic principles leave small unserviced areas between the hinterlands. Because this array is not realistic, central place theory proposes an adjustment to the idealized geometry of the urban system, bringing the low-order central places a little closer together, using hexagonal hinterlands, and introducing a competitive zone at the margins of each hinterland (Figure 3.17). This hexagonal pattern is the basic building block of central place theory.

Higher-order central places will also offer all of the lower-order goods and services, and it is assumed that consumers will use the nearest central place that offers the goods or services they require. As a result, a *nested hierarchy* of central places can be constructed, with marked discontinuities in the population size and packages of services offered at each level of the hierarchy.

Different configurations of nested hierarchies of hexagonal hinterlands can be derived from the application of varying assumptions about economic

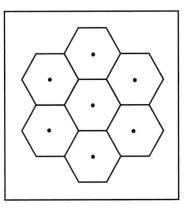

FIGURE 3.17   Hexagonal market areas around central places.

organization. Christaller was most interested in an economic environment dominated by marketing principles. These principles entail the important assumption that merchants' (and all other service providers') profits will be maximized when the number of central places is kept to a minimum. This produces the configuration of central places shown in Figure 3.18—the $k = 3$ solution, three being the number of hexagonal hinterlands of the smallest size category that are contained within the hinterland of the next-largest category of central places, while there are three times as many central places at each successively lower tier in the urban hierarchy.

Another possible form of economic organization is one that is based on principles of *consumer welfare*, the central concern of August Lösch.[5] Lösch found evidence of his spatial theory around the cities of Indianapolis and Toledo, and tests and elaborations of his and Christaller's solutions have produced an extensive literature. The urban landscapes of the upper Midwest, in particular, have tended to exhibit the size and spacing characteristics of Christaller's $k = 3$ framework, although—as in other case studies—there is little evidence of a stepped hierarchical structure. Rather, the size distribution of central places in regional subsystems that have been studied is best described as a continuum, which at least fits in with what we know about the rank-size characteristics of urban systems. Explanations for the absence of stepped hierarchies include spatial variations in population density or in purchasing power, uneven terrain, uneven access to transportation, and the tendency for people to combine trips for high-order and low-order items. Such explanations have, in turn, led to a body of research on *consumer behavior* in relation to the service packages offered in different central places. This research produced some valuable insights into the geography of retailing and consumers' travel behavior, helped to refine concepts of spatial interaction, and introduced urban analysts to some powerful analytical techniques; but it did not add a great deal to our ability to theorize the development of the hierarchical and spatial characteristics of systems of cities.

Beyond Consumer Hinterlands

One of the main shortcomings of central place theory is that, by focusing on central place functions, it fails to address some other important influences on the size and spacing of cities—manufacturing, for example, and

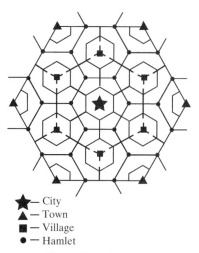

FIGURE 3.18   Christaller's $k = 3$ central place system: the geometry of settlement according to his "marketing" principle.

the long-distance trading functions (e.g., wholesaling) that are essential to the development of most urban functions, whether they stem from the agricultural, industrial, or service sector of the economy. Many central place functions, even in the early nineteenth century, depended on a long chain of production and exchange that involved numerous steps, from harvesting or mining through processing, packaging, shipment, and storage to retail distribution. The products offered for sale in Albany, Buffalo, and Cleveland, for example, would have originated from all corners of the United States, from many parts of Europe, and from a few other parts of the globe. Today, the internationalization of the economy means that almost every purchase we make is the result of a complex geography of international specialization and trade involving vast quantities of products whose origins range from makeshift kitchen workshops in the least developed countries to the giant electronics factories of South Korea and the perfumeries of Grasse (in Provence, France).

Another important shortcoming of central place theory is that it provides only a *static* model of settlement patterns—a model that is unresponsive to changes in population densities, consumer spending power, transportation technologies, or communications systems such as the Internet and also a model that "paints a picture of a world that somehow springs into existence fully formed, with no historical antecedents."[6]

THE ORGANIZATION OF INDUSTRY (1875–1920)

As **industrial capitalism** gathered momentum and transport and communications networks became more efficient, the urban system became increasingly consolidated, with more pronounced patterns of economic specialization and greater integration between component

parts. On the railroads the innovation of steel rails to replace iron ones made it possible to carry heavier loads at higher speeds (which encouraged the concentration of industry in larger cities) and provided the opportunity to standardize the railroad gauge throughout the country (which encouraged interurban trade and specialization). At the same time, the railroad system extended dramatically. The total length of railroad tracks increased from 30,000 miles in 1860 to 160,000 miles in 1890, allowing towns such as Birmingham, Jacksonville, Memphis, and Houston to emerge as central places with regional status and allowing the revival of colonial towns such as Savannah and Charleston. Across the Great Plains, the railroad companies, in collaboration with grain elevator companies and lumber dealers, literally dictated the pattern and design of lower-order central places. Desperate for traffic, they deliberately encouraged the urbanization of the West through propaganda that extolled the virtues of territory that earlier had been marked on maps as "The Great American Desert."

This period also saw some important demographic changes. Falling death rates (the result of better diets, improvements in sanitation and public health, and the introduction of scientific medicine) generated a steady **natural increase**. Meanwhile, from 1890 to 1910 over 12 million immigrants arrived in the United States, accounting for about one-third of the country's growth. It was not long, however, before pressure from labor unions (concerned about the effects of immigration on wages and unemployment) and from nationalistic groups such as the American Protective Association and the Daughters of the American Revolution (concerned about the impact of immigration on American values and society) led to restrictive legislation. The 1921 Immigration Act imposed an overall ceiling of 250,000 immigrants a year, with quotas for each country of origin, based on the existing composition of the white population. In 1924 another Immigration Act reduced the ceiling to 150,000 immigrants a year.

The overall effect of these demographic changes was the fueling of the growth of existing cities, the infilling of settled areas, and the colonization of the few remaining frontier regions. The emergence of small towns that accompanied the commercialization of agriculture in the West, the exploitation of major mineral deposits (copper in Montana, lead and zinc in Missouri, and iron ore around Lake Superior), and the escalating demand for coal helped to more than double the total number of urban places between 1870 and 1920. At the top of the urban hierarchy, New York and its adjacent counties in 1920 had 4.75 million residents. Boston, Chicago, Philadelphia, and Pittsburgh each had in the region of 1.5 million, while the next tier of cities, with populations in the region of 500,000, now included several cities from the West (Los Angeles, San Francisco, Seattle), Southwest (Dallas), and Midwest (Kansas City, Milwaukee, Minneapolis-St. Paul, St. Louis) as well as from the Northeast quadrant of the country (Baltimore, Cincinnati, Providence).

The most important feature of this period, however, was the way in which the logic of *industrial location*

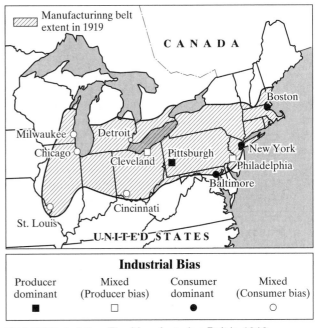

FIGURE 3.19   The Manufacturing Belt in 1919.

influenced patterns of urban growth and development. As we have seen, the framework of the Manufacturing Belt was already in place by 1870. What happened during the late nineteenth century was that the territorial expansion of the Manufacturing Belt was halted and its cities became more closely integrated with one another. By 1920 the Manufacturing Belt had become a national economic heartland (Figure 3.19), its larger cities blossoming into great metropolitan centers and its smaller cities providing the settings for highly specialized and very profitable manufacturing industries. Outside the Manufacturing Belt, cities existed to serve farms. Inside, farms existed to serve cities.

In overall terms, *the consolidation of the Manufacturing Belt was the result of the operation of the principle of initial advantage at a regional scale*. With its large markets, well-developed transport networks, and access to excellent coal reserves, it was ideally placed to take advantage of the general upsurge in demand for consumer goods, the increased efficiency made possible by the telegraph system, the postal services, and the rationalization of the banking system, and the reduced energy costs brought about by the advent of thermoelectric generating stations.

The net effects were twofold:

1. *Local specialization became geared to national rather than local markets.* Brewers in Milwaukee and St. Louis, for example, were able to take advantage of the integrated railroad system and of innovations such as refrigerated rail cars and mechanized production techniques. Similarly, men's clothing and fruit and vegetable canning

emerged as specialties in Baltimore; musical instrument manufacture and men's clothing in Boston; meat packing, furniture manufacture, and printing and publishing in Chicago; coach-building and furniture manufacture in Cincinnati; glass and iron and steel in Pittsburgh; and textile manufacture in Philadelphia. In smaller cities specialization came to be much more pronounced, as in the production of foundry and machine-shop products in Albany–Troy, Fort Wayne, and Paterson; iron and steel and coach-building in Columbus; furniture in Grand Rapids; distilled liquors in Peoria; agricultural machinery in Racine and Moline; and boots and shoes in Worcester.

2. *This specialization provided the basis for increased commodity flows between the towns and cities of the Manufacturing Belt, thus binding the region more tightly together.* The region's financial and commercial linkages and technical expertise made it particularly attractive to new industrial activities with large or national markets; this in turn stifled the chances of cities in other regions achieving comparable levels of industrialization. This does not mean, of course, that other cities did not become industrialized. Rather, it was the *scale* and *intensity* of industrialization that differed. Cities in other regions were able to develop an array of locally oriented manufacturers, together with some nationally oriented

activities based on particular local advantages or raw materials; but they were rarely able to attract manufacturers of mass-produced goods for the national market.

UNDERSTANDING AND ANALYZING UNEVEN URBAN DEVELOPMENT

One of the most important and distinctive characteristics of urban systems under industrial capitalism is their **uneven development**. Since the late nineteenth century, the basic expression of this unevenness in the United States has been the contrast between the highly urbanized, highly industrialized, and (until recently) relatively affluent "core" that corresponds to the Manufacturing Belt and the "periphery" that encompasses the rest of the country. This **core-periphery pattern** of unevenness is also apparent at more detailed scales. Within the "core," there is an important contrast between the extreme urban concentration of *Megalopolis* (the corridor of interpenetrating urbanization that stretches from Boston through New York and Philadelphia to Baltimore and Washington, D.C.—see Figure 3.20)[7] and parts of upper New York state, Pennsylvania, Ohio, and Indiana that are characterized by small town settings. Within the "periphery" there are still greater contrasts: between Seattle–Tacoma and Portland and the rest of the Pacific Northwest, for example; between Los Angeles and the Bay Area and the rest of the West

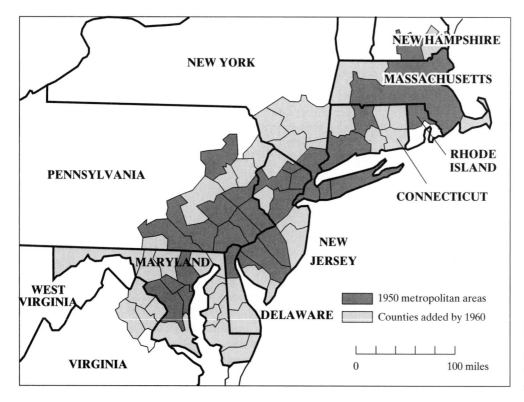

FIGURE 3.20
"Megalopolis," the extended region of metropolitan development identified by Jean Gottman in 1960.

and Southwest; between Minneapolis–St. Paul and the rest of the Upper Midwest; and so on.

Urban-Industrial Growth as a Self-Propelling Process

As we have seen, the early establishment and initial advantage enjoyed by gateway ports, entrepôts, and inland gateways go some way toward explaining this unevenness. In addition, the location of brand-new industrial cities—power sites, mining towns, transportation centers, and heavy manufacturing towns—can be seen as contributing to a decidedly uneven pattern of spatial development because of the inherently uneven geographical distribution of energy sources, raw materials, and so on. Agglomeration economies and localization economies have been identified as important elements in *reinforcing* this uneven urban growth, but we must also recognize the dynamics of local economic development that sustain the uneven processes of urbanization.

Urban-industrial development can be seen as a self-propelling growth process. Given a certain amount of original employment growth in a particular industry, a number of interrelated effects can be discerned:

1. **Backward linkages** occur as new companies arrive in order to provide the growing industry with components, supplies, specialized services, or facilities.

2. **Forward linkages** occur as new companies arrive that take the finished products of the growing industry and (depending on the nature of the products) use them as inputs to their own processing or assembly, finishing, packaging, or distribution. Together with the initial growth, the growth in these linked industries (both backward and forward) helps to create a threshold of activity large enough to attract *ancillary industries* and activities—maintenance, repair, recycling, security, and so on.

3. The existence of these interrelated industries establishes a pool of labor with skills and experience that are attractive to still more companies.

4. The linkages between all these industries help to promote interaction between professional and technical personnel and makes it possible to sustain research and development (R&D) facilities, research institutes, and so on, thus increasing the likelihood of new inventions and innovations that further stimulate growth.

Collectively, these four localization economies constitute the *primary* **multiplier effect**.

5. Meanwhile, the spiral of growth in the initial industry, the linked industries, and ancillary industries is associated with additional growth effects through a *secondary* multiplier effect. This effect stems from the growth of population associated with the expansion of employment: the employees themselves plus their families. Their presence creates a demand for housing, utilities, physical infrastructure, retailing, personal services, and so on—all of which generate additional jobs. This expansion, in turn, helps to create threshold populations large enough to attract more, higher-order services and activities.

6. Last—but by no means least—the overall growth in employment implies an expansion of the local tax base, which enables the provision of improved public utilities, roads, schools, health services, recreational amenities, and so on, all of which intensify agglomeration economies and so enhance the competitiveness of the setting in attracting further rounds of investment.

This whole self-propelling spiral of growth is known as the process of **cumulative causation** (Figure 3.21). Seen in broader context, cumulative causation will produce negative consequences for some other cities and regions: outmigration of people and outflows of capital, together with a downward spiral of events that mirrors the upward spiral of places with an initial advantage. These negative consequences are known as **backwash effects** or *polarization effects*.

Modifying Processes

The self-propelling processes involved in cumulative causation thus seem to provide a powerful explanation for the very uneven patterns and rates of urban growth that accompanied the organization of industry in the late nineteenth century. As they stand, though, they beg several important questions. Why did the big cities of the era not grow ever larger in relation to the rest of the urban system (something that would have been reflected by a change in the slope of the **rank-size** distribution)? How were urban "newcomers" such as San Diego able to develop into major metropolitan areas? And how did a city like Detroit shift from a vigorous spiral of cumulative causation in the 1920s to a track record of several decades of disinvestment and depopulation by the early twenty-first century?

So-called **spread** (or **trickle-down**) **effects** in a core-periphery pattern can provide a partial answer. These are the result of economic growth in peripheral cities and regions that is induced by high levels of demand in the core. The classic example is the growth of agricultural machinery production in peripheral cities as a consequence of the need to feed ever-larger and more affluent populations in core cities. Such effects have never been associated with major shifts in

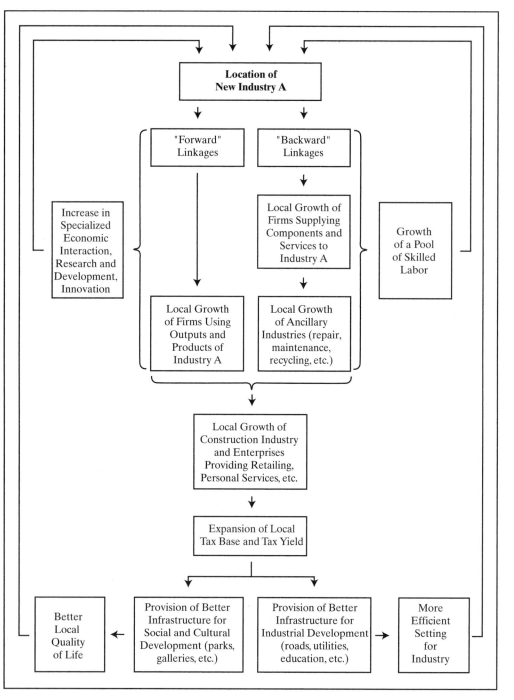

FIGURE 3.21   The self-propelling process of urban-industrial growth.

patterns of urban development, but they do begin to help us modify the rigid determinism of cumulative causation.

It is also possible for peripheral cities to develop economic strength through a process of **import substitution**, whereby local companies produce for their own market area goods and services that had previously been provided by companies in core settings. Although some things are hard to copy because of the constraints of climate or the limitations of natural resource endowment, local entrepreneurs can copy many products and services, thus capturing local capital, increasing local

employment opportunities, intensifying the use of local resources, and generating profits for further local investment. The amount of growth achieved through import substitution greatly depends, however, on the ability of local entrepreneurs to adopt appropriate innovations; without that ability, their businesses would not be competitive. As we have already seen, the diffusion of innovations is a function of urban hierarchy effects and distance ("neighborhood") effects, so that import substitution does not offer a very powerful explanation for instability within urban systems.

FIGURE 3.22
The archetypal landscape of urban-industrial development: Ensley, Alabama.

Another modifying factor that must be acknowledged here consists of the **agglomeration diseconomies** that can arise after urban growth reaches a certain point. These include the costs to companies of inflated land prices, traffic congestion, crowded port and railroad facilities, increasing unit costs of solid waste disposal, and higher taxes that are levied in order to support services and amenities that were previously considered unnecessary—traffic police, city planners, and metropolitan art galleries, for example. Companies simply pass on many diseconomies to consumers in the form of higher prices. In addition, there are agglomeration diseconomies that impinge directly on city residents rather than on companies: noise, air pollution, increased crime, commuting expenses, and housing costs, for example.

The fundamental reasons for the instability of uneven urban development, however, are to be found in the longer-term shifts in technology and labor processes, cyclical changes, and evolutionary development of capitalism outlined in Chapter 1. The technological innovations associated with successive **Kondratiev** upswings generate new industries that are not yet tied down by enormous investments in factories or tied into existing industrial complexes. Combined with innovations in transportation and communications, this creates *windows of locational opportunity* that can result in new industrial complexes, with small towns or cities growing into dominant metropolitan areas through new cycles of cumulative causation. This expansion, in turn, tilts the balance of comparative advantage and distorts previous patterns of demand and supply, prompting further redeployment of economic activity of many kinds.

Even more important are the consequent shifts in the relative profitability of "old" industries in core cities vis-à-vis the new industries in rapidly growing cities. As soon as the differential is large enough, some *disinvestment* will take place within core cities, leading to **deindustrialization** there; meanwhile, the capital is invested in the new ventures that are located elsewhere. This process is often referred to as the **creative destruction** inherent to the dynamics of capitalism. But the process does not stop there. If the deindustrialization of the old core cities is severe enough, the relative cost of their land, labor, and infrastructure might decline to the point where substantial flows of new investment are attracted. As a result, there is what Neil Smith calls a geographical "seesaw" movement of investment capital, which always

> strives to move from developed to underdeveloped space, then back to developed space which, because of its interim deprivation of capital, is now underdeveloped, and so on.[8]

As a result, as we will see in the remainder of this chapter and in Chapter 4, the evolution of the urban system reflects this seesaw movement, with the expansion of cities in the "growth periphery" of the Sunbelt and the deindustrialization and subsequent redevelopment of Manufacturing Belt cities. Smith also points out that, since this seesaw movement depends a great deal on the mobility of capital, the greatest instability and unevenness is likely to be seen at the intrametropolitan scale, where capital is most mobile. We will return to this topic in Chapter 13.

EARLY FORDISM, THE AUTOMOBILE ERA, SUBURBAN INFILL, AND THE GREAT DEPRESSION (1920–1945)

The mass production of the internal combustion engine with the introduction of **Fordism** brought decisive changes to the balance of forces that had governed economic and urban development before World War I. Automobiles began to compete with transit systems, initiating a phase of suburban infill (see Chapter 6). The availability of tractors enabled farmers to work still larger acreages with fewer farm hands, thus initiating a revolution in farm size and stimulating the urbanization process by releasing rural workers. Particularly significant in the 1920s was the migration of blacks from the rural South to northern industrial cities. Following the "Great Migration" of 1916–1918 that saw blacks induced to northern industrial cities because of labor shortages during World War I, this new wave of migration was in response to a number of factors: the mechanization of southern agriculture, discrimination and organized violence against blacks in the South, and labor shortages in northern cities caused by the reduced levels of immigration after the immigration acts of 1921 and 1924.

The result was a decisive restructuring of the country's social geography that included the creation of first-generation incipient black ghettos in northern cities. Automobiles and trucks, meanwhile, enabled rural residents to travel farther to market or to shop (or to rely on mail-order deliveries), thus leading to the stagnation or decline of many small towns. Another consequence of the transition from steam to internal combustion was the relative decline of coal and railroad centers and, conversely, the rapid growth of oilfield cities like Oklahoma City and Bakersfield. At the top of the urban hierarchy, air passenger transport reinforced the centralization of business management functions in a few key cities, while the development of highway networks induced the development of hundreds of satellite cities and dormitory towns, reinforcing the process of **regional decentralization** in which the entire structure of American cities was to be reorganized (Chapter 5).

The increased levels of economic integration and more sophisticated corporate organization that characterized the early twentieth century were also to have important implications for patterns of urban development. As competition to maintain profit levels intensified, strong companies began to buy out their competitors—the process of **horizontal economic integration**. The chances of new companies being successful were then reduced, as the larger companies were able to draw on economies of scale to edge out smaller competitors by temporary price cutting. Meanwhile, larger companies captured further profits

FIGURE 3.23   The Model T assembly line in Ford's Highland Park plant. Ford's revolutionary approach to industrial production, together with the revolutionary effects of the automobile itself, were to have far-reaching effects on urbanization.

through the processes of **vertical economic integration** (taking over the companies that provide their inputs or those that purchase their outputs, or both) and **diagonal economic integration** (buying highly profitable companies whose activities are completely unrelated). Because of a flurry of company mergers in the 1920s, just 1 percent of all companies accounted for nearly one-half of the country's productive capacity. *The long-term significance of this shift in corporate organization would be that the fortunes of many cities would no longer rest on the decisions of owners with local ties.*

A CRITICAL TURN FOR URBANIZATION: THE DEPRESSION AND MACROECONOMIC MANAGEMENT

Before the effects of this shift could be registered, however, the economy as a whole had moved into the downward side of the long wave. In October 1929 the stock market collapsed, marking the turning point to recession after a period of Type B economic growth (see p. 13). This time, though, problems of imbalance between demand and supply were unprecedented in their intensity, and every sector of the economy was involved in the recession. Economic historians differ in their explanation for this phenomenon, but they generally agree that the basic reason was an abnormal sensitivity within the economy.[9] Agriculture had become a victim of its own success, mechanization having boosted productivity so much that the selling price of a bushel of wheat fell from over $1.80 in 1920 to less than 50 cents in 1930. The Smoot-Hawley Tariff Act of 1930 sought to protect U.S. agriculture from cheap imports but succeeded only in contributing to a recession in world trade. Industrial and financial markets, meanwhile, had become unstable as a result of stock market speculation and complex corporate takeovers. Fearful of inflation, the Hoover administration placed tight restrictions on credit, which further dampened demand for industrial products.

The net result was that the unemployment rate rose from around 3 percent of the labor force in 1929 to more than 25 percent by 1933. The unevenness of urban-industrial growth in previous years was mirrored in the pattern of decline, so that the intensity of the Depression was greatest in cities that had come to specialize in manufactured goods or components. The rate of unemployment in some specialized manufacturing towns, for example, exceeded 60 percent; in large, heavily industrialized cities like Detroit and Pittsburgh the unemployment rate in the spring of 1933 was about 50 percent; while in large but more economically diversified cities such as Philadelphia and Seattle it was between 20 and 30 percent. Meanwhile, rural-to-urban migration streams came to a temporary halt, slowing urban population growth dramatically.

FIGURE 3.24   Unemployed workers lining up for jobs at a dockyard in the 1930s. The unprecedented severity of the Great Depression led to a much greater involvement by the federal government in macroeconomic management and in urban and regional policy.

The fundamental economic problem of the 1930s was the long-wave downswing, during which deflationary depression saw a collapse of both foreign and domestic markets. The subsequent situation, where surplus labor, surplus productive capacity, and surplus capital existed side by side, was an acute **overaccumulation crisis** (described in Chapter 1). Too much capital had been generated, in aggregate, relative to the available opportunities for its profitable investment. Viewed from a slightly different angle, the same circumstances can be interpreted as *underconsumption*: insufficient demand in relation to the capacity of the economic system. Overaccumulation-underconsumption had occurred before, but never on such a scale as during the Depression. In previous episodes the problem had been resolved by the redirection or "switching" of increased amounts of capital (a market response, not a switch that was consciously planned at any level) into the form of fixed capital within what is sometimes called the *secondary circuit* of the economy's **circuits of capital investment**.

At this point it should be explained that there are two main forms of fixed capital. First, there is fixed capital in the form of machinery and equipment that is enclosed within the production process (e.g., lathes, cranes) or in the form of offices, factories, warehouses, and the like that provide the framework for production. Second, there is fixed capital that is embedded within long-term consumption (e.g., refrigerators, washing machines) and in the form of the built environment (mainly housing) that provides a framework for consumption. In periods of overaccumulation, investment in some of these fixed assets—particularly those involving real estate and the built environment—becomes relatively attractive. As such investment takes place, the problem of overaccumulation-underconsumption is alleviated. This is a point that we will take up again, for it clearly has some very important implications for our understanding of the development of the built environment of cities.

In the early 1930s, however, the secondary circuit could not absorb the surplus capital. The crisis led to the emergence of an *increased level of government intervention in the economy*, using taxes (1) to channel capital into the *tertiary circuit* (investment in science and technology and in "social overhead" such as education, health, and welfare programs; the former in order to maintain the economy's productivity and the latter in order to maintain and reproduce a healthy and productive workforce) and (2) to manipulate consumer demand in order to manage the economy. Public expenditures were increasingly used to stimulate demand or, through public employment, to absorb surplus labor. This approach to macroeconomic management is now generally known as **Keynesianism** (after the doctrine of British economist John Maynard Keynes).

One of the most important long-term consequences of the Great Depression, therefore, was that the ideology of free enterprise, which had been the unchallenged basis for economic and urban development up to the 1930s, was severely shaken. Although President Hoover had felt morally obliged not to intervene in the "natural" workings of the business system, a majority of American voters preferred the idea of government intervention to provide relief, reform, and recovery, electing Franklin Roosevelt to the presidency in 1932 on a platform that was a precursor of Keynesian economic management. The progressive interventionism embodied in his New Deal programs not only brought the first real experiments in city and regional planning but also introduced a general expansion of government involvement in urban development. At the same time, the government's involvement in macroeconomic management led to its mediation of a social pact between organized labor and big business: a more structured approach to the labor process and to wage bargaining that was geared to productivity, with workers' welfare guaranteed by Social Security. This approach heralded the zenith of **organized capitalism** referred to in Chapter 1.

We will see in Chapter 16 that the government has come to play a major role in influencing the dynamics of urban development; and in Chapter 17 we will review the changing effects of government-sponsored planning and policy making on patterns of urban development.

FOLLOW UP

1. Be sure that you understand the following key terms:

   agglomeration economies and diseconomies

   backward and forward linkages

   central places

   circuits of capital investment

   comparative advantage

   creative destruction

   cumulative causation

   diagonal, horizontal, and vertical economic integration

   entrepôt

   external economies

   hinterland

   import substitution

   initial advantage

   Keynesianism

   localization economies

   Manufacturing Belt

   Megalopolis

   mercantile model (Vance)

   multiplier effects

   primate cities and primacy

   rank-size rule

   spread and backwash effects

   urbanization process

2. On page 53 it is suggested that "the impact of each phase of economic development was felt in different ways and to different degrees in different cities." Can you say why? Can you think of examples that would illustrate the point?

3. Construct a graph of population growth from 1840 to 1940 for a U.S. town or city that you know. The best source of data will be the website and publications of the U.S. Census Bureau. When you have drawn the graph (by hand or by using computer software such as Microsoft Excel), annotate it, suggesting reasons for any fluctuations in the city's rate of growth.

4. What is the economic history of the U.S. city or town that you live in (or is of most interest to you)? Find out what it was like during each of the epochs of urbanization described in this chapter.

5. Work on your *portfolio*. Research some examples of power-site towns, mining towns, transportation towns, and heavy manufacturing towns, for example (p. 63). Find out how and why they became established and what has happened to them during the twentieth and early twenty-first centuries.

   Think about how you can make the aspects of urbanization described in this chapter meaningful to you personally. If you live in the United States, perhaps your great-grandparents or great-great-grandparents were among the earlier waves of immigrants who contributed to American urban growth. Where and how did they arrive? What did they do, and what cities did they and their descendants live in? You might even be able to add sound recordings of people or city life to your portfolio.

Find some quotations that help to highlight some of the ideas in the chapter or that give expression to some of your reactions to them. In addition to the World Wide Web, a very useful resource is *The City:*

A Dictionary of Quotable Thoughts on Cities and Urban Life, by James A. Clapp (New Brunswick, N.J.: Center for Urban Policy Research, 1984).

6. If you have time, read a novel that is evocative of urban life and urban change during part of the period covered by this chapter. There are hundreds to choose from, including *McTeague*, by Frank Norris (New York: Fawcett, 1960) and *A Tree Grows in Brooklyn*, by Betty Smith (Philadelphia: Blakiston, 1943).

KEY SOURCES AND SUGGESTED READING

The outline of U.S. economic development in this chapter draws from the more detailed account given in "The economic organization of U.S. space," pp. 111–49 in *The United States: A Contemporary Human Geography*, P. L. Knox *et al.*, New York: Wiley, 1988. On the analysis of the structural characteristics of urban systems, see *The Structure of Urban Systems*, John U. Marshall, Toronto: University of Toronto Press, 1989. Other useful sources include:

Abbot, C. 1992. Urban America. pp. 110–28 in *Making America*, ed. L. S. Luedtke. Chapel Hill, N.C.: University of North Carolina Press.

Berry, B. J. L. 1988. *The Geography of Market Centers and Retail Distribution*. Englewood Cliffs, N.J.: Prentice Hall.

Borchert, J. 1967. American Metropolitan Evolution. *Geographical Review* 57: 301–32.

Conzen, M. P. 1981. The American Urban System in the Nineteenth Century, pp. 295–347 in *Geography and the Urban Environment*, ed. R. J. Johnston and D. T. Herbert, vol. 4. Chichester, UK: Wiley.

Harvey, D. 1989. *The Urban Experience*. Baltimore, Md.: Johns Hopkins University Press.

Malecki, E. 1997. *Technology and Economic Development: The Dynamics of Local, Regional, and National Competition*, 2nd ed. Essex, UK: Longman.

Meyer, D. R. 1983. Emergence of the Manufacturing Belt: An Interpretation. *Journal of Historical Geography* 9: 145–74.

Mills, E. S., and J. F. McDonald, eds. 1992. *Sources of Metropolitan Growth*. New Brunswick, N.J.: Center for Urban Policy Research.

Pred, A. R. 1966. *The Spatial Dynamics of U.S. Urban-Industrial Growth, 1800–1914*. Cambridge, Mass.: MIT Press.

Roberts, G., and J. Steadman. 1999. *American Cities and Technology*. London: Routledge.

Storper, M., and R. Walker 1989. *The Capitalist Imperative: Territory, Technology and Industrial Growth*. New York: Blackwell.

Sui, D. Z., and J. O. Wheeler. 1993. The Location of Office Space in the Metropolitan Service Economy of the United States. *Professional Geographer* 45: 33–43.

Vance, J. E., Jr. 1970. *The Merchant's World: The Geography of Wholesaling*. Englewood Cliffs, N.J.: Prentice Hall.

Weinstein, B. L., and R. E. Firestine. 1985. *Regional Growth and Decline in the United States*, 2nd ed. New York: Praeger.

RELATED WEBSITES

American Memory, Library of Congress:
http://memory.loc.gov/ammem/aapchtml/
This website contains From Slavery to Freedom: The African-American Pamphlet Collection, 1822–1909 comprising nearly 400 pamphlets from the Library of Congress Rare Book and Special Collections Division. These pamphlets, published between 1822 and 1909, are the work of African-American authors and others who wrote about slavery, African colonization, emancipation, reconstruction, and related topics.

The History Net: *http://history.about.com/*
This website offers articles and hyperlinks related to the history of twentieth-century and earlier cities.

Internet Modern History Sourcebook:
http://www.fordham.edu/halsall/mod/modsbook14.html
This searchable online bibliography (maintained by Fordham University) offers a variety of resources on modern history related to American cities.

H-Urban: *http://www.h-net.msu.edu/~urban/weblinks/index.htm*
H-Urban is an international electronic discussion network that was established in 1993 at the University of Illinois at Chicago (UIC) as a forum for scholars of urban history. The H-Urban website offers numerous hyperlinks to wide-ranging sources of information on American cities, including women and social movements in the United States from 1830 to 1930, a Philadelphia timeline from 1646–1899, and the 1906 San Francisco earthquake.

4

URBAN SYSTEMS
IN TRANSITION

By the end of World War II, the U.S. and Western European economies had begun to enter a substantially different phase in terms of what they produced and how and where they produced it. This phase is sometimes referred to as advanced capitalism (or disorganized capitalism). Fundamental to this change was a decrease in the proportion of the workforce involved in manufacturing and a marked increase in service-related employment. Agriculture and manufacturing continued to increase their productivity, using improved technologies and organizational practices to employ less labor but produce more. This increased productivity, together with greater specialization in both agriculture and industry, generated new opportunities for employment in distribution services (transportation, communications, utilities, and wholesaling) and producer services (services such as marketing, advertising, administration, finance, and insurance that enable companies to maintain their specialized roles). The prosperity associated with this phase of economic development, meanwhile, allowed increasing numbers of households to spend a larger proportion of their incomes on consumer services related to personal services, eating out, home improvement, and recreation.

CHAPTER PREVIEW

This chapter deals with the changes that have occurred to the U.S. and European urban systems as a result of the transition to an advanced form of capitalism (disorganized capitalism) that is increasingly global in its scope. As in Chapter 3, we trace urban-system change in terms of epochs of urbanization that were distinctive in their patterns of urban growth and patterns of interaction between cities.

For the United States, three epochs are identified. The first, spanning 1945–1972, corresponded with a period of postwar economic recovery and growth. We will see how the principles and processes of urban change identified in Chapter 3 resulted in new outcomes as they operated through the new industries and technologies of advanced capitalism.

The second epoch, between 1972 and 1983, was a brief but important period that corresponded with a phase of economic crisis and reorganization. This was reflected within the **urban system** in terms of differential economic stress as the crisis and reorganization transformed the political economy of urbanization, creating the preconditions for the next and most recent epoch of urban-system change.

Since 1983, the U.S. urban system has been experiencing yet another phase of change, dominated this time by the effects of high-tech economic development and new telecommunications technologies. We will see how this has resulted in a functional reorganization of the hierarchy of U.S. cities, how it has come to be reflected in variations in the quality of life, and how it has influenced—and been influenced by—recent demographic and social changes. Finally, we will examine the implications of all these developments in terms of concepts and theories of urban system change.

FREEWAYS, REGIONAL DECENTRALIZATION, AND METROPOLITAN CONSOLIDATION (1945–1972)

As in previous epochs of development in the United States, between 1945 and 1972 changes in transport and communications technologies strongly influenced the outcomes of the structural economic change to **advanced capitalism** involving **Fordist** mass production. In this first phase the most significant developments in this respect were the construction of the interstate highway system (Figure 4.1) and the growth of a network of regional and subregional airports capable of handling large passenger jets (Figure 4.2). The net effects of changes in employment and transportation on the structure of the urban system can be summarized in terms of two apparently contradictory (but in fact interrelated) outcomes: **regional decentralization** and **metropolitan consolidation**. Increased accessibility, coupled with the attractions of cheaper land, lower taxes, lower energy costs, local boosterism, and cheaper and less-militant labor, allowed cities in the South and West to grow rapidly.

The metropolitan areas of the "Sunbelt" were particularly attractive to labor-intensive manufacturing processes and to the high-growth, high-tech sectors of the fourth **Kondratiev** upswing: electronics, aerospace, and petrochemicals. These industries could establish themselves in Sunbelt cities and have the cities and their infrastructure conform to them, rather than the other way around. Meanwhile, it seemed that the benefits of **initial advantage** and **cumulative causation** in the Manufacturing Belt were now outweighed by the effects of **agglomeration diseconomies**, the increasing costs of now-inefficient infrastructure, and the relatively low productivity of dated machinery. The outcome was a decentralization of jobs and population within the urban system, away from the system "core" toward metropolitan centers in the periphery.

But the reorganization of U.S. business that led to the regional decentralization of the economy also made for a certain amount of metropolitan consolidation. Two key economic functions—headquarters offices and research and development (R&D) laboratories—tended to become increasingly localized in larger metropolitan areas. This localization was mainly in response to (1) corporate reorganization following mergers and acquisitions and (2) the differential growth and changing pattern of economic specialization within the urban system. The pattern of **control centers**—cities with a high proportion of corporate headquarters—changed to reflect the growth of cities like Atlanta, Houston (Figure 4.3), Los Angeles, and Dallas as regional business centers and the relative decline of Manufacturing Belt cities like Cincinnati, Detroit, Philadelphia, and Pittsburgh.

At the same time, there was a consolidation of corporate headquarters locations in New York and

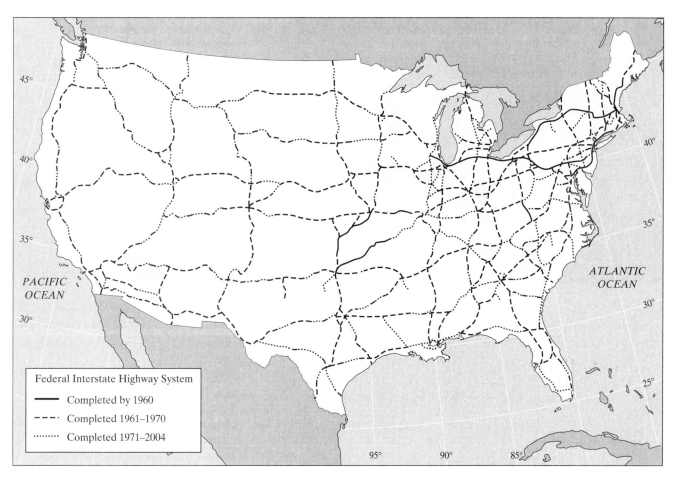

FIGURE 4.1   The growth of the interstate highway system.

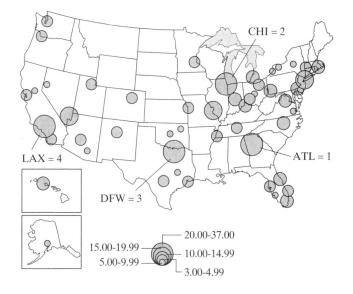

FIGURE 4.2   Air passenger traffic in 2002.

Chicago. By 1970 about 25 percent of the 500 largest industrial corporations had their headquarters in New York, as did the same proportion of the 300 largest corporations in banking, insurance, retailing, transportation, and utilities. The map of Fortune 500 corporate headquarters in 2002 reflects the continuing dominance of New York and Chicago as corporate control centers (Figure 4.4). Such dominance reflects "a process of cumulative and mutual reinforcement between relatively accessible locations and relatively effective entrepreneurship."[1] In other words, New York and Chicago were able to consolidate their importance as economic control points because of their reserves of entrepreneurial talent, the array of support services that they could offer, and their national and regional accessibility through the interstate highway and commercial airline networks. Meanwhile, localization is evident in the growing number and percentage of corporate headquarters in larger growth centers outside

FIGURE 4.3
The reorganization of U.S. business following World War II that led to the regional decentralization of the economy also made for a certain amount of metropolitan consolidation that promoted the growth of regional business centers like Houston.

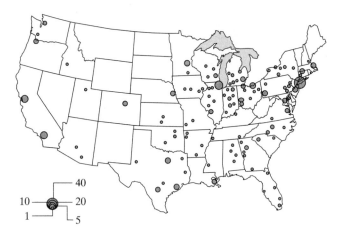

FIGURE 4.4   The location of Fortune 500 corporate headquarters in 2002. The size of the dots is proportional to the number of corporate headquarters in each city.

New York and Chicago—most notably the Sunbelt cities of Houston and Atlanta—that offered strong locational and entrepreneurial assets (Figure 4.4).[2]

R&D facilities, another high-income, high-growth economic activity, had traditionally been dispersed, their locational pattern tied to that of parent industries.

With the emergence of superlarge conglomerate companies under advanced capitalism, R&D activities became much more critical to corporate success. The increasingly high stakes involved made it vital to avoid market saturation, and companies could do that only by developing new products and improving old ones through R&D. Much long-range corporate R&D gravitated to central laboratories in headquarters complexes, where intraorganizational interaction could be fostered. In companies where independent divisions produced different product lines, however, R&D laboratories tended to be located in separate divisional laboratories. Both arrangements led to a consolidation of R&D activity in larger metropolitan areas. The most striking development (though not as important in terms of numbers of jobs) was the localization of some R&D activity in a few "innovation centers"—places with strong university research facilities, a strong federal scientific presence, and a range of cultural and recreational amenities of the kind that made them attractive to highly qualified scientific personnel. The best-known examples are Silicon Valley between Palo Alto and Santa Clara, California, the Route 128 area around Boston, and the Research Triangle area of North Carolina.

ECONOMIC CRISIS AND NEO-FORDIST URBAN RESTRUCTURING (1972–1983)

The Fordist period of postwar economic prosperity came to an abrupt end at the beginning of the 1970s in the United States and Western Europe. In October 1973 an Arab-Israeli war led Arab countries to impose an embargo on shipments of oil to the United States and other Western countries, and, soon afterward,

OPEC (the Organization of Petroleum Exporting Countries) quadrupled oil prices. The subsequent shock to the economic system reinforced a number of long-term structural economic problems, with the result that the U.S. economy was plunged into another episode of **stagflation**.

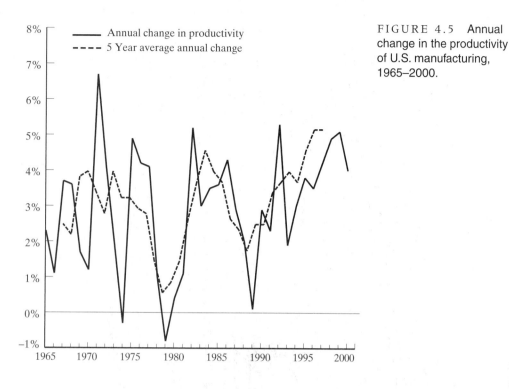

FIGURE 4.5 Annual change in the productivity of U.S. manufacturing, 1965–2000.

During the 1960s productivity in U.S. manufacturing had averaged between 2 and 3 percent growth per year. By the late 1970s productivity had fallen to less than 1 percent annual growth (Figure 4.5). At the end of the 1970s the average U.S. family had only 7 percent more real purchasing power than it had had a decade before. Between 1970 and 1983 the average weekly wages of U.S. workers fell, in real terms, from $375 to $365.[3] Between the late 1970s and the mid-1990s, the average weekly wages for production workers continued to fall, in real terms (Figure 4.6). Unemployment, having remained steady at around

4.5 percent until the early 1970s, reached 8.5 percent in 1974 and a peak of almost 10 percent in 1982. In 1971 the U.S. economy recorded its first trade deficit of the twentieth century; since then it has only twice shown a positive balance of trade with the rest of the world, in 1973 and 1975. By 1983 the United States had slipped below Switzerland, Japan, Norway, Sweden, and Finland in the international league table of per capita gross domestic product (GDP).

The fourfold rise in oil prices in 1973 as a result of the OPEC cartel is widely believed to have been responsible for initiating this downturn. The truth, however, is

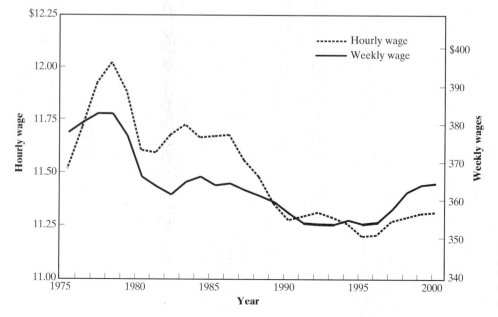

FIGURE 4.6 Changes in average hourly and weekly wages of U.S. production workers, 1975–2000 (in constant 1982 dollars).

that the sudden rise in energy prices only brought to a head a series of longer-term trends that were already working against the U.S. economy: the results of the latest long-wave downswing. This time, falling demand and rising inflation (stagflation) were aggravated by both the increasing penetration of Japanese and European companies into American markets and increased competition from **newly industrializing countries** or **NICs** (such as Taiwan, Mexico, Hong Kong, and South Korea) with significantly lower labor costs. At the same time, there was a shift to the new **technology systems** of the fifth **Kondratiev cycle**. As with previous epochal transformations, many of the new industries, based on new technologies and exploiting new resources, needed new settings. *Cities that had been the chief settings for the "old" industries of the previous Kondratiev cycle, meanwhile, experienced the brunt of economic and social dislocation.*

The result was a period of economic crisis that reverberated around the urban system while corporate America attempted to come to grips with the new situation. Within a decade corporate restructuring and redeployment had accelerated the decentralization of jobs from the older industrial neighborhoods of central cities (particularly in the Manufacturing Belt) and had stimulated the growth of new production spaces in Sunbelt cities, where land, labor, and local taxes were all cheaper.

The political repercussions of recession, meanwhile, led to a decisive shift away from the **Keynesian** approach to macroeconomic management and to a significant change in attitudes toward metropolitan management. Although the deepening economic recession after 1973 accentuated the vulnerability of more and more households, recession also made it difficult, both politically and economically, to finance the welfare programs that had expanded dramatically in the 1960s (see p. 412). Taxpayer "revolts," led by Proposition 13 in California in 1978 and Proposition $2\frac{1}{2}$ in Massachusetts in 1980, were followed by electoral victories by politicians who had argued that Keynesianism not only had generated unreasonably high levels of taxation, a bloated class of government workers, and disincentives for ordinary people to work and save, but also that it might have fostered soft attitudes toward problem groups in society. With the installation of Ronald Reagan as president in 1981, the federal government began a major round of *retrenchment*. Federal responsibilities were decentralized and many spheres of economic activity involving federal oversight were *deregulated* (see Chapter 17). Federal outlays on urban-related programs declined by about a third, in real terms, between 1978 and 1984, while many city governments, facing a fiscal crisis intensified by the withdrawal of federal funds (see p. 476), drastically reduced municipal workforces and cut back on public services.

URBAN DISTRESS

The stagflation crisis that began in 1973 lasted in the United States until 1982. Domestically, the symptoms of the crisis could be seen in falling rates of profit, plant closures, rising unemployment, and increasing poverty. High energy costs, high interest rates, foreign competition, and depressed levels of consumer demand drove many smaller companies out of business. Many larger companies, facing the same problems, responded by reorganizing their production processes, eliminating the duplication of activities among existing facilities, rearranging the division of tasks among them, closing down or trimming back activities in high-cost locations, investing in new facilities in low-cost locations, and diversifying into nonmanufacturing activities.

The economic sector that was hit hardest by the crisis was the traditional manufacturing sector, and the cities that were hit hardest were specialized manufacturing centers. Already suffering from post-World War II regional decentralization, the Manufacturing Belt fell into an accelerated decline that has been characterized as **deindustrialization** (Figure 4.8). In Youngstown, Ohio, which became emblematic of deindustrialization, the closure of the Campbell Steel Works in 1977 eliminated more than 10,000 jobs overnight. During the 1970s, Detroit lost more than 166,000 jobs (a net loss of nearly 30 percent of its 1970 employment base), while nearby Flint lost nearly 16,000 jobs (a 23 percent loss).

Deindustrialization on this scale involved a mutually reinforcing series of problems: a sort of cumulative

FIGURE 4.7   Rustbelt decline. Closed and abandoned stores in Gary, Indiana.

FIGURE 4.8 A sign and padlocked gates symbolize deindustrialization. The photograph shows a General Motors plant in Flint, Michigan, that was closed in the 1980s.

causation in reverse (Figure 4.9). Plant closures led to job losses in ancillary industries, and these in turn led to recession in retailing and personal services. Unemployment and **underemployment** led to lower incomes and increased levels of poverty, which in turn led to outmigration. Decreased prosperity within communities came to be reflected in aging housing and infrastructure and was reflected in weakened tax bases, which meant that city governments were unable to maintain or improve public services or amenities. In some cases the loss of tax revenues, combined with

increasing needs in terms of infrastructure maintenance and the provision of services and amenities, led to fiscal crises that threatened to bankrupt cities (see Chapter 16). All this amounted to a discouraging environment for investment, at least until wages and land prices were driven so low that some kind of **comparative advantage** might be recaptured. Meanwhile, many cities became acutely distressed.

Figure 4.10 shows the pattern of distress across the urban system (cities of 100,000 or more) at the end of the decade of crisis in the fiscal year 1982.

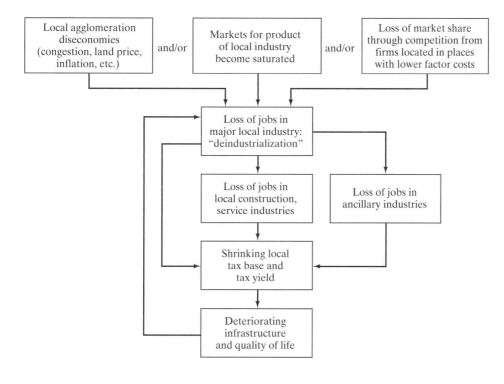

FIGURE 4.9 Deindustrialization and the downward spiral of economic decline.

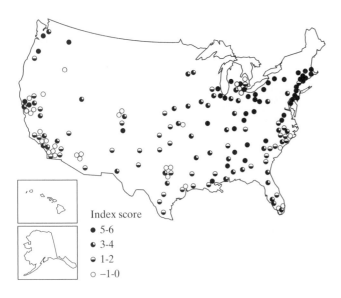

Index score
- ● 5-6
- ◕ 3-4
- ◔ 1-2
- ○ −1-0

FIGURE 4.10   Urban distress in the early 1980s. This index (based on rates of population growth and levels of household incomes, employment in manufacturing and retailing, poverty, older housing, and unemployment) was used by the federal government in allocating funds under the Urban Development Action Grant Program (see Chapter 17).

The index of distress is the one that the U.S. Department of Housing and Urban Development used at the time to determine eligibility for federal aid to cities (under the Urban Development Action Grant or UDAG program). It was based on a set of *minimum* standards relating to population growth, household incomes, and manufacturing and retail employment, and *maximum* standards in relation to the incidence of poverty, pre-1940 housing, and unemployment. Cities scored one point for each standard not met but lost a point if the incidence of poverty was less than half the national norm. The index thus ranged from −1 (for the most prosperous cities) to 6 (for the most distressed). The map clearly illustrates the distress that came to be associated with Manufacturing Belt cities, though it should be noted that high levels of distress were also found in the South (including Atlanta, Birmingham, Chattanooga, Knoxville, New Orleans, and Roanoke) and the West (including Oakland, Portland, Spokane, and Tacoma). It should

also be noted that a few of the most prosperous cities were located within the Manufacturing Belt, where the likes of Livonia, Sterling Heights, and Warren (all suburbs of Detroit) reflected the decentralization of jobs, population, and prosperity at the metropolitan scale (see Chapter 6).

It would be misleading, however, to portray the decade from 1973 to 1982 as a period of unmitigated economic gloom. Some sectors of the economy were able to prosper, while the quadrupling of oil prices in 1973 and the doubling of oil prices in 1979 enabled oil and petrochemical companies to generate vast profits. Yet, with much of the economy unable to offer attractive prospects for investment, the net result was a classic **overaccumulation crisis**, with surplus labor, surplus money capital, and idle productive capacity existing side by side. As we have seen (Chapter 3), this kind of market imbalance can temporarily be resolved within the economy's **circuits of capital investment** if capital investment is switched from the *primary* circuit of industrial production to the *secondary* circuit of capital assets or the *tertiary* circuit of education or technology research. This diversion seems to be what happened during the 1970s and early 1980s, though the evidence is by no means clear and analysts disagree about the nature of the switching mechanisms involved.

The important point, in the present context, is that investments in the secondary and tertiary circuits were to have profound effects on urban development in the United States. For, although large quantities of money capital were switched "offshore" (including, incidentally, massive loans to less developed countries that soon found that they could not service the debt), the principal target in the secondary circuit was speculative commercial development (especially high-rise office towers) in American metropolitan areas. The principal target for investment in the tertiary circuit, meanwhile, was investment in technology research, which helped to achieve a major restructuring of the economy after 1983. Investment in the secondary circuit is discussed in Chapter 11, which deals with changes in cities' built environment. The restructuring of the economy, which prompted some important changes in the urban system, is examined here.

NEO-FORDIST ECONOMIC RESTRUCTURING AND THE EMERGENCE OF "INFORMATIONAL" CITIES (1983–PRESENT)

As in previous epochs of economic and urban development, new technologies have been of critical importance in facilitating the neo-Fordist restructuring of the U.S. and European economies. Since the 1970s three kinds of "permissive" or "enabling" technologies

have been significant:

- *Production process technologies* such as electronically controlled assembly lines, automated machine tools, robotics, computerized sewing systems, and so on have increased the separability and flexibility of production processes. It is now easier for companies to take advantage of spatial variations in the costs of land and labor.

- *Transaction technologies*, and computer-based just-in-time inventory control systems in particular, have also increased companies' locational and organizational flexibility, allowing materials, components, or information to be purchased as and when needed and eliminating the necessity for large buffer stocks.

- *Circulation technologies* such as communications satellites, fiber-optic networks, microwave communications, e-mail, and wide-bodied jets have reduced the time and cost of distribution, bringing a wider geographic market within the range of an increasing range of business activities.

We can see the effects of these technologies in facilitating two of the three main processes that have been identified[4] as being central to the restructuring of the U.S. economy:

1. *A transformation of the relationship between corporate capital and labor*, with corporations recapturing the initiative over wage rates and conditions that had been lost in the Keynesian era of "organized" capitalism. Production process technologies played a major role in this transformation, since their ability to substitute machines for workers put labor into a very weak bargaining position. Circulation technologies were also important, since they gave large corporations the "global reach" that provided access to cheaper, nonunion labor. The decline of traditional manufacturing industries—the established base of organized labor—also played an important part in weakening labor unions, as did the restructuring of urban labor markets to take in growing numbers of women and new groups of immigrants (see below) who were less disposed to join labor unions.

2. *The emergence of new roles for the state and the public sector*, shifting their emphasis away from "collective consumption" (schools, hospitals, community services) toward (a) public-private cooperation, (b) deregulation, (c) increased support for high-technology R&D, and (d) increased spending on defense.

3. *The development of new international and intermetropolitan **divisions of labor***, as larger corporations exploited the locational and organizational flexibility made possible by new circulation and transaction technologies. It is important to emphasize that the "time-space compression" of these new technologies has, paradoxically, heightened the importance of geography. The reduction of spatial barriers has

had the effect of greatly magnifying the significance of small differences between local land and labor markets, since new technologies enable such differences to be quickly (if temporarily) exploited. As a result, the urban system is becoming a continuously variable geometry of labor, capital, management, production, and consumption. Meanwhile, the *pace* of change has been accelerating, raising some important questions about what might happen to the social stability of cities, civic loyalties, and peoples' sense of place (Chapter 17).

One of the most important aspects of the new inter-metropolitan division of labor concerns the location of professional and business services whose growth has been a fundamental part of the shift to advanced capitalism. Between 1976 and 1986 the number of jobs in professional and business services in the United States more than doubled, from 2.7 million to 5.6 million; by 1993 this number had more than doubled again to 11.5 million, and by 2002 it had grown to 16.0 million. Although much of this growth occurred in larger metropolitan areas, the greatest proportional growth has been in midsized metropolitan areas. Much of this growth has been localized and specialized. Boston, for example, has become a major center for computer and data-processing services, engineering services, and R&D laboratories. Washington, D.C., has a concentration of jobs in management consulting services; San Jose has a concentration in personnel supply services, R&D laboratories, and computer and data-processing services; Raleigh–Durham, Austin, and Orlando have concentrations of service jobs related to high-technology industries; and Huntsville, Colorado Springs, and Norfolk have concentrations of R&D labs and private professional and business services related to military activity.[5]

There is also an important geographical dimension to the contribution of these jobs to the overall employment base. Of the metropolitan areas with a greater-than-average share of their employment accounted for by jobs in professional and business services in 2001, fewer than 40 percent were in the Manufacturing Belt and the Midwest (Figure 4.11). Florida alone accounted for 18 percent of metropolitan areas with above average employment in professional and business services, with another 11 percent in the larger metropolitan areas on the West Coast and the remaining 45 percent stretching in a southerly band from California to the Carolinas.

The overall result of this restructuring, it is argued, is a new, "informational" mode of development in place of the industrial mode of development. Thus:

> In the industrial mode of development, the source of increasing surplus lies in the introduction of new energy sources and in the quality of the use of such energy. In the informational mode of development, . . .

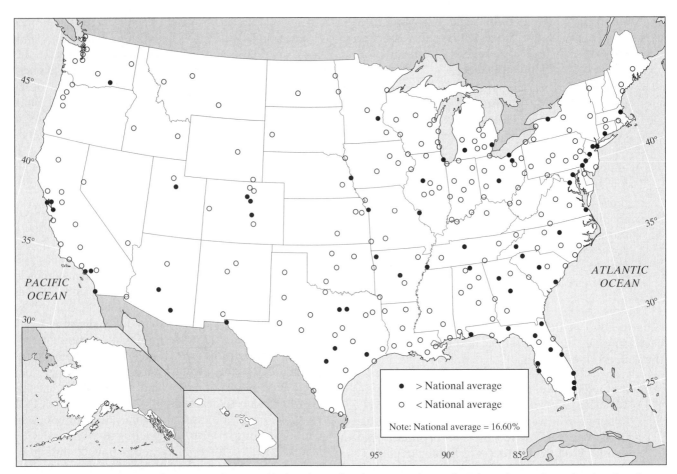

FIGURE 4.11   Metropolitan areas above and below the national average share of total employees in professional and business services (NAICS 54–56), 2001.

the source of productivity lies in the quality of knowl-edge . . . [which] mobilizes the generation of new knowledge as the key source of productivity through its impact on the other elements of the production process and their relationship.[6]

It follows that the critical dimension of economic restructuring is the ability of companies to acquire and use knowledge and information. Hence the notion of an urban system based on a "space of flows" rather than the space of **relative locations**.[7] What matters, in other words, is not so much the physical distance between cities, between companies, between factories and shopping malls, but the degree to which they are hooked in, via telecommunications systems, to the flow of ideas, capital, and people. New transaction and circulation technologies have also been important in shaping the flow of information (Figure 4.12).

The effectiveness of the new transaction and cir-culation technologies in supporting an informational mode of production has occurred partly because the new dominance of corporate capital over labor (achieved largely, as we have seen, as a result of the application of new rounds of production process

technologies and circulation technologies) has given companies the freedom to reorganize, redeploy, and relocate their resources. Analysis of the origins and destinations of Federal Express overnight letters, packages, and boxes provides an example of the ge-ography of this "space of flows."[8] Figure 4.13 shows a clear hierarchy of flows. The largest flows from one metropolitan area to another can be considered as "first-order" flows. These are depicted in Figure 4.13a, which illustrates very clearly New York's role

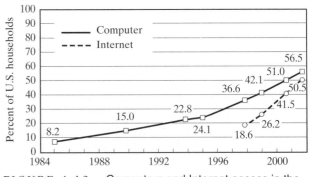

FIGURE 4.12   Computers and Internet access in the home, 1984–2001.

Box 4.1   The Digital Divide and Splintering Urbanism in the United States

Since the early 1980s the United States has experienced more than a fivefold increase in the proportion of households with one or more computers in the home, while the current rate of growth in Internet access is two million new users per month (Figure 4.12).[9] The "most wired" cities in 2001 were Portland and Seattle, followed closely by San Francisco and Boston—where nearly 70 percent of households had access to the Internet from a home computer.[10] E-mail and e-commerce have become part of many people's daily routine.

Despite the phenomenal increase in the use of computers and the Internet, a **digital divide** persists—a gap in opportunities among different segments of the population and different places to access advanced information and communication technologies for a variety of activities. More highly educated and more affluent adults are most likely to have home computers and use the Internet. In 2000, for example, 88 percent of higher-income U.S. households (with incomes of $75,000 or more) had at least one computer, and 79 percent had a least one household member who accessed the Internet at home. Among households with incomes below $25,000, only 28 percent had a computer, and 19 percent had Internet access. Within metropolitan areas in 2000, households in suburban locations were most likely to have a computer (58 percent) or Internet access (48 percent); in contrast, only 46 percent of households in central cities had a computer, and just 38 percent had Internet access. Nonmetropolitan households were least likely to have a computer or Internet access (42 percent and 32 percent, respectively). And although more school-age children use computers at school than have access to them at home, white children are more likely to have home computer

access or use the Internet than black or Hispanic children.[11]

So despite the initial promises that advanced information and communication technologies could benefit everyone everywhere:

> geography still matters ... electronic systems simultaneously reflect and transform existing topographies of class, gender, and ethnicity, creating and recreating hierarchies of places mirrored in the spatial architecture of computer networks. Far from eliminating differences among places, systems such as the Internet allow their differences to be exploited ... often reinforcing existing relations of wealth and power.[12]

The uneven diffusion of information and communications technologies is quite stark in U.S. cities, as new urban landscapes of innovation, economic development, and cultural transformation are forged at the same time that social and economic inequalities are intensified. The resulting splintering urbanism is visible as clusters and enclaves of "superconnected" people, firms, and institutions, often in the immediate vicinity of large numbers of people with nonexistent or rudimentary access to information and communications technologies.[13] The main concern is that a continued lack of access to these technologies will prevent many people from benefiting from the new knowledge-based economy—thereby widening existing socioeconomic and spatial differences in well-being within and among cities.[14] The policy responses that have been most effective in addressing this issue have been those that have sought to improve access to information and communication technologies within cities, while at the same time attempting to narrow existing educational and income disparities.

---

as *the* national information center. In addition, there are regional flows that focus on Dallas-Fort Worth, Houston, Atlanta, Miami, and Chicago. These cities, together with Washington, D.C., dominate the second-order flows (i.e., the second-highest flow volume from a given metropolitan area), which also underscore the strong ties between New York and centers in the West and between Los Angeles and the centers of the East (Figure 4.13b). Los Angeles dominates third-order flows (Figure 4.13c). The pattern of fourth-order flows (Figure 4.13d) is striking because of the absence of any pronounced regional hierarchy. Rather, a large number of metropolitan centers are involved in cross-national flows, emphasizing the

point that it is not so much the physical distance between cities that is important for the **informational city** as it is to be hooked into the "space of flows."

GLOBALIZATION, SPLINTERING URBANISM, AND UNEVEN URBAN DEVELOPMENT

The world's major cities and metropolitan regions are pivotal settings for the processes involved in economic, cultural, and political **globalization** (see Chapter 1). As a result, they reflect these processes in their own patterns of urban development and change. Central to this strong relationship between urbanization and

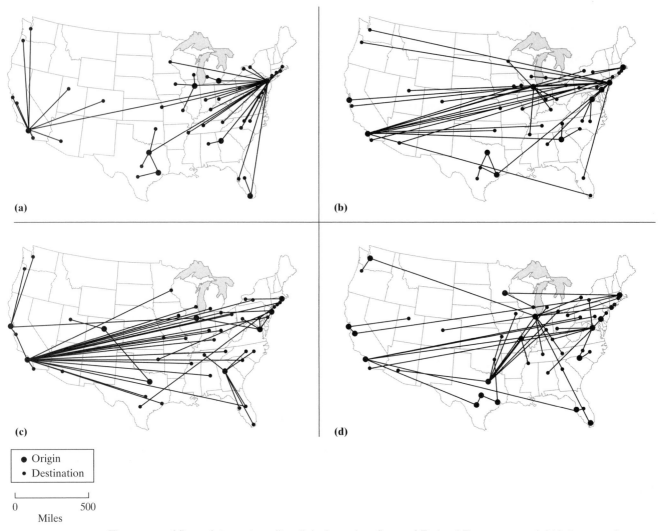

(a)

(b)

(c)

(d)

● Origin
• Destination

0   500
   Miles

FIGURE 4.13   The space of flows. Intermetropolitan links based on flows of Federal Express overnight letters, packages and boxes: a, first-order links; b, second-order; c, third-order, d, fourth-order.

globalization are networked infrastructures of transportation, information, and communications technologies, such as telephone systems, satellite television, computer networks, electronic commerce, and business-to-business Internet services. According to UN-Habitat (the United Nations Centre for Human Settlements),[15] information and communications technologies are intensifying global urbanization in three main ways:

1. They allow specialist urban centers, with their high value-added services and high-tech manufacturing, to extend their powers, markets, and control to ever-more distant regional, national, international, and global spheres of influence.

2. The growing speed, complexity, and riskiness of innovation in a global economy require a concentration of technological infrastructure and an associated knowledgeable technology-oriented culture in order to sustain competitiveness.

3. Demand for information and communications technologies is overwhelmingly driven by the growth of metropolitan markets. World cities (see pp. 92–94), especially, are of disproportionate importance in driving innovation and investment in networked infrastructures of information and communications technologies—due to their cultures of modernization, concentrations of capital, relatively high average disposable personal incomes, and concentrations of internationally oriented companies and institutions.

In contrast to the infrastructure networks of earlier technology systems that underpinned previous phases of urbanization, these information and communications technologies are not locally owned, operated, and regulated. Rather, they are designed, financed, and operated by **transnational corporations** to global market standards. Detached from local processes of urban development, these critical networked infrastructures are highly uneven in their impact, selectively serving only certain neighborhoods, certain cities, and certain kinds of metropolitan settings. Geographers Stephen Graham and Simon Marvin have called this important new tendency **splintering urbanism**.[16]

Splintering urbanism is characterized by an intense geographical differentiation, with individual cities and parts of cities engaged in different—and rapidly changing—ways in ever-broadening and increasingly complex circuits of economic and technological exchange. Figure 4.14 shows some of the processes associated with new telecommunications systems that can lead to social and spatial polarization within cities. The uneven evolution of networks of information and communications technologies is forging new and dynamic landscapes of innovation, economic development, and cultural transformation, while at the same time intensifying social and economic inequalities within and between cities.

Enclaves of 'superconnected' people, firms, and institutions, with their increasingly broadband connections to elsewhere via the Internet, mobile phones and satellite TVs and their easy access to information services, often exist cheek-by-jowl with much larger

numbers of people with at most rudimentary access to modern communications technologies and electronic information.[17]

The chief beneficiaries of splintering urbanism are world cities and major regional metropolitan centers, particularly in **core countries** where large corporations are able to undertake the necessary capital investments in new networked infrastructures and national governments and entrepreneurial urban agencies are able to subsidize and facilitate them. In Chapter 6 we identify several specific kinds of urban settings within cities that are most directly associated with this dimension of globalization (p. 158).

ECONOMIC FUNCTIONS WITHIN THE URBAN HIERARCHY

Successive epochs of economic and urban development have produced urban hierarchies that are highly differentiated in terms of economic functions. Whereas the dominance of large traditional manufacturing centers characterized the urban systems from the turn of the twentieth century to the 1950s in the

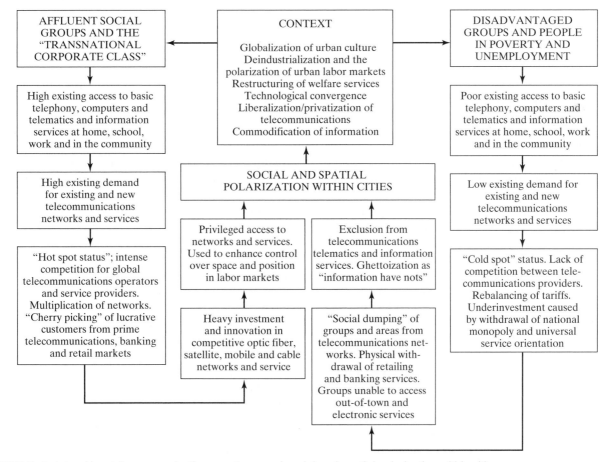

FIGURE 4.14   New telecommunications systems and social and spatial polarization within cities.

United States and Western Europe, the pivotal role of metropolitan areas specializing in service industries, particularly producer services, characterizes the urban systems under advanced capitalism and the current "informational" mode of development:

> In this new system, dominance is defined increasingly by the capacity of these service centers to organize and expand production on a system wide basis—more and more international in dimension—and increasingly less by their ability to get the goods out locally, as was the case in the past.[18]

What is remarkable about this transformation is that a relatively large number of these new service-oriented centers are large traditional manufacturing centers of the previous epoch that have been fundamentally transformed in terms of economic structure. More recently in the United States, a growing number of newer Sunbelt (or "trade and defense perimeter") cities have reached a dominant status. Some analyses[19] suggest that the urban hierarchy exhibits a four-tiered structure in terms of economic functions:

1. world cities
2. regional control centers
3. specialized producer-service centers
4. dependent centers

World Cities

At the top of the hierarchy, some of the largest metropolises (but not necessarily the very largest) have become so closely integrated within the *global* economic system that their economic activities—and with them many of their social, political and cultural attributes—qualify them as world cities. The term **world city** originated in 1915 with Patrick Geddes, one of the leading figures in the evolution of city planning (see Chapter 17). Geddes used the term to highlight the primacy of the few cities in which a disproportionate amount of the world's business was conducted. With the increasing interdependence of a world economy that is now much more transnational than in Geddes' day, the term has been adopted to highlight cities that are at the center of the "space of flows"—cities that are control points for transnational production and marketing and for international finance.

There are three world cities that, because of the importance of their financial markets and associated business services, stand out as being in a class of their own: London, New York, and Tokyo. These are the cities that dominate Europe, the Americas, and Asia, respectively, while being closely bound together through their involvement in the organization and management of the global system of finance and production (Figures 4.15, 4.17).

FIGURE 4.15   The New York Stock Exchange: an important element in New York's position as a dominant world city.

Below these three in the global urban system are a few other fully fledged world cities: Chicago and Los Angeles in the United States; Paris, Brussels, Frankfurt, and Zurich in Europe; São Paulo in South America; and Singapore and Hong Kong in Asia (Figure 4.16). Finally, there are about 23 more cities that qualify as "secondary" world cities: cities that complete the international space of flows between governments, major corporations, stock exchanges, futures exchanges, securities markets, commodity markets, major banks, and international organizations. There are four U.S. cities in this category: Houston, Miami, San Francisco (Figure 4.18), and Washington, D.C. Toronto in Canada is also a member of this category. Europe's "secondary" world cities are Berlin, Rotterdam, Vienna, Milan, and Madrid. Moscow also belongs in this category. The problems of gathering reliable and up-to-date data on cities' international activities, however, make it impossible to be definitive about world city status. The list given here and shown in Figure 4.16 is, therefore, to be taken as illustrative: a product of what we know and what we think we know about the world economy.

Similarly, it is impossible to be definitive about whether, and in what ways, world cities differ from

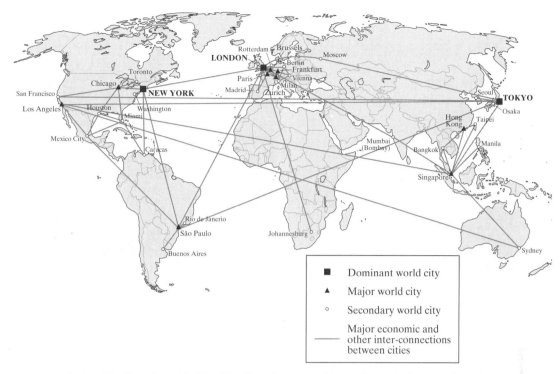

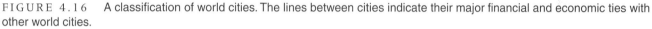

FIGURE 4.16   A classification of world cities. The lines between cities indicate their major financial and economic ties with other world cities.

other cities in terms of attributes such as social organization, culture, politics, and so on. Nevertheless, a conventional wisdom has emerged among urban theorists that they *are* different or at least are *becoming* different. Among the distinctive features of world cities in this context are:

1. Increasing control over the production and transmission of news, information, and culture.
2. Marked economic and social polarization and intense spatial segregation involving:
   • a growing international elite, dominated by a transnational producer-service group (in law,

FIGURE 4.17   America's dominant world city: New York.

FIGURE 4.18   San Francisco: a secondary world city.

FIGURE 4.19   Seattle: a regional nodal center.

banking, insurance, business services, accounting, engineering, advertising, and so on);

- pronounced inner-city **gentrification** and redevelopment for luxury use;
- a large **informal economy**; and
- a large and growing group of multiply disadvantaged people.

3. Massive concentrations of new immigrant groups.
4. Heightened political conflict over issues of urban growth and management and intensified race and class conflict.

Regional Control Centers

Regional control centers are characterized by large concentrations of national and regional headquarters of large corporations, well-developed banking facilities, dense networks of producer-service companies (e.g., insurance, accounting, advertising, legal counsel, public relations, R&D), and concentrations of important educational, medical, and public-sector institutions. Most of them are also important centers for the wholesale distribution of manufactured goods. In the United States, they include 19 *regional* **nodal centers** (such as Seattle: Figure 4.19) and 16 *subregional nodal centers* (such as Richmond: Figure 4.20). They are listed in Table 4.1.

Specialized Producer-Service Centers

The service specialization of the third-tier cities, specialized producer-service centers, tends to be more narrowly defined, while they depend on world cities and nodal centers for high-order **producer services** such as banking and advertising. Most are specialized in management and technical production for well-defined

industries (e.g., office equipment in Rochester or semiconductors in San Jose in the United States) and are therefore characterized by concentrations of the headquarters offices, R&D facilities, and technically oriented production plants of large companies in those industries. These *functional nodal centers* are found within the old U.S. Manufacturing Belt (examples include Detroit, Milwaukee, Pittsburgh, and Toledo) and increasingly in

TABLE 4.1

Regional Command and Control Centers

Regional nodal centers	*Subregional nodal centers*
Atlanta	Birmingham
Baltimore	Charlotte
Boston	Des Moines
Cincinnati	Jackson (Mississippi)
Cleveland	Jacksonville
Columbus	Little Rock
Dallas	Memphis
Denver	Mobile
Houston	Nashville
Indianapolis	Oklahoma City
Kansas City	Omaha
Miami	Salt Lake City
Minneapolis	Shreveport
New Orleans	Spokane
Philadelphia	Syracuse
Phoenix	Richmond
Portland	
St. Louis	
Seattle	

Source: *T. J. Noyelle and T. M. Stanback, Jr.,* The Economic Transformation of American Cities, *Totowa, N.J.: Rowman & Allanheld, 1984, Table 4.2, pp. 55–56.*

FIGURE 4.20   Richmond: a subregional nodal center.

FIGURE 4.21   Las Vegas, Nevada: a dependent resort center that experienced phenomenal growth in the 1990s—it grew at a rate of 83.3 percent—and increased its metropolitan population from 852,737 to 1,563,282 to become the thirty-second largest metropolitan area in the United States in 2000 (up from fifty-second place in 1990).

the South and West (examples include Orlando and San Jose). A second group of specialized producer-service centers consists of *government/education centers*, mostly state capitals that also have one or more large universities (U.S. examples include Albany, Austin, Madison, Raleigh–Durham, and Lansing).

Dependent Centers

Most fourth-tier cities, dependent centers, are smaller, though there are several important exceptions (including San Diego, Buffalo, Albuquerque, and Tampa in the United States). What these cities have in common is that their fortunes depend largely on the decisions made in regional, national, and global control centers. Four subcategories can be identified:

1. *Traditional manufacturing centers* such as Buffalo, Chattanooga, Erie, and Rockford. Nearly all of them are located in the old U.S. Manufacturing Belt.

2. *Industrial/military centers* such as Huntsville, Newport News, and San Diego. Most of them are in the South and West.

3. *Mining/industrial centers* such as Duluth and Charleston.

4. *Resort/retirement/residential centers* such as Albuquerque, Fort Lauderdale, Orlando, and Las Vegas (Figure 4.21). Most of these cities are also in the South and West.

BOX 4.2   VARIATIONS IN THE QUALITY OF LIFE WITHIN THE U.S. URBAN SYSTEM

Economic history and urban history have left some cities with crumbling infrastructure, few amenities, and large populations of vulnerable and disadvantaged people, while others enjoy not only sound infrastructure, a broad range of first-class amenities, and a relatively prosperous population, but also a balmy climate and attractive terrain. Even so, it is difficult to gauge overall differences in the quality of life. Apart from the problems of gathering data on all of the relevant dimensions of people's quality of life, there is an important conceptual problem in that some people—and, more important, some specific *groups* of people (based on age, ethnicity, income, education, occupation, and so on)—see things differently than others. For many younger people, for

example, the quality of a city's educational facilities is of particular importance in determining their quality of life. But for the many families without children, or whose children are grown, local educational facilities are at most a marginal consideration. Also, by no means does every person place much value on the amenities of "high" culture: art galleries, theaters, opera houses, and so on.

For these reasons it does not always make sense to calculate composite indexes of places' overall quality of life: There are too many hidden trade-offs. Rather, it is more useful to think in terms of the different *packages of attributes* that places offer. We are, nevertheless, still confronted with the important question of just which attributes to consider and

(Continued)

Box 4.2   Variations in the Quality of Life Within the U.S. Urban System   (Continued)

with the (sometimes very tricky) technical problem of how best to measure them. Are spectator sports, for example, important enough to include in the list of attributes to be considered? And, if so, how are they to be measured: By the number of different sports? By the seating capacity of stadiums and arenas? By the win–loss record of the city's teams? And how should an NCAA Division I football team count in comparison with an NFL team? Few economic, social, cultural or environmental indicators are unambiguous or definitive, and so we must be cautious in interpreting the results of quality-of-life studies.

One widely used source of data is the *Places Rated Almanac*,[20] which ranks 354 metropolitan areas in the United States and Canada in terms of nine components on the basis of points allocated for a set of criteria for each that are indexed against the North American average:

1. Cost of living (state and local taxes, home mortgage, utilities, food, health care, transportation, and recreation expenses).

2. Transportation (supply of public transit and the typical time it takes to get to work and back; connectivity with other metropolitan areas via national highways, scheduled air service, and passenger rail service; and relative location to all other metropolitan areas).

3. Job growth (percentage increase in new jobs projected by 2005; projected number of new jobs created, 2000–2005; number of new jobs with above average pay).

4. Education (school support, library popularity, college town, college options).

5. Climate (winter mildness, summer mildness, hazardousness, seasonal variation).

6. Crime (violent crime [murder, robbery, and aggravated assault] rate, property crime [burglary, larceny-theft, motor vehicle theft] rate x 0.1).

7. The arts (number of art museums, annual museum attendance, per capita museum attendance, annual ballet performances, touring artist bookings, opera performances, professional theater performances, and symphony performances).

8. Health care (office-based physicians in general and family practice, medical specialists, and surgeons; accredited short-term, general hospital beds; and hospitals with physician-teaching programs certified by the American Medical Association [AMA] or Association of Canadian Teaching Hospitals).

9. Recreation (amusement and theme parks, aquariums, auto racing, college sports, gambling, golf courses, good restaurants, movie theater screens, professional sports, protected recreation areas, skiing, water area, zoos).

Based on their scores for the different components, metropolitan areas can be grouped according to similarities in the "package" of these nine components. For example, a recent college graduate just entering the job market and not yet in a long-term relationship might be interested in that group of metropolitan regions that have the best scores for job outlook and recreation. The five metropolitan areas with the highest average scores in these two categories are Riverside–San Bernardino, Calif.; Orlando, Fla.; Las Vegas, Nev.–Ariz.; Phoenix-Mesa, Ariz.; and Salt Lake City–Ogden, Utah. An older person who is considering retirement might want to identify locations with a mild climate, low cost of living, adequate health care facilities, and a low crime rate. The five metropolitan areas with the highest average scores for these four categories are Johnson City–Kingsport–Bristol, Tenn.–Va.; Roanoke, Va.; Huntington–Ashland, W.Va.–Ky.–Ohio; Augusta–Aiken, Ga.–S.C.; and Asheville, N.C.

Box 4.3   Contemporary European Urbanization[21]

Although there is no standard European-wide definition for exactly what qualifies as a city, the majority of Europeans live and work in urban areas. Europe's urban population of approximately 390 million represents about 12 percent of the world's urban population or about 38 percent of the urban population of the world's developed countries. Although Europe as a whole is about 74 percent urban, wide variation is found from country to country, ranging from lows near 40 percent in such countries as Albania or Bosnia and Herzegovina, between 90 and 100 percent in countries such as Belgium, the Netherlands, and the United Kingdom, to virtually 100 percent in the microstates, such as Monaco (Table 4.2). These

(Continued)

Box 4.3   CONTEMPORARY EUROPEAN URBANIZATION   (*Continued*)

TABLE 4.2

Urban Population in Europe,[1] 2001

	Population (thousands)		Percent urban
	Urban	Total	
Western Europe			
Austria	5,444	8,075	67.4
Belgium	9,997	10,264	97.4
France	44,903	59,453	75.5
Germany	71,948	82,007	87.7
Luxembourg	406	442	91.9
Netherlands	14,272	15,930	89.6
Switzerland	4,826	7,170	67.3
Northern Europe			
Denmark	4,538	5,333	85.1
Estonia	955	1,377	69.4
Finland	3,031	5,178	58.5
Iceland	261	281	92.7
Ireland	2,276	3,841	59.3
Latvia	1,437	2,406	59.8
Lithuania	2,532	3,689	68.6
Norway	3,365	4,488	75.0
Sweden	7,358	8,833	83.3
United Kingdom	53,313	59,542	89.5
Southern Europe			
Albania	1,351	3,145	42.9
Bosnia and Herzegovina	1,764	4,067	43.4
Croatia	2,706	4,655	58.1
Greece	6,408	10,623	60.3
Italy	38,565	57,503	67.1
Malta	357	392	91.2
Portugal	6,601	10,033	65.8
Slovenia	975	1,985	49.1
Spain	31,073	39,921	77.8
Macedonia	1,213	2,044	59.4
Yugoslavia (Serbia and Montenegro)	5,446	10,538	51.7
Eastern Europe			
Bulgaria	5,303	7,867	67.4
Czech Republic	7,647	10,260	74.5
Hungary	6,428	9,917	64.8
Poland	24,123	38,577	62.5
Romania	12,363	22,388	55.2
Slovakia	3,111	5,403	57.6

[1] See United Nations web site (*http://www.un.org*) for urban population data for smaller countries, such as Monaco, Liechtenstein, etc.

Source: *United Nations,* World Urbanization Prospects: The 2001 Revision, *New York: Department of Economic and Social Affairs, Population Division, Table A-1 (Available at: http://www.un.org/esa/population/publications/wup2001/WUP2001report.htm).*

(*Continued*)

Box 4.3 Contemporary European Urbanization   (Continued)

figures reflect the close correspondence between the urbanization of a country and its economic, political, and even physical geography.

The empirical observation of rank-size rule holds for Belgium, Germany, Italy, Norway, and Switzerland. In contrast, the national urban systems with primate cities, often national capitals, include Paris, Budapest, Vienna, Reykjavik, Dublin, Athens, and Sofia. As each country developed its own national capital and urban system, however, few cities grew to the size and status of world cities on a par with New York and Tokyo. Of course, London is one of the world's largest metropolitan regions and, within Europe, Paris is next. Although these two cities reached the one million mark by 1800 and 1850 respectively, quite a few European cities have grown to a million or more inhabitants since 1900. Historically, population movement into cities has been the most important component of urban growth, especially during industrialization. This form of internal migration, though, has largely ceased, especially within Western Europe. As the birth rates in most countries have also fallen considerably since the Baby Boom, European cities are now among the slowest growing in the world, averaging just 0.11 percent growth a year.

Nevertheless, the cities of Europe have expanded outward as a result of transportation and communications improvements. In some instances the burgeoning neighboring towns and cities have coalesced into massive polycentric metropolises on a par with the Boston–Washington axis in the United States and the Tokyo–Osaka axis in Japan—especially the extensive urbanization connecting London, Birmingham, Liverpool, Manchester, Leeds, and Newcastle in England, but also the Rhine-Ruhr metropolitan region of Germany and the Randstad, a densely populated horseshoe-shaped metropolitan region in the western Netherlands (see Chapter 6).

Although criticized for being overly simplistic and static, a **core-periphery model** is often used to describe the spatial pattern of urban development in Europe (Figure 4.22). The dominance of the core cities is based on their superior endowment of those factors that determine the location of economic activity, such as accessibility to markets. The largest cities at the core are connected by the most advanced transportation and communications systems. Labor force quality and government policies make the core the most attractive area for major modern companies, institutions, and industries.

Cities in the core and periphery are linked in a symbiotic if unequal relationship. Core cities prosper and maintain their economic dominance at the expense of the periphery by capturing flows of migrants, taxes, and investments in cutting-edge industries such as high-technology manufacturing and in command and control functions like the headquarters of transnational corporations. Peripheral cities have more limited potential for economic development and attract tourists, innovations, and investment in branch plants from core locations. In between are the semiperipheral cities that have economic links with both the core and the periphery.

An early start gave some cities an initial advantage that led to **localization economies** involving clusters of particular industries. As the birthplace of the Industrial Revolution, the English Midlands formed the nucleus of an early industrial core of economic growth. This core extended as far as Paris to the south and the Ruhr cities of Düsseldorf and Dortmund to the east. The periphery included Ireland, western and southern France, Portugal, Spain, Italy, Albania, Greece, Bulgaria, and Romania.

Over time, changes in the locational preferences of companies and industries have affected the core-periphery pattern of development. From the early 1970s the old industrial core cities began to experience the severe economic and social impacts of deindustrialization. The shift from labor-intensive freight handling to **containerization** and bulk transportation methods devastated employment levels in traditional port cities like Liverpool and Rotterdam. Well-paid heavy industry jobs were replaced with either low-paying services employment or well-paid high technology and information services positions. This squeezing out of the middle-income groups has made the urban population more polarized economically and by residential location.

The core has now shifted to the south and east to areas of modern industrial growth. New core cities include Frankfurt, Stuttgart, and Munich in southern Germany, Zürich and Geneva in Switzerland, Milan and Turin in northern Italy, and Lyon in southern France (Figure 4.22). London and Paris have managed to retain their historic importance because of their size and established positions as major national and international cities in the urban hierarchy. The continued economic strength of the core is reinforced by the considerable political control that accompanies the role of the largest core cities as major centers of international decision-making.

(Continued)

BOX 4.3 CONTEMPORARY EUROPEAN URBANIZATION (*Continued*)

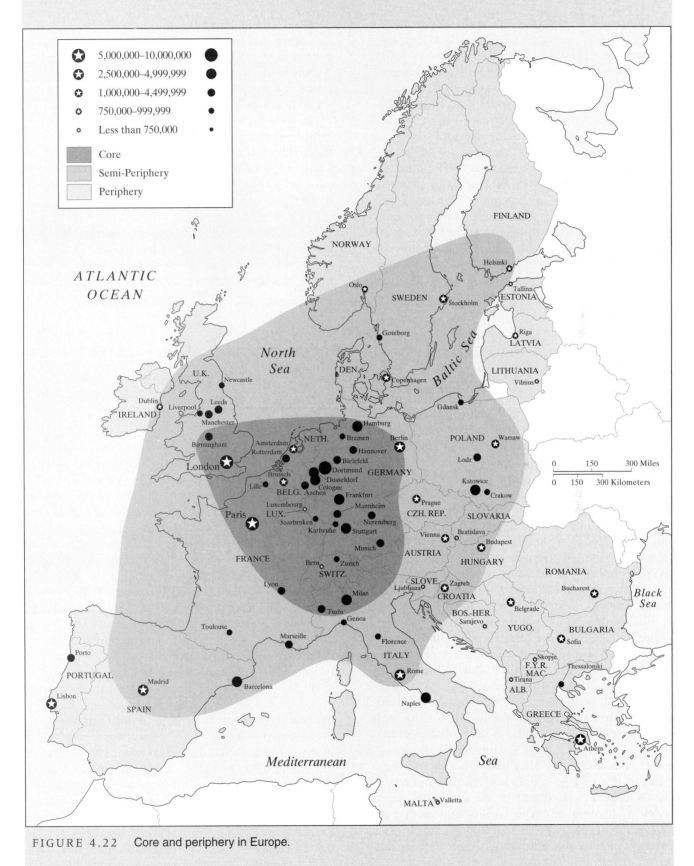

FIGURE 4.22 Core and periphery in Europe.

> ### Box 4.3 Contemporary European Urbanization   *(Continued)*
>
> Certainly, European urban areas are part of networks of cities that operate at a number of spatial scales. Since 1989 cities on either side of the former Iron Curtain have become more interconnected in the "new" European economy. Cities like Berlin, Prague, Vienna, and Ljubljana, formerly at the edge of Western Europe, have been well placed to take advantage of increasing trade between east and west and have emerged as vital links at the heart of a reunited Europe. Since the early 1990s, improved air, rail, and road transportation linkages connecting the major urban centers in both the east and the west have been laying the foundation for the complete integration of the European urban system. Meanwhile, the increasing economic and political integration among the member countries of the European Union (EU)—involving the removal of national
>
> barriers to the free movement of people, goods, information, and capital investment—continues to exert a strong influence on the development of the European urban system. Membership in the EU might better enable the former socialist cities to address their pressing social and economic problems.
>
> At some point serious consideration might be given to shifting some of the administrative functions of the EU from Brussels, Luxembourg, and Strasbourg to cities further east. Although London, Paris, Frankfurt, Brussels, and Milan will continue to dominate as major financial, political, and cultural centers, Berlin, Vienna, Warsaw, Prague, and Budapest, which have already emerged as important European centers, will surely shift the center of gravity of the core-periphery model further east as the twenty-first century progresses.

DEMOGRAPHIC AND SOCIAL CHANGE IN THE POSTWAR URBAN SYSTEMS

Parallel with all this economic change, some important demographic and social changes were under way. As they have interacted with economic changes, some important consequences for the pattern of urbanization have emerged. One of the most striking demographic events was the maturing of the so-called *Baby Boom* generation. The boom had its origins in markedly increased birth rates after 1945, when Americans and Europeans were—literally—catching up with one another during the relative stability of the immediate postwar period. Subsequently, the Baby Boom was sustained by the affluence of the 1950s and early 1960s, which encouraged people to marry younger, have children earlier, and have larger families. The boom ended rather abruptly in 1964, as the availability of the contraceptive pill made birth control and family planning much more effective. The economic recession of the 1970s, followed by the increasingly materialistic culture of the 1980s (see Chapter 12), resulted in delayed marriages, deferred childbearing, and smaller family sizes. The Baby Boom was over. Meanwhile, the availability of cheap and efficient contraceptives played an important role in the radical changes that took place in social attitudes in the mid to late 1960s. As attitudes toward sexual behavior changed, so too did people's attitudes to divorce and the family, and with these changes in attitudes there soon followed changes in patterns of household formation that were to have direct effects on patterns of demand for housing

space and all kinds of urban services, from schools to hospices.

The Baby Boomers

At the center of these social changes were the early cohorts of the Baby Boom generation, people who had been born in the late 1940s and who had reached college age in the mid-1960s. Their formative experiences had taken place in the context of the postwar economic boom, and now, as young adults, they were presented with the new freedoms of sex, drugs, and rock 'n' roll. This freedom, in turn, helped to set in motion a widespread rebellion against the apparent complacency and rigidity of what J. K. Galbraith called "The Affluent Society"[22]—a rebellion that was given added focus by civil rights issues and dissent over U.S. involvement in Vietnam in the late 1960s. The result was a *counterculture* movement with radical political agendas and a strong collectivist approach to the pursuit of freedom and self-realization.

In the 1970s in the United States, these cohorts found themselves on the labor market and in housing markets, where their large numbers caused unprecedented competition. Wages stood still while house prices ballooned. Meanwhile, new patterns of household formation associated with changed social attitudes—more single-adult households, smaller families, and more female-headed families (Figure 4.23)—produced *more* households and so even more competition for housing space.[23] In 1975 in the United States, average owners could meet the mortgage payments on a median-priced existing home

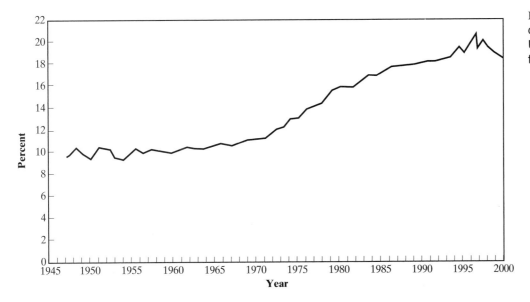

FIGURE 4.23 The changing percentage of U.S. families headed by females, 1947–2000.

with about 21 percent of their income. By 1981 the same house took about 36 percent of their income (Table 4.3). Meanwhile, these new patterns of household formation, coupled with greater numbers of women entering the labor force (another product of the revolutionary social changes of the 1960s) and an intensifying economic recession, led to even greater competition in labor markets.

Between 1973 and 1983 alone, the real median income of all U.S. households headed by someone under 35 fell by nearly 11 percent.[24] One of the first casualties was the progressive, collective radicalism of the late 1960s. Liberalism gave way to conservatism; a survival mentality eclipsed exploration and liberation; and philosophies of self-realization came to be directed inward, fostering a more selfish, narcissistic, and materialistic culture. The urban outcomes of this shift will echo through the pages of much of the rest of this book. What was most important in terms of the complexion of the overall urban system was that Baby Boomers, like previous generations who had lived through periods of economic restructuring, began voting with their feet, moving out of cities where the economic prospects were least attractive. Table 4.4 shows that all ten of the strongest "magnets" for adult Baby Boomers in the United States during the 1980s were Sunbelt metropolitan areas with expansive high-tech and defense-oriented economies, while all ten of the chief "losers" were in the old Manufacturing Belt.

Now, at the beginning of the twenty-first century, the 76 million Baby Boomers in the United States have reached the "nesting" stage of the lifecycle—between their early 40s and late 50s—the generally more stable, lower-migration, middle-aged, prime earning years. Table 4.5 shows the U.S. metropolitan areas with the greatest shares of Baby Boomers in 2000.

Among metros with the largest Boomer shares are several areas known for attracting "yuppie" Baby Boomers in the late 1970s and 1980s—including some in Seattle, San Francisco, Atlanta and Washington, D.C. Many of these territories that attracted the Boomers have matured along with their Boomer residents in terms of their cultural attractions and amenities. The housing prices have generally escalated as Boomer incomes rose. Several other metros not often associated with Boomers are on the list. Over the past decade, cities such as Denver and Nashville, Tenn., have attracted Boomers to their fast-growing job markets. Metros such as Minneapolis and Colorado Springs, Colo., have retained or enticed Boomers with their cultural or natural attractions ... the Boomer nests are areas in which Boomers still have some dominance in market share, political clout and in the general social and cultural scene. Their influence is likely to persist in these locales, since most are likely to remain during their retirement years.[25]

The Post-Boomers/Young Adults

The Post-Boomers/Young Adults, born between 1966 and 1980 and numbering about 59 million in the United States, have become another important demographic group as they have established their own households while their disposable income has grown.[26] This group has entered the housing market during a period of improving housing affordability for owners since the mid-1980s as falling mortgage interest rates have helped to offset rising home prices (Table 4.3). Still in their high migration years, the Post-Boomers include a significant proportion of new immigrants to the United States. This often independent, multicultural, and entrepreneurial group—of 20 to 34 year olds in 2000—has its own metropolitan distribution (Table 4.6). In contrast to the Baby Boomers in 2000

TABLE 4.3

Income and Housing Costs of Homeowners in the United States, 1975–2002 (in Constant 2002 Dollars)

Year	Pre-tax monthly income	Median existing home sales price	Mortgage rate (%)	Pre-tax mortgage payment	Percent of income to meet mortgage payment on median priced home (%)
1975	4,069	117,595	8.9	845	20.8
1976	4,045	119,746	8.9	857	21.2
1977	4,058	124,648	8.8	888	21.9
1978	4,101	132,482	9.4	991	24.2
1979	4,107	133,718	10.6	1,109	27.0
1980	3,857	127,843	12.5	1,224	31.7
1981	3,746	122,122	14.4	1,336	35.7
1982	3,751	118,418	14.7	1,325	35.3
1983	3,836	118,556	12.3	1,119	29.2
1984	3,937	118,155	12.0	1,093	27.8
1985	4,041	119,808	11.2	1,041	25.8
1986	4,184	125,806	9.8	976	23.3
1987	4,210	129,840	9.0	936	22.2
1988	4,233	132,392	9.0	957	22.6
1989	4,290	134,157	9.8	1,043	24.3
1990	4,163	131,450	9.7	1,016	24.4
1991	4,101	128,519	9.1	937	22.8
1992	4,070	128,116	7.8	832	20.5
1993	4,036	127,020	6.9	755	18.7
1994	4,077	127,076	7.3	785	19.3
1995	4,115	127,560	7.7	818	19.9
1996	4,184	128,811	7.6	817	19.5
1997	4,280	130,834	7.5	825	19.3
1998	4,407	135,592	7.0	809	18.4
1999	4,505	140,016	7.1	850	18.9
2000	4,457	145,670	7.9	949	21.3
2001	4,368	153,193	6.9	912	20.9
2002	4,383	161,043	6.4	910	20.8

Source: *Joint Center for Housing Studies of Harvard University,* The State of the Nation's Housing 2003, *(2003), Table A-1, Available at: http://www.jchs.harvard.edu/publications/markets/son2003.pdf.*

(Table 4.5), the U.S. metropolitan areas with the greatest shares of Post-Boomers are in the faster growing Sun Belt regions (Austin, Raleigh–Durham, Salt Lake City, and Dallas–Fort Worth), places with large numbers of recent immigrants (Salinas and Dan Diego), and cities, including those in the Midwest, with large universities (Madison, Lansing, and Columbus).

Given the relatively high migration rates and the significant numbers of recent immigrants among the Post-Boomers, it is useful to look also at the U.S. metropolitan areas that experienced the highest growth of this group since 1990 (Table 4.7). Only 6 of the 20 largest "magnets" for Post-Boomer growth during the 1990s also appear on the list of metropolitan areas with the greatest shares of Baby Boomers in 2000 (Atlanta, Denver, Nashville, Seattle, San Francisco, and

Colorado Springs). The metropolitan "magnets" that are attracting the Post-Boomers are generally less expensive and offer better job prospects than the Boomer "nests." The result is a spatial generational divide between the Baby Boomers and the Post-Boomers in which the Post-Boomers are more heavily concentrated in the South and West, with a more limited presence in the Midwest and East.[27]

The Elderly

Another group with an important mobile element is the elderly population. Between 1970 and 2000, the elderly population of the United States (people over 65 years of age) increased from 20.1 million to 35.0 million, and from 9.2 percent of the total population

TABLE 4.4

U.S. Metropolitan Areas with the Biggest Gains and Losses of Baby Boomers, 1980–1990

Magnets

Rank	Top 10 metropolitan areas	Percent growth in Baby Boomers, 1980–1990	Baby boomers as percent of total, 1990
1.	Orlando, Fla.	51.6	35.1
2.	Fort Worth–Arlington, Tex.	34.5	36.1
3.	Atlanta, Ga.	33.0	37.7
4.	Santa Rosa–Petaluma, Calif.	30.2	35.1
5.	Vallejo–Fairfield–Napa, Calif.	29.1	35.3
6.	Dallas, Tex.	28.3	37.5
7.	Manchester–Nashua, N.H.	24.2	36.3
8.	Reno, N.V.	23.4	36.3
9.	Portsmouth–Dover–Rochester, N.H.	23.3	36.4
10.	Seattle, Wash.	22.1	37.6

Losers

Rank	Bottom 10 metropolitan areas	Percent decline in Baby Boomers, 1980–1990	Baby boomers as percent of total, 1990
1.	Provo–Orem, Utah	−27.7	25.2
2.	Davenport–Rock Island–Moline, Iowa–Ill.	−20.2	30.7
3.	Peoria, Ill.	−20.1	30.1
4.	Huntington–Ashland, W.Va.–Ky.–Ohio	−17.4	29.2
5.	Gary–Hammond, Ind.	−16.9	30.9
6.	Youngstown–Warren, Ohio	−16.8	29.4
7.	Erie, Pa.	−15.8	29.9
8.	Beaumont–Port Arthur, Tex.	−15.5	30.1
9.	Pittsburgh, Pa.	−13.1	30.6
10.	Buffalo, N.Y.	−12.4	30.8

Source: *W. H. Frey, "Boomer Magnets,"* American Demographics *(March 1992): 36.*

TABLE 4.5

U.S. Large Metropolitan Areas with the Greatest Shares of Baby Boomers, 2000

Rank	Top 12 metropolitan areas	Share of total population (%)
1.	Seattle–Tacoma–Bremerton, Wash. CMSA	32.1
2.	Washington–Baltimore, D.C.–Md.–Va.–W.Va. CMSA	32.0
3.	Richmond–Petersburg, Va. MSA	31.7
4.	Denver–Boulder–Greeley, Colo. CMSA	31.6
5.	San Francisco–Oakland–San Jose, Calif. CMSA	31.5
6.	Minneapolis–St. Paul, Minn.–Wis. MSA	31.5
7.	Atlanta, Ga. MSA	31.3
8.	Hartford, Conn. NECMA	31.2
9.	Colorado Springs, Colo. MSA	31.0
10.	Louisville, Ky.–Ind. MSA	31.0
11.	Boston–Worcester–Lawrence, Mass.–N.H.–Maine–Conn. CMSA	30.9
12.	Nashville, Tenn. MSA	30.9

Source: *W. H. Frey, "Boomer Havens and Young Adult Magnets,"* American Demographics *(March 2001): 23.*

TABLE 4.6

Large U.S. Metropolitan Areas with the Greatest Shares of Post-Boomers/Young Adults, 2000

Rank	Top 12 metropolitan areas	Share of total population (%)
1.	Austin–San Marcos, Tex. MSA	28.2
2.	Madison, Wis. MSA	26.3
3.	Lexington, Ky. MSA	26.0
4.	Raleigh–Durham–Chapel Hill, N.C. MSA	25.9
5.	Salt Lake City–Ogden, Utah MSA	24.5
6.	Atlanta, Ga. MSA	24.5
7.	Dallas–Fort Worth, Tex. CMSA	24.0
8.	San Diego, Calif. MSA	24.0
9.	Lansing–East Lansing, Mich. MSA	23.9
10.	Columbus, Ohio MSA	23.7
11.	Salinas, Calif. MSA	23.6
12.	Denver–Boulder–Greeley, Colo. CMSA	23.5

Source: *W. H. Frey, "Boomer Havens and Young Adult Magnets,"* American Demographics *(March 2001): 23.*

TABLE 4.7

Large U.S. Metropolitan Areas with the Biggest Increases in Post-Boomers/Young Adults, 1990–2000

Rank	Top 20 "magnets"	Post-boomer growth, 1990–2000 (%)[1]
1.	Las Vegas, Nev.–Ariz. MSA	206
2.	Austin–San Marcos, Tex. MSA	165
3.	Phoenix–Mesa, Ariz. MSA	156
4.	Raleigh–Durham–Chapel Hill, N.C. MSA	156
5.	Atlanta, Ga. MSA	155
6.	Denver–Boulder–Greeley, Colo. CMSA	150
7.	Boise City, Idaho MSA	148
8.	Dallas–Fort Worth, Tex. CMSA	142
9.	Orlando, Fla. MSA	140
10.	West Palm Beach–Boca Raton, Fla. MSA	139
11.	Charlotte–Gastonia–Rock Hill, N.C.–S.C. MSA	138
12.	Portland–Salem, Ore.–Wash. CMSA	137
13.	Miami–Fort Lauderdale, Fla. CMSA	133
14.	Nashville, Tenn. MSA	133
15.	Seattle–Tacoma–Bremerton, Wash. CMSA	131
16.	San Francisco–Oakland–San Jose, Calif. CMSA	131
17.	Houston–Galveston–Brazoria, Tex. CMSA	127
18.	Fort Myers–Cape Coral, Fla. MSA	127
19.	Colorado Springs, Colo. MSA	125
20.	Lexington, Ky. MSA	124

[1] 20–34 year olds in 2000 as a percentage of 10–24 year olds in 1999.

Source: *W. H. Frey, "Boomer Havens and Young Adult Magnets,"* American Demographics *(March 2001): 24.*

to 12.4 percent. The elderly population's retirement years have been relatively affluent compared with those of their immediate predecessors, thanks mainly to significant improvements in both private-sector and public-sector pension programs (improvements that had been achieved during the prosperity of the 1960s). The financial security thus enjoyed by many (but by no means all) of the elderly, combined with windfall gains for homeowners that resulted from house price inflation, enabled large numbers of them to relocate to "places of reward and repose" in pursuit of new lifestyles in amenity-rich environments. Table 4.8 shows the top metropolitan magnets for the elderly during the 1990s, where an unprecedented kind of urbanization—one often fueled by elderly immigration—is producing some cities where 25 percent of the population is of retirement age (compared with the nationwide figure of just under 13 percent), with all

kinds of ramifications for local economic development, politics, and quality of life.

Within [the] Southern and Western states, selected metropolitan areas experienced exceptionally high rates of senior growth during the '90s. Las Vegas leads the pack with a growth rate of 86 percent. Yet a group of senior metro "hot spots" has much smaller populations. This list includes Naples, Fla.; Myrtle Beach, S.C.; Las Cruces, N.M.; and Flagstaff, Ariz., among other smaller metro areas with top quality amenities that have attracted seniors in droves. Other larger metro areas that have shown sharp rises in the over-65 set include "New South" urban centers like Houston, Dallas and Atlanta, along with counterparts in the "New West" like Phoenix and Denver.

Metro areas with a university environment like Austin, Texas, and Raleigh–Durham, N.C., were also attractive to seniors. In fact, the few Northern metro

TABLE 4.8

U.S. Metropolitan Areas with the Biggest Increases in Elderly Population, 1990–2000

	Population aged 65+	
Elderly "Magnets"	*65+ population as percent of total population in 2000*	*Percent change in 65+ population 1990–2000 as percent of 65+ population in 1990*
Major Metropolitan Areas[1]		
1. Las Vegas, Nev.–Ariz. MSA	11.8	+86.2
2. Phoenix–Mesa, Ariz. MSA	11.9	+38.0
3. Austin–San Marcos, Tex. MSA	7.3	+37.3
4. Houston–Galveston–Brazoria, Tex. CMSA	7.7	+31.8
5. Atlanta, Ga. MSA	7.6	+30.8
6. Orlando, Fla. MSA	12.4	+28.8
7. Sacramento–Yolo, Calif. CMSA	11.3	+27.8
8. Raleigh–Durham–Chapel Hill, N.C. MSA	8.6	+25.8
9. Denver–Boulder–Greeley, Colo. CMSA	8.9	+25.8
10. Dallas–Fort Worth, Tex. CMSA	8.1	+25.1
Smaller Metropolitan Areas[2]		
1. Naples, Fla. MSA	24.5	+77.9
2. Anchorage, Alaska MSA	5.5	+72.5
3. Myrtle Beach, S.C. MSA	15.0	+61.7
4. Las Cruces, N.M. MSA	10.6	+55.7
5. Fort Walton Beach, Fla. MSA	12.1	+55.1
6. Ocala, Fla. MSA	24.5	+47.0
7. Flagstaff, Ariz.–Utah MSA	7.5	+46.3
8. Wilmington, N.C. MSA	14.1	+45.7
9. McAllen–Edinburg–Mission, Tex. MSA	9.7	+43.8
10. Melbourne–Titusville–Palm Bay, Fla. MSA	19.9	+42.6

[1] Major metropolitan areas have 2000 total populations exceeding 1 million.

[2] Smaller metropolitan areas have 2000 total populations of less than 1 million.

Source: *W. H. Frey, "Seniors in Suburbia,"* American Demographics *(November 2001):19, Table 1; US Census Bureau, Geographic Comparison Table, GCT-PS. Age and Sex: 2000,* Census 2000 Summary File (SF 1) 100-Percent Data, available at American FactFinder: http://factfinder.census.gov/

areas with higher than average senior rates include university towns: State College, Pa.; Iowa City, Iowa; Bloomington, Ind.; and Madison, Wis. The cultural amenities available in these places are especially attractive to today's elderly, who have more leisure time, and are better educated than those in previous generations. These areas are likely to increase in population as the Boomers ascend to seniorhood.[28]

The New Immigrants

Another demographic trend with important implications for patterns of U.S. urbanization has been the increasing number of immigrants. One of the main reasons for the increase was the abolition in 1965 of the ceiling and quota systems that had been introduced by the immigration acts of the 1920s (see p. 69). Without taking into account illegal immigrants (estimated at 7 million in 2000), net immigration accounted for nearly one-third of the population growth within the urban system during the 1980s and 1990s. In marked contrast to earlier waves of immigration, the dominant nationalities among the arrivals since the 1960s have been from Latin America and Asia rather than Europe (Figure 4.24). In 2002 just over 42 percent of the 1,063,732 immigrants who were granted legal permanent resident status were from Latin America, and a further 31 percent were from Asia. Only 17 percent were from Europe.

At the same time, many of the descendants of the earlier generations of immigrants who had come from Europe were moving out of the cores of older metropolitan areas as they established themselves in American society. As a result, the ethnic composition of American cities has changed. Cambodian, Colombian, Cuban, Dominican, Haitian, Jamaican, Korean, and Vietnamese neighborhoods are now elements of the social mosaic of many American cities, replacing or supplementing the older ethnic residential concentrations of past waves of immigrants. There are some significant differences, however, in the destinations of these new immigrants. New York has traditionally been the most powerful single magnet, the metropolitan area being nominated as the intended place of residence by the largest number of immigrants each year until 2000. But since 2001 the Los Angeles metropolitan area has attracted more immigrants (Table 4.9). Other important magnets now include metropolitan areas on the West Coast and in the Southwest and Florida—places that historically attracted relatively few immigrants. The impact of recent immigration, measured in terms of immigrant arrivals as a percentage of the total population, is now greatest in Miami, while New York, Chicago, Washington, D.C., and Boston are the only traditional destinations among the ten metropolitan areas with the highest rates of immigration (Table 4.9). In short, recent streams of immigration are highly differentiated and have enormous impacts. This is a topic we will discuss in more detail in Chapter 12.

CONCEPTS OF URBAN SYSTEM CHANGE: COUNTERURBANIZATION, DEURBANIZATION, AND REURBANIZATION

The results of 30 years of economic crisis and restructuring, intersecting with various elements of demographic and social change, have been reflected in some strikingly differential patterns of growth and decline across the U.S. urban system. Two dramatic reversals of longstanding trends in the pattern of urbanization dominated the 1970s. The growth of larger metropolitan areas, which had always been greater than that of any other category of settlement, slowed dramatically—and in

TABLE 4.9

U.S. Metropolitan Areas with the Greatest Number of Immigrants, 1988 and 2002

	Metropolitan area	1988	Metropolitan area	2002
1.	New York, N.Y.–N.J.–Conn. CMSA	134,556	Los Angeles–Long Beach, Calif. PMSA	108,613
2.	Los Angeles, Calif. CMSA	110,450	New York, N.Y. PMSA	91,275
3.	Miami, Fla. CMSA	45,600	Chicago, Ill. PMSA	43,810
4.	San Francisco, Calif. CMSA	41,586	Miami, Fla. PMSA	40,832
5.	Chicago, Ill.–Ind.–Wis. CMSA	25,947	Washington, D.C.–Md.–Va.–W.Va. PMSA	38,468
6.	Washington, D.C.–Md.–Va. MSA	18,032	Houston, Tex. PMSA	28,225
7.	Boston, Miss.–N.H. CMSA	15,461	San Jose, Calif. PMSA	27,431
8.	San Diego, Calif. MSA	13,351	Orange County, Calif. PMSA	25,806
9.	Houston, Tex. CMSA	11,779	San Diego, Calif. MSA	22,484
10.	Philadelphia, Pa.–N.J.–Del.–Md. CMSA	10,543	Boston–Lawrence, Mass. PMSA	21,535

Note: Immigrants admitted to the United States, FY 1988, classed by intended area of residence.
Source: *W. H. Frey, "Metropolitan America: Beyond the Transition,"* Population Bulletin *45, 2 (1990): 21; U.S. Department of Homeland Security, 2002* Yearbook of Immigration Statistics, *Office of Immigration Statistics, October 2003, Washington, D.C., Table C, p. 9.*

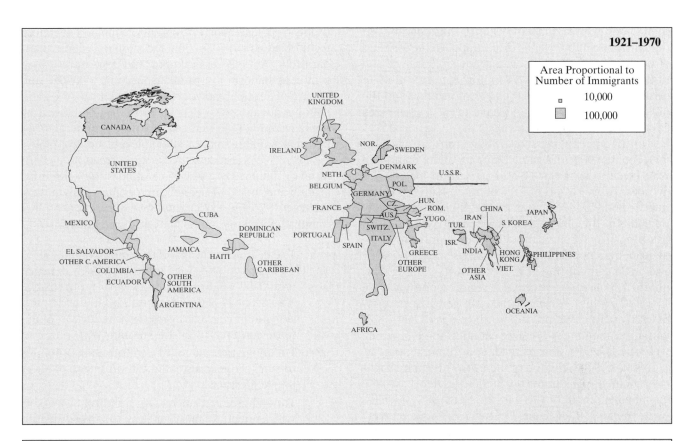

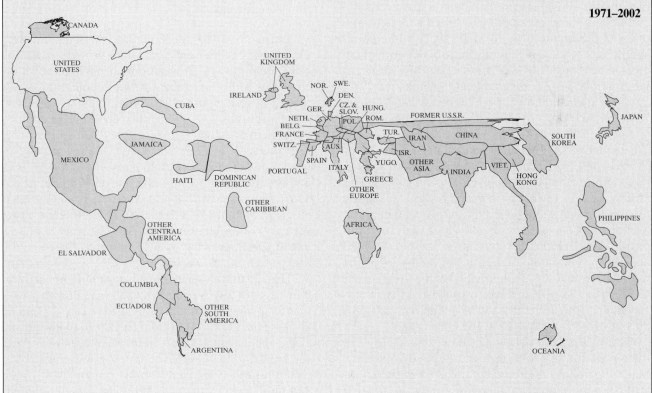

FIGURE 4.24   Legal immigration to the United States: countries of origin for 1921–1970 and 1971–2002.

some cases growth suddenly changed into decline. At the same time, there was a "turnaround" in the fortunes of nonmetropolitan areas, with many small towns and rural areas gaining population for the first time in decades. Together, the metropolitan slowdown and the nonmetropolitan turnaround have been characterized as **counterurbanization**.[29]

During the 1970s about 80 percent of all nonmetropolitan counties gained in population, compared with levels of between 40 and 50 percent in previous decades. Every region of the country experienced this turnaround; though it was barely discernible in New England, in the South, Southwest, and West it was of stunning proportions. Throughout the South the nonmetropolitan population had grown at rates between 1 and 3 percent during the 1960s; during the 1970s, nonmetropolitan growth in the region ran at levels of 15 to 20 percent. Strong population gains were registered in counties within commuting range of metropolitan areas and in more remote counties, some of which were specialized resort, recreation, or retirement areas.

Meanwhile, at the top of the urban hierarchy, eight metropolitan areas of over 1 million residents actually lost population, all of them in the Northeast and Midwest: Buffalo, Cleveland, Detroit, Milwaukee, New York, Philadelphia, Pittsburgh, and St. Louis. Metropolitan areas of similar size in the South and West did not decline, but their rate of growth slowed strikingly, and many of their central city cores did actually lose population. In contrast, small to medium-sized metropolitan areas in the South and West grew vigorously—faster, in fact, than the surrounding nonmetropolitan areas.

One explanation for the counterurbanization of the 1970s is that it was a consequence of improvements that had been made to rural electrification, new communications networks, improved water supplies, and better TV reception that made small towns and rural settings more attractive to both employers and individuals. This idea, in turn, is related to a second broad explanatory theme: corporate reorganization and decentralization. Nonmetropolitan settings offered, along with their improved infrastructure and accessibility, inexpensive land and cheap, nonunion labor. Thus at least part of the rural turnaround was a continuation of the trends of the 1960s, when large companies consolidated their managerial and R&D functions and decentralized their productive capacity to new plants in Sunbelt cities.

In the 1970s this decentralization extended to nonmetropolitan areas (and to less developed countries). Many of the jobs and workers involved in the counterurbanization process, therefore, were in **branch plants**—subsidiary factories responsible for standardized, routine production or assembly. This development conforms neatly with the **product-cycle model** of industrial location, which postulates that products evolve through stages that successively require (1) fewer inputs of skilled labor, entrepreneurship, and advanced technologies (as production techniques become standardized) and (2) less-affluent and less-sophisticated local populations (as products become cheaper and more familiar to broader markets). As a result, the comparative advantage in making a product shifts successively from metropolitan core areas to peripheral cities, nonmetropolitan areas, and less-developed economies.

There were also several *contingent reasons* for counterurbanization. The peak years of both the rural turnaround and the slowdown in metropolitan growth were between 1973 and 1975, when several events conspired to interrupt the traditional patterns of urbanization:

- The economic recession that followed the OPEC oil embargo and price increase, which deterred would-be migrants to cities with an industrial base. In cities with defense-related industries, this slowdown was aggravated by a reduction in defense contracts after the Vietnam War.
- The movement of counterculture Baby Boomers to rural areas and small towns in search of alternative lifestyles.
- The retirement of increasing numbers of workers with enough financial security to enable them to move out of metropolitan areas to resort or retirement areas.

During the 1980s the larger U.S. metropolitan areas once more experienced population growth, while the most rapid growth among nonmetropolitan areas occurred in counties adjacent to metropolitan areas. Of the 20 largest metropolitan areas, only two—Detroit and Pittsburgh—lost population. Los Angeles grew by more than 3 million people in the 1980s (1,150 new arrivals each and every work day!), Dallas grew by almost 1 million, San Francisco grew by nearly 900,000, and the Atlanta, Houston, Miami, New York, Phoenix, San Diego, and Washington, D.C. metropolitan areas all grew by more than 500,000. Orlando grew faster than any other large (million-plus) metropolitan area in the 1980s, growing by 53.3 percent to become the thirty-sixth largest metropolitan area in the country.

This urbanization exhibited a pattern of bicoastal growth, with towns and cities of all sizes, together with many surrounding nonmetropolitan areas, expanding all along the Pacific Coast from San Diego to Seattle and along the Atlantic Coast from Miami to New York (Figure 4.26(a)). Most of the fastest-growing small and medium-sized cities were located on the West Coast, and most were exurban communities (chiefly residential, but not contiguous to central cities or their suburbs) that had recently become **incorporated** as cities. It is telling that by 1990 they had overtaken in size some of the key **central places** (e.g., Roanoke, Virginia and Columbia, South Carolina) and traditional manufacturing

Box 4.4   The Deurbanization Scenario in the United States?

The unprecedented population shifts involved in counterurbanization in the 1970s were accompanied by other demographic shifts. Patterns of household formation changed toward the end of the 1960s in response to changing social attitudes. Putting these changes together with population changes suggested a scenario of urban decline that is captured by the model of **deurbanization** (Table 4.10). According to this scenario, the traditional sequence of urbanization—with high growth in both population and households in metropolitan areas, followed first by moderate population growth and continuing high growth in the number of households and then by only moderate growth in both population and households—is followed by three stages of deurbanization. The first is a stage of population decline but moderate growth in households—the situation realized by the counter-urbanization of the 1970s. The second is a stage of declining numbers of both population and households, a stage that might be anticipated as a result of the continuation of some of the trends that underpin counterurbanization, together with the decline in households that can be expected as the last of the Baby Boom generation matures. The third is a stage of "land decline"—the physical shrinkage of metropolitan areas as reduced numbers of people and households are reflected in patterns of land use.

Fortunately, such a scenario is not inevitable. Looking back at the twentieth century, it seems that the metropolitan slowdown of the 1970s has played itself out (though changes in Census Bureau definitions make it impossible to be certain—the changing definition of "metropolitan" by the U.S. Census Bureau (see Chapter 1) meant that while metropolitan areas certainly grew as a result of births and inmigration, they also grew from territorial expansion as new areas achieved metropolitan status and additional counties were added to existing metropolitan areas). Overall, from 1910 to 2000, the U.S. metropolitan population grew by nearly 200 million people, with the largest increase, 33.3 million (14 percent), occurring between 1990 and 2000.[30] Looking at the entire period from 1940 to 2000, the proportion (if not the number) of people living in central cities remained relatively stable, while the suburbs continued to grow substantially. By 2000 half of the U.S. population lived in the suburbs of metropolitan areas (Figure 4.25). At the same time, the nonmetropolitan growth rate in the 1990s was 10 percent—fueled by the growth of those counties adjacent to metropolitan areas. Nonmetropolitan population decline has been concentrated in the country's interior states—in a band of counties across the Great Plains area stretching from the Mexican border to the Canadian border—that have borne the brunt of the "rolling recession" associated with successive downturns in labor-intensive traditional manufacturing activity, agriculture, and energy-related and extractive industries.

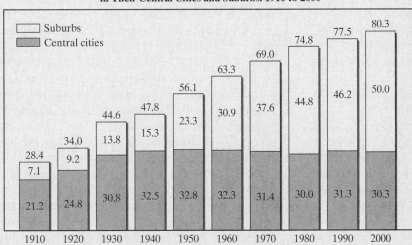

Percent of Total Population Living in Metropolitan Areas and in Their Central Cities and Suburbs: 1910 to 2000

FIGURE 4.25   Percent of total U.S. population living in metropolitan areas (and their central cities and suburbs): 1910 to 2000.

(Continued)

BOX 4.4   THE DEURBANIZATION SCENARIO IN THE UNITED STATES?   (Continued)

TABLE 4.10

Population and Household Change in a Changing Urban System

City growth characteristics	Stage of deurbanization	Residential decline	Vacant land	Service characteristics (metropolitan system)
High population/ high household	—	None	None	Booming demand for all services. Overcrowded schools.
Moderate population/ high household	—	None	None	High demand for services. New schools. Hospital shortages.
Moderate population/ moderate household	—	None	None	Stable demand for most services.
Negative population/ moderate household	Stage 1 (population decline)	None	None	Empty schools. Oversupply of hospitals. Varied retail conditions.
Negative population/ negative household	Stage 2 (household decline)	Some	Some	Declining demand for most services.
	Stage 3 (land decline)	Major	Substantial	Oversupply of schools and hospitals. Commercial blight.

Source: Modified from R. Sinclair, "Household Change, Space Demands and Spatial Patterns at Different Levels of the Urban System," in The Changing Geography of Urban Systems, ed. L. Bourne, et al. Servicio de Publicaciones de la Universidad de Navarra, Spain, 1989, Table 1, pp. 231–32.

centers (e.g., Youngstown, Ohio) of earlier epochs of urbanization. Two cities that reported populations of more than 100,000 in the 1990 census were not even incorporated in 1980: Santa Clarita and Moreno Valley, California. Both typify the exurban growth that occurred around the country's major metropolitan areas during the 1980s. The contrast between the traditional Manufacturing Belt and the Sunbelt consequently had been overlain by the new contrast between the interior and the "trade and defense perimeter," where the new, informational mode of development had found its fullest expression by the 1980s (Figure 4.26(a)).

During the 1990s the number of people living in metropolitan areas grew by 14 percent, so that by 2000, four out of every five people in the United States were living in cities or suburbs. The 1990s also saw continued growth in nonmetropolitan areas, especially in those counties adjacent to metropolitan areas.

The larger metropolitan areas continued to grow in the 1990s. By 2000 more than 50 percent of the U.S. population was living in a metropolitan area with a population of at least 1 million. Of these 50 largest metropolitan areas, only Pittsburgh and Buffalo lost population during the 1990s. Las Vegas grew fastest—at a rate of 83.3 percent—and increased its metropolitan population to 1,563,282 from 852,737, to become the thirty-second largest metropolitan area in the country (up from fifty-second place in 1990) (Figure 4.26(b)).

The 20 largest metropolitan areas all grew, and New York, Los Angeles, and Chicago remained the top three by population size. Los Angeles grew by 1.8 million, New York by 1.6 million, and Dallas, Atlanta, and Phoenix by just over 1 million. The Houston, Chicago, Washington, D.C., San Francisco, Miami, Denver, and Seattle metropolitan areas all grew by more than 500,000. In the top 20 group, Phoenix grew faster than any other large (million-plus) metropolitan area in the 1990s—by 45.3 percent—to become the fourteenth largest metropolitan area in the country (up from twentieth in 1990).

In the 1990s most of the fastest-growing small and medium-sized metropolitan areas (with fewer than 1 million people in 2000) were not located along the West Coast (Figure 4.26(b)). The generally more rapid growth of metropolitan areas in the South and West at the beginning of the twentieth-first century represents a continuation of the corporate reorganization and decentralization trends that began the 1960s amid increasing globalization.

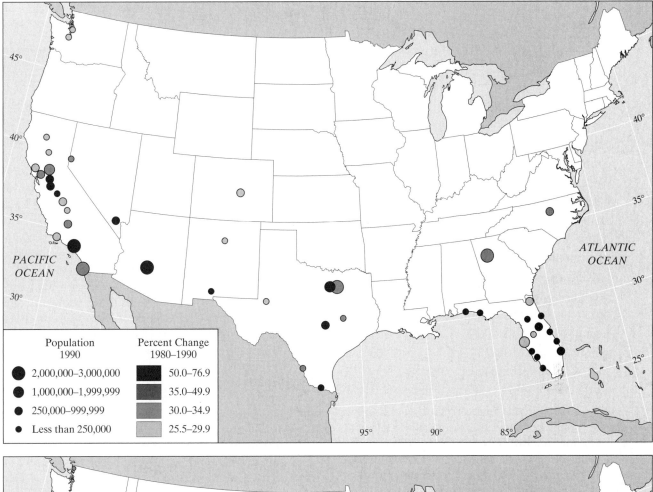

Population
1990

● 2,000,000–3,000,000
● 1,000,000–1,999,999
● 250,000–999,999
• Less than 250,000

Percent Change
1980–1990

■ 50.0–76.9
■ 35.0–49.9
■ 30.0–34.9
□ 25.5–29.9

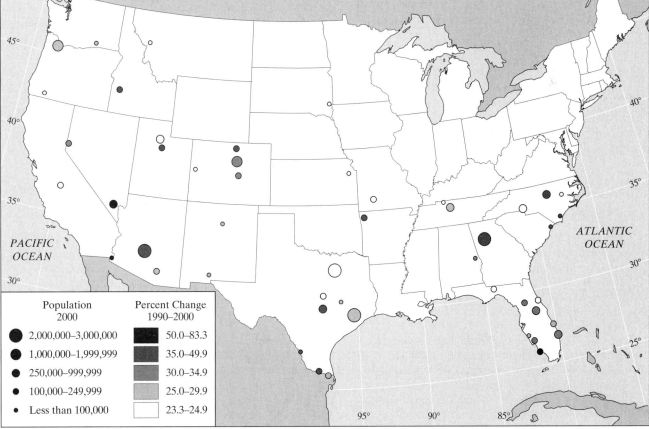

Population
2000

● 2,000,000–3,000,000
● 1,000,000–1,999,999
● 250,000–999,999
● 100,000–249,999
• Less than 100,000

Percent Change
1990–2000

■ 50.0–83.3
■ 35.0–49.9
■ 30.0–34.9
■ 25.0–29.9
□ 23.3–24.9

FIGURE 4.26   The 50 fastest-growing metropolitan areas in the 1980s and 1990s.

FOLLOW UP

1. Be sure that you understand the following key terms:

   collective consumption
   control centers
   core-periphery model
   counterurbanization
   deurbanization
   Digital Divide
   enabling technologies
   informational cities
   metropolitan consolidation
   nodal centers
   product-cycle model
   production process technologies
   regional decentralization
   splintering urbanism
   transactional technologies
   world cities

2. Watch a movie. Economic restructuring and urban change might not seem promising as the basis for a good movie, but *Roger and Me* (directed by Michael Moore; Warner Bros., 1989) is well worth watching and will bring to life the meaning of deindustrialization, corporate restructuring, and urban distress.

3. Find out all you can about the employment structure of the city or town in the United States that you live in (or is of interest to you). Use census data available from the U.S. Census Bureau's website to construct an industry-by-industry profile of employment and summarize the data in a graph or table. What do the data suggest in relation to the city's function within the urban hierarchy?

4. Consider the extent to which your daily life depends on economic linkages between cities around the country and around the world. Look, for example, at the labels on all the clothes you are wearing. How many different places were involved in their production? How did they get to you? How much was a product of your local economy?

5. *Portfolio* time. You can research several topics quite easily. Population, employment, and unemployment data are available in most libraries and on the Internet and provide a good opportunity to devise maps, charts, and graphs that update and amplify the changing experience of different cities. The same topics are also the subject of frequent reports on the websites of various government agencies (the U.S. Bureau of the Census and the U.S. Department of Labor, in particular). Magazines like *American Demographics* are good sources of up-to-date information on population changes. You should also monitor major newspapers and business magazines for stories that relate to the topics you cover in your basic reading. By accessing one of the online information retrieval services like LexisNexis (*http://www.lexisnexis.com/*), you hook into the "space of flows" that you are researching!

   You should also check recent paper or online issues of academic journals (such as *Urban Geography, Urban Studies, Cities, Environment & Planning A, Environment & Planning D (Society and Space), Progress in Human Geography*, and the *Annals of the Association of American Geographers*) for new research findings and interpretations of urban change.

KEY SOURCES AND SUGGESTED READING

Berube, A., and B. Forman. 2002. *Living on the Edge: Decentralization within Cities in the 1990s.* Washington, D.C.: The Brookings Institution.

Bluestone, B., and B. Harrison. 1982. *The Deindustrialization of America*. New York: Basic Books.

Castells, M. 1996. *The Rise of the Network Society*. Malden, Mass.: Blackwell.

Downs, A. 1994. *New Visions for Metropolitan America*. Washington, D.C.: The Brookings Institution.

Frey, W. H. 2003. *Metropolitan Magnets for International and Domestic Migrants*. Washington, D.C.: The Brookings Institution, Center on Urban and Metropolitan Policy (*http://www.brookings.edu/es/urban/publications/ 200310_Frey.htm*).

Glaeser, E., and J. M. Shapiro. 2001. *City Growth and the 2000 Census: Which Places Grew, and Why*. Washington, D.C.: The Brookings Institution (*http://www.brook.edu/es/ urban/projects/census/whygrowthexsum.htm*).

Graham, S., and S. Marvin. 2001. *Splintering Urbanism*. New York: Routledge.

Jacobs, B. 1992. *Fractured Cities: Capitalism, Community and Empowerment in Britain and America*. Andover, N.Y.: Routledge, Chapman, and Hall.

Joint Center for Housing Studies of Harvard University 2002. *The State of the Nation's Housing 2002*. Cambridge, Mass.: Harvard University, Table A-3 (*http://www.jchs.harvard.edu/publications/markets/ Son2002.pdf*).

Katz, B., and R. Lang, eds. 2003. *Redefining Urban and Suburban America: Evidence from Census 2000.* Washington, D.C.: The Brookings Institution.

Kephart, G. 1991. Economic Restructuring, Population Redistribution, and Migration in the U.S. pp. 12–34 in *Urban Life in Transition*, eds. M. Gottdiener and C. G. Pickvance, Urban Affairs Annual Reviews No. 39. Newbury Park, Calif.: Sage.

Knox, P., and P. Taylor, eds. 1995. *World Cities in a World System*. Cambridge, UK: Cambridge University Press.

Logan, J. R. 2002. *Regional Divisions Dampen '90s Prosperity: New Census Data Show Economic Gains Vary by Region.* Albany, N.Y.: Lewis Mumford Center for Comparative Urban and Regional Research, University at Albany.

McCarthy, L., and D. Danta. 2003. Cities of Europe. pp. 168–221 in *Cities of the World: World Regional Urban Development*, eds. S. D. Brunn, J. F. Williams, and D. J. Zeigler, 3rd ed. Lanham, Md: Rowman & Littlefield.

Noyelle, T. J., and T. M. Stanback, Jr. 1984. *The Economic Transformation of American Cities*. Totowa, N.J.: Rowman & Allanheld.

Rodwin, L., and H. Sazanami. 1989. *Deindustrialization and Regional Economic Transformation: The Experience of the United States*. Boston, Mass.: Unwin Hyman.

Sui, D. Z., and J. O. Wheeler. 1993. The Location of Office Space in the Metropolitan Service Economy of the United States. *Professional Geographer* 45: 33–43.

Townsend, A. M. 2001. The Internet and the Rise of the New Network Cities, 1969–1999. *Environment and Planning B: Planning and Design* 28: 39–58.

Weinstein, B. L., and R. E. Firestine. 1985. *Regional Growth and Decline in the United States*. 2nd ed. New York: Praeger.

Wilson, E., ed. 1997. *Globalization and the Changing U.S. City*. Thousand Oaks, Calif.: Sage.

RELATED WEBSITES

The Digital Divide Network: *http://www.digitaldividenetwork.org/* This Benton Foundation website (in partnership with funding agencies such as the Ford Foundation and Annie E. Casey Foundation) provides research and data on the urban digital divide from a variety of perspectives.

The Centre for Urban History at the University of Leicester: *http://www.le.ac.uk/ur/index.html* The Centre for Urban History contains research and hyperlinks related to European urban history.

Globalization and World Cities (GaWC) Study Group and Network: *http://www.lboro.ac.uk/gawc/* The GaWC website is an excellent source of information on world cities. This site also includes a Teaching Gateway page containing data and maps that can be used in student project work on world cities.

The Brookings Institution: *http://www.brook.edu/* The Brookings Institution is an independent, nonpartisan organization devoted to research, analysis, education, and publication focused on public policy issues related to economics, foreign policy, and governance. Its website offers publications, for example, on urban and regional demographic change in the United States based on analyses of the decennial censuses.

5

THE FOUNDATIONS
OF URBAN FORM
AND LAND USE

Just as we can view the urban system as both the result of and the framework for processes of economic, demographic, and social change, so too can we view the spatial form and organization of cities themselves. Changing patterns of urban form and land use must, like the changing urban system, be understood in relation to the rhythms of the economy. In the transition from mercantile towns to industrial cities and, now, to metropolitan complexes, urbanization has evolved from compact but very heterogeneous settings through horizontally separate sectors and zones of different land uses and on to multinodal frameworks that contain unexpected juxtapositions of land uses in suburban "cities" and in redeveloped central city areas.

CHAPTER PREVIEW

The evolution of urban form stems not only from the history of economic development, migration, and immigration described in earlier chapters but also from the interaction of economic and demographic growth phases with changes in social structure and lifestyles, innovations in building materials and construction techniques, innovations in urban transportation, and changes in the legal framework of land ownership, land use law, and land use policy.

Within each phase of urban development, many consider innovations in transport systems to have been the single most significant determinant of urban form and land use. Transport systems controlled the density and areal extent of urban development and gave expression to the pent-up energies of successive phases of economic and social change. In this chapter we will see the importance of early transportation technologies in shaping the foundations of urban form and land use.

We begin with the "mercantile city" (i.e., pre-1840), where we will see how the character of economic development and the lack of effective intraurban transportation gave rise to very compact cities with distinctive patterns of land use that had a great deal in common with "preindustrial" cities elsewhere. As urbanization gathered pace with the onset of industrialization (1840–1875), rapid growth, the catalytic effects of railroads, and the absence of controls on land use produced a second distinctive epoch of city building, the product of which was the early industrial city.

This phase in turn provided the foundations— literally—for the industrial cities that characterized the period of urban development between 1875 and 1920. As we will see, the "industrial city" was based on a highly specialized and very distinctive pattern of land use that reflected the dual processes of industrial location and social segregation. In the section on the industrial city, we will see how land use zoning laws and transit systems gave particular shape and character to urban form. We also discuss attempts that have been made to make sense of the specialized land use patterns of the industrial city using concepts of rent and referring to a model of urban land use that is based on widely occurring patterns of land values.

THE MERCANTILE CITY (BEFORE 1840)

Despite the steady growth in city population during the first half of the nineteenth century, American cities remained physically compact. Even the largest cities rarely extended beyond two miles from their downtown (the average distance that a person can walk in half an hour). As a result of this compactness, land use in these *walking cities* was very mixed, with workshops, warehouses, stores, banks, and offices interspersed among housing.

By the 1840s this compactness had also led to extremely high densities—75 to 100 people per acre— and chronic problems of overcrowding, with an average of two families to every dwelling. In river ports the development of ferries and bridges that opened up new river frontage provided temporary relief of this crowding. In the 1820s and 1830s, for example, New York expanded across the Hudson, Philadelphia across the Schuylkill, Pittsburgh across the Allegheny and the Monongahela, Cleveland across the Cuyahoga, and Boston across the Charles.

THE PEDESTRIAN CITY

Apart from such extensions, however, mercantile cities were severely constrained by the lack of fast but inexpensive forms of transportation. Most people walked from place to place, and most goods were moved by hand cart or horse cart. The result was a loose intermingling of activities in these *pedestrian cities*. Because most towns and cities of any size were seaports or river ports, the hub of economic activity was the waterfront, dominated by merchants' offices, warehouses, wharves, and workshops. Clustered nearby were hotels, churches, stores, and public buildings, together with the homes of prominent families. Housing for artisans, storekeepers, and laborers edged into vacant spaces between them and extended to the outer edge of town, where commercial activities that needed extra space (e.g., textile mills), large quantities of stream water (e.g., breweries), or that were particularly noxious (e.g., slaughterhouses, tanneries) tended to locate.

There was little separation between home and workplace. Factory owners often built their homes right next to their factories; artisans and storekeepers lived above or behind their workshop or store; laborers and service workers lived off alleyways and in lofts; servants lived in the upper floors of their masters' houses; and, in Southern cities, slaves lived in compounds behind the main house. As cities increased in population and density, enclaves of specialization did begin to emerge around clusters of workshops and factories and around concentrations of ethnic groups. Boston's North End, for example, became a distinctive Irish quarter in the early years of the nineteenth century. Nevertheless, the mercantile city was so small in compass that short distances separated poor from rich and artisan from laborer, and no enclave was large enough to be thought of as a separate district of specialized character.

MODELS OF THE MERCANTILE CITY

This scenario bears some resemblance to the generalizations derived from attempts to portray the distinguishing characteristics of walking cities, most of which were heavily influenced by European examples. The best-known and most influential is Sjoberg's model of the *preindustrial city*,[1] in which the social pyramid—a large group of outcasts and laborers, a smaller group of artisans, and an even smaller elite group—is reflected in the city's spatial pattern, with the elite clustered around key religious, political, and administrative institutions at the center of the city, artisans loosely clustered by occupation around this core, and laboring groups located at the outer edge of the city, grouped here and there into neighborhood clusters according to ethnicity (Figure 5.1).

This generalization has been questioned by James Vance, Jr.,[2] who gives much greater emphasis to the *mosaic of occupational subdistricts* and downplays the extent of a fringe of low-income laborers, placing them instead as lodgers scattered within these various functional subdistricts.

What is important, however, is not so much these details as the overall structure. Both Sjoberg and Vance describe preindustrial cities as having

- a central core dominated socially by the residences of an elite group.
- a number of occupationally distinctive but socially mixed quarters.
- a residual population of the very poor living in the back alleys and on the fringes of the city.

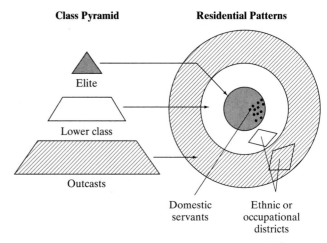

Class Pyramid   **Residential Patterns**

Elite

Lower class

Outcasts

Domestic servants   Ethnic or occupational districts

FIGURE 5.1   The relationship between class structure and place of residence in the preindustrial city (after Sjoberg).

- everything physical at a human scale: a walking city in which home and work were tightly bound by the organization of work into patriarchal and familial groupings.

These features, common to preindustrial cities over much of the world, could be found in American cities of the early nineteenth century: Charleston, S. C., for example (the sixth-largest city in the United States until 1830).[3] Beyond these broad parallels, however, the truth is that American cities exhibited all kinds of exceptions to any set of generalizations. They were young, still shaped by local, idiosyncratic, and contingent factors. They were growing rapidly and consequently had no time to "shake down" into the classic preindustrial patterns exhibited by the cities of medieval Europe and feudal Asia. And before they could do so, they were overtaken by the revolutionary forces of industrialization.

THE EARLY INDUSTRIAL CITY (1840–1875)

The impact of the Industrial Revolution on American cities was dramatic. But before the new regime came to be clearly reflected in the form and spatial organization of cities, there was an interlude of volatility, a transitional period during which the preconditions for change gained momentum and important precedents were established.

Probably the most fundamental change to emerge with the onset of industrialization was the intense competition that developed for the best and most accessible sites for the factories and the warehouses, shops, and offices that depended on them. Meanwhile, industrialization generated three new social groups: an industrial business elite, a white-collar group of managers and office workers who were needed to run the businesses, and a blue-collar group of factory workers who were needed to operate the machinery. It was not long before these social groups found themselves in competition for the best and most accessible sites for housing. The competition was fierce. Social status, newly ascribed in terms of money, became synonymous with ability to pay rent, so that a person's address came to acquire unprecedented importance.

Ultimately, these changes would turn the city inside out, with specialized industrial and commercial uses claiming most of the central area, the rich exchanging their central locations for the peripheral location of the poor, and occupational clustering giving way to residential differentiation by income.[4] But it took businesses, households, landowners, and city governments time to sort themselves out into these new patterns. They had to do so, moreover, in an unregulated environment and within a legal framework in which land ownership was treated as a civil liberty rather than as an economic or social resource. Most of all, the growth had to be accommodated without any major technological advances in intraurban transportation. For a whole generation cities were brimming with new arrivals and jammed with new factories, warehouses, and institutions, with no adequate means of circulation that might have helped to rationalize the organization of space (Figure 5.2). Meanwhile, crucial precedents were being set in terms of land law and the land development process.

URBANIZATION AND THE PUBLIC INTEREST

The early colonists from England had brought with them a strong desire to avoid the rigidity and inequality of feudal land laws. They found a continent with almost unlimited possibilities for land ownership (Native Americans notwithstanding), and so they established a simple approach to land law that was based on individuals' civil liberties. Any person would be able to own land and in return would be obliged to pay taxes. There were to be no constraints on who might buy the land, who might inherit it, or for what purpose it might be used. The Northwest Ordinances of 1784–1787 and the federal Constitution ratified and codified these principles, adding (through the Fifth and Fourteenth Amendments) the principle that no government agency should seize any private property except under due process of law (eminent domain) and then only for legitimate public purpose and with just compensation.

Nevertheless, it should be recognized that treating land ownership as a civil liberty has had some profound consequences for the nature of urban development. When public lands came to be disposed of in the nineteenth century, much of the country was delivered up to the control of private speculators:

> Those who had private capital, or who were bankers or agents for Eastern money, had an enormous advantage. Such men could purchase a whole valley, a promising townsite, whatever they wished. They could then sell part of it cheaply to settlers and retain large tracts and await the substantial price rise that would follow in the wake of the development of adjacent land.[5]

The influence of speculators was reinforced by their role as moneylenders to pioneers, many of whom, although able to pay cash for their land, did not have enough for capital improvements. As a result, the lending of money gave speculators a strong local control over land owned by others. A further consequence was the manipulation by speculators of the locations of key investments, both public and private. The concentration of land ownership and influence over land use decisions in the hands of speculators "settled a vast class of conservative interest across the

FIGURE 5.2   The early industrial city: Boston. This photograph shows Quincy Market and Faneuil Hall in the heart of the city, where industrial, retail, wholesale, commercial, and residential land uses were crammed together.

nation: It installed men who saw the duty of government to be the defense of private property in every town in North America."[6]

The onset of industrialization began to show up some of the disadvantages of enshrining property rights as a civil liberty. With no controls over land use and little regulation of the sale of land, urban development became a free-for-all. The result was that, although individual freedoms were protected, cities provided less stable settings than they might have been for businesses, less convenient settings than they could have been for their residents, and less healthy settings than they should have been for everyone.[7] As we will see in Chapter 17, these considerations eventually led to the acceptance of city planning in the United States.

INSTRUMENTS OF CHANGE: HORSECARS AND RAILROADS

The appearance of all kinds of industrial land uses, together with the arrival of thousands of immigrants and their subsequent emergence as a massive and threatening proletariat, soon gave the upper and upper-middle socioeconomic groups a strong motivation for moving permanently as far as possible away from them. (The wealthy had already been sending their families away to the countryside for the uncomfortable and epidemic-prone summer months.) Similarly, the intensifying stresses and conflicts of city life repelled many of the new class of affluent white-collar workers.

There was, therefore, a strong demand for the first forms of public transport to provide an opportunity to escape. A few families, wealthy enough to be able to afford private carriages or hire hackney carriages, had moved out to exclusive exurban settings at the first signs of industrial squalor. But it was not possible for others to join them until the development of horse-drawn omnibus systems, horsecars (light rail systems drawn by horses), and short-haul passenger railroad routes.

Horsecars

The first horse-drawn omnibus service was established by Abraham Brower in New York in 1829, picking up and setting down passengers by request along the length of Broadway for a flat fare of 12 cents. By the 1840s there were several hundred omnibuses operating in New York, and transportation companies had also appeared in Baltimore, Boston, Brooklyn, New Orleans, Philadelphia, Pittsburgh, St. Louis, and Washington, D.C. Meanwhile, John Mason had successfully established the first horsecar system in lower Manhattan in 1832. In 1854 Paris became the first European city with a horsecar system, by which time horsecar systems had become widespread in the United States (Figure 5.3), allowing people to commute up to three or four miles to downtown workplaces within a reasonable time (30 to 45 minutes). The cost of such travel, however, meant that only modest numbers of better-off families were able to move to the **horsecar suburbs** that sprang up at the edge of the walking city

FIGURE 5.3
A mule-drawn "horsecar" photographed in Washington, Georgia, in 1905.

around the extremes of the radial horsecar and omnibus routes. Ordinary wage earners received little more than a dollar a day; most round-trip fares were between 15 and 25 cents.

Railroads

The railroads have already been identified as a major influence on the development of the **urban system**. Although they were principally developed for interurban transport, they also had some striking effects on the structure and organization of land use within cities.

- First, they provided the catalyst for the reorganization of land uses in city centers. The railroad companies needed large, flat tracts of land at or near the city center in order to accommodate passenger stations and all the necessary infrastructure: locomotive sheds, shunting yards, and, often, goods-handling facilities. They also had the economic (and political) power to secure such sites.

- Once they were developed and tracks laid to them, the railroad facilities had the effect of realigning the pattern of central-area land uses, laying the foundation for modern **CBDs** (**central business districts**). Because of the inevitable funneling of large numbers of people through the railway stations, nearby locations became prime sites for hotels, restaurants, saloons and larger stores (Figure 5.4). And, because of the arrival and departure of bulk goods (usually from a different area of the terminus), nearby locations became prime sites for warehouses and the offices and storehouses of wholesale distributors.

- At about the same time, a critical new building technology was introduced: iron girders with curtain walls. Together with passenger elevators (another innovation of the 1850s) this new technology facilitated the beginnings of high-rise construction (see Chapter 10). The net result was the location within the embryonic CBD of the few land uses that could afford the costs of such technology: hotels and office buildings for insurance companies and publishing houses, for example.

- Within the central urban area the railroad infrastructure brought a radical change to the physical environment. As the celebrated critic Lewis Mumford observed, cities at this time

    consisted of the shattered fragments of land, with odd shapes and inconsequential streets and avenues, left over between the factories, the freight yards and dump heaps. In lieu of any kind of over-all municipal regulation or planning, the

FIGURE 5.4   The new downtown nexus: railroad station and railroad hotel. This example is St. Pancras in London, opened in 1874 as a main line railway terminus with a built-in 250-room hotel.

railroad itself was called upon to define the character and project the limits of the town. . . . [T]he railroad carried into the heart of the city not merely noise and soot but the industrial plants and the debased housing that alone could thrive in the environment it produced. Only the hypnotism of a new invention, in an age uncritically enamored of new inventions, could have prompted this wanton immolation under the wheels of the puffing Juggernaut. Every mistake in urban design that could be made was made by the new railroad engineers, for whom the movement of trains was more important than the human objects they achieved by that movement.[8]

Already, therefore, the consequences of treating land narrowly in terms of the rights of individual owners (as opposed to broader notions of collective good or environmental quality) were being inscribed into the structure of cities.

- Outside the central area the mainline tracks ran in swathes that radiated from the city center. Where they passed through stretches of broad flat land, they attracted stockyards, factories, and warehouses and repelled all but the lowest grade of housing. In this way, pronounced linear industrial belts were established. Because of the amount of fixed capital tied up in them, these industrial belts served as fixed elements around which the future development of cities had to take place.

- Elsewhere, the shadow of the railroad condemned smaller tracts of land to permanent dereliction or blight. Between junctions and crisscrossing branch lines were left no-man's-lands of junkyards and abandoned lots, while the viaducts that were often needed to carry the tracks into the heart of cities harbored beneath them an assortment of makeshift and unsavory activities. Even where the tracks simply passed through residential neighborhoods, the railroad brought its polarizing influence. From mid-century the social geography of industrial cities began to be regrouped around the "right" and "wrong" sides of the tracks.

- The railroads' greatest influence on land use, however, was in their promotion of commuter traffic. The railroad companies recognized the potential demand among more affluent middle-income households for a means of escape from the increasingly degraded inner-city environment that the railroads themselves had helped to create. The speed of steam trains meant that commuter stations and residential subdivisions could be established in the countryside beyond the city, creating beads of **exurban development** along the approaches to the bigger cities.

Boston was the first American city around which such a pattern developed, but New York, Philadelphia, and Chicago soon followed it (in the 1850s) and the pattern had spread to most other large cities by the 1860s. Considerable numbers of people were involved. About 20 percent of Boston's business classes traveled to work by train in 1848 from exurbs located between 12 and 15 miles from the central area. Ten years later 40 trains were running each day between downtown Philadelphia and the exurb of Germantown, and a similar number of trains were carrying the 5,000 or so commuters into Chicago from Evanston.[9] But the critical feature of the railroad exurbs was not their size so much as their exclusivity. Their cost—10 to 15 cents for a one-way ticket—made them accessible at first only to successful merchants, industrialists, doctors, lawyers, and other nouveau riches.

Like the omnibus and the horsecar, therefore, the railroads siphoned off from the city the affluent and the educated, depositing them in narrow nodes of development that extended only a comfortable stroll from the station or stop. Because commuters had to walk to the station, lot sizes in railroad exurbs tended to be fairly modest (8,000–10,000 square feet), allowing for a sufficient number of dwellings to make the station or stop profitable.

All this, of course, had a profound effect on the rhythm and pattern of urban life. The sheer speed and capacity of railway transport also changed the quality of metropolitan life. The railroads allowed for the rapid distribution of highly perishable foodstuffs—milk, vegetables, fish, and meat—and so facilitated marked improvements in the diet of city dwellers. This also led to the beginning of the standardization of diet and of patterns of consumption generally. In the same way, the bulk transport of building materials on the railways began to introduce a certain sameness to city buildings across broad regions. The railways' need for organization and punctuality also led to the standardization of time. In Britain, Manchester's municipal council decided in 1847 to adjust its clocks to London time; by 1852, every city had done the same, and clocks with standard time became prominent in every railway station.

THE INDUSTRIAL CITY (1875–1920)

It is no exaggeration to describe the changes of the industrial period as revolutionary. During this period, cities began to grow dramatically in population and territorial extent (Figure 5.5) while simultaneously experiencing unprecedented specialization and differentiation among land uses. More sophisticated technologies, powered by electricity and employing mechanized, in-line production techniques, brought to cities the logic of **economies of scale**, **agglomeration economies**, and the **division of labor**.

The options for urban development were dramatically expanded with the emergence of new technologies: gaslight, electrification, steel-frame skyscraper construction, and elevators. These changes, in turn, brought unprecedented opportunities for entrepreneurs to accumulate personal wealth, introduced the mass consumption of all kinds of new products, and financed the creation of a modern infrastructure of roads, tunnels, bridges, parks, schools, sewers, and street lighting. As a result, the organization of the economy, of society, and of urban space was radically transformed.

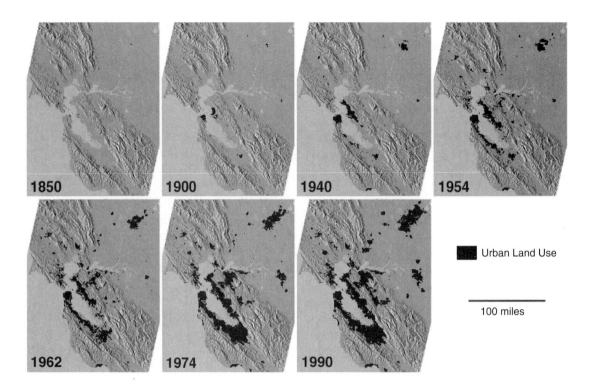

FIGURE 5.5   The growth of the built-up areas of the San Francisco Bay Area.

ECONOMIC SPECIALIZATION AND THE REORGANIZATION OF URBAN SPACE

The pressures of increased population densities, increased economic specialization, increased competition for space, and the increasing scale of economic activities meant that urban land use became highly specialized. Users of land became spatially segregated by their ability to pay for the most attractive locations—whether for industry, business, or homes. As economic growth brought more industry, businesses, and people into urban confines, both the extent and intensity of different land uses increased, bringing acute competition and conflict over land.

Factories were the linchpin of the new land use system, taking first pick of the available sites and dominating every aspect of spatial organization. For many of the pillars of industrialization—textiles, chemicals, iron and steel—the premium locations were near the waterfront, where they could secure the water needed to supply steam boilers, to cool hot surfaces, to make the necessary chemical solutions and dyes, and, above all, to dump soluble and suspendable forms of waste.

Around the factories, speculators built housing for the factory workers. This was something of a departure from the earlier phases of manufacturing, when factory owners built a good deal of housing for their workers. With large-scale industrialization in larger urban settings, workers were expected to buy housing with their wages. The result was the **generalized housing market**:[10] the creation of a mass market in rental housing, disconnected from any direct linkages between employment and tenancy.

The quality of this housing was dictated by the wages of the workers, who found themselves the least competitive of all land users. Because most workers were also unable to afford the available means of intraurban transportation on a regular basis, the developers of workers' housing were left with little choice: the construction of cramped and spartan dwellings in the leftover spaces between the factories, dumps, yards, and railroads. "To pay attention to such matters as dirt, noise, vibration, was accounted an effeminate delicacy."[11]

In a competitive market, builders and developers had to minimize costs and prices. This they achieved by using pattern-book standardization linked to economies of scale and by keeping lot sizes to a minimum. The consequence, in terms of the reorganization of urban land use, was the proliferation of pinched land divisions, narrow alleyways, solid blocks without parks or playgrounds, and narrow, inflexible street patterns (Figures 5.6, 5.7).

Between the first-pick of the factories and the last-pick of workers' housing, the reorganization of urban land uses involved the simultaneous operation of two tendencies: **centrifugal** (outward) and **centripetal** (inward) **movement**. Affluent professional and white-collar households, together with their attendant retail services, were drawn to suburban and exurban locations, creating centrifugal forces in the reorganization of land use. Most factories, warehouses, offices, hotels, and specialized retail activities, in contrast, were drawn to specialized, centrally located sites, creating centripetal forces with slightly different foci.

Some factories, however, needed so much land that they opted for peripheral, **greenfield sites**, as with the steel plants at Gary, Indiana (built in 1905) and Sparrow's Point outside Baltimore (built in 1913). Large-scale industrial developments of this sort triggered a new kind of suburbanization, involving skilled, unionized working-class households. *From this point on, American suburbanization encompassed a progressively broader cross section of groups, with the result that there began to emerge a differentiation of suburban neighborhoods according to occupation and income.*

FRAMING THE CITY: NETWORKED INFRASTRUCTURES

The networks of water, sewage, gas, electricity, telegraph, telephone, and streetcar and subway lines that were developed during the Industrial era were

FIGURE 5.6   A New York alleyway ("Bandit's Roost"), photographed by Jacob Riis in 1888.

fundamental to processes of urban change. As cities struggled to grow and to adapt to new industries, new institutions, and new ways of life, so they also continuously made and remade their infrastructure and their built environment, applying the logic of modernization and industrialization to urbanization itself. How this played out was in part a function of new technologies, in part a function of the financial capability to provide

FIGURE 5.7   Wash day in New York's tenement district.

them, and in part a function of highly uneven power struggles and social and political biases.

In the process, the whole culture of cities was transformed. One aspect of this was the sense of progress and optimism that came to be associated with urbanization in the Industrial era. Cities themselves became "electropolises," and new streets, bridges, telephone networks, gasworks, and streetcar systems became emblematic of the sense of progressive modernization (Figure 5.8). Lighting was just one of a series of radical changes to urban life. At the beginning of the nineteenth century, gas lighting created a new urban nightlife as well as a means for securing greater law and order. By the end of the nineteenth century, electrically lit shopping arcades had become the "dream spaces" of the modern city, icons of progress and cultural transformation.

A second aspect of this transformation was that cities themselves came to be regarded as complex "machines" that could be rationally organized as a unitary system. Engineers, sanitarians, and reformers drew on this notion in campaigning for improved networks of infrastructure to bring order, rationality, and sanitation to the troubled and apparently chaotic industrial metropolis.

One of the most influential early precedents was set in Paris by Napoleon III, who presided over a comprehensive program of urban redevelopment and monumental urban design. The work was carried out by Baron Georges Haussmann between 1853 and 1870. Haussmann demolished large sections of medieval Paris to make way for broad, new, tree-lined avenues (Figure 5.9), with numerous public open spaces and monuments. In doing so, he made

FIGURE 5.9   Haussmannization in Paris, 1853–70. The comprehensive program of urban redevelopment and monumental urban design involved demolishing large sections of medieval Paris to make way for numerous public open spaces and monuments, with broad, new, tree-lined avenues, such as the boulevards leading to the Arc de Triomphe.

the city not only more efficient (wide boulevards meant better flows of traffic) and a better place to live (parks and gardens allowed more fresh air and sunlight in a crowded city and were held to be a "civilizing" influence), but also safer from revolutionary politics (wide boulevards were hard to barricade and facilitated troop movement and cannonfire; monuments and statues helped to instill a sense of pride and identity).

Haussmann's approach heralded a technocratically minded, comprehensive approach to urban governance and planning that quickly spread elsewhere, notably to Barcelona, Berlin, Chicago, Cologne, Dresden, and London. In the newer cities of midwestern and western North America, rectilinear grids of streets became the norm for organizing networks of streets and utilities, making for an unprecedented openness of urban form. Streets in industrial metropolises everywhere, previously used only for local access, for informal economic activities, and for social life, were now seen primarily as part of the circulatory system of a functioning metropolitan "machine." Their size and layout were standardized; they were paved, regulated, and managed. Their construction also provided for the introduction of a subterranean system of channels, chambers, shafts, and water and sewage pipes that helped to "deodorize" and sanitize the city (Figure 5.10).

FIGURE 5.8   The "electropolis." The novelty and energy that electric light brought to cities is captured in this painting of *The Bowery at Night* by W. Louis Sonntag Jr., in 1895.

FIGURE 5.10   Workmen repairing a sewer under Fleet Street in London. The sewers carried 87 million gallons of water daily in 1854.

THE EMERGENCE OF LAND USE ZONING LAWS

Not surprisingly, the dramatic reorganization of urban space within the industrial city created some serious conflicts and tensions—problems that were all the more acute in the United States because of the civil liberties enshrined in private land ownership. It did not take long before some groups mobilized against the land users they perceived as threats, hazards, or nuisances. Their object was the legal exclusion of particular land uses from certain parts of the city. The outcome was **land use zoning**, a regulatory mechanism that was to become a key instrument in consolidating the specialization and segregation characteristic of land use in American cities.

Land use zoning had its roots in discrimination. In San Francisco discrimination against the Chinese was focused on the laundries that by the 1880s had spread well beyond Chinatown into the city's other ethnic neighborhoods, where they operated as social centers for the many Chinese domestic servants who were scattered around town. The white population, regarding the laundries has havens for "undesirables," sought to close more than 300 of them in 1886 by declaring that they were nuisances and fire hazards. But in the federal courts the case of *Yick Wo v. Hopkins, Sheriff* struck down the statute because of the way it gave arbitrary power of racial discrimination to a board of supervisors. Upon this decision, the city of Modesto promptly came up with a simple ploy to satisfy the requirements of the Fourteenth Amendment: the division of the city into two zones, one excluding laundries and another permitting them.

Similar nuisance-zone statutes, directed variously against laundries, brothels, pool halls, dance halls, livery stables, and slaughterhouses, were quickly adopted in other West Coast cities. Los Angeles was zoned into three districts: one residence-only, one industry-only, and one where a limited range of industries was permitted among residential uses.

The principle having been established, the prototype land use zoning ordinance was written in New York. The catalyst this time was discrimination against Jewish garment manufacturers, who had begun to encroach upon the territory of Fifth Avenue luxury stores from their base on the Lower East Side. Fearing that increased traffic and the presence of thousands of low-wage workers would deter wealthy customers, the Fifth Avenue Association mobilized a coalition of regulatory interest groups that sponsored zoning ordinances. These ordinances were designed to stabilize the broad pattern of land uses by encouraging specialization and uniformity at the micro scale. The New York zoning ordinance, drafted by lawyer Edward Bassett and passed in 1916, was based on the premise that *restrictions on land use are constitutional because they enable city governments to carry out their duties of protecting the health, safety, morals, and general welfare of their citizens.*

This philosophy, clearly, was a significant modification of the sanctity of land ownership as a civil right, though it did not by any means signal a retreat. The law itself was spelled out by way of a map, on which all private land was assigned a particular functional category for future land uses (it did not attempt, however, to weed out nonconforming uses already in existence). These categories were then listed in detail, including any restrictions such as the height, number of floors, overall size, and general characteristics of buildings, the overall density of population, the minimum amount of public open space, and so on.

Existing land users and speculators alike appreciated the advantages of such an approach. For existing land users, the law brought security from the threat of intrusive new activities. For speculators, the law brought a welcome degree of stability and predictability to the business of land development. No longer would the developer of middle-income housing have to worry about tumbling prices resulting from the establishment of a noisy saloon on an adjacent lot. So compelling were these advantages that within 10 years of the enactment of the New York law more than 500 other cities had adopted similar ordinances. The main effect (which did not fully take hold until after 1920) was to stabilize patterns of land use in the industrial city and to defuse much of the tension surrounding urban growth and spatial reorganization. Later, the ancestry of zoning began to tell, and it became a major means of discrimination against social groups perceived for one reason or another as undesirable.

At the start, however, land use zoning was unable to cope with the fundamental tension generated by the growth of industrial cities: the unease of the increasingly numerous and more affluent middle-income groups at being cooped up in the city along with the factories, railroads, warehouses, and "huddled masses" of "ordinary" people. Suddenly, in 1888, this tension was relieved by the introduction of a simple but radically effective innovation in urban transportation: the electric streetcar, or trolley.

THE SUBURBAN EXPLOSION: STREETCAR SUBURBS

The first successful streetcar system was established by Frank Sprague, who had left Thomas Edison to form his own company and experiment with electrically powered omnibus systems. Drawing on the experience of numerous attempts to solve the urban transportation system—including electrified trolley systems in Cleveland, Baltimore, East Orange (N. J.), Montgomery (Ala.), and, in particular, the Siemens system developed in Germany—Sprague perfected an electrically driven version of the horsecar, powered by way of overhead cables, that opened for business in Richmond, Virginia, in the spring of 1888.

The innovation was adopted the following year by Henry Whitney, who had established in Boston an integrated horsecar system that covered the entire city. By electrifying the system, Whitney could extend the radial lines, since streetcars could travel much faster than horsecars. In Boston the total length of track in 1887 had been just over 200 miles; by 1904 it

was almost 450 miles. Within five years of Sprague's success at Richmond, more than 200 other cities had adopted his system. By 1902 there were 22,000 miles of streetcar tracks in American cities.

By making it feasible to travel up to 10 miles from the downtown area in 30 minutes or so, the streetcar greatly increased the territory available for residential development and opened the way for the growth of **streetcar suburbs**. So much land became accessible at once that the price of land was kept down, thus ensuring inexpensive suburban lots. Affordable land, combined with the cheaper operating costs per passenger-mile of the streetcar (because of their large carrying capacity and the efficiency of electric power), ensured that developers of streetcar suburbs instantly found a market in the solid middle-income groups, and there was a rush of speculative suburban sprawl as developers and streetcar operators worked together to shape the massive increment of urban growth (Figure 5.11). Around the fringes of Chicago, for example, some 80,000 new residential lots were recorded between 1890 and 1920 (though many of these lots were not built on for some time afterward).

As well as enabling the growth of suburbanization, the streetcar influenced urban form in several other ways:

1. The sudden departure of thousands of middle-income households from the old core gave much-needed scope for the reassignment of space to nonresidential uses in the downtown.

2. The ease and cheapness of travel on streetcars led suburban housewives to use them on shopping trips outside rush hours, thus helping to sustain a specialized, high-order shopping area in the downtown.

3. The development of cross-town and circumferential streetcar lines created a series of junctions and interchanges around which grew small nuclei of commercial development. This signaled the end of the simple city structure with a single focal point.

4. Larger nodes of development emerged around the termini of the radial streetcar routes. Distinctive attractions often seeded these nodes. To counter the lack of commuters on weekends and holidays, streetcar companies promoted the idea of outings to amusement parks, beaches, picnic grounds, and even cemeteries at such times. These sites were often located near streetcar termini (partly as a result of sponsorship by the nexus of business interests involved with streetcar development), and they soon attracted a variety of ancillary services such as restaurants and convenience stores.

5. The streetcar facilitated the growth of exurban settlements and satellite townships that were

FIGURE 5.11   A streetcar in the Richmond district of San Francisco in 1909. In this new "streetcar suburb" the intersection of streetcar l ines shown here had already attracted a large commercial building containing offices and stores.

joined to the main city by railroad but were too small to sustain an economic base of their own until they acquired a streetcar system, thus serving a larger pool of local workers and customers. It was then possible to develop an industrial base and attract substantial clusters of shops and offices. In this way the streetcar sowed the seeds for the subsequent evolution of a *metropolitan* form of urbanization.

Rapid Transit

In a few larger cities the congestion created by numerous streetcars converging on the downtown area prompted the development of electrified rail systems in structures separated from streets—on elevated tracks or in tunnels (Figure 5.12). These *rapid transit* systems could travel faster than streetcars while carrying more passengers, and their effect was to stretch

FIGURE 5.12   New York's elevated railway at Coenties Slip in the early 1800s.

the fingers of urban development farther into the countryside, reinforcing and extending the radial form of the city.

Underground lines were pioneered in London, where the Metropolitan subway line was opened in 1863 and the Inner Circle line was completed in 1884, linking up the main above-ground railway termini (Figure 5.13). In the United States, Henry Whitney once again led the way, opening 1.6 miles of subway underneath Tremont Street in Boston's downtown in 1897. His success (over 50 million passengers were carried in the first year of operation) prompted the development of the New York subway system, which opened in 1904. The start-up costs of subway systems were extraordinary, however. Whitney's line cost nearly $150,000 for every 100 yards of track. Apart from a combined elevated and subway system in Philadelphia that opened in 1908, there were no further subway projects in the United States until the 1940s, when one opened in Chicago. Instead, elevated electric railways—the "els"—were the preferred solution in Boston, Brooklyn, Chicago, Kansas City, New York, and Philadelphia, where they were installed around the turn of the twentieth century.

Mass Transport and Real Estate Development

In every city there was a very close relationship between the development of mass transportation systems and suburban real estate development. In some cities the same organization undertook transit lines and real estate development—the streetcar suburb of Shaker Heights in Cleveland is an example. Transit lines often preceded land development, and in many cases the transit service was expected to lose money in the early days. But substantial increases in land value, because of access to transit lines, would more than make up for the early investment.

The link between transit and real estate development was particularly strong in West Coast suburban developments. In northern California, F. M. Smith bought and consolidated the trolley lines in San Francisco's East Bay and purchased 13,000 acres of land for development in the Oakland and Berkeley areas in the 1900s. In southern California, Henry Huntington, a founder of the Southern Pacific railroad, developed the Pacific Electric Interurban Transit Company in the Los Angeles area. He bought up the land along his routes and laid out suburban developments, at the same time avoiding competitors' land holdings unless he was made a partner. Among Huntingdon's partners was Harry Chandler, the largest developer in Los Angeles. In the first decade of the twentieth century, Chandler bought 47,500 acres in the San Fernando Valley, an area about the size of the city of Baltimore, and the Pacific Electric extended their lines into the valley. A $25-million water project, paid for by the city of Los Angeles, supplied the development with water after a vigorous campaign for the water project, led by Chandler's father-in-law, Harrington Grey Otis, publisher of the *Los Angeles Times*. Later Chandler bought the 300,000 acre (468 square mile) Tejon Ranch in Los Angeles and Kern counties, land that is still controlled by Chandler interests.

FIGURE 5.13 The world's first underground railway, in London: Baker Street Station, Metropolitan Railway, 1863.

For half a century, until approximately 1920, public transportation inexorably influenced the form and growth of the city. Transit lines formed the arteries along which the population moved and real estate development occurred. Transit stops became the nodes of commerce and entertainment, and the radial lines all converged at a hub of the metropolitan area. Residential neighborhoods grew up along the transit lines and small commercial districts were developed wherever lines intersected or branched.

URBAN STRUCTURE IN THE INDUSTRIAL ERA

To contemporary observers, the changes wrought by the industrial era were shocking. Never before had there been such visible growth, such intensive building and rebuilding, such novel arrangements and re-arrangements of urban form and function. As we look back, we can acknowledge these changes as having established the physical template for the development of the modern city. In this context three dimensions are of particular importance: the growth of CBDs, the assignment of locations according to principles of rent, and the emergence of distinctive patterns of residential segregation.

CENTRAL BUSINESS DISTRICTS

It was the CBD that gave visual expression to the growth and dynamics of the industrial city. The CBD became a symbol of progress, modernity, and affluence. In the United States one could read the precise status of any particular city from the size and degree of differentiation of its CBD. The CBD became the hub of economic, social, political, and cultural life in American cities, and the organization of functions within the CBD set the framework for the next 75 years of downtown development.

Department Stores and Shopping Districts

The location of railroad stations and the subsequent confluence of transit lines had shaped the early development of CBDs, and retailing activities dominated the land uses within CBDs. The more specialized, higher-order, and exclusive retailers could generate sufficient profits to outbid any other potential user of downtown space. Because most shoppers were reluctant (as they are today) to walk much more than 400 yards from their starting point, the location decisions of downtown retailers were tightly circumscribed. The result was a distinct and compact shopping area, a zone perhaps 300 yards in radius, in which retailing was practically the only activity along street frontages at ground level.

Anchoring this zone, at the very center, were grand, multistory *department stores*: Macy's, Bloomingdale's, and Lord and Taylor's in New York; Jordan March and Filene's in Boston; Woodward and Lothrop's in Washington, D.C.; Wanamaker's and Strawbridge and Clothier's in Philadelphia; Carson Pirie Scott and Marshall Field's in Chicago; I. Magnin's in San Francisco; Bullock's in Los Angeles; and so on (Figure 5.14). These stores were landmarks, and they needed to be: Only in prominent locations at the very center of things was the conflux of potential customers great enough to ensure enough daily business to carry the enormous overheads required to stock and display in proper style and quantity the best that industrialization had to offer. The classic location was a corner site at a major intersection, but the fundamental requirement was proximity to railroad termini and, crucially, the downtown stops and transfer points of streetcar lines.

Other stores were by no means located at random within the shopping zone of the CBD. Stores that were so specialized that there was only one of their kind in the city could afford to let their customers come to them. They could therefore locate almost anywhere within the shopping area, or even just beyond it, seeking out a cheaper site and having to pay little attention to their neighbors.

FIGURE 5.14   Strawbridge and Clothier's department store, Philadelphia, photographed in the early 1900s.

Most stores, however, had to consider carefully the microdetails of location. Then, as now, a significant part of consumer behavior was comparison shopping: Some people, "born to shop," seem to enjoy it for its own sake. But an equally important part of consumer behavior is based on the minimization of effort. It makes sense, therefore, for particular types of specialized stores to cluster together: men's outfitters, for example (Figure 5.15). In addition, shopkeepers had to be sensitive to the functional relationships between their businesses and those of others. It thus made sense for shoe stores and stores selling mass-market women's fashions to locate near department stores to catch the "overspill" of shoppers. Exclusive, high-fashion stores, in contrast, need to cluster together but, equally, need to be away from the great press of shoppers near the department stores. New York's Fifth Avenue, without a subway or an elevated train in the 1920s, is a good example of how relative inaccessibility allowed exclusivity to be maintained.

Downtown Office Districts

In the largest cities the pedestrian limitations of shoppers created a series of "walking zones" that constituted specialized retail subareas. Similarly, large cities could support specialized clusters of high-order services such as banks, law offices, or medical facilities. The office district (or subdistrict) was usually adjacent to the shopping district for it, too, had a functional dependence on the focus of transportation at routes that brought office workers on commuter trains and streetcars and out-of-town business visitors by long-distance trains.

It was the office district that gave the CBD its most prominent landmarks, with taller and taller office buildings. The earliest, biggest office buildings were those that were purpose-built for insurance companies and publishing houses. In addition to their need for personal intercommunication among large numbers of employees, these organizations found skyscrapers a valuable means of advertising. But with the emergence of the CBD as the organizational hub of a larger and increasingly complex array of urban functions, it did not take speculators long to realize the value of central office space to small companies. Speculatively built office blocks provided an "address" for smaller office functions, which could also share in the splendor of marbled lobbies and concierge services that such settings provided (Figure 5.16).

Warehouse Zones

In the early industrial city wholesaling and retailing had gone hand in hand within the CBD—to the extent, in some instances, of occupying the same building. The *scale* of operations in the industrial city meant that this could no longer continue. Wholesalers increasingly had to supply not only downtown

FIGURE 5.15   The early CBD. This stretch of Kearny Street (just off Market Street, the main commercial thoroughfare) in San Francisco shows the early expression (in 1894) of clustering and functional interdependence. Along these two blocks were several men's tailors and outfitters, together with a haberdasher and shirtmaker, a store selling men's furnishings and shirts, a hatter, a watch store, a cigar store, and a men's club. It would be nice to think that the florist at the corner of Kearny and Bush was part of the functional grouping, providing the finishing touch for the gentleman's shopping trip.

FIGURE 5.16　Speculative development and specialized land use: This 15-story medical arts building at 490 Post Street in the heart of San Francisco's CBD dates from the early 1900s, when central business districts began to be reorganized into functional subdistricts.

FIGURE 5.17　Philadelphia City Hall: civic pride symbolized in bricks and mortar.

merchants but also merchants in outlying nodes and in satellite towns and smaller cities farther afield. At the same time, they needed to have access to railroad freight yards for the in-shipment of carloads of goods, and they needed to be near the handling yards of express shipping companies like American Express and Wells Fargo that had sprung up to put together carloads of smaller quantities of outgoing goods. The result was a distinctive warehouse zone, located on the edge of the CBD next to the freight yards. Two very different kinds of wholesaling tended to remain embedded in the CBD itself, however: (1) low-volume, high-value goods such as diamonds and jewelry and (2) high-volume, perishable produce such as fruit and vegetables, fish, and meat, each sold separately from central locations accessible to restaurant chefs and the owners of small food stores.

City Halls and Civic Pride

Another element in the growth of CBDs was that of the *city hall* and its associated functions. It was convenient for the offices of city bureaucrats and officials to be at the hub of things: convenient both for them and for their constituents, given that the CBD was the single most accessible place in the city. But the main reason for their location within the CBD was not accessibility or efficiency but symbolism. City governments, with new powers and responsibilities, needed to be identified with the hub of things. Their image was the city's image. A similar logic dictated that city halls and associated civic buildings be the equals—or betters—of the department stores and office buildings that were the landmarks of the commercial districts (Figure 5.17).

In addition, the civic pride invested in city halls dictated that they be more imposing and more splendid than those of rival cities, with landscaped malls, parks, and statuary to show them off to the best advantage. The result was another distinctive walking zone, dominated by the city hall itself and containing the main library, the central post office, and the courthouse, museum, assembly hall, opera house, and art gallery that made up the tableau of civic amenities. In between were the offices of lawyers and the buildings of vocational schools and colleges, together with the premises of others who traded on the stream of workers and visitors frequenting the area: lunch counters, bars, tobacco and newsstands, and so on.

The Spatial Organization of CBDs

From these observations we can put together a number of generalizations about the organization of land use within the CBD. The overall spatial structure of CBDs tends to be dominated by a high-density *core* that contains the retail, office, entertainment, and civic zones and a lower-density *frame* that contains zones of warehousing, educational facilities, hotels, medical services, and an intermixture of specialized shops (antiquarian bookstores, for example) and services (e.g., acupuncturists, picture framers) that have neither the

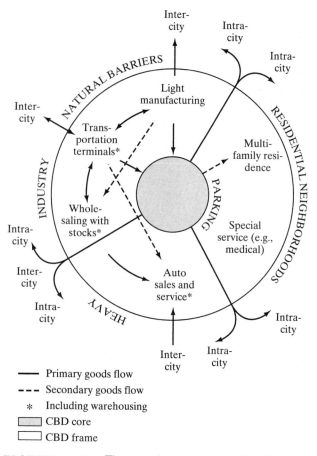

Primary goods flow
--- Secondary goods flow
* Including warehousing
CBD core
CBD frame

FIGURE 5.18   The core–frame concept of the CBD.

FIGURE 5.19   Land use specialization in the CBD's zone of discard: X-rated businesses are typical of this niche in the changing urban fabric.

functional linkages nor the potential profitability to justify locations within the core (Figure 5.18).

Nevertheless, constant changes in the relative fortunes of different kinds of activities, changes in the transportation infrastructure, and the inevitable aging and obsolescence of buildings, make the CBD very dynamic. The core tends to expand toward the higher-status, newer elements in the frame, creating a *zone of assimilation*, and to move away from lower-status, older elements, creating a *zone of discard*, which often forms a niche for another kind of economic specialization: vice and X-rated businesses (Figure 5.19). The frame itself overlaps with older structures and residential districts in a **zone in transition** ripe for conversion to CBD uses.

LAND VALUES AND URBAN LAND USE

The spatial organization of the CBD points to the importance of accessibility and relative location in understanding patterns of land values and land use. In New York in the mid-1920s the peak land values were $22,000 per foot of street frontage on Broadway; only half a mile away land values were less than $3,000 per foot.[12] Comparable gradients could be found in every large city (Figure 5.20). In 1903 the economist Robert Hurd made the point in a much-quoted statement:

> Since value depends on economic rent, and rent on location, and location on convenience, and convenience on nearness, we may eliminate the intermediate steps and say that value depends on nearness.[13]

Hurd's reference to economic rent (sometimes referred to as ground rent or "location rent") requires some elaboration, since the concept has become fundamental to attempts to theorize patterns of urban land use.

First, it should be clear that economic rent, or ground rent, does not correspond to the popular usage of "rent" (more property called "contract rent"). **Ground rent** is the surplus paid (to land owners, in this instance) above the minimum amount that would be necessary to use the land at all. It is often defined, in practice, as the total revenue that can be generated by a particular activity on a particular parcel of land, less the total production and transportation costs associated with that same parcel of land. (Not all activities are run on business lines, however. Households, for example, along with public agencies, do not evaluate their success in terms of profits but in terms of satisfaction, disposable income, efficiency, or some other criterion. The term *utility* is, therefore, often used in place of *revenue*.) It follows that ground rent reflects the utility of a particular site for a particular activity. Furthermore, it is likely that for each site there will be one activity that generates the highest possible economic rent: the *highest and best use* for

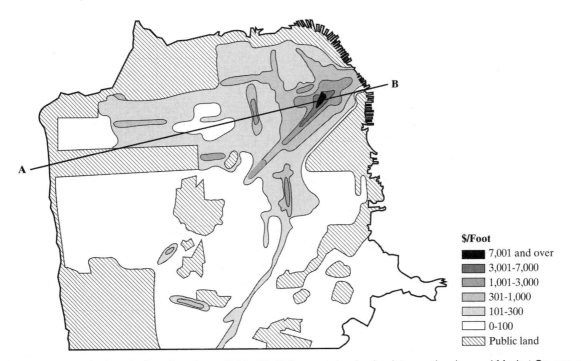

FIGURE 5.20   Land values in San Francisco, 1926–27. At the peak land value intersection (around Market Square and the intersection of Powell and Market Streets), land values were in excess of $7,000 per foot, falling away very rapidly to under $3,000 per foot within a few blocks. Elsewhere in the city, relatively high land values were restricted to narrow corridors that correspond to major shopping thoroughfares.

that site. Land use should, therefore, respond to land values in a predictable fashion. Activities that can derive the greatest utility from a given site should outbid all other activities for that site (see the Box entitled "The Bid-Rent Theory of Urban Land Use").

Sectors and Zones

In practice, land use in industrial cities did conform to certain general patterns. The classic study of land use in American cities of the Industrial era was undertaken by Homer Hoyt,[14] who made a comparative study of patterns of rental values in 142 U.S. cities on behalf of the **Federal Housing Administration**. Hoyt developed a **sector model** of urban land use that was based on a number of generalizations derived from his study of patterns of rental values from 1878 to 1928. The generalizations were these:

- Rent, and therefore socioeconomic status, varied within cities primarily by radial sectors (wedges).
- The highest rents were to be found in a single sector that often extended out continuously from the CBD.
- Intermediate rents, associated with middle-income neighborhoods, were commonly found in sectors on either side of the high-rent, high-status sector.

- Low rents, associated with working class and low-income housing, were usually found on the side of the city opposite the high-rent sector.
- Over time, the high-rent sectors tended to:
  - grow outward along major transportation routes;
  - extend along ridges of high ground, free from the risk of flooding and with panoramic views; and
  - be drawn toward the homes of community leaders.

Putting these generalizations together, Hoyt came up with a generalized model of urban land use (Figure 5.22). The main point about this model is the **relative location** of the different sectors. Hoyt argued that corridors of industry and warehousing will always tend to be surrounded on both sides by sectors of working-class housing, while middle-income housing will always tend to act as a buffer between the industrial/working-class half of the city and the city's main sector of elite neighborhoods.

The key to the dynamics that produced these patterns was, for Hoyt, the behavior of affluent households. Once the CBD has been established and corridors of industrial development laid out, affluent households will always have first pick of the most desirable sites: away from industry and the press of humanity in the

BOX 5.1   THE BID-RENT THEORY OF URBAN LAND USE

The idea that companies and households compete for space in a way that maximizes their utility is the basis for a neoclassical economic model of urban land use developed by William Alonso.[15] Like all approaches in **neoclassical economics**, **bid-rent theory** begins with some basic assumptions:

- The city is located on a flat, featureless plain, so that all parcels of land are uniform apart from their relative distances from one another.

- Transport is ubiquitous, with transport costs being a direct function of the linear distance between places.

- The city has a CBD that contains all the employment opportunities and is located in the geographical center of the city, the single most accessible point.

- The city is inhabited by rational people whose behavior is geared to maximizing profits, utility, and satisfaction while minimizing effort. This assumption also implies a situation of *perfect free-market competition*, with no restrictions on land or buildings and no buyers or sellers with sufficient resources to affect the overall state of the market by their own activities.

Given these assumptions, the argument proceeds as follows. For all types of land use, the most central sites will be the most attractive. Because of the geometry of the city there are relatively few central sites in relation to the total space available. As a result, competition for central sites will be intense, and the prices offered for them will be higher than those offered for less central sites. Different types of land users will place different financial evaluations on the utility of centrality, depending on their particular schedule of expected income and expenditures. It is logical, for example, to expect some offices, banks, hotels, and other commercial establishments to be able and willing to outbid households for central sites because the extra income accruing to a central location through increased trade is likely to outweigh the savings in commuting costs obtained at the same site by a household (or even by several households living on the same site).

Each type of land user can thus be thought of as having a distinctive bid-rent curve that reflects the prices that that type is prepared to pay for sites at different distances from the CBD (Figure 5.21). Juxtaposing the bid-rent curves of different users shows that the users with the steeper curves capture the more central sites, while those with shallower curves—residential users and manufacturing and distribution facilities (that require a large site near a freeway intersection) in this example—are left with the peripheral sites. This locational equilibrium is reflected in a simple pattern of concentric zones.

When we plot bid-rent curves for households of different income groups, an important relationship is exposed: Those with higher incomes will have steeper bid-rent curves and so end up nearer to the city center, while the lowest-income groups will end up on the periphery. This outcome is counterintuitive, given what we have seen about the suburbanization of the middle-income groups. Alonso's major contribution here was to point to the need to make additional assumptions about household behavior: That different households will bid for different-sized lots in different locations *according to their relative preferences for living space versus the utility of accessibility*. Residential bid-rent curves are thus seen as a *trade-off* between living space and commuting costs. Alonso's model is, therefore, sometimes referred to as the "trade-off model" of urban land use.

In practice, high-income households are best able to afford the recurring costs of commuting and tend to trade off those costs for extra living space. Although this fits the commonly observed pattern of high-income households consuming relatively large amounts of land on the urban fringe, however, it poses an apparent paradox: low-income households are left with nothing but the more expensive inner-city sites. This paradox is resolved, however, by the rational behavior of low-income households: Forced to consume land in high-cost locations, they minimize household costs by living at higher densities.

The apparent paradox of patterns of residential land use having been resolved, some of the more restrictive assumptions of bid-rent theory can be relaxed and new variables introduced. By allowing for secondary centers of employment and shopping in a multinodal metropolitan setting and introducing the influence of transit routes, for example, it is possible to obtain a much more realistic projection of contemporary urban structure (see Chapter 6). Figure 5.21 indicates, however, that although there are secondary peaks surrounding freeway intersections, for example, the bid-rent curves continue to show an overall decline with distance from the CBD.

(Continued)

Box 5.1   The Bid-Rent Theory of Urban Land Use   *(Continued)*

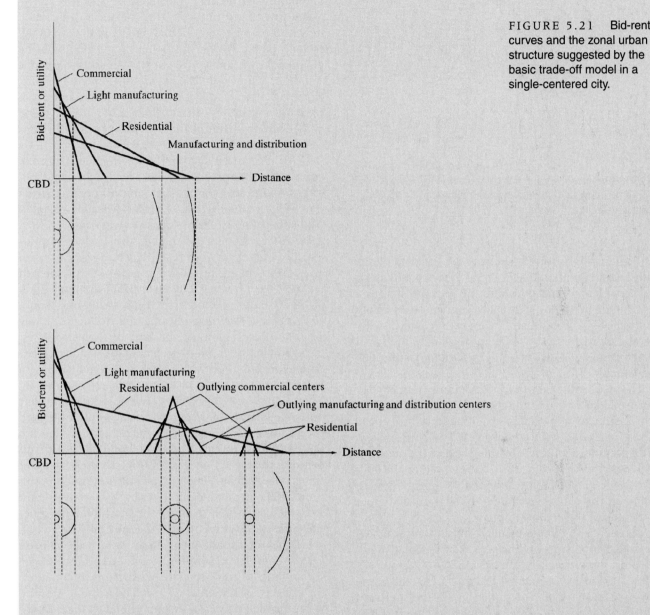

FIGURE 5.21   Bid-rent curves and the zonal urban structure suggested by the basic trade-off model in a single-centered city.

There are, however, some severe limitations to the usefulness of bid-rent theory, mostly because neoclassical economic assumptions about perfect competition and people's rationality do not hold up very well in the real world. People's behavior is, in reality, conditioned by habit, convention, deference, ignorance, and, sometimes, downright irrationality. Their search for a place to live or to set up a business is constrained by time as well as money, by imperfect information, and by the effects of monopolistic elements (big corporate land users like the railway companies in the nineteenth century; big landholders like the speculative developers and the banks). Land and property markets, meanwhile, are constrained by regulatory and fiscal controls, by the intervention of special interest groups such as preservationists and tenants' associations, and by the actions of key power holders and decision makers such as developers, realtors, and loan officers.

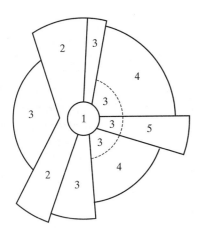

FIGURE 5.22   Homer Hoyt's model of urban structure. 1–CBD; 2–wholesaling and light manufacturing; 3–lower-income residential; 4–middle-income residential; 5–upper-income residential.

CBD and on high ground free from the risk of flooding. With urban growth, the high-status area expands along the axes of transportation lines, in response first to a desire among the most affluent to combine accessibility with suburban or exurban living and, second, to a desire among the almost-as-affluent to acquire the social cachet of living in the same neighborhood as the rich and famous.

The result was a *generational shift* in the residence of upper-middle-income households. Each new generation of affluent households would build or buy at the current true edge of the city: as far as possible from less-fortunate households and within sight of the country-side. When their parents died, they would retain their new home (in the more fashionable neighborhood), selling off the old family home—in all probability to a household of slightly lower status.

Middle-class, middle-income housing, meanwhile, is established in surrounding sectors by speculative developers, who recognize the desirability of "good" addresses among prospective customers. At the same time, an important consequence of the outward growth of each sector is the banding of each sector into *zones* of different age, style, and condition.

Filtering and Vacancy Chains

The most critical part of this overall dynamic is a mechanism of neighborhood development and change implied by Hoyt's model. The basis of this mechanism is the chain of residential moves initiated by the construction of new homes for the affluent: the **filtering** of households up the housing scale and the consequent filtering of houses down the social scale. As a new home is custom built for a family at the top of the socioeconomic ladder, the home they leave might be purchased by a family of slightly lower socio-economic standing, whose old home, in turn, might be purchased by a third family of still lesser means, and so on, creating a **vacancy chain** that, other things being equal, can go on until the last vacancy is a broken-down old place that nobody wants.

Hoyt recognized that, for the filtering mechanism of growth and change to work, there have to be some special stimuli to lead the most affluent to new construction: Why, in other words, should they leave well-built, well-situated homes? Hoyt's answer was *housing obsolescence*. For the affluent, he argued, several kinds of obsolescence can trigger a desire for new housing. *Functional obsolescence* might result from advances in kitchen technology or heating systems or from the innovation of new luxury features such as swimming pools. *Obsolescence of form* might result from trends in family size and organization. At the time of Hoyt's analyses, for example, the trend away from large families, combined with increases in the real cost of maintaining a household of domestic servants, gardeners, and so on, was beginning to make large, free-standing mansions outmoded. *Locational obsolescence* might result from land use changes in the immediate neighborhood that are seen as intrusive: increased levels of residential development, the arrival of mass transit, the appearance of stores, for example. Finally, *style obsolescence* might result from shifts in architectural styles and conventions, something to which many of the most affluent (especially the newly affluent) are especially sensitive.

Hoyt's interpretation of urban land use has been very influential in pointing to the importance of the social dimensions of spatial organization. In addition, the concept of filtering has provided a useful starting point for the analysis of neighborhood change. As we will see, however (Chapter 13), urban housing markets are in fact much more complex than Hoyt allowed. In any case, the scale, form, and dynamics of cities were about to be revised yet again by a new round of economic and technological change, as we describe in the next chapter.

FOLLOW UP

1. Be sure that you understand the following key terms:

   bid-rent theory
   CBD core and frame

   centripetal and centrifugal movement
   exurban development
   filtering
   generalization of housing

generational shift

ground rent

land use zoning

pedestrian city

preindustrial city

sector model (Hoyt)

streetcar suburbs

vacancy chain

zone of assimilation

zone of discard

zone in transition

2. Consider the statement on p. 118 that "Treating land ownership as a civil liberty has had some profound consequences for the nature of urban development." What other approaches might a society take to the issue of land ownership? Would American cities have developed any differently if another approach had been adopted before industrialization took root?

3. Consult the zoning ordinances and land use zoning map for your town or city. What conclusions can you draw about the city's objectives? Which of the models and concepts discussed in this chapter are most useful in understanding the overall pattern of land use shown in the map?

4. Construct a sketch map of the downtown area of any city with which you are familiar. How well does the land use pattern conform to the idealized pattern described in Figure 5.18?

5. Attend to your *portfolio*. You should continue to look for supplementary materials along the lines suggested in previous chapters. Since the topics covered in this chapter are focused on local-scale, historical events, you might wish to consider drawing on old newspaper accounts. Copies of nineteenth-century and early twentieth-century newspapers might not be available online but are often available on microfilm or microfiche in large libraries, and it is usually possible to obtain a photocopy of a particular story. You can then include that material in your portfolio with your own observations on how the story is related to themes in the text (or to your own perspective).

KEY SOURCES AND SUGGESTED READING

Fishman, R. 1987. *Bourgeois Utopias*. New York: Basic Books.

Goodman, D., and C. Chant. eds. 1999. *European Cities and Technology*. London: Routledge.

Gowans, A. 1986. *The Comfortable House: North American Suburban Architecture 1890–1930*. Cambridge, Mass.: MIT Press.

Graham, S., and S. Marvin. 2001. *Splintering Urbanism*. New York: Routledge.

Jackson, K. T. 1985. *Crabgrass Frontier: The Suburbanization of the United States*. New York: Oxford University Press.

Kostof, S. 1992. *The City Assembled: Elements of Urban Form Through History*. Waltham, Mass.: Little, Brown.

Roberts, G. K., and P. Steadman. 1999. *American Cities and Technology*. London: Routledge.

Vance, J. E., Jr., 1990. *The Continuing City: Urban Morphology in Western Civilization*. Baltimore, Md.: Johns Hopkins University Press.

Warner, S. B., Jr. 1972. *The Urban Wilderness: A History of the American City*. New York: Harper and Row.

RELATED WEBSITES

David Rumsey Map Collection: *http://www.davidrumsey.com/* This online historical map collection at the University of California–Berkeley offers over 10,000 free downloadable maps. The collection largely contains maps of North and South America from the eighteenth to the twentieth century. There are wonderful user-friendly ways to view and analyse the maps, including the simple "Insight Browser" as well as the more sophisticated "GIS Browser" that allows multiple historical maps of cities such as Boston, New York, San Francisco, and Washington to be overlaid in order to trace the growth and change of individual cities through time.

UNESCO: *http://whc.unesco.org/sites/cities.htm* The United Nations Educational, Scientific and Cultural Organization's World Heritage List includes information and images of cities and towns of global importance from around the world, including the Historic District of Québec and Old Town Lunenburg in Canada and La Fortaleza and San Juan Historic Site in Puerto Rico.

Trolley Cars Dot Com: *http://www.trolleycars.com/* This website contains well-illustrated information about trolley cars with hyperlinks to trolley car museums in North America and other mass transit (streetcars, trams, etc.) websites.

Geo-Images: *http://geogweb.berkeley.edu/GeoImages/BainCalif/Urban.html* This University of California–Berkeley website contains images of urban land uses, including urban transit, urban waterways, urban parks, and urban sprawl.

6

CHANGING METROPOLITAN FORM

In the 1920s the commercial development of the internal combustion engine unleashed the social, economic, and political forces that have given physical shape to the contemporary metropolis. Automobiles, trucks, and aircraft—along with electricity and the telephone—helped to recast the imprint of urbanization, producing sprawling, multinodal metropolitan settings with complex patterns of land use. Just as downtown areas had established their unrivaled dominance and internal functional specialization as described in Chapter 5, a new logic of transportation and location led to the decentralization of many of their retailing, wholesaling, manufacturing, and office functions, leaving them more specialized and less dominant. And just as the suburbs radiating out along transit lines had redrawn the social map, a new form of suburbanization materialized, bringing much lower densities, much greater social segregation, and a much greater variety of suburban land uses. Meanwhile, the automobiles that triggered these new developments also undermined the old mass transit and pedestrian circulation systems on which the land use patterns and functional organization of the industrial city had been based. All these changes, in turn, brought new challenges to understanding the development of urban form and land use: new concepts and theories, modified ideas, and alternative models.

CHAPTER PREVIEW

Looking back at the changes that have occurred to urban form and land use since the first widespread availability of trucks and automobiles, we can identify three broad phases:

1. Suburban infill, as automobiles began to compete with transit systems (1920–1945);

2. Suburban sprawl and economic decentralization, after automobiles became the dominant form of transportation, served by new roads and freeways (1945–1972); and

3. Splintering urbanism (1973–present), consisting of a highly fragmented matrix of lower-density land uses, a product of the limitations of automobiles, and the possibilities of new communications technologies.

We need to see the contemporary metropolis as a composite of these most recent phases of development,

grafted in turn onto the surviving legacies of the mercantile city, the early industrial city, and the industrial city that were described in Chapter 5. In this chapter we review the principal aspects of change during each of these recent phases, paying particular attention to the changing patterns of land use that have come to define the physical framework of contemporary urbanization. As in previous chapters, we address the conceptual and theoretical issues that emerge in the light of major shifts in urban outcomes. In this context we will see that the spatial reorganization of cities triggered by the increasing use of trucks and ownership of automobiles led to the development of new models of metropolitan form, and how the locational flexibility made possible by new digital technologies has led to some entirely new ways of conceptualizing urban form.

SUBURBAN INFILL (1920–1945)

American cities were the pioneers and exemplars of automobile-based suburbanization. According to the records, there were 4 motor vehicles in use in the United States in 1894, 16 in 1896, 8,000 in 1900, almost 470,000 in 1910, over 9 million in 1920, and nearly 27 million in 1930 (Figure 6.1). This first spurt of growth in automobile production and ownership corresponds to the period during which automobiles were being substituted for horse-drawn and electric-powered intraurban transport. The rate of growth decreased in the

1930s and 1940s as the process of substitution was completed, reinforced by the dampening effects of the Great Depression and World War II.

FORDISM

Over the period as a whole, Henry Ford's vision of mass production, coupled with mass consumption, came to fruition in the automobile industry—a precursor of the

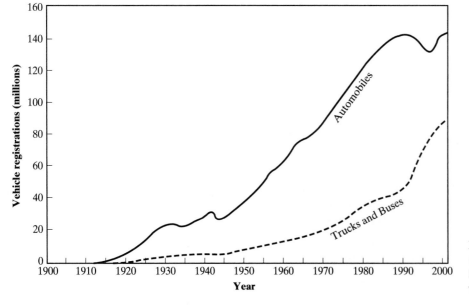

FIGURE 6.1   Motor vehicle registrations, 1990–2001 (includes both private and publicly owned vehicles).

Fordism that characterized entire economies in the subsequent period. Ford's vision was made possible by a combination of lower prices and higher wages and was paid for by higher productivity squeezed from **economies of scale**, assembly-line production, and "scientific" management (**Taylorism**). One result was that *the price of automobiles fell significantly*, making automobile ownership possible for a widening spectrum of society and drawing more of the middle-middle and lower-middle socioeconomic groups into the infill suburbs between the streetcar corridors and railroad exurbs. The Model T, which in 1909 sold for $950 (the equivalent of 22 months pay for the average worker), cost less than $300 (the equivalent of less than three months pay) in 1925, when Ford's factories were producing 9,000 cars each working day: one every 10 seconds.

PAVING THE WAY FOR SUBURBANIZATION

Automobiles, however, are of little use without good roads. It is therefore essential to acknowledge the coalition of special interests that emerged to support the building and improvement of roads. The first push for good roads came in the 1890s, from bicyclists, the Post Office, and farmers' Granges—associations that wanted to promote better access to markets. This was consolidated a decade later when the automobile, oil, rubber, and construction industries were joined by the Good Roads Association, a confederation of urban merchants and industrialists who believed that better highways would improve business.[1] Their campaigns resulted in the 1916 Federal Aid Roads Act, which required every state, as a condition of federal aid, to establish a state highway department to plan, build, and maintain interurban highways. A second act in 1921 provided additional funds with the objective of integrating the long-distance road network.

By the 1930s federal funds were being designated for road construction and improvement *within* cities. For some time before that, however, local coalitions had been able to persuade local governments to step up their road-building and road-improvement programs. State governments began to levy a gasoline tax in order to pay for road-building—the very first being Oregon's gasoline tax of one cent per gallon in 1919. Street improvement and highway construction were already the second largest item (after education) in local government budgets. Chicago, for example, improved more than 100 miles of streets between 1915 and 1930 at a cost of over $1 million per mile.[2] By 1940 about half of the country's 3 million miles of roads, urban and interurban, had acquired smooth hardtop surfaces. Parking space, however, remained an outstanding problem (Figure 6.2).

FIGURE 6.2   With the emergence of mass ownership of automobiles, North American cities faced a new pressure on land use in industrial and commercial districts: space for parking lots. This example is from Montreal in the late 1930s.

Parkways

Some of the most imaginative and progressive initiatives were taken around New York City, guided by master planner Robert Moses. Moses developed the idea of the **parkway** as a limited-access recreational highway. The germ of the idea had been contained in Frederick Law Olmsted's design for New York's Central Park in 1858, and the first full-scale parkway in the United States was William K. Vanderbilt's Long Island Motor Parkway, begun in 1906 and completed in 1911. Moses saw that parkways had great appeal in the age of the automobile and that they could improve environmental quality in existing built-up areas by replacing blighted areas with sinuous, airy green corridors.

Beginning with the Bronx River Parkway (1906–1923), the Hutchinson River Parkway (1928), and the Saw Mill Parkway (1929), Moses supervised the construction of over 100 miles of four-lane, limited-access, and lavishly landscaped parkways that ran through Westchester County to the north and through Long Island to the east, giving New York automobile owners access to ocean beaches that were much less crowded than those served by subway and streetcar. Moses also managed the Triborough Bridge and Tunnel Authority, which enabled him to link the parkway system together and to make it accessible to residents of Manhattan and the Bronx.

During the 1920s and 1930s many other cities added stretches of parkway and built bridges and tunnels that brought automobiles to previously inaccessible and undeveloped tracts. Among the more notable examples were Chicago's Wacker Drive north of the Loop (the **CBD**), the Bay Bridge and Golden Gate Bridge in San Francisco, and the Benjamin Franklin Bridge in Philadelphia. At the same time, engineers in every city were kept busy with the less glamorous infrastructure of suburban extension: smaller bridges, sewer lines, and utilities.

The Decline of Mass Transit

The success of what the boosters called "automobility" was directly responsible for a downward spiral in the fortunes of transit systems. New automobile suburbs lured commuters away from the high-density corridors near transit stops. They also captured the attention of developers, thus ending the marriage of convenience in which development companies had subsidized transit companies. As shops, offices, wholesaling, and industry moved out to the automobile suburbs in order to take advantage of growing markets and cheaper land, the transit systems, still rigidly focused on the downtown, were unable to serve the consequent crosstown and intersuburban commuting flows and so lost further ground.

Meanwhile, increasing numbers of automobile owners substituted Sunday drives for weekend streetcar outings. The transit companies' consequent loss of revenue forced a reduction in maintenance, the abandonment of less-profitable routes, cutbacks in the frequency of weekday services, and increases in fares—all of which pushed many more commuters to turn to the automobile. More vehicular traffic in the streets, in turn, slowed streetcars, increasing their operating costs and prompting still larger numbers of passengers to abandon the trolleys in favor of automobile commuting. By the 1920s about 30 percent of the people entering the CBDs of older, more congested cities did so by car; in newer cities west of the Mississippi, the proportion was nearer 60 percent.

It has also been suggested that auto–oil–rubber corporate interests conspired in the hope of hastening the demise of transit systems:

> The way it worked was that General Motors, Firestone Tire and Standard Oil of California and some other big companies, depending on the location of the target, would arrange financing for an outfit called National City Lines, which cozied up to city councils and city commissioners and bought up transit systems like La.'s. Then they would junk or sell the electric cars and pry up the rails for scrap and beautiful modern buses would be substituted, buses made by General Motors and running on Firestone Tires and burning Standard's gas.[3]

During the 1930s General Motors (GM) created a holding company through which GM and other auto-related companies channeled funds to buy up streetcar systems in 45 U.S. cities.[4] In the late 1960s GM was convicted in a Chicago federal court of having conspired to destroy electric transit and to convert trolley systems to diesel buses, whose production GM monopolized. The corporate executives involved each received a $1 fine.

It must be acknowledged, however, that the main reason for the demise of streetcar systems was the fact that their fixed routes could not cope with changing metropolitan form and patterns of land use. Buses were able to go around congestion that stopped the trolley in its tracks, and they were able to reach new residential subdivisions and peripheral workplaces on hard-surfaced streets whose costs were borne not by the transit companies but by the whole population (through a mixture of federal, state, and local taxes).

PATTERNS OF SUBURBAN GROWTH

The early investment in automobiles, roads, and suburban infill is widely credited with rescuing the U.S. economy from a period of stagnation and igniting the boom of the Roaring Twenties. The locus of the boom was the suburbs. In the 1920s, for the first time, suburbs grew faster than central cities—much faster. While central cities grew 19 percent, adding 5 million new residents, the suburbs grew 39 percent, adding over 4 million residents.

Automobile Suburbs

At first, new automobile suburbs appeared as simple accretions to existing, transit-dependent suburban corridors, and they functioned as such, depending on the central city for employment and shopping. But because developers no longer had to consider the constraints imposed by people having to walk to transit stops, the form of suburban development was very different from the start. By the standards of the time, building lots were much larger (even though the houses were no bigger and were often less substantial), resulting in much lower densities. Many developers dispensed with sidewalks altogether, partly to save money and partly to emphasize the exclusivity of the neighborhoods to the automobile-owning classes. In some cases grid layouts gave way to curvilinear street patterns that minimized the number of junctions and were considered to lend a distinctive and upscale appearance to the neighborhood. In every American city these infill suburbs survive as a distinctive element of urban form: what are now considered to be relatively high-density suburbs with relatively small houses and few neighborhood amenities (Figure 6.3). Most have now filtered down the social scale to become working-class suburbs.

FIGURE 6.3 Infill suburbs survive as a distinctive element of urban form in every American city. These relatively high-density suburbs have relatively small houses and few neighborhood amenities. Most, like this one in New Jersey, have now filtered down the social scale to become working-class suburbs.

Just as these new suburbs were beginning to expand, their form was also influenced by the U.S. Supreme Court in its landmark case on **land use zoning** law: *Village of Euclid, Ohio v. Ambler Realty Co.* (1926). Ruling in favor of the municipality's right to prevent a property owner from using land for purposes other than for which it had been zoned, the Court established the power of local governments to "abate a nuisance." The latter, the Court ruled, could be defined very broadly to include anything affecting the general welfare of a residential area. As a result, zoning quickly came to be used to exclude not only undesirable land uses from residential areas but also (by establishing large minimum lot or dwelling sizes, for example) undesirable *people*.

The consequences for suburban development were far-reaching. Growing suburban areas rushed to become **incorporated** as municipalities in order to be able to control the pace and nature of growth. Speculative developers, reassured by the stability conferred on the land market by these jurisdictions' zoning maps, were emboldened to lay out larger and larger subdivisions, copying, as far as possible, Fordist techniques of mass production for mass consumption.

As these automobile suburbs were laid out, the overall pattern of urban development, no longer tied to the star-shaped corridors of transit lines, reverted to a more symmetrical shape (Figure 6.4). In these new suburbs lot sizes were larger and densities were lower. The average size of a building lot rose from about 3,000 square feet in streetcar suburbs to about 5,000 square feet in automobile suburbs; residential densities fell from about 20,000 people per square mile in streetcar suburbs to about 10,000 per square mile in automobile suburbs.

Planned Suburbs

Decentralization on this scale brought unprecedented opportunities and challenges. The chief *opportunities* were to establish new patterns of development that incorporated retail, commercial, industrial, and recreational land uses in addition to housing. The chief

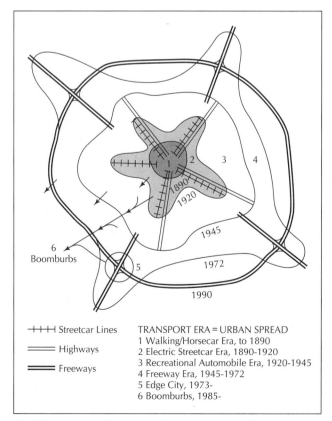

FIGURE 6.4   The form of the automobile city.

challenges were to avoid unnecessary traffic congestion and to preserve environmental quality. Although most suburban infill was totally unplanned and entirely without any conscious recognition of these challenges and opportunities beyond the immediate implications for profitability, there were several critical experiments and innovations that were to become models for subsequent phases of urbanization.

In fact, the first experiments in creating planned suburban settings were those that had been designed for the upper-middle-income families that developers had sought to bring to outlying railroad stations and streetcar termini before the onset of the automobile suburbs. Among the best known are Llewellyn Park in West Orange, N.J. (1853); Chestnut Hill in Philadelphia (1854); Lake Forest, Ill. (1856); Riverside, southwest of Chicago (1869); Roland Park, Baltimore (1891); and Forest Hills Gardens, New York (1909). They were precursors for the idea of secluded, affluent neighborhoods that were self-contained at least in terms of lower-order services, but they were compact, and their layout and design were unsuited to the automobile.

In the mid-1920s there were several attempts to plan communities that would be based on "automobility." The best-known of these are Palos Verdes Estates (Los Angeles), Shaker Heights (Cleveland), River Oaks (Houston), the Country Club District (Kansas City), and two "master suburbs" in Florida: Coral Gables (Figure 6.5) and Boca Raton. Although each had its own innovations, they were all built by private developers for profit, with an upper-middle-class market in mind. They were characterized by very low densities (for the time) of about three dwellings per acre, by high-quality landscaping, by the inclusion of recreational facilities (golf courses, in particular), public gardens, and plazas, in addition to shopping amenities, and by detailed deed covenants aimed at preserving the character and appearance of the entire community.

In retrospect, the Country Club District was perhaps the single most influential of these. It was the creation of developer Jesse Clyde Nichols, who later founded the Urban Land Institute, an independent research organization concerned with urban land use and development from the developers' point of view. Nichols had been impressed by the European Garden City movement and by the City Beautiful movement in the United States (see Chapter 10), and he was determined to put together a project large enough to sustain a self-contained community. Nichols had to be concerned with profitability, however, so that from the start he set out to create a setting that would appeal to the most lucrative section of the residential market: the top end. It took him 14 years to acquire the land he needed, before embarking, in 1922, on the construction of 6,000 homes and 160 apartment buildings that eventually housed over 35,000 residents.

The centerpiece of the development was Country Club Plaza (Figure 6.6), the world's first automobile-based shopping center. Built in Spanish colonial

FIGURE 6.5 A planned suburb: Coral Gables, Miami.

FIGURE 6.6
Country Club Plaza,
Kansas City.

revival style (inspired by the Spanish colonial revival style of the 1915 Pan-Pacific Exposition in San Diego), it featured waterfalls, fountains, flowers, and expensive landscaping, with extensive parking lots behind ornamental brick walls. Nichols carefully controlled the composition of businesses in the Plaza through leasing policies that brought upscale retail stores to the first floor of the development and lawyers, physicians, and accountants to the offices on the second floor. Similarly, the residential sections of the Country Club District were carefully landscaped and controlled. Densities were kept low, streets were curvilinear, trees were preserved wherever possible, and houses were set back some distance from the street, with driveways and garages. All sales were subject to racially restrictive deed covenants, and all purchasers were required to join the homeowners' association, the purpose of which was to ensure lawn care and to supervise the general upkeep and tidiness of streets and open spaces. It was a commercial success from the very start and, despite the obvious elitism of the whole venture, it also attracted critical acclaim from the builders, developers, and planners who came from across the United States to view the shape of the future.

In 1923 a small group of intellectuals founded the *Regional Planning Association of America (RPAA)*. Among the founders were Lewis Mumford and several architects

and planners—including Clarence Stein and Henry Wright. Although RPAA members were mostly interested in big-picture, long-term scenarios for urbanization and regional development, they did manage, through the agency of Alexander Bing—a New York real estate developer and founding member—to create two planned communities. The first, Sunnyside Gardens (built between 1924 and 1928), was an undeveloped inner-city site in Queens only five miles from Manhattan. Here, Clarence Stein and Henry Wright designed big traffic-free superblocks that enabled the creation of vast interior garden spaces.

The second was Radburn (started in 1928), 15 miles from Manhattan in the borough of Fair Lawn, N.J., where Stein and Wright were able to release the Sunnyside superblock principle from the rigid grid of the inner city. Traffic was channeled through a hierarchy of roads, so that most residential areas could be kept virtually free of traffic. Pedestrians and cyclists were given their own paths that crossed traffic arteries under rustic bridges; housing was clustered cozily around irregular-shaped open spaces (Figures 6.7 and 6.8). The RPAA had hoped to draw in a socially mixed group of residents, but Radburn's attractiveness quickly ensured that it became a commuter suburb for upper-middle-class families, whereupon realtors took it upon themselves to keep out Jews and blacks in order to maintain property prices.[5] Nevertheless,

FIGURE 6.7
Radburn, N.J., designed
by Clarence Stein and
Henry Wright. It embodies
many of the ideas
promoted by the RPAA.

Radburn, like the Country Club District, became an influential landmark in the history of urban design.

Suburbanization and Federal Policy

Suburbanization, like all other forms of economic and urban development, came to a shuddering halt with the onset of the Great Depression. Between 1928 and 1933 residential construction activity in the United States fell by 95 percent. In the same period a million nonfarm households lost their property through foreclosure. By the spring of 1933 half of all home mortgages were technically in default, and every day saw a thousand new foreclosures.[6] The Hoover administration responded by creating a credit reserve for mortgage lenders (the **Federal Home Loan Bank Act** of 1932) and a fund for making loans to nonprofit corporations formed to build or upgrade housing for low-income families (the Emergency Relief and Construction Act of 1932). Neither of these initiatives was framed effectively, however. If they had any value at all, it was as precedents that acknowledged the need for the federal government to play a role in stabilizing and protecting home ownership and housing quality.

The Roosevelt administration built on these precedents in ways that not only helped revive the expansion of automobile suburbs but also had lasting impacts on the nature of urbanization in the United States. The Home Owners Loan Corporation (HOLC), created in 1933, helped to stop the slide by refinancing tens of thousands of mortgages that were in danger of foreclosure and by establishing low-interest

mortgages that allowed former owners to recover the homes that they had lost through foreclosure. The following year, in an attempt to reduce unemployment by stimulating construction (a labor-intensive industry), Roosevelt created the **Federal Housing Administration (FHA)**.

This initiative was a critical part of the shift to **Keynesian** macroeconomic management (see Chapter 3) and led to the development of the so-called **Keynesian suburb**. The FHA was given responsibility for stimulating construction by the private sector, and it chose to do so by stabilizing the mortgage market and facilitating sound home financing on reasonable terms. The FHA achieved this goal not by lending money but by *insuring* mortgage loans made by private institutions for home construction or purchase. The insurance gave banks and savings and loan associations the confidence to disburse more mortgages and allowed them to charge one or two percentage points less than before, thus stimulating demand for home ownership. FHA guarantees also stimulated demand as a result of terms that required smaller down payments and extended repayment periods (which meant lower monthly repayments). The FHA established and enforced minimum standards for housing financed by its guaranteed loans, thus helping to eliminate shoddy suburban construction.

At the time, many saw federal support for home ownership as a means of defending the property system. By giving as many people as possible a stake in the system and financing their stake by way of long-term loans, conservatives argued, greater social and political stability would be achieved. The

FIGURE 6.8   Part of the proposed plan for Radburn, N.J. In an effort to promote "community" and minimize the intrusive effects of automobiles, houses face onto common walkways; streets give access to the rear of the houses.

widespread debt encumbrance represented by mortgage repayments would commit home "owners" to oppose any changes to the social and economic structure of society that might endanger the value of their property or make it more difficult for them to pay off their debt.

The immediate effect of federal policy was to reignite suburban growth. Whereas housing starts had fallen to fewer than 100,000 in 1933, the number of new homes started in 1937 was 332,000, and in 1941 it was 619,000 (Figure 6.9). New Deal administrators saw a further opportunity: to plan suburban

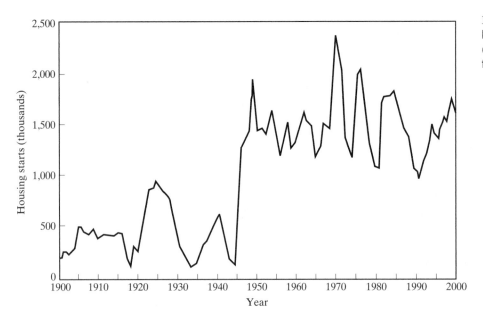

development, drawing people from central cities in order to make room there for slum clearance and re-development. In 1935 Roosevelt created the **Resettlement Administration** with just these objectives, appointing Rexford Guy Tugwell as its head. Drawing directly on the design of Radburn, Tugwell envisaged some 3,000 **greenbelt cities** that would contain government-sponsored low-cost housing and promptly drew up a list of 25 cities on which to start. Funds were allocated for only eight, however, and Congress, under strong pressure from the private development industry, whittled this number down to five. Two of the five were blocked by local legal action. The remaining three were built: Greendale, southwest of Milwaukee (Figure 6.10); Greenhills, near Cincinnati; and Greenbelt, just north of Washington, D.C. But in 1938 the Resettlement Administration was abolished and after World War II all three were sold off to non-profit corporations, to be swallowed up soon afterward in the continuing sprawl of automobile suburbs.

Suburbanization of Commerce and Industry

The decentralization of commerce and industry was slower than the decentralization of households during the early years of the infill era. As Figure 6.1 shows, the number of trucks and buses did not really begin to increase significantly until after World War II. But once a substantial and highly mobile population was in place in the suburbs, the logic of commercial and industrial location changed decisively. Suburbanites represented a large and very affluent market

FIGURE 6.10
Greendale, Wis., one of the three "greenbelt" towns built by the Resettlement Administration after the Great Depression.

for everything from convenience items to expensive clothes and furniture. At the same time, they represented a highly qualified workforce.

Retailing was one of the first commercial activities to diffuse to the new automobile suburbs. There were few attempts to replicate the success of Country Club Plaza (1922), partly because of the huge amounts of capital required. By 1945 there were only seven other such centers in the entire United States: the Upper Darby Center in West Philadelphia (1927); Suburban Square in Ardmore, Pa. (1928); Highland Park Shopping Village, outside Dallas (1931); River Oaks in Houston (1937); Hampton Village in St. Louis (1941); Colony in Toledo (1944); and Shirlington in Arlington, south of Washington, D.C. (1944). Rather, automobile-related retailing took the form of a hierarchy of retail settings, each with a distinctive range of services and each with a different market area. The automobile, in short, enabled a sort of central place system to develop within central places.

Some *office* development decentralized from downtown districts to outlying business centers and along principal business thoroughfares. Automobile travel made these locations attractive to all sorts of business services and to smaller corporate offices. Meanwhile, smaller offices for professionals such as physicians, dentists, and attorneys serving local populations, diffused to neighborhood business streets.

During the 1920s and 1930s there were also some shifts in *wholesaling*. Distributors whose market was dominated by in-town buyers were increasingly tempted to move to locations with cheaper rents. For most wholesalers, such locations had to be reasonably close to the traditional railroad freight yard warehouse district because of their reliance on rail cars for incoming goods. For some (such as department store warehouses and warehouses dealing in local manufactures), however, trucking made it feasible to relocate to the cheapest sites on the very edge of the city.

It was the decentralization of *industry*, however, that was the most significant change induced by trucking. By the 1920s there were strong but pent-up pressures for industrial decentralization. One such pressure was the increasing cost of doing business in congested industrial districts and corridors, where rents were high. In addition, the advent of electrical power made multi-story factories (which had been designed to maximize the efficiency of steam-powered machinery) obsolete. Using new assembly-line techniques made possible by electrical power demanded a horizontal arrangement of industrial space, and this required unprecedented amounts of land. Trucks enabled factories to move out to cheaper, peripheral land, while automobiles and buses enabled their workers to get to them.

By the late 1930s almost no new factories were opening in traditional central factory districts, while many of the older factories were being closed down as their owners invested in newer, more efficient plants in the suburbs. Thus began the first round of *inner-city* **deindustrialization** and restructuring. Factory districts, starved of investment and with some buildings 50 years old or older, became shabby and semiderelict. Nearby working-class housing of similar vintage degenerated into slums, their problems suddenly intensified by the Great Depression. And on the edge of the central city the first-tier **streetcar suburbs**, having been made both socially and functionally obsolete by the automobile, began to tumble down the social ladder, blurring the established order of the city.

Within the CBD the civic and entertainment zones were the least affected by these changes (though decentralization did attract away some "fringe" activities such as speakeasies, gambling dens, and brothels to roadhouses along the urban fringe). The decentralization of offices and shops caused CBD office and retail zones to become appreciably more specialized, catering increasingly to the upper end of the market with high-order goods and services. Department stores, in contrast, sought to maintain their magnetism by diversifying into all kinds of activities:

> The department store became a zoo (Bloomingdale's and Wanamaker's in New York had enormous pet stores), a botanical garden (floral shops, miniature conservatories, roof gardens), a restaurant (some of the major stores had lavish restaurants bigger than any other in their cities), a barber shop, a butcher shop, a museum (gift and art shops, art exhibits), a world's fair, a library, a post office, a beauty parlor.[7]

Hedging their bets, some department store companies opened new branches in the CBD frame (see Chapter 5), where they could offer free parking to customers. Sears and Montgomery Ward, in Chicago and New York, were the first to adopt this strategy, and other stores in virtually every other city soon followed them. Because car driving was seen as a male preserve, however, these new stores carried a limited range of items, dominated by household goods, automotive accessories, and building supplies. Downtown stores, in turn, came to emphasize haberdashery, millinery, cosmetics, and women's fashions.

New Patterns of Land Use

It was in response to all this spatial reorganization that Chauncy Harris and Edward Ullman set out, in 1945, a generalized model that attempted to describe the outcomes in terms of the spatial organization of land uses.[8] This model, generally referred to as the **multiple-nuclei model**, is a schematic representation of the

relative locations of major categories of land use (Figure 6.11), based on the evident proliferation of commercial and industrial nodes beyond the CBD. Harris and Ullman argued that new, automobile-based suburban nodes of commercial and industrial activity were not arranged in any predictable fashion, except in relation to surrounding land uses. They might develop around a government center, a university, a transit stop, or a highway intersection; but if they were office and retailing centers they would attract middle-income residential development, whereas if they were industrial centers they would attract working-class residential development.

Underpinning the overall pattern was a fundamental set of functional relationships, given expression as a result of the mobility conferred by automobiles and trucks and by the locational flexibility conferred by telephones and electricity. Fundamental to these relationships were the mutual attractiveness of certain groups of activities and the tendency for some land uses to be repelled by others. The result is an irregular-shaped patchwork of land uses across which there is a loose functional order. Looking back, we can see that Harris and Ullman were remarkably prescient: The multiple-nuclei city was the embryo form of uniquely American manifestations of contemporary urbanization—metropolitan sprawl with new "production spaces" and "edge cities."

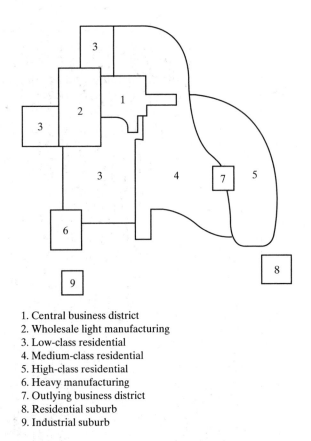

1. Central business district
2. Wholesale light manufacturing
3. Low-class residential
4. Medium-class residential
5. High-class residential
6. Heavy manufacturing
7. Outlying business district
8. Residential suburb
9. Industrial suburb

FIGURE 6.11   The multiple-nuclei model of metropolitan form.

FREEWAYS AND METROPOLITAN SPRAWL (1945–1972)

After 1945 there began a second spurt of growth in automobile ownership in the United States. From just under 26 million in 1945, the number of automobiles on the roads jumped to more than 52 million in 1955 and just over 97 million by 1972 (Figure 6.1). Over the same period, the rate of substitution of surfaced for unsurfaced roads matched this growth almost exactly, while the number of people per automobile fell from five to two. Advances in automotive engineering in the 1930s had effected a dramatic improvement in power-to-weight ratios, and the average top speed of low-priced automobiles had also increased—from less than 50 miles per hour in the 1920s to more than 80 miles per hour in the 1950s.

The result was a dramatic spurt in suburban growth. The 1950s was the decade of the greatest-ever growth in suburban population. While central cities in the United States grew by 6 million people (11.6 percent), suburban counties added 19 million people (45.9 percent). In almost every metropolitan area the ring of suburban counties grew much faster than the central city (or cities). Population statistics did not always reflect this growth, however, because of the tremendous amount of **annexation** undertaken by central cities in the South and West. Corpus Christi, Dallas, Houston, Oklahoma City, Phoenix, San Diego, and San Jose each annexed over 100 square miles of territory from surrounding counties during the 1950s and 1960s, thus capturing the demographic growth of the suburbs.

THE PRECONDITIONS FOR SPRAWL

Following Peter Hall,[9] we can recognize four preconditions for the emergence of this unprecedented suburban sprawl.

1. The principle of land use zoning, established in the *Euclid v. Ambler* case in 1926 (see p. 143), that allowed for the production of uniform residential tracts with stable property values.
2. The backlog of unfulfilled demand for housing from the Depression and war years, combined with the postwar baby boom. During the war there had been a moratorium on new construction, so that by 1945 there was an accumulated

backlog of between three and four million dwellings. After the war the resumption of domestic life by 16 million men and women who had been in the armed services led to a sudden increase in the rate of household formation and an increase in birth rates.

3. The cheap, long-term home financing that the HOLC and the FHA had initiated during the New Deal. In 1944 the **Servicemen's Readjustment Act** (the "**GI Bill**") created the Veterans Administration, one of the major goals of which was to facilitate home ownership for returning veterans. It did so through a program of mortgage insurance along the lines of the FHA, whose own lending powers were massively increased under the terms of the 1949 Housing Act.

4. New and improved roads. The Federal Aid Highway Act (1956) authorized 41,000 miles of limited-access highway to be built, at more than $1 million per mile, with 90 percent of the funding coming from a Highway Trust Fund, established through excise taxes on vehicles, gasoline, and tires. Every major city was to be linked into the system, while the city's internal traffic was to be restructured—often according to a hub-and-spoke model, the outer rim of the "wheel" being a ring road, or interstate "beltway" (Figures 6.12, 6.13), that made outlying locations more accessible, particularly where there were intersections with major radial "spokes."

These preconditions for sprawl encouraged developers to leapfrog ahead of the suburban frontier, building large shopping malls at highly accessible sites along new freeway corridors and cleverly using these shopping meccas as marketing tools for vast new residential communities. Industrial parks were attracted to the new highway and freeway corridors, catering to light industry, and distribution/warehouse operations, while offices were attracted to prestigious, highly visible sites in landscaped, campuslike settings near freeway intersections.

THE FORDIST SUBURB

The Fordist period was characterized by a spectacular increase in housing construction. From a low point of just over 90,000 housing starts in 1933 and a prewar average of about 350,000 starts per year, construction jumped to almost two million by 1950 and has averaged more than 1.5 million ever since. Between 1945 and 1972 the FHA helped nearly 11 million households to own houses, and the overall level of home ownership increased from around 45 percent to nearly 65 percent. Nearly all of this growth was accomplished

FIGURE 6.12   Circumferential beltways like this one around Washington, D.C. (I-495) were important in establishing the framework for metropolitan sprawl in the 1960s and 1970s.

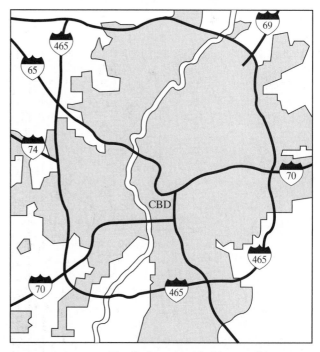

FIGURE 6.13   The "hub-and-spoke" layout of interstate highways in Indianapolis.

without the benefit of any kind of overall planning or metropolitan management, thus sowing the seeds of a number of problems and conflicts that we will examine in later chapters.

The only consistent element of planning came from developers, and their chief goal, of course, was profitability. The strategy deployed by most developers was based on Fordism: mass production for mass consumption. This strategy required the pursuit of economies of scale, the standardization of products, and the perfection of prefabrication technology. The developers' equivalent to Henry Ford's assembly line was **balloon-frame construction** (Figure 6.14). Instead of using heavy beams and corner posts held together by mortise and tenon joints, balloon framing used standardized, machine-cut 2-by-4-inch studs nailed together by inexpensive, machine-cut nails. By spreading the stress over a large number of light boards, the balloon frame had strength and stability far beyond its insubstantial appearance. Using standardized components and requiring only semiskilled labor, balloon framing was around 40 percent less expensive than traditional methods.

The most popular form of house for the new suburbs was a single-story structure with a low-slung roof, a carport or a garage, and large windows. The influential *Ladies Home Journal* had popularized this type of bungalow dwelling from as early as the turn of the century. Standardized by developers in the United States in the 1920s, the bungalow had been designed originally by British engineers for the tropical climate of colonial Bengal. It was low and well-ventilated and

FIGURE 6.14   Balloon framing. This inexpensive and efficient system of construction helped to revolutionize the residential construction industry.

opened out to the garden. It could be grand or modest; using balloon framing, it was the design that made it possible for the working classes to have a "dream" home (Figure 6.15).

Taking advantage of pattern-book bungalow designs and balloon-framing techniques, developers began to encircle every city of any significance with huge, sprawling subdivisions. One of the largest was the Lakewood Park complex (Figure 6.16), built to accommodate more than 100,000 people on 16 square miles of brushland south of Los Angeles.

FIGURE 6.15
Developers heavily marketed suburbia as the setting for the "American dream" of private home ownership.

FIGURE 6.16   Lakewood Park, Calif., an enormous project of 17,150 homes. Work began in February 1950; by the end of the year, families were moving in at the rate of 25 each day.

Without doubt the most famous was the original *Levittown* on Long Island, begun in 1948 by Abraham Levitt and his sons William and Alfred. They were the first large-scale developers to deploy the assembly-line approach in residential development. Combined with the use of innovative materials, new tools, standardized designs, low prices (the original Levittown Cape Cod design sold for $100 down and $57 per month) and, finally, slick marketing, this approach unrolled more than 17,000 homes onto the suburban fringe in short order (Figure 6.17). This was morphogenesis on a heroic scale. It marked the end of urban development as a process dominated by fine-grained accretion

and infill and the beginning of a process dominated by the mass production of coarse chunks of suburban and exurban fabric.

So pervasive did the large-sale suburban subdivision become that cultural geographer Donald Meinig nominated the stereotypical "California Suburbia" as one of three landscapes symbolic of America itself, part of "the iconography of nationhood" (the other two were the New England Village and the Main Street of Middle America). In his words, California Suburbia was characterized by

> low, wide-spreading, single-story houses standing on broad lots fronted by open, perfect green lawns; the most prominent feature of the house is the two-car garage opening onto a broad driveway, connecting to the broad curving street (with no sidewalks, for pedestrians are unknown and unwanted) which leads to the great freeways on which these affluent nuclear families can be carried swiftly and effortlessly in air-conditioned comfort to surfing or skiing, golfing, boating, or country-clubbing, as well as to the great shopping plazas and to drive-in facilities catering to every need and whim.[10]

In the economic boom of the 1950s and 1960s it was not long before Fordist suburbs encircled every large city, with landscapes that have contributed a great deal to the distinctiveness of North American urbanization. Designed for automobile travel, the Fordist suburb is characterized by drive-in and drive-through establishments for all kinds of services: drive-through banks and fast-food outlets, drive-up windows to drop off and pick up laundry, drive-in theaters, motels, gasoline service stations, and minimarts. The suburban shops and shopping centers of the 1950s and 1960s not only represented distinctive new morphological elements in

FIGURE 6.17   Levittown, Long Island. Formerly potato fields, the site was developed in 1947 to house 45,000 people.

FIGURE 6.18   "Hamburger Row." This example is along U.S. Highway 412 in Springdale, a town north of Fayetteville in northeast Arkansas. The classic strip in Norman, Okla., was described by Grady Clay in his book *Close Up: How to Read the American City* (New York: Praeger, 1974).

their own right but also brought with them acres and acres of parking space that imposed an unrelieved sterility to the suburban landscape. To accommodate this parking space, buildings had to be set back so far from the road that signs became larger and more outlandish in an attempt to catch the motorist's eye. The net result is the classic strip (Figure 6.18).

There were a few exceptions to Fordist suburban development, and their successes in designing automobile-oriented environments without unwanted, unaesthetic side effects were to become influential examples to a subsequent generation of developers. In the Chicago area communities such as Park Forest, Elk Grove, and Oak Brook were distinctive for having generous extra measures of green space within an otherwise standard 1950s format of curvilinear streets, cul-de-sacs, and shopping centers. In the Washington, D.C., area, developers returned to the greenbelt cities model, refining it to suit upper-middle-class markets of the 1960s in their plans for Columbia (Maryland) and Reston (Virginia: Figure 6.19). Both were organized into "villages" of 10,000 to 15,000 people, and each had a town center, recreational facilities, and schools. Meadows and woods separated the villages from one another and from a system of open space corridors containing walkways and bicycle paths.

Meanwhile, in California there emerged a rather different recipe, featuring recreational amenities such as country clubs, golf courses, parks, ballfields, man-made boating lakes, swimming pools, and equestrian

areas scattered among vast developments. Each phase of these developments was planned as a "neighborhood" with its own school and commercial subcenter, along with a package of amenities. Together, the neighborhoods constituted a complex of subdivisions with a distinctive, recreation-oriented environment. Examples include Irvine (carved out of the Irvine

FIGURE 6.19   Part of one of the residential "villages" in Reston, a privately developed new town in the Virginia suburbs of Washington, D.C., that became a model for many of the exclusive "master-planned" communities built around the fringes of every large metropolitan area during the 1980s and 1990s.

Ranch), Valencia (on the site of the Newhall Ranch), and Mission Viejo (on the O'Neill Ranch).

SUBURBAN PRODUCTION AND CONSUMPTION SPACES

Continuing the logic of decentralization that began during the prewar infill era, freeways, bigger and faster trucks, and larger factories and warehouses called for a new economic geography that only larger and less expensive parcels of suburban land could accommodate. Although heavy industry was usually tied to older outlying industrial satellites with access to rail sidings, most industry had become "footloose," able to set up wherever local authorities could be persuaded to zone land for industry. Corridors of land along major highways quickly became one common setting for footloose industries; the area around the junction of major highways became another; industrial parks set up by developers a third (Figure 6.20); and the area around airports a fourth.

Geographer Allen Scott has described the production space of the metropolis as a mosaic of such settings, each element in the mosaic being composed of particular *kinds* and *mixtures* of industrial land use: factories of different sorts, warehousing, and related offices, with retailing and services catering to the industrial workforce.[11] The largest clusters of industry are further characterized by a network of interplant linkages that bind many of the land users together in an *industrial complex,* such as the aerospace and electronics complex in Orange County, outside Los Angeles.

FIGURE 6.20   An industrial park in a suburb of Paris, France.

Along with decentralized population and industry came retailing and office functions. Freeways encouraged the development of a catalytic new element: integrated shopping centers. Freeways provided the degree of accessibility necessary to support large new department stores that could anchor shopping malls with a variety of specialized, higher-order retail stores. By 1957 some 2,000 shopping centers were scattered around U.S. metropolitan areas. This figure grew to 8,240 by 1965, 12,170 by 1970, and over 22,000 by 1980. Branches of major department stores anchored the largest of these, filled out with branches of chain stores catering to the middle-middle classes and rounded off with small independent stores or franchises catering to the needs and tastes of the class of residents in the surrounding realm. By the mid-1980s about 55 percent of all retail sales (excluding motor vehicles and gasoline) were accounted for by shopping centers.[12]

CENTRAL CITY LAND USE

The other side of all this suburbanization was the relative decline of central cities and the further reorganization of CBDs. Not only was metropolitan production space being decentralized, the whole structure of the economy was shifting away from the older manufacturing industries that had traditionally been located in central areas. Between 1953 and 1970 New York City lost 206,000 manufacturing jobs, Philadelphia lost 102,000, St. Louis lost 61,000, Boston lost 30,000, and Baltimore lost 25,000. Similar losses were recorded in central cities around the country, along with losses in retailing and wholesaling (New York and Philadelphia each lost 26,000 retailing and wholesaling jobs, for example, while Boston lost 21,000, St. Louis lost 14,000, and Baltimore lost 5,000). These losses were often more than balanced by gains in white-collar service jobs, but because white-collar workers tended to prefer living in the suburbs and commuting to work rather than living in one of the aging central city neighborhoods, the population of central cities began to thin out.

This shift in the employment base and population density of central cities resulted in some radical changes in land use and morphology. Old factories and warehouses were abandoned or demolished; railroad tracks and freight yards were torn up; and slum housing was cleared. Meanwhile, new office blocks were built. But the construction often did not take place on the site of demolished or abandoned areas. The locational requirements of white-collar services were different from those of traditional industries. Office blocks

FIGURE 6.21   Inner-city dereliction can introduce a contagious "blight" to surrounding areas, making investors reluctant to redevelop, and even leading to disinvestment by both the private and the public sector.

crowded into the CBD in skyscrapers, leaving industrial districts in a sorry state of deterioration. Slum clearance and abandoned housing (Figure 6.21) added to the empty and desolate appearance of large tracts of central cities.

Two other factors added to the sense of despoliation in central cities. Freeway systems, as they penetrated central cities in fulfillment of the hub-and-spoke designs of highway engineers, required 200 to 300-foot rights-of-way that took the line of least (commercial) resistance, carving through low-rent neighborhoods and blighting open spaces, parks, and river margins (Figure 6.22). Then, at the end of the 1960s, riots and civil disorder (see Chapter 15) literally laid waste to large sections of some inner cities. The result was that it became common for people to talk and write about American cities as "doughnut cities"—empty holes surrounded by a ring of suburbanization.

Like all such characterizations, the doughnut city idea was a gross exaggeration, capturing only the most dramatic and symptomatic element of change. A great deal of the fabric of central cities was relatively unaltered during the 1950s and 1960s while CBDs, though suffering a decline in their relative importance, continued as major hubs of retail and commercial activity. There were, nevertheless, some significant adjustments. In the CBD frame nearest to higher-status suburbs, upscale apartment houses and hotels began to replace large old mansions, drawing with them the zone of assimilation of the CBD core. As the CBD crept slowly in this direction, so in the zone of discard older hotels degenerated into rooming houses, and stores

FIGURE 6.22   The intrusive effects of interstate highways on downtown areas are clearly shown in this photograph of downtown Seattle.

changed hands. Military surplus stores, pawnshops, and wig stores replaced small tailors and clothiers; stores selling cheap novelties, fire-damaged goods, and pop records replaced housewares and dry goods stores.

In more general terms, the form of the Fordist city can be seen as an extension and outgrowth of the industrial city (Figure 6.23), with a CBD that has been penetrated by industrial as well as international elements; inner-city districts that are subject to urban renewal (Chapter 16) and **gentrification** (Chapter 13); and an extended metropolitan fringe containing new production spaces and the commercial nuclei of emerging "edge cities" (see below, pp. 158–162).

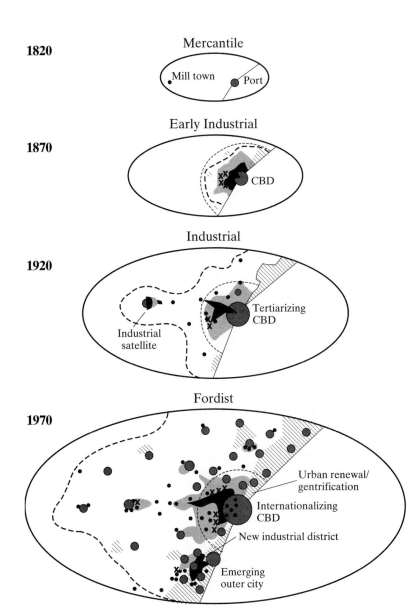

Land use zones

- ▰ commercial/financial
- ▰ Industrial
- ▰ Working class residential
- ▨ Elite residential
- ------ Central city boundary
- --- Built up area boundary
- x ˣ x Minority ghettoes

FIGURE 6.23   The evolution of urban form in the United States: from Mercantile to Fordist cities.

NEO-FORDIST DEVELOPMENT (1973–PRESENT)

Beginning in the mid-1970s, economic **globalization** and the influence of new digital telecommunications technologies brought a transition to **neo-Fordist** regimes of production (see Chapter 4). Individual cities and parts of cities became engaged in different—and rapidly changing—ways in ever-broadening and increasingly complex circuits of economic and technological exchange. Traditional patterns of urbanization began to be overwritten by a very new dynamic dominated by enclaves of superconnected people, companies, and institutions, with their increasingly broadband connections to elsewhere via the Internet, mobile phones, and satellite TVs and their easy access to information services. The uneven evolution of networks of information and communications technologies began to forge new urban landscapes of innovation, economic development, and cultural transformation while at the same time intensifying social and economic inequalities within cities, resulting in what Stephen Graham and Simon Marvin have termed **splintering urbanism**:[13]

SPLINTERING URBANISM

We can identify several distinctive kinds of urban settings that are the product of splintering urbanism:

- Enclaves of international banking, finance, and business services in world cities and major regional centers. Examples include the business districts in Lower Manhattan, the City of London, Frankfurt, Hong Kong, and Kuala Lumpur.
- Enclaves of Internet and digital multimedia technology development, mostly in world cities in the developed countries. Examples include "Multimedia Gulch" in the SOMA (South of Market Street) district of downtown San Francisco and New York's "Silicon Alley" (just south of 41st Street in Manhattan).
- Technopoles and clusters of high-tech industrial innovation. These have emerged in campus-like suburban settings around world cities in developed countries (as in London, Paris, and Berlin); in new and renewed industrial regions within developed countries (as in southern California, Baden-Württemburg in Germany, and Rhône-Alps in France); and in emerging high-tech production and innovation spaces in **newly industrializing countries** or **NICs** (as in Bangalore, India, and the Multimedia Super Corridor south of Kuala Lumpur, Malaysia).

- Places configured for **foreign direct investment** in manufacturing, with customized infrastructure, expedited development approval processes, tax concessions, and, in some cases, exceptions to labor and environmental regulations. Such places have emerged in economically depressed regions of developed countries (including Northern England and parts of the U.S. Manufacturing Belt), but are mostly found in or near major cities in less developed countries (as in the Brazilian cities of Porto Alegre and Paraná, which have attracted foreign-owned automobile plants).
- Enclaves of back-office spaces, data-processing, e-commerce, and call centers. These have emerged in older industrial cities within developed countries (e.g. Roanoke, USA, and Sunderland, England) and in many cities within newly industrializing countries, most notably in India, the Philippines, and the Caribbean.
- Spaces customized as "logistics zones." Airports, ports, **export processing zones**: enclaves in major cities around the world within which the precise and rapid movement of goods, freight, and people are coordinated, managed, and synchronized between various transport modes.

Connected to one another through a complex dynamic of flows, these urban spaces and settings are key elements in the spatial articulation of economic globalization. They are embedded within regions and metropolitan areas whose economic foundations derive from earlier **technology systems** and whose social and cultural fabric derives from more traditional bases. The result is that the local effects of splintering urbanism are transforming traditional patterns of land use and spatial organization in many parts of the world.

THE POLYCENTRIC METROPOLIS

North American cities were the first to break away from traditional patterns. Geographer Pierce Lewis coined the term "galactic metropolis" to capture the disjointed and decentralized urban landscapes of late twentieth-century North America.[14] The galactic metropolis is fragmented and multinodal, with mixed densities and unexpected juxtapositions of form and function. It is characterized by **"edge cities"**—suburban hubs of shops and offices that sometimes overshadow the old

FIGURE 6.24 The neo-Fordist galactic metropolis.

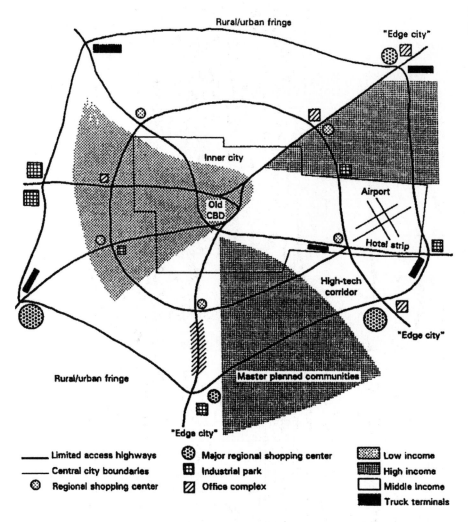

Rural/urban fringe

"Edge city"

Inner city

Airport

Old CBD

Hotel strip

High-tech corridor

"Edge city"

Rural/urban fringe

Master planned communities

"Edge city"

———— Limited access highways ⊗ Major regional shopping center ▨ Low income

———— Central city boundaries ⊞ Industrial park ▦ High income

⊗ Regional shopping center ▨ Office complex ☐ Middle income

▬ Truck terminals

downtown. Edge cities are nodal concentrations of shopping and office space situated on the fringes of metropolitan areas, typically near major highway intersections (Figure 6.24).

The result is a polycentric metropolitan structure that now has variants around the world. Geographer Peter Hall has identified six common types of nodes within the polycentric metropolis:[15]

- The traditional downtown center, based on walking distances and served by a radial transportation system. The hub of the traditional metropolis, it has become the setting for the oldest informational services: banking, insurance, and government. Examples include the City of London, Châtelet-Les Halles (Paris), Lower Manhattan (New York), and Maronouchi/Otemachi (Tokyo).
- Newer business centers, often developing in an old prestigious residential quarter and serving as a setting for newer services such as corporate

headquarters, the media, advertising, public relations, and design. Examples include London's West End, the 16th arrondissment in Paris, Midtown Manhattan, and Akasaki/Roppongi (Tokyo).
- Internal edge cities, resulting from pressure for space in traditional centers and speculative development in nearby obsolescent industrial or transportation sites. Examples include London's Docklands, La Défense (Paris: Figure 6.25), and Shinjuku (Tokyo).
- External edge cities, often located on an axis with a major airport, sometimes adjacent to a high-speed train station, always linked to an urban freeway system. Examples include Washington's Dulles corridor (Figure 6.26), London's Heathrow district, the O'Hare area (Chicago), Schipol (Amsterdam), and Arlanda (Stockholm). Some of these edge cities have grown so quickly that they have not been incorporated as separate jurisdictions and do not have official names that

FIGURE 6.25
La Défense, Paris,
an example of an
"internal edge city."

show up on street maps or in guide books, post office directories, or government statistical tables. In many respects they are **stealth cities**, invisible administratively and politically, without their own chambers of commerce, libraries, town halls, public squares, or courthouses. They are, nevertheless, very real, showing up clearly both on the skyline and on plots of land values.

• Outermost edge city complexes for back offices and R&D operations, typically near major

train stations 20 to 30 miles from the main core. Examples include Reading (outside London), St. Quentin-en-Yvelines (Paris), Greenwich, Connecticut (outside New York), and Shin-Yokohama (Tokyo: Figure 6.27).

• Specialized subcenters, usually for education, entertainment, and sporting complexes and exhibition and convention centers. These take a great variety of forms and locations. Some are on reclaimed or recycled land close to the traditional core; some are older centers, formerly separate and independent, that have become progressively embedded in the wider metropolitan area.

The largest of the world's polycentric metropolises have become "100-mile cities"—metropolitan regions that are literally 100 miles or so across, consisting of a loose coalition of **urban realms**, economic subregions that are bound together through urban freeways (Figures 6.28, 6.29, 6.30). Each urban realm tends to function semi-independently, with a broad mix of land uses and populations of between 175,000 and 250,000. Each realm has retail, commercial, and residential subareas, as well as a commercial and retailing node that functions as a high-order **central place** for the majority of local residents. As a result, most residents of metropolitan areas do not have much to do with the central core except for occasional trips to major sporting events, large concerts, and so on. The sprawling suburbs of the fast-growing realms consist for the most part of a generic landscape of freeways, commercial strips, condominium complexes, office parks, and single-family subdivisions (see the Box entitled "Boomburbs and 'Generica'").

FIGURE 6.26   Part of the corridor of development that runs between Tysons Corner and Dulles airport, in the Washington D.C. metropolitan area, an example of an "external edge city."

FIGURE 6.27 The Shin–Yokohama district in Tokyo's metropolitan fringe, an example of an "outermost edge city."

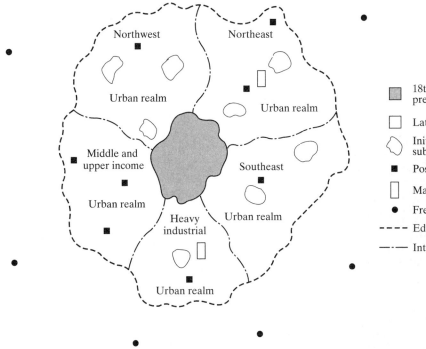

FIGURE 6.28   The concept of "urban realms" within a metropolitan framework.

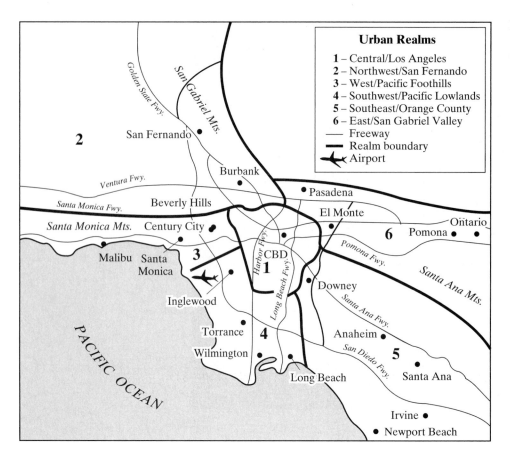

FIGURE 6.29 Urban realms in the Los Angeles metropolitan area.

Urban Realms
1 – Central/Los Angeles
2 – Northwest/San Fernando
3 – West/Pacific Foothills
4 – Southwest/Pacific Lowlands
5 – Southeast/Orange County
6 – East/San Gabriel Valley
— Freeway
▬ Realm boundary
✈ Airport

THE END OF "SUBURBIA"

The scale and organization of metropolitan areas have reached the point where urban functions have been dispersed across decentralized landscapes that are not suburban in the traditional sense. Meanwhile, many central cities are now "central" only in the most limited geographical sense, their economic and demographic importance having been eclipsed by

FIGURE 6.30 Part of the freeway system that feeds the "100-mile city" of Los Angeles.

surrounding urban realms. Hence historian Robert Fishman's announcement of the end of "suburbia."[16]

The other side of the same coin is that central cities are no longer exclusively "urban." Central cities have experienced a continued deindustrialization and decentralization of jobs and population but also a selective reinvestment that has reversed the downward spiral of some neighborhoods' decline, conserved and preserved some of the remaining fragments of past eras, and introduced residential and mixed-use developments to some areas that were formerly declining. CBDs have experienced a selective recentralization of economic activity that has brought a renaissance of urbanity and a rush of speculative building.

At the heart of recent land use changes in central cities has been the growth of skilled jobs associated with advanced business services (advertising, banking, insurance, design, etc.) that have replaced the semi-skilled manufacturing jobs lost through deindustrialization and the decentralization of retailing. These employment changes and the associated changes in class composition are dealt with in greater detail in Chapter 12. For the moment, we should note that it has been the advanced business service companies that have filled the office space created by the building booms of the 1980s and 1990s, and it has been their employees who have fueled the processes of gentrification (see p. 370) and historic preservation (p. 267).

Box 6.1   Boomburbs and "Generica"

Between 1985 and 2000, when the population of U.S. metropolitan areas increased by 17 percent, about 25 million acres of farmland and open space (roughly the size of Indiana) was developed around these metropolitan areas: a 47 percent increase in developed land.

The touchstones of the polycentric metropolis are **boomburbs**, the fastest-growing suburban jurisdictions in the United States, typically located along the interstate beltways that ring large metropolitan areas in the western United States. Boomburbs do not resemble traditional central cities or older satellite cities. Although they possess most elements found in cities, such as housing, retailing, entertainment, and offices, they are not typically patterned in a traditional urban form. Boomburbs almost always lack, for example, a dense business core. They can thus be seen as distinct from traditional cities not so much in their function but in their low density and loosely configured spatial structure.

Robert Lang of Virginia Tech's Metropolitan Institute defines boomburbs as places with more than 100,000 residents that are not the largest city in their metropolitan areas, and which have maintained double-digit rates of population growth in recent decades. While boomburbs may be found throughout the United States, they occur mostly in the Southwest, with almost half in California alone (Figure 6.31). In 2000, there were 53 boomburbs in the United States, the most populous of which was Mesa, Ariz., with 396,375 residents—bigger than such traditional large cities as Minneapolis (382,618), Miami (362,470), and St. Louis (348,189). Arlington, Tex.—the second-biggest boomburb, with 332,969 residents—fell just behind Pittsburgh (334,536) and just ahead of Cincinnati (331,285). Even such smaller boomburbs as Chandler, Ariz., and Henderson, Nev., with 176,581 and 175,381 residents respectively, surpassed older mid-size cities such as Knoxville,

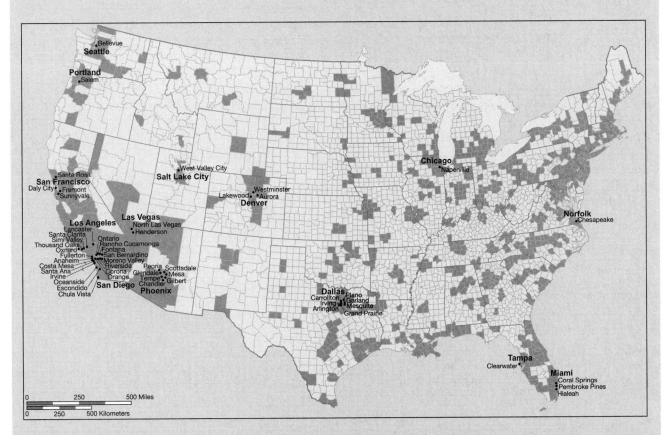

FIGURE 6.31   America's boomburbs. This map shows the distribution of the 53 places with more than 100,000 residents in 2000 that were not the largest city in their metropolitan areas and that maintained double-digit rates of population growth in the 1980s and 1990s.

(Continued)

BOX 6.1   BOOMBURBS AND "GENERICA"   (Continued)

Tenn. (173,890), Providence, R.I. (173, 618), and
Worcester, Mass. (172,648).

In these boomburbs the scale of the development industry is now such that urbanization occurs in large increments, cutting—and within
months filling—swathes of rural land with residential subdivisions, condominium complexes,
bleak access roads, strip malls, parking lots, office
parks, big-box stores, and, most characteristically,
strip development. Beyond these strips lie subdivisions dominated by large-lot, single-family homes
(Figure 6.32). Most of the architecture and urban
design is without merit, adding up to what architect Rem Koolhaas has called "Generica." Following the logic of a fast return on investment and
flexibility in use, most commercial structures are
simple boxes, while economies of scale dictate a
cookie-cutter approach for all but the most upscale
residential subdivisions—where "monster homes,"
"starter castles," and "McMansions" take over as
the norm.

FIGURE 6.32   Boomburb development. The pace of
growth in America's boomburbs forces developers to
constantly search for new tracts of land suitable for
suburban development. This photograph shows a new
subdivision in North Las Vegas, Nev. Note the exclusionary
walls and the broad boulevard.

BOX 6.2   JAPANESE CITIES: TOKYO AND THE TOKAIDO MEGALOPOLIS[17]

Japan's phenomenal economic growth in the
decades following World War II stimulated dramatic
urban development. Between 1950 and 1970 the
percentage of people living in cities with a population of 50,000 or more rose from 33 to 64 percent,
while the overall urban population reached 72 percent. As the population in cities exploded, the number and size of larger cities increased dramatically.
Since 1970 the urban population has continued to
increase, but more slowly, reaching 78 percent by
the late 1990s.

A distinguishing feature of Japan's urban landscape is the concentration of its major cities into a
relatively small portion of its already small land
area. The Tokaido Megalopolis contains 44 percent
of Japan's population of 126 million. This **megalopolis** comprises the three urban industrial regions of Keihin (Tokyo–Yokohama), with more than
30 million people; Hanshin (Osaka–Kobe–Kyoto),

with more than 16 million; and Chukyo (Nagoya),
at nearly 9 million (Figure 6.33).

Japan exhibits a classic **core-periphery model**
of development: a core capital region centered on
the primate city of Tokyo and the peripheral regions
elsewhere in Japan. The rapid urban growth from
the late 1950s through the early 1970s saw a shift in
population from the rural areas to the big cities in
general and to Tokyo in particular. Since then, rural-
to-urban migration has been increasingly toward
Tokyo at the expense of the rest of the country.

The plans for Tokyo's urban development that
were drawn up immediately after World War II
were not all implemented. Instead, the city experienced largely haphazard growth that produced
congestion and a disorganized city layout. In general, urban growth has been concentrated around
key subcenters, such as Shibuya, Shinjuku (an "internal" edge city that is now the city government

(Continued)

BOX 6.2   JAPANESE CITIES: TOKYO AND THE TOKAIDO MEGALOPOLIS   *(Continued)*

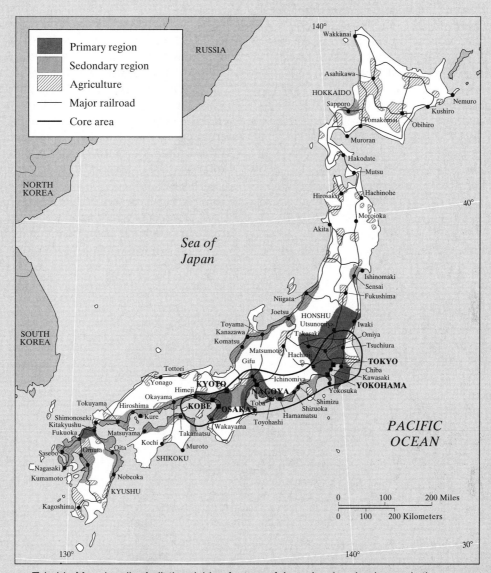

FIGURE 6.33   Tokaido Megalopolis. A distinguishing feature of Japan's urban landscape is the concentration of its major cities into a relatively small portion of its already small land area. The Tokaido Megalopolis contains 44 percent of Japan's population of 126 million. It comprises three major urban industrial regions centered on Tokyo–Yokohama, Osaka–Kobe–Kyoto, and Nagoya.

headquarters), and along the key transportation arteries (rail and expressway) radiating outward from the Chiyoda District, the old historic core of the Imperial Palace.

In contrast to many U.S. cities, Tokyo's CBD has maintained its corporate primacy. This has occurred despite the decentralization of middle-income workers to peripheral areas in search of affordable housing

as land costs spiraled upward in the 1980s in the lead-up to the recession of the 1990s. The workers who commute to the central city each day use one of the busiest and most crowded mass transit systems in the world—the average Tokyo commute is a 2-hour journey in each direction. Compared to U.S. cities, however, the residential pattern in Tokyo exhibits fewer serious disparities along racial or socioeconomic lines.

Box 6.3   Australian Edge Cities?

The majority of people in Australia live within a relatively narrow band about 125 miles wide along the east, southeast, and southwest coastal areas. Away from the coast, settlement is very sparse, especially in the desert areas of the outback. Although smaller than **megacities** elsewhere in the world, such as Los Angeles, Jakarta, and Mexico City, Australia's largest cities—Sydney, Melbourne, Brisbane, Perth, and Adelaide—have grown to become so-called "mega-metropolitan regions" (Figure 6.34).[18]

The powerful role of edge city development—in reshaping the edge of the urban landscape of cities like Los Angeles and Chicago in the United States and London and Amsterdam in Europe—has not been matched to the same extent in Australia.[19] Certainly, large upscale master-planned communities and office and science park developments are found toward the edge of Australia's largest cities. But these developments are not extensive enough spatially as yet to be characterized as "external" or "outermost" edge cities (see pp. 159–160).[20]

Meanwhile, although office and retail activity has been decentralizing to the suburbs since the 1950s, Australian central cities have never suffered the same level of disinvestment, infrastructure deterioration, and racial segregation as those in the United States. In 1995, for example, Australian central cities still contained about one-third of Australia's finance, insurance, business, and property services. Meanwhile, the recent economic restructuring toward producer services and other services such as tourism and health has tended to reinforce the corporate primacy of Australian CBDs.

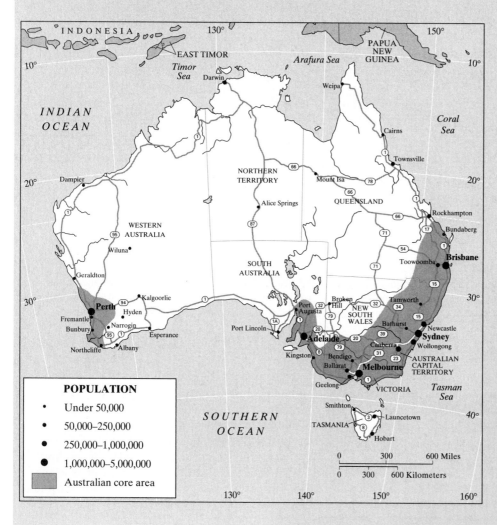

FIGURE 6.34
Australia's "mega-metropolitan regions." The majority of people in Australia live within a core region comprising a relatively narrow band about 125 miles wide along the east, southeast, and southwest coastal areas containing the "mega-metropolitan regions" of Brisbane, Sydney, Melbourne, Adelaide, and Perth.

(Continued)

Box 6.3   Australian Edge Cities?   *(Continued)*

The lack of urban freeway construction and peripheral ring roads has worked against the development of edge cities in Australia. Likewise, despite suburbanization, the radial orientation of the transportation networks has helped maintain central city vitality and facilitated recent urban revitalization. In contrast to the United States, the strong regulatory tradition in Australian cities has also helped restrict the development of edge cities. State government planning agencies in Australia have strictly managed land release for urban expansion and the staging of corridor development. In addition, by specifying the location of "district centers" and assigning land for this purpose, state governments have curbed speculative business development beyond the edge of the built-up area.

That is not to say that neo-Fordist economic restructuring has not had an impact on Australian cities. In contrast to the experience in the United States, however, the brunt of deindustrialization in Australia has been quite selective in that it has most adversely affected the postwar industrial suburbs. Particularly hard hit have been the middle and outer suburbs in Sydney, Melbourne, and Adelaide that formerly contained both the traditional manufacturing jobs and the public housing estates that were built for factory workers and their families. These public housing estates, and tracts of low-rent housing surrounding public resettlement hostels, now contain a significant proportion of recent immigrants, many of whom arrived in Australia from Southeast Asia as "boat people" and political refugees. Nevertheless, the level of ethnic residential segregation in Australia is not as extreme as in U.S. central cities such as Detroit and Milwaukee.

Like the United States, however, Australia has experienced a polarization of occupations and incomes that has been associated with neo-Fordist economic restructuring. Unlike the trend in U.S. cities, however, the growth of business and professional services jobs in Australian cities has been associated with a particularly strong centralization of wealth. High levels of housing reinvestment associated with the gentrification of the inner suburbs have been a constant feature of Australian cities during the last three decades (Figure 6.35).

Since the 1970s, as Australian cities have become increasingly integrated into regional and global networks of cities, the national and state governments have adopted neoliberal strategies to revitalize the old industrial landscapes and waterfront areas of the city centers (see Chapter 17). The national Better Cities program, for example, incorporated strategic investment in urban infrastructure and other projects with the goal of enabling cities like Sydney to capture a greater share of the financial flows in the Pacific region and move up in the league table of World Cities. As in the case of the neoliberal policies in the United States and United Kingdom, the Better Cities public-private partnerships have promoted upscale central city revitalization projects—luxury downtown apartments and condominiums, festival marketplaces, and so on—at the expense of affordable housing.

FIGURE 6.35   Gentrification in The Rocks, Sydney. The Rocks is the site of Australia's first European Settlement. High levels of housing reinvestment associated with gentrification have been a consistent feature in central parts of Australian cities during the last three decades.

FIGURE 6.36   South Street Seaport in New York City, one of a number of high-profile "specialty marketplace" projects designed to attract both residents and tourists by utilizing waterfront locations and historic buildings to create settings in which people can shop, promenade, and safely engage in the timeless pursuit of people watching.

Between them these changes have transformed many of the townscapes of central cities. Where they have been insufficient to rejuvenate the most acute cases of deindustrialization and retail decentralization, **civic entrepreneurialism** has stepped in to engineer the appearance of large, set-piece mixed-use developments designed to bolster the city's image and pull in other forms of development. The best-known examples include Baltimore's Harbor Place, Miami's Bayside, Riverwalk in New Orleans, Riverfront in Savannah, Quincy Market in Boston, Pioneer Square in Seattle, and South Street Seaport in New York (Figure 6.36).

Even some of the largest and most spectacular of these developments have encountered financial difficulties, however, partly because of the inherent high risks of such projects, and partly because intense competition between cities for investment dollars has contributed to over-building. Meanwhile, the underlying shift toward a service-based and information-based economy has created numerous land use conflicts and social tensions in central cities. These are topics that are examined in some detail in subsequent chapters on politics (Chapter 16) and planning (Chapter 17).

FOLLOW UP

1. Be sure that you understand the following key terms:

   annexation
   civic entrepreneurialism
   edge city
   galactic metropolis
   greenbelt cities
   high-tech corridor
   incorporation
   industrial complex
   Keynesian suburbs
   multiple-nuclei model
   parkway
   production space
   retail-strip corridor
   secondary mortgage markets
   urban realms

2. Be sure that you understand the significance of the following in relation to patterns of urban change:

   Country Club District
   Federal Aid Highway Act (1956)
   Federal Home Loan Bank Act (1932)
   Housing Act (1949)
   RPAA
   1926 *Euclid v. Ambler*
   Federal Aid Roads Act (1916)
   FHA
   Levittown
   Resettlement Administration
   Servicemen's Readjustment Act (1944)

3. Construct a graph that shows the changing population between 1940 and 2000 of the central city and suburban counties of your nearest metropolitan area. Annotate the graph to suggest reasons for the trends that you find.

4. Update your *portfolio*. The material in this chapter lends itself especially well to maps of all kinds. For example, you might prepare a hand-drawn, GIS, or computer cartography map of the city you know best, showing major industrial spaces, shopping centers, shopping strips, suburban downtowns, major highways, and, if it is big enough, urban realms. Another example: make a map of just part of the city, showing the elements of neo-Fordist development. Be sure to make use of other sources in your library and online: newspapers, magazines, specialized journals, reports, aerial photos, and even satellite images. Try to give the portfolio a balanced content as well as a distinctive flavor.

KEY SOURCES AND SUGGESTED READING

Backhaus, G., and J. Murungi, eds. 2002. *Transformations of Urban and Suburban Landscapes: Perspectives from Philosophy, Geography, and Architecture.* Lanham, Md.: Lexington Books.

Brooks, D. 2002. Patio Man and the Sprawl People. *Weekly Standard*: 19–29.

Cuthbert, A. R. ed. 2003. *Designing Cities: Critical Readings in Urban Design.* Oxford, UK.: Blackwell.

Ellin, N. 1995. *Postmodern Urbanism.* Oxford, UK.: Blackwell.

Fishman, R. 2002. *Urban Utopias in the Twentieth Century.* Cambridge, Mass.: MIT Press.

Fishman, R. 1997. Cities After the End of Cities. *Harvard Design Magazine*: Winter/Spring.

Ford, L. R. 2003. *America's New Downtowns: Revitalization or Reinvention?* Baltimore, Md.: Johns Hopkins University Press.

Ford, L. R. 1994. *Cities and Buildings: Skyscrapers, Skid Rows, and Suburbs.* Baltimore, Md.: Johns Hopkins University Press.

Goodman, D., and C. Chant. eds. 1999. *European Cities and Technology.* London: Routledge.

Gowans, A. 1986. *The Comfortable House: North American Suburban Architecture 1890–1930.* Cambridge, Mass.: MIT Press.

Graham, S., and S. Marvin. 2001. *Splintering Urbanism.* New York: Routledge.

Hall, P. 2002. *Cities of Tomorrow.* 3rd ed., Oxford, UK.: Basil Blackwell.

Harris, R. 1996. *Unplanned Suburbs: Toronto's American Tragedy 1900 to 1950.* Baltimore, Md.: Johns Hopkins University Press.

Jackson, K. T. 1985. *Crabgrass Frontier: The Suburbanization of the United States.* New York: Oxford University Press.

Kay, J. H. 1997. *Asphalt Nation.* New York: Crown Publishers.

Kostof, S. 1992. *The City Assembled: Elements of Urban Form Through History.* Waltham, Mass.: Little, Brown.

Lang, R. E., and P. A. Simmons. 2001. *'Boomburbs': The Emergence of Large, Fast-Growing Suburban Cities in the United States.* Census Note 06, Washington, D.C.: Fannie Mae Foundation (*http://www.fanniemaefoundation.org/programs/census_notes.shtml*).

Langdon, P. 1994. *A Better Place to Live: Reshaping the American Suburb.* Amherst, Mass.: University of Massachusetts Press.

Lewis, P. G. 1996. *Shaping Suburbia.* Pittsburgh, Pa.: University of Pittsburgh Press.

Lewis, R. D. 1998. Running Rings Around the City. North American Industrial Suburbs 1850–1950 in eds. R. Harris and P. Larkham, *Changing Suburbs.* London: Chapman and Hall.

Roberts, G. K., and P. Steadman. 1999. *American Cities and Technology.* London: Routledge.

Soja, E. 2000. *Postmetropolis.* Oxford, UK.: Blackwell.

Sudjic, D. 1999. *The 100-mile City.* New York: Andre Deutsch.

Szold, T., and A. Carbonell, eds. 2002. *Smart Growth: Form and Consequences.* Cambridge, Mass.: Lincoln Institute of Land Policy.

Wiewel, W. 2002. *Suburban Sprawl: Private Decisions and Public Policy.* Armonk, N.Y.: M. E. Sharpe.

RELATED WEBSITES

Metropolitan Institute at Virginia Tech: *http://www.mi.vt.edu/* The Metropolitan Institute's research on urban and metropolitan affairs is available on its website, including publications on suburbanization, boomburbs, edge cities, and world cities.

Levittown Anniversary Committee: *http://www.levittownpa.org/* In celebration of the town's 50-year anniversary (1952–2002), the Levittown Anniversary Committee's website offers a history of Levittown that includes photos from the 1950s and 1960s.

Virtual Cities Directory: *http://www.intoronto.com/cities/home.html* The Virtual Cities Directory is a collection of links to 3-D representations of real cities around the world including Boston, Helsinki, London, New York, San Francisco, Sydney, Tokyo, Toronto, Vancouver, and Warsaw.

Innovative Transportation Technologies: *http://faculty.Washington.edu/~jbs/itrans/* This University of Washington website contains information and photographs of innovative urban transportation technologies including suspended mass transit systems, monorails, maglevs, and automated highway systems.

7

URBANIZATION IN THE LESS DEVELOPED COUNTRIES

From a geographical perspective, the most significant aspect of recent world urbanization is the dramatic difference in trends and projections between the core industrial regions comprising developed countries like the United States, the United Kingdom, Japan, and Australia and the semi-peripheral and peripheral regions comprising the less developed countries of Latin America, Africa, and Asia.[1] In 1950 two-thirds of the world's city dwellers were concentrated in the more developed countries. Since then the world's urban population has tripled, with the bulk of the growth in the less developed countries. As we saw in Chapter 2, economic development and industrialization in the core countries depended greatly on the exploitation of peripheral regions. Inevitably, the international division of labor that underpins this relationship fundamentally influenced the patterns and processes of urbanization in the periphery. Although variations in internal and external factors produced different urbanization experiences in each country, a pressing problem today for many less developed countries is a process of overurbanization in which cities are growing more rapidly than the jobs and housing that they can sustain.

CHAPTER PREVIEW

This chapter examines urbanization in the less developed countries in global and historical context. A look at current trends and projections shows urban population increasing at twice the general population growth rate in the less developed countries.[2] The factors promoting this kind of urban growth vary within and between the different parts of the world. In sharp contrast to the experience of the world's **core regions**, where urbanization was largely an outcome of economic growth, urbanization in **peripheral regions** has resulted from demographic growth that preceded economic development. We consider some of the attempts made during the last 50 years to disentangle the relationship between urbanization and economic development and to explain "underdevelopment" in the less developed countries.

Colonization and the expansion of trade around the world allowed Europeans to influence the world's economies and societies. Although the various peripheral regions were at different levels of urbanization on the eve of the European encounters, colonization and the Industrial Revolution created unprecedented concentrations of people in cities that were connected in networks and hierarchies of interdependence around the world. We use a sequence of five phases of colonial urbanization as a framework for examining how this process and its impacts changed over time in different parts of the world. We consider, in turn, mercantile colonialism, industrial colonialism, late colonialism, early independence, and neocolonialism. This brings us to a pressing problem for many cities in the less developed countries today: a process of overurbanization in which cities grow more rapidly than the jobs and housing that they can sustain.

URBANIZATION TRENDS AND PROJECTIONS: THE LESS DEVELOPED COUNTRIES IN GLOBAL CONTEXT

The world's urban population reached 2.9 billion in 2000 and is expected to rise to 5 billion by 2030. According to United Nations estimates, almost half of the world's population is now urbanized.[3] Of the major world regions, North America is the most urbanized, with over 77 percent of the people living in the towns and cities of the United States and Canada. Africa and Asia are the least urbanized—less than 40 percent of the people are city dwellers (Figure 7.1).

To put these figures into historical perspective, in 1950 less than 30 percent of the world's population was urbanized. In that year only 83 metropolitan areas had a million or more people, and just 8 topped the five million mark. By 2000, there were 372 metropolitan areas of a million or more, with 40 containing over 5 million people. Looking ahead to 2015, there will be about 475 cities with a population of a million plus, including about 58 with over five million. In fact, the predictions for the near future are for virtually all the world's population growth to occur in urban areas. By 2030, for example, more than 60 percent of the world's population is expected to be urbanized.

From a geographical perspective, the most significant aspect of recent world urbanization is the incredible difference in trends and projections between the core industrial regions comprising developed countries like the United States, United Kingdom, Japan, and Australia and the semi-peripheral and peripheral regions comprising the less developed

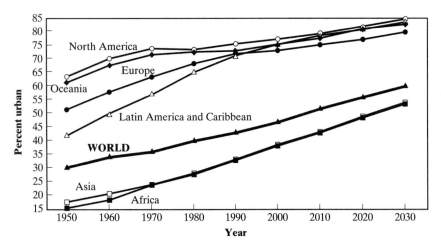

FIGURE 7.1   Urbanization by major world region, 1950–2030.

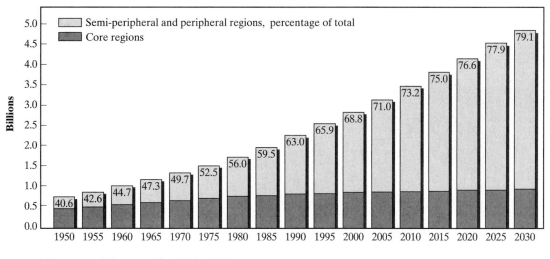

FIGURE 7.2   Urban population growth, 1950–2030.

countries of Latin America, Africa, and Asia (see the Box entitled "Core, Semi-Periphery, and Periphery in the World-System"). In 1950 just under 60 percent of the world's urban population was concentrated in the more developed countries. Since then the world's urban population has tripled, with the bulk of the growth in the world's periphery. In fact, almost all of the population increase between now and 2030 will be absorbed by the less developed countries, whose urban population is expected to

rise from almost 2 billion in 2000 to nearly 4 billion or almost 80 percent of the world's urban population by 2030 (Figure 7.2).

Of the world's 30 largest metropolitan areas in 1950, 21 were in the **core countries** of Europe and North America. It took only until 1980 for the situation to be reversed, with 19 of the 30 largest metropolitan areas located in the less developed countries. By 2015 all but 5 of the 30 largest metropolitan areas will be outside the core regions. What is more, the

BOX 7.1   CORE, SEMI-PERIPHERY, AND PERIPHERY IN THE WORLD-SYSTEM

With the end of the Cold War and economic growth in newly industrializing countries, or NICs, like South Korea and Taiwan, the designation of "Third World" is no longer useful for distinguishing the combined group of less developed countries from either the capitalist, developed countries like the United States and the United Kingdom (the "First World") or the former communist countries in what was the Soviet bloc (the "Second World").[4]

More helpful is Immanuel Wallerstein's world-system conceptualization[5] that sees the entire world economy as an evolving market system comprising an economic hierarchy of regions—a **core**, a **semi-periphery**, and a **periphery**. The labels "core" and "periphery" refer to the dominant *processes* operating at particular levels in the hierarchy. Core processes are characterized by economic relations that incorporate relatively high wages, advanced technologies, and a diversified production mix. Periphery processes

involve low wages, more rudimentary technologies, and a less diversified production mix. The label "semi-periphery" refers to places in which there is a combination of both sets of processes.

Figure 7.3 is an attempt to portray the current composition of countries in each of these three categories, using national figures for total and per capita **gross domestic product** or **GDP**. Countries with high scores on both indicators are likely to have politically strong states, large internal markets, and predominantly high-wage, capital-intensive production—all, theoretically, defining characteristics of core status. These are the so-called developed countries.

Conversely, countries with a low national economic output and a low per capita GDP are likely to have weak states and predominantly low-wage, labor-intensive production—characteristic of peripheral status. The category of countries with intermediate

(Continued)

BOX 7.1   CORE, SEMI-PERIPHERY, AND PERIPHERY IN THE WORLD-SYSTEM   *(Continued)*

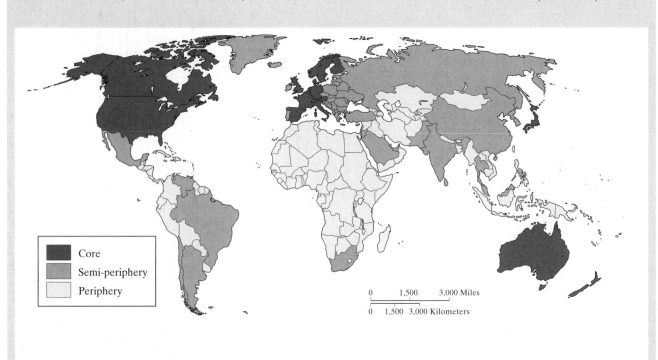

FIGURE 7.3   The world system: core, semi-periphery, and periphery.

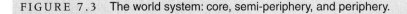

scores for total and per capita GDP—characteristic of semi-peripheral status—contains quite a diverse group. The countries with intermediate levels of living include resource-exporting countries such as Saudi Arabia and South Africa, recently industrialized or "old" NICs like Mexico, Brazil, Hong Kong, and Singapore, as well as the newly industrializing or "new" NICs such as Malaysia and Thailand. The semi-periphery group of countries also includes China, India, poorer European countries like Greece and Portugal, and the formerly socialist countries of Eastern Europe and Russia. The peripheral and semi-peripheral countries together comprise the so-called less developed countries.

This kind of categorization is based on rather sweeping assumptions, and the allocation of individual countries to particular categories is inevitably somewhat arbitrary. Certainly, significant variations

exist within and between the countries and regions of the world. Although economic *and* locational peripherality may coincide, for example, in parts of Sub-Saharan Africa, not all countries in the less developed regions of the world exhibit low levels of GDP. Oil-rich Saudi Arabia, Libya, and Venezuela, for example, together with one or two small export-processing locations like Hong Kong and Singapore, have per capita incomes comparable to some of the less affluent developed countries.

What is important is that this division of the world into a core, a semi-periphery, and a periphery represents more than an alternative classification of countries: it is a reflection of a particular conception of the dynamics of the world economy. As such, it is subject to change in response to changes in the processes and in the economic fortunes of different countries in the world system.

number of people living in the largest metropolitan regions is rising rapidly (Table 7.1)

Asia provides some dramatic examples of this trend. From a region of villages, Asia is quickly becoming a region of towns and cities. Between 1950 and 1995, for example, its urban population increased nearly fourfold to nearly 1.2 billion people. By 2030

over 50 percent of Asia's population is expected to be living in urban areas (Figure 7.1).

Nowhere is the trend toward rapid urbanization more pronounced than in China. For decades the communist government imposed strict controls on where people could live because it feared the transformative and liberating effects of cities. By tying people's

TABLE 7.1

Trading Places on the Top 30 List: The World's Largest Metropolitan Areas, Ranked by Population Size (Millions)

1950	Population	1980	Population	2015	Population
New York, USA	12.3	Tokyo, Japan	21.9	Tokyo, Japan	27.2
London, UK	8.7	New York, USA	15.6	Dhaka, Bangladesh	22.8
Tokyo, Japan	6.9	Mexico City, Mexico	13.9	Mumbai, India	22.6
Paris, France	5.4	São Paulo, Brazil	12.1	São Paulo, Brazil	21.2
Moscow, Russia	5.4	Shanghai, China	11.7	Delhi, India	20.9
Shanghai, China	5.3	Osaka, Japan	10.0	Mexico City, Mexico	20.4
Rhine-Ruhr North (Essen), Germany	5.3	Buenos Aires, Argentina	9.9	New York, USA	17.9
Buenos Aires, Argentina	5.0	Los Angeles, USA	9.5	Jakarta, Indonesia	17.3
Chicago, USA	4.9	Kolkata, India	9.0	Kolkata, India	16.7
Kolkata (Calcutta), India	4.4	Beijing, China	9.0	Karachi, Pakistan	16.2
Osaka, Japan	4.1	Paris, France	8.7	Lagos, Nigeria	16.0
Los Angeles, USA	4.0	Rio de Janeiro, Brazil	8.7	Los Angeles, USA	14.5
Beijing, China	3.9	Seoul, South Korea	8.3	Shanghai, China	13.6
Milan, Italy	3.6	Moscow, Russia	8.2	Buenos Aires, Argentina	13.2
Berlin, Germany	3.3	Mumbai, India	8.0	Manila, Philippines	12.6
Mexico City, Mexico	3.1	London, UK	7.8	Beijing, China	11.7
Philadelphia, USA	2.9	Tianjin, China	7.7	Rio de Janeiro, Brazil	11.5
St. Petersburg, Russia	2.9	Cairo, Egypt	6.9	Cairo, Egypt	11.5
Mumbai (Bombay), India	2.9	Chicago, USA	6.8	Istanbul, Turkey	11.4
Rio de Janeiro, Brazil	2.9	Rhine-Ruhr North, Germany	6.7	Osaka, Japan	11.0
Detroit, USA	2.8	Jakarta, Indonesia	6.4	Tianjin, China	10.3
Naples, Italy	2.8	Manila, Philippines	6.0	Seoul, South Korea	9.9
Manchester, UK	2.5	Delhi, India	5.5	Kinshasa, Dem. Rep. Congo	9.9
São Paulo, Brazil	2.4	Milan, Italy	5.4	Paris, France	9.9
Cairo, Egypt	2.4	Tehran, Iran	5.4	Bangkok, Thailand	9.8
Tianjin, China	2.4	Karachi, Pakistan	5.0	Lima, Peru	9.4
Birmingham, UK	2.3	Bangkok, Thailand	4.8	Bogotá, Columbia	9.0
Frankfurt, Germany	2.3	St. Petersburg, Russia	4.7	Lahore, Pakistan	8.7
Boston, USA	2.2	Hong Kong	4.5	Bangalore, India	8.4
Hamburg, Germany	2.2	Lima, Peru	4.4	Tehran, Iran	8.2
Total	**121.5**	**Total**	**252.5**	**Total**	**423.7**

Source: *United Nations,* World Urbanization Prospects: The 2001 Revision, *New York: Department of Economic and Social Affairs, Population Division, 2002, Table A-11 (http://www.un.org/esa/population/publications/wup2001/WUP2001report.htm)*

jobs, school admission, and even the right to buy food to the places where they were registered to live, the government made it almost impossible for rural residents to migrate to towns or cities. As recently as 1985, more than 70 percent of China's one billion people still lived in the countryside.

China is now rapidly making up for lost time. Having decided that towns and cities can be engines of economic growth within a communist system, the Chinese government not only relaxed residency laws but drafted plans to establish over 430 new cities. Between 1981 and 2001 the number of people living in cities in China more than doubled, from 162 million to 385 million, and the number of cities with a population of half a million or more increased from 16 to 92.

In the world's core countries, levels of urbanization are high and have been for some time (Figure 7.4). Belgium, the Netherlands, and the United Kingdom are 90 percent or more urbanized, while Australia, Canada, Denmark, France, Germany, Japan,

New Zealand, Spain, Sweden, and the United States are all more than 75 percent urban. Despite low *rates* of urbanization, especially when compared to the less developed countries, the urban population in the core countries is still expected to rise to 83 percent by 2030.

Levels of urbanization are also very high in many of the world's semi-peripheral countries (Figure 7.4). Brazil, Mexico, Taiwan, Singapore, and South Korea are all at least 75 percent urbanized. And unlike the core countries, their rates of urban growth are also high.

In the peripheral countries rates of urbanization are even higher, and the forecasts are for unprecedented urban growth and size. Jakarta, Indonesia, for example, grew from 1.4 million to 11 million between 1950 and 2000 and is expected to reach 17.3 million by 2015. Likewise, Lagos, Nigeria, a city of less than 300,000 in 1950, reached 8.7 million in 2000 and is projected to have a population of 16 million by 2015. Dhaka (Bangladesh), Mumbai (Bombay, India), São Paulo (Brazil), Delhi (India), and Mexico City, are all

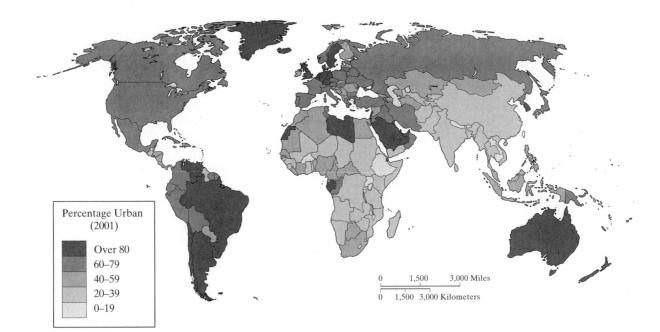

FIGURE 7.4   World urbanization, 2001.

projected to have populations in excess of 20 million by 2015 (Table 7.1).

Many of the largest cities are growing at annual rates of between 4 and 7 percent. The doubling time of a city's population is the time needed at current growth rates for it to double in size. At the higher rate of 7 percent, the populations of the largest cities will double in 10 years; at the lower rate of 4 percent they will double in population in 17 years. To put the situation in numerical terms, metropolitan areas like Mexico City and São Paulo are adding half a million people to their populations each year: nearly 10,000 every week, even with deaths and out-migrants. It took London 190 years to grow from half a million to 10 million; New York took 140 years. In contrast, Mexico City, São Paulo, Buenos Aires, Kolkata (Calcutta), Rio de Janeiro, Seoul, and Mumbai all took less than 75 years to grow from half a million to 10 million inhabitants.

FACTORS PROMOTING URBAN GROWTH

The factors promoting this kind of urban growth vary within and between different parts of the world. In sharp contrast to the experience of the world's core regions, where urbanization was largely an outcome of economic growth, urbanization in peripheral regions has resulted from demographic growth that preceded economic development.

The more rapid decline in death rates compared to birth rates—part of a **Demographic Transition** in peripheral regions—is a fairly recent trend that has generated large increases in population well in advance of any significant levels of industrialization or rural economic development. In rural regions this has produced fast-growing populations in places that face increasing problems with agricultural development.

Rural-to-urban migration is a common response as many impoverished rural residents migrate to the larger towns and cities in search of a better life. They are driven by the desire for employment and the prospect of access to schools, health clinics, piped water, and the kinds of public facilities and services that are often unavailable in rural regions. Overall, the metropolises of the periphery have absorbed four out of five of the 1.2 billion city dwellers added to the world's population since 1970.

Rural migrants have poured into cities out of desperation and hope, rather than being drawn by actual jobs and opportunities. Because of the disproportionate number of teenagers and young adults in these migration streams, an important additional component of urban growth has followed—exceptionally high rates of **natural increase** of the population. In most peripheral countries the rate of natural increase in cities exceeds that of net in-migration. On average, about 60 percent of urban population growth in peripheral countries is attributable to natural increase.

Political and environmental circumstances can also promote urban growth. Wars in Liberia and Sierra Leone caused hundreds of thousands of refugees to flee to their capitals, Monrovia and Freetown. In Mauritania, Niger, and other countries along the southern edge of the Sahara, deforestation and overgrazing in conjunction with government inaction have forced people to move to the cities as the expanding desert has overtaken whole villages.

THEORIES OF URBANIZATION AND ECONOMIC DEVELOPMENT

Historically, an association has existed between urbanization and economic development: Countries with higher levels of urbanization tend to have higher levels of economic development (Figure 7.5). What is not as clear is the direction of causality—the extent to which economic development promotes urbanization or urbanization promotes economic development. In the developed countries, although urbanization was largely an outcome of economic development, it was a reciprocal arrangement in which the urbanization that was driven by economic growth in turn stimulated further economic development (see Figure 1.4). The various attempts made during the last 50 years to disentangle this relationship and explain urbanization and "underdevelopment" in the less developed countries can be grouped into three categories.

MODERNIZATION THEORIES:
THE DEVELOPMENTAL APPROACH

In the 1950s ideas about the development of the less developed countries were based on extrapolations of the European experience. This developmental approach prescribed an economic transition along a continuum of progress from a "traditional" rural society

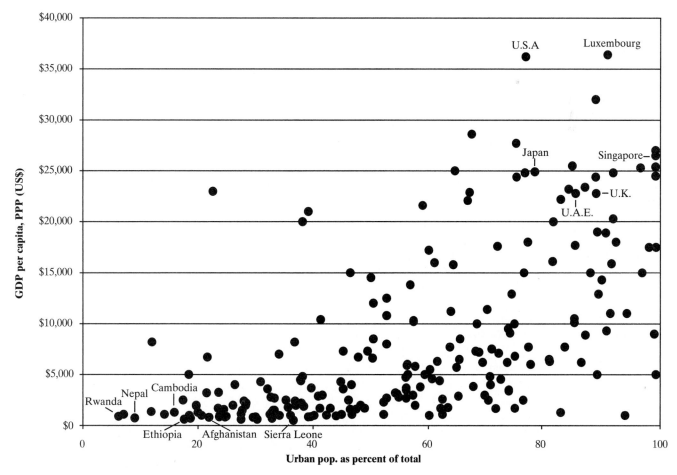

FIGURE 7.5 Urbanization and economic development, 2000.

toward a "modern" urban industrialized one. Models, such as Rostow's stages of economic growth, informed the developmental approach.[6] This model shows the five successive stages through which less developed countries must pass in order to achieve economic convergence with the **core economies** (Figure 7.6).

Similarly, Myrdal's notion of **cumulative causation**[7] saw the growth of the peripheral regions following the patterns of urbanization experienced by the core region of Europe during the Industrial Revolution (see Chapter 2). Economic growth in one region would trigger strong demand for food, consumer goods, and other manufactures that local producers could not satisfy. This demand would create the opportunity for investors in peripheral regions to establish a local capacity to meet the demand; entrepreneurs would take advantage of the cheaper land and labor in these peripheral regions. If strong enough, these **spread effects** could enable peripheral regions to develop their own upward spiral of cumulative causation (Figure 7.7).

Myrdal's influential model was followed by others who used a similar logic. Hirshman's model described **trickle-down effects**.[8] Perroux highlighted the importance of the *propulsive industries* that are charac-

teristic of regions with high rates of economic growth,[9] such as the textile industry in England during the Industrial Revolution. As the propulsive industry grows, it attracts other related industries, generating a set of **agglomeration economies**. A **growth pole** is formed and an urban growth center develops. These ideas are shown spatially in Friedmann's **core-periphery model**.[10] The model shows an urban core of economic advantage and growth, surrounded by nearby agricultural areas that are in the process of development due to their proximity to the core, and a distant stagnant or declining periphery (Figure 7.8).

Although these kinds of development models have informed public policy and practice in the past, they are now regarded as too simplistic. They perpetuate the myth of "developmentalism," that all countries and regions—despite differences in their political, cultural, technological and other characteristics—are on the same economic growth trajectory to become "modern" urban industrialized societies. A major weakness of developmentalism is that it fails to appreciate that the prospects for late starters are different from those of places that enjoyed an earlier **initial advantage** free from effective competition and limiting precedents.

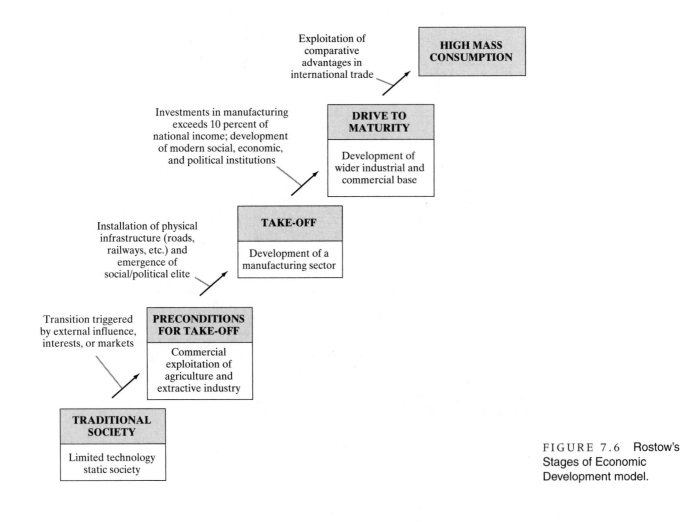

FIGURE 7.6 Rostow's Stages of Economic Development model.

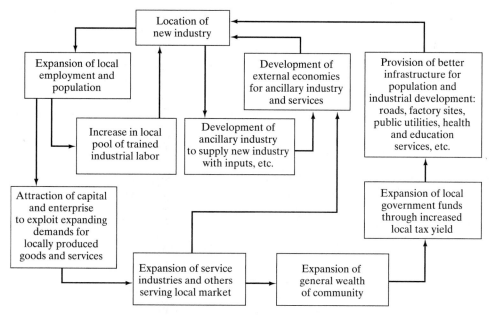

FIGURE 7.7 Myrdal's model of cumulative causation.

The less developed cities and countries must compete in a crowded field and overcome barriers that were created by the success of some of the early starters in the core regions. Most problematic for the modernization theories, though, was the indisputable evidence that the pattern of urbanization in the less developed countries was not following that of the core regions—urban growth was not producing the expected boost in economic development.

URBAN BIAS AND *UNDERDEVELOPMENT*

In 1977, Michael Lipton coined the term *urban bias* in his book *Why Poor People Stay Poor: Urban Bias in World Development*.[11] Urban bias describes how the urban-based elite who hold power in some less developed countries tends to implement policies that allocate resources for the benefit of cities. By concentrating resources in the urban areas, urbanization rates accelerate

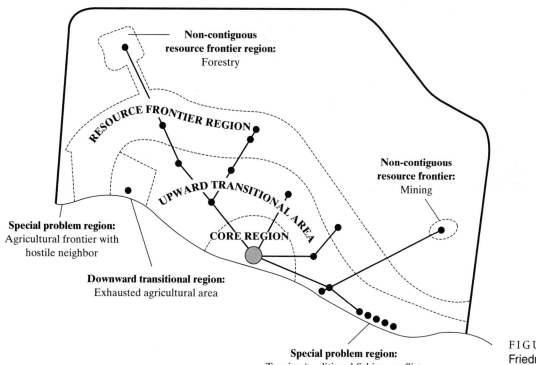

FIGURE 7.8
Friedmann's core-periphery model.

and overall national economic development is impaired; urban–rural inequalities intensify because most of the poor people still live in rural areas despite often massive rural-to-urban migration.

Although influential at the time, this notion of a urban–rural divide is now viewed as a simplistic generalization that, at best, describes only those countries experiencing very rapid urbanization. As was the case for modernization theories, the urban bias idea has been faulted for neglecting the international constraints on economic development in the peripheral regions.

An avalanche of critical writings argued that the prosperity of the developed countries *depended* on *under*development in other parts of the world. The less developed countries, their role in the world-system already established (and firmly controlled by the economic and military power of the more developed countries), could not "follow" the historical experience of developed countries. In fact, the global system of unequal trade, exploitation of labor, and profit extraction guaranteed that the less developed countries would become more, rather than less, impoverished.

Writers like André Gunder Frank rejected the idea that underdevelopment was the result of geographical isolation or a failure to embrace Western technology, investment, and values. Instead, underdevelopment stemmed directly from the unequal nature of the interrelationships between the developed and less developed parts of the world. Figure 7.9 shows the system as described by Frank:[12]

as a photograph of the world taken at a point of time, this model of a world metropolis (today the United States) and its governing class, and its national and international satellites and their leaders—national satellites like the Southern states of the United States, and international satellites like São Paulo. Since São Paulo is a national metropolis in its own right, the model consists further of its satellites: the provincial metropolises, like Recife or Belo Horizonte, and their regional and local satellites in turn. That is, taking a photograph of a slice of the world we get a whole chain of metropolises and satellites, which runs from the world metropolis down to the hacienda or rural merchant who are satellites of the local commercial metropolitan center but who in their turn have peasants as their satellites. If we

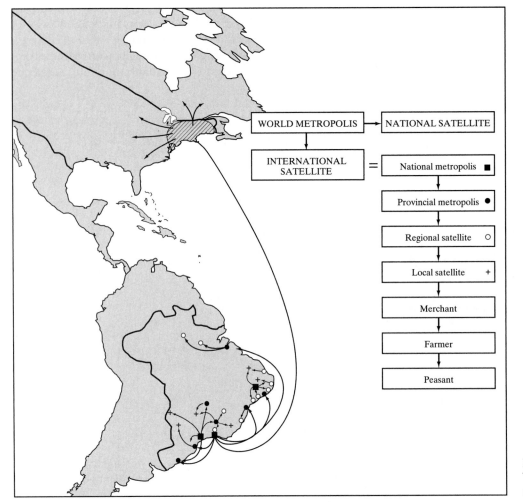

FIGURE 7.9
The Frank model of dependency.

take a photograph of the world as a whole, we get a whole series of such constellations of metropolises and satellites.

Certainly the unequal structure of the world economy and the kinds of monopoly power of a metropolis over its satellites have changed over time—for example, with the switch from **merchant capitalism** to **industrial capitalism** in the nineteenth century or following political independence for former colonies. But the transfer of wealth from satellites to metropolis has continued to fuel growth in some places at the expense of others.

Frank's approach is an example of *dependency theory*, which has been very influential in explaining global patterns of development and underdevelopment. Dependency theory states, essentially, that development and underdevelopment are the reverse sides of the same global process: Independent development is impossible because development in one location requires underdevelopment in another.

Immanuel Wallerstein's *world-system theory*[13] addressed certain criticisms of dependency theory: that dependency theory ignored differences in the characteristics and processes among the less developed countries and focused too much on how these countries are "locked into" a position of dependence. According to the world-system perspective, the entire world economy is an evolving economic system with a hierarchy of countries comprising a core, semi-periphery, and periphery (see the Box entitled "Core,

Semi-Periphery, and Periphery in the World-System," including Figure 7.2). Core countries take advantage of their position in the world economy to exploit the peripheral and semi-peripheral countries. The semi-peripheral countries also exploit peripheral countries, while in turn being exploited by core countries. What is important about this conceptualization for the less developed countries is that the composition of this hierarchy is variable, with movement possible in either direction, including from periphery to semi-periphery and semi-periphery to core.

Looking specifically at cities, the model of peripheral urbanization used a political economic approach[14] to extend the dependency and world-system perspectives to the national **urban systems** of less developed countries (see the Box entitled "A Model of Peripheral Urbanization"). This six-stage model describes how the extension of the global economic system to the less developed countries generated a strong process of urbanization.[15] But like dependency theory, this model has been criticized for being too deterministic; it implies that the incorporation of the less developed countries into the world economy makes the associated problems of urban development identified in the model seem inevitable. David A. Smith, for one, has stressed that:

> structural similarity between nations or regions in the global hierarchy may lead to parallel patterns of urban growth. But this will not always be the case... The real challenge is to identify the ways in which social

BOX 7.2   A MODEL OF PERIPHERAL URBANIZATION

The model of peripheral urbanization captures how the global system of production and trade generates a strong process of urbanization that affects individual cities and systems of cities in the less developed countries.[16]

1. Rural-to-urban migration increases as "traditional" forms of agriculture are disrupted by the introduction of commercial agriculture, by fiscal taxes on the rural population, and by competitive pressures on craft industries, initially from cheap imports and later from products by national manufacturers.

2. Production in the rural areas by national and foreign businesses promotes the development of major transportation and market centers and the rapid expansion of national capitals and major port cities.

3. The growth of manufacturing concentrates production even more within the largest cities,

stimulates the expansion of a national state bureaucracy to encourage the process of industrialization, and leads to the concentration of high-income groups in the major centers.

4. Workers move to the largest cities in search of employment, and their labor and spending support further economic expansion.

5. The state supports industrial expansion and the system is maintained through the provision of physical infrastructure in the main urban centers and social services to selected groups.

6. As development accelerates, private investment begins to spread outward to avoid the rising land prices, labor costs, and traffic congestion in the central city. The state may encourage this process of deconcentration with measures to encourage metropolitan decentralization.

relations and outcomes on the local level are linked to macrostructural processes, including those of the global political economy.[17]

NEW MODELS FROM THE LESS DEVELOPED COUNTRIES: OPPORTUNITIES FOR DEVELOPMENT

Significant departures from earlier thinking characterize contemporary perspectives on urbanization and economic development.[18]

- There has been a shift away from the dependency theory focus on the international constraints on development in the peripheral regions. The emphasis is now on the internal opportunities for development, using models, assumptions, and objectives that have been devised within the less developed countries themselves. This shift has been associated with an increasing recognition of the importance of local context for development and of the potential benefits of the economic and cultural interrelationships that already exist within the less developed countries.

- The related renewed interest in the role of indigenous social and cultural institutions and in the structure of the economic relations within less developed countries has led to a resurgence of concern for gender divisions in terms of how these affect and are affected by economic development.

- The relationship been the environment and development within the context of **sustainable urban development** for the less developed countries and at a global scale is attracting increasing attention. This represents a shift away from conceptualizing development as economic growth and "modernization" until the 1970s, and from then until the mid-1980s of assigning limited importance to the environmental context in dependency theory.

A HISTORICAL PERSPECTIVE ON COLONIAL URBANIZATION

Economic development and industrialization in the core countries depended greatly on the exploitation of peripheral regions. Inevitably, the **international division of labor** that this relationship put in place fundamentally influenced the patterns and processes of urbanization in the periphery. The colonial powers established **gateway cities** in peripheral regions. In an effort to establish economic and political control over continental interiors, colonial cities were created as centers of administration, political control, and commerce. A **colonial city** is one that a colonial power deliberately established or developed as an administrative or commercial center.

One type of colonial city was the new city that was "planted" in a location where no significant urban settlement had previously existed. Such cities were laid out expressly to fulfill colonial functions, with ceremonial spaces, offices, and depots for colonial traders, plantation representatives, and government officials; barracks for a garrison of soldiers; and housing for colonists. As these cities grew, housing and commercial land uses were added for the local people who were drawn to the city by employment opportunities as service workers, such as servants, clerks, or porters. Examples of pure colonial cities were the original settlements of Mumbai (Bombay), Kolkata (Calcutta), Ho Chi Minh City (Saigon), Hong Kong, Jakarta, Manila, and Nairobi.

The other type of colonial city had colonial functions grafted onto an existing settlement in order to take advantage of a good site and a ready supply of labor. Examples include Mexico City, Shanghai, Tunis, and Delhi. In these cities the colonial imprint is most visible at the center of the city in the formal squares and public spaces, the layout of avenues, and the presence of colonial architecture and monuments. This architecture includes churches, city halls, and railway stations; the palaces of governors and archbishops; and the houses of wealthy traders, colonial administrators, and landowners.

The colonial legacy can also be seen in the planning and building regulations of many cities. Colonial planning regulations were usually the same ones used for the colonizing country itself. Because they were based on Western concepts, these regulations were often inappropriate for colonial contexts. Most colonial building codes, for example, used Western models of a small family home in a residential neighborhood at some distance from the workplace. This is at odds with the needs of large, extended families whose members work in a busy domestic economy in family businesses that are traditionally integrated with the residential setting. Colonial planning, with its **gridiron** street layouts, **land use zoning** regulations that do not allow for mixed land uses, and building codes designed for European climates, ignored the specific needs and cultural preferences of local communities (see the Box entitled "Delhi: The Evolution of an Imperial City").

King[19] offered a conceptual framework to facilitate analysis of different colonial cities at different times

Box 7.3　Delhi: The Evolution of an Imperial City[20]

THE PRECOLONIAL CITY

When the British arrived, Delhi was still the Mogul capital of North India, but it was in the hands of rebels and its population had fallen to about 150,000. Almost everyone lived within the 2.5 square miles enclosed by the 5.5-mile-long city wall. At the heart of the city were the political, religious, and commercial focal points: the Royal Palace, the Jama Mosque, and Chandni Chowk (Figure 7.10). The rest of the city was a preindustrial core of narrow, twisting lanes and mixed land uses—not unlike medieval European cities.

THE PERIOD OF COEXISTENCE, 1803–1857

In its early colonial years, Delhi was merely a district military post for the Punjab region. There were no more than a few hundred Europeans and, because the situation was stable, the military forces were located to the northwest of the city. Most of the British lived in the city itself, near the Royal Palace (containing the

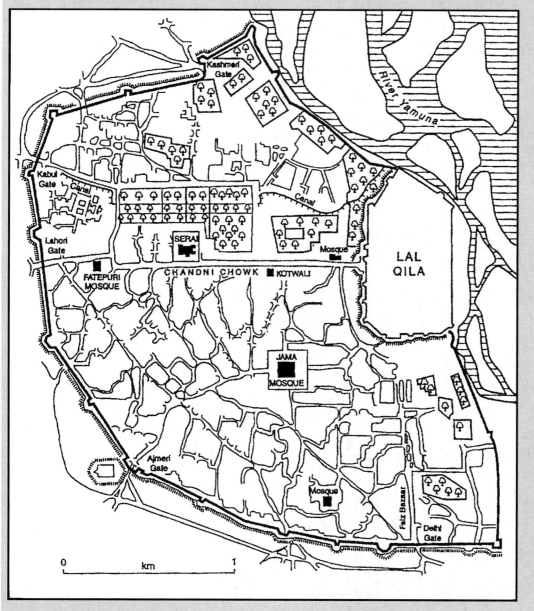

FIGURE 7.10
Mogul residential patterns in Shahjahanabad (Old Delhi).

(Continued)

BOX 7.3 DELHI: THE EVOLUTION OF AN IMPERIAL CITY   *(Continued)*

puppet Mogul emperor) formerly occupied by the Mogul aristocracy. Western technology was limited, and most of the British lived similarly to the indigenous elite. There was little social contact between the different groups, but little conflict either.

COLONIAL CONSOLIDATION, 1857–1911

The pressures of colonial expansion erupted in India with the mutiny of 1857 in which the puppet Mogul emperor was implicated and dethroned. The consequent sharpening of military control resulted in a more forceful imposition of political power and cultural values. The Royal Palace became Delhi Fort, and a free-fire zone, 550 yards wide, was created around it and the city walls. Military forces occupied the northern third of the city and civilians were moved out to the civil lines to the north of the walls where several physically imposing buildings were constructed as institutional symbols of power. The

indigenous population was thus confined to the remaining area of the old city that, although crowded, was still functionally sound.

Culturally, however, Old Delhi and its population became increasingly isolated as the British withdrew to a distinct area to the north, separated by the military zone, police lines, newly constructed gardens, and the railways. Poor communications had been blamed for many of the problems of the mutiny and by 1911 eight separate railways had been linked to the city, further eroding the area available to a growing indigenous population. As a result, nearly 500,000 Indians were crammed into 1.5 square miles of the old city, while a few thousand British enjoyed the relatively open spaces of the northern districts.

IMPERIAL DELHI, 1911–1947

Railway technology enabled Delhi to be developed as the new centralized capital of a consolidated India—the jewel in King George V's imperial crown.

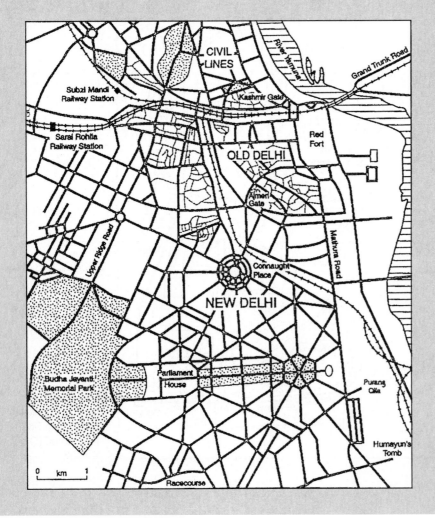

FIGURE 7.11   New Delhi: the social morphology of imperialism.

(Continued)

BOX 7.3   DELHI: THE EVOLUTION OF AN IMPERIAL CITY   *(Continued)*

In 1921 the site of the city was moved south to a new location. New Delhi was planned on a vast scale by Edwin Lutyens and Herbert Baker, and as early as 1931 it covered 30 square miles. New technological developments—the automobile and the telephone—allowed such a large area to be interconnected in theory. In practice, these technologies spread slowly and erratically, and for many years communications depended on people on foot or on bicycles.

Although the new capital had acquired several important processing industries, including flour milling, there was no provision for manufacturing growth. But the most extraordinary example of colonial influence on urban planning was the minute residential stratification of New Delhi into rigid spatial categories based on the infinitely complex combination of Indian, British, military, civil, and colonial ranking systems. This resulted in several specifically designated zones (Figure 7.11) within which there was further stratification by size of house and garden and by orientation in relation to the most important building, Government House.

The old city received some improvements to the water supply and drainage systems but, as in most colonial cities, urban planning was primarily reserved for the expatriate zones. With time, Old Delhi became even more physically and socially isolated by more gardens and a spacious new central business district. Further in-migration led to massive overcrowding and rapidly growing squatter settlements to the west and east of the old city. Not surprisingly, by World War II death rates in Old Delhi were five times higher than in the new capital.

Independence did little to change this contrast; 10 years later, Old Delhi still contained 60 per cent of the city's population at an average density of 16,000 people per square mile. But this was but a foretaste of what came later as the capital's population quadrupled to its present size of almost 8 million. Clearly, the problems of inadequate services that still plague contemporary planners in Old Delhi have their roots firmly in the residential segregation established during its colonial past.

within the context of their internal and external relationships (Figure 7.12). Because the role and function of colonial cities can vary depending on the scale of analysis, it can be helpful to use a number of spatial scales for examining colonial cities:

1. *City:* The city itself—its internal dynamics, functions, and form.
2. *Region:* The city within the context of its immediate region—its regional production systems, trading relations, settlement patterns, transportation networks, and labor movements.
3. *Colonized Society or Territory:* The city in relation to the colonized society or territory—in terms of changes in social stratification, cultural attitudes, and so on.
4. *Metropolitan Power:* The city within the context of its interactions with the colonial metropolitan power—through trade, capital investment, and colonial policies.
5. *Colonial Empire:* The city within the context of its assigned role within the empire—as an administrative center, commercial port, or transportation hub.
6. *World Economic System:* The city and its place in the evolving world economy and global system of cities.

Internal and external factors produced a different urbanization experience for every country with a colonial history. Timing was important—colonial control was imposed at, and for very different, periods of time in various parts of the world (Figure 7.13). Regional differences in ethnicity and culture, kinds and levels of urbanization, social, political, and economic systems, environmental conditions, and level of technology all affected how colonial urbanization played itself out. The motivation for colonial expansion into a particular part of the world was also a factor.

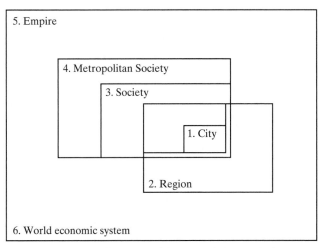

FIGURE 7.12   Conceptual framework for analyzing colonial cities.

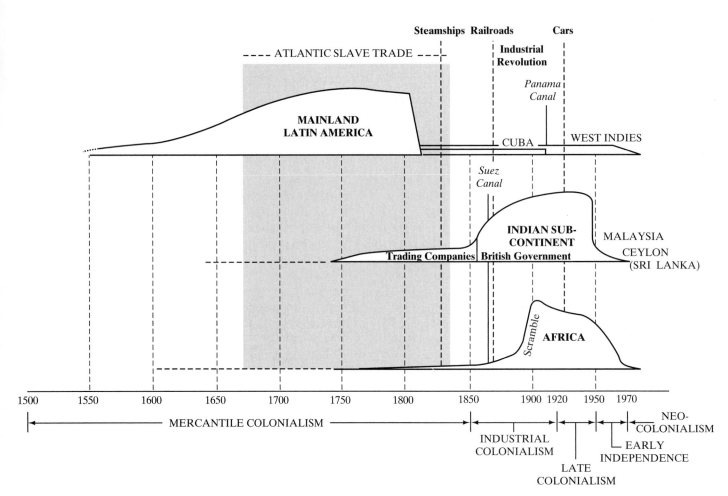

FIGURE 7.13   Comparative periods of colonial rule and phases of colonial urbanization.

Simon[21] theorized about the determinants and evolving nature of the relationships between the colonizers and the colonized. He identified ten general determinants of colonial and post-colonial urban form:

1. The motives for colonization—for example, trade (mercantilism), agricultural settlement, or strategic acquisition.
2. The nature of precolonial settlement—for example, isolated villages or permanent urban centers.
3. The nature of imperial or colonial settlement—for example, imperial control required military security with little if any permanent settlement; colonialism could involve significant levels of permanent settlement.
4. Relationships between the colonizers and the indigenous population—for example, extermination (Australia and the United States), assimilation (Hispanic America after the initial conquest), or some intermediate form of accommodation (much of Africa).

5. The presence or absence of indigenous towns—for example, where such centers did exist, they were destroyed, ignored, added to, or incorporated within a new planned city. Where there were no preexisting centers, new colonial cities were established, sometimes for the colonists alone, sometimes for colonists and indigenous people in separate sections of the city, and sometimes for all groups without formalized segregation.
6. The nature of the anticolonial struggle, the means by which independence was ultimately gained, and the degree to which the new leadership identified with existing administrative centers, especially the capital city, that symbolized both the colonial past and the achievement of liberation and independence.
7. The extent to which the ex-colonial elite retained economic dominance or were supplemented or replaced by skilled expatriates.
8. Policies pursued by the new national elite with respect to national integration, ethnic and class conflict, and the nature of the country's insertion into the world economy.

9. The functioning of the economy, including government policies affecting the various sectors: private, state, and **informal** sectors.

10. The extent of urban legislative change under capitalist expansion policies or some model of socialist centralization.

Although not every region had the same experience, Drakakis-Smith's sequence of five phases of colonial urbanization[22] is a useful framework for examining how this process and its impacts changed over time in different parts of the world (Figure 7.13):

- Mercantile colonialism
- Industrial colonialism
- Late colonialism
- Early independence
- Neocolonialism

We begin with a brief review of the state of urbanization at the eve of the European encounters.

INDIGENOUS URBANIZATION AT THE EVE OF THE EUROPEAN ENCOUNTERS

The Europeans were not the first to create city-based empires (see Chapter 2). In fact, while Western Europe was in a period of urban stagnation during the **Dark Ages**, urbanization was going strong in other parts of the world. Cities and city systems of great size and political, cultural, economic, and technological importance thrived in what are now the less developed countries of the world *before* the arrival of the Europeans.

Islamic influence and Muslim culture dominated the cities of Southwest Asia from the seventh century A.D. In Africa between the eleventh and sixteenth centuries, Arab influence extended in a band across the north and south of the Sahara and along the east coast. The precolonial kingdoms of West Africa, including Mali, and the cities of Timbuktu, Jenne, and Gao, thrived on long-distance trans-Saharan caravan trade (Figure 7.14). Further south, cities like Mogadishu, Mombasa, Zanzibar, Bulawayo, and Great Zimbabwe

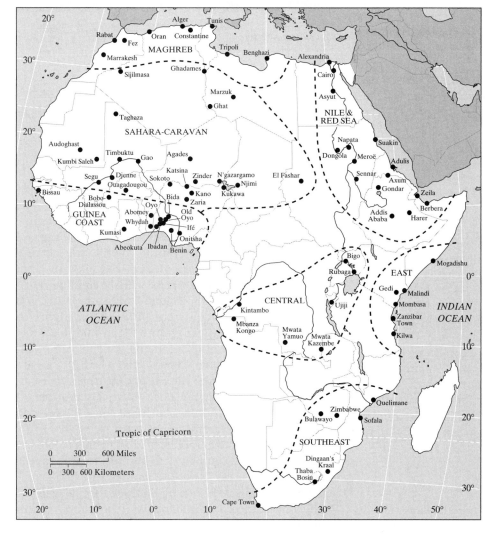

FIGURE 7.14 Historical centers of urbanization in Africa.

FIGURE 7.15   Great Zimbabwe, the capital of the Rozvi Mutapa Empire, flourished between the fourth and ninth centuries.

FIGURE 7.16   Aztec Temple of the Sun at Teotihuacan to the north of Mexico City.

were important centers of commercial and cultural exchange, religion, and learning (Figure 7.15).

In Asia urban civilizations, such as the Ming Dynasty during the fifteenth and sixteenth centuries in China, flourished. In Southeast Asia the urban centers that developed as early as the first century A.D. were usually either inland sacred cities with ritual rulers or coastal or riverine centers—some with up to 100,000 inhabitants—that thrived on their long-distance trade connections.

In Mesoamerica the Aztec urban system in the highlands of central Mexico was centered on Tenochtitlán, the capital city from 1325 A.D. The towering, brightly colored pyramids and palaces impressed the Spanish when they arrived in 1519. With between 100,000 and 300,000 inhabitants, at that time the city was larger than Lisbon or Seville. The Aztecs had a highly stratified society that included a ruler, soldiers, priests, skilled craftsmen such as gold and metalworkers, merchants, and large numbers of agricultural peasants using irrigated farming and terraced fields (Figure 7.16). Further south along the Andes between central Chile and Colombia, the Inca Empire reached its height in the fifteenth century A.D. The capital city, Cuzco, had impressive monuments and a population of between 100,000 and 300,000 people. The Inca Empire was organized around a system of 170 administrative centers connected by an elaborate road system that was used to transport the military and the agricultural produce from the irrigated and terraced field farming.

MERCANTILE COLONIALISM

Individual entrepreneurs made initial forays outside Europe in search of riches like gold and silver.[23]

Later, attention turned to commodities that were valued within the European trading system, such as spices, silk, and sugar. There was no extensive overseas European settlement because *mercantile colonialism* was based on private companies rather than state enterprises. These companies could afford to locate only a limited number of permanent representatives within the existing coastal centers. As a result, the local trading and collection networks were retained and incorporated into the new European trading systems. The nature of mercantile colonialism did vary in response to the different local contexts. In Latin America the earliest contacts were incredibly detrimental to the populations and cities of the Aztec and Inca empires. This contrasts with the experience in Asia where the Chinese empire did not permit direct European contact in the trade of valuable commodities of Chinese origin.

As profits increased the Europeans began to establish a more extensive presence. Company representatives used troops to take control of local trading and collection networks and to protect warehouses. Later, the demand for commodities of dependable quality forced European companies to become involved in the production process itself. The various East India companies that operated out of European states, such as England, France, and Holland, were prominent in this process.

Overall, the mercantile colonial period had only a limited impact on individual cities and systems of cities. Europeans were usually confined to small areas of existing cities that were already organized into ethnic or occupational districts. Colonial vernacular architectural styles were European in function but local in design and materials. No new urban hierarchies were created, and settlements of purely colonial origin, such as Lima, Manila, or Cape Town, were the exceptions.

By about 1800, diminishing European interest in overseas business activity resulted in a transitional period in the nature of colonial urbanization as

- The Napoleonic wars tied up many of the adventurers and some venture capital in Europe.
- The shift from trade to production increased the cost of colonial activities for individual companies, forcing some, most notably the British, French, and Dutch East India Companies, into liquidation and government takeover.
- Greater profits could be made from the Industrial Revolution in Europe.

INDUSTRIAL COLONIALISM

By the 1870s, European investment was again flowing overseas in response to the enormous demand for raw materials and food for the growing urban workforce of the Industrial Revolution in Europe. With government involvement needed to acquire territory and organize production, colonial influence began to have a profound impact on the cities and urban systems outside Europe.

Within colonial cities functional and residential segregation intensified. Although manufacturing was limited to prevent competition with European exports, a large commercial and service sector served the trading and consumer needs of the colonial power. Functional specialization was based on class and ethnicity. Europeans and their institutions dominated foreign trade, expatriate non-Europeans controlled local assembly and distribution, and the indigenous population was involved only in local production, and then under expatriate supervision. This functional specialization reinforced earlier ethnic and occupational residential segregation. Many European districts were separated from non-European ones by physical barriers such as railway lines, parade grounds, police barracks, and racecourses. When combined with the existing social stratification of the indigenous communities, this produced extremely complex patterns of residential segregation (see the Box entitled "Delhi: The Evolution of an Imperial City").

Industrial colonialism so affected the urban systems in some countries, such as Malaysia, that new urban hierarchies were created. Africa saw a general reorientation of urban economic activity from the interior trading routes to the new coastal ports (see the Box entitled "Generalized Model of the Coevolution of Transportation Networks and Urban Hierarchies in a Colonial Context"). Control of production and distribution gave these cities a crucial role in an evolving world economy and international division of labor that supported the growth of the European industrial economy in the nineteenth and early twentieth centuries. But the concentration of economic and political power in certain cities at the expense of others established the foundations for urban primacy that we see today in many less developed countries.

LATE COLONIALISM

The First and Second World Wars and an intervening economic recession in Europe caused erratic demand

---

BOX 7.4   GENERALIZED MODEL OF THE COEVOLUTION OF TRANSPORTATION NETWORKS AND URBAN HIERARCHIES IN A COLONIAL CONTEXT

Based on their work on Ghana and Nigeria, Taaffe, Morrill, and Gould[24] produced a generalized model to describe the main phases in the coevolution of the transportation network and urban hierarchy in a colonial context in Africa (Figure 7.17). The colonial transportation network developed around road and railway lines of "penetration" from the coast into the interior in order to

- Establish political and military control by connecting interior regions with coastal administrative centers.

- Reach and exploit areas rich in minerals.
- Access areas of agricultural produce for export.

The sequence in the model is best seen as a process rather than as a series of discrete historical stages—at any point in time, a transportation network and urban system may show evidence of all phases.

(a) *Scattered ports:* The early stages of *mercantile colonialism* involve a scattering of small ports and trading posts along the coast. There is

(Continued)

BOX 7.4   GENERALIZED MODEL OF THE COEVOLUTION OF TRANSPORTATION NETWORKS AND URBAN HIERARCHIES IN A COLONIAL CONTEXT *(Continued)*

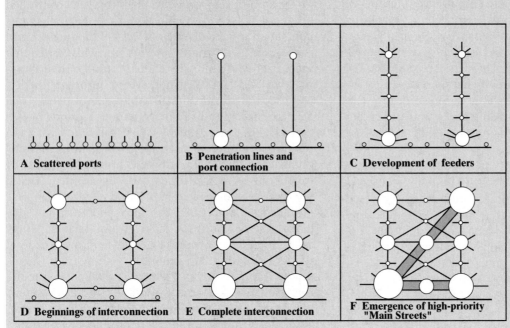

A Scattered ports

B Penetration lines and port connection

C Development of feeders

D Beginnings of interconnection

E Complete interconnection

F Emergence of high-priority "Main Streets"

FIGURE 7.17
The Taaffe, Morrill, and Gould model of the evolution of transportation networks and urban hierarchies.

little lateral interconnection and each port has an extremely limited **hinterland**.

(b) *Penetration lines and port connection:* With industrial colonialism, the construction of major lines of penetration reduces the cost of transportation into the interior for some ports. Markets expand at these ports and at their interior centers, and port concentration begins.

(c) *Development of feeders:* Feeder routes focus on the major ports and interior centers, and these ports enlarge their hinterlands at the expense of adjacent smaller ports.

(d) *Beginnings of interconnection:* As feeder development continues, small nodes develop along the main lines of penetration that become focal points for feeder networks of their own. Interior concentration begins as these nodes draw from the hinterlands of the smaller nodes on either side.

(e) *Complete interconnection:* As the feeder networks continue to develop around the ports, interior centers, and main nodes, some of the larger feeders begin to link up. Lateral interconnection continues until all the ports, interior centers, and main nodes are linked.

(f) *Emergence of high-priority "Main Streets":* A set of high-priority national routes—with the best rail schedules, widest paved roads, and so on—emerges, and the process of concentration is repeated, but at a higher level, as the largest centers connected by these "Main Streets" grow at the expense of others.

Although helpful in conceptualizing the evolution of the transportation and urban systems in a colonial context, this kind of development model suffers from many of the weaknesses identified for modernization theories such as Rostow's stage model[25] and Myrdal's cumulative causation notion.[26] Since independence, instead of spread effects, the initial advantage of the former coastal capital cities has allowed them to maintain a significant **competitive advantage** over other urban centers. Despite decades of independence in some countries, the restructuring of the inherited colonial transportation and urban networks has not yet occurred. This has been the experience even in countries where new capital cities were deliberately established in the interior in an effort to promote a more balanced urban system (e.g., Dodoma replaced Dar-es-Salaam in Tanzania, Abuja replaced Lagos in Nigeria, Yamoussoukro replaced Abidjan in Côte d'Ivoire, Brasilia replaced Rio de Janeiro in Brazil, and Islamabad replaced Karachi in Pakistan).

for **primary products** from the colonies. In an effort to ensure continued profitability, most regions sought to improve efficiency through **economies of scale—** land reforms and mechanization—that forced out smaller producers and landholders. This fuelled rapid rural-to-urban migration that generated more workers than the slower growth in domestic service and factory jobs in the cities could accommodate. A planning process dominated by Europeans and their interests allowed uncontrolled **squatter settlements** to develop; at the same time, the high point of colonial planning and architecture was producing new and re-designed European districts based on the *Garden City* concept (see Chapter 17) and impressive institutional structures like city halls, universities, and banks in the downtown areas (Figure 7.18) (see the Box entitled "Delhi: The Evolution of an Imperial City"). An accelerated in-migration into colonial cities of blue-collar and white-collar workers attempting to escape the recession in Europe accompanied the rural-to-urban migration. This expatriate influx made it increasingly difficult for the slowly growing group of educated indigenous residents to break into middle-income occupations in either administration or commerce.

EARLY INDEPENDENCE

The 1950s and 1960s saw independence spread rapidly throughout most of Asia and Africa. After the colonial powers departed, there was an influx of indigenous people into the cities in search of jobs in the administrative and commercial sectors from which they had been excluded. During the early years of independence,

these job opportunities were limited because of the continuation of European control of commercial companies and the sluggish demand for primary products from a shattered postwar Europe.

Ironically, a major problem in Europe was a shortage of unskilled labor. As a result, unemployed workers from the former **colonial cities** migrated to Western Europe in search of work. Initially, the migrants came from former colonies to former metropolitan powers (e.g., Indians to Britain or Algerians to France), but this migration soon spread to many other poor countries, especially those around the Mediterranean Sea (e.g., Turkey). The workers were abundant, nonunionized, cheap, and, being easily threatened with deportation, compliant. During the 1950s and 1960s these migrant workers represented a lucrative bonus for European industrialists and their governments and economies. The governments of the sending countries encouraged this labor migration because it helped slow urban population growth, increase foreign exchange revenues through the remittances sent home, and, they hoped, train some of their workers.

The build-up of workers in Europe was rapid and highly concentrated. By the late 1960s West Germany and France together had some six million foreign workers, concentrated in the industrial cities in the most menial jobs. The sending countries gained few benefits relative to their losses: large numbers of their younger and most trainable workers gone; few migrants receiving useful skills or training; and the remittances used mostly by the returning migrants to finance small consumer businesses in the larger cities, accentuating already serious urban problems. By the 1970s a growing recession in Europe was forcing

FIGURE 7.18 City Hall and downtown square in Cape Town, South Africa, an example of colonial architecture and urban design.

most countries to tighten their labor immigration laws and reduce the inflow of migrant workers.

During the period of early independence, the economic situation in most less developed countries saw little improvement. Expatriate firms and the same commercial and trading relations from the colonial period continued to dominate these new countries—they exported inexpensive primary products and imported expensive manufactured ones. Socially within cities, the major change was the emergence of a large segment of poor people who were unable to secure paid employment and so could not afford adequate housing, education, and healthcare for themselves and their families. These households looked inward for their survival to create an **informal sector** in which meager incomes are earned and then spent in a wide variety of illegal and quasi-legal economic activities (Figure 7.19) (see Chapter 9).

NEOCOLONIALISM

During the late 1960s and early 1970s there was a dramatic change in the way that workers from the less developed countries were integrated into the world economic system. The **new international division of labor** involved **transnational corporations** from the developed countries shifting the labor-intensive parts of the production process to cities in less developed countries. This was done for a number of reasons:

- The cost of production had risen in the cities of the developed countries due to the rising cost of wages, rents, and imported raw materials, combined with declining productivity and increasing environmental regulations.

FIGURE 7.19   A woman selling American cigarettes and candy is part of the informal sector of the economy in Saigon, Vietnam.

- Steady rural-to-urban migration kept the cost of labor down in the cities of the less developed countries.
- A large informal sector, representing a reserve army of labor, depressed demands for wage increases.
- Advances in technology allowed the separation of production from management. E-mail, satellite links, and **containerization** made it possible for the labor-intensive parts of the production

FIGURE 7.20   Container port in Valparaiso, Chile. Automated cranes load and unload cargo containers between the ships and the flat beds of nearby trains and trucks.

process to be located in the cities of less developed countries (Figure 7.20) at the same time that the head offices of companies could remain in the largest cities of the developed countries.

- International agencies and national governments supported the new international division of labor because the new jobs created in the growing cities in the less developed countries could help promote economic and political stability.

The new international division of labor has had a varied and complex impact on the cities in the less developed countries.

- Rapid economic growth has been highly selective. Only a relatively small number of **newly industrializing countries**, or **NICs**, like South Korea and Taiwan, initially experienced rapid industrial growth. Following rising labor costs in these early NICs, transnational corporations looked for new supplies of cheap labor, and new industrial producers emerged, including Malaysia and Thailand.
- Cities have received the bulk of this **foreign direct investment** or **FDI**. This has induced further rural-to-urban migration and exacerbated existing urban problems.
- The effect on social class formation has been considerable. The new waged workforce is relatively conservative, while the informal sector has continued to grow, raising concerns about urban instability.
- In contrast to the early years of independence, women have been incorporated into the urban workforce at unprecedented rates.

These changes have taken place in most cities in the less developed countries with the support of the national

FIGURE 7.21   A garment factory in the Kaohsiung Export Processing Zone in Kaohsiung Habor in South Taiwan that combines the features of a free trade zone with those of an industrial district.

governments. The state's role in taking over the management of capital cities from local governments has been crucial in facilitating the penetration of foreign direct investment. In an effort to encourage economic growth, many governments borrowed heavily to modernize their cities, adding new airports, conference centers, and **free trade zones** (Figure 7.21). Already in debt to international banks and agencies, many governments have had to adopt economic adjustment programs dictated by the **World Bank** or the **International Monetary Fund** (**IMF**) as a precondition for further loans. An important part of these structural adjustments has been a reduction in government spending in areas such as social welfare. These cutbacks have affected the population in the cities most, and have worsened the already existing shortfall in the provision of basic urban services.

OVERURBANIZATION AND MEGACITIES

The problem today for many cities in the less developed countries is not a lack of urban growth, but rather a process of **overurbanization** in which cities and their populations grow more rapidly than the jobs and housing that they can sustain. Unprecedented rates of urbanization have been associated with the growth of **megacities**. Megacities can be distinguished by a very obvious characteristic—sheer size. Most have a population of 10 million or more. Examples of megacities include Jakarta, São Paulo, Bangkok, Beijing, Cairo, Kolkata (Calcutta), Dhaka (Bangladesh), Lagos, Manila, Mexico City, New Delhi, Shanghai, and Tehran (Figure 7.22). Each has more inhabitants than 100 of the countries that are members of the United Nations.

These megacities are usually characterized by primacy and a high degree of centrality within their

national urban systems. **Primacy** and primate cities occur when the population of the largest city in an urban system is disproportionately large in relation to the second-largest and third-largest cities in that system (see the discussion of the **rank-size rule** in Chapter 3). **Centrality** refers to the functional dominance of cities within an urban system. Cities that have a disproportionately large share of national economic, political, and cultural activities have a high degree of centrality within their urban system. This combination of size and functional centrality often causes megacities in different countries to have more in common with one another than with the smaller metropolitan areas and cities within their own countries.

Although most megacities do not function as **world cities**, they do perform an important intermediate role

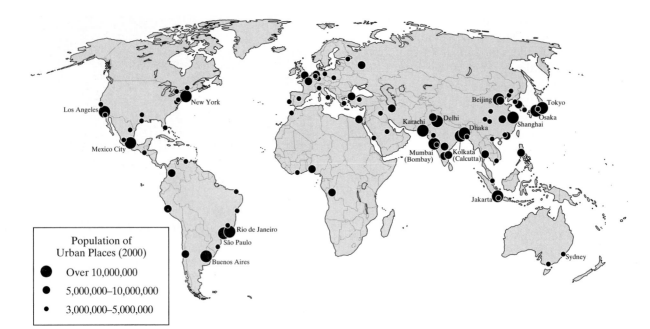

FIGURE 7.22   Major urban agglomerations of the world, 2000.

between the upper tiers of the system of world cities and the provincial towns and villages. Not only do the megacities link the local and provincial economies into the world economy, but they also represent important points of contact between the formal and informal sectors of the urban economy.

But critical networked transportation, information, and communications infrastructures are highly uneven in their extent, selectively serving only certain kinds of regions and metropolitan settings. They are embedded within regions and metropolitan areas whose economic foundations derive from earlier **technology systems** and whose social and cultural fabric derives from more traditional bases. In the next chapter we will see that the close relationship between **globalization** and urbanization means that the traditional patterns of urban land use and spatial organization in many of the less developed countries are being transformed by the local effects of **splintering urbanism**.[27] As we will see in Chapter 9, these local effects include squatter settlements in megacities that often are associated with severe problems of social disorganization and environmental degradation. Nevertheless, many neighborhoods have been able to develop self-help networks that have formed the basis of community within often overwhelmingly poor and crowded cities.

FOLLOW UP

1. Be sure that you understand the following key terms:

   centrality
   colonialism
   colonial cities
   containerization
   core, semi-peripheral, and peripheral countries
   Demographic Transition
   dependency theory
   economies of scale
   free trade zones
   megacities
   modernization theory
   newly industrializing countries (NICs)

   transnational corporation (TNC)
   underdevelopment
   urban bias
   world-systems theory

2. If you have access to a video library, you should arrange to watch Episode 3 in the *Americas* video series, produced by WGBH Boston and Central Television Enterprises for Channel 4, UK (1993). This video, *Continent on the Move: Migration and Urbanization*, examines rural-to-urban migration in Mexico within the context of the underlying processes driving this movement of people to the cities and the associated social and economic problems that have overwhelmed the resources of city governments.

3. Use the figures in Table 7.1 to make a map that shows the 30 largest metropolitan regions around the world in 2015. Make a list of the most important factors that you think are responsible for the spatial distribution shown on your map. Categorize these factors into those operating mainly at an intrametropolitan or intranational scale and those that are triggered by larger global processes. Then think about some of the social, economic, political, and environmental implications of what you have identified for the urban residents and city governments of these metropolitan regions.

4. Pick a large metropolitan region with a colonial history that interests you, perhaps Lagos (Nigeria), São Paulo (Brazil), or Jakarta (Indonesia). Do some research in the library and on the World Wide Web to find historical information that you can use to fill in something for your metropolitan region under these headings: Mercantile Colonialism; Industrial Colonialism; Late Colonialism; Early Independence; and Neocolonialism. Think about some of the reasons why what you find out about the history of your metropolitan region might differ from that of other large metropolitan regions in other parts of the less developed world.

5. Work on your *portfolio*. You might consider spending some time going through the United Nations Human Settlements Program (UN-Habitat) website (at *http://www.unchs.org/*) to find supplementary information about some of the processes and outcomes of urbanization in the less developed countries of the world that you are least familiar with. It might also be helpful for you to find maps and data that help you to consider how and why the urban experience of the less developed countries has been different from that of the developed countries.

KEY SOURCES AND SUGGESTED READING

Drakakis-Smith, D. 2000. *Third World Cities*, 2nd ed. London: Routledge.

Friedmann, J. 1992. The End of the Third World. *Third World Planning Review* 14: iv–viii.

Gilbert, A. 1998. *The Latin American City*. 2nd ed. London: Latin American Bureau.

Gilbert, A. and J. Gugler. 1992. *Cities, Poverty and Development: Urbanization in the Third World*. 2nd ed. Oxford, UK: Oxford University Press.

Ginsburg, N., B. Koppel, and T. G. McGee, eds. 1991. *The Extended Metropolis: Settlement Transition in Asia*. Honolulu, Hawaii: University of Hawaii Press.

Gugler, J., ed. 1997. *Cities in the Developing World: Issues, Theory, and Policy*. Oxford, UK: Oxford University Press.

King, A. D. 1990. *Urbanism, Colonialism, and the World Economy: Cultural and Spatial Foundations of the World Urban System*. London: Routledge.

Lowder, S. 1986. *The Geography of Third World Cities*. Totowa, N.J.: Barnes & Noble Books.

Potter, R. B., and S. Lloyd-Evans. 1998. *The City in the Developing World*. Harlow, UK: Longman.

Simon, D. 1992. *Cities, Capital and Development: African Cities in the World Economy*. London: Belhaven Press.

Smith, D. A. 1996. *Third World Cities in Global Perspective: The Political Economy of Uneven Urbanization*. Boulder, Colo.: Westview Press.

United Nations Centre for Human Settlements (Habitat). 2001. *The State of the World's Cities Report, 2001*. Nairobi: UN Habitat (*http://www.unchs.org/Istanbul+5/statereport.htm*).

United Nations Centre for Human Settlements (Habitat). 2001. *Cities in a Globalizing World: Global Report on Human Settlements 2001*. London: Earthscan Publications.

RELATED WEBSITES

United Nations Human Settlement Programme (UN-Habitat): *http://www.unchs.org/*
UN-Habitat, the United Nations agency for human settlements, attempts to promote socially and environmentally sustainable towns and cities with the goal of adequate shelter for urban residents. The UN-Habitat website contains information and publications related to its two major worldwide campaigns: the Global Campaign on Urban Governance and the Global Campaign for Secure Tenure. It also offers its annual "State of the World's Cities Report."

The Megacities Foundation: *http://www.megacities.nl/foundation_fr.html*
The Megacities Foundation in the Netherlands was established as a result of an initiative by UNESCO to focus attention on the problems of explosively growing megacities around the world. The Foundation website contains publications and other information related to megacities, their growth, and associated problems.

United Nations and Decolonisation: *http://www.un.org/Depts/dpi/decolonization/main.htm*
This United Nations website provides information on decolonisation for countries that have gained their independence and joined the United Nations since World War Two.

The United Nations Population Fund: *http://www.unfpa.org/*
The UNFPA website contains information and downloadable publications on a variety of topics (poverty, human rights, and gender equality, to name a few) related to urbanization and migration in less developed countries.

8

URBAN FORM AND LAND USE IN THE LESS DEVELOPED COUNTRIES

A variety of historic and contemporary processes—including colonialism, rural-to-urban migration, and overurbanization—have shaped the cities of less developed countries. Although global influences are generating certain similarities in urban structure worldwide, the patterns of land use and functional organization in cities in different regions remain distinct, reflecting variations in such factors as historical legacies, levels of technology, and environmental and cultural influences. Meanwhile, the local effects of splintering urbanism are now transforming these historical patterns of urban land use and spatial organization. Economic and cultural globalization is generating new urban landscapes of innovation, economic development, and cultural transformation. A major problem is that the uneven diffusion of the networked infrastructures of information and communications technologies has intensified economic and social inequalities not only between urban residents in the more and less developed parts of the world, but also among city dwellers in the less developed countries themselves.

CHAPTER PREVIEW

In the less developed countries, as in other parts of the world, competition for territory and location strongly shapes urban form and land use. In general, different categories of land users—commercial and industrial, as well as residential—compete for the most convenient and accessible locations in cities. This chapter examines the various patterns of urban form and land use in Latin American, African, Islamic, South Asian, Southeast Asian, and East Asian cities. We draw on some well-known descriptive urban models—simplifications of reality—to establish useful snapshots of the major differences among these regions. Of course, these rather static generalizations cannot capture all the outcomes of the historic and contemporary social, cultural, economic, political, technological, and environmental processes affecting cities.

Since the earliest cities (described in Chapter 2), internal urban structure has varied a great deal from one region of the world to another in response to the influence of factors such as history, culture, environmental conditions, and the different roles of individual cities within the **world-system**. Nevertheless, the urban form and land use of many cities in the less developed countries bear the imprint of certain general demographic, cultural, economic, and political processes (described in Chapter 7).

Demographic processes such as **rural-to-urban migration** have helped shape urban form and land

use within less developed countries. As we will see in Chapter 9, overurbanization in the largest cities has led to a large **informal sector** of the economy, as recent migrants and others who cannot find regularly paid work resort to earning a living in jobs that are not regulated by the state. The resulting pronounced *dualism*, or juxtaposition in geographic space of the formal and informal sectors of the economy, is quite evident in the form and land use of cities.

For example, there has been a "quartering" of cities into spatially partitioned, compartmentalized residential enclaves. Luxury homes and apartment complexes correspond with an often dynamic formal sector that offers well paid jobs and opportunities; these contrast sharply with the slums and **squatter settlements** of people working in the informal sector who are disadvantaged by a lack of formal education and training and the often rigid **divisions of labor** shaped by gender, race, and ethnicity. In addition, the imprint of colonialism remains evident, often at the center of the city in the formal squares and public spaces, the layout of avenues, and the presence of colonial architecture and monuments. The colonial legacy, combined with the general processes of urban demographic, cultural, economic, and political change, continue to influence the internal structure and land use at urban and metropolitan scales within less developed countries.

PATTERNS OF URBAN FORM AND LAND USE

LATIN AMERICAN CITIES

The internal structure of many Latin American towns today shows evidence of a colonial past—always most visible in what is now the urban core. Spanish colonial towns of the 17th century generally were located and planned in compliance with royal edicts—the so-called "Laws of the Indes"—which contained design features reflecting the legacy of the Roman Empire in Spain itself. This city planning legislation dictated a **gridiron** pattern of square or rectangular blocks subdivided into long narrow lots along similarly narrow streets. A central plaza (market square) was surrounded by the important buildings—the Roman Catholic church, town hall, hospital, governor's palace, and commercial and retail arcades (Figure 8.1). The Spanish settlers lived near the center of the city, with the wealthy closest to the plaza and residents of middle or lower income in neighborhoods (*barrios*) further out. The indigenous population was located at the periphery along with undesirable land uses such as slaughterhouses and cemeteries.

As an urban society, the Spanish used towns as social control mechanisms in the pursuit of their three goals: "God, Glory, and Gold." The Roman Catholic Church wanted to convert as many of the indigenous people as possible, and forcibly relocating them to the cities made this task easier. The towns facilitated the expansion of empire, and the indigenous people—as forced laborers in the mines and fields—could more easily be recruited and controlled in an urban setting because the gridiron street pattern allowed uprisings to be put down swiftly.

In contrast, the towns of Portuguese origin in what is now Brazil evolved without any royal guidelines. Some urban centers grew up around forts or along important transportation routes. Terrain and geology influenced the layout of the mining towns. Mission settlements typically had a large rectangular plaza (*praca*) with a church. Port cities usually had a linear layout because of their location between the waterfront and surrounding hills. The lower city (*cidade baixa*) housed the port and main market activities. The upper, hilly portion of the city (*cidade alta*) was fortified

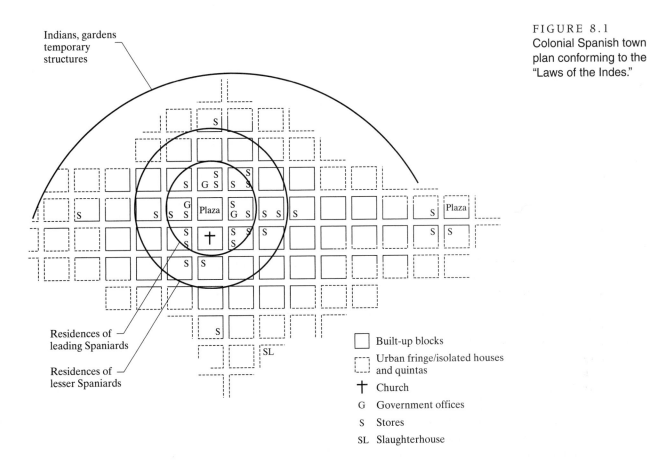

FIGURE 8.1
Colonial Spanish town
plan conforming to the
"Laws of the Indes."

Indians, gardens
temporary
structures

Residences of
leading Spaniards

Residences of
lesser Spaniards

☐ Built-up blocks

┌--┐
└--┘ Urban fringe/isolated houses
 and quintas

✝ Church

G Government offices

S Stores

SL Slaughterhouse

and contained church and government buildings and wealthier residential neighborhoods.[1]

Although the original plaza remains important as a historic or tourist attraction in larger cities and might retain many of its commercial and administrative functions in smaller towns and cities, contemporary Latin American cities exhibit certain additional characteristic features of urban form and land use. The model in Figure 8.2 shows a commercial spine and adjacent elite residential sectors surrounded by a series of encircling zones in which residential quality decreases with distance from the **central business district (CBD)** in what has been described as an **inverse concentric zone** pattern. The principal elements are as follows:[2]

- The downtown comprises a CBD and a market that reflect how many Latin American downtowns have modern self-contained commercial and retail districts that are now quite separate from the more traditional mixed market districts with their small, street-oriented businesses. The dominance of the CBD reflects a road and public transportation system focused on the downtown and the presence of a large and relatively affluent population living in central city neighborhoods.

- An extension of the CBD—a commercial/industrial "spine" surrounded by elite residential

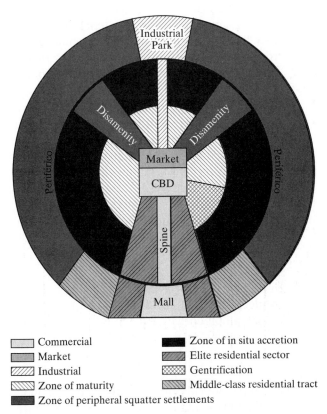

☐ Commercial

☐ Market

▨ Industrial

▨ Zone of maturity

■ Zone of peripheral squatter settlements

■ Zone of in situ accretion

▨ Elite residential sector

▨ Gentrification

▨ Middle-class residential tract

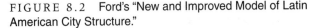

FIGURE 8.2   Ford's "New and Improved Model of Latin American City Structure."

neighborhoods—stretches toward the periphery along the main transportation artery. This sector has the best urban services and most of the high-end locations outside the CBD: tree-lined boulevards, golf courses, parks, restaurants, and office buildings. A suburban shopping center at the periphery of the commercial spine might compete with the downtown.

- A separate industrial sector, along a railroad or main highway, terminates in a suburban industrial park with modern single-story factories, warehouses, and distribution facilities.

- A zone of maturity surrounding the CBD comprises stable middle-income neighborhoods that are well served by paved streets, lighting, public transportation, schools, water, and sewerage. Newer middle-income housing tracts are also located adjacent to the elite sector. There is a small area of **gentrification** in some attractive historic neighborhoods within the zone of maturity near the CBD and elite sectors.

- Further out is a zone of *in situ* accretion (literally, in place addition) containing lower-income neighborhoods that show signs of transition toward becoming part of the zone of maturity. Because currency fluctuations and devaluations can reduce the value of a bank account overnight, many low-income homeowners instead spend their savings on building materials to enlarge their homes in order to generate rental income from a lodger. The zone is in a constant state of construction, and many houses have half-finished rooms or second stories.

- Around the edge of the built-up area is a zone of peripheral squatter settlements housing impoverished recent immigrants to the city. This zone has the worst quality housing, with most shacks made from scavenged materials such as timber and corrugated iron. The zone is almost completely without urban services.

- Sectors of disamenity—along polluted rivers and industrial corridors—run outward from the core. Squatter settlements in these sectors tie into those at the periphery.

- In larger cities a peripheral ring road (*periférico*) connects the shopping mall and industrial park, but development is restricted by the difficulties of expanding infrastructure and upgrading the zone of peripheral squatter settlements.

It is particularly difficult, however, to generalize about urban form and land use in contemporary Latin American cities because land use regulations are weak and often completely disregarded. Figure 8.3 shows a model of land use that combines typical commercial, residential, and industrial patterns within a broad overview that reflects the disregard for **land use zoning** regulations in Latin America.[3] Some important features highlighted by this diagrammatic representation are:

- Locations where the informal sector of the economy dominates.
- Widely dispersed individual commercial establishments, such as small grocery stores, specialty stores (furniture, clothing, housewares), and restaurants, reflecting the fact that most people do not own an automobile.
- Strip malls along major highways.
- Industrial activities in individual factories and small plants throughout the city.
- Low-income housing covering a large part of the city, including slums in less desirable locations near the downtown, especially near the old industrial areas.

In contrast to Latin American cities further south, many towns in northern Mexico have experienced phenomenal growth stimulated by **foreign direct investment**, (FDI), from U.S. firms since the early 1990s and Mexico's entry into the North American Free Trade Agreement (NAFTA) and before that as a result of the Border Industrialization (**Maquiladora**) Program. Since 1965 this program has allowed many U.S. firms to import materials and components duty free into Mexico if the goods then produced by the low-cost Mexican workers are reexported to the United States.

This imprint of **globalization** is captured in Arreola and Curtis's model of the Mexican border city shown in Figure 8.4.[4] Note the following features:

- The pre-1950 "traditional" urban core is truncated along its northern edge by the international border.
- The CBD contains a small tourist district.
- Interconnected major commercial strips radiate out from the CBD, including a higher-order spine that supports regional shopping centers.
- Industrial activities—that have shifted progressively toward the periphery of the built-up area—are located near the railroad and main highways.
- The maquiladoras are close to the U.S. border and the airport.
- Residential quality declines generally with distance from the CBD in an inverse concentric zone pattern—from primarily upper-middle and middle-income neighborhoods to lower-middle and lower-income areas.
- At the periphery of the city adjacent to the major highways are comparatively small squatter settlements that are growing up in an irregular pattern away from their points of origin.
- Exceptions to this pattern include inner-city slums, which fan out from the CBD in a sector

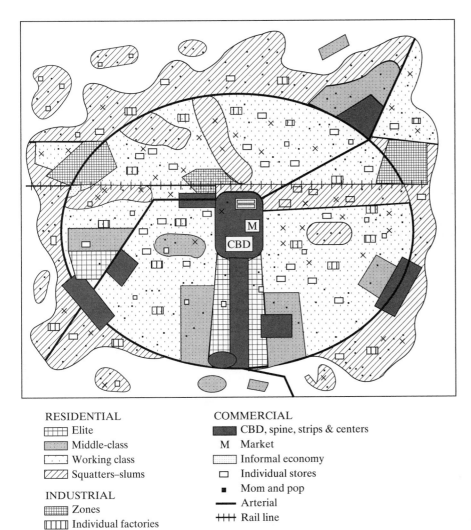

FIGURE 8.3 Crowley's model of major land uses in Latin American cities.

RESIDENTIAL
Elite
Middle-class
Working class
Squatters–slums

INDUSTRIAL
Zones
Individual factories
× Small plants

COMMERCIAL
CBD, spine, strips & centers
M Market
Informal economy
Individual stores
Mom and pop
— Arterial
+++ Rail line

of disamenity between the railroad line and the border.

- On the opposite side of the city is a prestigious sector separated from other neighborhoods by the spine. The elite residential areas within it occupy limited space, are not connected to each other, and are pushing steadily outward toward the sector's periphery.

- Newer upper-middle and middle-income housing surrounds the elite residential areas, with some also in outlying suburban developments.

- Residential tracts of public housing are located near the industrial areas (especially the maquiladoras).

A distinguishing feature of recent urbanization in parts of Latin America has been the extension of some **megacities** beyond their city and metropolitan boundaries to form **extended metropolitan regions**.[5] In Brazil "polygonized development" reflects a relative decline in the formerly overwhelming dominance of the city of São Paulo as a result of a relatively "concentrated decentralization" pattern of new economic activity

that is locally concentrated in surrounding smaller cities and towns (Figure 8.5).[6] New economic activity has been concentrated in the geographical polygon that now surrounds São Paulo and encompasses Belo Horizonte, Uberlândia, Londrina, Maringá, Porto Alegre, Florianópolis, and São José dos Campos. The main **growth poles** of high technology in Brazil have developed within this new industrial polygon.

A number of factors have led to this expansion of São Paulo into an extended metropolitan region and have worked against a broader deconcentration of new investment throughout Brazil. These include:

1. **Agglomeration diseconomies** in the metropolitan area of São Paulo and growing **agglomeration economies** in nearby urban centers

2. Government intervention in the region surrounding São Paulo, including public investment in infrastructure and fiscal incentives to companies

3. The extraordinary regional concentration of income and research resources in São Paulo that has restricted development to the immediate region

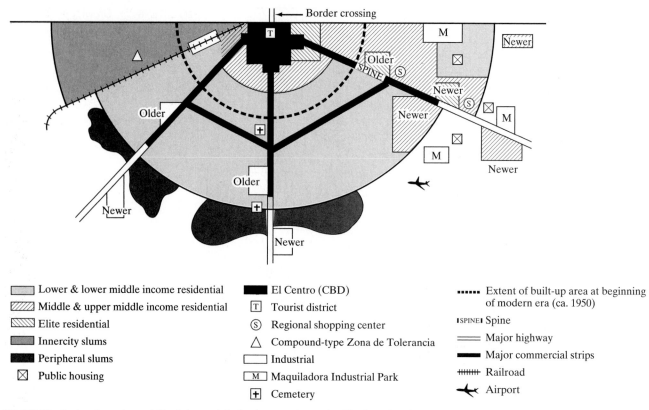

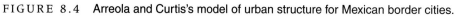

Lower & lower middle income residential

Middle & upper middle income residential

Elite residential

Innercity slums

Peripheral slums

⊠   Public housing

■   El Centro (CBD)

T   Tourist district

Ⓢ   Regional shopping center

△   Compound-type Zona de Tolerancia

▢   Industrial

M   Maquiladora Industrial Park

⊞   Cemetery

•••••   Extent of built-up area at beginning of modern era (ca. 1950)

ⅠSPINEⅠ Spine

═══   Major highway

▬▬▬   Major commercial strips

⧓⧓⧓   Railroad

✈   Airport

FIGURE 8.4   Arreola and Curtis's model of urban structure for Mexican border cities.

FIGURE 8.5   São Paulo's extended metropolitan region.

AFRICAN CITIES

Although there have been attempts to create a generalized model of the urban structure of African cities,[7] the varied histories and rapid changes in urban structure since independence in African countries make it more realistic to think in terms of six categories of African cities.[8]

1. *The indigenous (native) city:* Most towns established before major European contact were administrative centers of indigenous power that also had craft and trade functions. These include the Yoruba cities of southwest Nigeria that date back to at least the tenth century. The larger towns were walled and contained central palaces and courtyards. Some, like Ibadan, had populations exceeding 50,000 people by the time Europeans arrived and were subsequently profoundly influenced by colonial rule. Further east, Addis Ababa grew to become the largest indigenous city after Emperor Menelik made it his capital in 1886; although Ethiopia retained its political independence, the city was still influenced significantly by European contact.

2. *The Islamic city:* Islamic cities in Africa—including Kano in predominantly Muslim northern Nigeria in the west, Dar-es-Salaam to the south in Tanzania, and Merca in Somalia to the east—were founded by African people or invaders who brought Islamic influences (Figure 8.6). These cities flourished as the capitals of empires, as religious centers, and as marketplaces at the end of long-distance trans-Saharan caravan routes. They contained features characteristic of Islamic cities in the Middle East and other parts of Asia, including the main covered bazaars or street markets (*suqs*), mosques, a citadel (fortress), and public baths.

FIGURE 8.6   The Central Mosque of Kano, Nigeria, the oldest city in West Africa.

3. *The colonial (administrative) city:* Many cities in Africa date from the late nineteenth and early twentieth centuries and are colonial in origin. These cities were founded for administrative and trade purposes rather than as settlements for Europeans. Many, such as Dakar (Senegal) and Freetown (Sierra Leone), were ports that served as the main points of contact between the colonial powers and the local population. The colonial transportation network of road and railway lines connected the interior regions with these coastal trade centers; this allowed the Europeans to maintain political and military control and to extract minerals and agricultural produce from the interior areas for export (see the Box entitled "Generalized Model of the Co-evolution of Transportation Networks and Urban Hierarchies in a Colonial Context" in Chapter 7). These cities were characterized by quite sharp functional and residential segregation that was imposed by the colonial powers.

4. *The "European" city:* The "European" city is a special category of **colonial city**. It is a true "colonial" city in the original sense of a colony as a place of permanent new settlement. Cities such as Harare (formerly Salisbury) in Zimbabwe, Lusaka in Zambia, and Nairobi in Kenya were established primarily as places for Europeans to live and to provide urban services for permanent European settlers in the surrounding rural areas. They were designed as replicas of towns in Europe and reflected European town planning ideas. Although their main functions were administration and trade, manufacturing to meet the needs of the European settlers was allowed. The Europeans imposed such sharp functional and residential segregation that the Africans who lived and worked in these cities were viewed as temporary residents. The most extreme expression of the "European" city was the apartheid city of South Africa (see the Box entitled "South African Cities").

5. *The dual city:* The dual city represents a juxtaposition of two or more of the indigenous, Islamic, colonial, or "European" categories of cities. Kano, for example, combines both an ancient walled Muslim city surrounded by a new area of colonial origin. Greater Khartoum comprises the indigenous city of Omdurman on the west bank of the Nile and the colonial city of Khartoum on the opposite bank. These dual cities are interdependent but physically separate—each has a full range of urban functions and each developed with a certain degree of independence.

6. *The hybrid city:* Indigenous and foreign elements are integrated in the hybrid city. Since independence,

BOX 8.1   SOUTH AFRICAN CITIES

There was racial segregation in the "European" cities of South Africa well before the formal introduction of apartheid. European colonists imposed functional and residential segregation in cities like Durban, but there were also multiracial neighborhoods, especially in working-class areas, that were segregated by income. The preapartheid "Segregation City" comprised a number of elements (Figure 8.7(a)):[10]

- A dominant white central business district (CBD)
- Secondary and peripheral Indian or Chinese CBDs, or both
- Industrial districts, owned by whites but constituting common employment space
- A white residential core with suburban extensions along sectors of desirable amenities, strongly differentiated by socioeconomic status
- Indian, Coloured (mixed ancestry), and Chinese enclaves in older centrally located residential areas
- An African working zone of single quarters in barracks and compounds
- African, Indian, and Coloured peripheral settlements in small townships or larger concentrations of squatter settlements and privately developed neighborhoods of very mixed quality
- African domestic quarters, widely distributed throughout the city
- Mixed zones that represented black diffusion into white residential areas

The introduction of apartheid transformed the racial segregation of the "European" city from informal practice to strict rule. The apartheid system had the explicit goal of separate racial development. The 1950 Group Areas Act required the spatial separation of the officially defined racial groups (Whites, Asians, Coloureds, and Blacks). Different areas—separated by clear natural divides such as greenbelts or physical barriers such as railways, roadways, and industrial strips or, if necessary, buffer zones of vacant land—were designed for each group (Figure 8.7(b)). The sectoral design and radial transportation routes allowed separate growth at the periphery of each sector while minimizing contact and interaction across sectors. Ten "homelands" (*Bantustans*) comprising only about 14 percent of the area of the country—mostly barren land—were designated for black people, who comprised 70 percent of the population. The 1952 "Pass Laws" gave black males temporary work permits that allowed them to work in the cities as needed, but they had to return to their designated homeland each year.

For nearly 40 years the white minority used apartheid to control the black majority, using fundamentally geographical means to achieve the spatial separation of races within cities, within the countryside, within buildings, in employment, and by marriage. The decentralization of industry to the homelands, though encouraged, was not successful. Instead, black settlements grew up in homeland townships close to white urban areas. Soweto near Johannesburg is a well-known example. These settlements gave the white-controled cities access to nearby cheap labor without requiring the authorities to pay for housing, infrastructure, and services for the black population. The ensuing protests against apartheid were quickly and ruthlessly suppressed. African National Congress leaders like Nelson Mandela were jailed, and activists such as Steven Biko were killed.

The international pressure that contributed to the end of apartheid included sports boycotts, a forced

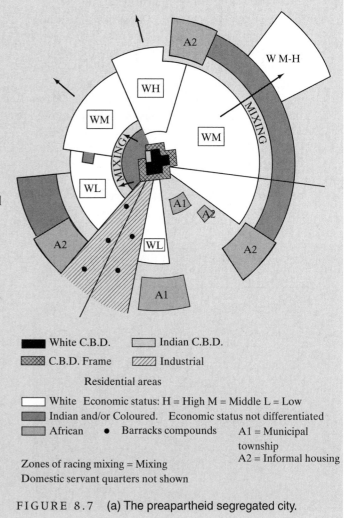

■ White C.B.D.		▢ Indian C.B.D.	
▨ C.B.D. Frame		▨ Industrial	

Residential areas

▢ White Economic status: H = High M = Middle L = Low
▨ Indian and/or Coloured. Economic status not differentiated
▨ African ● Barracks compounds A1 = Municipal township
A2 = Informal housing

Zones of racing mixing = Mixing
Domestic servant quarters not shown

FIGURE 8.7   (a) The preapartheid segregated city.

(Continued)

Box 8.1   South African Cities   (*Continued*)

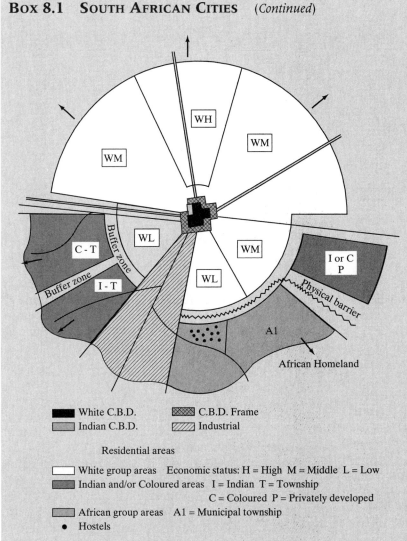

FIGURE 8.7   (b) The apartheid city in South Africa.

Legend:

- ■ White C.B.D.
- ▨ C.B.D. Frame
- ▢ Indian C.B.D.
- ▨ Industrial

Residential areas

- ▢ White group areas   Economic status: H = High M = Middle L = Low
- ▨ Indian and/or Coloured areas   I = Indian T = Township
-   C = Coloured P = Privately developed
- ▨ African group areas   A1 = Municipal township
- ● Hostels

Economic status of Black Group Areas not differentiated
Domestic servant quarters not shown

withdrawal from the British Commonwealth, trade sanctions, and voluntary investment bans by major transnational corporations. A number of white South Africans were also vocal in their opposition to the system. The 1990s saw the end of apartheid in South Africa. Nelson Mandela was freed from jail, and President F. W. de Klerk agreed to political power sharing between blacks and whites. In 1994, South Africa held the first election in its history in which blacks were allowed to vote. Nelson Mandela was elected the first black president.

Despite efforts to promote greater integration, the shadow of apartheid planning is expected to remain visible in the geography of South African cities for decades.[11] The white elite still occupies the affluent neighborhoods, while vast numbers of poor blacks live in peripheral squatter settlements. Much progress

remains to be made in addressing apartheid's legacy and in reducing the deep social, economic, and spatial divisions between blacks and whites.

Some observers are optimistic about South Africa's urban future—that postapartheid planners and politicians can work closely with the communities they serve, attempting to meet their priorities despite limited resources and enormous needs and making urban amenities and services available to all. The major resource of South African cities is the people. Despite their poverty, they have shown resilience and resourcefulness in the face of oppressive urban policies in the past. These qualities can be harnessed, not just for survival, as so often was the case during apartheid, but for the steady improvement of South African cities and the opportunities they offer for all the people of South Africa.[12]

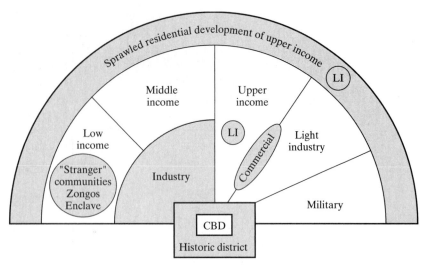

FIGURE 8.8   A model of a sub-Saharan African City.

most African cities have become hybrids as their indigenous, Islamic, colonial, and "European" elements have become more integrated. Accra, Ghana, is a good example of this kind of city.[9] It was an indigenous city, founded prior to the sixteenth century, that became a colonial administrative center in the late nineteenth century. It combines indigenous, colonial, and modern development. Aryeetey-Attoh's model of the internal structure of Accra shows an irregular pattern of ethnic and religious enclaves superimposed upon residential sectors of different income level (Figure 8.8). Unplanned peripheral sprawl of high-income development accompanied Accra's rapid expansion. This recent trend contrasts with earlier descriptions of an inverse concentric zone pattern for African cities in which the wealthy and middle-income groups congregated close to the government center to avoid the hazards of a time-consuming commute, while the poor occupied the peripheral locations.

A characteristic of many African cities is the existence, within the same municipal boundary, of two enclaves: a poor highly marginalized population and a wealthy expatriate elite.[13] The urban elite is connected to the international economy; poorer people are tied into the informal sector of the urban economy. Greater integration is difficult because the African formal sector has not been able to disengage from its unequal political, economic, and cultural links with other parts of the world and make urbanization work for the benefit of African cities and their residents. Most observers believe that it is unlikely that people in the informal sector can make the transition to the more prosperous formal sector of the urban economy. Modern technology, which is relatively easily available to those in the formal sector, is virtually inaccessible to most people in the informal sector. This situation presents particular

problems for the larger cities that are experiencing **overurbanization**. These include sub-Saharan Africa's only megacity, Lagos (Nigeria), and other rapidly growing **primate cities**, such as Addis Ababa (Ethiopia), Harare (Zimbabwe), and Nairobi (Kenya).

ISLAMIC CITIES

Islamic cities are found in the Arabian Peninsula and the Middle East—the heartland of the Islamic Empire under the prophet Muhammad (A.D. 570–632)—and in those regions into which Islam spread, including parts of Africa, South-Central Asia, and Indonesia. Many elements of the traditional Islamic city can be found in towns and cities as far away as Seville, Granada, and Córdoba in southern Spain (the western extent of Islam), Kano in northern Nigeria and Dar-es-Salaam in Tanzania (the southern extent), and Davao in the Philippines (the eastern extent).

Islamic cities are good examples of how cultural values, economic necessities, and environmental conditions are reflected in the form and land use of cities (Figure 8.9). The fundamentals of the layout and design of the traditional Islamic city are so intimately connected with Islamic cultural values that they are found in the *Qur'an* (the Koran), the holy book of Islam. Although urban growth does not have to conform to any master plan, certain basic regulations and principles are intended to ensure Islam's emphasis on personal privacy and virtue, on communal well-being, and on the inner essence of things, rather than on outward appearance.

Privacy is central in the construction of the Islamic city. According to Islamic values, women must be protected from the gaze of unrelated men. Traditionally, doors must not face each other across a minor street, and windows must be small, narrow, and above eye level. Dead-end streets are used to restrict the number of people approaching the homes, and angled entrances prevent intrusive glances. Larger homes are

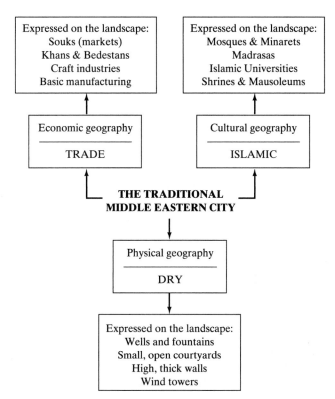

```
┌──────────────────────────┐   ┌──────────────────────────┐
│  Expressed on the        │   │  Expressed on the        │
│  landscape:              │   │  landscape:              │
│  Souks (markets)         │   │  Mosques & Minarets      │
│  Khans & Bedestans       │   │  Madrasas                │
│  Craft industries        │   │  Islamic Universities    │
│  Basic manufacturing     │   │  Shrines & Mausoleums    │
└──────────────────────────┘   └──────────────────────────┘
            ↑                               ↑
┌──────────────────────────┐   ┌──────────────────────────┐
│  Economic geography      │   │  Cultural geography      │
│  ──────────────────      │   │  ──────────────────      │
│         TRADE            │   │        ISLAMIC           │
└──────────────────────────┘   └──────────────────────────┘
            ↑                               ↑
          THE TRADITIONAL
          MIDDLE EASTERN CITY
                    ↓
        ┌──────────────────────────┐
        │   Physical geography     │
        │   ──────────────────     │
        │          DRY             │
        └──────────────────────────┘
                    ↓
        ┌──────────────────────────┐
        │  Expressed on the        │
        │  landscape:              │
        │  Wells and fountains     │
        │  Small, open courtyards  │
        │  High, thick walls       │
        │  Wind towers             │
        └──────────────────────────┘
```

FIGURE 8.9   The traditional Middle Eastern city.

built around courtyards that provide an interior and private focus for domestic life.

The rights of others are also given strong emphasis. The Qur'an specifies an obligation to neighborly consideration and cooperation that traditionally is interpreted as applying to a minimum radius of 40 houses. In traditional designs parapets surround the roofs to prevent views of the homes of neighbors and drainage channels are steered away from adjacent houses. Because most Islamic cities are in hot, dry climates, these basic principles of urban design have evolved in conjunction with certain practical responses to intense sunlight and heat. Narrow winding streets help to maximize shade, as do latticework on windows and the courtyard design of residential areas. In the Persian/Arabian Gulf and Iran, local architectural styles include air ducts and roof funnels with adjustable shutters that allow dust-free drafts to penetrate the houses. The overall effect is a compact, cellular urban structure within which it is possible to maintain a high degree of privacy.

The traditional Islamic city grew up at desert oasis sites, and prospered from trade connections. The model of the internal structure of the Islamic metropolis in Figure 8.10 reflects several important features:[14]

- At the heart of the city is the walled *Kasbah*, or citadel (fortress). Its often ornate gates open onto the palace buildings, baths, barracks, small mosque, and shops.
- Surrounding the Kasbah is the *medina*, or old city.
- The medina usually had a wall with several watchtowers for defense. City gates controlled access, allowing strangers to be scrutinized and merchants to be taxed.
- Until the eleventh century, the absence of corporate bodies in a society composed of rulers and subjects meant that there was no need for public buildings and communal meeting places.[15] After that time mosques began to proliferate, and the *Jami*—the city's principal mosque—became the

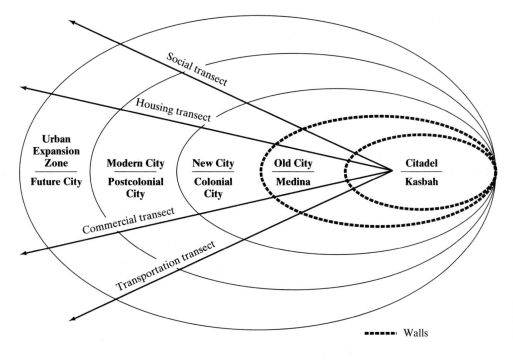

FIGURE 8.10   Internal structure of the Middle Eastern metropolis.

------ Walls

dominant feature of the traditional Islamic city. Located centrally, the mosque complex is not only a center of worship, but is also a hub for education and a broad range of welfare functions. As cities grew, new, smaller mosques were built toward the periphery, each out of earshot of the call to prayer from the Jami and one another.

- The covered bazaars or street markets (*suqs*) nearest the Jami typically specialize in the cleanest and most prestigious goods, such as books, perfumes, prayer mats, and modern consumer goods. Those nearer the city gates specialize in bulkier and less valuable goods such as basic foodstuffs, building materials, textiles, leather goods, and pots and pans. Within the suqs, each profession and line of business has its own alley.

- The residential population in the old city tended to be concentrated in distinctive quarters or *ahya'*, many of which had gates of their own. There were different quarters for workers with similar occupations, for Jews, Europeans, Christian sects, and ethnic groups, and for people of different village, tribal, or regional origin. Both rich and poor lived together in each quarter.

- In the colonial era a new city developed outside the old city walls (Figure 8.10). The traditional elements of urban form—mosques, public baths, and small shops—were incorporated but augmented with the modern amenities and architectural styles of the colonizers—new government buildings, wide boulevards, hotels, department stores, and corporate offices.

- Like cities everywhere, contemporary Islamic cities bear the imprint of globalization despite the fact that Islamic culture is self-consciously resistant to many aspects of Western-based culture. The new city that developed outside the old city walls during the colonial era became surrounded by the modern, postcolonial city (Figure 8.10); as this happened, courtyard homes all but disappeared, their place taken by apartment blocks, with high-income and single-family units increasingly common. The postcolonial city became the zone of international hotels, department chain stores, modern skyscrapers and office blocks, modern universities, factories, and highways. The postcolonial city has also become the location for squatter settlements that are now home to recent rural immigrants.

- An urban expansion zone in which small villages are undergoing urbanization has grown up beyond the modern city (Figure 8.10). New industrial estates, an international airport, modest new housing tracts, or, in the case of Egypt and Saudi Arabia, new cities, are being added. In the richest countries experiencing growing automobile ownership, this can also be a zone of uncontrolled urban sprawl.

The contemporary Islamic city has become an interlocking set of zones—the citadel, old city, new city, modern city, and urban expansion zone—that show changes in urban form and function that are patterned in time. Zeigler[16] describes a series of different transects running out for the center of the city through each of these zones.

1. *Social Transect:* The old city has become socially marginalized as wealthy residents have moved out. The modern city is attracting most of the new housing investment and the best urban services. Even foreign tourists stay in hotels in the modern city and make only brief visits to the old city in air-conditioned buses.

2. *Housing Transect:* The old city usually comprises traditional two-story courtyard houses. The modern city contains multistory apartment blocks—now the most characteristic feature of the residential landscape of the contemporary Islamic city.

3. *Commercial Transect:* The old city still contains fully functional suqs, traditional industries, and small-scale, family-owned craft businesses. The modern city offers department stores as well as international and national franchise businesses.

4. *Transportation Transect:* The old city contains narrow congested streets of pedestrians, donkey carts, and taxis. The new city reflects the proliferation of private automobiles and associated amenities, such as gas stations and parking spaces.

A feature of recent urbanization has been the entirely new group of Islamic cities that has grown up around the modern oil and gas fields. These are best exemplified by the cities of the United Arab Emirates, Kuwait, Saudi Arabia, Bahrain, and Qatar (see the Box entitled "Contemporary Islamic Cities and the Imprints of Globalization").

Another characteristic of recent urbanization is the development of increasingly integrated urban regions containing a number of Islamic cities. These **megalopolises** include one that stretches along a new superhighway across the desert from Cairo to Alexandria and another, anchored by Istanbul and Bursa, that forms Turkey's Marmara megalopolis.

BOX 8.2   CONTEMPORARY ISLAMIC CITIES AND THE IMPRINTS OF GLOBALIZATION

An entirely new group of cities has grown up since the 1960s around the modern oil and gas fields of countries like the United Arab Emirates, Kuwait, Saudi Arabia, Bahrain, and Qatar.[17] Petrodollars financed this city building.

These cities often comprise two zones:

- A tiny core anchored by a fort and perhaps what remains of a traditional wharf area or high-income neighborhood. Despite its small size, the layout and architectural style of the core provides a source of local identity.

- Surrounding the core is a modern city of high-rise office buildings, shopping malls, golf courses, apartment complexes, sprawling sub-urbs, and mosques. Oil wealth has paid for the world's most creative architects to design cities that blend modern buildings with traditional themes. One surprising and very visible element of these new urban landscapes is how green they are. Figuratively, the oil is turned into water and the water into green space.[18]

The oil that fueled this urban growth has also ignited rural-to-urban migration. As the rural population has become urbanized, nomadism as a way of life has been virtually eliminated. The economic magnetism of cities like Dubai, Kuwait, and Abu Dhabi has attracted unskilled workers from as far away as India, Pakistan, Sri Lanka, and the Philippines and skilled workers from Europe, the United States, and other parts of the Arab world.

In contrast to the industrialization experience of many cities in Europe and North America, industrial production in these oil cities is limited to local craft industries and manufacturing-in-transit at free ports such as Dubai's Jebel Ali Free Zone and Airport Free Zone. The port cities of the Persian/Arabian Gulf have become **entrepôts** supplying Central Asia with cars, electronics, and other high-end goods. Amid concerns about overspecialization in petroleum-related activities, some cities have begun to establish a diversified economic base that will allow them to survive in the world economy when the oil reserves are gone. For example, as banking and trading centers, cities like Dubai, Abu Dhabi, and Manama have become increasingly important transactional nodes in the economic systems reshaping the region.

SOUTH ASIAN CITIES

Two main forces—colonial and traditional—have influenced the urban form of South Asian cities. Dutt captures these two forces in his colonial-based and bazaar-based models.[19]

The *colonial-based* city model incorporates land use patterns found in colonial cities in other parts of the world but also contains certain regional characteristics that reflect the colonial functions specific to the Indian subcontinent (Figure 8.11). The development of colonial cities in South Asia typically involved the following:

- The port facility often became the nucleus of the colonial city. The need for trade and military reinforcements required a waterfront location accessible to oceangoing ships.

- A walled fort beside the port functioned as a military outpost. If the port contained factories that processed agricultural raw materials for export, this formed the initial basis for colonial trade.

- Beyond the fort, the native town—overcrowded, unplanned, inadequately serviced—housed the local workers who serviced the fort and the colonial administration.

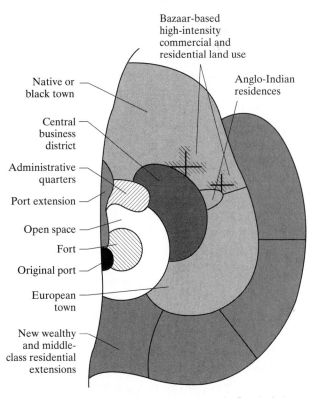

FIGURE 8.11   The colonial-based city in South Asia.

- A Western-style CBD adjacent to the fort and native town contained the government, administrative, commercial, retail, and public buildings.
- The European town grew in a different area or direction from the native town. It was low-density with spacious bungalows, elegant apartment houses, planned boulevards, and good urban services and recreational facilities.
- Between the fort and European town, an extensive open space (*maidan*) was reserved for military parades and European recreational activities, such as cricket and horse racing.
- Between the native and European town, Anglo-Indian colonies contained the Christian offspring of European-Indian marriages who were never fully accepted by either community.
- Since independence the colonial city has been extended into reclaimed lowland or undeveloped areas to create new living space, particularly for the native elite.

Bazaar-based cities were part of a thriving **urban system** in South Asia before colonialism, and they contain features that predate the colonial period (Figure 8.12). Their urban form today reflects their roles as early centers of trade and commerce, administration, or religious pilgrimage. Typically:

- There is a retail business concentration at the main crossroads. Around this intersection—known as a *chowk* in northern India—are the houses of the wealthy and the merchants who live above or behind their shops and warehouses.
- The bazaar contains a variety of land uses, although commercial activities dominate. With the bulk of family income spent on basic necessities—food, clothing, and shelter—most streets contain retail outlets for foodstuffs and clothes; sidewalk vendors are everywhere. As the bazaar evolved, functional separation of retail and wholesale businesses occurred. There are specific areas for retail traders—textile shops and tailors, grain and bread shops, jewelry stores and pawnshops, and vendors of perishable goods (vegetables, meat, and fish) that people buy daily because most do not own a refrigerator. Public or nonprofit inns provide inexpensive accommodation in smaller towns; hotel accommodation in the larger towns and cities reflects Western influence.
- Surrounding this inner core is a zone of wealthy residences, with separate housing for domestic servants, cleaners, shop assistants, and porters.
- The homes of the poor comprise a third zone in which limited demand for land keeps rents low.
- Beyond this third zone, the native elite and middle-income groups settled in what had been the

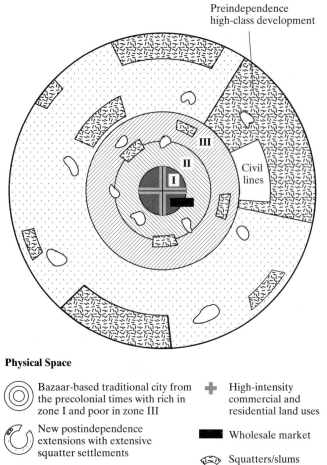

Physical Space

⊚ Bazaar-based traditional city from the precolonial times with rich in zone I and poor in zone III

◔ New postindependence extensions with extensive squatter settlements

✚ Chowk, or crossroads

✚ High-intensity commercial and residential land uses

▬ Wholesale market

▱ Squatters/slums

Cultural Space

◯ Religious and linguistic clusters and Untouchables

FIGURE 8.12   The bazaar-based city in South Asia.

"civil lines" in the colonial period—zones containing civil functions, such as the courthouse, police headquarters, jail, hospital, and public library.

- Ethnic, religious, linguistic, and caste neighborhoods formed in specific areas depending on the timing of settlement and the availability of land; the "**Untouchables**" always live at the periphery of the city, the site of most squatter settlements.
- As in the colonial cities, many traditional bazaar-based cities were extended as the city grew during the twentieth century. Governmental, semigovernmental, cooperative, and private agencies built planned high-income developments, especially for the native elite.

The *planned city* represents a third category of South Asian urban form. Ancient cities in the Indian subcontinent that were planned, such as Mohenjo

Daro and Harappa, did not survive (see Chapter 2). Other planned cities from the precolonial, colonial, and independence periods formed the nuclei for later development. An example of a precolonial planned city is Jaipur, which was founded in 1727 by the Maharajah of Jaipur (Figure 8.13). A city that was planned in the British colonial period is Jamshedpur, which was designed as a company town around India's first steel mill (Figure 8.14).

As they grew, the colonial-based, traditional bazaar-based, and planned cities developed hybrid elements. Colonial functional demands affected the traditional bazaar city layout; traditional elements were added to the colonial cities. Many bazaar and colonial cities have new planned extensions at their peripheries. The original cores of the planned cities were modified over time and surrounded by later unplanned traditional developments and semiplanned modern extensions. The traditional bazaar often thrived alongside the modern CBD, even in former colonial centers like Kolkata (Calcutta) and Mumbai (Bombay) that have since become megacities. In South Asia, megacities such as these have grown so far beyond their city and metropolitan boundaries that they now form extended metropolitan regions.

FIGURE 8.14   Jamshedpur, India, an example of a city that was planned in the British colonial period, which was designed as a company town around India's first steel mill. This photo of the Tata Iron and Steel Co. works was taken in 1952.

Although many cities in South Asia are only marginally connected to the world economy, some are beginning to tap into the service employment opportunities associated with foreign direct investment. In India, cities like Bangalore and Hyderbad have focused on providing educated English speakers and good telecommunications infrastructures to attract **transnational corporations**. As a result, India has become a global center for software development and back-office processing (accounting, medical transcription, payroll management, maintaining legal databases, and processing insurance claims and credit card applications). In a world economy, India has been able to also attract call center business (telemarketing and help desk support) by offering these services for 30 to 40 percent less than countries that have traditionally provided these services (such as Ireland, Canada, Australia, and even Hong Kong and the Philippines). And although the first companies to set up back-office processing or call centers in India were from Europe and the United States (e.g., Swissair, British Airways, General Electric, and American Express), Indian businesses now also offer outsourcing opportunities to transnational corporations (Figure 8.15).

FIGURE 8.13   Jaipur, India, an example of a precolonial planned city, which was founded in 1727 by the Maharajah of Jaipur. This is the old section of the city, known as the Pink City, seen from the Jantar Mantar, an astronomical observatory also built by the Maharajah. The City Palace is at center.

SOUTHEAST ASIAN CITIES

The earliest cities in Southeast Asia originated as a result of the diffusion of Indian and Chinese influences into the region beginning in the first century A.D.[20] Two main types of cities developed: the sacred city and the trading city.

Sacred cities were inland agrarian capitals that were centers of spiritual authority. They contained great

FIGURE 8.15   Indian employees at a call center in Bangalore, India provide service support to international customers.

temple complexes—such as Angkor Wat in Angkor, Cambodia—for the spiritual rulers. These sacred cities were located and designed according to cosmological principles and were frequently relocated on the advice of court astrologers. The fortunes of these cities rose and fell depending on agricultural productivity and their rulers' military successes.

In contrast, *trading cities* were bustling coastal or river ports that were part of a complex international trading network. The largest contained between 50,000 and 100,000 people. The ethnic elite lived inside the walls of these cities while the native and foreign traders lived outside.

The arrival of Europeans resulted in the establishment of colonial cities and systems of cities in Southeast Asia. The early centuries of European contact produced an embryonic colonial urban network of existing and new **gateway cities** to facilitate control of Asian trade.[21] In the nineteenth century the need for markets and raw materials for the rapidly industrializing countries of Western Europe prompted a shift in colonial emphasis toward Southeast Asia. This fueled the growth of a larger network of colonial urban settlements that included port cities, administrative centers, and mining and market towns, linked together and to areas of production by an extensive transportation system.

The colonial port cities were oriented primarily toward Europe. Virtually all were located on sites of existing settlements—Saigon (Ho Chi Minh City, Vietnam) was built near the Chinese village of Cholon, Singapore was built in the vicinity of a Malay fishing village, and Batavia (Jakarta, Indonesia) was located at the site of an existing trading center.

Colonial cities were typically laid out using a gridiron street plan. The commercial and administrative functions were concentrated at the urban core. The

exclusive residential districts for the Europeans were located outside the central "urban villages" (*kampungs*) that contained different ethnic groups. In some cities elements of the precolonial urban form—religious, military, or court functions at the core and the markets and merchants at the periphery—have survived, as in the old walled fort in Manila and the Shwe Dagon Pagoda and Buddhist monasteries in Yangon (Rangoon).[22]

Aside from their considerable size and growth rates, a major characteristic of the larger colonial port cities was the tremendous mixing of economic activities and land uses. (Figure 8.16):[23]

- The port zone—the center of economic activity in the colonial era—has retained its importance in the postcolonial period.
- Instead of a single CBD, there are separate civic and commercial/retail concentrations:
  - A government zone.
  - A Western commercial zone with foreign-owned banks, office towers, department stores, and modern hotels.
  - One or more high-density "alien" (often ethnic Chinese or Indian) commercial zones, containing small businesses—jewelers, tailors, chemists, and so on—in two-story shops and warehouses where the goods are manufactured and sold and where the business owner lives.

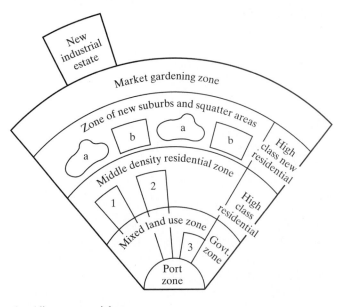

1 = Alien commercial zone
2 = Alien commercial zone
3 = Western commercial zone
a = Squatter areas
b = Suburbs

FIGURE 8.16   The Southeast Asian city.

- In addition to the colonial-era elite residential neighborhoods near the government zone, new high-income suburbs have been built to accommodate recent growth.

- Similarly, in addition to the middle-density neighborhoods (*kampungs*) of different ethnic groups closer to the urban core, there are now new middle-income suburbs.

- Squatter settlements are found in zones of disamenity throughout the city (e.g., along polluted rivers) and at the edge of the built-up area.

- A peripheral zone of intensive market gardening supplies fresh produce to the city's markets.

- New industrial estates have also been built at the urban periphery to attract transnational corporations and promote job growth.

This diversity remains evident in the urban form of larger postcolonial port cities in Southeast Asia, as in the example of Indonesian port cities (Figure 8.17).[24] Their principal elements are as follows:

- *Port-colonial city zone:* The original port and colonial features remain major components of urban form. Many port-related activities persist even when new, larger facilities have been constructed elsewhere. In some cities the area that was the Dutch colonial city is no longer functionally important and is now a historic preservation district.

- *Chinese commercial zone:* Much of the city's business is conducted in the Chinese commercial zone that occupies parts of the colonial city and the mixed commercial zone. This area contains

traditional high-density two-story shops over which the owners live and new shopping plazas with discount appliance stores.

- *Mixed commercial zone:* Leading from the waterfront, the Chinese commercial zone gradually merges into a spine of mixed commercial development that is ethnically, functionally, and architecturally diverse. This zone is likely to be the actual economic heart of the city. It contains a variety of commercial activities, from open traditional markets with Javanese rice sellers and Chinese jewelers to modern shopping malls with Pizza Huts and McDonalds.

- *International commercial zone:* Although Western in origin, this zone is best described as international because Japanese investment has financed much of it. This zone runs along the main monumental boulevard and contains skyscraper offices, upscale shopping malls, international hotels, and convention centers.

- *Government zone:* The government zone is relatively distant from the colonial city and contains preindependence and postindependence areas with high-rise offices and other buildings set in spacious gardens and parks.

- *Elite residential zone:* The elite residential neighborhoods developed initially near the government zone. More recently, high-income suburbs and gated communities have been built in areas along modern highways that offer urban services and luxury shopping.

- *Middle-income suburbs:* Although the majority of the growing Indonesian middle-income group live in gradually upgrading inner-city *kampungs*, demand for housing has exceeded supply, even in the new planned suburbs that have been built near the ring roads, suburban industrial parks, shopping centers, and universities.

- *Industrial zones:* New suburban industrial parks, port facilities, and satellite cities have been built since the 1970s as a result of government efforts to diversify the economy by attracting foreign direct investment in traditional industries such as textiles and food processing and in new industries such as automobile assembly and electronics.

- *Kampungs:* These are unplanned low-income and middle-income "urban villages." Inner-city kampungs are typically old, high-density areas with serious environmental problems such as flooding. Mid-city kampungs are less densely populated, better serviced, and in better condition, having benefited from government-supported kampung improvement schemes. The rural kampungs, some of which are still agricultural in character, are both peripheral and relatively

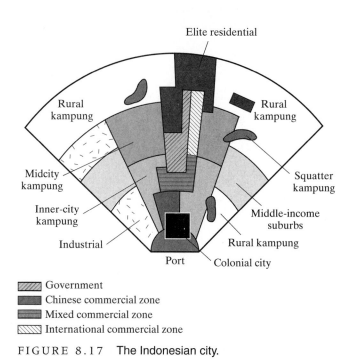

Government

Chinese commercial zone

Mixed commercial zone

International commercial zone

FIGURE 8.17   The Indonesian city.

self-contained. Temporary squatter kampungs are located in areas of disamenity (such as along polluted canals or railroad tracks) throughout the city except in the elite zones.

As in other parts of Asia and in Latin America, a distinguishing feature of recent urbanization in Southeast Asia has been the extension of megacities beyond their city and metropolitan boundaries to form extended metropolitan regions. Megacities like Jakarta, Bangkok, and Manila now lie at the epicenter of extended metropolitan regions.

> Extended metropolitan development tends to produce an amorphous and amoebic-like spatial form, with no set boundaries or geographic extent and long regional peripheries, their radii sometimes stretching 75 to 100 km from the urban core. The entire territory—comprising the central city, the developments within the transportation corridors, the satellite towns and other projects in the peri-urban fringe, and the outer zones—is emerging as a single, economically integrated 'mega-urban region,' or 'extended metropolitan region.' Within this territory are a large number of individual jurisdictions, both urban and rural, each with its own administrative machinery, laws, and regulations. No single authority is responsible for overall planning or management.[25]

The extended metropolitan region of Jakarta is called Jabotabek, which gets its name by combining the first letters of each jurisdiction that comprises the larger region (Jakarta, Bogor, Tangerang, and Bekasi) (Figure 8.18). Jabotabek contains more than 20 million people and thousands of factories that together represent 80 percent of Indonesia's industrial employment.

Surrounding these megacities are "peri-urban" suburban fringes that are within commuting distance of the urban core. Further out are what McGee calls *desakotas*—whose name comes from combining the Bahasa Indonesian words for village (*desa*) and town (*kota*)—to capture the intensive mix of agricultural and nonagricultural land uses in areas at the urban periphery and along the transportation corridors connecting the major cities (Figure 8.19).[26] Six features characterize the *desakota* areas:

1. A large population engaged in smallholder rice cultivation
2. An increasing amount of nonagricultural activities in areas that were previously largely agricultural
3. Intense movements of people and goods using relatively cheap transportation, such as two-stroke motorbikes, buses, and trucks
4. Intense mixing of land uses, including agriculture, cottage industries, industrial estates, and suburban residential developments
5. Increasing participation of females in nonagricultural labor
6. "Invisible" or "gray" areas of informal sector activities where government regulations may not apply or are difficult to enforce

The evolution of high-density rural areas of intensive agriculture close to the major urban centers establishes the preconditions for the development of extended metropolitan regions in Asia.[27] These preconditions include:

1. Sizeable pools of labor that are available for relocating industries
2. Significant improvements in transportation and communications that have increased the accessibility of regions close to the urban core and in

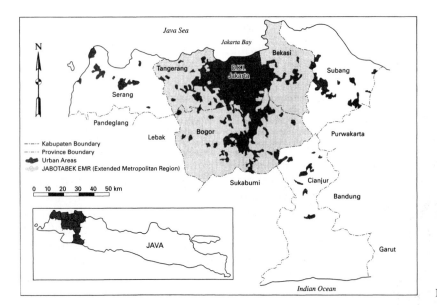

FIGURE 8.18   Jabotabek, Indonesia.

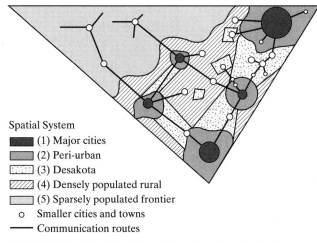

Spatial System
- ■ (1) Major cities
- ▨ (2) Peri-urban
- ▦ (3) Desakota
- ▧ (4) Densely populated rural
- ▤ (5) Sparsely populated frontier
- ○ Smaller cities and towns
- —— Communication routes

FIGURE 8.19 Spatial configuration of extended metropolitan regions in Asia.

some cases led to corridor development between two major urban centers

3. Areas within the extended metropolitan regions that are attractive to decentralizing activities, such as labor-intensive industry and housing

Although this kind of urbanization appears to repeat the experience of the developed countries during the nineteenth and early twentieth centuries, there are some important differences:[28] The distinction between rural and urban activities is breaking down faster in Asia than it did in the developed countries; considerable advances in technology—facilitating the flow of goods, people, and capital—have allowed extended metropolitan regions to develop; and rapid industrialization and economic growth have focused primarily on the extended metropolitan regions and particularly on the peripheral parts of these regions.

EAST ASIAN CITIES

Since World War II, the distinctive differences that have emerged in the internal structure of East Asian cities reflect the influence of the socialist tradition in communist China, North Korea, and Mongolia versus the capitalist system in South Korea, Taiwan, Hong Kong, and Macao.

The internal structure of the free market cities reflects the high levels of industrialization and urbanization that come with being tied into the world economy. Their urban form and land uses reflect the operation of the capitalist economic system, including private ownership of property, the importance of private investment decisions, relatively high standards of living, increasing reliance on the private automobile, and at times significant stratification by socioeconomic background.[29]

In contrast, the planning of socialist cities in China and Mongolia between the late 1940s and the late 1970s, and still today in North Korea, reflects the communist ideal of a centralized and highly standardized form of social organization that integrated a classless society with state economic planning based on national and local self-reliance.[30] Changes to the internal structure of cities associated with their transformation to a socialist model after 1949 involved:

1. Replacing the retail concentration at the core with political–cultural–administrative functions
2. Standardizing housing
3. Adopting the *self-contained neighborhood unit* concept

Maoist policies created a peculiar mix of old and new in the urban form and land use of Chinese cities. Dutt et al.'s model of a traditional Chinese city with post-1949 socialist developments captures this mix (Figure 8.20).[31] The surviving elements of the traditional Chinese city reflect that the original design was based on cosmological principles of urban planning. This dictated a normally square city shape with a surrounding series of walls and moats, three gates in each city wall, and three streets within the walls that ran east–west and north–south. Government buildings and the palace of the imperial household were very important in the traditional Chinese city and were located at the core. Religious and commercial land uses were secondary. Residential land uses were differentiated based on social status or occupation.

Post-1949 changes include the addition of a number of long, wide boulevards running in an east–west direction through the city that established the basis for the city street system. Streets were named for revolutionary heroes and events. The streets cut the city into grids of varying sizes, many of which comprise self-contained neighborhood units—large rectangular parcels of housing and other uses surrounded by enclosing walls. The buildings are arranged in rows and columns and divided into subcompounds based on their residential, office, service, or other functions. Most buildings share the same uniform, box-like character. Near the city center there is a huge square for mass gatherings that contains a monument such as a revolutionary hero's sculpture. Government compounds line both sides of the major boulevards toward the center of the city, along with party committee compounds, revolutionary history exhibition halls, commercial centers, and entertainment complexes. Large factories are located at the periphery, and the universities are concentrated in one area of the urban outskirts (Figure 8.20).

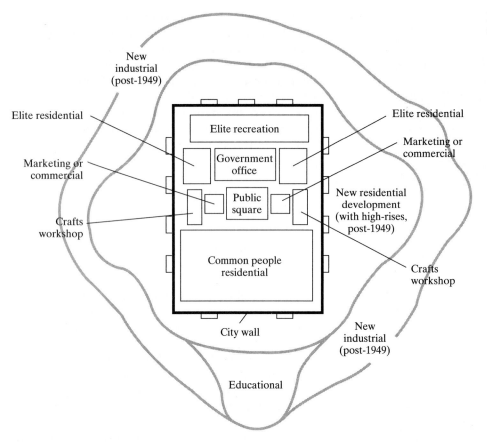

FIGURE 8.20   Traditional Chinese city model with post-1949 socialist developments.

This transformation of traditional Chinese cities into socialist manufacturing and administrative centers is also captured in Lo's model, which, however, pays less attention to the details of the old city and more to the surrounding areas (Figure 8.21).[32] The inner self-contained neighborhood units are shown divided into multiple polygons to reflect a sense of irregularity in this older part of the city. As in the Dutt et al. model, new factory workers' housing estates comprise neighborhoods that form a distinct outer ring that also contains large-scale factories. Lo additionally shows a zone of vegetable cultivation toward the periphery, that itself is surrounded by a zone of food grains and industrial crops. These agricultural zones reflect the socialist ideals of self-reliance and the integration of rural and urban areas. Large cities, including Beijing, expanded in a series of concentric zones as shown in this model.[33]

In general in East Asia, colonial penetration was more limited than in other parts of Asia. Even so, European and Japanese contact had a strong influence on Chinese port cities like Shanghai and Tianjin. The

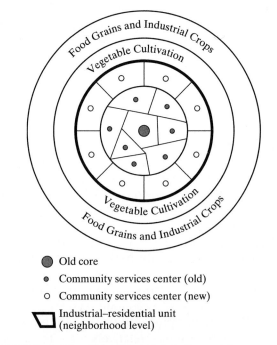

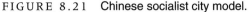

FIGURE 8.21   Chinese socialist city model.

Treaty of Nanjing in 1842 ceded the island of Hong Kong to Britain and gave the British the right to reside in a small number of Chinese port cities. Subsequent refinements to this treaty—the so-called *Open Door policy*—gave other foreign powers the same rights as the British. By 1911, after more treaties, about 90 cities were open to foreigners—along the coast, up the Chang Jiang (Yangtze) valley, in northern China, and in Manchuria.[34]

After being opened up to foreigners, these colonial concession-based cities became the largest industrial, commercial, and transportation centers at that time in East Asia.[35] They contained three sections:

- *A concession zone:* Docks and military bases were established initially along coasts or rivers. Factories and warehouses were built later. With increased trade, commercial zones and elite Western residential districts were added.

- *A Chinese residential zone:* The majority of the Chinese population lived in the old city that had been designed using traditional Chinese principles of urban planning.

- *A buffer zone:* This area became a highly desirable location for the Chinese elite—senior Chinese officials, business people, and high-ranking employees of foreign companies.

After the defeat of Japan in 1945, the colonial era ended in East Asia except in (British) Hong Kong and (Portuguese) Macao. After that time the internal structure of the colonial concession-based

cities experienced significant changes as they were transformed to reflect socialist planning principles (Figure 8.22). The Western residential districts became offices or houses for the party bosses, new port facilities were added as part of industrial expansion, infill neighborhood units were built on available sites in the area surrounding the old city, and high-rise housing developments were constructed at the periphery.

In the reform (post-Mao) era since the late 1970s, free market forces and global capital investment have been allowed increasingly to dictate the course of urban development.[36] As has occurred in cities like Seoul in South Korea and Taipei in Taiwan, the internal structure of the larger cities in China has been modified to include:

- Ring roads to channel some of the increased traffic around cities.

- Satellite towns near larger cities to absorb new population growth.

- Renovation of older central city commercial and residential districts.

- Preservation districts in older parts of cities that provide a visual link to their presocialist past.

- An urban open space system to provide recreational opportunities for residents.

- Shiny new downtowns with modern high-rise offices and luxury hotels. Competition to have the world's tallest building has increased in Asia since the Petronas Towers in Malaysia's Kuala

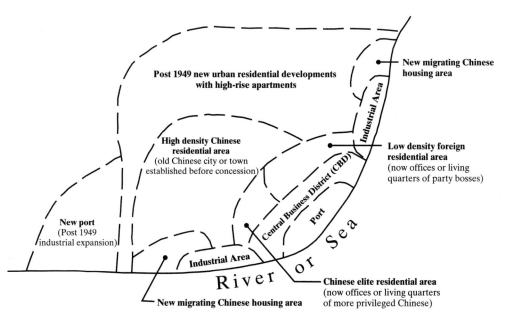

FIGURE 8.22 Colonial concession-based Chinese city model.

Lumpur took the title from the Sears Tower in Chicago. The Petronas Towers then lost the title in 2003 on completion of "Taipei 101." This 101-story building in Taiwan was officially opened in 2004 (Figure 8.23). It has offices, a shopping mall, an observation deck on the 91st floor, and the world's fastest elevators. "Taipei 101" is not expected to keep its title for long, though. Developers in China say that the Shanghai World Financial Center—scheduled for completion in 2007—will be taller.

The imprints of globalization such as these are most visible in the megacities and extended metropolitan regions of East Asia. Megacities like Shanghai and Beijing in China and Seoul in South Korea have experienced enormous transformations in their economies, societies, and internal structure as their governments and businesses have made concerted efforts to take advantage of their increasing integration into the world economy (see the Box entitled "Shanghai and the Pudong New Area—The 'Dragon Head' of China's Economy?"). As in other parts of Asia and in Latin America, the megacities in East Asia are also growing beyond their city and metropolitan boundaries to form extended metropolitan regions (see the Box entitled "Hong Kong's Extended Metropolitan Region: Foreign Direct Investment and Regional Development in the Pearl River Delta Region").

FIGURE 8.23   The Taipei 101 tower looms over the city of Taipei in Taiwan. This 101-storey building became the world's tallest office building when it opened in 2004. It has offices, a shopping mall, an observation deck on the 91st floor, and the world's fastest elevators.

BOX 8.3   SHANGHAI AND THE PUDONG NEW AREA—THE "DRAGON HEAD" OF CHINA'S ECONOMY?

The megacity of Shanghai, with about 15 million residents, is China's largest city. The entire extended metropolitan region of 17 city districts and 3 rural counties is estimated to contain 17 million people. It is located on the Huangpu River near the mouth of the Chang Jiang (Yangtze) River (Figure 8.24).

At the time of the Communist revolution of 1949, Shanghai had a population of well over a million, was one of the three largest manufacturing centers in the world, and was the busiest international port in Asia. Its growth was fueled by its role as the primary Asian entrepôt for trade in opium, silk, and tea. The British were the first to develop Shanghai in this way at the end of the Opium Wars in 1842. By 1847 the French had arrived, and by 1895 large sections of the city had been divided into autonomous international settlements that were not under Chinese law. The centerpiece of the city was the Bund, the riverfront development of

monumental neoclassical buildings containing the major international banks and trading houses.

Just as the foreigners had drained the wealth out of the city, now the Communist government tapped into its economic strength, but to redress regional imbalances in prosperity across China. Shanghai was heavily taxed—75 percent or more of its revenues went to the national government—with very little reinvested in the city. As a result, Shanghai lost its vibrancy and became rundown despite industrialization and growth under the Communists.

Dramatic changes came in the 1980s when the central government recognized the need for massive investments in Shanghai's infrastructure. Designated as one of 14 "open cities," the Shanghai Economic Zone and several Economic and Technological Development Zones were established as attractive locations for international investment. In addition, about 200 square miles of low-density farm, industrial, and

(Continued)

BOX 8.3   SHANGHAI AND THE PUDONG NEW AREA—THE "DRAGON HEAD" OF CHINA'S ECONOMY?   *(Continued)*

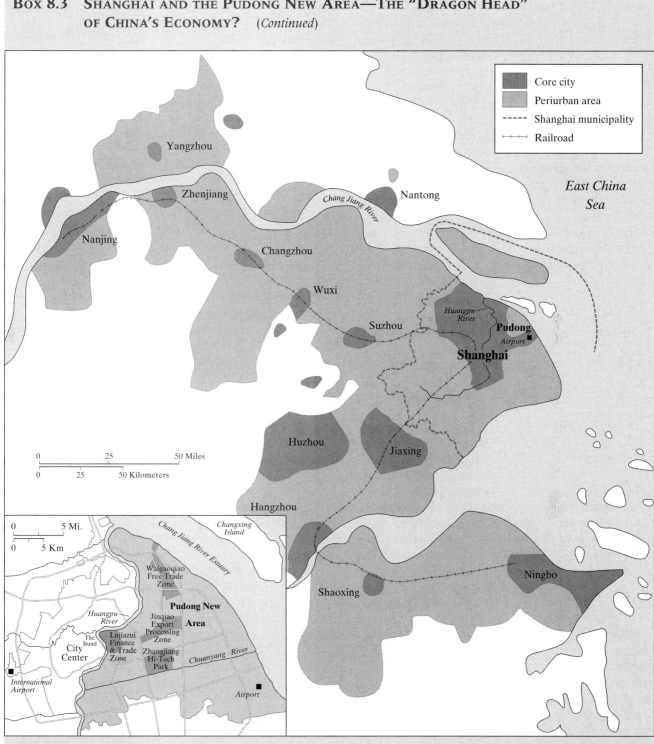

FIGURE 8.24   The Shanghai extended metropolitan region.

residential land to the east of the Huangpu River was designated as the Pudong New Area (literally, "east of the (Huang)pu (*Pudong Xinchu*)") (Figure 8.24). Shanghai was to be the "dragon head"—the economic engine that would stimulate development along the entire length of the Chang Jiang valley—the "dragon body."[37]

The official designation of Pudong as a national development project in 1990 was an acknowledgment of the strategic importance of Shanghai for China's

(Continued)

BOX 8.3   SHANGHAI AND THE PUDONG NEW AREA—THE "DRAGON HEAD" OF CHINA'S ECONOMY?   *(Continued)*

overall development. This designation allowed the city government to offer a comprehensive range of incentives, including tax breaks, to attract domestic and foreign companies and investment and to utilize central government funds for infrastructure and other expenditures.[38] Plans were drawn up for new bridges across the Huangpu River, inner and outer ring roads, subway lines, sewage works, and a second international airport. Pudong is now a multibillion dollar concentration of export-processing industries; office space in Lujiazui that serves as the city's new CBD and as a national financial center; scientific, research and educational facilities; and new residential communities.

Today, Shanghai is a city of extremes, with grinding poverty and vice alongside luxury apartment buildings and upscale department stores. Its phenomenal growth has not only reinforced is status as China's largest city but has also propelled it into the role of a **world city**, a key node in the flows of capital, goods, and information that underpin the global economy. Some observers are concerned that in order to sustain Shanghai's phenomenal growth, it is not enough to focus on the development of Pudong alone or to allow problems that have marred Pudong's development to persist.[39] These include:

- An excessive role for foreign investors and transnational corporations, raising concerns about minimal technology transfer and "demonstration effects" for local companies
- Rising wages that may deter some foreign companies from considering investing in Pudong
- Poor public transportation links for ordinary workers commuting to Pudong, despite huge public investments in infrastructure
- Inadequate coordination between the separate administrative authorities responsible for the Bund and the Lujiazui financial district, leading to wasteful competition within Shanghai for foreign financial institutions

BOX 8.4   HONG KONG'S EXTENDED METROPOLITAN REGION: FOREIGN DIRECT INVESTMENT AND REGIONAL DEVELOPMENT IN THE PEARL RIVER DELTA

The Pearl River Delta in Guangdong Province is one of the fastest growing urban regions in the world (Figure 8.25). Anchored by the major metropolitan centers of Hong Kong, Macao, and Guangzhou, the Pearl River Delta is an extended metropolitan region of about 50 million people. Since the liberal economic reforms initiated with the post-Mao "Opening and Reform" in 1978, the Chinese government has fostered this region as one of its "engines" of capitalist growth. With this reversal of its antiforeign investment stance, China has become the most important destination for foreign direct investment among the less developed countries, and the Pearl River Delta has become the country's most intensive region of foreign direct investment, economic development, and urbanization.

Within the Pearl River Delta, Hong Kong, a British colony until 1997, is a metropolis of 7.4 million people with a world-class manufacturing, trading, and financial base. It is the world's largest container port, the ninth-largest banking center, the seventh-largest center for foreign exchange trade, and the tenth-largest trading economy. In recognition of Hong Kong's role as a capitalist economic dynamo, the Chinese government set it up as a Special Administrative District with its own currency, legislature, and legal system, based on the principle of "one country, two systems."

Hong Kong's success prompted the Chinese government to create two of its first Special Economic Zones (SEZs) in nearby Shenzhen and Zhuhai (Figure 8.25). These were established as **export processing zones** that offered cheap labor, land, and tax breaks to attract transnational corporations and their investment, technology, and management practices. Many Hong Kong businesses shifted their labor-intensive manufacturing or contracted out their assembly-line work to Chinese subcontractors in this new "back factory." Hong Kong now acts as the "front shop," concentrating on marketing, design, purchasing raw materials, inventory control, management and technical supervision, and financial arrangements. Sit and Yang[40] describe this so-called "front shop, back factory" (*Qiandian Houchang*) model as a new cross-border joint-venture system that links the Pearl River Delta into the world economy (Figure 8.26). The entire delta region was subsequently designated an Open Economic Area, in which local governments, individual businesses, and farm households enjoy a high degree of independence in economic decision-making.

(Continued)

BOX 8.4 HONG KONG'S EXTENDED METROPOLITAN REGION: FOREIGN DIRECT INVESTMENT AND REGIONAL DEVELOPMENT IN THE PEARL RIVER DELTA *(Continued)*

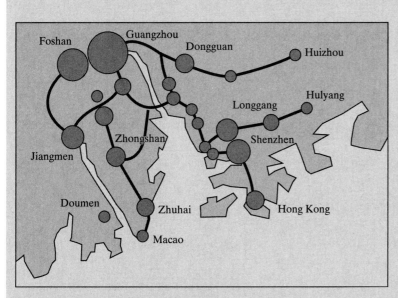

FIGURE 8.25 The Pearl River Delta extended metropolitan region.

The relaxation of central government controls allowed the region's growing rural population to migrate to urban areas in search of assembly-line jobs or to stay and diversify their rice-paddy cultivation into more profitable agricultural activities such as market gardening, livestock husbandry, or fisheries. Economic freedom also allowed rural industrialization, involving mostly low-tech, small-scale, labor-intensive industries.

The triangular area between Hong Kong, Macao, and Guangzhou has emerged as a particularly important economic zone due to its relatively cheap land and labor and the enormous investments by regional and local governments in the transportation and communications infrastructures to meet the needs of local and international companies. New infrastructure includes major airports, high-speed toll roads, satellite ground stations, port installations, light rail networks, and new water management systems. These, in turn, have attracted business and technology parks, financial centers, and resort complexes in a loose-knit sprawl of urban development. The result is a distinctive extended metropolitan region in which numerous small towns play an increasingly important role in fostering the process of urbanization, with an intense mixture of agricultural and nonagricultural activities and an intimate interaction between urban and rural areas.

As China's most "open" area for foreign investment, this new pattern of exo(genous)-urbanization in the Pearl River Delta is quite different from the endo(genous)-urbanization (or urbanization driven by intranational or regional forces) that took place during the pre-1978 period and that is still prevalent in other parts of China.[41] This foreign-investment-induced urbanization, or exo-urbanization, is characterized by:

- Intensive inflows of foreign direct investment (and from Hong Kong) that have driven the region's transformation into a globally integrated export-oriented industrial economy

- The predominance of labor-intensive assembly manufacturing that has led to large-scale rural-to-urban migration of largely unskilled females and to industrial and urban expansion at the expense of agriculture

- Export oriented, "trade-creating" foreign investment that has generated large-scale cross-border flows of people, goods, and information between the Pearl River Delta, Hong Kong, and Macao, and promoted a pattern of border urbanization

- Small-scale, processing-type manufacturing associated with the predominantly export-oriented (rather than domestic) investment from Hong Kong that has been biased towards locations outside major cities, generating rapid growth in smaller urban centers like Jiangmen, Zhongshan, Nanhai, Shunde, Dongguan, and Huzhou (Figure 8.25)

Although the spatial pattern of Hong Kong's extended metropolitan region appears similar to that in other parts of Asia, the exo-urbanization in the Pearl River Delta region involves some important

(Continued)

Box 8.4   Hong Kong's Extended Metropolitan Region: Foreign Direct Investment and Regional Development in The Pearl River Delta   (*Continued*)

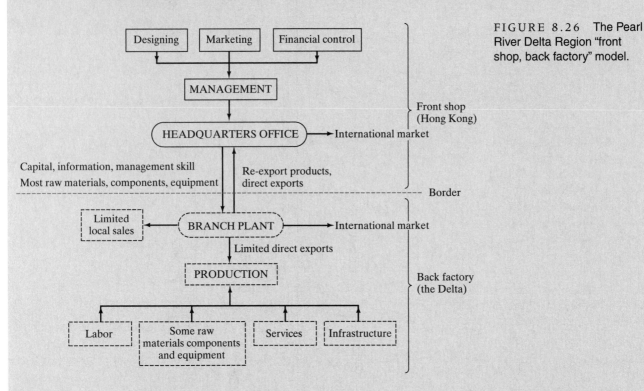

FIGURE 8.26   The Pearl River Delta Region "front shop, back factory" model.

differences. The "urbanization from below" process that has fueled the growth of the smaller urban centers and rural areas in the delta region has not been the result of decentralization of investment from the core city of Guangzhou. In fact, in contrast to the explosive growth of megacities like Jakarta and Bangkok, the regional primate city of Guangzhou has experienced relative decline. Spatially, the pattern of urbanization in the Pearl River Delta region may have more in common with the "polygonized development" that has been identified for Latin America. This has involved a relatively concentrated dispersal pattern that is locally concentrated in the smaller cities and towns surrounding megacities like São Paulo.[42]

Today the Pearl River Delta is a thriving export-processing platform. Although enjoying double-digit annual economic growth during the past two decades, concerns have been raised about its long-term sustainability. One issue is that the region's growth is dependent on foreign direct investment, something that can be extremely volatile in its source, volume, and destination within and between countries. Other concerns relate to the pressures of economic development and urbanization on the environment and agriculture in the delta region.

FOLLOW UP

1. Be sure that you understand the following key terms:

   apartheid
   Desakota
   exo-urbanization
   export processing zone or EPZ
   extended metropolitan region or EMR
   foreign direct investment or FDI
   "front shop, back factory" model
   growth pole

   informal sector of the economy
   Laws of the Indes
   maquiladora
   overurbanization
   rural-to-urban migration
   self-contained neighborhood units (in China)
   squatter settlements
   transnational corporation or TNC

2. If you have time, read a novel that is evocative of urban life in one of the less developed regions of

the world. There are hundreds to choose from, including bestsellers about Kolkata in India, such as *City of Joy* by Dominique LaPierre (New York: Warner Books, 1966), which became a Hollywood movie, or the popular classic, *Calcutta*, by Geoffrey Moorehouse (New York: Holt, 1983).

3. Try to get a copy of W. K. Crowley's article "Order and Disorder—A Model of Latin American Urban Land Use" in D. E. Turbeville III (ed.), 1995, *Yearbook of the Association of Pacific Coast Geographers*, 57, pp. 9–31, Corvallis: Oregon State University Press. This article might be available in your university library either electronically or in paper format or if not, through interlibrary loan. Compare Crowley's composite model (p. 201) to Ford's 1996 model that is described in this chapter (p. 199). What are some of the main differences, and what factors do you think might account for these differences?

4. Use your library or the World Wide Web to find maps that were produced at different points in time to chronicle the growth of Lagos in Nigeria as it grew to become the first sub-Saharan African megacity. Examine the extent and direction of growth and think about some of the most important social, economic, technological, and political processes that are driving this growth.

5. Find a map of Jakarta and consider how and why Ford's model of an Indonesian city does or does not fit this megacity.

6. Work on your *portfolio*. Make a diagrammatic and annotated entry that describes each of the models of urban structure in the less developed countries that were covered in this chapter. Make a list of the advantages and disadvantages of using models such as these to help our understanding of the internal structure of cities in the less developed countries.

KEY SOURCES AND SUGGESTED READING

Aryeetey-Attoh, S. 2003. Urban Geography of Sub-Saharan Africa. pp. 254–97 in *Geography of Sub-Saharan Africa*, ed. S. Aryeetey-Attoh, 2nd ed. Upper Saddle River, N.J.: Pearson Education.

Brunn, S. D., J. F. Williams, and D. J. Zeigler, eds. 2003. *Cities of the World: World Regional Urban Development*, 3rd ed. Lanham, Md.: Rowman & Littlefield.

Crowley, W. K. 1995. Order and Disorder—A Model of Latin American Urban Land Use. pp. 9–31 in *Yearbook of the Association of Pacific Coast Geographers*, 57, ed. D. E. Turbeville III. Corvallis, Ore.: Oregon State University Press.

Diniz, C. C. 1994. Polygonized Development in Brazil: Neither Decentralization nor Continued Polarization. *International Journal of Urban and Regional Research* 18: 293–314.

Dutt, A. K., Y. Xie, F. J. Costa, and Z. Yang. 1994. City Forms of China and India in Global Perspective, pp. 25–51 in *The Asian City: Processes of Development*, *Characteristics and Planning*, Dordrecht, The Netherlands: Kluwer Academic Publishers.

Ford, L. R. 1996. A New and Improved Model of Latin American City Structure. *The Geographical Review* 86: 437–40.

Ginsburg, N., B. Koppel, and T. G. McGee, eds. 1991. *The Extended Metropolis: Settlement Transition in Asia*. Honolulu, Hawaii: University of Hawaii Press.

Lemon, A., ed. 1991. *Homes Apart: South Africa's Segregated Cities*. London: Paul Chapman Publishing Ltd.

McGee, T. G. 1967. *The Southeast Asian City: A Social Geography of the Primate Cities of South East Asia*. New York: Frederick A. Praeger, Publishers.

Olds, C. 2001. *Globalization and Urban Change: Capital, Culture, and Pacific Rim Mega-Projects*. Oxford, UK: Oxford University Press.

Sit, V. F. S., and C. Yang. 1997. Foreign-Investment-Induced Exo-Urbanisation in the Pearl River Delta, China. *Urban Studies* 34: 647–77.

RELATED WEBSITES

Internet Resources on Cities in Latin America:
http://darkwing.uoregon.edu/~caguirre/urban.html
This website, by Professor Carlos Aguirre, Department of History, University of Oregon, offers a wealth of hyperlinks to useful sources of information on Latin American cities, including The Megacities Project and a Virtual Guide of Mexico City.

Urban China Research Network:
http://mumford.cas.albany.edu/chinanet/index.asp
Established in 1999, the Urban China Research Network, a joint venture between the Lewis Mumford Center and the Center for Social and Demographic Analysis (CSDA), is a virtual research center with the goal of strengthening contacts and collaboration among scholars throughout the world with interests in urbanization in contemporary China. The website contains research ranging from urban

morphology to the dynamics of rural-to-urban migration to issues concerning housing, crime, and changing family structure within China.

Islamic History:
http://instruct1.cit.cornell.edu/Courses/nes257/nes.html
This class website for an Islamic History course at Cornell University by Professor David Powers contains numerous hyperlinks to useful sites related to Islamic cities. Click, for example, on the hyperlink to an Index of Images from Articles on Islamic Cities (*http://instruct1.cit.cornell.edu/Courses/nes257/cities.html*) to access a wealth of maps, aerial photos, and sketches of Islamic cities.

Digital South Asia Library: *http://dsal.uchicago.edu/*
The Digital South Asia Library at the University of Chicago offers a rich collection of material on South Asian cities, including photographs, maps, and statistics.

9

URBAN PROBLEMS
AND RESPONSES
IN THE LESS DEVELOPED
COUNTRIES

Rapid urbanization in many less developed countries has created a host of problems that weaken the role of cities as engines of economic growth. Pervasive poverty, inadequate housing, lack of urban services, transportation problems, and environmental degradation all contribute to dreadful living conditions for many urban dwellers. Recent progress has been made in responding to these problems within the context of sustainable urban development. Government involvement is important because this is the point where urban sustainability as a concept overlaps with urban management as a practical process. Government efforts can also help to establish democratic institutions and a participatory planning process that facilitate the involvement of impoverished local communities—those disproportionately affected by the problems of urbanization.

CHAPTER PREVIEW

This chapter is not intended to provide an exhaustive inventory of urban problems and the responses to them in the less developed countries. Instead, the goal is to show how the processes involved in urbanization are inherently problematic and how space and place often play key roles. We focus on five major sets of problems—poverty, inadequate housing, lack of urban services, transportation problems, and environmental degradation.

The close relationship between globalization and urbanization means that the local effects of **splintering urbanism** are transforming traditional patterns of spatial organization. These local effects are associated with acute problems of social disorganization and environmental degradation in less developed countries. The problems are visible in the extensive areas of slum and squatter settlements in **megacities**, high rates of unemployment and underemployment, and an ever-growing informal sector of the economy in which people seek economic survival.

If present trends continue, an increasing number of the largest settlements in less developed countries are likely to face these problems of urbanization. Certainly, cities in both the **newly industrializing countries (NICs)** and the oil-rich countries face serious difficulties. Islamic culture and urban design principles, for example, have not always been able to cope with the pressures of contemporary rates of urbanization. But this chapter focuses on the poorest countries of Africa, Latin America, and Asia that must confront the most severe urban problems.

Many communities have been able to establish self-help networks and organizations—with and without the help of government agencies and nongovernmental organizations (NGOs)—that have formed the basis of community cohesion within the often overwhelmingly poor and crowded cities. There is also increasing recognition that responses to the problems of urbanization must be framed within the context of the concept of sustainable urban development; nevertheless, the intensity of urban problems swamps all efforts to respond.

Part of the difficulty arises from the "globalization paradox" faced by cities. Cities must act as collective units in order to respond to international competition and compete in the world economy, but cities also face growing internal social and economic fragmentation that hampers their capacity to build coalitions, mobilize resources, and develop good governance structures.

During the last decade four changes in urban governance have emerged that have attempted to address this "globalization paradox": decentralization and formal government reforms, participation of local communities in urban policy-making and implementation, multilevel governance and public–private partnerships, and process-oriented and area-based policies. As we will see, although some of these changes are not new, the rationale for introducing them is.

URBAN PROBLEMS

POVERTY

The number of people subsisting below the international poverty line of US$1 a day rose from 1.2 billion in 1987 to 1.5 billion in 2000 and is projected to rise to 1.9 billion by 2015.[1] Poverty is also increasingly becoming an urban phenomenon. In countries such as Bangladesh, El Salvador, Gambia, Guatemala, Haiti, and Honduras, more than 50 percent of the urban populations live below their respective national poverty lines.

In less developed countries, a combination of internal and external factors influences the economies of cities and shapes their labor markets. Within national economies, the urban and rural sectors interact through exchanges of people, goods, services, money, and information (Figure 9.1). Many cities are experiencing unprecedented rates of growth that are driven by rural **"push" factors**—overpopulation and the lack of employment opportunities—rather than urban **"pull" factors**—prospective jobs and a better quality of life. With high rates of **natural increase** and many more people moving to the cities than jobs available, the net result is widespread unemployment and underemployment. **Underemployment** occurs when people work less than full time even though they would prefer to work more. Underemployment is difficult to measure accurately, but estimates range from 30 to 50 percent of the employed workforce in the less developed countries.

The urban and rural sectors of the economy, separately and together, are also linked through trade and other exchanges into the world economy (Figure 9.1). Through the **new international division of labor**, towns and cities in different parts of the less developed world play a key, if often unequal, role in international economic flows that link provincial centers into their national urban network and, ultimately, into the hierarchy of **world cities** and the global economy.

Based on the assumptions underlying modernization theories (see Chapter 7), in the 1970s many

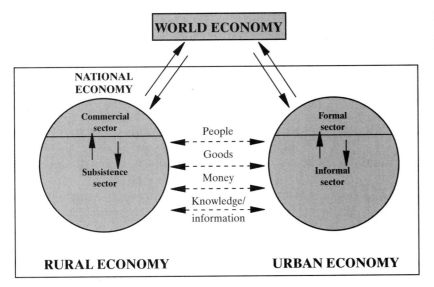

FIGURE 9.1   Spatial and sectoral interactions involving the urban economies in less developed countries.

governments adopted policies to promote urban economic development, including fiscal and other incentives to attract **foreign direct investment**. Public officials were attempting to take advantage of the new international division of labor—to attract some of the international capital that was gravitating toward the greater profit opportunities offered by lower-cost locations in the less developed countries.[2]

Cheap labor is certainly an important factor in the locational decision-making of **transnational corporations**. But the demand of the global market for quality products means that other factors—educated and trainable workers, a good-quality transportation network, and reliable services such as water and electricity—are also important. This meant that the poorest countries in sub-Saharan Africa and elsewhere in the less developed world were bypassed. The relatively few locations that could offer attractive conditions in the 1970s experienced rapid growth, most notably the four so-called Asian Tigers (Hong Kong, Singapore, South Korea, and Taiwan), as well as Mexico and Brazil. Since then, international investment has promoted growth in other countries, such as Malaysia and Thailand. With time, international investment has shifted to the larger urban centers, reinforcing existing investment patterns and aggravating already serious problems of urban **primacy** and **uneven development** in the less developed countries.

Dependence on international capital and global economic trends has introduced instability into the labor market and further fragmented both the urban and the rural sectors of the economy into large informal and smaller formal components (Figure 9.1). **Overurbanization** in the largest cities has led to a large informal sector because people who cannot find regularly paid work resort to earning a living in jobs that are not regulated by the state. The resulting pronounced *dualism*, or juxtaposition in geographic space of the formal and informal sectors of the economy, is quite evident in the built environment: Modern high-rise office and apartment complexes and luxury homes correspond with an often dynamic formal sector that offers well-paid jobs and opportunities; these contrast sharply with the slums and **squatter settlements** of people working in the informal sector who are disadvantaged by a lack of education, formal training, recognized qualifications, and often rigid **divisions of labor** shaped by gender, race, and ethnicity (Figure 9.2).[3]

But the informal sector of the economy is ambiguous—it encompasses wealth and poverty, productivity and efficiency, exploitation and liberation.[4] Four main kinds of **informal sector** activities attract different kinds of workers (Figure 9.3).

1. **Subsistence activities** in which goods and services, such as clothing and repairs, are primarily for household consumption rather than for profit

2. Small-scale producers and retailers, often self-employed, who produce or sell to earn an income—the typical workers in this category are street traders, artisans, and food vendors

3. Petty capitalists who see greater opportunities for profit in the informal sector where they can avoid taxes and legal employment regulations, such as minimum wages and safe working conditions

4. Criminal and socially undesirable activities, including drug dealing, smuggling, theft, extortion, and prostitution

FIGURE 9.2   A squatter settlement in Lagos, Nigeria.

In many cities more than one-third of the population is engaged in the informal sector; in some cities this figure is more than two-thirds (Figure 9.4). Across Africa, the International Labor Office has estimated that informal-sector work is growing 10 times faster than formal-sector employment. Although informal jobs—selling souvenirs, driving pedicabs, or dressmaking—might seem marginal from the point of view of the world economy, they support more than a billion people around the world.

In many less developed countries the labor force of the unregulated informal sector includes the world's most vulnerable workers—women and children. In environments of extreme poverty, every family member must contribute something. Industries in the formal sector often take advantage of this situation and the fact that labor standards are nearly impossible to enforce in the informal sector. Many companies farm out their production using subcontracting schemes that are based, not in factories, but in home settings that use child workers. Table 9.1 lists a combination of factors that creates an unfavorable position for women in both the informal and formal sectors of urban economies in the less developed countries.

Urban geographers recognize that the formal and informal sectors are interconnected—the informal sector represents an important resource for the formal sector of urban economies. The informal sector provides a huge range of cheap goods and services that reduce the cost of living for employees in the formal sector, allowing employers to keep wages low. Although this arrangement does not contribute to urban economic growth or help alleviate poverty, it does keep many companies competitive within the global economic system. For export-oriented businesses, in particular, the informal sector provides a considerable indirect production subsidy. And while this subsidy is often passed on to consumers in the developed countries in the form of lower prices, the poorest households in the cities of the least developed countries are forced to resort to increasingly drastic strategies for coping with worsening poverty (Table 9.2).

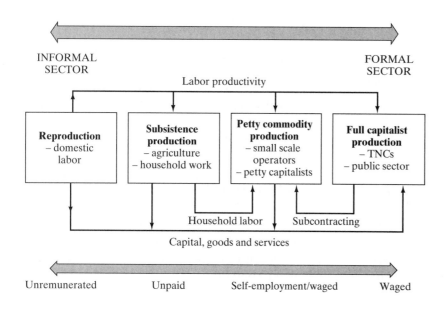

FIGURE 9.3   The informal–formal sector continuum.

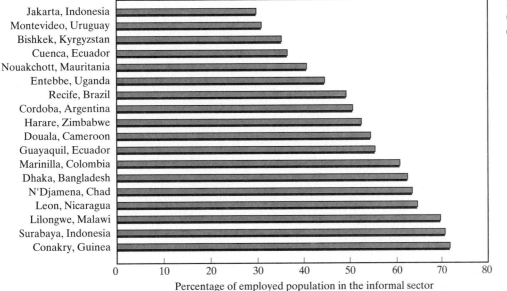

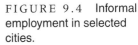

FIGURE 9.4   Informal employment in selected cities.

Percentage of employed population in the informal sector

TABLE 9.1

Factors Producing an Unfavorable Position for Women in the Urban Labor Force in Less Developed Countries

1. Women's reproductive role, particularly in relation to domestic work
2. Colonial introduction of the notion of the male breadwinner and female housewife
3. Orthodox economic perception of women as having relatively low levels of skills, aptitudes, and/or education
4. Women seen as more "passive" workers, less likely to organize and resist exploitation by employers
5. Gender stereotyping of women as dexterous and able to undertake repetitive tasks

Source: *R. B. Potter and S. Lloyd-Evans,* The City in the Developing World, *Harlow, UK: Longman, 1998, pp. 166–67.*

INADEQUATE HOUSING

Social polarization is a dimension of urbanization that is intensified by globalization and splintering urbanism. The United Nations Human Settlements Programme (UN-Habitat) has identified social polarization as an indirect but crucial determinant of contemporary patterns of segregation of people and land uses around the world—the "quartering" of cities into spatially partitioned, compartmentalized residential enclaves. In less developed countries, these separate "quarters" include protected enclaves of the wealthy; middle-income neighborhoods; low-income neighborhoods; ethnic enclaves; and "excluded ghettos" (see Chapter 6).

TABLE 9.2

Household Strategies for Coping with Worsening Urban Poverty

Increase resources

- Add more household members to the labor force
- Set up new businesses
- Grow own food
- Increase scavenging of items for use or resale; increase foraging for wild foods
- Rent out rooms to tenants

Limit consumption

- Reduce or eliminate consumption of items such as new clothes, meat, or "luxury" food and drink
- Buy cheaper food and secondhand clothes
- Withdraw children from school
- Delay medical treatment
- Postpone repair or replacement of household equipment
- Defer house repairs or improvements
- Reduce social events, including visits to rural homes

Change household composition

- Postpone or stop having children
- Increase household size, with women and married children, to keep wage earners in the family and enable older women to work
- Migration

Source: *C. Rakodi. Poverty lines or household strategies? A review of conceptual issues in the study of household poverty.* Habitat International *19(4), 1995, p. 418.*

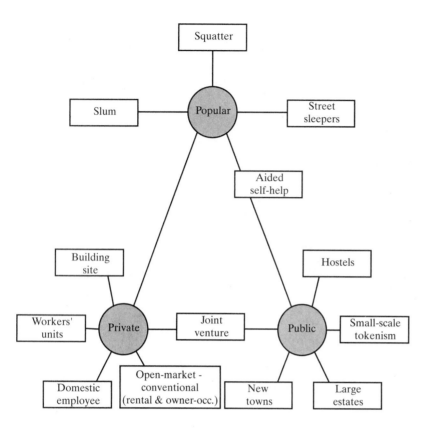

FIGURE 9.5 Typology of low-cost urban housing supply in less developed countries.

The typology[5] of urban housing supply shown in Figure 9.5 describes several kinds of housing options in the cities of less developed countries. Private housing (such as owner-occupied, rental, and employer-provided) and public units (in government-built or subsidized dwellings) comprise a relatively small proportion of the total housing supply. The "popular" housing category (that includes the slums and squatter settlements of the "excluded ghettos") contains the majority of urban residents—the very poor, the never employed, the permanently unemployed, and the homeless.

The United Nations has estimated that more than one billion of the world's urban residents live in inadequate housing, mostly in the slums and squatter settlements in the less developed countries (Figure 9.6).

FIGURE 9.6   A squatter settlement built by poverty-stricken people along a huge water pipe in New Delhi, India.

More than half of the housing in these countries is substandard, with one-quarter being makeshift structures and more than one-third not complying with local building regulations.[6]

The informal labor market, then, is directly paralleled in the slums and squatter settlements: With so few jobs offering regular wages, most people cannot afford rent or mortgage payments for decent housing. Unemployment, underemployment, and poverty result in overcrowding. In situations where overurbanization has overtaken the supply of cheap housing—outstripping the capacity of builders and governments to provide new and affordable homes—the inevitable outcomes are slums and squatter settlements that offer precarious shelter at best.

Whereas slums are, at least technically, legal permanent homes that have become dreadfully substandard over time, squatter settlements contain nonpermanent makeshift housing that is built illegally on land that people desperate for shelter neither own nor rent. Such housing is constructed on unpaved streets in the least desirable locations—derelict sites, poorly drained land, even cemeteries and waste dumps—usually with open sewers and no basic services like electricity or running water. Shacks are cobbled together from any material available—corrugated iron, planks, mud, thatch, tar paper, and cardboard. In Chile squatter settlements are called *callampas*, meaning "mushroom cities"; in Turkey, *gecekindu*—literally, "built after dusk and before dawn." In India they are called *bustees;* in Brazil, *favelas;* and in Argentina, simply *villas miserias.* In the worst cases, overcrowding, inadequate sanitation, and lack of maintenance promote extremely high levels of illness and infant mortality.

Faced with the growth of these squatter settlements, many governments initially responded by eradicating them. Encouraged by Western housing experts and development economists, many cities tried to stamp out this kind of unintended urbanization through large-scale eviction and clearance programs. In dozens of cities—including Jakarta (Indonesia), Caracas (Venezuela), Lagos (Nigeria), and Kolkata (India; formerly Calcutta)—hundreds of thousands of people in slums and squatter settlements were ordered out on short notice and their homes bulldozed to make way for public works, land speculation, luxury housing, urban renewal, and, on occasion, to improve the appearance of cities for visitors and tourists. Seoul, South Korea, has probably had the most forced evictions of any city in the world. Since 1966, as part of a sustained government clean-up campaign, millions of people have been forced out of housing that they owned or rented. Beginning in 1983, the government evicted nearly 750,000 people as part of a beautification program to prepare for the 1988 Olympic Games. Most cities cannot afford to build new low-income housing to replace the demolished neighborhoods.

Displaced households have no option but to create new squatter settlements elsewhere in the city. The futility of slum clearance has led to a widespread reevaluation of such policies; this housing is now seen as a rational response to poverty. Slum and squatter neighborhoods can provide affordable shelter as well as reception areas that might offer supportive community networks and informal employment opportunities for rural migrants. Recognizing the positive functions of this housing and the many self-help improvement efforts by residents, city governments have begun to support them instead of sending in police and municipal workers with bulldozers.[7] But there are limits to community-based efforts, and there are many squatter settlements where self-help organizations do not emerge.

LACK OF URBAN SERVICES

The influence of the informal labor market is felt in the inability of cities to provide basic services (Figure 9.7).

FIGURE 9.7   A garbage-strewn alley and water channel in Cairo, Egypt, caused by a lack of basic urban services and illegal dumping.

Because the informal sector yields no tax revenues, there are limited municipal funds for providing adequate health and educational services or for maintaining a clean, safe environment.

Water supply is an acute problem for many cities. Definitions of what constitutes an adequate amount of safe drinking water vary from country to country. Although many governments classify the existence of a water tap within about 300 feet of a house as "adequate," such a tap does not guarantee enough water for good health for nearby households. In many cities residents must wait in long lines to fill even one bucket of water because communal taps often function for only a few hours every day. In Rajkot, a city of 600,000 people in India, piped water routinely runs for only 20 minutes each day. The **World Bank** has estimated that only about 70 percent of urban residents in the less developed countries have access to an adequate source of water. Hundreds of millions of urban dwellers have no alternative but to use contaminated water—or at least water whose quality is not guaranteed. Only a small minority, typically the residents of the most affluent neighborhoods, has water piped into their homes.

Sewage services are just as bad. According to World Bank estimates, only about 40 percent of urban residents in less developed countries are connected to sewers. About 90 percent of these sewers discharge their waste untreated into a nearby river, a lake, or the sea. São Paulo has over 1,000 miles of open sewers; raw sewage from the city's slums drains into the Billings reservoir, a major source of the city's drinking water. Jakarta has no waterborne sewage system at all. Septic tanks serve about one-quarter of the city's population; others must use pit latrines, cesspools, and roadside ditches. A survey of over 3,000 towns and cities in India found that only 8 had full sewage treatment facilities and another 209 had partial treatment facilities. Each day the Yamuna River picks up 53 million gallons of untreated sewage and more than 5 million gallons of industrial effluents as it passes through Delhi.

In many cities, including Bangkok, Karachi, Dar es Salaam, and Bogotá, less than one-third of all garbage and solid waste is collected and removed—the rest is partially recycled informally, tipped into gullies, canals, or rivers, or simply left to rot. These problems provide employment opportunities within the informal sector. Street vendors, who get their water from private tanker and borehole operators, sell water from 2-gallon or 4-gallon cans. They typically charge 5 to 10 times the local rate set by public water utilities; in some cities they charge 60 to 100 times as much. Similarly, many cities have informal-sector mechanisms for sewage disposal. In some Asian cities, for example, handcart operators remove human waste at night. The problem with this arrangement is that this waste is often disposed of improperly and eventually pollutes the rivers or lakes from which the urban poor draw their water.

In NICs such as Singapore, South Korea, and Taiwan, an educated labor force was an important factor in economic growth.[8] Yet access to education remains a problem in many countries, especially for rural residents, because the centers of education are concentrated in cities. Investments in education have benefited middle-income groups more than the poor and men more than women. The bias against women varies from country to country but is most marked in Islamic cities and in the poorest households.

Urbanization does not necessarily bring improved health because physical well-being and poverty are inextricably intertwined.[9] The United Nations has identified a number of problematic issues specific to urban health.[10]

- Urban health patterns are different from those in rural areas because urban populations are leading the *epidemiological transition:* a shift from communicable to noncommunicable diseases. The World Health Organization has predicted that by the third decade of the 21st century, the social and economic changes associated with urbanization will make depression, heart attacks, and traffic accidents the leading healthcare burdens in the cities of less developed countries, as opposed to respiratory diseases, diarrhea, and early childhood conditions now.

- Some communicable health problems, including HIV/AIDS, still dominate urban areas. In fact, urbanization has been paralleled by increasing HIV/AIDS cases in cities, particularly in sub-Saharan Africa (see the Box entitled "HIV/AIDS in Sub-Saharan African Cities"). Within cities, children are particularly susceptible to communicable diseases, including acute respiratory infections.

- Poor urban residents suffer from the health-related problems of rural areas, such as inadequate diet and poor sanitation, in addition to being exposed to health problems associated with factors specific to cities, such as stress arising from overcrowding and poor working conditions.

- The health burden of the urban poor can only be fully understood within the context of overall inequities within cities.

Box 9.1   HIV/AIDS in Sub-Saharan African Cities

Of the 42 million people worldwide who were estimated to be living with HIV/AIDS in 2003, more than 29 million were in sub-Saharan Africa.[11] Despite having only about 10 percent of the world's population, this region is home to more than 70 percent of the infected people. One in five adults now has HIV. AIDS has already claimed 15 million lives and orphaned more than 12 million children there. In the worst-hit countries, like Botswana and Swaziland, AIDS is expected to wipe out half the teenagers and plunge life expectancy to age 30 by as early as 2010.

Infection rates are high in cities because the concentration of large numbers of people at high density increases the speed of transmission of HIV. Prostitution, multiple sexual partners, and teenage pregnancy are more common in cities than in rural areas. In addition to this devastating human tragedy, the spread of HIV/AIDS adds to already overwhelming economic problems and drains limited public resources in sub-Saharan African cities in a number of ways:[12]

- Productivity is reduced because AIDS deaths decrease the number of experienced workers, especially those in their most productive years.
- International competitiveness is hurt by higher production costs because of a shortage of skilled workers.
- Employment creation is slower due to lower government revenues and reduced private savings.
- Higher public expenditures are needed for monitoring, prevention, and health care.

Research suggests that there is a link between rapid urbanization amid poverty and the urban character of the HIV/AIDS problem. The African dilemma is particularly troublesome because, although 28 percent of urban residents live in poverty in the less developed countries as a whole, more than 41 percent do so in Africa.[13] In some cities—Kinshasa, Lagos, and Nairobi, to name three—between 60 and 80 percent of residents live in slums and squatter settlements.

A recent study in Nairobi, Kenya, found that poor slum dwellers start sexual intercourse at an earlier age, have more sexual partners, and are less likely than other city residents to know about or adopt condom use or other preventative measures against contracting HIV/AIDS.[14] Three features of slum conditions help socialize children into very early sexual intercourse: poverty, a social context that is accepting of prostitution, and household arrangements that do not allow privacy, causing children to think that sexual intercourse is an activity that they too can participate in at a very early age.

The long-term deteriorating economic conditions in sub-Saharan Africa in general, and in cities in particular, have increased the likelihood that women, especially adolescents, will engage in risky sexual behavior for economic survival despite the risk of contracting and spreading HIV/AIDS. Desperation can push some women to rely on sexual relations to supplement the household income to cover the cost of rent, schooling, and other basic necessities; many of these women maximize the number of sexual partners in an effort to increase their economic security.

The Nairobi study highlighted the need to treat people in squatter settlements as a low-income group that is particularly vulnerable to contracting HIV/AIDS because of poor access to health services, family planning facilities, education, and basic amenities due to their relative geographic isolation, low incomes, and illegal or informal residence. The results of the study suggest that it might be difficult to improve the reproductive health status of residents in slums and squatter settlements without also improving their economic and living conditions.

TRANSPORTATION PROBLEMS

Even though city governments in less developed countries typically spend nearly all their budgets on transportation infrastructure in a race to keep up with population growth, conditions are bad and rapidly getting worse. Cities in less developed countries have always been congested, but in recent years, congestion has turned into near gridlock (Figure 9.8).

In megacities the availability and use of private automobiles have increased sharply. Not only are there now more people and traffic, but also the changing spatial organization of these cities has increased the need for transportation. Traditional patterns of land use have been superseded by the agglomerating tendencies inherent in modern industry and the segregating tendencies inherent in contemporary societies. The greatest single change has been the separation of

FIGURE 9.8   Traffic on Ratchadmari Road in Bangkok, Thailand. Roads in the larger cities of less developed countries have always been congested, but in recent years, congestion has turned into near gridlock.

home from work, which has caused a significant increase in commuting.

The general trend has been for increasing automobile dependence and decreasing urban density. Figure 9.9 shows the relationship between city form and dominant transportation system in three kinds of cities:[15]

1. The traditional "walking" or "pedestrian" city with densities of 250–500 persons per acre.
2. The "transit" city with densities of 175–250 persons per acre.
3. The "automobile dependent" city with densities of 25–50 persons per acre.

The trend away from traditional "walking" cities and toward "automobile dependent" cities in the less developed countries has generated unprecedented traffic problems. Despite some innovative responses, transportation systems in many cities are breaking down, with poorly maintained roads, traffic jams, long delays at intersections, and frequent accidents. Many governments have invested in expensive new freeways and street-widening schemes. Because the new roads tend to focus on city centers—still the settings for the majority of jobs and most services and amenities—they ultimately fail, emptying vehicles into a congested and chaotic mixture of motorized traffic, bicycles, animal-drawn vehicles, and hand-drawn carts. Bangkok and Mexico City have some of the worst traffic problems in the world (see the Box entitled "How Rationing Can Backfire: The 'Day Without a Car' Regulation in Mexico City"). In Bangkok, the 15-mile trip from Don Muang Airport to the city center can take 3 hours. The

costs of these traffic backups are enormous. The annual costs of traffic delays in Bangkok alone have been estimated at $272 million—the equivalent of more than 2 percent of Thailand's gross domestic product.

ENVIRONMENTAL DEGRADATION

Due to pressing problems of poverty, poor housing, and inadequate service and transportation infrastructures, cities in less developed countries are unable to devote many resources to environmental issues (Figure 9.11). Because of the speed of population growth, these problems are escalating rapidly. Industrial and human wastes pile up in lakes and lagoons and pollute long stretches of rivers, estuaries, and coastal zones. Chemicals leaching from uncontrolled dump sites pollute groundwater. The demand for timber and domestic fuels are denuding forests near many cities. This environmental degradation is, of course, directly linked to human health. People living in such environments have much higher rates of diarrhea, respiratory infections, and tuberculosis and much shorter life expectancies than people living in the surrounding rural communities. Children in squatter settlements may be 50 times as likely to die before the age of five as those born in affluent countries.

Air pollution has escalated to hazardous levels in many cities. With the growth of a modern industrial sector and automobile ownership, but without enforceable pollution and vehicle emissions regulations, tons of lead, sulfur oxides, fluorides, carbon monoxide, nitrogen oxides, petrochemical oxidants, and other

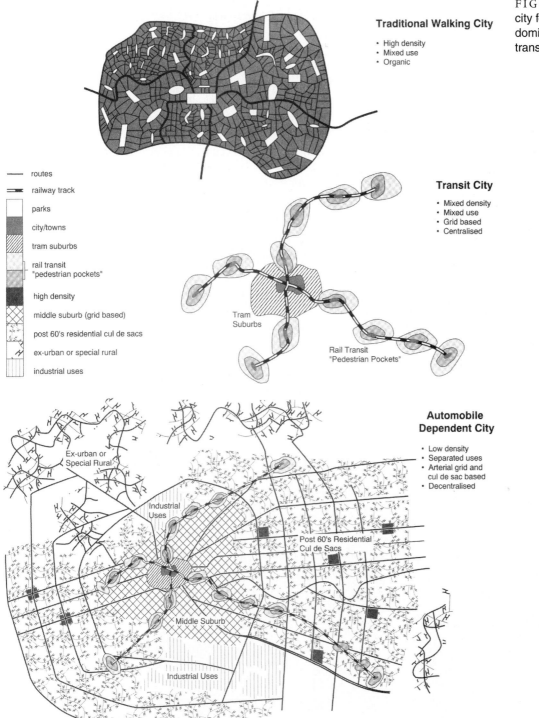

Traditional Walking City

- High density
- Mixed use
- Organic

Transit City

- Mixed density
- Mixed use
- Grid based
- Centralised

Automobile Dependent City

- Low density
- Separated uses
- Arterial grid and cul de sac based
- Decentralised

routes
railway track
parks
city/towns
tram suburbs
rail transit "pedestrian pockets"
high density
middle suburb (grid based)
post 60's residential cul de sacs
ex-urban or special rural
industrial uses

Tram Suburbs

Rail Transit "Pedestrian Pockets"

Ex-urban or Special Rural

Industrial Uses

Post 60's Residential Cul de Sacs

Middle Suburb

Industrial Uses

toxic chemicals are pumped into the atmosphere every day in large cities. The burning of charcoal, wood, and kerosene for fuel and cooking in low-income neighborhoods also contributes significantly to poor air quality. The United Nations has estimated that more than 1.1 billion people worldwide live in urban areas where air pollution exceeds healthful levels.

A United Nations study of 20 megacities found that all had at least one major pollutant at levels exceeding World Health Organization (WHO) guidelines. Fourteen of the twenty had two major pollutants exceeding WHO guidelines, and seven had three. Such pollution is not only unpleasant, but dangerous. In Bangkok, where air pollution is almost as

Box 9.2   How Rationing Can Backfire: The "Day Without a Car" Regulation in Mexico City[16]

In Mexico City daily traffic backups can extend for more than 60 miles (Figure 9.10). Associated air pollution is not only unpleasant, but dangerous. Lead and sulphur dioxide concentrations are two to four times above World Health Organization (WHO) guidelines; the city exceeds national ozone levels on more than half the days of the year; and 7 in 10 newborns have dangerously high levels of lead in their bloodstream.

In an effort to stem the use of automobiles, Mexico City's government introduced a regulation in 1989 that banned the use of each automobile on a specific day of the week. The "Day Without a Car" regulation—*Hoy no circula* ("this one does not circulate today")—prohibited automobiles with license plate numbers ending in 0 or 1 from being driven on Mondays, 2 or 3 from being driven on Tuesdays, and so on. The regulation was controversial. Proponents believed that it was reasonable to expect automobile owners to contribute to easing traffic congestion and air pollution problems. Those who opposed the regulation argued that it

was unfair because some people could avoid the ban more easily than others.

Ultimately, the regulation was counterproductive and a failure—it caused higher congestion and pollution levels because people who could afford a second automobile purchased one and effectively circumvented the ban on their first vehicle. Total car use and traffic congestion in Mexico City actually increased due to the regulation.[17] Even though a second car was purchased primarily to replace a first car on its driving ban day, total car use increased in households with multiple drivers because a second car was now available. The regulation led to more rather than less pollution because, in addition to higher car use causing more emissions, people tended to buy older second cars with lower technical standards. Vehicle ownership data indicates that in the years before the regulation was introduced, Mexico City usually exported an average of 74,000 used vehicles to the rest of the country. In the years following the regulation, this trend was reversed as Mexico City began to import 85,000 vehicles annually.

FIGURE 9.10
A major street in the Miguel Hidalgo area of Mexico City is clogged with traffic and smog during the morning rush hour.

severe as in Mexico City, it has been estimated that the pall of dust and smoke blanketing the city causes more than 1,400 deaths each year. Traffic and pollution-linked illnesses cost $3.1 billion annually in lost productivity in Bangkok.

Proximity to industrial facilities—often because poor people need to live near their place of employment—poses another set of risks. In 1984 a major accident at the Union Carbide chemical factory in Bhopal, India, caused 2,988 deaths and more than 100,000 injuries, mostly among the residents in the nearby squatter settlements. The interrelationships between poor housing conditions, poverty, and urban environmental problems reflect a number of factors:

- A greater incidence of sickness and death is closely associated with housing that is crowded, poorly built, located in unsafe areas (subject to natural and other hazards), and inadequately serviced by water facilities, sewage treatment, and garbage disposal.

- Poor quality housing is often makeshift and temporary, dilapidated, poorly maintained, and fire-prone.

- Low incomes combined with rising land and property prices result in overcrowding and homelessness.

- Insecure and illegal land tenures come with eviction and demolition, which remove any motivation to invest in home improvements.

- The poorest members of society are the most disenfranchised and the least able to articulate their concerns about environmental and other problems.[18]

FIGURE 9.11   Garbage floats behind the sampans of the Chao Praya River in Bangkok, Thailand.

RESPONSES TO THE PROBLEMS OF URBANIZATION

SUSTAINABLE URBAN DEVELOPMENT

The relationship between urban economic development and environmental conditions is attracting increasing attention within the context of sustainable development. There is now greater recognition that the responses to the problems of urbanization must be framed within the context of the concept of **sustainable urban development**. Responses to the urbanization problems in the less developed countries must address several interconnected components of sustainable urban development (Figure 9.12):[19]

- *Economy:* The most sustainable economic activities are those that integrate the external role of cities within the regional, national, and world economies with the needs of the labor force and the revenue requirements of city governments. Jobs in the informal sector of the economy have a role, but government employment policies are also necessary. Addressing poverty is fundamental not only for the well-being of urban residents but also for the prosperity of the cities themselves.

- *Environment:* The so-called *Brown Agenda*[20] involves addressing urban environmental problems such as pollution, land degradation, and hazardous living and working conditions. Some of these sustainability issues, such as water and air quality, are citywide and are primarily the responsibility of government planning efforts. Other issues, involving renewable and especially nonrenewable resources, can involve individuals and households to a greater extent. This could include complying with environmental regulations and supporting poverty-alleviating policies for improving the economic situation of the poor, thereby easing some of the pressures on the urban environment.

- *Society:* The basic needs of society include a range of issues, from shelter to food to education to healthcare. Government has a role in meeting citywide needs, such as schools and hospitals, as well as in improving human rights. Opportunities for genuine community involvement and self-help programs are also important.

- *Demographic situation:* Exceptionally high rates of natural increase of the population within cities, combined with often massive **rural-to-urban migration**, not only have overwhelmed the resources of city governments, but also have intensified and complicated the social and ethnic dimensions of urbanization, threatening the sustainability of much urban development.

- *Political sphere:* The role of government is an important dimension of urban sustainability. Governments can help to establish democratic institutions and a planning process that allows the participation of all members of society, including

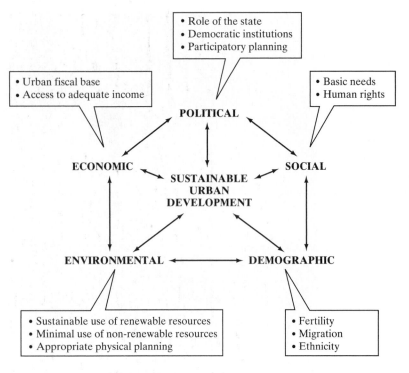

FIGURE 9.12   The components of sustainable urban development.

the poorest people. The political arena is also the point at which urban sustainability, as a concept, overlaps with urban management as a practical process.

THE "GLOBALIZATION PARADOX" AND RECENT CHANGES IN URBAN GOVERNANCE

The United Nations[21] has drawn attention to the "globalization paradox" faced by cities in the less developed countries as they try to achieve sustainable urban planning and management. The increased competition and fragmentation associated with **globalization** and splintering urbanism are having contradictory effects on cities. A city must act as a collective unit in order to respond to international competition and compete effectively in the world economy while being less able to rely on higher levels of government for assistance. At the same time, growing social and economic fragmentation hampers the capacity of cities to build coalitions, mobilize resources, and develop good governance structures. This predicament is particularly detrimental when urban change is dramatic and requires enhanced decision-making capacities—as in many of the cities and metropolitan regions in less developed countries.

During the last decade, four changes in urban governance have emerged in attempts to address the "globalization paradox" in the less developed countries.[22] The relative importance of these changes varies among countries, and they do not necessarily herald the end of

more traditional systems of governance. Some of the changes, such as decentralization and formal government reforms, have been around for some time. What is new is the rationale for introducing them.

1. *Decentralization and formal government reforms:* Although certain issues that affect cities, such as rural poverty and rural-to-urban migration, are best addressed at the national scale, many urban problems are better tackled closer to the source. This principle underlies the trend toward decentralizing power to regional and local levels of government. Decentralization has included strengthening the powers of city governments in African countries, such as South Africa, and in many Latin American countries.

Regional programs—such as the Support System for Training for Local Municipal Development in Latin America (SACDEL)—support the decentralization efforts. Funding and support comes from the national governments of Latin American countries such as Colombia, Costa Rica, Ecuador, and Peru, as well as the Economic Development Institute of the World Bank and UN-Habitat, to name a few. The program operates from a regional center in Quito, Ecuador, and provides training and assistance to public and private training agencies that are charged with enhancing the local capacity of municipal governments in Latin America.[23]

There have also been attempts at formal government reform at the metropolitan level—including establishing new government structures and agencies with *metropolitan-wide* responsibilities for strategic planning, economic development, urban services, and, more recently, environmental protection. The rationale for these reforms includes long-established goals of efficiently and cost-effectively providing urban services and infrastructure. More recent reasons for reform include the need to devise and implement policies that better address metropolitan-wide problems of environmental degradation, social exclusion, crime, and violence.

The United Nations maintains that strategies that create new governmental forums can increase the voice of marginalized groups, particularly in cities where ethnic minorities are geographically concentrated. The Popular Participation Law in Bolivia, for example, created municipal councils in which representatives from the Quechua (descendants of the Incas) and Aymará (pre-Inca ethnic group) minorities play a role in allocating resources within cities. In India a constitutional amendment was passed in 1993 that reserved seats for women in local government. Mandating the representation of women in new governmental forums is a good first step; but the empowerment of women in many less developed countries is often still constrained by traditional gender relations.

2. *Participation of local communities in urban policy-making and implementation:* The emphasis is now on the internal opportunities for development, using models, assumptions, and objectives that have been devised within the less developed countries themselves. This shift has been associated with an increasing recognition of the importance of local context for development and growing support for participatory approaches that give citizens a stronger voice in urban policy-making and implementation (see the Box entitled "Different Levels of Participation in Urban Policy-Making and Implementation").

The greater involvement of urban residents—notably through participation by community-based organizations—is becoming more common in urban planning. By helping to legitimize policy-making, resident participation in devising and implementing policies—by both men and women—can make public policy implementation more efficient and responsive to the needs of the community (Figure 9.13). This is particularly important in cities where, because of the inability of local public institutions to provide basic urban

FIGURE 9.13   Workers on a community self-help sanitation project examine the pipes for a new sewer.

services, people have had to organize themselves to undertake self-help water and sanitation projects (see the Box entitled "Urban Social Movements and the Role of Women").

When local citizens in Dar es Salaam, Tanzania, found that the government was economically and technically unable to improve the water supply, they looked for affordable ways to do it themselves.[24] Local residents formed organizations with elected members, such as the Kijiotoyama Development Community (KIJCO). Negotiations between KIJCO and the city government resulted in a Community Infrastructure Program through which the residents received assistance in constructing two deep wells.

Orangi, a squatter settlement of about 1 million people in Karachi, Pakistan, had no public sanitation system.[25] A local organization called the Orangi Pilot Project helped the community to organize into small groups of households surrounding each lane. The Pilot Project helped survey the site, draw up plans, and prepare cost estimates. The residents collected money to pay for the sewer installation. As more households became involved, the local government began to provide some financial support. In the first eight years of the program, the residents administered the construction of about 69,000 latrines and 4,500 sewerage lines.

3. *Multilevel governance and public–private partnerships:* Multilevel governance involves a set of joint practices, including partnership and co-funding. Decentralization has increased the

overlap of responsibilities, creating the need to involve different levels of government in urban policy-making. In many cities multilevel governance involves not only public institutions but also private companies, nonprofit agencies, and NGOs. The private sector is often involved because of privatization policies, as when the water systems were privatized in many African countries. Community-based organizations may be involved because they have legitimacy in representing the people and first-hand knowledge of local problems. NGOs too can play an important role because they are knowledgeable about program management or because they have assumed certain responsibilities from the state. Multilevel governance often comes about for practical reasons, but it is increasingly considered a new way of policy-making and implementation.

BOX 9.3   DIFFERENT LEVELS OF PARTICIPATION IN URBAN POLICY-MAKING AND IMPLEMENTATION [26]

Although greater community participation is now accepted as an important component of urban policy-making and implementation, the quality of actual citizen participation varies within and between cities. The continuum of participation runs from a pretense of participation and "passive" situations in which people are simply told what is going to happen, all the way up to "self-mobilization," in which people in an area work together to develop and implement their own initiatives.

Manipulation and decoration	People are involved in a pretense of participation, for example, by including unelected community representatives on official boards but without giving them any power in the decision-making process.
Passive participation	People are told what is going to happen without their input and with no power to change decisions.
Consultation and information-giving	People's views are sought through a consultation process that aims to elicit their priorities and needs. But the process is undertaken by external agencies that define the information-gathering process and control the analysis in which the problem is defined and the solutions devised. The people are given no decision-making powers, and the outside agencies make no commitment to respond to the citizen-defined priorities.
Participation for material incentives	People participate, but only in implementation and only in response to the desire for material incentives (such as contributing labor to a project in return for wages, or building a house as part of a "self-help" project as a condition for receiving the land and services).
Functional participation	An external agency sees participation as a way to achieve its project goals, especially reducing costs through free citizen labor and management. People might participate by forming groups to meet the agency's predetermined objectives. At worst, people are simply coopted to serve the goals of an external agency that accord little with their own. At best, this limited form of participation has brought real benefits to some communities.
Interactive participation	An external agency launches a project in conjunction with local people who have often already identified their needs and initiated contact with the agency themselves. Participation is seen as a citizen's right, not just a means to achieve a project's goals. People participate in joint analysis, action plan development, and the formation or strengthening of institutions for implementation and management. The citizens have considerable influence in determining how available resources are used.
Self-mobilization	A project is initiated by the people themselves, who establish contacts with external agencies in order to secure the resources and technical advice they need. But the people retain control over how the resources are used. Within poor neighborhoods, this is less project-oriented and more part of a process through which people organize to get things done and to negotiate with external agencies for support in doing so on an ongoing basis.

BOX 9.4   URBAN SOCIAL MOVEMENTS AND THE ROLE OF WOMEN

In many cities in less developed countries, poor residents are joining self-help efforts to improve their living conditions.[27] When community-based organizations form broad-based associations, the possibility of an **urban social movement** arises. An urban social movement can be defined as a social organization with a territorially based identity that strives for emancipation through collective action. The concept of emancipation here involves satisfying basic needs (including housing and urban services), developing a respectful attitude toward the environment, lack of discrimination (on the basis of race, ethnicity, religion, gender, socioeconomic status, or residential location), and gaining access to urban policy-making and implementation.

Theorists like Manuel Castells believe that a basic characteristic of a true urban social movement is that it seeks societal change.[28] Others, like Schuurman and van Naerssen,[29] argue that the poor in the less developed countries might not be able to achieve societal reform because they cannot always overcome the political and social constraints that limit their efforts. Nevertheless, urban social movements often represent the only mechanism for improving the living conditions of poor people and are becoming more prevalent in the less developed countries. Because women comprise a large percentage of the poor, they have high participation rates in urban social movements in many countries.

In São Paulo, Brazil, for example, many urban social movements emerged during the 1970s. This was a period of intense industrial growth and dramatic population increase that put enormous pressures on already inadequate urban services, such as housing, water, sanitation, transportation, and health care.[30] The state of São Paulo was unable to cope with the huge influx of rural workers. Most of the new immigrants were forced to settle at the urban periphery in squatter settlements. These settlements became the main political bases for the urban social movements.

Women have been able to become politically active in urban social movements in less developed countries in a way that has not been possible in the conventional political arena of political parties and trade unions. The women are already part of the community networks that the urban social movements use to organize residents. In São Paulo women use a number of approaches based on community networking:

- *House meetings (reuniões de casa):* The small-scale and nonhierarchical format of informal meetings in women's homes facilitate the establishment of working relationships and greater solidarity by allowing all participants, especially those with little experience in political organizing, to express their views in a less intimidating environment than typically found at the larger general meetings of the movement.

- *Communication and publicity:* Although all urban social movements use written material to publicize their activities to the general population, within the movements themselves, direct personal contact is more effective, especially when literacy rates are low. Women take advantage of their personal contacts within the community to gain support, publicize their goals, and encourage people to participate.

- *"Spontaneous" political action:* "Spontaneous" political action describes protests with no identifiable organizational source. A breakdown in negotiations or the failure of government to keep its promises might provide the impetus. Unlike organized political demonstrations, the movement chooses only a time and place, usually at very short notice, and then spreads the word through the community network; how many people show up and what actually happens during the mobilization is generally unplanned. In Brazil, "sacking" (looting) is the most common form of "spontaneous" political action. Undertaken almost exclusively by women, a small group decides to ransack a local grocery store, picks the time and place, and spreads the word through the community network. "Sacking" involves a general "code of practice" in which only basic foodstuffs (beans, rice, and oil) are taken, with only rare incidents of violence or destruction of property.

- *Community research:* Often finding themselves outmaneuvered during negotiations due to government use of official data and statistics, urban social movements have implemented new political techniques, including community research. In addition to the kinds of qualitative research that they were already using, such as sharing personal experiences, the women began to collect basic statistics on conditions and opinions using simple questionnaires. They then use the findings to reinforce their arguments with government officials, while also identifying the community's priorities for action.

For example, a partnership of residents, national and local government agencies, and NGOs came together to set up the Masese Women's Self Help Project to upgrade the "Masese slum" in Jinja, Uganda. The project's Housing and Human Settlement Upgrading Programme established a settlement and credit plan that enabled women to get secure land tenure and construction materials for housing. The project also developed a community infrastructure that offers employment, health, and education services.[31]

4. *Process-oriented and area-based policies:* Urban governance today involves a system in which a variety of participants work in partnership with government agencies. This complicates policy-making. Policy-making now involves coalitions and compromises; it also requires discussion and debate, which, in turn, depend on appropriate negotiation procedures. This is a new way of reaching decisions in which policy content is, at least in part, a function of the decision-making process itself. These new forms of collective action can no longer take place at the central government level.

Area-based regulation has replaced national regulation because of the state's inability to solve problems and address issues at lower levels of government and the need to integrate the diverse elements of effective public policies.

The extent to which these changes in governance address the "globalization paradox" and represent the kind of policy-making and implementation that move cities in the less developed countries toward sustainable urban development can be judged using certain basic criteria.[32] These include the quality of life of urban residents, including levels of poverty, social exclusion, and human rights; the scale of nonrenewable resource use and waste recycling; and the scale and nature of renewable resource use, including freshwater resources.

The cities and governments in the less developed countries—especially the poorest ones—face a daunting set of problems with often limited resources. As we will see in Chapter 15, there is little comfort in the fact that, despite significantly higher levels of prosperity, problems of urbanization persist in the cities of developed countries like the United States.

FOLLOW UP

1. Be sure that you understand the following key terms:

   Brown Agenda
   decentralization of government power
   dualism
   epidemiological transition
   "globalization paradox"
   natural increase
   new international division of labor (NIDL)
   push and pull factors
   slums
   squatter settlements
   sustainable urban development
   underemployment
   urban social movement

2. If you have access to a computer, collect some digital images from the World Wide Web—jpeg format is best—that document some of the acute housing, traffic, and pollution problems in the cities of the less developed countries. If you have access to the software, display the images with accompanying captions in a short Microsoft PowerPoint presentation[©].

3. Because of their rapid growth and high underemployment, it is among the peripheral metropolises of the world—Mexico City (Mexico), São Paulo (Brazil), Lagos (Nigeria), Mumbai (India; formerly Bombay), Dhaka (Pakistan), Jakarta (Indonesia), Karachi (Pakistan), and Manila (the Philippines)—that we can find contenders for the title of "shock city" of the early twenty-first century. Using library and World Wide Web sources, document the most remarkable and disturbing economic, social, and cultural changes for one of these megacities. Think about how and why the processes and outcomes you identify for this "shock city" in a less developed country today are similar to and different from those in Manchester (UK) and Chicago in the nineteenth century.

4. Work on your *portfolio*. Collect official figures that document some urban problems in less developed countries that particularly interest you. The United Nations (UN-Habitat) has economic and social indicators for selected cities in recent editions of its annual *Global Report on Human Settlements* that you can find in the library or on the UN's website (at *http://www.unhabitat.org/*).

KEY SOURCES AND SUGGESTED READING

Anzorena, J., J. Bolnick, S. Boonyabancha, Y. Cabannes, A. Hardoy, A. Hasan, C. Levy, D. Mitlin, D. Murphy, S. Patel, M. Saborido, D. Satterthwaite, and A. Stein. 1998. Reducing Urban Poverty: Some Lessons From Experience. *Environment and Urbanization* 10: 167–86.

Brunn, S. D., J. F. Williams, and D. J. Zeigler, eds. 2003. *Cities of the World: World Regional Urban Development*, 3rd ed. Lanham, Md.: Rowman & Littlefield.

Drakakis-Smith, D. 2000. *Third World Cities*, 2nd ed. London: Routledge.

Drakakis-Smith, D. 1995. Third World Cities: Sustainable Urban Development, 1. *Urban Studies* 32: 659–77.

Friedmann, J. 1992. *Empowerment: The Politics of Alternative Development*. Cambridge, Mass.: Blackwell.

Potter, R. B., and S. Lloyd-Evans. 1998. *The City in the Developing World*. Harlow, UK: Longman.

Rakodi, C. 1995. Poverty Lines or Household Strategies? A Review of Conceptual Issues in the Study of Household Poverty. *Habitat International* 19: 407–26.

Schuurman, F., and T. van Naerssen. 1989. *Urban Social Movements in the Third World*. London: Routledge.

United Nations Centre for Human Settlements (HABITAT). 2001. *Cities in a Globalizing World: Global Report on Human Settlements 2001*. London: Earthscan Publications Ltd.

United Nations Centre for Human Settlements (HABITAT). 2001. *The State of the World's Cities Report, 2001*. Nairobi: UN Habitat (*http://www.unchs.org/Istanbul+5/statereport.htm*).

Williams, S. W. 1997. "The Brown Agenda": Urban Environmental Problems and Policies in the Developing World. *Geography* 82: 17–26.

RELATED WEBSITES

World Resources Institute: *http://www.wri.org/*
This website contains research and best practice on global environmental problems and includes *EarthTrends: The Environmental Information Portal* (*http://earthtrends.wri.org*), which offers data and maps on environmental, social, and economic trends related to cities in less developed countries.

UN Environmental Program (UNEP): *http://www.grida.no/*
This UN website contains information and hyperlinks for environmental issues of importance to urban and metropolitan areas around the world. Each issue of its magazine, "Our Planet" (*http://www.ourplanet.com*), focuses on a specific theme, such as water and sanitation, hazardous waste, or globalization, poverty, trade, and the environment.

United Nations Development Programme: *http://www.undp.org/*
The UNDP website contains excellent information on economic development issues affecting cities in less developed countries. It offers publications and other resources on issues such as poverty, democratic governance, energy and the environment, and HIV/AIDS.

The International Labour Organization: *http://www.ilo.org/*
This UN agency offers research and data on labor issues around the world, including information about work standards and child workers. The ILO also has a separate site devoted to export processing zones (*http://www.ilo.org/public/english/dialogue/govlab/legrel/tc/epz/*).

10

THE CITY AS TEXT: ARCHITECTURE AND URBAN DESIGN

The design of the built environment is what gives expression, meaning, and identity to the broad sweep of forces involved in urbanization. It provides cues for all kinds of behavior. It is symbolic of all sorts of political, social, and cultural forces. As a result, each group of structures of a given period and type tends to be a carrier of the zeitgeist, or "spirit of its time." Any city, therefore, can be "read" as a multilayered "text," a narrative of signs and symbols. If we think in this way of the city as a text, the built environment becomes a biography of urbanization. The text metaphor can be developed in several ways. It is possible, for example, to analyze particular settings in terms of their "grammar"—the way that buildings and spaces take on meaning in relation to one another. Similarly, it is possible to search for meanings that are conveyed by means of the graphic equivalent of allegory. Perhaps the most illuminating way to develop the metaphor, however, is to look for the equivalent of key pages or passages in the narrative: settings that are emblematic of the overall political economy, or fragments of the built environment that mark turning points in the story, or significant events or relationships. Yet in reading any text carefully we must also take note of the subtext, "reading between the lines" in order to understand the full story.

CHAPTER PREVIEW

The purpose of this chapter is to show how architecture and urban design are linked into the dynamics of urban change and spatial organization. Most of the chapter is a review of successive phases of design that have left a distinctive legacy on contemporary cityscapes. First, however, we will establish the importance of architecture and design in more general terms—as they relate to the political economy of urbanization. We also note that architectural styles are not simply imposed, layer on layer, onto metropolitan form; they are diffused differentially through time and space, creating distinctive morphological subareas.

We will see that with the advent of the industrial city, architecture and urban design developed some especially significant relationships with the urban political economy. The interdependencies between society and

design shifted rapidly to accommodate new ideas and new economic, technological, social, cultural, and political forces. We give particular attention here to three successive styles: the Arcadian Classicism of the early Industrial era, the Beaux Arts style adopted by the City Beautiful movement, and the early skyscraper style.

By the end of the Industrial era the most influential style of all—Modernism—had begun to emerge. We will review Modern architecture and urban design in some detail, since it has left a strong imprint on contemporary urbanization. As we will see, however, reactions to Modernist designs, together with new political-economic circumstances, prompted a good deal of recent city building and rebuilding, undertaken in the form of postmodern designs or through a strategy of historic preservation.

ARCHITECTURE AND THE DYNAMICS OF URBAN CHANGE

As Lewis Mumford put it in 1938, "in the state of building at any period one may discover, in legible script, the complicated process and changes that are taking place within civilization itself."[1] Thirty years later, sociologist Ruth Glass was to characterize the city as "a mirror . . . of history, class structure and culture."[2] Both comments echo the Marxian perspective on the built environment: that it is part of the "superstructure" of the dominant mode of production, reflecting the underlying relationships, tensions, and contradictions in society.

Thus we can see downtown landscapes as emblematic not only of the rise of great cities but also of corporate power, whose fortunes have been so closely interwoven with the process of urbanization. The landscapes of civic architecture are another common element in the urban text. We can see that the lavish city halls of the nineteenth century were built to show that a city had "arrived"; that the Beaux Arts libraries and museums of the early twentieth century were built to establish a moral order; that the anonymous Modernist office blocks of the mid-twentieth century were built to house the growing civic bureaucracy in appropriately functional settings; and that the festival settings of recent decades have been built to reassert the image of the city, to attract economic development and, some would argue, to mask the inequality and decay of the inner city.

ARCHITECTURE AND EXCHANGE VALUE

But architecture, like other dimensions of the superstructure, not only reflects underlying economic and

institutional structures, it also serves as one of the means through which they are sustained, protected from opposing forces, and legitimized. In this context, one of the most obvious roles for architecture is in helping to stimulate consumption through *product differentiation* aimed at tapping particular market segments.

If the capitalist economy is to be successfully maintained, it is vital for urbanization to facilitate the manipulation of demand not only to increase consumption but also to ensure the constant circulation of capital. The architect, by virtue of the prestige and mystique socially accorded to creativity, adds exchange value to a building through his or her decisions about design, "so that the label 'architect designed' confers a presumption of quality even though, like the emperor's clothes, this quality may not be apparent to the observer."[3]

ARCHITECTURE AND THE CIRCULATION OF CAPITAL

The professional ideology and career structure that reward innovation and the ability to feel the pulse of fashion also serves to promote the circulation of capital. Without a steady supply of new fashions in domestic architecture, the **filtering** mechanism on which the whole homeowner market is based would slow down to a level unacceptable not only to developers but also to financial institutions.

As Homer Hoyt recognized (see p. 133), one way in which the dynamics of urban growth are sustained is through the cachet of fashionable design and state-of-the-art technology that induces affluent households to move from comfortable and prestigious homes into newer ones; hence the rapid succession of themes that can be revived and "rereleased," much like the contrived revivals of *haute couture*. In some cities new housing for upper-income groups is now promoted through annual exhibits of "this year's designs," echoing the automobile industry's old strategy of carefully planned obsolescence in design. The same interpretation of the role of architecture can be applied over a longer time horizon. Here, the emphasis is on the way that major shifts in design—such as the shift from Modernism to postmodernism—help to resolve problems of underconsumption during **over-accumulation crisis** periods.

ARCHITECTURE AND LEGITIMATION

Another role for architecture in sustaining the superstructure is that of legitimation. A major theme in the literature on critical architectural history is the way that architecture has repeatedly veiled and obscured the realities of economic and social relations. The physical arrangement and appearance of the built environment can help to suggest stability amid change (or vice versa), to create order amid uncertainty, and to make the social order appear natural and permanent.

Part of this effect is achieved through what political scientist Harold Lasswell calls the "signature of power."[4] It is manifest in two ways: (1) through majestic displays of power inherent in urban design and (2) through a "strategy of admiration," aimed at diverting the audience with spectacular and histrionic design effects. It must be recognized, however, that it may not always be desirable to display power. Legitimation may, therefore, involve modest or low-profile architectural motifs. Conversely, it is by no means only "high" architecture that sustains the social order. The everyday settings of workplace and neighborhood also help to structure and reproduce class relations—part of the "sociospatial dialectic" that we will explore in Chapter 14.

Meaning and Symbolism

Any consideration of these connotative roles of architecture, however, soon runs into complex issues of meaning and symbolism. When we move down from high-level generalizations about socioeconomic processes and urban design, we find that people often endow buildings with meanings in ways that can be highly individualistic and often independent of their class or power. If architecture communicates different things to different people, or groups of people, we have to look more closely at questions of communication

by whom, to *what audience*, to *what purpose*, and with *what results*.

The first distinction to make here is that between the *intended* meaning of architecture (on the part of architects and their clients) and its *perceived* meaning as interpreted by others. Thus:

> An environment becomes a social symbol when it is intended or perceived as a representative of someone or of some social group, when social meaning plays an influential role in relation to its other functions. There are thus two sides to the coin of social symbolism, (a) when an environmental action is intended by its proponent to convey social meanings, that is, as a symbolic action and (b) when the environment is perceived as a social symbol, whether this was intended or not.[5]

Sometimes, however, both sides of the same coin can be seen at once. Lasswell's "signatures of power," for example, may serve to reassure the rich, strong, and self-confident and reinforce feelings of deference among the poor and the weak; but the same symbolism may provoke and radicalize *some* of the poor and the weak. *The point is that much of the social meaning of architecture depends on the audience.* Meanwhile, of course, designers' and developers' preconceptions of the audience(s) will help to determine the kinds of messages that are sent in the first place.

ARCHITECTURE VERSUS "MERE BUILDING"

We count on the built environment in several significant ways. Depending on our tastes, we look for the built environment to be functional and attractive—not only to have a sense of identity and continuity but also to contain variety and to provide a sense of security. Yet for the most part the structures that constitute the built environment were not designed with such considerations. A great deal of the built environment is not "designed" at all, in any sense that architects would recognize. The builders of many homes and smaller retail, commercial, and industrial structures simply work from pattern books, copy or hybridize other buildings, or use prefabricated technology; hence the snooty distinction made by British architectural critic Sir Nicholas Pevsner between architecture and "mere building."

For the perpetrators of mere building, the prime considerations have to do with costs and marketability. Because the markets concerned are generally mass markets of one kind or another, aesthetics and design tend to finish a poor second in their calculations. For the creators of architecture, in contrast, their product aspires to be Art (or at the very least a contribution to the R&D division of the culture industry). As architects' professional turf came increasingly under threat from engineers and other building specialists after World War I, it was art, aesthetics, and design that secured the profession's status and legitimacy. As a result, architecture has often been uncoupled—both in practice and analysis—from the larger built environment. The profession tends to rate its members on their artistic achievements; influential glossy trade magazines have always stressed the aesthetic over the practical; and schools of architecture have consistently instilled in their students an ethic of avant-gardism.

Nevertheless, we cannot afford to dismiss High Architecture as the product of, or for, a self-referential elite. It is High Architecture that not only punctuates the built environment with major public and commercial structures but also sets the style for others to follow: journeyman architects and, eventually, the perpetrators of "mere building." Furthermore, these linkages have become increasingly influential as society has become more affluent and better educated, thus increasing the number of "cultivated" consumers. Our task, therefore, is to establish some guideposts in terms of the evolution of major architectural styles and design themes that have emerged in each of the major phases of urbanization.

THE STYLE OF PRODUCTION/THE PRODUCTION OF STYLE

The early industrial city of the mid-nineteenth century created unprecedented challenges for architects. Like other artists, they found themselves having to come to terms intellectually with industrialization, modernization, and, indeed, urbanization itself. Meanwhile, the tremendous increase in the scale and pace of change in cities required their new buildings to fulfill all kinds of new economic, social, and cultural imperatives. In the turmoil, it was some time before distinctive new styles emerged and even longer before any consensus could be reached on what was "good" architecture for the industrial city.

Meanwhile, the architects commissioned to build the new public buildings, factories, and mansions of the period resorted to a variety of borrowed themes and revivals. One of the most pronounced underlying themes was a reaction against industrialization, modernization, and urbanization that found expression in Romanticism and Classicism. As a result, state capitols and city halls ended up looking like Roman forums, railway stations like medieval cathedrals, offices like Renaissance palazzi, banks and college buildings like Greek temples, and hotels like Jacobean mansions (Figure 10.1).

ARCADIAN CLASSICISM AND THE "MIDDLE LANDSCAPE"

Although this reactionary Romanticism and Classicism heavily influenced the general Western intellectual climate of the mid-nineteenth century, American design professionals were especially influenced by the literary culture of the so-called **American Renaissance**. This was rooted in the works of Ralph Waldo Emerson and Henry David Thoreau and propagated by Walt Whitman and Herman Melville, whose objective was to confront the future rather than turn away from it; to define ideals appropriate to the time. Emerson had drawn, in the 1830s, on the European Romantics' notion of the pastoral ideal in arguing for settlements that incorporated the benefits of both city and country.[6] Thoreau, a disciple of Emerson's, popularized the idea of Nature as a spiritual wellspring for city dwellers in his book *Walden* (1854).[7]

By the time of the Industrial era, Americans had come to think of their relationship with Nature and the Great Outdoors as something distinctively "American." The historian Frederick Jackson Turner advanced the influential idea that the "frontier experience" was the single most significant factor in determining the American character,[8] and it became broadly understood that "access to undefiled, bountiful and sublime Nature is what accounts for the virtue and special good fortune of Americans."[9]

FIGURE 10.1   Neoclassical designs, such as the one used for a First National Bank building in Alabama, reflected a reaction during the early years of the industrial city against industrialization, modernization, and urbanization.

This response came from a perspective that was also influenced by a disdain for the values and ethics of the commerce and industry that were transforming urban society. The principles of industrial economic development were associated with corruption, exploitation, and moral degeneration. At the same time, there was an abiding fear of the social and physical consequences of this exploitation and degeneration: the alienated "mob" and its squalid and unhealthy neighborhoods that were seen as a threat to physical well-being and social order in the industrial city.

Against this backdrop, the American Renaissance took Nature as a fundamental spiritual wellspring, defining the ideal as a setting in which man and Nature had achieved a state of balance—what landscape architect Leo Marx has described as a "middle landscape" of pastoral and picturesque settings.[10] At the same time, the intellectuals of the period emphasized the moral superiority of domesticity and the virtues of republicanism and sanitary reform. This attitude led to a vision of ideal urban landscapes that combined the morality attributed to Nature with the enriching and refining influences of cultural, political, and social institutions. Progressive intellectuals like Andrew Jackson Downing advocated a program of "popular refinement" involving the creation of a whole series of institutions and settings such as public libraries, galleries, museums, and parks, in order to bring out the best in "ordinary" people.

Public Parks

The first manifestations of these ideals came in the arcadian landscapes of "rural" cemeteries like Mount Auburn in Cambridge, Laurel Hill in Philadelphia, and Green-Wood in Brooklyn. They were not only a response to the crowded and distasteful conditions of the unplanned and unregulated burial grounds in cities but also an attempt to express religious and social ideals. Religious ideals were expressed through monumental portals and sculptures, while social ideals were expressed in spaciousness and naturalistic scenery.

These cemeteries were influential because they inspired the creation of public parks as a means of carrying these ideals into the heart of the city, where, it was hoped, they might provide a civilizing, spiritually uplifting, and socially instructive setting. Very quickly, parks became one of the major issues in the struggle for political and institutional reform. For the present, however, we must focus on the design aspects of the park movement, and here the most influential figure was Frederick Law Olmsted, Sr.

Olmsted saw his work as serving the psychological and social needs of city residents to have access to a naturalistic landscape, a secluded escape from the dirt and noise of the city, a place for leisure and recreation, and an environment that would foster restraint and decorum. New York City's Central Park was Olmsted's first opportunity to achieve those goals on any scale. In 1858 he and Calvert Vaux won the competition for the design of the park, which was completed in 1862. Their design included a succession of specific areas for sport, recreation, and culture, all embedded in a picturesque landscape.

The park was integrated with the city by means of four avenues laid out with an elaborate system of independent traffic lanes, bridges, and underpasses that were designed not to interrupt the continuity of the landscape. The result was widely acclaimed, and Olmsted went on to design park projects in other cities (including Boston, Brooklyn, Buffalo, Chicago, Detroit, Milwaukee, Newark, Philadelphia, and San Francisco) and campuses for the University of California at Berkeley and Columbia University in New York, as well as collaborating again with Vaux on the Romantic-styled railway suburb of Riverside near Chicago.

Architects like Vaux and Downing, working within the genre of Arcadian Classicism, designed "picturesque" buildings (Figure 10.2) that were meant to unfold to the viewer one at a time as they were approached along the contrived curves of their site plans. Similarly, the interior spaces were asymmetrical, supposedly promoting feelings of freedom and discovery as visitors moved from room to room. Their facades were drawn from an eclectic range of decorative and picturesque styles, the most popular of which were neo-Gothic, Tuscan, and Moorish.

Commercial Confusion

Meanwhile, the growth of cities was overwhelmingly dominated by industrialists, speculators, and developers whose contribution to the built environment was guided by laissez-faire principles rather than progressive ideals. With few controls over design and construction, the result was a combination of the prosaic and the bizarre. Where appearance did not matter much to city makers, anonymous and cheap-looking buildings appeared. Where appearance did matter, buildings competed with one another through sheer size and exuberance. As art historian Barbara Rubin has pointed out, the "culture industry" was relatively new and there was no generally accepted standard of good taste in architectural style.

The city's new commercial and industrial elite, anxious literally to make its mark on the urban scene, was especially vulnerable to this uncertainty. Striving for an image of personal achievement and individuality and intent on monumentalizing their success for the benefit of future generations, this group commissioned great edifices of stone and brick from prominent architects, "only to find photographs of their new corporate headquarters illustrating 'Architectural Aberrations,' a regular feature in the professional journal *Architectural Record*."[11] This confusion prevailed until the 1890s, when the canons of good taste

FIGURE 10.2   A design for a villa in the "Tuscan" style, from Calvert Vaux's *Villas and Cottages*, 1857.

derived from the Beaux Arts style came to North America in the form of the City Beautiful movement.

BEAUX ARTS AND THE CITY BEAUTIFUL

The Beaux Arts movement took its name from L'Ecole des Beaux Arts in Paris, where, from the mid-nineteenth century, architects were trained to draw on Classical, Renaissance, and Baroque styles, synthesizing them in new buildings that might blend artfully with the significant older buildings that dominated European city centers. Though North American cities had no buildings of comparable vintage that needed to be complemented in this way, the Beaux Arts style provided a convenient packaging of High Culture that promised to resolve confusion and uncertainty about how to build the industrial city. The idiom was showcased by Daniel Burnham's neoclassical architecture for the World's Columbian Exposition in Chicago in 1893 (Figure 10.3). The temporary structures of the exposition showed what might be done: high culture could be married with American Arcadian Classicism.

FIGURE 10.3
The Beaux Arts style. The World's Columbian Exposition in Chicago, 1893, designed by Daniel Burnham.

A few years later, in 1899, this marriage produced a short-lived but widely influential movement of its own: the City Beautiful movement.[12] The thrust of the movement was decisively toward the role of the built environment as an uplifting and civilizing influence. The preferred architectural style was neoclassical, accompanied by matching statuary, monuments, and triumphal arches—all, if possible, laid out like Burnham's **White City** at the Chicago Exposition, with uniform building heights and imposing avenues with dramatic perspectives.

In 1901, Burnham collaborated with several others (including Frederick Law Olmsted, Jr.) on the McMillan Plan for Washington, D.C. (named after Senator James McMillan, Chairman of the Senate Committee on the District of Columbia). The purpose of the McMillan Plan was to rescue the Mall area from the neglected and unfinished framework derived from Pierre Charles L'Enfant's original plan of 1791. The centerpiece of the new plan was the redeveloped Mall and Federal Triangle, with neoclassical buildings along the Mall, a terminal memorial (the Lincoln Memorial), a pantheon (the Jefferson Monument), the Memorial Bridge, and a water basin (Figure 10.4).

Although the scheme was not completed until 1922, the plans and sketches provided enough publicity to ensure the immediate future of the City Beautiful movement. Burnham went on to draw up plans for Cleveland (1902), San Francisco (1905), and Chicago (1909) before his death in 1912. These plans were also very influential, though little of the San Francisco plan was actually realized. Meanwhile, John Olmsted (F. L. Olmsted, Sr.'s stepson) devised a plan for Seattle; others worked on schemes for Kansas City, Denver, and Harrisburg, Pennsylvania. In cities already too densely built up to accommodate a City Beautiful plan, the movement's themes found expression in a variety of inspirational statues, monuments, columns, and arches.

The success of the movement helped to foster two key developments in architecture and urban design. The first was professionalization. In 1897, for example, the American Park and Outdoor Art Association was founded, followed by the American Society of Landscape Architects in 1899 and the National Playground Association in 1906; courses in landscape architecture were introduced at Harvard University in 1900. The second development was an intensification of discourse and critical analysis on the aesthetics of the built environment. The City Beautiful movement had brought architecture and urban design to the forefront of public debate, provoking an intense critique of bourgeois art, architecture, and urban design that, in turn, helped energize the emerging design professions.

It is clear, looking back, that the City Beautiful movement was an explicit and rather authoritarian attempt to create moral and social order in the face of urbanization processes that seemed to threaten disorder and instability (see Chapter 17). It was successful because it allowed private enterprise to function more efficiently while symbolizing a noble idealism that was endorsed by the pedigree of Beaux Arts neoclassicism. "Over and beyond the good intentions of its leading figures and technicians, the City Beautiful movement perfectly fulfilled its true function of matching maximum planning with maximum speculation."[13]

FIGURE 10.4   The City Beautiful. This aerial view of Washington, D.C. shows the Mall and Federal Triangle, with neoclassical buildings along the Mall, a terminal memorial (the Lincoln Memorial) in the lower foreground, aligned with the Washington Monument and the Capitol, with a pantheon (the Jefferson Monument) and a water basin to the right.

Its success did not last long, however, because the dynamics of urbanization were changing even as Burnham, Olmsted, and their followers were prescribing remedies to old ailments. Streetcars and electric power had begun to turn cities inside out, the severe recession of 1893–1894 had shaken up relations between classes and introduced profound changes to the social and political organization of the city, and by the time of Burnham's death in 1912, automobiles were beginning to make their mark. In this new context monumentality was seen as impractical, while the movement's total lack of concern for housing was seen as elitist. The immediate cause of the movement's demise, however, was that it was completely unable to accommodate the raw energy and exuberance of architects and clients who wanted to build skyscrapers.

THE AMERICAN WAY: SKYSCRAPERS

Even as Burnham was drawing up his plans for the Chicago Exposition, the skyscraper had established its

FIGURE 10.5   The Flatiron Building in New York shortly after completion. Designed by Daniel Burnham and built in 1902, it is a fine example of a Beaux Arts skyscraper.

presence in the city. Indeed, Burnham himself had worked on the 16-story Monadnock Building in Chicago (completed in 1891) and derived most of his income from designing high-rise structures like the famous Flatiron Building in New York (1902: Figure 10.5). Chicago had in fact become the seedbed for skyscraper development because the technical preconditions for its development had coincided with the need to rebuild the center of the city after the fire of 1871. The first of these preconditions was fulfilled as early as the 1850s, when Elisha Graves Otis perfected the passenger elevator. In the same decade architect James Bogardus used wrought iron beams in designing a six-story building for Harper's, the publisher, in New York. But it was not until the late 1860s, with the construction of the Equitable Insurance Company Building (also in New York), that an iron-case framework rather than masonry was successfully used to carry the dead weight of a building.

Freed from the crushing weight of masonry that had to be spread in thick walls and across a wide base, buildings were suddenly allowed to rise to unprecedented heights on relatively small lots (Figure 10.6). Given the enabling technology of elevators and iron-cage construction, skyscrapers found their fundamental economic rationale in the inflated land values and increasingly specialized patterns of land use in the **central business district (CBD)** of the late nineteenth century. Even so, it took another technological innovation—the telephone—to make skyscrapers practicable settings for businesses. The telephone allowed

FIGURE 10.6   American cityscapes were revolutionized beginning in the early decades of the twentieth century by the ability of iron-cage frameworks to carry the dead weight of tall buildings.

businesses to dispense with human messengers, who would have clogged the elevators of tall buildings.

It was the awesome spectacle of such structures, however, that led to their first appearance in urban landscapes. Businesses—newspapers and insurance companies, in particular—saw the immediate benefits of advertising their presence, their success, and their dependability through highly visible, symbolic structures. From the start, therefore, the skyscraper form was less of an architectural style than it was an expression of economics and advertising (Figure 10.7).[14]

Nevertheless, the new form did provide a challenge for designers in that its format was so novel. The first architect to conceive a distinctive style for skyscraper construction was Louis Sullivan, working in Chicago in conjunction with his partner Dankmar Adler. Sullivan, influenced by Arcadian Classicism in his belief that nature should be made manifest through structure and ornamentation in art and architecture, was the author of the now-famous dictum, "form follows function." The Guaranty Building (Figure 10.8; later to become the Prudential Building), built in Buffalo in 1895, is generally reckoned to be not only the best of Sullivan's skyscraper designs but also the inspiration for hundreds of look-alikes

FIGURE 10.8   The Guaranty Building, built in Buffalo in 1895, was a distinctive contribution to the skyscraper form. Designed by Louis Sullivan, it soon became the model for hundreds of look-alike buildings in other large cities.

FIGURE 10.7   American Gothic. Early skyscrapers in Chicago, showing the highly decorated form that was intended to be awesomely spectacular.

produced by other architects in other cities around the turn of the century. Its most notable feature is considered to be the way it honestly expressed the steel framing, with columns of windows (emphasized by recessed terra-cotta transoms) rising between brick-clad piers that appear to be supported by a two-story stone base.

As skyscrapers proliferated and grew taller, they began to present problems. Within the canyons of larger CBDs, sunlight was cut off and air circulated poorly. As a result, some cities enacted **land use zoning** ordinances that required **setbacks**: Stepping back the upper stories of a building, beyond a certain height, in order to allow light and air to reach the street (Figure 10.9). This restriction resulted in the characteristic wedding-cake profile of interwar skyscrapers. By the 1920s skyscrapers had come to dominate the skyline of most large American cities, while even smaller cities had a centerpiece of 20–30 stories. By this time, however, cityscapes were beginning to reflect the influence of Modernism.

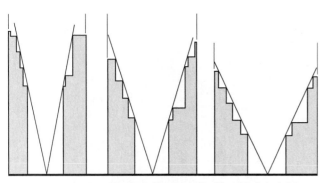

FIGURE 10.9   Zoning diagrams for New York height districts, 1916. These diagrams described the maximum permitted building height, in relation to the width of the street. Above a certain level, the mass of the building had to step back one foot for each additional three feet in height, in order to allow air and sunlight to street level. These laws produced skyscrapers that were stepped in form—a characteristic element in today's cityscape in midtown Manhattan.

MODERNISM: ARCHITECTURE AS SOCIAL REDEMPTION

The roots of the Modern Movement are deep and tangled. They can be traced to several independent reactions to the conspicuous consumption of bourgeois Victorian society and to the richness of conventional tastes in the decorative arts. These reactions, however, collided with one another, provoked further reactions, and fused into a Modern Movement with an avant-garde culture riven by dozens of twists, turns, contradictions, and controversies. Running through these tangled roots was a fundamental shift in aesthetic values that had a profound effect on the appearance and physical arrangement of urban landscapes. Stated simply, this shift rested on the conviction that the best design is pure, simple, and timeless; that good design should be functional; and that through good design it is possible to create the matrix for new, progressive social conditions.

Arts and Crafts and Art Nouveau

The prologue to the Modern Movement was written in the 1890s and early 1900s by the Arts and Crafts movement in England, by the Art Nouveau movement (called *Jugendstil* in central Europe) in France, and by Frank Lloyd Wright (Figure 10.10) in the United States.

The **Arts and Crafts movement**, led by William Morris, was a reaction to the extravagance of Victorian taste. Morris wanted to bring artists and craftsmen together in order to synthesize honest, simple, and popular forms. In architecture this meant drawing heavily on vernacular themes, something that proved

FIGURE 10.10   America's most celebrated architect, Frank Lloyd Wright. Although his visions for urban design were never fulfilled, his influence on domestic architectural style has been inestimable.

convenient and attractive to the planners of the first garden cities (see p. 144). The **Art Nouveau movement** emphasized sensuous shapes and organic themes as an alternative to Victorian extravagance. Art Nouveau style became widely used in graphic art, but in architecture it was limited to the embellishment of buildings rather than being reflected in their form. The terra-cotta transoms on Sullivan's Guaranty Building, for example, were decorated in an intricate Art Nouveau style.

The Early Modernists

Meanwhile, Frank Lloyd Wright, who had worked for a spell in Sullivan's office, had introduced a radically different style to residential architecture. Wright seems to have been influenced by the Japanese pavilion at the Chicago Exposition. His houses were long and low, with a geometry that owed nothing to Romanticism or neoclassicism. They were low-slung buildings with a massive central chimney, a hip roof with deep overhangs, and no attics or basements, which Wright declared "unwholesome" (Figure 10.11). This style is generally known as Wright's Prairie Style, after the pattern-book design he published in 1901 in the *Ladies' Home Journal* that he called "A Home in a Prairie Town."

Wright's departures from earlier styles were mild, however, in comparison with the revolutionary ideas that emerged as artists not only struggled to free themselves from the canons of bourgeois taste but also struggled to come to grips with the age of machinery, technology, and unprecedented speed. Modernism, in pursuit of progress, strove to break with history.

Many critics have set the turning point at 1907, the year Picasso painted *Les Demoiselles d'Avignon*, thus initiating the abstractionist challenge to representational art. Art critic Robert Hughes has suggested that the Cubism of Picasso and Braque was at least in part an attempt to capture the experience of constantly altering landscapes as seen from a moving train or automobile.[15] Architects took their cue from this abstractionism, seeking to design buildings appropriate to the needs and experiences of the machine age, translating the angularity of Cubism into built form.

Several future-oriented schools of thought had already emerged in response to the challenge of the machine age. One was the **Secessionists**, headed by Austrian architect Adolf Loos, whose principal motivation was the elimination of all "useless" ornamentation

FIGURE 10.11   The Robie House in Chicago, designed by Frank Lloyd Wright. For this house, one of the famous private residences in the world, Wright's distinctive horizontal emphasis, known as the Prairie Style, was adapted to a narrow city lot. The influence of the Prairie Style has echoed throughout suburbia ever since.

from architecture. Another was the **Deutscher Werkbund**, led by Peter Behrens (architect and chief designer for the German electrical manufacturer AEG). The Werkbund was a loose coalition of artists and craft firms founded in 1906 to reform the relationship between artists and industry, their guiding principle being that quantity and quality should complement each other. Behrens saw industrialization as the manifest destiny of the German nation, and his factory designs were for muscular temples to technological power. Other members of the Werkbund (including architects Bruno Taut and Walter Gropius), meanwhile, emphasized the need to overcome the alienating aspects of traditional society by leaving behind all the architectural refinements and symbolism of the old order and replacing them with a no-nonsense style.

The Bauhaus and the Modern Movement

Such sentiments were to become central to the Modern Movement in architecture as it developed in the 1920s. In general, the Modernists developed the idea of the importance of redesigning entire cities in the name of technological progress and social democracy. One subset, the **Futurists**, wanted nothing at all to do with the past. Led by Filippo Marinetti and represented in the field of architecture and urban design by Antonio Sant'Elia, they sought to provoke social and institutional change through cities that were to be stages for permanent revolt, with huge and spectacular edifices that were at once monuments to the masses and to technology. Sant'Elia's drawings of *La Citta Nuova* (1914) showed massive concrete buildings with soaring towers and huge parapets, enormous generating stations, and huge factories and airfields, the like of which were not to appear in real urban landscapes for another 50 years.

It was from these roots that there emerged in the 1920s a Modernism that was founded in the idea of architecture and design as agents of social redemption. Through industrialized production, modern materials and functional design, architecture (as opposed to mere building) could be produced inexpensively, become available to all, and thus improve the physical, social, moral, and aesthetic condition of cities.

It was the **Bauhaus School** that did more than any other single movement in promoting this ideal. Founded by Walter Gropius in Weimar in 1919 and later (1926) relocated to Dessau, the Bauhaus became the "refinery of . . . European avant-gardes,"[16] setting the standards for modern design in everything from teapots to workers' housing. The unifying themes were simplicity of line, plain surfaces, and suitability for mass production. The socialist ideology that was clearly embodied in the school's approach did not sit well, however, with Adolf Hitler's fascism, and in 1933 the school closed.

Many of the leading practitioners of the Bauhaus moved to the United States, and it was at this point that Modern architecture and urban design began to appear in North American urban landscapes. The decisive event was an exhibition at the Museum of Modern Art in New York City, organized in 1932 by two American architects, Philip Johnson and Henry-Russell Hitchcock. The exhibition included examples of work by Ludwig Mies van der Rohe (the last director of the Bauhaus and author of the famous dictum "Less is more") and many other European Modernists. More in hope than as a reflection of reality, the exhibition was called *The International Style*. It did not take long, however, for the Modernism of the International Style to become truly international as the principal language of design:

> Both Walter Gropius and Mies van der Rohe ended up in America, where, after World War II, their design influence would be almost unquestioningly accepted by the executives of great business corporations and institutions anxious to install themselves in buildings which symbolized progress and prosperity.[17]

Le Corbusier

The international impact of Modernism on urban design was sharpened by the energy and self-promotion of another key figure: the French-based Swiss architect Charles-Édouard Jeanneret (Figure 10.12), who is better known by his pseudonym, *Le Corbusier* ("crowlike"). Le Corbusier worked in the offices of early Modernist Auguste Perret in Paris and briefly with Behrens in Berlin before returning to France to work through his ideas on art, architecture, and **urbanism**. By 1919 he had conceived his basic formula for the relationship between architecture and cities: both should be a "machine for living."

In 1922 he published his ideas on the principles of urban design in a book entitled *La Ville Contemporaine*. The key, he argued, was to reduce the congestion of city centers by increasing their density—by building upwards, in other words. High-density, high-rise city cores, he pointed out, would leave plenty of space for wide avenues to carry automobile traffic and for green space for recreation (Figure 10.13).

La Ville Contemporaine was conceived as a class-segregated city, with the best-located, most spacious and best-appointed tower blocks reserved for elite

FIGURE 10.12   Le Corbusier, an ardent advocate of the need for architecture to embrace the message of the machine age.

cadres of industrialists, scientists, and artists. Blue-collar workers were to have smaller garden apartments located in satellite units at some distance from the central cultural and entertainment complex. His plan for Paris, the Plan Voisin, reflected this strategy without any concessions to the existing city or to its inhabitants. The Plan Voisin called for 18 700-foot towers that would have required the demolition of most of historic Paris north of the Seine (Figure 10.14). Inside the towers the apartments ("cells," as he called them) were to be uniform, with the same standard furniture.

Not surprisingly, such a radical, totalitarian plan attracted a lot of attention and a great deal of opposition. Le Corbusier claimed not to be able to understand the opposition, choosing instead to regard his own ideas as beyond the grasp of ordinary citizens. After the power elite failed to support his ideas and the Great Depression took away the ability of industrialists to back him, he revised his ideas on urban design. In *La Ville Radieuse* (1933) he adopted a rather different stance, arguing that everyone should live in giant collective apartment blocks called *unités*, with a minimum of interior space.

Although his reformulated prescriptions came no nearer to being realized at the time than his earlier

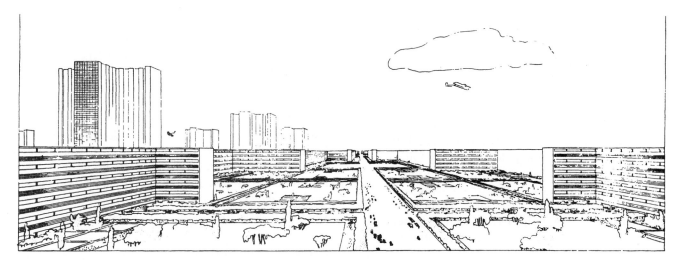

FIGURE 10.13   Illustration for *La Ville Contemporaine*, by Le Corbusier.

ones, they were to become widely influential in urban design circles. They were incorporated into the discourse of CIAM (the Congrés Internationaux d'Architecture Moderne), an international association of avant-garde architects formed in 1928 to promote Modernist architecture, and they found expression in the *Athens Charter*, a document published by CIAM in 1943 that set out the ideological basis of Modernist urban design. But the reasons for Corbusier's influence have as much to do with professional dynamics and avant-gardism as with aesthetics or functionality. The heroic scale of his ideas and his sheer irrepressibility drew admiration from architects and urban designers who wanted leadership and recognition, while his

willingness to confront the automobile era drew admiration from technocrats.

From this admiration grew a conventional wisdom that was centered on the need to modernize cities through ruthless redevelopment, tearing out their centers and replacing them with high-rise housing linked by intrusive freeways. This impulse, indeed, was what began to materialize at an alarming rate in American cities in the 1960s (see Chapter 17), prompting urban geographer Sir Peter Hall to observe that "The evil that Le Corbusier did lives after him; the good perhaps interred with his books, which are seldom read for the simple reason that most are unreadable."[18] Hall's criticism of Corbusier

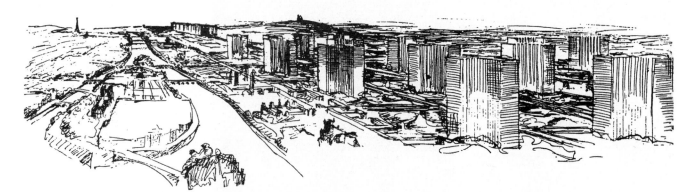

FIGURE 10.14   Illustration for the Plan Voisin, by Le Corbusier.

and his followers is not directed at his designs as much as at the mindless bureaucratic arrogance and political naïveté that led the designs to be inserted to an urban process to which they were not actually suited.

But Le Corbusier's work and influence were not finished with *La Ville Radieuse*. After World War II he forsook the heroic scale of urban design in favor of individual structures. His work ranged from expressionist designs like his famous chapel at Ronchamp in France (built 1950–1955; interpreted variously as resembling a nun's three-cornered hat, hands held in prayer, or the prow of a ship) to Cubist-inspired angular concrete buildings. The most celebrated example of the latter is an apartment block in Marseilles—*L'Unité d'Habitation* (Figure 10.15)—that carried through his formula of a "machine for living" with its integrated community services, daycare facilities, and shops.

It was the style of L'Unité, however, that proved to be Corbusier's second legacy to mid-twentieth century urbanization. The poured concrete sections and panels, textured and sculpted with recessed windows and balconies, not only struck the Modernist chord but also proved to be relatively inexpensive to construct and amenable to prefabrication.

It was not long before cities everywhere were being recast in the image of L'Unité. Urban geographer Edward Relph, describing the evolution of contemporary urban landscapes, wrote:

> The dominant lines can be made vertical or horizontal or into a grid; the colours of the glass and metal can be changed; the proportions of the windows can be varied; the shape of the building can be altered to fit the site. . . . [B]ut . . . [t]he result is that most Modernist buildings are almost indistinguishable one from another. They merge

both in perception and in memory into a confusion of pale tones, sharp angles and cereal box forms, standing on end downtown and lying flat in the suburbs.[19]

This pervasive anonymity eventually contributed, as we will see, to a strong reaction on the part of corporate owners, developers, the general public, and, indeed, many professional architects. We should note, however, that it was not so much Corbusian design per se that was at the root of the problem. Rather, it was a question of the cheapness of the execution of Corbusian forms, coupled with their imposition in thoughtless (and sometimes ruthless) ways. We should also note that within architectural and design circles, Corbusier has been elevated alongside Wright, Gropius, and Mies van der Rohe as the embodiment of artistic genius.

An American Response

While Corbusier was developing his ideas, Frank Lloyd Wright was pursuing the possibility of an American response to the challenge of machine age, automobile-based urbanization. His vision was for a "Usonian" (a word-play on U.S. own) future. It was a vision that was heavily influenced by the individualism and naturalism of Jefferson, Thoreau, and Emerson. Wright, like other American intellectuals of the automobile era, saw the popularity of the automobile as a threat to the special good fortune of America.

Wright hoped that, through planned metropolitan decentralization, cities might be spared from the sprawl and congestion threatened by the automobile. From this position it was only a short step to the promise of the automobile as the means of gaining access—for some, at least—to the blessings of Nature. In this way,

FIGURE 10.15   L'Unité d'Habitation, Corbusier's influential apartment block, a "machine for living" that became the model for countless Modern buildings throughout the world. The Society for the General Esthetics of France called it "an eyesore and a public nuisance."

two powerful American traits—a pioneer affinity with the outdoors and a love of the automobile—might be reconciled.

Frank Lloyd Wright became America's home-grown architectural hero by pursuing this very formula. **Usonia** was to be the physical framework for a new American way of life where "wage-slaves" would be emancipated from the trap of congested but expensive cities, living instead in "Prairie Style" homes in a semirural setting. Where Le Corbusier wanted to increase densities and build tall, Wright wanted to decrease densities and spread out. Where Le Corbusier wanted people to live in identical cells, Wright wanted them to live in differentiated and individualized homes that were designed to be in harmony with their natural surroundings. His ideal city, conceived in the mid-1920s but not written up until some 20 years later,[20] was **Broadacre City** and it was to be built on the basis of two new technologies: the automobile and mass-production building technology using high-pressure concrete, plywood, and plastic.

The inaccessibility inherent to the large lots and low densities of Broadacre City was to be conquered by a network of landscaped **parkways** and freeways, with the focal point of semirural neighborhoods being provided by huge public service stations, architectural centerpieces that would provide a broad range of low-order goods and services. Much to Wright's frustration and annoyance, Broadacre City drew both widespread attention and comprehensive criticism. "For his pains, he was attacked by almost everyone: for naïveté, for architectural determinism, for encouraging suburbanization, for wasteful use of resources, for lack of urbanity, above all for being insufficiently collective in his philosophy."[21]

Meanwhile, however, some of Wright's designs for individual buildings—such as administration buildings—for example, the administration building for the S.C. Johnson Wax Company in Racine, Wis., and the Kaufmann House ("Fallingwater": Figure 10.16) in Bear Run, Pa.—had established him as an architect of first rank. It was his messianic role that was unsuccessful. He never gave up his Usonian vision—indeed, he came to identify himself completely with his own myth—but after World War II he concentrated more on the design of individual buildings in a new phase, marked by the use of curvilinear forms, arches, and spirals (the Solomon R. Guggenheim Museum in New York (Figure 10.17) is the best-known example), that itself was very influential in introducing a Modernist **Expressionism** into American urban landscapes.

The International Style and Late Modernism

Concurrent with the postwar Modernism of Le Corbusier and Wright, and equally influential, was the

FIGURE 10.16   Fallingwater, Frank Lloyd Wright's most celebrated attempt to harmonize architecture with nature. The stream actually runs through the house.

continuation of the International Style associated with Mies van der Rohe. His dictum, "Less is more," was violated by the sculpted concrete forms favored by Corbusian designers and by any form of Expressionism. Mies himself had pointed to an alternative interpretation of Modernist ideals in 1938 in his designs for the campus of the Illinois Institute of Technology in Chicago. There, all the buildings were low-slung stark steel-and-glass cubes arranged rigidly at right angles to one another (Figure 10.18).

Downtown settings required a different solution, first suggested in 1951 by Gordon Bunshaft of Skidmore, Owings and Merrill in his design for Lever House on New York's Park Avenue. By setting back the base of Lever House, Bunshaft was able to have a steel-and-glass skyscraper with pure form: no wedding-cake setbacks in the building itself. The space left at the foot of the building, meanwhile, could be used as a corporate plaza. A few years later, Mies himself designed an even sleeker steel-and-glass skyscraper, the Seagram Building (Figure 10.19), just across Park Avenue from Lever House.

Between them these two buildings immediately set the preferred image for corporate skyscrapers. Their imprint can be seen on the downtown landscape of any large American city. It was not until later in the 1960s, however, that technological improvements

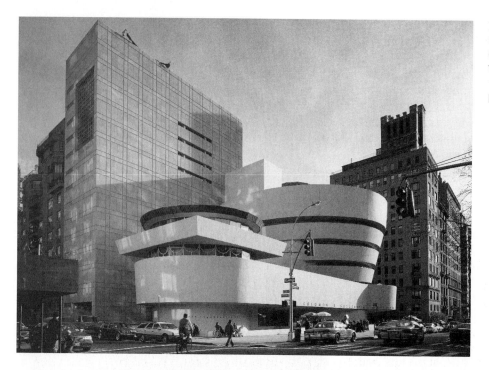

FIGURE 10.17 The Guggenheim Museum in New York City. Designed by Frank Lloyd Wright and built between 1943 and 1957, this is an example of Modernist Expressionism.

made it possible to have glass facades without having to pay the price of uncomfortably high levels of solar gain. At about the same time, new techniques for mounting glass made it possible to have continuous glass surfaces. Putting the two together resulted in "glass-box" architecture, the ultimate expression of "Less is more," buildings that disappear in the reflection of their own surroundings (Figure 10.20). Because of the relatively low cost of cladding buildings in glass, together with the spectacular effect, this "aviator

sunglasses" style became very popular for corporate buildings of all kinds in the 1970s and 1980s.

By this time, however, the execution of Modernism had begun to drift somewhat from its founding ideology, while variants such as the mirror-glass box represented an evolution that was captured by architecture critics in the label Late Modern. Among the more commonly occurring of these variants were windowless blank boxes of brick or concrete (made possible by sophisticated lighting and air-handling

FIGURE 10.18 S. R. Crown Hall, on the campus of the Illinois Institute of Technology, designed by Mies van der Rohe and built in 1945–1946.

FIGURE 10.19   The Seagram Building, on the other side of Park Avenue from Lever House: Mies van der Rohe and Philip Johnson, architects, 1956–1958. This building is important not only for its success in articulating Modernist principles of form but also for its innovative granite plaza that provided the skyscraper form with a new sense of grandeur and prestige.

FIGURE 10.20   The ultimate face of Modernist architecture: buildings that disappear in the reflection of their own surroundings—in this case, the glass facade of a modern skyscraper shows the reflection of the Empire State Building.

technologies—Figure 10.21); buildings that dramatized their structural framework with exposed girders or cantilevers; and buildings that emphasized technology, with exposed ducts, pipes, elevators, and railings (Figure 10.22).

The Critique of Modernism

According to some commentators, the symbolic death of Modern architecture took place at 3:32 P.M. on July 15, 1972, with the dynamiting of the Pruitt-Igoe project, a group of 33 public housing apartment blocks in St. Louis (Figure 10.23). Seventeen years before, the architect for the project, Minoru Yamasaki, had won an award from the American Institute of Architects. Good as Pruitt-Igoe looked on the drawing board, it turned out to be unlivable. It was the tenants themselves who suggested that their homes be dynamited.

Although Modern architecture has never actually died, the demolition of Pruitt-Igoe did draw attention to a crisis in architecture and urban design that had first been articulated in 1961 by Jane Jacobs in her famous book *The Death and Life of Great American Cities*.[22] Jacobs reasoned that the design professions had taken away the life and vitality of cities, tearing out their sclerotic hearts only to replace them with a "great blight of Dullness" in the form of Corbusian tower blocks.

A decade later, as Pruitt-Igoe was being prepared for demolition, Oscar Newman published an equally influential attack on Modernist housing. He argued that Modern architecture had been too preoccupied with form, with architecture as sculpture, and insufficiently attentive to people's need for functional, defensible spaces.[23] Specifically, he suggested that much of the petty crime, vandalism, mugging, and burglary in modern housing projects was related to a weakening of community life and a withdrawal of local social order caused by the inability of residents to identify with, or exert any control over, the space beyond their own front door. Sociability, in short, had been "designed out" by Modernists, along with color, variety, and ornamentation.

FIGURE 10.21   Late Modernism. Salk Institute, La Jolla, California. Architect Louis Kahn.

FIGURE 10.22   Late Modernism. Lloyds Building, London. Architect Richard Rogers.

In fact, there have been many worthy heirs to the early Modernists, architects whose contributions to the built environment in cities around the world have been both positive and influential. Among these can be included the work of Louis Kahn, Richard Meier (Figure 10.24), I. M. Pei, Cesar Pelli, Norman Foster, Richard Rogers, Mario Botta, Tadao Ando, Christian Portzamparc (Figure 10.25), Renzo Piano, Rem Koolhaas, Jacques Herzog, and Pierre DeMeuron (Figure 10.26). But although their designs were successful refinements of the clarity of form intrinsic to early Modernist design, they all more or less abandoned the idea of any social purpose to Modernism. This was highlighted by the abstract theorizing of iconoclastic "art compound" architects like Peter Eisenman and Bernard Tschumi, the fantasy architecture of celebrity architects like Frank Gehry and Zaha Hadid, and the downright elitism of high-profile commercial architects like Robert Stern, Charles Gwathmey, and Charles Moore. As a result, Modernist design in general came increasingly to be regarded as elitist and dysfunctional. It was in this climate that postmodern architecture began to prosper when the post-1973 recession burst the bubble of postwar growth. When economic prosperity returned a few years later, it was with a very different cultural sensibility: one of consumer materialism rather than one of radical progressivism. As a result, the idea of a futuristic, Modernist Utopia became increasingly anachronistic.

FIGURE 10.23   The demolition of prize-winning apartment blocks in the Pruitt-Igoe project in St. Louis has come to symbolize the public's loss of faith in Modernist urban design.

THE POSTMODERN INTERLUDE

The explicit beginnings of postmodern architecture stem from the work of architect Robert Venturi, who openly challenged Mies's aphorism. "Less is a *bore*," argued Venturi; hybrid elements should displace Modernist purity, witty and ironic references should be included in design, and complexity and contradiction should have a place. In short, architects should "Learn from Las Vegas."[24] Venturi was really arguing for an architecture appropriate to a new phase of urbanization:

In the Venturi cosmology, the people could no longer be thought of in terms of the industrial proletariat, the workers with raised fists, engorged brachial arteries, and necks wider than their heads. . . . The people were now the "middle-middle" class, as Venturi called them. . . . They were the "sprawling" masses, as opposed to the huddled ones.[25]

What the middle-middle classes wanted most, it transpired, was to live in comfortable and affluent-looking houses in neighborhoods with human scale and

FIGURE 10.24   The Getty Center, Los Angeles, California. Architect Richard Meier.

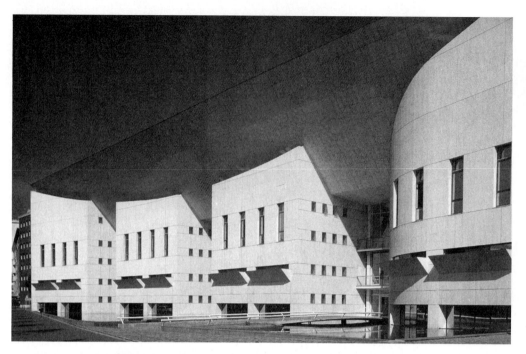

FIGURE 10.25
Cite de la Musique in Paris, France. Architect Christian Portzamparc designed this music school.

decorative detail and to work and shop in grand and spectacular settings. This vision was not lost on developers with upscale businesses and consumers in mind: They soon realized that although it would cost more to build a "rich" building, it would sell or rent more quickly—and often at a premium—because it can project the appropriate "look." The shift in attitudes was soon reflected in the pages of trade magazines.

In contrast to the abstract formalism of Modern architecture, postmodern buildings were to be scenographic, decorative, and full of signs and symbols. Postmodern architecture is by definition wide-ranging and eclectic. One of its principal characteristics is "double coding," combining Modernist styling or materials with something else—usually historical or vernacular motifs. This mixing allows the deployment of the symbolism of everything from historicism and revivalism to metaphysical references and eclectic pastiche. Postmodern architectural style is the style of styles (Figure 10.27). It was this self-conscious stylishness that made it attractive to the middle classes and developers.

In the 1980s and 1990s commercial and residential townscapes throughout metropolitan areas were suffused with the hallmarks of postmodern design. New single-family homes were built in revival styles or historicist modes, while shoebox stores and offices are peppered with arches, atria, columns, keystones, semicircular windows, and cornices; and office villages and shopping centers came to resemble period stage settings—"lite" architecture, the built environment's equivalent of easy-listening music, lite beer, and lo-cal snacks. Meanwhile, celebrity architects like Michael Graves and Philippe Starck were featured in advertisements for a variety of consumer products and lent their names to lines of "designer" housewares.

Postmodernism and Neo-Fordism

Postmodern design was sustained by a new alliance of taste and capital. Instead of an alliance between a liberal elite and public capital or between a cultural avant-garde and corporate capital, it developed through an alliance between representatives of the status-oriented and consumption-oriented society of the 1980s and 1990s and the managers of flexible capital.

FIGURE 10.26   Goetz modern art gallery in Munich, Germany. Architects Jacques Herzog and Pierre DeMeuron.

FIGURE 10.27   Postmodern style. (a) The AT&T Building in midtown Manhattan, a Postmodern skyscraper designed by Philip Johnson, formerly a leading exponent of Modernist design. The building is now occupied by Sony Music. (b) The glass elephant building in Maximilian Park, Hamm, Germany. Metaphorical references are one of the many wide-ranging characteristics of Postmodern architecture.

In terms of the broader context of urbanization, postmodern architecture, along with postmodern culture and philosophy, can thus be interpreted as having emerged in tandem with the development of globalized, more flexible forms of capitalist enterprise. David Harvey argues that, although this transition and the critique of Modernism had been under way for some time, it was not until the international economic crisis of 1973 that the relationship between art and society was sufficiently shaken to allow postmodernism to became both accepted and institutionalized as the "cultural clothing" of neo-Fordism.[26] As we will see in Chapter 11, the real estate development industry was quick to adopt postmodern design in order to ensure product differentiation and maximize exchange values. At the same time, the new conservatism and materialism of Western society that emerged in the 1980s and 1990s (Chapter 4) produced a marketplace eager for these packaged products.

Packaged Landscapes

Within the broader cultural shift associated with neo-Fordism, the symbolic properties of places and material possessions have assumed unprecedented importance, with places becoming objects of consumption in themselves. Meanwhile, as producers of all kinds of material goods have sought out profitable niche markets, packaging has become a key marketing strategy. These trends have found expression in the built environment of most cities in Western countries and in many of the major cities of the rest of the world.

The preferred architectural styling for much of this has been postmodern. Having emerged as a deliberate reaction to the perceived shortcomings of Modern design, its emphasis on decoration and self-conscious stylishness has made it a very convenient form of packaging for the new global consumer culture. It is geared to a cosmopolitan market, and it draws quite deliberately on a mixture of elements from different places and times. In many ways it has become the transnational style for the more affluent communities of the world's cities.

The result is a mosaic of packaged developments and mega-projects: refurbished heritage and cultural zones, waterfront redevelopments, campuses and technopoles for high-tech industry, and airport complexes, as well as business centers, condominium complexes, master-planned communities, and shopping malls. The developers of business parks bundle office space with day care centers, fitness centers, and integrated retail and entertainment spaces, surrounded by lush landscaping and perhaps a nine-hole golf course. Shopping malls are themed and packaged with movie theaters and dining, exhibit, and performance spaces. Condominium complexes are packaged with high-bandwidth Internet access, enhanced telephone services, movies on demand, party centers, pools, and fitness centers. Private, master-planned residential communities are packaged with security systems, concierge services, bike trails, "town" centers, and even elementary schools.

Marketing packaged landscapes like these has been very successful in metropolitan regions around the world. In Istanbul, Turkey, communities in the **edge cities** of Esenkent and Bogazköy are modeled on American suburbia. In Manila in the Philippines, private master-planned communities are packaged with privatized infrastructure networks of roads, drainage, water supplies, power, and telecommunications, and designed and marketed as fragments of urban Europe, with names like "Brittany" and "Little Italy." In the United States some 47 million people—one in six of the total population—live in 230,000 privately planned residential communities, and half the new home sales in big cities are in these communities.

New (Retro) Urbanism

New Urbanism is a school of thought within urban design circles that advocates one particular kind of privately planned residential community: "neotraditional communities" that seek to reproduce the simple, communitarian, white-picket fence ambience, appearance, and serenity of small-town, walkable neighborhoods (Figure 10.28). It is based on a prescriptive system of design that seeks to replicate an older way of making places through the adoption of land use zoning regulations that require traditional street grids, mixed-use zoning, careful regulation of building materials and massing, and an emphasis on creating pedestrian-friendly environments. It is conventionally championed as a response to urban sprawl, as an appropriate residential form for edge cities, as a sustainable setting that minimizes automobile use, and as a new urban form appropriate for new lifestyles.

But New Urbanism has been widely criticized as catering to pretentious materialism, social exclusion and fear, and as being based on a selective recall of urban history. Critics of New Urbanism argue that there is nothing either new or urban about it, and that the outcome at best is very bland rear-guard buildings; at worst, kitschy, Disneyesque architecture that is more about imageability than liveability. In contrast to the hype surrounding New Urbanism, it has succeeded only in producing a plethora of middle-class subdivisions that are isolated from their host communities

FIGURE 10.28   New Urbanism. Seaside, Fla., an upscale resort community designed by Andres Duany and Elizabeth Plater-Zyberk to embody "neo-traditional" urban design.

by private management and retro architecture that is too naively artificial to bear its own stylistic weight. New Urbanism's claims in terms of automobile use and sustainability are belied by actual commuting patterns and levels of SUV ownership.

New Urbanism relies on webs of "servitude regimes" (i.e., covenants, controls, and restrictions that regulate both the physical environment and social comportment) that serve to frame a distinctive physical and social environment that appeals both to producers and to consumers. For producers, a key mechanism is the residential community association, which is used to establish by-laws, a constitution, and an overarching servitude regime in order to ensure stability until the subdivision is completed and sold off. Developers thus become benevolent dictators, establishing a tyranny of taste that defines the "legal landscapes" of suburban America. For consumers, these servitude regimes offer a means of narrowing uncertainty, protecting equity values, and, above all, establishing the proscenia—the backdrop and "stage"—for their lifestyles.

Historic Preservation

Historic preservation has been a significant architectural movement during the past few decades in the United States.[27] It has saved, refurbished, and provided a new lease on life to downtown department stores, office blocks, hotels, government buildings, railroad stations, and warehouses (Figure 10.29). It is no accident that it has coincided with the emergence of postmodern design: It must be interpreted as drawing heavily on some of the same economic and sociocultural forces. Historic buildings and districts lend

both distinctiveness and identity; they also form an obvious linkage with that aspect of postmodern culture that emphasizes the past, the vernacular, and the decorative. In the United States the National Historic Preservation Act of 1966 established the National Register of Historic Places as the official list of national historic resources worthy of preservation.[28] In 1968 only about 1,000 properties and districts were listed on the National Register of Historic Places; in 1985 there were about 37,000 entries, and by 2004 there were nearly 77,000. Over the same period the number of communities with historic district commissions with protective powers increased from around 100 to well over 2,000.[29] Between 1981 and 1986 in particular, the availability of tax credits and accelerated depreciation benefits for investments in historic property significantly bolstered historic preservation in U.S. cities.

But in most cities historic preservation was initiated by citizen groups that had been produced by the "counterculture" of the late 1960s. Confronting developers and public agencies, they gained widespread support by tapping the anti-Modernist attitude that had developed among the public. One of the landmark cases was the victory by Don't Tear It Down, a Washington, D.C., preservation group that succeeded in stopping the demolition of the Old Post Office, a Romanesque Revival federal building on Pennsylvania Avenue that developers subsequently adapted to an upscale market hall. During the 1970s such issues became an important element in local politics (Chapter 16).

It did not take long before historic preservation became institutionalized, with city councils passing legislation and city planners making plans that reflected the shift in values among the electorate. Developers were

FIGURE 10.29 Head House Square, part of the Old Market Restoration project in Philadelphia. This historical district contains many well-preserved houses from the eighteenth and nineteenth centuries. Market House (on the right) dates to the eighteenth century.

FIGURE 10.30   Facade preservation has become a popular compromise between redevelopment and the conservation of entire structures. This new luxury condominium development in New York City rises behind an old facade after the decision was made to keep the old facade and construct the new building behind it.

also quick to respond. Instead of offering resistance, developers have taken the initiative, hiring architectural historians and conservation experts and seeking out older buildings ripe for reconditioning. Preservation groups, meanwhile, having established a secure power base, began to invite the newly enlightened developers to serve on their boards, to accept donations from the development industry, and to compromise with, rather than confront, redevelopment projects. The result has been that parts of some downtowns, particularly on the East Coast, now amount to tableaux

of preservation projects. In a graphic reflection of the compromises between developers and preservation groups, many of the elements in these tableaux are buildings whose facades have been saved but whose interior spaces have been entirely remodeled to accommodate contemporary office layouts, air conditioning, and wiring for new telecommunications systems (Figure 10.30)—a practice referred to irreverently as "facodomy" among Modernist designers. This practice is also prevalent in many European cities where historic preservation regulations tend to be much more stringent than in the United States.

DESIGN FOR DYSTOPIA

Within cities that contain intensifying concentrations of poor, minority households, growing numbers of homeless people, and increasing levels of crime, violence, and vandalism, security has become a "positional good" (i.e., a measure of social status). The casualties of economic and social change—the poor, the socially marginalized, and the disadvantaged—are perceived as an unnecessary and undesirable part of the contemporary landscape of planned communities, gallerias, malls, office plazas, festival marketplaces, and so on.

As a result, architecture and urban design have increasingly incorporated a combination of built-in electronic surveillance and carefully designed architectural policing of social boundaries (Figure 10.31). This is a direct reflection of the intensifying social polarization associated with **neo-Fordism**. Gated subdivisions are a form of privately planned residential

FIGURE 10.31
Security and surveillance systems have become an important component of urban design as more and more city spaces become exclusive and exclusionary, as in the case of this gated Los Angeles community in Brentwood, California.

community (some of which also employ New Urbanist designs—as if traditional neighborhoods were gated!). More than 7 million households in the United States live in developments behind walls and fences—about 4 million of them in communities where access is controlled by gates, entry codes, key cards, or security guards. Gated developments are more prevalent in Sunbelt metropolitan areas such as Dallas, Houston, Phoenix, and Los Angeles, but they are rapidly becoming popular in older metropolitan areas like New Orleans, Long Island, N.Y., Chicago, Atlanta, and Washington, D.C. A significant number of gated communities in the United States are, moreover, inhabited by minority households—especially Hispanics—with moderate incomes.

Fortress L.A.

In "Fortress L.A.," described by urban design critic Mike Davis,[30] affluent neighborhoods are themselves miniature fortresses, with encompassing walls, restricted entry points, privatized roadways, overlapping public and private police services, homes that bristle with alarm and surveillance systems, and lawns staked with signs that advise would-be trespassers of an armed response. In less-affluent neighborhoods, gates are replaced by random police checkpoints, homes have wrought-iron grilles over their windows and are surrounded by chain-link fences, and thousands of rooftops are painted with identifying street numbers for the benefit of the police helicopters.

Downtown, and in the commercial areas of edge cities, "public" spaces—donated by developers in order to secure zoning permissions and development rights—are monitored by cameras and security personnel. Street furniture and landscaping features convex "bum-proof" benches and sprinkler systems programmed to drench unsuspecting sleepers at random times during the night. Office builders maintain their distance from unwanted and inappropriate visitors by removing logos and names from street-level access ways and retreating behind mirror-glass facades. Retailers find safety in galleries or in developments that can be reached only by elevated pedestrian ways that are easily monitored.

Design for commercial facilities in less-affluent areas is epitomized by the Martin Luther King, Jr. Center in Watts, with its 8-foot-high wrought-iron fence, video cameras equipped with motion detectors, floodlighting, infrared beams, and a security observatory that contains the headquarters of the shopping center manager, the monitoring and communications center for private security personnel, and a substation of the Los Angeles Police Department.

All this paraphernalia has been translated into High Design through the architecture of Frank Gehry. Beginning in the 1960s, Gehry found ways of designing for upscale uses amid decaying neighborhoods. His "stealth houses" (Mike Davis's term) hid their luxurious interiors behind blank or proletarian facades. The frontage for his Danziger Studio in Hollywood, for example (Figure 10.32), is simply a massive gray wall, "treated with a rough finish to ensure that it would collect dust from passing traffic and weather into a simulacrum of nearby porn studios and garages."[31]

FIGURE 10.32 The Danziger House and Studio, Los Angeles. Architect Frank Gehry.

Later, encouraged by the postmodern penchant for irony, Gehry began to recycle elements of decaying and polarized urban landscape into his buildings—roughly cast concrete, for example, and chain-link fencing—that made his 1980s designs the equivalent of post-Punk, radical-chic clothing. Gehry transformed the security theme from a secondary dimension of design, where it had been low-profile and high-tech, to an explicit fortress style, high-profile and low-tech. Thus his Regional Branch Library in Hollywood looks as if it is built to withstand a siege, with 15-foot security walls of stucco-covered concrete block, antigraffiti barricades covered in ceramic tile, a sunken entrance, and stylized sentry boxes. We can read this architecture either as a mordantly witty parody or as a prologue to the text of 21st century urbanization.

FOLLOW UP

1. Be sure that you understand the following key terms:

    Art Nouveau movement

    Arts and Crafts movement

    Beaux Arts movement

    City Beautiful movement

    Expressionism

    Futurists

    International Style

    Late Modernism

    Modern Movement

    Postmodernism

    Romanticism

    Secessionists

    setbacks

    townscapes

2. Be sure that you understand the significance of the following in relation to the development of contemporary town- and cityscapes:

    American Renaissance

    Athens Charter

    Bauhaus School

    Broadacre City

    Chicago's White City

    Deutscher Werkbund

    passenger elevators

    Guaranty Building, Buffalo

    Laurel Hill, Philadelphia

    La Ville Radieuse

    legitimation

    "Less is more"

    L'Unité d'Habitation

    New Urbanism

    product differentiation

    Pruitt-Igoe

    Usonia

3. With what characteristics of architecture and urban design do you associate the following?

    Le Corbusier

    Frank Gehry

    Philip Johnson

    Frederick Law Olmsted, Sr.

    Mies van der Rohe

    Louis Sullivan

    Robert Venturi

    Frank Lloyd Wright

4. Select a large town, city, or metropolitan area you know well and attempt to "read" its biography through the "text" of its built environment. What districts and neighborhoods are distinctive in terms of their built environment, and why? Are there distinctive architectural landmarks or "signature" structures? What do they symbolize, and to whom? Are there town- or cityscapes that are typical of American towns or cities in general? Of the region? Of particular social groups? Are there settings that are unique to the place you have selected and, if so, what is it about them that makes them unique?

5. Work on your *portfolio*. The material in this chapter lends itself especially well to sketches, photographs, and video recordings. Be selective in the way you put these together—choose specific themes, or particular kinds of townscapes. Provide captions, annotations, or commentaries that explain or interpret them in relation to the overall "text" of the city.

In addition, you could select a landmark building in your nearest town or city, find out as much as you can about who commissioned it and who designed it; and then write a short essay or put together a video, Microsoft PowerPoint, or web-based presentation that explains the building in relation to its geographical and historical context. You could also investigate buildings and districts that are currently designated for historic preservation: How many are there? Where are they? What sorts of places are involved, in what parts of the city? If you live in the United States, you could start with the *National Register of Historic Places* (see the National Park Service website [*http://www.cr.nps.gov/nr/about.htm*] or try the historic preservation office of your state or city).

6. If you have time, read a paperback with a story that has as its backdrop some aspect of architecture and urban design covered in this chapter. There are many to choose from, including *The Devil in the White City: Murder, Magic, and Madness at the Fair that Changed America*, by Erik Larson (New York: Crown, 2003) and *The Celebration Chronicles: Life, Liberty, and the Pursuit of Property Value in Disney's New Town*, by Andrew Ross (New York: Balantine, 1999).

KEY SOURCES AND SUGGESTED READING

Blakeley, E. J., and M. G. Snyder. 1997. *Fortress America: Gated Communities in the United States*. Washington, D.C.: Brookings Institution Press.

Clay, G. 1974. *Close Up: How to Read the American City*. New York: Praeger.

Dear, M. 1999. *The Postmodern Urban Condition*. Oxford, UK: Blackwell.

Ford, L. R. 1994. *Cities and Buildings: Skyscrapers, Skid Rows, and Suburbs*. Baltimore, Md.: Johns Hopkins University Press.

Hall, P. 2002. *Cities of Tomorrow*, 3rd ed. New York: Basil Blackwell. See especially Chapters 6 and 7.

Hughes, R. 1980. *The Shock of the New*. New York: Knopf. See especially Chapter 4: "Trouble in Utopia."

Knox, P. L. 1987. The Social Production of the Built Environment. Architects, Architecture, and the Post-Modern City. *Progress in Human Geography* 11: 354–77.

Knox, P. L., ed. 1993. *The Restless Urban Landscape*. Englewood Cliffs, N.J.: Prentice Hall.

Larsson, L. O. 1984. Metropolis Architecture, pp. 191–220 in *Metropolis 1890–1940*, ed. A. Sutcliffe. Chicago, Ill.: University of Chicago Press.

Relph, E. 1987. *The Modern Urban Landscape*. London: Croom Helm.

Sanchez, T. W., and R. E. Lang. 2002. *Security versus Status: The Two Worlds of Gated Communities*. Census Note 02.02. Alexandria, Va.: Metropolitan Institute, Virginia Tech.

Scully, V. 1988. *American Architecture and Urbanism*, 2nd ed. New York: Henry Holt.

Soja, E. 2000. *Postmetropolis*. Oxford, UK: Blackwell.

Wiseman, C. 1998. *Shaping a Nation: Twentieth-Century American Architecture and Its Makers*. New York: W. W. Norton.

RELATED WEBSITES

Archpedia Architectural Encyclopedia:
http://www.archpedia.com/
The Archpedia site contains very useful and well-illustrated information on a variety of architectural styles (including Prairie, Modern, and postmodern) and famous architects (including Frank Lloyd Wright, Le Corbusier, Mies van der Rohe, Walter Gropius, Louis Kahn, I. M. Pei, Frank Gehry, Michael Graves, Cesar Pelli, Robert Venturi, Renzo Piano, Richard Rogers, Rem Koolhaas, Norman Foster, and many others).

The Great Buildings Collection:
http://www.greatbuildings.com/gbc.html
This website is a gateway to architecture from around the world and through history and includes 3-D models, photographic images, and architectural drawings, plus commentaries, bibliographies, and web links, for hundreds of buildings and well-known architects.

A Digital Archive of American Architecture:
http://www.bc.edu/bc_org/avp/cas/fnart/fa267/contents.html
This Boston College website contains images dating back to the seventeenth century in America and photos of prominent architects, building types and styles, and world's fairs (including the Columbian Exhibition of 1893 in Chicago).

U.S. National Park Service, National Register of Historic Places: *http://www.cr.nps.gov/nr/about.htm*
This website offers information on the National Historic Preservation Act of 1966 and the National Register of Historic Places.

11

THE URBAN
DEVELOPMENT PROCESS

Although architecture and urban design are important in contributing to the character of distinctive morphological subareas in cities, much of the decision making about *what kind* of structure gets built and *when* and *where* is in the hands not of architects and urban designers but of other "city makers" such as developers and politicians. It is useful to think of urban development as a process that involves a variety of "actors" or decision makers, each with rather different goals and motivations. As they interact with one another over specific development issues, they constitute an organizational framework for city building. These frameworks have been called "structures of building provision." It is through these frameworks, or structures of building provision, that the urban fabric is created and modified. The resulting process of urban development involves a complex ebb and flow of investment, disinvestment, and reinvestment.

CHAPTER PREVIEW

In this chapter we take a closer look at the processes and some of the principal actors involved in the production of the built environment. What makes the built environment especially interesting within the overall context of urbanization is that it reflects, through its very creation, the decisions of city makers such as landowners, financiers, developers, builders, politicians, and bureaucrats as well as members of the design professions. Townscapes must be seen as the culmination of land development processes that involve all these key actors within the framework of changing metropolitan form, land use, and architecture described in Chapters 5, 6, and 10. Understanding these processes requires us to identify the key actors, their motivations and objectives, their interpretations of market demand, and their relationships with one another. These, in turn, must be seen within the context of one key precondition for investment in the built environment: the expectation of a satisfactory rate of return on a prospective project. We begin, therefore, with a review of the relationships between property, location, rent, and investment.

PROPERTY, LOCATION, RENT, AND INVESTMENT

In general terms, investment in the built environment depends on:

- efficiently managing capital by mediating financial institutions, responding to prevailing rates of return in the different circuits of capital investment flows, and

- government intervention, acting on principles of **Keynesian** economic management or principles of social welfare and conflict resolution.

We must also recognize that investment in the built environment depends in part on the broader context of other investment opportunities. One dimension of this broader context is related to the role of the built environment in relation to the **circuits of capital investment** flows that maintain the circulation of the economy. We have already encountered one important aspect of this role: investment in the built environment as a response to the occasional **overaccumulation crises** of **stagflation**. At such times capital seems to "switch" from the primary circuit (investment in manufacturing production) into the secondary circuit (fixed capital for production and consumption) or the tertiary circuit (science, technology, and social infrastructures), thereby alleviating the consequences of underconsumption.

It was geographer David Harvey who first conceptualized these circuits and pointed to the central role of the built environment and the city-building process in the overall dynamics of capitalism.[1] Harvey has shown not only that investment in the built environment is critical at times of overaccumulation but also that it is constantly important as a precondition for successful capital accumulation. If the built environment is not renewed and extended, the economy will stagnate and social tensions will develop.

A second dimension of the broader context stems from the overall relationship between the supply of property, prevailing interest rates, and current rates of profit from investment in property. Using the example of investment in housing, we can specify four main ways in which rent interacts with market conditions to affect the flow of capital into the built environment:[2]

1. If interest rates in general are high relative to profits from house building, there will be a financial disincentive for the flow of capital into new construction, even if there is a shortage of housing stock. Continued housing shortages, however, will induce higher rents for existing dwellings, which in turn may inspire speculative investment in vacant land. Eventually, continued housing shortages will lead to higher profits from construction, thus leading to renewed investment in housing.

2. If interest rates in general are high relative to profits from house building but housing space is abundant, rents will be depressed and there will be a tendency for disinvestment. This trend will continue until the supply of housing is reduced to the point that rents rise, leading to an eventual upturn in investment in housing.

3. With low interest rates relative to profits from house building and a shortage of housing space, land acquisition and construction will boom until the backlog in demand is met. Typically, construction overshoots this point, leading to overbuilding, high vacancy rates (though not necessarily of the newly constructed housing), lower profits, and lower rents.

4. When interest rates are low relative to profits from house building, there will be a speculative boom even if there is an abundance of housing space. This boom may also encourage the upward revaluation (on paper, at least) of existing investments in housing.

Box 11.1   Concepts of Rent

The most general concept of rent in economics and urban studies is that of ground rent (sometimes referred to as location rent or economic rent). It is simply the net difference between the minimum amount that it would be necessary to pay for the particular use of a piece of property[3] (i.e., for the owner to meet costs) and the actual amount that is paid. This is the concept we encountered in Chapter 5 in relation to bid-rent theory.

Bid-rent theory, however, is based on a notion of perfect free-market competition that includes the assumption that property is fragmented among many owners, all (or at least most) of whom are "rational" in being motivated by short-term profit. In practice, markets are imperfect. On the supply side, factors other than short-term profit motivate the behavior of many owners. Property ownership brings social prestige and political influence, so it is often treated as a long-term investment. In addition, property ownership has always been dominated by a relative handful of people and institutions whose extensive holdings insulate them from the need to behave "rationally" in economic terms. These considerations help to explain the many inconsistencies in urban land use. They have also been identified as major barriers to economic development, the argument being that large-scale property owners have been content to live off ground rent, thus forfeiting the potential capital accumulation and productive investment from a more rational use of property investments.

Such considerations point to the importance of recognizing three major components of ground rent: absolute, differential, and monopoly rent.

- *Absolute rent* arises from the ability of property owners as a group to block any new investment in land or property until a base rent is paid. We can recognize four ways in which this occurs:[4] (a) the speculative holding of property; (b) the withholding of property from the market by cartel-like groups; (c) the need for investors to assemble usable "packages" of property when ownership is fragmented; and (d) government limitations on the size or location of investment (through zoning or conservation policies, for example).
- *Differential rent* is that part of ground rent that is attributable to the difference between (a) the economic returns to be expected from the most marginal property for any given use and (b) the economic returns for the same use on the particular property in question. Such differences are, in practice, largely attributable to relative location.
- *Monopoly rent* arises from the power of owners of particular *kinds* of property to extract higher rents. Thus, whereas absolute rent derives from the ability of property owners to charge for the mere availability of property of *any* use, monopoly rent is related to the exercise of a degree of monopoly power over a specific category of use. Owners of cheap inner-city apartments, for example, may be able to charge an element of monopoly rent because as a group of owners they have a virtual monopoly over a distinctive type of property. Cheap apartments might be built elsewhere in the city, but they would be very difficult to produce in the inner city because of the much higher differential rent in such locations. Low-income households living in inner-city apartment blocks must therefore contribute an element of monopoly rent (sometimes called "class-monopoly rent" because it reflects the relative economic power of property owners over other socioeconomic classes) to the overall ground rent realized by the apartment blocks.

Ground rent, it turns out, is not the straightforward function of location suggested by Hurd in 1903 (see p. 132).

FIGURE 11.1   The lease charged to the eventual occupiers of this office space will, primarily, be a function of the relative location of the office building.

A third dimension of the broader context of investment opportunities operates at the intrametropolitan scale and involves the existence of localized property submarkets. Although interest rates do not normally vary within metropolitan areas, other market conditions do. Demand for certain types of residential, commercial, retail, or industrial space varies sharply from one part of the metropolis to another. At the same time, different parts of the metropolis consist of property with different degrees of capital investment ("capitalization"). In some areas, new, highly capitalized property precludes most development initiatives. Around the fringes of the metropolis and in some isolated pockets within the metropolis are parcels of undeveloped property that are open to a variety of initiatives. And in between are older properties that are undercapitalized and devalued by age *where renewed investment could yield higher rents.*

Geographer Neil Smith described this last situation as representing a **rent gap**. In Chapter 13 we will see the importance of rent gaps in explaining processes of neighborhood change such as **gentrification**. For the moment, however, it is sufficient to note their role in the overall process of **uneven development** that is a fundamental characteristic of capitalist development at any scale. Investment always flows to locations with relative advantages in terms of cost and revenues where, other things being equal, rates of return will be highest. In the case of property development, capital moves away from locations where land and building costs are high to those where they are low; from locations where revenues are low to those where they are high; and from undercapitalized projects (older housing, outmoded shopping strips, decayed waterfronts, and so on) to projects that can command higher rents at the same location (condominiums, shopping malls, festival marketplaces, and the like). There is, therefore, a constant restlessness to the built environment, as both simultaneous and sequential processes of investment, disinvestment, and reinvestment take place.

A TYPOLOGY OF INVESTMENT IN LAND AND PROPERTY

The restlessness of the built environment has been complicated by the fact that some property owners do not behave "rationally." Rather than treating property purely as a financial asset, they may treat it in part as a source of social status, political influence, or "pocket money," thus failing to respond (or responding only belatedly) to changes in market conditions. Real estate analyst Anne Haila has developed a

TIME HORIZON OF INVESTMENT

		Present	Future
AIM OF INVESTMENT	Use *Agent* *Source* *Purpose* *Function*	**Bazaar** Serendipitous actor Own money Use value Consumption	**Organism** Planner Public revenues Restructuring Public coordination
	Exchange *Agent* *Source* *Purpose* *Function*	**Jungle** Dealer Productive sector Annual rent Investment switch	**Circus** Speculator Borrowed money Capital gain Intrinsic dynamic

FIGURE 11.2   A typology of investment in land and property.

typology of investment in land and property that takes such behavior into account (Figure 11.2). It is based on the *purpose* of the investment (i.e., for the use value of the land or property or for its future exchange value) and the *time horizon* in which the investment decision is made (present or future). As Haila explains, "A piece of land can be acquired and a building can be constructed for occupation (use), or for monetary return (exchange) the property yields when rented (annual rent) or when sold (capital gain, defined as the difference between the purchase price and the resale price). An investor acquiring real property and developing it can be oriented towards satisfying a present need or receiving a short-term revenue (present), or can expect to receive benefit or revenue in the long term (future)."[5]

Each of the four types of investment defined in this way is characterized by particular kinds of actors ("Agent" in Figure 11.2) with particular motives ("Purpose"), working with particular resources ("Source"), and contributing to a particular dimension of market dynamics ("Function"). Each of the four is also characterized by a distinctive contribution to urban spatial change: "bazaar," "jungle," "organism," and "circus."

BAZAAR: The agents are serendipitous actors. They acquire (occupy, buy or inherit) a property with no intention of speculation but, when selling the property, may find that it has increased in value. The actors are serendipitous in the sense that the profit they possibly receive is accidental. The source of investment is the surplus accumulated mostly by the owners or users. . . . The purpose of investment is use value. . . . The function of investment is the consumption of these use values. The consumption satisfies the need of the agents. The action of an agent is oriented to present use. The produced land-use pattern is like a bazaar. . . . Individuals with different tastes and skills produce rich and hetero-geneous space with many nuances and differences.

JUNGLE: The agent is a dealer. The source of investment is surplus produced in the productive (primary) sector. The main purpose of investment is annual rent. Dealers switch investment from the productive sector to the real estate sector. This happens when there is an overaccu-mulation crisis in the productive sector. The cycles of production determine the timing of investment. The function of investment is to switch overaccumulated capital to better use. Investment in the real estate sector is made only as the second-best alternative, when there are no profitable investment opportunities in produc-tion. Dealers' action is oriented primarily to the present exchange value. . . . The produced land-use pattern is like a jungle. The external laws of production cycles dic-tate the production and use of space. . . . With a turn in the cycle of production, investment is shifted away from the real estate sector and real estate becomes devalued. Because investment is not determined by the need for space, there is always a possibility of overdevelopment.

ORGANISM: The agent is a planner. Planners represent public authorities at different levels of government (state, district, city). They channel investment into pub-lic works (streets, parks) and construction projects (public housing, public buildings). They draw up plans and grant construction permits. They assemble land for redevelopment, coordinate land uses and carry out re-newal projects. The source of investment is public revenues. . . . The purpose of investment is to create conditions for the reproduction of labor power, eco-nomic restructuring and capital accumulation. The as-semblage of parcels of land, the coordination of land uses, investment in infrastructure, public services and renewal projects serve the needs of the economy. The action of an agent is oriented to future use, and the an-ticipated needs dictate the forms of activity.

CIRCUS: The agent is a speculator. Speculators seek gain by using the market for their self-interest. They try to an-ticipate the change of prices and to sell and buy in favor-able market situations. They can also try to manipulate the market by, for example, lobbying for or against plan-ning permissions. The source of investment is monies borrowed and gathered from different sources (savings of small investors who buy shares in real estate investment trusts (REITs); capital of firms, financial institutions and public institutions that form joint-stock companies for construction projects). The purpose of investment is capi-tal gain. Agents are mainly interested in land as a finan-cial asset, so land is treated as a financial asset. . . . The intrinsic laws of the real estate sector determine the low and high tides of flows of money, not the external laws of production cycles. The action of speculators is oriented to future exchange. . . . The produced land-use pattern is called a circus. A circus can be erected in one night. It is easy to remove and change. Its arena can be used for various purposes. . . . A circus is manipulated, used and

controlled by speculators for their self-interest. A circus is a place for struggle, competition and risk-taking. It is a place for large fortunes, but also for enormous losses.[6]

PROPERTY AS A FINANCIAL ASSET

For many observers it is the last model—the "circus"—that is currently most important in the urban develop-ment process. David Harvey has argued that in contemporary society rents from property are increas-ingly seen as in principle no different from the returns on investments in bonds, stocks, and shares:

> The money laid out is interest-bearing capital in every case. The land becomes a form of fictitious capital, and the land market functions simply as a particular branch—albeit with some special characteristics—of the circulation of interest-bearing capital. Under such con-ditions the land is treated as a pure financial asset which is bought and sold for the rent it yields.[7]

As a result, "interest-bearing capital circulates through land markets perpetually in search of enhanced future ground rents."[8] Although Harvey sees this change as part of the overall evolution of the economy, there are several specific trends that have been important in con-solidating the *tendency to treat property as a financial asset*:[9]

- The **globalization** of the economy has brought the property market to a wider sphere, in-creasing competition (particularly for office space and for high-status residential space), dri-ving up rents while increasing the degree of differentiation between rents in different lo-calities, and opening local property markets to a wider circle of investors (including interna-tional investors).

- The rapid reorganization of urban form and land use, described in Chapter 6, has increased competition in local property markets, thereby increasing the pressure for property owners to respond rationally to market conditions.

- The more sophisticated use of advertising and marketing in promoting property investments has helped both to attract a wider pool of in-vestors and to promote the idea of property as an "ordinary" asset or commodity.

- The loosening of planning and development controls (described in Chapter 17) has further facilitated investors' ability to treat property as an ordinary asset.

- The involvement of new kinds of investors (transnational corporations, pension funds, and

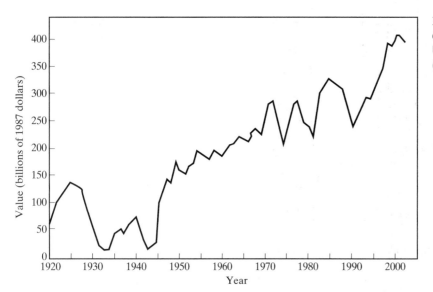

FIGURE 11.3   The value of new private construction put in place each year in the United States, 1920–2002, in constant (1987) dollars.

so on) and the emergence of new kinds of professionals (real estate managers and property investment analysts) have propagated a more calculating (rent-maximizing) attitude in property markets generally.

- The deregulation of financial markets in general, and of the savings and loan industry in particular (see p. 514), has removed many institutional and legal barriers between capital markets that were previously highly segmented. This deregulation has also increased competition within property markets, driven up rents, attracted a wider pool of investors, and facilitated their ability to treat property purely as a financial asset. As a result, there has been an overall tendency toward increasing investment in the built environment throughout the postwar period, cyclical fluctuations notwithstanding (Figure 11.3).

THE STRUCTURES OF BUILDING PROVISION

The conception of the structures of building provision is based on the observation that, in addition to general principles of demand and supply and universal theories of rent and investment, each project is the result of the actions of a variety of social agents and mediating institutions. The creation of the built environment, therefore, must be seen in terms of the functional linkages between specific sets of decision makers and institutions. Table 11.1 indicates the range of key decision makers, or "city makers," involved in the urban development process. Figure 11.4 shows how they fit into the structures of building provision for residential housing. In this section we will identify the most important of these city makers, paying particular attention to their motivations and constraints before going on to examine the overall functional linkages between them in the context of land development processes.

CITY MAKERS

In any given situation the creation of the built environment is the result of a variety of agents, all with their own objectives, motivations, resources, and constraints, and all connected with one another in several different ways. The suburban residential development process has been described as "a sort of three-dimensional spider web that can be moved by impact in any corner."[10] The same applies to the development process in the **CBD**, to the office development process, and to the process of developing industrial space—though the structure and composition of the web will vary in each case. To complicate things still further, each category of agents will involve many individuals. In any city of any size, for example, there will be hundreds of major landowners and dozens of developers and builders. Some agents will act for themselves within the web of the development process; others will be representing groups of people, large corporations, or public agencies. Some agents may play more than one role at a time. Landowners may be actively involved in subdividing and building, for example, while city governments may act as both regulators and entrepreneurs. Finally, all the agents within the web of the development process must operate within a locally and historically specific context of market conditions and political constraints.

TABLE 11.1

Urban Development: Decision Categories and Selected
Decision Makers in the United States

1. *Industrial and commercial location decisions:*
   Executives of industrial companies
   Executives of commercial companies
2. *Development decisions:*
   Executives of development companies
      (developers)
   Land speculators and landowners
   Apartment owners and landlords
3. *Financial decisions:*
   Commercial bankers
   Executives of savings and loan associations
      ("thrifts")
   Executives of insurance companies
   Executives of mortgage companies
   Executives of real estate investment
      trusts (REITs)
4. *Construction decisions:*
   Builders and developer-builders
   Executives of architectural and
      engineering firms
   Construction subcontractors
5. *Support decisions:*
   Chamber of Commerce executives
   Real estate brokers
   Executives of leasing companies
   Apartment management firms

Source: *J. Feagin and R. Parker,* Building American Cities: The Urban
Real Estate Game, *Englewood Cliffs, N.J.: Prentice Hall, 1991, Table 1.1.*

As long as we bear these caveats in mind, it is possible to outline the agents that are typically involved in the creation of the built environment. It is also possible to make some generalizations about the roles and objectives of each of the major groups within the web of the development process.[11]

Landowners

Landowners stand at the beginning of the chain of events involved in urban development. There are three major types of landowners, each with rather different perspectives:

- The first, *landed estates*, encompasses "old" money, whose ancestors were able to acquire large land holdings at an early stage in the history of development. For this group of landowners profitability is important, but it is often modified by social and historical ties. Decisions about the sale of land are made with an eye to the very long term.

- The second, *industrial land owners*, is dominated by commercial farmers, a group that is crucial to the land conversion process at the urban fringe. Their decision-making typically has to balance short-term financial considerations against longer-term lifestyle considerations. Maintaining an agrarian occupation and lifestyle often means paying higher taxes that result from the inflation of land values because of proximity to an expanding urban area. The alternative is to capture the rising value of land by selling some (or all) of it to speculators or developers. In response, many state and local governments in the United States have established tax schedules that favor agricultural uses.

- The third type is based on *financial ownership*, dominated by property companies and financial institutions such as insurance companies and pension funds. As we will see, the importance of financial institutions has grown rapidly, as savings and profits have been channeled into long-term investments and as large **transnational corporations** have diversified into property. Property companies, in contrast, are less concerned with the long-term appreciation of assets, focusing on the exploitation of urban land markets for short-term and medium-term profits.

Although each of these groups tends to behave in rather different ways within the web of city makers, all of them have influenced the outcome of the city building process in two broad ways: (1) through the size and spatial pattern of parcels of land that are delivered to speculators and developers and (2) through conditions that they may impose on the subsequent nature of development. In terms of the size and spatial pattern of land parcels, much of course depends on the overall pattern of land holdings. The large *ranchos* and mission lands around Los Angeles, for example, have formed the basis of extensive tracts of uniform suburban development, while in eastern cities, where the early pattern of land holdings was fragmented, development has been more piecemeal. Because of the structure of the tax system, however, it is often preferable for landowners to sell smaller parcels over an extended period. As a result, developers often arrange "installment contracts" with a variety of landowners in order to maintain a sufficient supply of building land. The outcome in terms of spatial patterns of development tends to be one of apparently random urban sprawl.

Because many landowners often sell only part of their holdings at a time, they have a strong interest in what happens to the land they sell. Any change in the use of sold-off parcels is likely to affect the future exchange value of remaining holdings. In the past it was

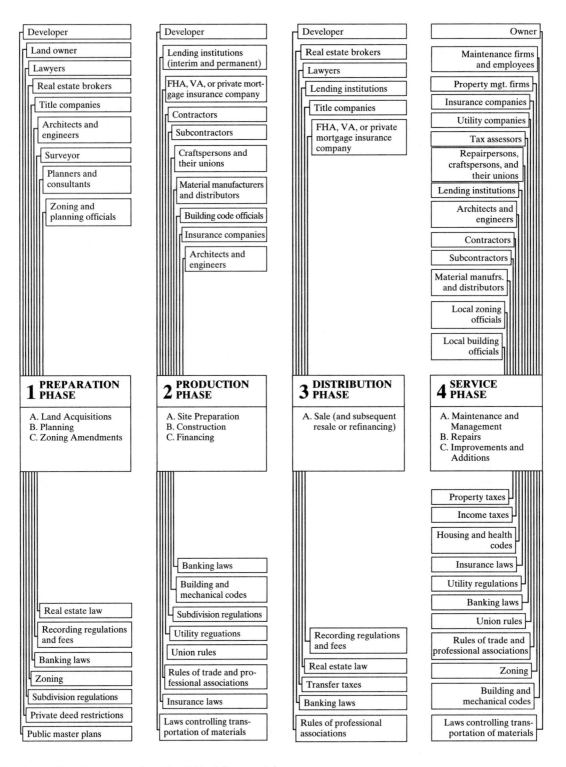

FIGURE 11.4   The structures of residential building provision.

very common for landowners to sell off parcels of land with contractual provisos—**restrictive covenants**—that limited the nature of subsequent development. Such covenants usually discriminated against low-income groups and socially undesirable land uses, sometimes in a very explicit way. With changed social attitudes and tougher laws against discrimination, restrictive covenants are somewhat less common but they have by no means disappeared. Rather, the practice has been to frame them obliquely, stipulating minimum lot sizes or residential densities, for example, and so ensuring development for more affluent users.

Speculators

Speculators seek to buy relatively low-priced land just before it begins to appreciate rapidly in value and to sell it just as it reaches a peak. Sociologists John Logan and Harvey Molotch have identified three very different kinds of speculators (or, as they call them, "place entrepreneurs").[12]

- The first is the *serendipitous entrepreneur*—someone who has inherited property or who has bought it with a particular use in mind and then finds that it would be more valuable sold or rented for some other use.
- The second is the *active entrepreneur*—the individual who hopes to anticipate changing patterns of land use and land values, buying and selling land accordingly. The prototypical active entrepreneur is a small or medium-scale investor: individuals (not corporations) who attempt to monitor the investments and disinvestments of bigger players, using local social networks to find out who is going to do what, when, and where.
- The third is the *structural speculator*—the bigger players who rely not merely on an ability to anticipate changing patterns but who also hope to influence or engineer change for their own benefit. "Their strategy is to create differential rents by influencing the larger arena of decision making that will determine local advantages."[13] They may attempt, for example, to influence the route of a freeway or the location of a rapid transit stop, to change the **land use zoning** map or the master plan, or to encourage public expenditure on particular amenities or services.

Developers

The principal role of developers is in deciding on the nature and form of new projects, platting large parcels of land into smaller lots, installing the infrastructure necessary for a particular use (e.g., streets, sewer and water mains, gas and electric lines), and selling the lots to builders. These activities generally fall under the descriptive label of "subdivision." Many development companies, however, have extended their activities well beyond the business of subdivision to include land assembly and speculation, design, construction, and marketing. Because it is the developers who must decide on the *type* of project to be undertaken on a particular site, they can fairly claim to be the single most important group of city makers.

The Development Process. Figure 11.5 shows the major steps involved in the overall development process.[14] The preliminary development activities consist of site selection, the conceptualization of the project (whether it is to be a residential subdivision, a private master-planned community, an office park, or whatever), and the consideration of the concept's feasibility. *Site selection and project conceptualization* stand together at the very beginning of the process, each influencing the other. This first step is clearly very important to the outcome of the city-building process, since the developers are inscribing their judgment and interpretation onto the landscape. Other things being equal, developers will opt for what is easiest to produce and what is the safest bet in terms of effective demand—the middle of the market. Only a few will have both the nerve to gamble on innovative projects and the ability to persuade financiers and customers that the potential outweighs the risks. In terms of residential development, this conservative approach translates into housing for the "typical" household (or, at least, the developer's idea of the typical household).

Through the 1960s and 1970s this approach resulted in a preponderance of three-bedroom single-family suburban housing, with little provision for atypical households—who were effectively excluded from new

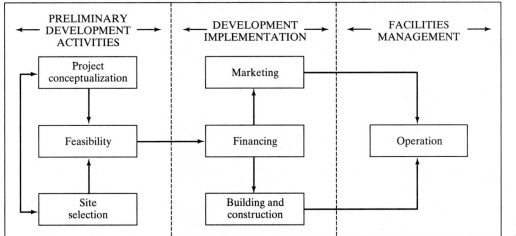

FIGURE 11.5 The development process.

suburban tracts. Only in the 1980s, when marketing consultants caught up with social shifts that had made the "typical" household a demographic minority, did developers begin to cater to affluent singles, divorcees, retirees, and "DINKs" (dual-income, no kids), adding luxury condominiums, townhouses, artists' lofts, and the like to their standard repertoire.

Among the developer's concerns in the site selection process will be the location, size, and cost of available land, the availability of utilities and municipal services, the possibility of special engineering or construction needs because of the site's physical properties, and the potential reaction of neighboring land users to the project (Figure 11.6). For most residential development, the overarching criterion is that of site costs, although the nature of the proposed project can introduce important qualitative considerations:

> Builders constructing lower-priced units are concerned primarily with minimizing costs. They seek to supply basic housing without frills; flat terrain with few trees is ideal. By contrast, builders of higher-priced houses look for sites with natural or social amenities. The costs of building on more rugged terrain and in denser vegetation are higher, but builders invariably find that the value of the completed house is increased by an even greater amount. When units will be sold for a higher price, builders and subdividers may create ponds out of marshes or add to the relief of an otherwise flat area by cutting streets lower and piling excess dirt up on the building sites. Characteristics that are limiting factors to the builder of lower-priced houses therefore are considered in a positive light in the trade-off calculations of the builders of higher-priced houses.[15]

For most commercial and industrial development, in contrast, the main criterion is the availability of sufficient land in an appropriate location; site costs are a secondary consideration. Indeed, as urban sprawl has accelerated and development companies have become larger, the whole question of the availability of land has increased in importance, even for residential developers. Some companies create **land banks**, partly as a speculative venture but mainly to ensure a supply of developable land (many of the parking lots on the edge of downtown areas, for example, are in fact held primarily for their speculative value rather than for their earning capacity as parking lots). Larger companies, with a compelling need to acquire land at a rapid rate (in order to keep their organization fully employed), search out and bid for suitable land before it has been put on the market (and before any thought has been given to project conceptualization): a tactic known in the trade as **bird-dogging**.

The final phase of predevelopment activities is that of determining *feasibility*. Typically, this phase requires coordination with local planners in order to check on compliance with zoning ordinances and legal codes, approaching community leaders in order to gauge reactions to the proposed project, undertaking detailed market analyses, drawing up alternative schematic designs ("schematics"), investigating any special technical issues arising from these schematics, and projecting costs and revenues for each of them.

Having completed the preliminary phase, the developer moves into implementation: financing, marketing, design, and construction. *Financing* involves convincing others of the project's feasibility. Typically, the developer, just like the would-be homeowner, must put down part of the cost: the developer's *equity*. The remainder is sought from a bank or from some other backer or consortium of backers—pension funds, insurance companies, and the like—who may themselves require certain changes in the nature of the project. The development industry is highly "leveraged," meaning that the developer's equity often works out to

FIGURE 11.6
Large-scale development of lower-priced housing depends on the availability of large tracts of cheap land, on which Fordist principles of mass production can be exploited. Flat terrain with few trees is ideal.

be a much smaller proportion of the overall cost than the homeowner's equity.

Marketing has become increasingly sophisticated as development projects have become larger. In addition to professional market research, this phase involves active promotion well in advance of the availability of the developer's product. If possible, finished space will be sold or leased in advance of construction. In extreme cases, where stakes are very high, developers may buy out the existing leases of prestige customers in order to allow those customers to move into the project and thus lure in other tenants.

The *design* and *construction* phase begins with detailed contract drawings being produced from the schematics, followed by a bidding process in which general contractors are invited to bid for various aspects of the engineering, construction, and landscaping. Speed of operation is essential during this phase. Interest has to be paid on construction loans based upon a balance outstanding that increases as the project proceeds. Any delays in producing revenue from a project can result in significant losses due to increased interest payments, particularly when such delays occur toward the latter stages of construction. Although it usually takes about 10 years for a typical mixed-use project to break even, the survival of a project depends on generating some revenues as quickly as possible simply to avoid foreclosure. Once the construction is completed, the developer has to search for and manage tenants, collect rents, generally maintain and administer the project, or sell to new owners. This is the *facilities management* stage of the process.

Developers' Specializations and Linkages. The actual development process will vary a good deal, of course, depending on the size and nature of the project and the resources and scope of operations of the development company. Many developers specialize in a certain type of project: suburban subdivisions, master-planned communities, office blocks, mixed-use developments, hotels, factories, and so on. And although some developers are active across several regional markets, many specialize in just one metropolitan market, or even a particular downtown or **urban realm**. In some cases a few developers have been able to dominate a very large proportion of the development activity in a metropolitan area through their connections with an equally small number of key speculators and builders. Around Los Angeles, for example, both sprawl and infill have been dominated since the late 1970s by what Mike Davis calls the "new Octopus" (the original Octopus having been the economic and urban development nexus controlled by the Southern Pacific Railroad[16] in the late nineteenth century). This new Octopus consists of three distinct but interlocking types of enterprise:[17]

- The successors to the great nineteenth-century *landed estates*. These include the Irvine Company, C. J. Segerstrom & Sons, Watson Land Company, and the Newhall Land and Farming Company.

- Between 15 and 20 *developers* or "community builders" (including Lewis Homes, Kaufman and Broad, the Lusk Company, the Koll Company, the Haagen Company, and Ahmanson Commercial

FIGURE 11.7 Donald Trump visits Trump World Tower in New York City during construction.

Company), who among them dominate the "starter home" and commercial and industrial development activity of the region.

- Owners of *land-intensive industries* who have found that their land holdings are among their best assets and who have gone on to become major developers in their own right. These land-intensive industries include aerospace (where Howard Hughes's move into property development continues with The Howard Hughes Corporation[18]), entertainment (MCA and Disney), energy (Chevron), and transportation (Union Pacific).

In general, however, surprisingly little is known about the linkages between developers and other key agents within local webs of the structures of building provision.

Builders

As we have seen, developers sometimes extend their operations to include building; more often than not, however, developers will subcontract to general building contractors. At the same time, many small and medium-sized building firms will undertake their own speculative land acquisition and development functions. Much depends, as with development companies, on the size and internal organization of the company. The typical large builder reduces costs through direct purchasing of materials by the truckload, the development of efficient subcontracting relationships, the retention of a specialized labor force, the use of federal financial aid and housing research, and the use of mass production methods on large parcels of land.

As a result, large builders inevitably are concerned almost exclusively with construction for mass-market suburban development. Medium-sized companies cannot afford to pay the interest on large parcels of developed land, so their preferred strategy is to maximize profits by building at high densities (condominiums, apartment blocks) or by catering to the high-profit luxury end of the market (where the mass-production orientation of the big companies is a handicap). This strategy leaves small companies to use their more detailed local knowledge to scavenge for "custom" building contracts and smaller infill opportunities, at which point they will assemble the necessary materials and labor and seek to build as quickly as possible, usually aiming at the market for larger, higher-quality dwellings in neighborhoods with an established social reputation.

Consumers

Consumers—households and commercial and industrial businesses—represent the demand side of the development process. Demand for space within the built environment is an important and complicated topic, and a full discussion is reserved for Chapters 12 and 13.

One point that needs to be made in the present context, however, is that consumer preferences and consumer behavior must be seen as having developed in a social context "that is fundamentally shaped by markets ruled by capitalists and top managers in industrial and development corporations. . . . The physical structure of production builds barriers and sets limits to individual choices. Moreover, citizen preferences are frequently created or manipulated by powerful investors and their associates working through advertising, public relations, and the mass media."[19]

It should also be stressed that people need not always react individually—as "consumers"—to the choices given to them as a result of business decision makers and managers. As we will see in Chapter 16, they may also affect the development process through citizen protests over specific development projects, through involvement in pro-growth, no-growth, or slow-growth politics, or through involvement in residents' associations.

Real Estate Agents, Financiers, and Other Facilitators

Real estate agents, financiers, and other professional facilitators are essential to the development process as facilitators, intermediaries, and specialized experts. A wide range of professionals is involved, including surveyors, market analysts, advertising companies, lawyers, title insurance companies, appraisers, property managers, engineers, ecologists, and geologists. The most important of these "exchange professionals," however, are mortgage financiers and realtors. Their activities go well beyond the actual *creation* of the built environment to encompass continuing processes of neighborhood change. As "gatekeepers" of this change, they are discussed at length in Chapter 13.

Government Agencies

The development industry forms a staple of local politics and the focus of a good deal of local policy. Indeed, local economic development in the United States since the early 1980s has been associated with an unprecedented form of **growth machine** politics and unprecedented partnerships between development companies and public agencies. As we will see in Chapter 17, city governments have increasingly shifted to a new culture of **civic entrepreneurialism** that draws heavily on public-private partnerships in which public resources and legal powers are joined with private interests in order to undertake development projects. This shift has fostered a speculative and piecemeal approach to the management of cities. Local governments (with additional funding leveraged from state and federal agencies) have subsidized projects such as downtown shopping malls, festival market places, new stadiums, theme parks, and conference centers because these

TABLE 11.2

Urban Development: Decision Categories and Selected
Government Decision Makers in the United States

1. *Actions on utility services, building codes, zoning, tax
   abatements, boosterism, and progrowth actions*:
   > Mayors and city council members
   > County governmental officials
   > Officials on local zoning and planning
   >   commissions
   > Officials in special local agencies
2. *Housing, redevelopment, and tax decisions*:
   > Members of Congress
   > U.S. Department of Housing and Urban
   >   Development (HUD) officials
   > Local and state government officials

Source: *J. Feagin and R. Parker,* Building American Cities: The Urban
Real Estate Game, *Englewood Cliffs, N.J.: Prentice Hall, 1991, Table 1.2.*

have been seen as having the capacity to enhance property values and generate retail turnover and employment growth (see Chapter 16). If successful, the reasoning goes, such projects can cast a beneficial glow over the whole city; but even in the face of poor economic performance they can be regarded as a kind of "loss leader" that may bolster the image of a city and attract other forms of development.

Development is much less of a political issue at the federal level but, as we have already seen with the examples of the interstate highway system and federally insured mortgages, federal policy can be a crucial factor in relation to urban development. As a result, many important development-related decisions are made by politicians and government officials at the federal as well as the local level (Table 11.2). In Chapters 16 and 17 we will explore several issues where the development process intersects with urban politics and policy. Meanwhile, it is worth noting that the United States is somewhat unusual in comparison with other countries in the degree to which regulation of the property development process is weak and decentralized—another direct legacy of the sanctification of private property under U.S. law (first discussed in Chapter 5). Moreover, as we saw in Chapter 6, many federal responsibilities were decentralized and many spheres of activity involving federal oversight were deregulated in the 1980s as part of the overall retrenchment and restructuring that marked the shift away from Keynesian economic management.

MARKET RESPONSES OF THE DEVELOPMENT INDUSTRY

Like most other industries, the development industry has undergone some radical changes in the past 30 years as part of the restructuring and redeployment that followed the stagflation of the mid-1970s. In parallel with producers in many other industries, developers and builders have found it advantageous to decrease their emphasis on **Fordist** strategies of mass production-mass consumption in favor of more flexible approaches aimed at exploiting profitable new market niches (see p. 12). In this section we illustrate some of these changes, noting how they have inscribed new elements into the built environment.

We can take the shopping center industry as a good illustration of the impact of consolidation and centralization within the development industry.[20] During the 1970s mergers and acquisitions led to the concentration of the major department stores within regional shopping malls under the ownership of a few corporations (including Allied Stores, Associated Dry Goods, Federated Department Stores, R. H. Macy & Co., and May Department Stores). Meanwhile, ownership of the malls themselves became consolidated in the hands of a few big companies (including the Hahn Company, the Rouse Company, Melvin Simon & Associates, and the Edward J. DeBartolo Corporation),[21] and before long the largest retailing organizations had established their own development companies. Sears, for example, established the Homart Development Company, May established May Centers, and Federated established Federated Stores Realty.[22]

During the 1980s and 1990s the pace of mergers and acquisitions increased, partly because underutilized real estate assets made department store companies attractive targets for leveraged buyouts and partly because development companies acquired department store companies. May Department Stores acquired Associated Dry Goods (which included Lord & Taylor and Kaufmann's) in 1986 and Filene's in 1988. Canadian-based Campeau Corporation purchased Allied (including the Bon Marché, Jordan Marsh, Maas Bros., and Stern's chains) in 1986 and Federated (including the Bloomingdale's, Burdines, Lazarus, and Rich's chains) in 1988. In 1990, saddled by debt resulting from the highly leveraged Campeau takeover of Federated, both Federated and Allied filed for bankruptcy and were reorganized as subsidiaries of Federated, a new holding company of Campeau. In 1994, Federated acquired R. H. Macy & Co. and became the largest department store retailer in the United States.

Centralization has also increased because real estate has become popular with institutional investors such as insurance companies and pension funds and as deregulation of financial markets has removed many of the institutional barriers that had separated investments in residential development from those in nonresidential development and in stocks and bonds. In particular, the "securitization"[23] of real estate assets has caused property financing to become linked more

BOX 11.2 BROWNFIELD REDEVELOPMENT[24]

Brownfields are abandoned or underutilized industrial and commercial facilities where real or perceived contamination complicates expansion or redevelopment. This definition, used by the U.S. Environmental Protection Agency (EPA) and Department of Housing and Urban Development (HUD), characterizes a tremendous number of properties as brownfields—because the severity of contamination is not specified and the environmental problems can be merely *suspected* as well as actually documented.

Many brownfields are found in central cities and industrial suburbs with a history of traditional manufacturing. These sites can be abandoned industrial and railroad facilities or manufacturing plants that are operating but show signs of pollution. Brownfields can also be small commercial or even residential lots with only suspected contamination. The most contaminated brownfields in the United States are the 1,300 or so Superfund sites on the National Priorities List. These sites contain waste with amounts of toxic chemicals, like lead or mercury, considered hazardous by the EPA. Most brownfields, though, have only low to medium levels of known or suspected contamination from ordinary waste like nonhazardous garbage.

Brownfield cleanup and redevelopment involve a wide range of citymakers. For landowners, speculators, developers, builders, real estate agents, and financiers, brownfields can offer opportunities for profit from a large underexploited source of land within established communities. The benefits for government agencies and households have focused traditionally on employment and tax base generation.

Brownfield redevelopment, however, presents a dual land-use policy challenge: reducing the barriers to private-sector redevelopment while connecting reuse to broader community goals. The first part of this challenge involves addressing the uncertainties and risks—involving cost and time—created for the private sector by four main issues: legal liability for contamination; lack of information about the level of contamination and uncertainties about cleanup standards; availability of funding for site investigation, cleanup, and redevelopment; and complicated regulatory requirements. The second part of this challenge involves connecting brownfield redevelopment and the reuse of these sites to wider community efforts—involving sustainable development and environmental justice—to achieve environmental and health protection, improved public safety, targeted jobs and training, central city revitalization, and reduced metropolitan sprawl.

During the last few decades, federal, state, and local governments in the United States have attempted to reduce the barriers to brownfield reuse for landowners, speculators, developers, and builders in a variety of ways, including legislative changes addressing landowner and lender legal liability, increased financial incentives, and efforts to improve interagency coordination at all levels of government. In comparison to the United States, the national governments in Europe typically are more involved in brownfield redevelopment because of their stronger role in urban policy and planning. In the United Kingdom, for example, the central government in 1998 set a target of 60 percent of all new housing to be built on previously developed land by 2008. The government offered favorable tax treatment to encourage brownfield redevelopment while introducing strict planning policies to protect **greenfield sites** at the periphery of cities from uncontrolled development. In addition, despite the different national regulatory frameworks across Europe, the involvement of the European Union (EU) has facilitated a more coordinated approach to environmental policymaking in general and to brownfield redevelopment in particular as part of government efforts to address the problem of contaminated sites in the cities across Europe.

directly to broader capital markets and, indeed, to international capital markets. Now, in the early twenty-first century, the shopping center industry is both highly concentrated and centralized, as are other branches of real estate and urban development. An "hourglass" structure has emerged in the real estate industry. At the top end are the large investors and developers like Trammell Crow Company, Jmb Realty Company, and MetLife Real Estate Investments (the real estate investment arm of Metropolitan Life Insurance Company); a few real estate players occupy the thin middle of the hourglass; and the remaining large number of real estate players at the bottom range from one-person brokerage companies to syndicators and small developers.[25] There are few competitive medium-sized companies because of the increased institutionalization of real estate development since the advent of the large, public real estate companies and REITs. Although it is easier to become a small niche player by focusing on those deals that are not large enough for the big-league institutional companies, the small developers that seek to grow by taking on larger projects eventually hit a wall due to competition from the large investors and developers.[26]

BOX 11.3   URBAN DEVELOPMENT IS LESS AND LESS A LOCAL ACTIVITY

Investment in the built environment has come increasingly to emphasize large-scale projects, and because of this trend urban development has, in turn, become more international and less and less a local activity. Take, for example, 30 St. Mary Axe, the first tall structure to be built in the heart of heritage-conscious London since the 1970s and the second-tallest skyscraper in the one-mile-square historic City of London, after Tower 42. This new €200-million, 40-story building in London's financial and insurance district is cone-shaped so that the wind will pass easily around it. Its unique silhouette has earned it a variety of nicknames, including the "Erotic Gherkin" and the "Towering Innuendo" (Figure 11.8).

The project involved a number of international companies and/or their subsidiaries (Figure 11.8). The developer and primary occupant is Swiss Re, a Swiss reinsurance company. The architect was Foster and Partners, a British company headed by Sir Norman Foster, whose designs include the Centre Pompidou in Paris and the Hong Kong and Shanghai Bank in Hong Kong. The general contractor was Skanska, a Swedish company. The structural engineers were Ove Arup & Partners, a British company, as was the mechanical and electrical engineer, Hilson Moran. The facade

supplier was Schmidlin, a Swiss company, and the facade maintenance system supplier was Lalesse Gevelliften, a Dutch company. The structural steel supplier was Victor Buyck-Hollandia, a joint venture by two companies—Belgian and Dutch. The steel dome was supplied by Waagner Biro, an Austrian company. The elevator supplier was KONE, a Finnish company, and the elevator engineers were Van Deusen & Associates, a company from New Jersey. Universal Builders Supply (UBS), a company from New York, was subcontracted to supply the tower and hoist erectors.

Certainly the scale of activity in the development industry has now reached the point where it is often international in its dimensions. Canadian-based Olympia and York, for example, had assets in 1991 of over $31 billion and annual revenues of nearly $9 billion, controlling more than 100 million square feet of office space in cities across North America, including trophy structures such as the World Financial Center (formerly Battery Park City) and the Park Avenue Atrium in New York, Exchange Place (Boston), Yerba Buena Gardens (San Francisco), and First Canadian Place (Toronto). When Olympia and York, by then the world's largest real estate conglomerate and developer, ran into serious financial

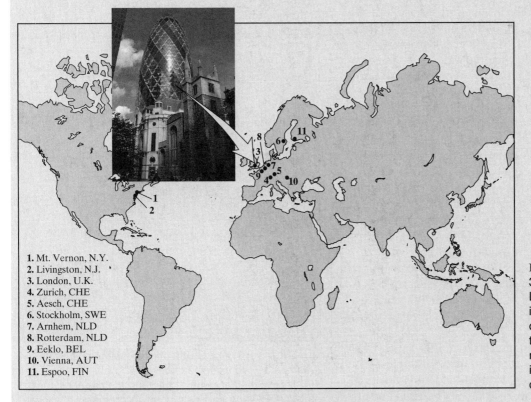

1. Mt. Vernon, N.Y.
2. Livingston, N.J.
3. London, U.K.
4. Zurich, CHE
5. Aesch, CHE
6. Stockholm, SWE
7. Arnhem, NLD
8. Rotterdam, NLD
9. Eeklo, BEL
10. Vienna, AUT
11. Espoo, FIN

FIGURE 11.8
30 St. Mary Axe in London, completed in 2004, is a reflection of how the scale of activity in the development industry is often international in its dimensions.

(Continued)

Box 11.3   Urban Development Is Less and Less a Local Activity   (Continued)

difficulties in 1992—most seriously, its inability to finance the $6.9-billion Canary Wharf project in the Docklands area of London, the largest commercial property development in Europe at that time—the reverberations were felt in stock exchanges and money markets around the world.

Since bankruptcy and reorganization, Olympia and York became O&Y Properties Inc., which became O&Y Properties Corp. (OYP) in 1997 when it took over Camdev, (the new name for Campeau Corporation following its own bankruptcy and reorganization after it overextended itself in the highly leveraged takeover in 1988 of Federated Department Stores (including the Bloomingdale's, Burdines, Lazarus, and Rich's chains in the United States)) (see p. 285). OYP then became the largest commercial and real estate services provider in Canada when it acquired

Enterprise Property Group in 1998. OYP's flagship properties include First Canadian Place in Toronto—lost in the Olympia and York bankruptcy and recovered from the banks for nearly $400 million in 1997.

The significance of these giant corporations lies in their potential for removing much of the debate and control over patterns of development from local arenas of municipal government, from the influence of local "growth machine" coalitions, and from the voices of neighborhood and environmental groups. Although large national and international development companies often work with local partners in joint ventures in order to exploit local networks of key contacts, design professionals, and construction companies, the larger, stronger partners tend to learn quickly, setting up local subsidiaries or simply gaining the experience to take on more projects on a full-ownership basis.

New Products

The development industry has responded to the need for greater flexibility through the pursuit of *product differentiation* and *niche marketing*. In the *commercial* sector product differentiation has resulted in a variety of new formats for hotels: luxury/full service, executive conference resorts, extended-stay (with kitchen and laundry facilities en suite), economy-only, and all-suite. Developers of office buildings responded to the changing business climate by producing self-consciously luxurious buildings (Figure 11.9). Developments for retailing have similarly seen different formats for different market segments: upscale downtown galleries and malls, for example (Figures 11.10 and 11.11), and "big box power centers" (community shopping centers anchored by a major tenant such as Wal-Mart, along with one or two other major retailers and a complementary mix of specialty retailers, restaurants, banks, or other **consumer services**). Another significant new "product line" for developers is the specialized mall: a medical mall, for example, that is crafted to provide busy, affluent consumers with one-stop shopping that offers physicians, counselors, therapists, medical laboratories, pharmacies, outpatient facilities, fitness centers, health food stores, and cafes.

Developers of business and industrial parks, meanwhile, are offering "flex-space": single-story structures with "designer" frontages, loading docks at the rear, and interior space that can be used for offices, R&D labs, storage, or manufacturing, in any ratio (Figure 11.12). Old product lines can also be

FIGURE 11.9   La Coline office development, Paris, France is a good example of developers' attempts to capture a distinctive market niche among upscale business and professional services. The building incorporates a wooden bridge that arches across an ornamental pool in a glass-ceilinged atrium.

FIGURE 11.10   San Francisco's Crocker Center Galleria, a typical example of attempts to reinsert upscale specialty retailing with climate-controlled and security-intensive settings in downtown locations.

master bedrooms, bathrooms with whirlpool tubs and saunas, exercise rooms, game rooms, libraries, "gourmet" kitchens, temperature-controlled wine rooms, multizone air conditioning systems, high-speed Category 5 voice/data telephone wiring, cable/DSS wiring, and stereo systems prewired to multiple locations. In addition to custom pools and spas, many properties include cottages for guests (Figure 11.13).

Large, privately planned communities also became popular with developers because they allow more flexibility in design and product type and enable developers to respond quickly to changing market demand (such as the growing number of older Baby Boomers seeking smaller homes with less maintenance).[27] The essential features of these communities are "a definable boundary; a consistent, but not necessarily uniform, character; overall control during the development process by a single development entity; private ownership of recreational amenities; and enforcement of covenants, conditions and restrictions by a master community association."[28] The direct descendants of the planned

"treated" in order to enhance flexibility within the market. Business and industrial parks have been repackaged as "planned corporate environments" with built-in daycare facilities, fitness centers, jogging trails, restaurants and convenience stores, lavish interior decor, and lush exterior landscaping.

In the *residential* sector some developers have "repositioned" themselves away from single-family "starter" homes to build more multifamily projects (that, like business parks, are packaged with services: in this case, security systems, concierge services, exercise facilities, bike trails, and so on) or more expensive homes for the "move-up" market, where the basic product—single-family suburban homes—is differentiated by features such as dramatic master bedroom–bathroom suites, marble floors, and signature landscaping.

At the very top end of the residential market are speculative homes differentiated by the most lavish "designer" features. New speculatively built homes just north of Orlando in Florida, for example, have elaborate

FIGURE 11.11   Grand Avenue Mall in Milwaukee, a typical example of attempts to revive the fortunes of downtown department stores and chain stores through large-scale remodeling that replicates the amenities of suburban malls in a downtown location.

FIGURE 11.12   "Flexspace": a new product line for developers in the 1980s, this concept combines "designer" office frontages with rear-area loading bays and interior space that can be used as office, industrial, or warehousing space, in any proportion.

FIGURE 11.13   The consumer boom and increased materialism beginning in the 1980s brought a rash of new speculative building at the top of the residential market, which is normally dominated by custom building such as this mansion on a canal in Fort Lauderdale in Florida.

communities of the 1960s (see p. 154), they are a result of an extreme form of product differentiation and carefully targeted niche marketing. By exploiting new and more flexible land use zoning regulations, developers can put together projects that are attractive to a very profitable sector of the residential market while retaining scope for flexibility in the composition and timing of the development.

Residents of such communities are offered sequestered settings with an extensive package of amenities that typically include a golf course, tennis courts, swimming pools, play areas, jogging courses, an auditorium, exercise rooms, a shopping center, a daycare center, and a security system symbolized by imposing gateways and operated by electronic card-key systems. Housing is typically a mixture of expensive single-family houses, upscale townhouses and condominiums, and smaller studios or apartments for young singles or the elderly, all in High Suburban style: mock-Tudor, mock-Georgian, neo-Colonial, Giant Cape Cod, and so on (Figure 11.14).

The entire ensemble is typically framed in a carefully landscaped setting that might contain a lake stocked with swans or a neoconservationist assemblage of remnant woodland, an artificial wetlands environment, and plantings of wild flowers. The landscape is completed by a parade of joggers in expensive warm-up suits and by busy delivery vans

bringing affordable luxuries ordered over the Internet from the mail-order branches of designer clothing and household furnishing companies. The very names of the communities are carefully selected in order to set a tone of distinction, heritage, and authenticity. Advertising imagery draws on the totemism of golf, equestrianism, and pastoral landscapes, while advertising copy leaves no doubts about the status and stylishness of the product.

Clearly the importance of design in the built environment is increasing. As one developer put it, "My buildings are a product. They are products like Scotch Tape is a product, or Saran Wrap. The packaging of that product is the first thing that people see. I am selling space and renting space and it has to be in a package that is attractive enough to be financially successful."[29] Such sentiments are by no means new, though they are perhaps felt more keenly nowadays. As we saw in Chapter 10, however, design is much more than packaging. It involves languages and ideologies that go well beyond the orbits of developers' worlds. It follows that we can "read" these "designer" neighborhoods as the product of our times, the carriers of our society's concern with materialism and social distinction. As we will see in Chapter 12, these themes have emerged very clearly in the residential patterns of American cities in particular.

FIGURE 11.14
Townhouses in a new, private, master-planned community. Note the variety that has been introduced to windows, doors, rooflines, and so on in an attempt to signal the distinctiveness of the project.

FOLLOW UP

1. Be sure that you understand the following key terms:

> absolute rent
>
> active entrepreneur
>
> bird-dogging
>
> brownfields
>
> differential rent
>
> land banks
>
> landed estates
>
> monopoly rent
>
> niche marketing
>
> rent gap
>
> restrictive covenants
>
> serendipitous entrepreneur
>
> structural speculator

2. It is instructive to follow the various stages of the development process for a single project. Two books that do this are *From the Ground Up*, by Douglas Frantz (New York: Henry Holt, 1991) and *Skyscraper*, by Karl Sabagh (New York: Penguin, 1989). The latter is the basis for a video series that you may find in your video library.

3. Consider the statement (p. 276) that "There is ... a constant restlessness to the built environment, as both simultaneous and sequential processes of investment, disinvestment, and reinvestment take place." Can you find examples from your nearest town or city that illustrate this?

4. Update your *portfolio* and gather more materials. A very good source in relation to the material covered in this chapter is the real estate section of any major metropolitan newspaper. You may be able to use the print edition of your own area's newspaper or go online to access newspapers from other places. Real estate sections usually appear once a week and often contain useful insights and illustrations of the activities and interactions of various city makers. Look for features that illustrate the treatment of property as a financial asset and the responses of the development industry to the opportunities presented by urban change. You may also find data on vacancy rates and on land and property prices that you can use to illustrate aspects of changing urban geography.

KEY SOURCES AND SUGGESTED READING

Ball, M. 1985. The Urban Rent Question. *Environment and Planning A* 17: 503–25.

Downs, A. 1985. *The Revolution in Real Estate Finance.* Washington, D.C.: The Brookings Institution.

Fainstein, S. 2001. *The City Builders: Property Development in New York and London: 1980–2000*, 2nd ed. Lawrence, Kan.: University Press of Kansas.

Feagin, J. R. and R. Parker. 2003. *Building American Cities: The Urban Real Estate Game* 2nd ed. Englewood Cliffs, N.J.: Prentice Hall.

Grigsby, W., M. Baratz, G. Galster, and D. MacLennan. 1987. The Dynamics of Neighborhood Change and Decline. *Progress in Planning* 29: 1–76.

Harvey, D. W. 1989. *The Urban Experience.* Baltimore, Md.: Johns Hopkins University Press. This book is an abridged and slightly modified version of two books that were first published in 1985 by Johns Hopkins

University Press: *The Urbanization of Capital* and *Consciousness and the Urban Experience.* See especially Chapters 1–3 in *The Urban Experience.*

Harvey, D. W. 1989. *The Condition of Postmodernity*, Oxford, UK: Basil Blackwell. See especially Chapter 4, "Postmodernism in the City: Architecture and Urban Design."

Knox, P. L., ed. 1993. *The Restless Urban Landscape*, Englewood Cliffs, N.J.: Prentice Hall. See especially Chapters 2, 3, and 6.

MacLaran, A., ed. 2003. *Making Space: Property Development and Urban Planning.* London: Edward Arnold.

Miles, M. E., G. Berens, and M. A. Weiss. 2000. *Real Estate Development: Principles and Process*, 3rd ed., Washington, D.C.: The Urban Land Institute.

Weiss, M. A. 1987. *The Rise of the Community Builders.* New York: Columbia University Press.

RELATED WEBSITES

The International Network for Urban Development: *http://www.wirehub.nl/~intainfo/00-intro/IntroFrames.htm* Established in Paris in 1974, INTA is a nonprofit international network that encourages the exchange of information, experience, and best practices on urban development and revitalization worldwide. This website includes information on INTA's activities in the areas of urban development and regeneration, economic promotion, urban governance, urban services and mobility, and social and cultural inclusion.

Urban Land Institute: *http://www.uli.org/* The ULI site includes research on real estate development, including development case studies for cities in the United States.

The Celebration Company: *http://www.celebrationfl.com/* This is the official website for the town of Celebration, just south of Orlando in Florida. The Celebration Company, a subsidiary of The Walt Disney Company, designed, developed, and manages this new town, which opened to residents in 1996. The website provides information and images of New Urbanism planning and Neo Vernacular architecture.

U.S. Environmental Protection Agency's Brownfields Cleanup and Redevelopment website: *http://www.epa.gov/swerosps/bf/index.html* This EPA website offers information on brownfields in the United States, including material on cleanup technology, standards, procedures, and costs, as well as a description of the most contaminated "Superfund" sites on the National Priorities List.

12

THE RESIDENTIAL KALEIDOSCOPE

In contrast to European cities, one of the most powerful images of American urbanization is that of a system of "melting pots." American cities, according to this view, have absorbed a fluctuating stream of migrants and immigrants, processing people with a tremendous diversity of backgrounds into a pluralistic society with a common language and a shared ideology of democratic free enterprise. The degree to which reality has borne out this image is a matter for some debate. It is clear, however, that a great deal of the dynamism of cities is related to the processes through which new arrivals interact with one another and with existing groups—striving for economic and social success while dealing with their own cultural identity. In these processes people sort themselves—and are sorted—into different parts of the urban fabric, resulting in mosaic patterns of neighborhood differentiation (Figure 12.1). Through simultaneous processes of segregation, assimilation, and resegregation, different parts of the city and different types of housing come to be characterized, for a time at least, by households of different socioeconomic status, families of different size and structure, and people of different racial and ethnic backgrounds.

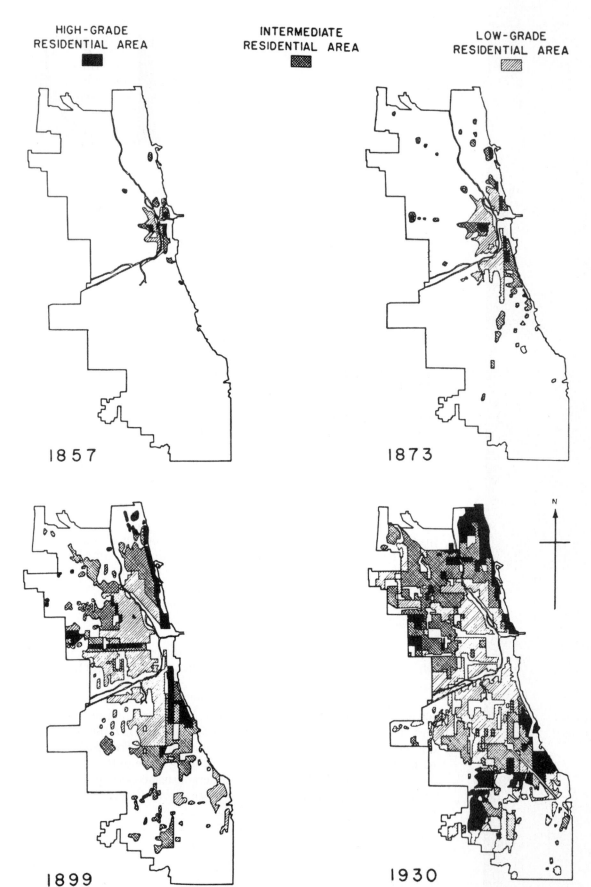

FIGURE 12.1   The evolution of socioeconomic areas in Chicago, 1857–1930.

CHAPTER PREVIEW

In this chapter we examine the "classic" arrangement of residential subareas in U.S. and European cities before going on to consider the rearrangements that have taken place as a result of economic reorganization and social change since the mid-1970s. We begin with an examination of the principal features of urban social interaction and residential segregation. In this context the central issues are the relationships between physical distance, social distance, and patterns of social interaction. As we will see, there are several very good reasons why physical distance and social distance continue to act as mutually reinforcing aspects of social interaction and residential segregation, not the least of which is territoriality.

Given these relationships, we will then see how and why the cornerstones of neighborhood differentiation in American cities are based on residential segregation in terms of social status, household types, ethnicity, and lifestyles. In contrast, in continental European cities ethnicity is generally not an independent dimension, partly because of the absence of substantial ethnic minorities (unlike the United States). Having established these "classic" dimensions and patterns of residential differentiation, we take a closer look at recent changes that reflect the new opportunities and constraints of the changing political economy and changing metropolitan form. Here, we will see how new social groups, new kinds of household organization, and new lifestyle orientations have been imprinted on the social map.

SOCIAL INTERACTION AND RESIDENTIAL SEGREGATION

The residential mosaic is what gives definition to cities. It is what gives character and life, flesh and blood to the skeleton of metropolitan form and land use. In particular, the residential mosaic is the framework for (and product of):

- patterns of friendship, relationships, marriage, and social interaction;
- the reproduction of distinctive life worlds and communities; and
- local politics.

Unlike a real mosaic, however, urban residential patterns are not set in cement. Although they do tend to reflect the inertia of self-perpetuating and self-reinforcing patterns, they also change, in the long run, in response to new economic, demographic, social, cultural, and political conditions. A better (though still imperfect) metaphor is therefore provided by the kaleidoscope, with its constant rearrangements of different fragments. Mapping these arrangements and rearrangements is a fundamental task of geographic analysis, providing descriptive models that are valuable in generating and testing hypotheses and theories concerning urbanization processes.

Patterns of urban residential differentiation stem, ultimately, from the dynamics of urban social interaction that develop within the structural frameworks of socioeconomic background, demographics, and culture (Figure 12.2). Inevitably, these dynamics are complex and multidimensional. Although we can think of most social interaction as being sustained by people's affinity for others with similar socioeconomic resources and cultural values ("people like us"), the sources of this affinity—lifestyle, age,

ethnicity, and so on—are interdependent and constantly shifting.

It is often helpful to distinguish between social interaction based on primary and secondary relationships:

- *Primary relationships* include those between kin—based on ties of loyalty and duty—and those between personal friends—based on ties of attraction and mutual interest.
- *Secondary relationships* are more purposive, involving individuals who group together to achieve particular ends. They are often divided into those in which there is some intrinsic satisfaction ("expressive" interaction) and those in which the interaction is merely a means of achieving some common goal ("instrumental"

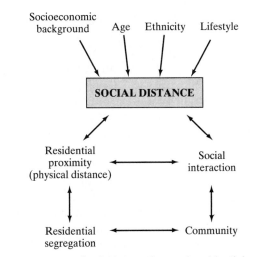

FIGURE 12.2 Social interaction and residential segregation.

interaction). Expressive interaction is typically organized around voluntary associations of various kinds: sports, hobby and social clubs, and volunteer groups. Instrumental interaction, on the other hand, normally takes place within the frameworks of business associations, trade unions, political parties, and pressure groups. Such a division reveals only part of the story, however, since a great deal of social interaction can be seen to be both expressive *and* instrumental: interaction that takes place within the framework of ethnic, religious, and cultural organizations, for example.

The degree of social interaction between particular groups of people in the city is a function of **social distance**, which can be conceptualized in terms of people's attitudes toward other groups. Short social distances are reflected in people's willingness to think of members of other groups as potential marriage partners, and progressively greater social distance is reflected in willingness to have members of another group as friends, neighbors, colleagues, fellow citizens, and (most distant) visitors from another city or country. This scale points to the close relationship between social distance and physical distance. The less the social distance, the greater the probable physical proximity between people—their *residential proximity*. Conversely, the greater the residential proximity, the greater the likelihood of social interaction.

This relationship between social interaction, social distance, and physical distance is not as simple as it may appear. The effects of social and physical distance are closely interwoven and difficult to isolate. It has been argued, however, that the influences of physical distance have been rapidly diminishing in the "shrinking world" of modern technology and mass communications.[1] Improvements in personal mobility and the advent of personal computers and new telecommunications networks like the Internet would seem to have released people from neighborhood ties. The middle and upper-middle income groups, who inhabit urban worlds without finite geographical borders, have become "cosmopolites," for whom distance is "elastic." Meanwhile, even "localities" have found themselves less bound to residential proximity, it is suggested, as better social services and increased economic security have made local support systems less important and as greater differentiation in the rhythms of daily life has made it more difficult for increasing numbers of people to participate in neighborhood life.[2]

Against such arguments we must recognize that distinctive residential neighborhoods still exist. Neighbors continue to supply much of the raw material for social life for many social groups, while for relatively immobile groups (such as the poor and the very elderly) the neighborhood represents virtually the only opportunity for social interaction. Even the more mobile are susceptible to chance local encounters and the subsequent social interaction that may follow; and most householders will establish some contact with neighbors from the purely instrumental point of view of mutual security.

There are in any case several good reasons why social distance and physical distance should continue to act as mutually reinforcing aspects of social interaction and residential segregation. Geographer Ron Johnston, noting that physical distance restricts any social interaction involving face-to-face contact for the simple reason that traveling takes time and costs money, goes on to make the basic case as follows:

> Lasting social contacts are based on common interests. These may be very specialized—a particular hobby, perhaps—but many are based on interests, notably the problems of child rearing, which reflect common values and lifestyles. These commonalities, in turn, reflect incomes, occupations and educational backgrounds. The more similar people are on these criteria, the more likely they are to have in common and the more mutual benefits they are likely to receive from frequent social contact—both formal (local societies, churches, etc.) and informal (coffee mornings, cocktail parties, bridge clubs, etc.). . . . [C]ongregation and segregation are obvious means to this end, ensuring that one's neighbors are potentially valuable social contacts, even if that potential is never realized.[3]

Sociologist Gerald Suttles has identified additional reasons for the persistence of residential segregation.[4] They include:

- Minimization of conflict between social groups with different values and attitudes.
- Maximization of the political voice and influence resulting from spatial clustering.
- Greater degree of social control (i.e., collective self-control and self-policing) that is possible with relatively homogeneous residential groupings.

TERRITORIALITY

Inherent to Suttles's interpretation of residential segregation is the idea of **territoriality**: the tendency for particular groups to attempt to establish some form of control, dominance, or exclusivity within a localized area. Group territoriality depends primarily on the logic of using space as a focus and symbol for group membership and identity and as a means of regulating social interaction.[5] In many ways it can be seen as the product of the transition from the rigid social order of the preindustrial city to the competitiveness and social mobility of industrial cities. In preindustrial society the relative stability of society and the rigidity of group membership enabled people effectively to maintain social distance through appearance and comportment—what sociologist Lyn Lofland, coining a dreadful piece of jargon, called "appearential ordering."[6]

In the more rapidly changing environment of the industrial city, with its constant arrivals of strangers from other places, appearances could be deceptive. The city became a stage; people could pose and masquerade, their background and social credentials unchallenged in the turmoil. Meanwhile, the more intense economic competition, the greater variety of people attracted to cities, and the emergence of new occupational groupings caused sensitivities to social distance to become even greater. With appearance no longer a reliable guide, spatial ordering—group territoriality—provided a means of establishing and maintaining social distance. The outcome can be seen most clearly in the demarcation of the "turf" of street gangs, in the segregation of ethnic groups antagonistic to one another (Figure 12.3), and in the exclusivity of wealthy residential enclaves. The very fluidity of modern urban social and economic life, however, makes it very difficult to maintain territoriality at such extreme levels.

THE FOUNDATIONS OF RESIDENTIAL SEGREGATION

Residential segregation is based primarily on four interrelated dimensions of society—social status, household type, ethnicity, and lifestyle—each of which influences people's perceptions of social distance. Before going on to examine the patterns and intensity of urban residential segregation, therefore, we first need to develop some understanding of the nature of these social cleavages.

Social Status

The structure and dynamics of social status are central not only to social interaction and residential segregation but also to the broader sweep of urban politics and urban change. At a purely empirical level, social status can be interpreted in terms of educational qualifications, occupation, and income.

At a more conceptual level, however, social status is usually taken to involve additional commonalities of values and culture, and it is more appropriate to talk in terms of social *class*. For most theorists these commonalities are rooted in economic organization and structure. Karl Marx, for example, saw class in terms of the fundamental division between the owners and controllers of property and capital (the bourgeoisie) and those who must sell their labor, however skilled (the proletariat). Max Weber, the second great theorist of social relations and social structure in modern urban settings, argued for a more refined categorization of class, based on the same criteria of control and ownership of property and capital but also recognizing differences (particularly within the huge proletariat) based on marketable skills and the consequent differences in access to consumer goods, living conditions, and personal experiences. The development of the ideas of Marx and Weber on social class by contemporary social theorists has resulted in some important distinctions that must be summarized very briefly here.[7]

First is the concept of **class structure**, which refers to the formal categorization of class positions in a society at any given time. It is based on the positions people hold within the division of labor and the framework of economic organization. The broad categories of class

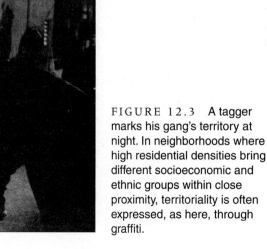

FIGURE 12.3 A tagger marks his gang's territory at night. In neighborhoods where high residential densities bring different socioeconomic and ethnic groups within close proximity, territoriality is often expressed, as here, through graffiti.

FIGURE 12.4   Many larger cities in the United States have retained elite residential neighborhoods in inner-city locations around enclaves of town mansions built by wealthy local industrialists in the nineteenth century. This example is from Grand Rapids, Mich.

structure are cast in terms of fairly heterogeneous groups: the "middle class," for example, consisting of a great variety of occupations. Narrower categories within the class structure ("the professions," for example) are sometimes referred to as **class fractions**.

Yet people are not always conscious of the contours of this formal categorization. Rather, "People *experience* class as it is built into their lives in particular ways and come to realize the force of class in and through the immediate circumstances they can experience and understand directly."[8] It is these experiences that constitute the process of **class formation**, resulting in conscious collectivities of people. Although many of these experiences are framed around class structure and class fractions, many are the result of other dimensions of what geographer David Harvey has termed **class structuration**. According to Harvey, these include (1) the division of labor that determines

FIGURE 12.5   Segregation by socioeconomic status is an attribute of "gentrified" neighborhoods—older residential areas that have been "invaded" and renovated by affluent households seeking the amenities and lifestyle of central-city locations. In Old Town Alexandria, Va., gentrification has established an elite area among older town houses that had previously comprised lower- and middle-income neighborhoods.

the formal class structure, (2) institutional barriers to social mobility, (3) the system of authority, and (4) the dominant consumption patterns of a particular time and place.[9]

Several factors are of direct relevance to patterns of social status, social interaction, and residential segregation in today's cities, including local marriage patterns and the social processes of socialization and stereotyping. By far the most important factor in reinforcing class structure, formation, and structuration is the educational system. Education is an important determinant of the skills that decide a person's starting position in the division of labor. The spatial organization of the American educational system is particularly important here. Most children attend a neighborhood school that is run by a locally elected school board that largely depends on local funding, usually through local property taxes. It follows that there are significant differences between school districts and between individual schools in the quality and type of education provided. Consequently, school catchment areas have become an important element in a self-perpetuating cycle of class structuration and residential segregation. "Good" schools will tend to be found in localities with high property values and with populations with high average levels of educational achievement who tend to elect progressive school boards. The success of these schools is a strong attraction for educated, affluent households with school-age children. As a result, affluent, middle-class neighborhoods with good schools tend to stay that way, constantly attracting a supply of members of this class fraction.

Economic competition, played out in the housing market, is the basic mechanism that ensures this stability, but it is supplemented by a local politics that is directed toward maintaining the quality of educational facilities and the exclusion of families whose finances and attitudes might threaten the educational environment. Given this stability, the school setting is an important element of **social reproduction** (see p. 491), providing (in this case) the skills that are the passport to higher-paid, higher-status occupations. At the other extreme, underfunded schools serving disadvantaged neighborhoods are a factor in the social reproduction of low-income households.

Marriage patterns also involve a circular chain of social reproduction and residential segregation. Studies of mate selection have shown that a high proportion of marriages are between people living within one or two miles of each other, with greater distances encompassing progressively fewer marriages.[10] This **distance-decay effect** reflects the role of spatial segregation in ensuring social reproduction through marriage: People tend to marry their equals in social status, and because neighbors tend to be social equals, they tend to find marriage partners from the neighborhood.

Family, school, and neighborhood settings are all important to the overall process of *socialization* by which people are introduced to group norms, attitudes, values, dress codes, speech styles, and standards of comportment. This process produces a local convergence that not only reinforces the process of social reproduction but also helps to give identity and character to the kaleidoscope of residential areas within cities. The process is so strong and pervasive that even adult newcomers to a neighborhood often seem to conform, after a while, to local norms. This phenomenon, known as the **neighborhood effect**, is difficult for researchers to track and quantify, but there is evidence that it operates in a wide variety of ways, influencing lifestyle preferences and voting patterns, for example (see also p. 388).

Finally, we should note that these self-reinforcing patterns of social interaction and residential segregation are further consolidated as a result of the *stereotyping* that is itself a consequence of spatial segregation. Because many people do not regularly mix with members of other social groups, they tend to create models of them—stereotypes—that are based on incomplete, second-hand knowledge and that often exaggerate certain attributes. Intergroup differences are thus exaggerated, increasing perceived social distances and sustaining the imperative for residential segregation.

Household Type

Households are the basic unit of residential organization. As such, they are very important to an understanding of residential segregation. In the United States, for example, although the traditional suburban household—consisting of the nuclear family unit: mother, father, and children—has by no means disappeared, other forms of household organization are fast outnumbering it. The single-parent family is the fastest growing of all household types, and one in every four households consists of a person living alone. Similar changes are occurring in the societies in most other developed countries in response to the social, economic, demographic, and technological processes discussed in Chapter 4. Other common household types in American cities are childless (or child-free) young heterosexual couples, gay couples, "empty-nesters" whose grown children have left home, families consisting of divorced and remarried parents and the children from their former marriages, and small groups of people, college students for example, living communally (and not necessarily involving sexual relations between any of the members).

Members of these different household types tend to share certain attributes, particularly in relation to housing needs and preferences. In addition to the necessity to match household size to housing space, each household type tends to have very specific housing

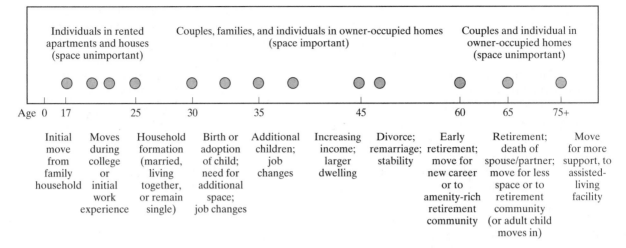

FIGURE 12.6   A household life cycle model of middle-income residential mobility and housing preferences.

needs and preferences. These similarities in residential needs and preferences lead to spatial congregation and greater within-group interaction, thus adding a further dimension to residential segregation.

One useful way of looking at the relationship between household types and residential segregation is through the process of the **household life cycle**. Figure 12.6 shows an idealized model of the life cycle of middle-income households, implying that there are recognizable stages, each with a distinctive household composition that is associated with particular space needs, specific preferences in terms of accessibility, type of housing tenure, locational setting, and different propensities to move. According to this scheme, the typical household life cycle begins with a brief stage based on single-person households or communal households of single persons who have left home to study or find work. In this stage the physical size of the dwelling is relatively unimportant, whereas accessibility to jobs and urban amenities is often a major consideration. The first year or two of living together or marriage can represent a continuation of this phase. In the event of the birth or adoption of a child, for couples and singles alike, the household's real and perceived needs change dramatically, reversing the relative importance of dwelling space and accessibility and placing a premium on safe, quiet residential environments. Over time, a series of adjusting moves will be made as household income increases.

By the time children are of junior-high or middle-school age, access to high quality neighborhood facilities—especially schools and social amenities—becomes increasingly significant. Meanwhile, the household's earning power is likely to have increased, partly through job promotions and if the second partner returns to the labor force after a period of homemaking. This increased income enables the household to move to a new suburban or exurban home. Divorce and remarriage result in more moves. Some middle-aged adults who experience divorce or lose a job may move in with their elderly, usually widowed parent—offering companionship and assistance in exchange for sharing the parent's home. At the "empty nester" stage, when a household's space requirements decrease considerably, neighborhood ties and sentimental attachments can prevent many couples or single parents from moving to smaller homes; housing professionals have recently begun to pay increasing attention to household desires to remain in a cherished home or valued neighborhood by refitting existing homes to meet changing household configurations. At the same time, the increasing financial independence of older people since the 1960s has been accompanied by increasing residential independence and associated housing options, such as assisted living accommodation or specialized retirement communities.[11] The final residential change for many people, however, can come at the stage when they move to live with a son or daughter or to institutionalized accommodation such as a hospice (Figures 12.7, 12.8).

This model fits with much of the available evidence about conventional residential mobility and urban structure (see Chapter 13). It should be stressed, however, that it is an idealized model that is *inherently family-centered* and *middle-income* in orientation. (For people who choose to remain single, for example, different assumptions have to be made. Similarly, purchasing a home cannot be assumed to be an option for low-income households.) The idea of a household life cycle therefore needs to be linked to intersecting cleavages of socioeconomic background, ethnicity, and lifestyle before we can chart the impact of household type on the residential kaleidoscope with any sensitivity.

Ethnicity

The term *ethnicity* covers any group that is primarily characterized by attributes of race, religion, nationality, or culture. Implicit in the use of the term is the idea

FIGURE 12.7 A circular neighborhood, part of Sun City, on the edge of the Phoenix metropolitan area, a specialized adult community—in this case, with an active, resort retirement lifestyle—whose 45,000 residents make it the largest retirement community in the country.

that such groups are minority groups whose presence in the city stems from a past or continuing stream of in-migration. In this sense, important ethnic groups in American cities include African Americans, Chinese, Italians, Jews, Mexicans, Puerto Ricans, and Vietnamese. The term **charter group** is sometimes used to describe the host society, the matrix in which ethnic groups find themselves. The charter group itself may not be ethnically homogeneous, but it is usually

dominated, culturally, if not always numerically, by a particular combination of race, religion, and national origin—as in the white, Anglo-Saxon charter group of many American cities (Figure 12.10).

What is clear enough from the available evidence, however, is that most minority groups tend to be highly segregated from the charter group (Figure 12.11). In addition, this segregation has been shown to be greater than might be anticipated from the socioeconomic

FIGURE 12.8 Residential segregation along demographic lines is illustrated here by the single-room occupancy (SRO) hotel, which is commonly dominated by single and elderly people. This example is from Seattle.

FIGURE 12.9   This photo of San Francisco's Chinatown shows part of an area that is completely and intensely ethnic in terms of residence, businesses, cultural institutions, and street life.

Box 12.1   The Social Construction of Race[12]

A dilemma facing urban researchers who want to map and analyze residential segregation and economic inequalities between different groups over time is that the most reliable sources of long-term data, such as the U.S. Census, categorize people by "race." Yet genetic research has not found any significant biological basis for differentiating people into separate racial groups. Rather, race is a **social construction** that artificially categorizes people into separate groups based on characteristics that include physical appearance (particularly skin color), ancestral heritage, and cultural history.

Historically, nineteenth-century Europeans classified the people in their colonies into an artificially created hierarchy of categories that placed themselves at the top. This social construction inevitably classified some people as inferior and often translated into the economic and social exploitation of these people. In the United States before the reforms associated with the civil rights movement, the social construction of race resulted in legal construction of laws that sought to separate people and treat them differently based on race.

Yet if race is a social construction, it does not explain a person's behavior or economic situation. Similarly, racism—and resultant discrimination—is not a uniform or invariable condition of human nature but, rather, consists of sets of attitudes that are rooted in societal perceptions as they relate to political culture and economic circumstances.

Consequently, the interaction of political culture and economic circumstances is particularly important to an understanding of residential segregation in societies, such as the United States and the United Kingdom, where institutional discrimination carries racism into the housing delivery system. Despite the reforms associated with the Civil Rights movements in these countries, institutional discrimination continues to permeate the legal framework, government policies (relating to, among others, urban renewal, public housing, and suburban development), urban **land use zoning** ordinances, and the practices of builders, landlords, bankers, insurance companies, appraisers, and real estate agents (see Chapter 13, pp. 364–370).

The impersonal web of exclusionary practices that results from this institutional discrimination has reinforced the racism and discrimination of individuals to the point where segregated housing has led to *de facto* segregated schools, shopping areas, and recreational facilities. This spatial segregation, in turn, serves to socially reproduce racism and to sustain economic inequalities between different groups within cities.

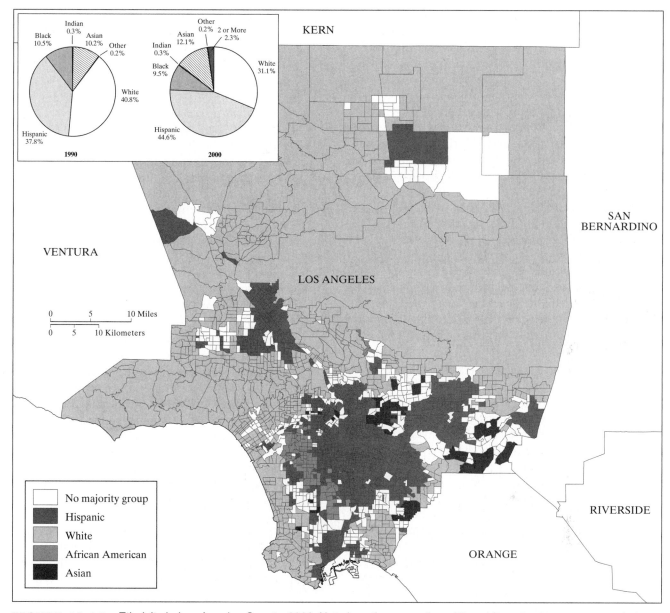

FIGURE 12.10   Ethnicity in Los Angeles County, 2000. Note how the proportion of the white—charter group—population has fallen over time to well below 50 percent and how little overlap there is in the major concentrations of Hispanic, Asian, and African-American populations.

status of the groups concerned. In other words, the low socioeconomic status of certain minority groups can only partially explain their high levels of residential segregation. The extent to which ethnic groups are spatially segregated from the charter group varies a great deal, depending on the degree to which the process of **assimilation** occurs and the amount of time that the process has been operating. Assimilation, however, is not simply the process of one culture being absorbed into another; both the charter and ethnic minority cultures are changed by assimilation through the creation of new hybrid forms of identity. In overall terms, however, the degree of assimilation is a function of the social distance between the charter

group and a particular ethnic group. It is, therefore, possible to think of ethnic groups as distinctive class fractions, subject to circular chains of social reproduction and residential segregation through socialization in family, school, and neighborhood settings and through stereotyping by other groups.

As a result, the rate and degree of assimilation of a minority group, and the spatial patterns of residential segregation, can be seen to depend on the intensifying effects of (1) external factors, including charter group prejudice, institutional discrimination in the housing market, and the structural effects of often low socioeconomic status, and (2) internal group cohesiveness associated with the desire of ethnic group members to maintain cultural identity. In fact, there may be differences not only in the degree and speed of the overall assimilation of different ethnic groups but also in the behavioral assimilation and structural assimilation of particular ethnic groups. **Behavioral assimilation** occurs when members of a minority ethnic group acquire the language, norms, and values of the charter group, thus becoming *acculturated* to the mainstream life of the city. **Structural assimilation**, on the other hand, refers to the diffusion of members of a minority ethnic group through the social and occupational strata of the charter group.

Given this distinction, we can see that **congregation**—the residential clustering of an ethnic minority through choice—fulfills several functions:[13]

- *Defensive functions*, whereby the existence of a territorial "heartland" helps to reduce the isolation and vulnerability of members of a particular ethnic minority. Such functions are particularly important when charter group discrimination is widespread and intense.

- *Support functions*, whereby ethnic enclaves serve as a "port of entry" to the city for new migrants and a haven for longer-term residents. By clustering together in a mutually supportive haven, members of the group are able to avoid the hostility and rejection of the charter group (and of other ethnic groups), exchanging insecurity and anxiety for familiarity and strength. The existence of ethnic institutions such as places of worship can be a particularly important source of both practical and spiritual support. In addition, most ethnic groups develop localized, informal self-help networks and welfare organizations. Finally, the existence of ethnic enclaves provides protected niches for ethnic enterprise (both legitimate and illegitimate) providing an expression of group solidarity as well as a means of social and economic advancement for successful entrepreneurs and a route by which workers can bypass the charter-group dominated labor market.

- *Cultural preservation functions*, whereby ethnic residential segregation helps to preserve and promote a distinctive cultural heritage. The distance-decay effect on the selection of marriage partners is particularly important here. In addition, the ethnic institutions and businesses supported by residential clustering are important elements of cultural solidarity. Territorial clustering is particularly important to ethnic groups for whom the observance of religious precepts relating to dietary laws, the preparation of food, and attendance at prayer and religious ceremonies are important.

- *"Attack" functions*, whereby the ethnic neighborhood serves as a base for action. This action is usually both peaceful and legitimate: Spatial concentrations often enable ethnic groups to gain representation within the institutional framework of urban politics, for example. Sometimes, however, ethnic heartlands are used as a convenient base for insurrectionary groups and urban guerrillas, who are able to "disappear" within the camouflage of a distinctive cultural milieu, protected by a silence resulting from a mixture of sympathy and intimidation.

We can also see that ethnic residential segregation might be expressed differently, depending on the importance of these functions combined with internal group cohesiveness and on the importance

FIGURE 12.11
The boundary between ethnic areas is sometimes very clear. A Chinese restaurant and an Italian restuarant on Grand Street mark the dividing line between Chinatown and Little Italy in New York City.

of charter group attitudes combined with institutional discrimination and structural effects associated with socioeconomic status. Three main types of ethnic segregation and congregation have been identified: colonies, enclaves, and ghettos.[14]

- *Colonies* are the result of situations in which a particular area of a city serves as a port-of-entry for an immigrant (or migrant) ethnic group. It is a temporary phenomenon, a base from which ethnic group members are culturally assimilated and spatially dispersed.

- *Enclaves* are ethnic concentrations that exist over several generations mainly because their inhabitants choose to congregate for functional reasons.

- *Ghettos* are ethnic concentrations that exist over several generations mainly because of the constraints of charter group attitudes and discrimination, which are often institutionalized through the operation of housing markets (see pp. 346–357).

Lifestyle

The intersecting cleavages of class, household type, and ethnicity, together with people's experience and their personal preferences and aspirations, lead to a variety of urban lifestyles. People pursuing similar lifestyles often cluster together, partly in order to fulfill the lifestyle itself and partly because affinity to people with similar lifestyle preferences tends to narrow perceived social distance, thus inducing the self-perpetuating chain of social reproduction and

residential segregation. The "classic" American lifestyles identified by sociologist Wendell Bell[15] were based on three stereotypes:

- *Familists*, who are home-centered and who tend to spend much of their spare time with their children. As a result, they are attracted to residential settings such as suburban neighborhoods that are dominated by other familists and are close to schools and parks.

- *Careerists*, who are attracted to prestige residential settings that are conveniently located in relation to workplaces or transportation nodes.

- *Consumerists*, whose preference is for the material benefits and amenities of cities, leading to a tendency to congregate in central locations close to clubs, theaters, art galleries, restaurants, and so on; or in amenity-rich planned suburban settings (Figure 12.12).

It is questionable, however, whether individuals or households can realistically be categorized this way. It is perhaps more accurate to think of most of them as subscribing to all three sets of preferences, but to different degrees and at different points in their life cycle (Figure 12.13). In addition, changing class structure, changing household types, changing social values, and increasing affluence have resulted in numerous additions to these "classic" lifestyles: Gay lifestyles, recreation-oriented and sport-oriented lifestyles, "singles" lifestyles (e.g., metrosexuals: Straight urban males who are in touch with their feminine side and not afraid to show it by cooking, having good grooming habits, and dressing in style),

FIGURE 12.12 Residential segregation according to lifestyle preferences has been made possible by the development of specialized, amenity-rich suburban communities such as the Wynmoor Village retirement community in Coconut Creek, Fla.

and the lifestyle of the active retirement community, for example. At the same time, it must be acknowledged that there remains a section of the population for whom "lifestyle" is experienced largely through television and the Internet, the home and the neighborhood being viewed more as havens from the outside world rather than as settings for the enactment of a favored lifestyle.

BOX 12.2   SOCIAL EXCLUSION AND MIGRANT WORKERS IN WEST EUROPEAN CITIES

The post-World War II revival of large-scale industrialization in Western Europe generated a strong demand for labor in the cities of more prosperous countries like France, Germany, and the United Kingdom. Tens of thousands of foreign-born migrant workers were brought in to fill low-wage assembly-line and service-sector jobs. These "guest" workers came from the countries of Mediterranean Europe, including Italy, Greece, Portugal, and Spain, and from former European colonies. West Germany attracted migrants from Turkey, Yugoslavia, Italy, and Greece; France brought in workers from Algeria, Tunisia, Morocco, Spain, and Portugal; while Britain drew on Commonwealth citizens from the Caribbean, India, Pakistan, and from Ireland.

The migrants were intended as temporary workers—a ready stream of labor that could be turned on and off as needed. The volume of incoming migrants, in fact, paralleled the highs and lows of the business cycle, with peaks in the mid-1960s and the early 1970s. By the mid-1970s, there were nearly 4 million foreign workers in West Germany, over 3.5 million in France, 1.5 million in the United Kingdom, nearly 1 million in Switzerland, and about half a million each in Belgium, the Netherlands, and Sweden. The onset of a deep economic recession in 1973, however, brought a dramatic check to the flow of migrants. Most governments imposed immigration restrictions. Some countries offered financial incentives to encourage return migration. Inducements were effective only in regions of severe unemployment, such as the Ruhr. Otherwise, the incentives were not enough to motivate migrants to return home to countries where employment opportunities were bleaker than in Europe.

Since the early 1970s, increasing globalization and deindustrialization, combined with the transformations following the fall of Communism in the late 1980s, have brought massive economic changes to Europe that have been associated with often high levels of long-term unemployment and increased occupational and social polarization. One of the key objectives of the European Union's 1992 Maastricht Treaty was greater social and economic cohesion among its member countries. The treaty's provisions, however, created two classes of immigrants within Europe: the 5 million E.U. migrants with the right to free movement among member countries; and the 10 million non-E.U. immigrants who, although long-settled, do not have citizenship of the E.U. country in which they reside or the right to move to other member countries.

Within each of these countries, the impact of the migrant workers has been localized in the larger metropolitan areas. More than one-third of the migrants in France are concentrated in the Paris region, where they represent over 15 percent of the population. Foreign-born residents make up 15 to 25 percent of the population in German cities like Frankfurt, Stuttgart, and Munich. More than half the population of Amsterdam is non-Dutch, being composed primarily of individuals from the former colonies of Indonesia and Suriname.

The migrants have had a very localized impact within these urban areas, reflecting the fact that they took over the inner-city areas and jobs left vacant because of the suburbanization of the upwardly mobile, native-born population. Because nearly all of the migrants were initially recruited to low-skilled, low-wage employment, they have inevitably found accommodation in the cheapest housing and most run-down neighborhoods. The position of migrants in the labor market reflects the objective of many to earn as much as possible as quickly as possible, the easiest strategy being to take on employment with an hourly wage where overtime and even a second job can be pursued. Most immigrants, however, have tended also to be kept at the foot of the socioeconomic ladder by a combination of social and institutional discrimination. The net result is reflected by the statistics for West Germany in the mid-1970s: Only 1 percent of migrant workers held nonmanual jobs. Meanwhile, the migrants' position in the housing market reflects their need for inexpensive accommodation. The localization of migrants in camps, factory hostels, immigrant hostels (*hotels meublés*), suburban **squatter settlements** (*bidonvilles* or *banlieue*), and inner-city tenements, however, has also been reinforced by bureaucratic restrictions and discrimination.

FIGURE 12.13   "Keynesian" suburbs like this one facilitated residential segregation by socioeconomic background (middle income), household type (married couples with young children), ethnicity (white), and lifestyle (family-centered). This aerial photo shows part of Levittown in 1948, shortly after this mass-produced suburb was completed on Long Island farmland just 25 miles east of Manhattan.

Box 12.3   How Segregated Are Urban Neighborhoods?

Segregation can be measured empirically in several ways. After a long debate among social scientists, the most generally used measure is the *index of dissimilarity*, a simple but effective statistic that is analogous to the Gini index of inequality that is widely used in economics. The index of dissimilarity is used to compare the spatial distribution of two groups (blacks and whites, for example), and it provides the answer to the question: What percentage of one group would have to move to a different locality in order to achieve a proportional distribution of the groups within each locality that is the same as their proportional distribution within the city as a whole? The formula for the index (D) can be written as:

$$D = \frac{1}{2} \sum_{i=1}^{k} |x_i - y_i|$$

where x_i and y_i represent the percentages of group members and others, respectively, in subarea i, the summation being over all the k territorial subunits within the city. This formula, multiplied by 100, yields an index with a theoretical range of values from 0 (no segregation) to 100 (complete segregation). The territorial units of measurement are usually census tracts, sometimes city blocks. It should be noted, however, that the index is very sensitive to variations in the size of the areal units (the smaller the mesh of areal units, the higher the resultant index of dissimilarity). A similar measure, the *index of segregation*, can be used to compare the spatial distribution of one group with *all* other groups (e.g., Asians and all other races). Both the index of segregation and the index of dissimilarity are measures of *absolute* levels of segregation. That is, they quantify the discrepancy between actual

(Continued)

BOX 12.3   HOW SEGREGATED ARE URBAN NEIGHBORHOODS?   (*Continued*)

TABLE 12.1

Changes in Black Residential Segregation[1] in the Most Segregated U.S. Metropolitan Areas as Measured by the Index of Dissimilarity, 1980–2000

2000 rank	Metropolitan area	Index of dissimilarity			Percent change	
		1980	*1990*	*2000*	*1990–2000*	*1980–2000*
1	Detroit, Mich.	87.4	87.4	84.6	–3.3	–3.3
2	Milwaukee, Wisc.	83.9	82.6	81.8	–1.0	–2.5
3	New York, N.Y.	81.2	81.3	81.0	–0.4	–0.2
4	Newark, N.J.	82.7	82.5	80.1	–2.9	–3.2
5	Chicago, Ill.	87.8	83.8	79.7	–4.8	–9.2
6	Cleveland, Ohio	85.4	82.4	76.8	–6.8	–10.1
7	Buffalo, N.Y.	80.1	80.0	76.6	–4.4	–4.4
8	Cincinnati, Ohio	78.1	76.1	73.9	–2.9	–5.3
9	St. Louis, Mo.	81.7	76.9	73.1	–5.0	–10.5
10	Nassau–Suffolk, N.Y.	76.7	76.1	73.0	–4.1	–4.8
11	Bergen–Passaic, N.J.	80.3	76.8	72.3	–5.9	–10.0
12	Philadelphia, Pa.	78.1	76.8	72.0	–6.2	–7.8
13	Indianapolis, Ind.	78.8	74.6	70.4	–5.5	–10.6
14	Miami, Fla.	78.5	69.0	69.4	+0.5	–11.5
15	Kansas City, Mo.	77.3	72.5	68.8	–5.1	–10.9
16	New Orleans, La.	69.8	67.9	68.4	+0.8	–1.9
17	Baltimore, Md.	74.4	71.3	67.5	–5.3	–9.3
18	Pittsburgh, Pa.	72.5	70.7	67.1	–5.1	–7.5
19	Los Angeles, Calif.	80.8	72.8	66.4	–8.9	–17.9
20	Houston, Tex.	75.4	66.4	66.3	–0.1	–12.1

[1] *Segregation: Blacks versus White Non-Hispanics.*

Source: *J. Iceland and D. H. Weinberg,* Racial and Ethnic Residential Segregation in the United States: 1980–2000, *Washington, D.C.: U.S. Department of Commerce, 2002, Tables 5-4 and 5-5.*

spatial patterns and those that would exist if groups were randomly intermixed. As a result, they do not capture differences in the proportional size of the groups. An index of dissimilarity of 90 between Asians and whites in a city may seem striking, but its significance is diminished if we learn that less than 1 percent of the city's population is Asian while 96 percent are white. Where researchers need to gauge the relative mixture of two groups across a set of localities, they use a different index, the *index of isolation* (sometimes called the index of dominance).[16]

For our purposes, the most widely used index, the index of dissimilarity, provides an effective measure of the degree of residential segregation in American cities. There is plenty of evidence of the residential segregation of ethnic groups. Indexes of dissimilarity across metropolitan areas at the census tract level in the United States range between 30 and 50 for most ethnic groups, with the higher levels of segregation being associated with more recent groups of migrants such as Mexicans and Vietnamese. A striking exception to this generalization is the very high degree of residential segregation experienced by African Americans. At the census tract level, indexes of dissimilarity between the African-American and the white non-Hispanic population in American metropolitan areas average about 64.

These data suggest that *the idea of American cities as melting pots may be something of a myth.* The metropolitan areas in the Midwest and Northeast are quite segregated (at 74.1 and 73.9 respectively). Segregation is highest in the largest metropolitan areas, with the top 10 most segregated metropolitan areas located in the Northeast-Midwest "Rust Belt" (Table 12.1). The U.S. Census Bureau averaged the ranks of the 43 largest metropolitan areas (with 1 million or more people in 1980 and at least

(*Continued*)

BOX 12.3   HOW SEGREGATED ARE URBAN NEIGHBORHOODS?   *(Continued)*

3 percent or 20,000 or more blacks) across five different indices of segregation, including the index of dissimilarity and the index of isolation, and found that in 2000 Milwaukee was the most segregated, followed by Detroit.[17]

Certainly, the residential segregation of the African-American population is currently at its lowest point since roughly 1920. And although there remains a large number of "hypersegregated" metropolitan areas, the 1990s continued a trend toward decreasing segregation that began in the 1970s.[18] Between 1980 and 1990, black segregation across all metropolitan area census tracts declined from 72.7 to 67.8; by 2000 it fell to 64.0. In 2000, the West was the most integrated region of the United States (55.9), followed by the South (58.1). Some of the more rapidly growing metropolitan areas in the West and South—Las Vegas, Phoenix, Austin, and Raleigh-Durham—have relatively low and declining levels of segregation (between 30 and 45).

These changes have been associated with the efforts of the civil rights movement since the 1960s and government action against discrimination in housing and lending. Although a 2002 study by the Urban Institute on behalf of the U.S. Department of Housing and Urban Development (HUD) found that 21.6 percent of black renters and 17.0 percent of black homebuyers encountered some form of housing discrimination, these figures represent a decline in discrimination since the previous HUD study in 1989 (from 26.4 percent for black renters and 29.0 percent for black homebuyers).[19] Meanwhile, the growth of black middle-income groups is reflected in census data that show an increase from 15 percent in 1960 to 36 percent in 2002 in the percentage of blacks living in the suburbs.

At the same time, however, the HUD study showed that **steering** on the basis of neighborhood racial composition increased significantly for black homebuyers between 1989 and 2000. Blacks were more likely to be steered by real estate agents to neighborhoods that were predominantly black compared to the neighborhoods recommended to comparable white homebuyers. Meanwhile, the many blacks who have not achieved middle-income status—and cannot afford to buy a home—remain concentrated in impoverished central city neighborhoods (see Figure 12.22).

INTERPRETATIONS OF RESIDENTIAL ECOLOGY

Contemporary patterns of residential segregation have developed on the basis of residential neighborhoods established in earlier growth phases, and the principal dimensions of urban residential structure were forged in the early decades of the twentieth century. As we have seen (Chapter 5), the emergence of the Industrial City brought about a radical transformation of urban space, resulting in a broad framework of *sectors* and *zones* of specialized land use. Within the older, central districts of cities, unprecedented numbers of migrants and immigrants from a wide spectrum of ethnic origins struggled to establish themselves. In the newer suburbs, meanwhile, the newly expanded middle-income groups sought to pursue the new trend of domesticity in quiet, secluded neighborhoods. Observing the resultant conflux of residential sorting and resorting, sociologists from the University of Chicago (among the most notable of whom were Robert Park, Ernest Burgess, and Roderick McKenzie) developed a theory of residential segregation and a model of urban residential structure that became benchmarks of urban theory.

THE CHICAGO SCHOOL: HUMAN ECOLOGY

Robert Park, having worked in Chicago's neighborhoods as a journalist before becoming chairman of the Department of Sociology at the University of Chicago, was struck by the distinctiveness of different neighborhoods: "A mosaic of little worlds that touch but do not interpenetrate."[20] Each of these little worlds, suggested Park and his colleagues, could be thought of as an *ecological* unit, a particular mix of people that had come to dominate a particular niche in the urban fabric. This interpretation led them to adopt a view of the city as a kind of social organism, with social interaction governed by a "struggle for existence."

This kind of biological analogy provided the Chicago School with an attractive general framework within which to set their detailed studies of the "natural histories" and "social worlds" of different groups. Just as in plant and animal communities, they concluded, order in human populations must emerge through the operation of "natural" process such as impersonal competition for territory and dominance. It should be borne in mind that these ideas were the product of an era when the appeal of **neoclassical economics** (with its emphasis on unfettered competition) was powerful, when the influence of Darwinism was relatively strong, and when the social sciences were struggling to establish a "scientific" respectability. Plant and animal science provided a rich source of ecological concepts and a graphic terminology with which to portray the social geography of the city.

The Chicago School saw social interaction as an expression of symbiosis in an overall social context that was dominated by competition for living space. The net result was a series of *natural areas* within which different groups were dominant. The Chicago School produced some very detailed and painstakingly researched studies of some of these "natural areas." One of the classics was Harvey Zorbaugh's *The Gold Coast and the Slum*,[21] a study of distinctive "natural areas" in Chicago's Near North Side. These included the "Gold Coast," a wealthy neighborhood adjacent to the lakeshore, and a slum area containing clusters of migrants and immigrants. Zorbaugh showed how the personalities of these different quarters were related to their physical attributes—the "habitat" they offered—as well as to the attributes and ways of life of their inhabitants.

The Human Ecologists of the Chicago School, however, did not see "natural areas" such as these as being permanently fixed. As the numerical strength and competitive (market) power of different groups altered and the relative attractiveness and suitability of different settings changed (through the physical and demographic changes imposed by the occupancy of particular groups with distinctive demographic mixes and ways of life), the ecological processes of *invasion* and *succession* (Figure 12.16) brought modifications to the pattern of "natural areas." Meanwhile, in addition to

the biotic dimension of social organizations (based on impersonal competition for territory and resources), Park and his colleagues recognized a cultural dimension, shaped by the consensus of social values.

Ernest Burgess brought these concepts together in his famous **concentric zone model** of residential differentiation and neighborhood change in Chicago.[22] Having delineated the location and extent of some 75 of Chicago's natural areas, Burgess interpreted them as falling into a series of concentric zones (Figure 12.17). This interpretation was reinforced by the conviction that as the city grew in population, economic competition and an increasingly specialized division of labor would lead to the complementary ecological processes of *centralization* and *decentralization*. Centralization occurred through the operation of **agglomeration economies** (see p. 65), while decentralization occurred as economic functions that lost out in central city competition removed themselves to peripheral settings. (This part of the argument is very similar to the principles that underpin **bid-rent theory**: see p. 134.)

The innermost zone, the **central business district (CBD)**, was above all the expression of the economic forces of centralization. It was surrounded by a **zone in transition**, comprising an inner belt of factories and warehouses surrounded by deteriorating residential neighborhoods. This deterioration was largely the product, according to Burgess, of the "invasion" of older residential areas by the burgeoning factories and warehouses generated by agglomeration around the CBD. This invasion and deterioration prompted existing residents to leave, thus providing housing opportunities for low-income migrants and immigrants. Surrounding the zone in transition was the *zone of independent workingmen's homes*, dominated by second-generation

FIGURE 12.14 Chicago's "Gold Coast" in the 1930s. A "natural area" investigated by the Chicago School.

FIGURE 12.15 A Chicago slum in the 1930s. Another kind of "natural area" investigated by the Chicago School.

migrants and immigrants who had established their economic and social status to the point where they had been successfully able to "invade" inner suburbs and consolidate themselves there in ethnic communities. Those who made it into the middle classes, or who were born into the middle class, lived in the *zone of better residences*, dominated by single-family homes with spacious yards; or in the *commuters' zone*, small towns and villages with a dormitory function but with little industry or employment of their own.

Although Burgess saw these broad zones as reflections of centralization, decentralization, and the differential economic competitive power of broad groups within society, he saw the "natural areas" that made up each zone (Chinatown, Little Sicily, and so on within the zone in transition, for example) as reflections of symbiotic relationships forged on the basis of language, culture, and race.

Criticisms of Human Ecology

The Chicago School became so influential and the zonal model so popular as a device for describing and explaining urban structure that it has become a benchmark in urban studies. It is also important to recognize, however, that the zonal model was derived from a particular set of circumstances: a city (Chicago) with a single dominant economic core and a metropolitan area that was growing rapidly through continuous streams of migration and immigration that produced a very heterogeneous population. It follows that the model can be applied only to cities where such circumstances exist. It is also important to know that the Chicago School version of Human Ecology came under heavy criticism during the late 1930s and 1940s.

Some of the most stringent criticisms were directed at the relative neglect of what Park and his colleagues

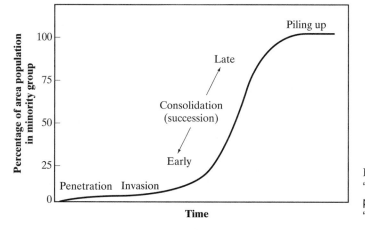

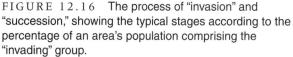

FIGURE 12.16 The process of "invasion" and "succession," showing the typical stages according to the percentage of an area's population comprising the "invading" group.

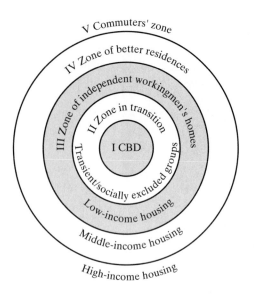

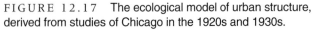

FIGURE 12.17   The ecological model of urban structure, derived from studies of Chicago in the 1920s and 1930s.

had identified (but failed to explore in sufficient detail) as the "cultural" dimension of social organization. Walter Firey, for example, criticized the Chicago School's approach because it overlooked the roles of sentiment and symbolism in people's behavior. He pointed to the evidence of social patterns in Boston where, although there were "vague concentric patterns," it was clear that the persistence of the status and social characteristics of distinctive neighborhoods such as Beacon Hill, the Common, and the Italian North End could be attributed in large part to the "irrational" and "sentimental" values attached to them by particular sections of

the population.[23] Firey's point was that social values could—and often did—override impersonal economic competition as the basis for social interaction and residential segregation. Other critics of the Chicago School pointed to the failure of its general structural concepts (such as the "natural area" and concentric zonation) to hold up under comparative examination. The most telling criticisms, however, were those that addressed the biotic analogies on which ecological theory had been developed. With similar analogies being used in an entirely different context to justify the geopolitics of Hitler's Nazi regime, social theories based on biotic analogies were suddenly revealed to be dangerously simplistic.

As a result, ecological ideas had to be reformulated, eliminating the crude mechanistic and biotic analogies. Following the early lead of sociologists such as Amos Hawley and Leo Schnore,[24] many analysts now find it useful to think simply in terms of an urban "ecology" consisting of discrete territories (such as "social areas" or "neighborhood types") containing relatively homogeneous populations with distinctive socioeconomic characteristics. One important development prompted by this reformulation was the exploration of the notion of urban ecology as a conceptual framework within which to portray the residential structure of the city through statistical analyses of socioeconomic data. "Factorial ecology," involving the use of factor analysis and the associated family of multivariate statistical techniques (including principal components analysis), became one of the most widely used techniques for tackling the complex question of measuring urban sociospatial differentiation.

Box 12.4   Factorial Ecology

Factor analysis—a powerful form of multivariate statistical analysis based on the methodological family that includes correlation and regression analysis—is a widely used approach for addressing the question of delineating the social and spatial dimensions of residential segregation. Factor analysis can help to identify what common patterns may exist across data that measure a broad range of attributes. The method is a summarizing technique that can identify groups of variables with similar patterns of spatial variation.[25]

The first result from studies of the factorial ecology of American cities suggested that the most important dimensions of residential differentiation were, in descending order of importance, socioeconomic status, family status or life cycle, and ethnicity—with spatial expressions associated respectively with sectoral, zonal, and clustered differentiation. One of the

striking aspects of such strides, beginning with a rash of studies in the 1960s and 1970s, was the consistency of these dimensions and their associated spatial patterns from one city to another, often over several decades and despite variations in the input variables used in the analyses.

A conventional wisdom emerged about the residential structure of the idealized North American city (Figure 12.18). Geographer Robert Murdie suggested that socioeconomic status, family status, and ethnicity represent the main dimensions of social space that, when superimposed on the city's physical template, serve to isolate areas of social homogeneity "in cells defined by the spider's web of the sectoral-zonal lattice."[26] As Murdie pointed out, the city's physical template should not be thought of as somehow independent of social space. Radial transportation routes, for example, are likely to govern

(Continued)

Box 12.4   Factorial Ecology   (*Continued*)

the positioning of sectors and to distort zonal patterns (as we saw in Chapter 5). Similarly, the configuration of both sectors and zones is likely to be influenced by specific patterns of land use and by patterns of urban growth.

In addition to (and sometimes instead of) the three "classic" dimensions represented in Murdie's model, it is quite common for factor-ecological analyses to identify *other* dimensions of residential differentiation. A comparative analysis of the factorial ecology of 21 U.S. metropolitan areas,[27] for example, underlined the diversity of patterns of race and ethnicity by region. African Americans were found to be most concentrated in metropolitan areas where they represented the largest minority group, and, in these settings, factor structures revealed particularly close interrelationships between race, female-headed households, and poverty. A comparative study of 24 Canadian cities found the widespread incidence of factors additional to the three identified by Murdie, including socioeconomic polarization ("impoverishment") and several factors related to family structure.[28]

With the dimensions of Murdie's idealized model being seen less clearly now than 35 years ago, it has

been suggested that these "classic" spatial patterns are changing. As we saw in Chapter 4, U.S. cities are undergoing fundamental economic transformations that have been accompanied by social, demographic, and technological changes. Advances in telecommunications, for example, have already begun to remove many of the traditional frictions of space, not only for households but also for economic activities. This is not to suggest that residential differentiation and segregation are disappearing or changing abruptly; but the recent economic and other changes that are affecting cities are being manifested in more complex ways and at a finer level of resolution than the "classic" sectors, zones, and clusters that have been associated traditionally with socioeconomic status, family status, and ethnicity.

Among the axes of differentiation that are likely to develop in the United States are:

- The emergence of *migrant* status as a potent source of differentiation.
- The reinforcement of *ethnic* differentiation with the arrival of new immigrant groups.
- The emergence of new dimensions of *occupational differentiation* related to the expansion of service jobs.
- The appearance of significant distinctions in the degree of *welfare dependency*.
- The relative increase in the importance of *poverty* and *substandard housing*.
- The increased social and spatial differentiation of *post-Boomers/young adults* and the *elderly*.[29]

Although the applicability of the "classic" three-factor model has been borne out by previous factor-ecological analyses of cities in other developed countries—not only Canada, but also Australia and New Zealand—the evidence from studies of European cities has always been less conclusive. Overall, residential differentiation in continental European cities tends to be dominated by a socioeconomic status dimension that is often associated with family status and the localization of self-employed workers (see Figure 12.19). Ethnicity, however, does not generally occur as an independent dimension, partly because of the absence of substantial ethnic minorities (unlike the United States). British cities exhibit a somewhat distinctive ecological structure, with the principal dimensions of the "classic" model being modified by the legacy of the construction and rental policies associated with the traditionally relatively large public housing sector.

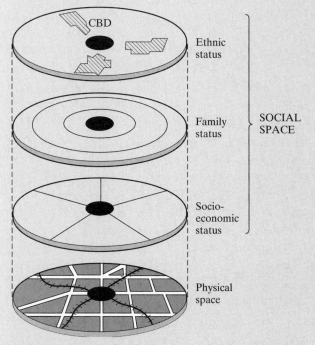

FIGURE 12.18   An idealized model of the residential structure of North American cities, based on the study of factorial ecology.

(*Continued*)

Box 12.4 Factorial Ecology *(Continued)*

Fundamental methodological issues remain as stumbling blocks in the pursuit of more sophisticated conceptual and empirical analyses of residential differentiation. Chief among them is the fact that most analyses rely almost exclusively on decennial censuses for their data. This means that research is often constrained by a lack of data covering a full range of socioeconomic characteristics, because many census bureaus have been wary of demanding information on sensitive subjects such as religion and have been unwilling to spend limited resources collecting information on complex characteristics such as people's lifestyle or activity patterns.

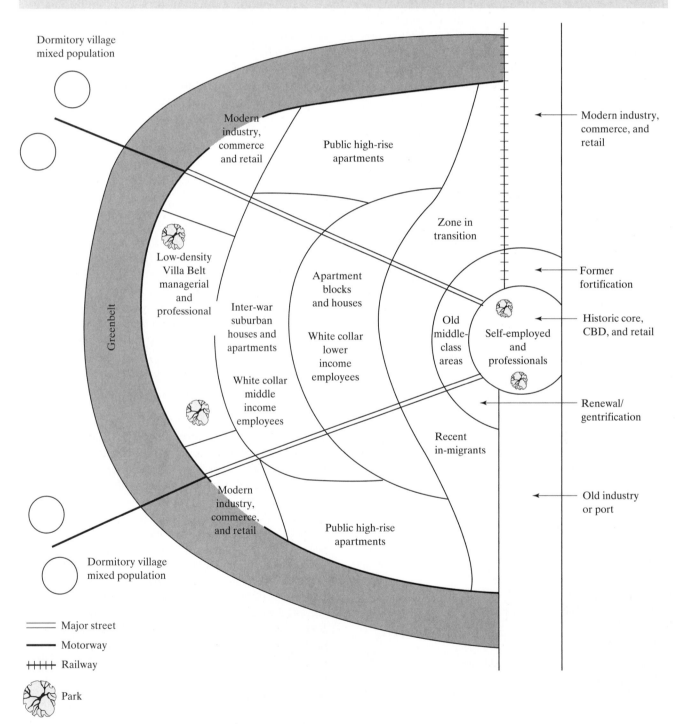

FIGURE 12.19   Model of Northwest European city structure.

BOX 12.5   RESIDENTIAL AND ECONOMIC STRUCTURE IN EUROPEAN CITIES[30]

The classic North American models of urban residential structure do not apply well to the diverse social patterns within European cities. A model with concentric circles of increasing socioeconomic status with distance from the core is most applicable to British cities. Mediterranean cities exhibit the **inverse concentric zone pattern** found in Latin America. In Mediterranean Europe the elite concentrate in central areas near major transportation arteries, while the very poor can be found in shantytowns in inadequately serviced parts of the periphery. In Europe it is the demographic pattern that fits a model of concentric circles because the number of persons per household usually increases with distance from the city center.

A **sector model** corresponds with the pattern of socioeconomic status where sectors containing different income groups radiate out from the city center. This social pattern reflects the historic preference of the wealthy to locate in pleasant axial areas, such as along monumental boulevards or upwind of pollution sources. The poorer residents were left with unattractive linear sectors of land along railway lines or zones of heavy industry. The multiple nuclei model fits the pattern of ethnic differentiation—different ethnic groups are concentrated either in nodes within the inner city or in high-rise public housing at the periphery.

Figure 12.19 shows a model of Northwestern European city structure. The preindustrial core contains the market square and historic structures such as a medieval cathedral and town hall. Apartment buildings contain upper-income and middle-income residents above small street-level shops and offices. The narrow winding streets extend out about a third of a mile. A few wider streets radiate out from the square and form a pedestrianized corridor that runs to the train station and contains national and international department stores, restaurants, and hotels. Skyscrapers are concentrated in the commercial and financial district. New downtown shopping malls or festival marketplaces are in refurbished historic buildings. Some of the old industrial and port areas contain new retail, commercial, and residential waterfront developments. Wealthy mansions grace the western part of the core.

Encircling the core are some **zones in transition** found in the concentric zone model. The area of the former city wall is a circular zone of nineteenth-century redevelopment. Upper-income residents have bought and gentrified some of the deteriorated middle-income housing, while other sections provide low-rent accommodation for students and poor immigrants.

Surrounding this area is another zone in transition—an old industrial zone with disused railway lines. In the 1950s and 1960s, new industrial plants, such as light engineering and food processing, replaced derelict old factories and warehousing. Low-income renters and owners live in the run-down nineteenth-century housing. Some houses have been refurbished or replaced with new units; foreign immigrants have added their cultural imprint in certain neighborhoods.

As in the concentric zone model, beyond this inner area is a zone of "workingmen's homes." This is a stable lower middle-income zone dating from the first half of the twentieth century. These **streetcar suburbs** contain apartment blocks and houses without garages and are anchored by a small shopping area, community center, library, and school. Outside this area are middle-income automobile suburbs, containing apartments and single-family houses with garages, that correspond with the zone of better residences in the concentric zone model. Further out are nodes containing the most exclusive neighborhoods, as might be expected in the **multiple-nuclei model**.

The estates of public high-rise apartments and new private middle-income and lower-middle-income "starter" homes at the urban periphery that lack basic amenities like shops and banks also correspond with the multiple-nuclei model. The periphery also contains commercial and industrial nodes, containing shopping malls, business and science parks, and high technology manufacturing.

Beginning with Vienna in 1904, many cities delineated a greenbelt at the edge of the built-up area within which development was prohibited. The greenbelt was intended to prevent urban sprawl and provide recreational space. Commuters live outside the greenbelt in dormitory villages and small towns that correspond with the commuters' zone in the concentric zone model. The airport and related activities such as hotels and modern manufacturing are located further out on a major freeway.

URBANIZATION AND CHANGING SOCIAL STRUCTURES

Many of the recent changes in the way urban analysts have conceptualized the residential structure of cities have evolved in tandem with the many profound changes that are occurring within cities themselves (such as increasing black suburbanization; see p. 320) as well as the widespread changes that are affecting cities from outside (including changes in the migration streams to U.S. cities that have been associated with increasing **globalization** or heightened national security concerns following the terrorist attacks of September 11, 2001). These changes point to the importance of bearing in mind the dynamics of urbanization. We have to see residential differentiation and urban ecologies in relation to the overall framework in which they are embedded. Geographer Wayne Davies has suggested that, historically, four major dimensions of residential differentiation have dominated cities everywhere—social rank, family status, ethnicity, and migration status—and that these have been combined in different ways in different types of society to produce varying social structures.[31]

In the earliest settlements family-related considerations dominated the social structure; prestige and status were ascribed primarily through kinship. With the beginnings of capitalist development, the division of labor intensified, a merchant class was added to the elite, and selective migration streams added to social and ethnic complexity. Davies suggests that these changes led to the creation of very different types of residential structure:

- In *preindustrial* cities the perpetuation of family kinship patterns and the continued importance of the established elite combined to produce a single axis of differentiation (social rank/family status), while the arrival of migrant groups of different ethnic origins created a second major axis of differentiation.

- In *colonial* cities located in previously settled areas, immigrants were politically and socially dominant, so that social rank, ethnicity, and migration status were collapsed into a single dimension of residential differentiation. Meanwhile, family status characteristics represented an independent dimension.

- In *immigrant* cities, the indigenous political elite remained dominant, while the age-, ethnic-, and sex-selective process of in-migration tended to overwhelm residential variations in family status.

- The onset of industrialization brought a great increase in occupational specialization, while income and wealth became more important as yardsticks of social prestige. As transport technologies made large-scale suburbanization possible, distinct patterns of differentiation emerged in terms of social rank and family status. Meanwhile, much greater heterogeneity in migration streams contributed to residential segregation in terms of race and ethnicity, thus completing the "classic" ecology of the *industrial* (modern) city.

- More recently, globalization, **deindustrialization**, and metropolitan decentralization have transformed this classic ecology, with the *informational city* exhibiting the much more fragmented and multidimensional ecology of **splintering urbanism**.

We now turn, therefore, to an examination of some important changes that have occurred in the relationships between socioeconomic background, household types, ethnicity, lifestyle, and residential segregation.

RECENT CHANGES TO THE FOUNDATIONS OF RESIDENTIAL SEGREGATION

The economic restructuring since the early 1970s has triggered a number of changes that have resulted in significant rearrangements of urban residential structure. At the heart of these rearrangements are three processes. The first is the *occupational polarization* that has resulted from sectoral shifts in the economy combined with the effects of corporate reorganization and redeployment, advances in robotics and automation, and the increasing participation of women in the labor force. The net effect has been an increase in the number of higher-paid jobs (in producer services, high-tech manufacturing, and the media), an increase in the number of low-paid jobs (in routine clerical positions, retail sales, fast food operations, and so on), and a decrease in middle-income jobs (skilled traditional manufacturing). These trends are reflected in the increasing inequality in family incomes in the United States since the 1960s (Figure 12.20). The new class fractions that emerged from this polarization at both ends of the socioeconomic ladder soon left their imprint on urban residential structure, while the thinning ranks of the traditional middle-income groups contributed to the breakup of solid middle-income suburban neighborhoods that previously had been the cornerstones of urban residential structure.

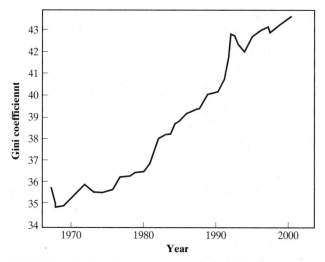

FIGURE 12.20 Annual changes in family income inequality in the United States, 1967–2001. The degree of inequality is measured by the Gini coefficient, which has a range of 0 to 100, with higher values representing greater inequality.

The second set of processes stems from the experience of the *Baby Boom generation* (see pp. 100–101), whose early cohorts found themselves having to cope with the transition from the social awareness, ethnic consciousness, and brotherly love of the counterculture years to the harsh economic realities of labor and housing markets that were suddenly depressed by the effects of **stagflation**. Attitudes and lifestyles changed, and residential preferences changed with them. Changed attitudes toward work, home, and family, meanwhile, were also reflected in changing patterns of household organization (increases in the proportion of single-person households, single-parent families and dual-income households with no children, for example), and these also came to be reflected in patterns of residential differentiation.

When the "back to downtown" movement started in the United States in the 1990s, most renters and buyers were young singles or childless couples; however, aging Baby Boomers—either retired or still working—are becoming a bigger part of the market. This has been associated with the trend away from "cocooning" (seeking to be holed up at home with family, using the home as an escape from life's hectic pace) and toward "hiving" (seeking frequent social interaction with friends, family, and community, using the home as a base for making those connections). These households are attracted to housing that is mixed with, adjacent to, within walking distance of, or connected by transit to recreational, cultural, entertainment, and work locations. The growing consumer interest in downtown housing has encouraged developers to provide luxury apartments and condominiums in new developments as well as in converted commercial, industrial, and warehouse buildings.[32]

The third set of processes stems from the experience of *ethnic minorities*. The arrival in American cities of new streams of immigrants has recharged the ethnic dimension of residential structure, recalibrated the scales of social distance, and led to the spatial reorganization of ethnic neighborhoods. It has done so differentially, however, because of the differential pattern of origins and destinations. As we saw in Chapter 4, legal immigration since the 1960s has not only been higher than at any time since the 1920s but it has also been very different in character from previous streams, with fewer Europeans and more people from Latin America, the Caribbean, and Asia. These new immigrants, moreover, have been highly localized in their impact—Mexicans locating disproportionately in Los Angeles, Chicago, San Diego, and Houston, for example; Puerto Ricans, Dominicans, Jamaicans, Chinese, and Indians in New York; Cubans, with growing numbers of Salvadorans and Nicaraguans, in Miami; Iranians and Salvadorans in Los Angeles; and Filipinos and Chinese in San Francisco.

Meanwhile, the experience of longer-established ethnic minorities has been closely tied to occupational polarization and to the sociocultural changes associated with the Baby Boom generation. For some, these occupational shifts and changed values facilitated behavioral or structural assimilation or both; for others, they resulted in social isolation and economic vulnerability. Black suburbanization, for example, took off in the 1960s with the growth of black middle-income groups (Figure 12.21). Between 1960 and 2002, the percentage of blacks living in U.S. suburbs rose from 15 to 36 percent (Figure 12.22). The metropolitan areas with the highest suburban black percentages are in the South. In Columbia and Charleston in South Carolina and in Atlanta, blacks comprise a quarter or more of the suburban population. Overall, however, black suburban representation remains low—exceeding 10 percent in only 24 of the 102 largest metropolitan areas.[33] And, for the United States as a whole, over half of all blacks still live in central cities—often concentrated in impoverished neighborhoods—compared with just over one fifth of all whites.

NEW DIVISIONS OF LABOR, NEW HOUSEHOLD TYPES, AND NEW LIFESTYLES

At the upper end of the socioeconomic ladder, the most significant changes have been associated with the growth of producer services occupations (involved in nonmaterial products): financial analysts, management consultants, personnel experts, designers, marketing experts, and so on. Such occupations have expanded

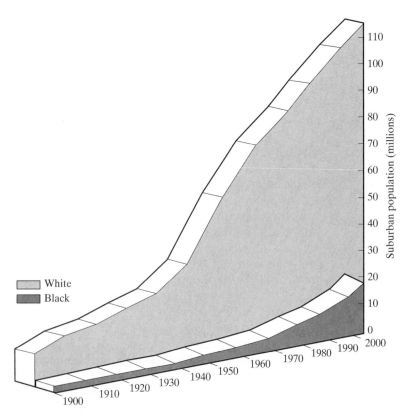

FIGURE 12.21   Black–white suburbanization trends in the United States, 1900–2002.

since the 1960s in the United States (Table 12.2), particularly in the larger metropolitan areas, where overall employment growth was relatively modest, and in the metropolitan areas of other developed countries—including Europe (Figure 12.25), Australia, and Japan.

At the other end of the socioeconomic ladder, the most significant changes have involved the creation of a *secondary labor market* characterized by low wages, poor working conditions, little or no job security, no benefits, and few prospects of promotion. This labor market has been the product of (1) the relative decrease in metropolitan labor markets of semiskilled and skilled traditional manufacturing jobs (the effect of regional and international decentralization associated with globalization and the **new international division of labor**) combined with (2) new production technologies involving computers and robotics and (3) the growth of unskilled service-sector jobs (such as flipping hamburgers, waiting on tables, stocking shelves).

The effects of these underlying trends have been compounded by some of the strategies associated with flexible (**neo-Fordist** as opposed to **Fordist**) production systems. It has been estimated that about one in every four jobs in the United States today is "flexible," in that it involves some form of temporary, part-time, or independent subcontracting employment relations.[34] Elaborate systems of subcontracting, for example, have driven down the wage rates of workers in the many small companies that bid for contract work from corporate giants. **Flexible production systems** in general and subcontracting in particular have also encouraged the use of temporary and part-time labor which, from the employer's point of view, has twin advantages. Employers can fine-tune their workforce according to market conditions, and they can realize substantial savings on the costs of pensions, health insurance, and unemployment insurance fund contributions (which must be paid for permanent, full-time workers). In contrast, such strategies leave the workers vulnerable to unemployment and poverty.

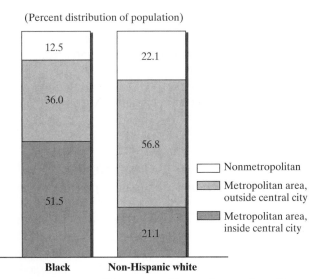

FIGURE 12.22   U.S. metropolitan and nonmetropolitan residence by race, 2002.

BOX 12.6 THE ETHNOBURB—A NEW SUBURBAN ETHNIC SETTLEMENT

Whereas the U.S. suburban population was only 18 percent minority in 1990, this figure had risen to 25 percent by 2000. Of course, these statistics mask significant variations both in the residential patterns of different minority groups and among different metropolitan areas. Variations in the suburban residential patterns of minorities reflect a variety of conditions that differ across metropolitan areas. These include city-suburban disparities in housing availability and costs, racial and ethnic discrimination, the relative mix and socioeconomic status of an area's minority groups, and the historic development of minority communities as shaped by the specific migration flows and residential patterns of the minority groups in each area.[35]

Compared to blacks or Hispanics, Asians are more likely to live in the suburbs than in central cities. Beginning in the 1960s, upwardly mobile Asians moved out of central city enclaves and ghettos in search of bigger houses, nicer neighborhoods, and better school districts. Meanwhile, the more recent Asian immigrants with higher educational attainment and professional jobs have moved directly into suburban areas. During the 1990s the number of Asians in U.S. suburbs grew by 84 percent (compared with 38 percent for blacks and 72 percent for Hispanics).

In the 102 largest metropolitan areas in 2000, nearly 55 percent of Asians were suburban residents (compared to only 35 percent of blacks and not quite 50 percent of Hispanics). At the same time, however, the share of suburban residents that is Asian is typically much smaller than the black or Hispanic shares. This is because Asians comprise a smaller proportion (5.1 percent in 2000) of the combined populations of the 102 largest metropolitan areas than either blacks (13.8 percent) or Hispanics (15.6 percent).

The majority of Asians are concentrated in a relatively small number of metropolitan areas. Those with the highest Asian suburban percentages are on the West Coast and in Hawaii. Despite their relatively small numbers, Asians comprise 17 percent of the suburban populations in the San Francisco and Oakland metropolitan areas, 23 percent in San Jose, and 48 percent in Honolulu.[36]

And although in the past, Chinatowns in major cities have symbolized the Asian presence in the United States, there are now as many as five distinct Asian national-origin groups in addition to the Chinese—Filipinos, Japanese, Indians, Koreans, and Vietnamese—along with a million more residents from other Asian countries such as Iran and Pakistan. Despite this diversity, the different groups are highly

localized in their impact. The New York metropolitan area, for example, has the largest number of Chinese and Indians and the second largest share of Koreans. Los Angeles has the most Filipinos and Koreans and the second largest number of Chinese and Japanese.[37] Meanwhile, the traditional Asian inner-city ethnic ghettos and enclaves no longer house the majority of Asian residents. Overall, more than half of Asians (58 percent) lived in the suburbs of metropolitan areas of all sizes in 2000, up from 53 percent in 1990.[38]

Geographer Wei Lei has proposed a new model of ethnic suburban settlement in large American metropolitan areas. In recent decades, under the influence of international geopolitical and global economic restructuring, shifting national immigration and trade policies, and local demographic, economic and political changes, a new type of ethnic area—the **ethnoburb**—has emerged (Figure 12.23).

Ethnoburbs are suburban ethnic clusters of residential areas and business districts. They are characterized by vibrant ethnic economies that depend on the presence of large numbers of local ethnic minority consumers; they also have strong ties to the global economy that reflects their role as outposts in the international economic system through business transactions, capital circulation, and flows of entrepreneurs and other workers. Ethnoburbs are multiethnic communities in which one ethnic minority group has a significant concentration but does not necessarily comprise a majority. Ethnoburbs function as a settlement type that replicates some features of an ethnic enclave and some features of a suburb lacking a single ethnic identity. Ethnoburbs coexist alongside the traditional inner-city ethnic ghettos and enclaves.[39]

Wei argues that the clear ethnic imprint on the suburban residential and business landscape sets the ethnoburb apart from the typical American suburb, in which members of ethnic minorities are more dispersed among the white charter group. Her case study analyses focused on the largest suburban Chinese concentration in the United States—the San Gabriel Valley in the eastern suburban part of Los Angeles County (Figure 12.24).[40] This Chinese ethnoburb evolved in the 1960s in the city of Monterey Park, a suburban bedroom community that was very accessible by the freeway system to the CBD and Chinatown. Since then, the suburbanization of Chinese people and businesses has spread not only throughout this western part of the San Gabriel Valley but also into the east, to suburban areas like Diamond Bar, Hacienda Heights, Rowland Heights, and Walnut (Figure 12.24).

(Continued)

Box 12.6   THE ETHNOBURB—A NEW SUBURBAN ETHNIC SETTLEMENT   (*Continued*)

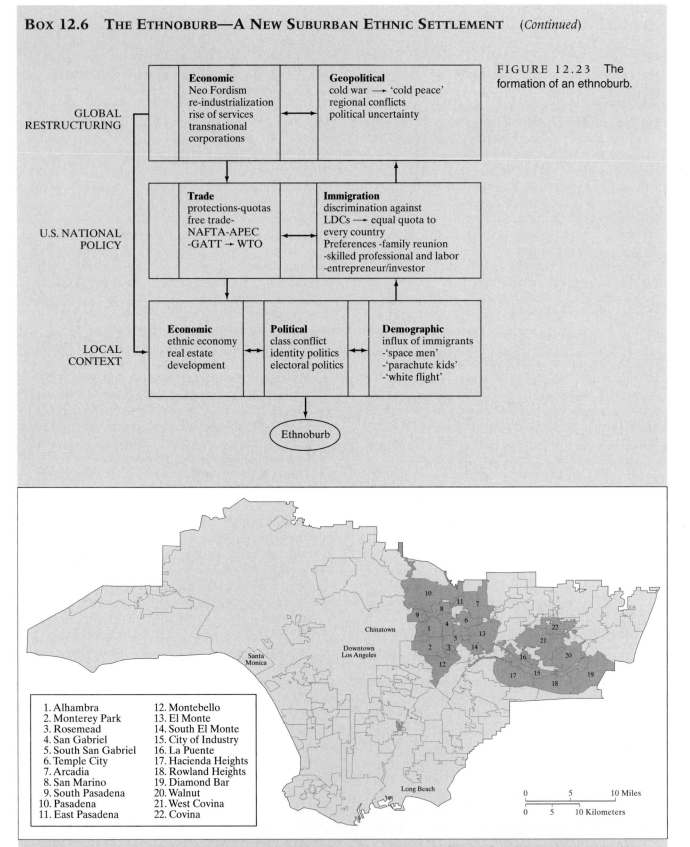

FIGURE 12.23   The formation of an ethnoburb.

GLOBAL RESTRUCTURING

Economic
Neo Fordism
re-industrialization
rise of services
transnational corporations

Geopolitical
cold war → 'cold peace'
regional conflicts
political uncertainty

U.S. NATIONAL POLICY

Trade
protections-quotas
free trade-
NAFTA-APEC
-GATT → WTO

Immigration
discrimination against
LDCs → equal quota to
every country
Preferences -family reunion
-skilled professional and labor
-entrepreneur/investor

LOCAL CONTEXT

Economic
ethnic economy
real estate
development

Political
class conflict
identity politics
electoral politics

Demographic
influx of immigrants
-'space men'
-'parachute kids'
-'white flight'

Ethnoburb

1. Alhambra
2. Monterey Park
3. Rosemead
4. San Gabriel
5. South San Gabriel
6. Temple City
7. Arcadia
8. San Marino
9. South Pasadena
10. Pasadena
11. East Pasadena
12. Montebello
13. El Monte
14. South El Monte
15. City of Industry
16. La Puente
17. Hacienda Heights
18. Rowland Heights
19. Diamond Bar
20. Walnut
21. West Covina
22. Covina

FIGURE 12.24   The San Gabriel Valley's ethnoburb in Southern Los Angeles County. Coded in the legend are cities and census-designated places that constitute the ethnoburb.

TABLE 12.2

Management, Professional, and Related Occupations—Total (in Thousands) and
Percent of Employed U.S. Metropolitan Residents, 1970–2000

	All MSAs/PMSAs		New York		Los Angeles	
	Total	*Percent*	*Total*	*Percent*	*Total*	*Percent*
1970	14,250	24.5%	854	25.4%	708	26.5%
1980	21,353	27.2%	1,027	29.5%	962	27.7%
1980	30,066	31.9%	1,355	34.9%	1,301	30.9%
2000	37,148	35.2%	1,483	38.1%	1,356	34.3%

	Chicago		Washington, D.C.		San Francisco	
	Total	*Percent*	*Total*	*Percent*	*Total*	*Percent*
1970	632	23.3%	419	34.8%	171	28.6%
1980	899	27.2%	664	38.8%	254	33.0%
1980	1,162	32.2%	1,031	44.1%	332	38.4%
2000	1,411	36.4%	1,238	48.6%	429	46.7%

Source: *U.S. Department of Housing and Urban Development,* State of the Cities Data Systems
(SOCDS) *(Available at: http://socds.huduser.org/).*

New Roles for Women

A particularly important aspect of these trends has been that a large number of part-time and temporary jobs have been filled by *women.* Whereas only about 36 percent of women of working age were actively engaged in the labor force in 1960, slightly more than 60 percent of such women were so engaged by 2000 (see Figure 14.21). This shift includes significant numbers of women who have been able to enter the workforce primarily as a result of changed social attitudes in the wake of the so-called Women's Liberation Movement of the 1960s. It is underpinned, however, by several other trends. Many companies have deliberately recruited women as a flexible source of relatively low-wage and poorly organized (in terms of union membership) labor. Many women have felt compelled to join the labor force in order to maintain household spending power in the face of price inflation. And many more households are now headed by women who must work in order to support dependent children. By

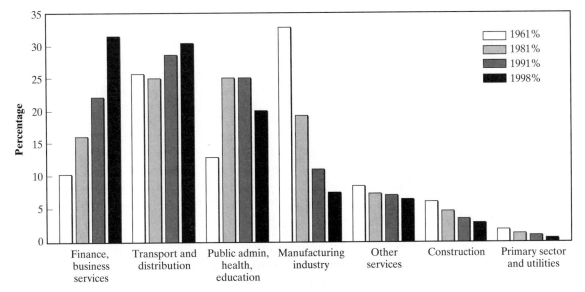

FIGURE 12.25   The changing employment structure of Greater London, 1961–1998.

2002 only 7 percent of all U.S. households consisted of a married couple with children in which only the husband worked.

New Patterns of Household Formation

Although the number of conventional two-parent families remained relatively stable, at about 26 million, married-couple families with children as a percentage of all households fell from 40 percent in 1970 to 25 percent by 2000. Meanwhile, single-mother families increased from 3 million to nearly 9 million (while the number of single-father families grew from 393,000 to 2 million). Overall in 2000, female-headed households (with and without children) comprised nearly 30 percent of all households.

For whites, the growth in divorce among couples with children was a major reason for the increase in single-parent households. The more liberal attitudes fostered by the counterculture movement of the 1960s and early 1970s and the general increase in female participation in the labor force, which has enabled more women to contemplate divorce, have meant that white single-mother families are more likely to result from marital disruption (50 percent were divorced in 2000) than an out-of-wedlock birth (30 percent were never married in 2000).[41]

The increase in single-parent households of African Americans from 1970 to 2000 was particularly striking, and the main reason was not increased divorce rates—black single mothers are least likely to be divorced (17 percent)—but an increase in the numbers of unwed teenage mothers. This increase was less a product of any sustained increased incidence of pregnancy among African-American teenage women and more a consequence of the demise of the "shotgun wedding"—black single mothers are most likely to be never married (65 percent)—in the face of changing social attitudes and, in particular, the diminishing probability that marriage would represent economic support or security. It was young African-American men whose wages, as a group, grew most slowly after 1973, and it was African-American men in general who bore the brunt of the economic and spatial restructuring of the post-1973 period. Figure 12.26 shows very clearly the somewhat erratic but nevertheless decisive decline in the ratio of African-American to white median family incomes between 1970 and 1992. And despite significant improvement during the 1990s, the ratio of 0.64 in 2000—reflecting a median family income of $33,676 for African-American and $53,029 for white families—captures the continued relatively lower levels of prosperity of many African-American households.

The pattern of household types was further diversified as counterculture sexual liberation movements

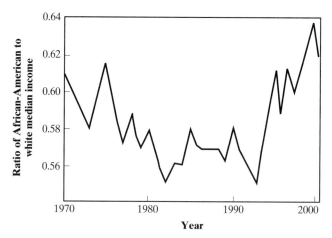

FIGURE 12.26   Ratio of African-American to white median family income in the United States, 1970–2001.

fostered the proliferation of households of unmarried heterosexual and homosexual couples and of communal households (including small groups of single people who have left home to study or find work). Meanwhile, increased longevity (among whites, at least) contributed to growing numbers of two-person elderly households; and declining fertility levels after the mid-1960s revolution in birth control technology increased the numbers of two-person young-adult households. In addition, the Baby Boomers maintained an obtrusive influence in the mix of household types over the years. Initially, they were the explanation for an increasing number of larger families with young children, then for an increase in the number of households of single young adults, then for an increase in the number of DINK households (dual-income, no kids), and, most recently, for an increase in the number of households consisting of couples with young children. *This diversification of household types has diminished the utility of the notion of the family life cycle as a basis for understanding residential segregation.* Rather, it is more appropriate to think in terms of a household life cycle consisting of a more broadly defined set of stages, each triggered by events such as entering and leaving school, entering the labor market, marriage, divorce, job change, and retirement, but with many different paths from childhood to retirement.

The diversification of household types has also resulted in large tracts of the city being dominated now by nonfamily households. In Milwaukee, Wis., for example, there is an easterly north-south axis in the heart of the metropolitan area with a concentration of households consisting of unrelated people living together (Figure 12.27). These households are concentrated in multifamily areas, mostly near the universities (to the northeast of the CBD surrounding the University of Wisconsin–Milwaukee and just west of

Increased Materialism and New Lifestyles

With the intersection of the maturing Baby Boom generation and the growth in high-paying **producer services** occupations (financial analysts, management consultants, marketing experts, and so on) in the 1970s, the cultural landscape changed significantly. A much greater emphasis on materialism and style came about, only a few years after confident predictions of a pronounced societal trend of declining materialism.[42] The Baby Boomers, whose formative experience was the postwar economic boom (see p. 100), provided the main basis for such predictions. Their rebellion against the apparent complacency of industrial progress and affluence was channeled into a countercultural movement with a collectivist approach to the exploration of freedom and self-realization through a variety of movements, including radical politics, drugs, and sexual liberation. The high-water mark of this "alienation generation"[43] was 1968, the year of sit-ins, student worker alliances, protest marches, general strikes, civil disorder, and riots. The failure of these events produced the **postmodern generation**, characterized by materialism and a pluralism of taste during the 1970s and 1980s. David Ley summarizes the transition as follows:

> The post-modern project is the project of the new cultural class, representatives of the arts and the soft professions who came to political awareness in the 1960s and were receptive to the oppositional ideas of the counter-culture. But through the 1970s and 1980s hippies have all too readily become yuppies, as the subjective philosophies of phenomenology and existentialism which opposed the impersonality of modernism in the 1960s and redeemed the individual have been directed inwards, and the celebration of meaning has often shifted subtly to the celebration of meaning of the self.[44]

These changes were part of a wider reconfiguration of sociocultural values that has proceeded in tandem with the reconfiguration of the political economy. We have already seen (Chapter 10) how postmodernism in architecture and urban design was an expression of materialistic values. Postmodernity is also manifest in many aspects of urban life, from art, literature, film, and music to commodity aesthetics. It represents a cultural "sensibility" that competes with the rationalistic, hierarchical, ascetic, and machismo themes of Modernity.

Postmodernity is, above all, pluralistic and consumption-oriented. What is particularly important in the present context is that this pluralism has provided the basis for highly differentiated patterns of consumption. This differentiation, in turn, has made for a sociocultural environment in which the emphasis is not so much on ownership and consumption per se

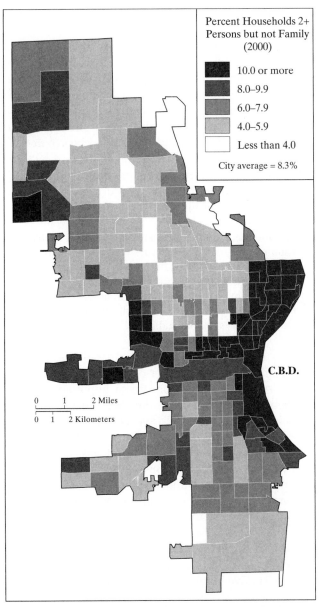

FIGURE 12.27   Non-family households with 2 or more persons as a percentage of all households in Milwaukee census tracts, 2000.

the CBD surrounding Marquette University). What is important about all these changes in the present context is that they have altered not only the composition of urban populations that represent the raw material for residential differentiation but that they have also, through their interactions with one another, modified the very foundations of social distance and neighborhood segregation. In particular, they have contributed to (1) increased residential segregation in terms of material culture and lifestyle and (2) the spatial isolation of "new" groups of vulnerable and disadvantaged households.

but on the possession of particular *combinations* of things and the *style* of consumption.

Meanwhile, the space-time compression of new communication technologies and the globalization and homogenization of consumer culture have fostered the perceived need for distinctiveness and identity: *"This society eliminates geographical distance only to reap distance internally in the form of spectacular separation."*[45] In this *society of the spectacle*, where emphasis is on appearances, the symbolic properties of urban settings and people's material possessions have assumed unprecedented importance. Social distinctions, previously marked by the ownership of particular kinds of consumer goods, now have to be established via the symbolism of "aestheticized" commodities.

In addition, the maturing Baby Boomers of the postmodern generation found themselves flooding labor and housing markets in the early 1970s, just as the economy was experiencing the worst recession since the 1930s. Wages stood still while food and house prices ballooned. Here is one personal testament to the consequent apostasy of the postmodern generation.

> In our 20s, my friends and I hardly cared. We finished college (paid for primarily by our parents), ate tofu, and hung Indian bedspreads in rented apartments. We were young: it was a lark. We scorned consumerism. But in our 30s, as we married or got sick of having apartments sold out from under us, we wanted nice things, we wanted houses.[46]

Yet economic circumstances did not permit a smooth transition to materialism during the 1980s, even for the college-educated middle-income groups:

> My friends dressed and ate well, but despite our expensive educations, most had only one or two elements of the dream we had all laughed at in our 20s and now could not attain. We had to choose between kids, houses and time. Those with new cars had no houses; those with houses, no children; those with children, no houses. A few lucky supercouples–a lawyer married to a doctor, say, with an annual combined income of more than $100,000–had everything but time.[47]

Unable to fulfill the American Dream, the postmodern generation saved less, borrowed more, deferred parenthood, comforted itself with the luxuries that were advertised in glossy magazines as symbols of style and distinctiveness, and generally surrendered to the hedonism of lives infused with extravagant details. The signs—literally—of this new materialism were everywhere in the 1980s: bumper stickers advising SHOP TIL YOU DROP or HE WHO DIES WITH THE MOST TOYS WINS; designer labels—Prada, Dolce & Gabbana, Louis Vuitton, Hugo Boss—deliberately left on clothes and accessories.

The culture of stylish materialism that developed in the 1980s also brought new patterns of social behavior. Barbara Ehrenreich argues that a cornerstone of the "yuppie strategy" was the determination of both men and women to find proven wage-earners as potential partners to marry or live with. The result, she suggests, was the consolidation of a **class fraction** characterized by high educational status and by very high household incomes. The importance of finding suitably qualified partners had, in turn, intensified the potency of material signifiers:

> . . . since bank accounts and resumés are not visible attributes, a myriad of other cues were required to sort the good prospects from the losers. Upscale spending patterns created the cultural space in which the financially well matched could find each other—far from the burger eaters and Bud drinkers and those unfortunate enough to wear unnatural fibers.[48]

We can see how such phenomena translate into the urban landscape in a number of ways. Because, for example, style and distinction are so important to consumption (and even to the very experience of shopping), the *retail* landscape became polarized. To avoid going out of business, department stores, for example, were faced with the choice of specializing at either end of the market—upscale for the wealthy or discount for the poor and thrifty. Bloomingdale's and Neiman-Marcus are at one extreme and Wal-mart and dollar stores are at the other. Undifferentiated chains, such as Korvette's and Gimbel's, with a middle-income consumer market, were forced to close. Sears and JC Penney anxiously attempted to "reposition" themselves to survive in the increasingly segmented market.[49]

The *residential* landscape began to reflect the increased importance of materialism and stylish distinction in a variety of ways, including the **gentrification** of increasing numbers of inner-city neighborhoods through the physical rehabilitation and social change associated with "invading" young professionals (see pp. 370–374) and the development of *private, master-planned communities* as suburban settings for recreation-oriented and consumption-oriented lifestyles (see p. 289). Increasingly, as lifestyles have proliferated within the different socioeconomic groups, household types, and ethnic groups, developers have provided a greater variety of residential settings (condominium apartments, townhouses, retirement communities, mater-planned communities), "packaging" them with keynote design themes and key lifestyle amenities. Thanks to the time-space compression of technological advances such as cable television and cheaper and more effective telecommunications such as the Internet, the resultant lifestyles are not localized and idiosyncratic but, rather, *replicated across the whole country*. Thanks to specialized relocation consultant services and the emergence of nationwide realtor networks such as ERA, Century 21, Coldwell Banker, and Red Carpet, people who can afford to can easily find the kind of lifestyle setting they are interested in and easily move from one to another. The result has been described as a *mosaic culture* that takes the form of "archipelagos" of similar lifestyle communities stretching from coast to coast, with "islands" in every major metropolitan area.[50]

The Spatial Isolation of the Vulnerable and Disadvantaged

At the bottom end of the socioeconomic ladder, the intersection of occupational shifts, immigration streams, and the cultural and demographic dynamics associated with the Baby Boom generation has resulted in some dramatic changes in the composition of vulnerable and disadvantaged households. During the 1970s, for example, single-parent families increased by 22.8 percent, the lone elderly increased by 11.3 percent, immigrants increased by 53.8 percent, and people on

---

BOX 12.7 "INCONSPICUOUS CONSUMPTION" DURING THE 1990S

Besides the decennial census, the Consumer Expenditure Survey (CEX) is perhaps the most important survey conducted by the U.S. Census Bureau that provides researchers and businesses with insights into consumer behavior. The survey reveals that during the 1990s Americans were not the "conspicuous consumers" that economist Thorstein Veblen described more than a century ago in *The Theory of the Leisure Class* (1899)—a label, in fact, that better fits 1980s spending and consumption patterns. As far as the 1990s are concerned, Veblen would be hard-pressed to find the abundance of material goods whose purchase he criticized as a primitive form of snobbery and self-doubt. Instead, as journalist Michael Weiss, who wrote the book *The Clustered World: How We Live, What We Buy, and What It All Means About Who We Are*,[51] puts it: "The 1990s was a decade of less flash and more cash for life's big and little necessities."[52] Compared to the 1980s, the 1990s was a more restrained decade, in which bargain-hunting consumers looked for opportunities to buy more for less. Certainly, household spending rose between 1990 and 2000—by 34 percent, outpacing a 28 percent increase in inflation—but consumers spent a larger share of their budgets on health care, housing, and transportation and less on new clothes, jewelry, eating out, and entertainment. Demographic trends played an important role in influencing consumer spending.

> In the 1990s, it was the Baby Boomers between the ages of 45 and 54—the peak earning years for Americans—who proved most influential. Over the course of the decade, these older Boomers increased their share of aggregate spending by 22 percent, to $1 trillion. Their dominant numbers helped increase total spending on health insurance and drugs (more aches

and pains), education (more kids in college) and housing (more homeownership than ever before).[53]

The category of spending that experienced the greatest decline was clothing.

> Among younger age groups, the drop in clothing expenditure reflects more casual attitudes toward fashion. In the 1990s, fashionistas deemed it acceptable to mix and match pricey fashions with discount rags . . . . It became OK to have a Gucci bag with a $20 raincoat from Target . . . .You could get your T-shirts at the Gap and cashmere sweaters from Bergdorf . . . . The bottom line for such eclecticism was lower clothing bills. That trend also reflects a values shift in which people are less brand-conscious and more time-pressed. Today, working couples who have less time to shop are looking for convenience and value when they go shopping . . . . Boomers don't care what pair of khakis they buy as long as it has the best price . . . . As for the Gen Ys, they'd rather have the coolest electronics than the coolest clothes . . . . Such changing priorities have helped discounters like Target and Wal-mart, while hurting specialty stores like Gap and Abercrombie & Fitch . . . . Apparel used to be about image. Now it's about function.[54]

During the coming decades, of course, that thinking may change as a result of the complicated interplay within cities of socioeconomic status, household types, ethnicity, and lifestyle. In less than a decade the oldest members of the Baby Boom generation will reach 65 and enjoy longer life expectancies than any previous generation. Although it is not certain how future consumption and housing choices will differ from today's, what is certain is that the sheer size of the Baby Boomers as a demographic group means that their choices will translate visibly into the urban landscape.

TABLE 12.3

Selected Socioeconomic and Demographic Indicators (Percent of Total) for U.S. Metropolitan Areas (MSAs/PMSAs), 1970–2000

	1970	1980	1990	2000
Single parent families with children	13.2	19.9	23.5	28.5
Poverty rate*	11.4	11.5	12.1	11.8
Low income families (national lowest 20% income bracket)	17.0	18.2	17.8	19.1
Persons aged 25 or more who graduated from high school	31.7	34.5	28.7	26.9
Unemployment rate	4.2	6.2	6.1	5.7
Immigrants (foreign born population)	5.7	7.4	9.5	13.0

* Poverty rate figures are for 1969, 1979, 1989, and 1999.

Source: *U.S. Department of Housing and Urban Development*, State of the Cities Data Systems (SOCDS) *(Available at: http://socds.huduser.org/)*.

work disability increased by 19.5 percent. At the same time, an increasing proportion of large families and single-parent families were *multiply disadvantaged*,[55] even though the overall incidence of multiply disadvantaged households decreased.

Although the 1990s saw some improvement, the changes in the composition of vulnerable and disadvantaged households within U.S. metropolitan areas since the 1960s have not been encouraging (Table 12.3). The trends that contributed to these changes (occupational shifts, corporate reorganization and redeployment, and so on) have also contributed to shifts within the **urban system**, so that there has been a great deal of variability *among* and *within* metropolitan areas in the changing composition of their vulnerable and disadvantaged populations. A

comparison of socioeconomic and demographic changes within and between the Detroit, Phoenix, and Los Angeles metropolitan areas, their central cities, and suburbs between 1970 and 2000 is illustrative of this variability (Table 12.4).

An important aspect of these changes is the **feminization of poverty** that has occurred, particularly among the African-American population (Figure 12.28). Female-headed households increased from just over 20 percent of all households below the poverty level in 1960 to 50 percent in 2002. As a result, one in six children in the United States are now being raised in poverty. A factor in this trend was the increase during the 1980s in teenage single mothers in communities where economic restructuring had left high levels of unemployment. Other contributory factors

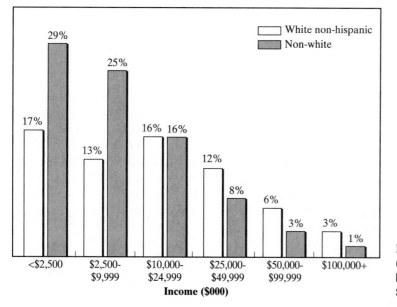

FIGURE 12.28 Female-headed families (no spouse present) as a percentage of all family households, by race and income group, United States, 2001.

TABLE 12.4

Selected Socioeconomic and Demographic Indicators for Detroit, Los Angeles, and Phoenix, 1970–2000

	Central City				Suburbs				PMSA			
	1970	1980	1990	2000	1970	1980	1990	2000	1970	1980	1990	2000
Detroit, Mich.												
Population (millions (rounded))	1.5	1.2	1.0	1.0	2.8	3.0	3.0	3.3	4.5	4.4	4.3	4.4
African-Americans (%)	43.7	62.6	75.5	81.2	3.6	3.4	4.3	6.2	18.0	20.2	22.0	22.8
Single parent families with children (% of families)	23.4	45.6	61.0	63.1	8.3	14.7	17.6	21.9	12.9	23.0	28.8	31.0
Low income families* (national lowest 20% income bracket)	19.7	30.2	40.5	36.2	8.3	10.1	10.7	10.7	12.4	15.6	17.9	16.2
Persons aged 25 or more who graduated from high school (%)	28.1	31.5	27.7	30.0	36.5	39.1	31.2	29.1	33.3	36.9	30.3	29.3
Unemployment rate	7.2	18.5	19.7	13.8	4.7	9.4	6.0	4.0	5.7	11.7	8.9	5.9
Recent immigrants (foreign born population) (%)	8.0	5.7	3.4	4.8	6.2	6.6	6.0	7.9	6.9	6.4	5.5	7.5
Los Angeles, Calif.												
Population (millions (rounded))	2.8	3.0	3.5	3.7	3.7	4.0	4.7	5.1	7.0	7.5	8.9	9.5
African Americans (%)	17.9	16.7	13.2	10.9	6.4	9.2	8.4	7.7	10.8	12.4	10.7	9.5
Single parent families with children (% of families)	21.4	29.3	29.9	32.4	14.7	22.2	23.0	26.7	17.5	25.3	26.1	29.6
Low income families* (national lowest 20% income bracket)	18.5	23.4	24.2	31.0	13.4	16.6	15.9	20.9	15.7	19.5	19.3	25.2
Persons aged 25 or more who graduated from high school (%)	31.1	28.0	19.2	17.4	34.0	31.9	21.8	19.9	32.7	30.2	20.7	18.8
Unemployment rate	6.9	6.8	8.3	9.3	5.6	5.5	6.7	7.3	6.2	6.0	7.3	8.2
Recent immigrants (foreign born population) (%)	14.5	27.1	38.4	40.9	9.2	19.6	29.9	34.2	11.2	22.3	32.7	36.2
Phoenix, Ariz.												
Population (millions (rounded))	0.6	0.8	1.0	1.3	0.3	0.5	0.7	1.2	1.0	1.6	2.2	3.3
African-Americans (%)	4.8	4.7	5.1	4.8	1.2	2.0	2.2	2.7	3.4	3.1	3.4	3.5
Single parent families with children (% of families)	14.4	20.2	26.4	32.0	11.4	14.6	21.9	24.9	12.9	17.9	23.9	28.5
Low income families* (national lowest 20% income bracket)	17.2	17.5	20.3	22.1	25.7	20.8	19.9	16.0	18.6	17.7	18.7	18.0
Persons aged 25 or more who graduated from high school (%)	32.4	35.4	25.5	22.9	28.7	34.6	28.0	25.5	31.9	34.7	25.7	23.5
Unemployment rate	3.8	5.5	6.6	5.6	4.5	5.9	6.4	4.6	3.9	5.4	6.1	4.9
Recent immigrants (foreign born population) (%)	4.0	5.7	8.6	19.5	4.4	5.6	6.4	9.9	3.9	5.4	7.2	14.1

* Figures are for 1969, 1979, 1989, and 1999.

Source: U.S. Department of Housing and Urban Development, State of the Cities Data Systems (SOCDS) (Available at: http://socds.huduser.org/); U.S. Bureau of the Census, County and City Data Book, 1972. Washington, D.C.: U.S. Government Printing Office, 1973.

include the persistence of a gender gap in wages, inequitable division of assets at the time of divorce, and the ineffective implementation of child support agreements.

It is important to bear in mind that, just as these changes were beginning to be felt, the **new conservatism** of the post-1973 period resulted in policy shifts that raised taxes and dismantled much of the welfare safety net. In 1988 the poorest tenth of American households were paying 20 percent more of their incomes in federal taxes than they had been a decade earlier, whereas the richest tenth were paying almost 20 percent less. Between 1979 and 2000 the average after-tax income of the bottom fifth of households increased by 8.7 percent (compared to 68.3 percent for the top fifth) (Table 12.5). During that time the share of after-tax income of the bottom fifth of households fell from 6.8 percent to 4.9 percent (while that of the top fifth rose from 42.4 percent to 51.3 percent) (Figure 12.29). Unemployment benefits have been reduced, employment and training programs cut, AFDC (Aid to Families with Dependent Children) benefits held down, eligibility requirements for work disability benefits tightened,

TABLE 12.5

Average After-Tax Income of U.S. Households, by Income Group (in 2000 Dollars)

Income category	1979	1989	2000	Percent change 1979–2000	Dollar change 1979–2000
Lowest fifth	12,600	12,100	13,700	8.7	1,100
Second fifth	25,600	25,100	29,000	13.3	3,400
Middle fifth	36,400	37,500	41,900	15.1	5,500
Fourth fifth	47,700	51,800	59,200	24.1	11,500
Highest fifth	84,000	108,000	141,400	68.3	57,400
81st–95th percentile	65,300	75,800	88,700	35.9	23,400
96th–99th percentile	103,600	129,800	158,600	53.1	55,000
Top 1%	286,300	506,500	862,700	201.3	576,400

Source: *R. Greenstein and I. Shapiro,* The New, Definitive CBO Data on Income and Tax Trends. *Washington, D.C.: Center on Budget and Policy Priorities, 2003, Table 2 (Available at: http://www.centeronbudget.org/9-23-03tax.htm).*

and federal aid to economically distressed cities and neighborhoods virtually abolished (see Chapter 17 for a full review).

Since 1974, Congress has amended the Fair Labor Standards Act several times to increase the federal hourly minimum wage—from $2.00 to $5.15. But because the minimum wage is not indexed for inflation—and during that time, the consumer price index rose by over 270 percent—the value of the increase in the minimum wage has been lost. In fact, the real (inflation-adjusted) value of the minimum wage is now far less than it was in 1974 (Table 12.6).

Sociologist-planner Manuel Castells argues that the net result has been the increasing exclusion of the most marginal and vulnerable groups in society,[56] while sociologist William J. Wilson has advanced the concept of an *underclass* of people who have become socially isolated and economically detached from the

TABLE 12.6

Increases in Federal Hourly Minimum Wage and Consumer Price Index, 1974–2003

Effective date	Minimum wage (actual dollars)	Minimum wage (2003 dollars)	Consumer price index (annual average)
May 1, 1974	2.00	7.46	49.3
January 1, 1975	2.10	7.18	53.8
January 1, 1976	2.30	7.44	56.9
January 1, 1978	2.65	7.48	65.2
January 1, 1979	2.90	7.35	72.6
January 1, 1980	3.10	6.92	82.4
January 1, 1981	3.35	6.78	90.9
April 1, 1990	3.50	4.93	130.7
April 1, 1991	4.25	5.74	136.2
October 1, 1996	4.75	5.57	156.9
September 1, 1997	5.15	5.90	160.5
2003	5.15	5.15	183.8

Source: *U.S. Department of Labor,* History of Federal Minimum Wage Rates Under the Fair Labor Standards Act, 1938–1996 *(Available at: http://www.dol.gov/esa/minwage/chart.htm); Federal Reserve Bank of Minneapolis,* Consumer Price Index, 1913- *(Available at: http://minneapolisfed.org/Research/data/us/calc/hist1913.cfm).*

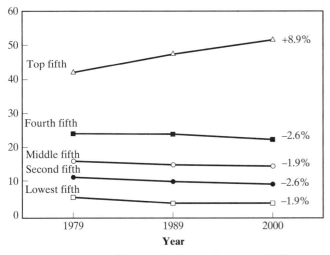

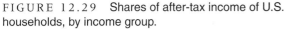

FIGURE 12.29   Shares of after-tax income of U.S. households, by income group.

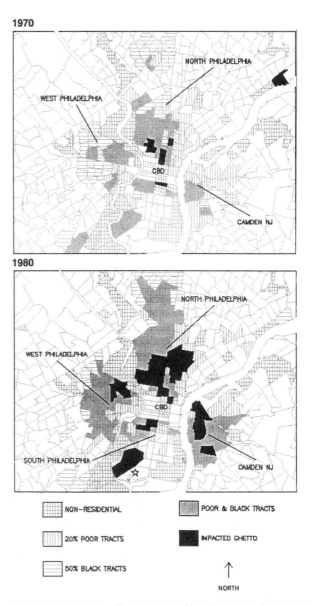

FIGURE 12.30   The growth of the "impacted" ghetto in Philadelphia.

labor force: mostly African Americans, with a disproportionate share of female-headed households that have become dependent on welfare benefits (see also p. 424).[57]

Most striking among the landscapes of the excluded are *impacted ghettos*—spatially isolated concentrations of the very poor (Figure 12.30), usually (though not always) dominated by African Americans and often denuded of community leaders and containing very high concentrations of female-headed households struggling to survive in downgraded environments that also serve as refuges for criminals. Less visible, but more decisively excluded, are the "landscapes of despair" inhabited by the homeless:[58] microspaces that range from the vest-pocket parks of downtown Los Angeles, the city square in San Francisco, and the federal area of the District of Columbia to the anonymous alleyways and park benches of every big city (see Chapter 15).

THE NEW MOSAIC: ATTEMPTING TO IDENTIFY URBAN "LIFESTYLE" COMMUNITIES

The reorganization of world markets has expanded the consumption functions of mature urban economies, creating new jobs and new spaces of consumption. Many of these jobs are low-paying jobs in stores, restaurants, hotels and domestic and personal services. While many of the new consumption spaces rely on a high level of skill and knowledge, and provide cultural products of beauty, originality and complexity, others are standardized, trivial and oriented toward predictability and profit. At the same time, individual men and women express their complex social identities by combining markers of gender, ethnicity, social class and—for want of a better word—cultural style. Many of these markers are created in, and diffused from, cities: on the streets, in advertising offices and photography studios, on MTV. Many of the people who create these markers live in cities, too. They are artists, new media designers, feminists, gays, single parents and immigrants ... urban cultural diversity holds a curious and yet wondrously creative mirror to the paradox of polarization: while cities become more like other places, they continue to attract the extremes of poor, migrant and footloose urban populations and the very rich. Their ability to forge 'urban' lifestyles continues to be the city's most important product.[59]

Given that these changes are still very much under way, it is premature to attempt to generalize about their imprint on an urban scale. Efforts have been made, however, to generate lifestyle profiles that describe different neighborhoods and consumer groups. Often through the use of Geographic Information Systems (GIS) marketing applications, geodemographics—cluster analyses that combine census data with market research information—have identified neighborhood types or "clusters" for dozens of countries, from the United States and Canada to Europe, South Africa, Australia, and Japan. In North America alone there are an estimated 15,000 companies, nonprofit groups, and politicians that use clusters as part of their targeted marketing strategies.

Table 12.7 lists 62 U.S. *lifestyle clusters*, categorized into 15 social groupings, that Claritas Inc. has identified for use in targeted marketing and that are included in the book *The Clustered World: How We Live, What We Buy, and What It All Means About Who We Are*, by journalist Michael Weiss.[60] Claritas Inc. generated these "clusters" using its PRIZM "segmentation" system and available socioeconomic statistics (such as census data, state income tax records and driver's license files, local property files, business phone listings, motor vehicle records, and federal and state mortgage lending data).[61]

The use of socioeconomic data to identify these lifestyle clusters, of course, captures the continuing importance of the *socioeconomic status* dimension in residential differentiation. An examination of the descriptive labels given to each "cluster" shows that the household life cycle also continues to be reflected in residential differentiation. The "apprentice years" of the life cycle are represented by several different lifestyle clusters, including "Towns and Gowns" (college-town singles in 10+ unit rental housing) and "Young Influentials" (upwardly mobile singles and couples in single-family housing). The "achievement years" are represented by "Blue Blood Estates" ("super-rich" suburban households in single-family housing) and "Kids and Cul-de-Sacs" ("upscale" suburban households in single-family housing). The "harvest years" are represented by "Money and Brains" ("sophisticated" urban-fringe couples in single-family housing) and "Pools and Patios" (established empty-nesters in single-family housing). Finally,

the "retirement years" are represented by "Gray Power" (affluent married and single retirees in Sunbelt cities in owner-occupied single-unit and multiunit housing) and "Sunset City Blues" (empty-nesters in aging industrial cities in single-family housing). *Ethnicity* also continues to be reflected in residential differentiation, as indicated by the lifestyle communities labeled "American Dreams" or "Hispanic Mix." Cutting across all of these, however, is the differentiation in terms of *lifestyles* that is brought out by the marketing data.

In the United States marketers spend an estimated $300 million each year on cluster analyses as a tool in targeting direct-mail campaigns, selecting sites for new retail facilities, and profiling the behavior of the country's more than 100 million households.[62] The marketing strategies are based on the assumption, as declared on the Claritas Inc. website, that:

> People living in the same neighborhoods tend to have similar lifestyles, proving the old adage that "birds of a feather flock together" still holds true. To a large extent, you are where you live![63]

According to Claritas Inc., the lifestyle traits of the residents in its "Blue Blood Estates" category, for example, include taking a golf vacation, reading *Fortune* magazine, and driving an Acura SUV. As Weiss describes it, "It's a world of million-dollar homes and manicured lawns, high-end cars and exclusive private schools . . . a cluster of such desired addresses as Scarsdale, New York; Winnetka, Illinois; and Potomac, Maryland . . . one in ten residents is a multimillionaire . . . . The residents of Blue Blood Estates are the well-heeled Americans who enjoy the embellishments of money—lavish homes, expensive clothes, luxury cars, and private club memberships—in a gilded suburban setting with others who share the same values."[64]

In contrast, and although Weiss acknowledges that the "Inner Cities" cluster currently is undergoing somewhat of a renaissance, he begins his description of this category as the "poorest of urban lifestyles, Inner Cities suffers its share of bad karma. High unemployment, low educational levels, and a high percentage of residents on public assistance combine to create a desolate economic landscape . . . . The reputation of Inner Cities has been traditionally the worst of any cluster: a desolate place with high levels of crime, drugs, prostitution, and homeless people. In the South Bronx, New York, home of the largest public housing project in the nation, residents in the late '70s and early '80s figured the best solution was simply to burn it down." Similarly, Claritas Inc. describes this "most economically challenged urban segment . . . as a transient world of young, ethnically diverse singles and

TABLE 12.7

Lifestyle "Clusters" as the Basis for Target Marketing in the United States

Clusters	Demographic snapshots	Social groupings
Blue Blood Estates	Elite super-rich families	Elite Suburbs
Winner's Circle	Executive suburban families	
Executive Suites	Upscale white-collar couples	
Pools & Patios	Established empty-nesters	
Kids & Cul-de-Sacs	Upscale suburban families	
Urban Gold Coast	Elite urban singles and couples	Urban Uptown
Money & Brains	Sophisticated town-house couples	
Young Literati	Upscale urban singles and couples	
American Dreams	Established urban immigrant families	
Bohemian Mix	Bohemian singles and couples	
Second City Elite	Upscale executive families	Second City Society
Upward Bound	Young upscale white-collar families	
Gray Power	Affluent retirees in Sunbelt cities	
Country Squires	Elite exurban families	Landed Gentry
God's Country	Executive exurban families	
Big Fish, Small Pond	Small-town executive families	
Greenbelt Families	Young middle-class town families	
Young Influentials	Upwardly mobile singles and couples	The Affluentials
New Empty Nests	Upscale suburban fringe couples	
Boomers and Babies	Young white-collar suburban families	
Suburban Sprawl	Young suburban town-house couples	
Blue-Chip Blues	Upscale blue-collar families	
Upstarts & Seniors	Middle-income empty-nesters	Inner Suburbs
New Beginnings	Young mobile city singles	
Mobility Blues	Young blue-collar/service families	
Gray Collars	Aging couples in inner suburbia	
Urban Achievers	Midlevel white-collar urban couples	Urban Midscale
Big City Blend	Middle-income immigrant families	
Old Yankee Rows	Empty-nest middle-class families	
Mid-City Mix	African-American singles and families	
Latino America	Hispanic middle-class families	
Middleburg Managers	Midlevel white-collar couples	Second City Centers
Boomtown Singles	Middle-income young singles	
Starter Families	Young middle-class families	
Sunset City Blues	Empty-nesters in aging industrial cities	
Towns & Gowns	College town singles	
New Homesteaders	Young middle-class families	Exurban Blues
Middle America	Midscale families in midsize towns	
Red, White, & Blues	Small-town blue-collar families	
Military Quarters	GIs & surrounding off-base families	
Big Sky Families	Midscale couples, kids, and farmland	Country Families
New Eco-Topia	Rural white- or blue-collar and farm families	
River City, USA	Middle-class rural families	
Shotguns & Pickups	Rural blue-collar workers and families	
Single City Blues	Ethnically mixed urban singles	Urban Cores
Hispanic Mix	Urban Hispanic singles and families	
Inner Cities	Inner-city solo-parent families	

(Continued)

TABLE 12.7   *Continued*

Clusters	Demographic snapshots	Social groupings
Smalltown Downtown	Older renters and young families	Second City Blues
Hometown Retired	Low-income, older singles & couples	
Family Scramble	Low-income Hispanic families	
Southside City	African-American service workers	
Golden Ponds	Retirement town seniors	Working Towns
Rural Industria	Low-income blue-collar families	
Norma Rae-Ville	Young families, biracial mill towns	
Mines & Mills	Older families, mines and mill towns	
Agri-Business	Rural farm town and ranch families	The Heartlanders
Grain Belt	Farm owners and tenants	
Blue Highways	Moderate blue-collar/farm families	Rustic Living
Rustic Elders	Low-income, older, rural couples	
Back Country Folks	Remote rural/town families	
Scrub Pine Flats	Older African-American farm families	
Hard Scrabble	Older families in poor, isolated areas	

Source: *M. J. Weiss,* The Clustered World: How We Live, What We Buy, and What It All Means About Who We Are. *Boston: Little, Brown and Company, 2000, pp. 12–13.*

single parents. Home values are low—about half the national average—and even then less than a quarter of residents can afford to own real estate. Typically, the commercial base of Mom-and-Pop stores is struggling and in need of a renaissance." The lifestyle traits of the residents of this cluster category are listed by Claritas Inc. as buying gospel music, shopping at Foot Locker, reading *Ebony* magazine, and watching the *Steve Harvey Show* on TV (Figure 12.31).

One potential drawback of this kind of cluster analysis is that it can result in stereotypical and polarized descriptions of different groups within metropolitan areas. Compare the labels for some of the different clusters in Table 12.7—notice the contrast, for example, between labels like "Winner's Circle" and "Money and Brains" versus "Shotguns & Pickups" and "Back Country Folks." This kind of cluster analysis is particularly problematical if it promotes stereotyping that causes some neighborhoods to be **labeled** as poor prospects for certain kinds of economic or other activities (as in **redlining**, where mortgage-lending agencies refused to lend in certain areas—see Chapter 13, p. 367). The concern is that the "You Are Where You Live" assumption of Claritas Inc. may become a self-fulfilling prophecy if companies target customers based on the cluster descriptions.[65]

The gap between "urban legends"—such as, "no one works in inner-city neighborhoods"—and urban realities is wide, and often the private marketing data that people rely on to provide accurate neighborhood information actually exacerbate these myths. These marketing data have several problems. They are often based on infrequently updated Census information that undercounts central city residents; they make misleading generalizations (e.g., crime statistics are not based on the number of crimes committed in the neighborhood but estimated from the "type" of people living there); and they fail to review local data for trend analysis, thereby missing many positive developments in cities. The data have serious repercussions for cities, influencing the location and product decisions of businesses; government policies on welfare, housing, and bus routes; and even prospective students' choice of university.[66]

The main point to repeat here is that lifestyle communities are complex and multidimensional—and not amenable to all-encompassing descriptions. The actual diversity within particular city blocks and even within individual households, of course, mark an important break from the idea of a generalized pattern of residential differentiation as depicted in Figure 12.18; but they follow from the regional and functional differentiation that is inherent to the economic and demographic trends that we have seen are influencing the cities and metropolitan areas in the United States.

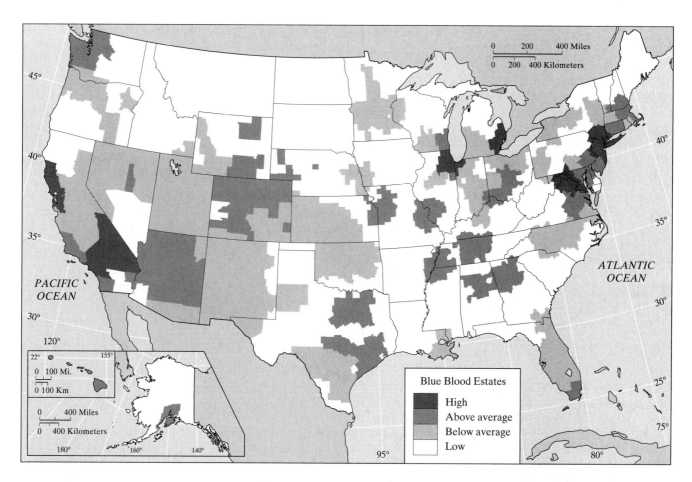

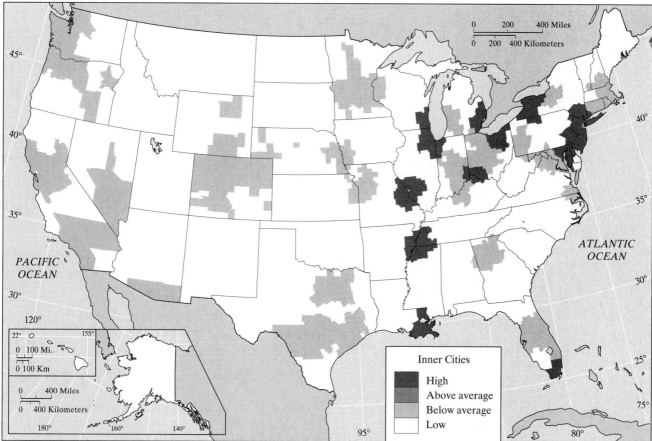

FIGURE 12.31   "Blue Blood Estates" and "Inner Cities" targeted marketing lifestyle clusters.

Box 12.8   GIS Marketing Applications Help Starbucks to Brew Up Better Locational Analyses[67]

Like other companies in the retail and food business, Starbucks Coffee Co. uses GIS—Geographic Information Systems—in its locational analysis when determining where to locate a new store (Figure 12.32). In CBDs, Starbucks aims to take advantage of high volumes of pedestrian traffic. Figure 12.33 shows a GIS map of walk times in the vicinity of one coffee shop. This kind of analysis—based on sales forecasting data—helps retail outlets like Starbucks to explore the potential of specific locations so that they can make sound site-selection decisions. Maps like this one can help site-selection teams reduce the risks inherent in attempting to identify a good location for a new store by improving the chances that the new location will cater to a high-volume and responsive customer base.

In addition to incorporating existing government-generated socioeconomic statistics, such as census data, ZIP codes, highway data, and city street maps, maps such as this can also contain data on sales and customer characteristics that the company generates from so-called store intercept studies. Through the application of GIS technologies, companies like Starbucks are also able to uncover spatial patterns, such as residence and consumer spending, that they can then use to help implement more effective marketing campaigns.

But the potential for invasion of privacy through the coupling of different sorts of data sets is ever present. For example, when a customer pays by credit card or check in a U.S. grocery store that uses an electronic inventory system, a record of the purchase can be connected to the customer's name. This could allow marketers to target that customer based on information on sales and manufacturers' coupons. With many Americans shopping at large grocery stores that offer more than just food items, marketers can potentially access information about the customer's use of prescription drugs, alcohol consumption, magazine preferences, and DVD rentals. So although GIS and the enormous data sets that support it open up new opportunities for marketing, the possibility also exists for the erosion of people's privacy.

FIGURE 12.32   High volumes of pedestrian traffic are important to the success of coffee shops like Starbucks—because coffee drinkers tend to stop in on an impulse.

(Continued)

BOX 12.8   GIS MARKETING APPLICATIONS HELP STARBUCKS TO BREW UP BETTER
LOCATIONAL ANALYSES   (*Continued*)

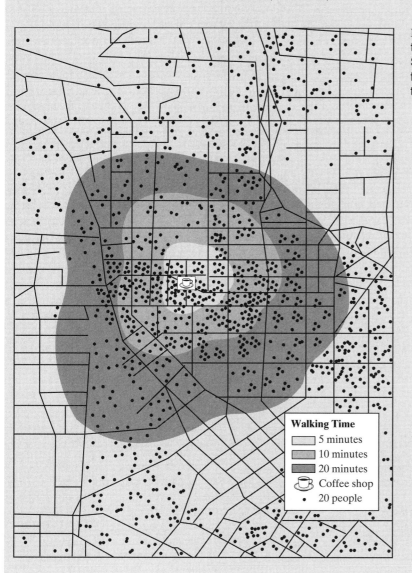

FIGURE 12.33   Walk times to different trade areas in a CBD—maps like this help Starbucks to understand the hinterland area from which an individual coffee shop is likely to draw its customers.

Walking Time
- 5 minutes
- 10 minutes
- 20 minutes
- ☕ Coffee shop
- • 20 people

FOLLOW UP

1. Be sure that you understand the following key terms:

acculturation

behavioral assimilation

charter group

class formation

class fractions

class structuration

class structure

congregation

distance decay effect

ethnoburbs

factorial ecology

family life cycle (or life course)

feminization of poverty

gentrification

Human Ecology

impacted ghetto

index of dissimilarity

lifestyle communities

neighborhood effects

postmodern generation

primary relationships

secondary relationships

social distance

socialization

social reproduction

society of the spectacle

stereotyping

structural assimilation

territoriality

underclass

2. Compare and contrast the socioeconomic and demographic indicators for the Detroit, Los Angeles, and Phoenix metropolitan areas and their central cities and suburbs in Table 12.4. Use your library or surf the Web to try to identify particular trends and factors that can help explain some of the differences and similarities that you identified.

3. Consider the lifestyle clusters described on pages 331–335. Can you recognize any of these categories in a city that you know? What are some of the problems associated with urban analyses based on

geodemographics that may result in stereotypical characterizations of residents and neighborhoods? What may be missed by concentrating on people's income, consumption patterns, and material possessions? What kinds of attributes do *you* think are most important in describing people and their neighborhoods?

4. Consider the observation that "The city became a stage; people could pose and masquerade, their background and social credentials unchallenged in the turmoil" (p. 299). Is this aspect of cities exciting or threatening to you (or perhaps both)? Why?

5. Work on your *portfolio*. The topics covered in this chapter provide a good opportunity for you to personalize your work by writing about your own experience in relation to social networks, territoriality, discrimination, lifestyles, and neighborhoods, and relating it to the concepts and generalizations that you have read about. You will also be able to use features and news stories from local newspapers to illustrate different aspects of sociospatial segregation. Remember too that you can summarize, quote, reproduce, or analyze material from the suggested reading and the World Wide Web. Look for maps and charts that illustrate key points.

KEY SOURCES AND SUGGESTED READING

Bullard, R. D., J. E. Grigsby, and C. Lee, eds. 1994. *Residential Apartheid: The American Legacy*. Los Angeles, Calif.: CAAS Publications.

Davies, W. K. D. 1984. *Factorial Ecology*. Aldershot, UK: Gower Press.

Emerson, M. O., G. Yancey, and K. J. Chai. 2001. Does Race Matter in Residential Segregation? Exploring the Preferences of White Americans. *American Sociological Review* 66:922–35.

Frey, W. H. 2002. *Metro Magnets for Minorities and Whites: Melting Pots, the New Sunbelt, and the Heartland*. Ann Arbor, Mich.: Population Studies Center at the Institute for Social Research, University of Michigan.

Glaeser, E. L. and J. L. Vigdor. 2001. *Racial Segregation in the 2000 Census: Promising News*. Washington, D.C.: The Brookings Institution.

Iceland, J. and D. H. Weinberg. 2002. *Racial and Ethnic Residential Segregation in the United States: 1980–2000*. Washington, D.C.: U.S. Department of Commerce.

Knox, P. and S. Pinch. 2000. *Urban Social Geography*. Harlow, UK: Pearson Education, 4th ed.

Logan, J. R. 2002. *Separate and Unequal: The Neighborhood Gap for Blacks and Hispanics in Metropolitan America*. Albany, N.Y.: Lewis Mumford Center for Comparative Urban and Regional Research (Available at: *http://mumford1.dyndns.org/cen2000/SepUneq/SUReport/Separate_and_Unequal.pdf*).

Madanipour, A., G. Cars, and J. Allen, eds. 1998. *Social Exclusion in European Cities: Processes, Experiences and Responses*. London: Jessica Kingsley Publishers.

Massey, D. S. and N. A. Denton. 1993. *American Apartheid: Segregation and the Making of the Underclass*. Cambridge, Mass.: Harvard University Press.

Sandercock, L. 2003. *Cosmopolis II: Mongrel Cities in the 21st Century*. London: Continuum.

Smelser, N. J., W. J. Wilson, and F. Mitchell, eds. 2001. *America Becoming: Racial Trends and their Consequences*. Washington, D.C.: National Academy Press.

Ward, D. 1989. *Poverty, Ethnicity and the American City, 1840–1925*. New York: Cambridge University Press.

Zukin, S. 1998. Urban lifestyles: Diversity and standardization in spaces of consumption. *Urban Studies* 35:825–39.

RELATED WEBSITES

The Virtual Geography Department Project, The University of Texas at Austin: *http://www.uncc.edu/hscampbe/landuse/a-intro/intro.html* The Virtual Geography Department Project site offers curriculum materials and classroom exercises from geography departments in a number of universities in the United States and elsewhere. See, for example, the "Urban Land Use: Residential Patterns and Change" module that introduces theories and models of urban land use and offers an exercise involving applying geographic concepts relevant to urban land use.

Lewis Mumford Center for Comparative Urban and Regional Research: *http://www.albany.edu/mumford/* The Lewis Mumford Center at the University at Albany undertakes comparative and historical urban research. Its website offers publications and reports on a variety of topics, including metropolitan racial and ethnic change, residential and school segregation, and inner-city poverty.

International Labour Organization: *http://www.ilo.org/* The International Labour Organization (ILO) is a United Nations agency that seeks to promote social justice and internationally recognized human and labor rights. The ILO formulates international labor standards that set minimum standards of basic labor rights. Its website offers information and publications on issues related to migrant workers in cities in Europe and other parts of the world.

Claritas Inc.: *http://www.claritas.com/* Claritas is a marketing information resources company that offers cluster analysis to help marketers and others target consumers and other businesses. Visitors to the website can enter a 5-digit U.S. ZIP code and see the *lifestyle clusters* that the Claritas PRIZM "segmentation" system has identified for that ZIP code area. You can enter your own ZIP code (if you live in the United States) or a ZIP code for a U.S. city that you know and assess the benefits and drawbacks of a cluster analysis system like PRIZM for yourself. How accurately do the results portray your own ZIP code area?

13

HOW NEIGHBORHOODS CHANGE

We have seen in previous chapters that in many parts of the world the overall form of larger metropolitan areas has evolved from a relatively straightforward, monocentric structure to a "galactic" sprawl with a polycentric structure—fragmented and multinodal—characterized by surrounding urban realms containing new urban spaces and settings that include edge cities and boomburbs. Embedded in this framework of splintering urbanism is the kaleidoscope of residential neighborhoods, arranged in terms of intersecting cleavages of socioeconomic background, household type, ethnicity, and lifestyle. If we look at these neighborhoods closely, we can see that each one is the product of a steady flux of change: additions and replacements, abandonments and demolitions, births and deaths, arrivals and departures that collectively carry each neighborhood in a certain direction at a particular speed. In this chapter we examine this dynamism in detail, emphasizing the operation of housing markets and the role of residential mobility in shaping and reshaping neighborhoods.

CHAPTER PREVIEW

This chapter examines how and why neighborhoods change. The first task is to clarify the different components of neighborhood change: the aging of the physical environment, the aging of residents, and the movement of households into and out of the neighborhood. We will see that each of these components exhibits a different periodicity and that the overall effect can be conceptualized in terms of neighborhood life cycles.

Having established these fundamental dynamics, the next task is to review another important dynamic: the changing pattern of housing tenure that links broad shifts in the political economy to the context of local housing markets. At this point an important question is raised: Why has public housing been so limited in U.S. cities in comparison with urban housing markets in other developed countries such as the United Kingdom? Another important issue raised here concerns the existence of spatial submarkets in which the dynamics of neighborhood change are played out.

The chief actors in these submarkets are of course individual households, whose decisions about where to live and whether or not to move are the subject of the middle part of the chapter. In this section we will see how similarities in the behavior of particular types of households (in terms of socioeconomic background, household type, ethnicity, and so on) can be linked in causal terms to processes of neighborhood change.

Yet, although we can consider households' patterns of demand within spatial submarkets to be the fundamental motor of neighborhood change, the actions of key "gatekeepers" such as real estate agents and mortgage financiers also influence the outcomes. As we will see, they are sometimes the agents of social and racial discrimination and bias.

The final section of the chapter draws on all these ideas—neighborhood life cycles, housing submarkets, household behavior, and social gatekeepers—in illustrating the nature of neighborhood transformation in relation to gentrification, involving social changes and physical upgrading in older, inner-city residential neighborhoods.

NEIGHBORHOOD CHANGE

Several components of neighborhood dynamism can be isolated, although, as Figure 13.1 suggests, they are all highly interdependent, each influencing—and being influenced by—the others. The most obvious and straightforward aspect of neighborhood change is that of the *physical deterioration* of the housing stock. The rate of physical deterioration of any neighborhood is chiefly a function of two factors: the quality of initial construction and the level of subsequent maintenance.

Both of these are in turn related to the socioeconomic backgrounds and lifestyles of the occupants. Although it is not uncommon for fragments of the urban fabric to last for 100 years or more, 50 to 60 years can be considered to be a reasonable life expectancy in most circumstances in the United States. In general terms, each subdivision or tract of new housing can be thought of as describing a **depreciation curve** over time as it ages (Figure 13.2). It should be acknowledged,

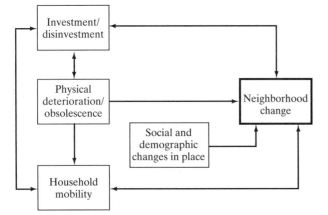

FIGURE 13.1   The principal determinants of neighborhood change.

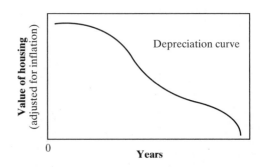

FIGURE 13.2   A hypothetical depreciation curve for a tract of suburban housing.

however, that such curves are averages that mask local unevenness in physical deterioration. Housing of the same age and initial quality wears unevenly because of variations in maintenance and improvements and the localized effects of road works, redevelopment, conversion, abandonment, fire damage, and so on. This unevenness in the rate of depreciation sets up an important precondition for neighborhood social change.

Closely related to physical deterioration is the structural and technological *obsolescence* of the housing stock. Both, in turn, are clearly a function of the needs and expectations of the occupants (and of the potential occupants) of the housing. Structural obsolescence occurs when the nature of the housing becomes unsuited to contemporary needs. Obsolescence does not necessarily bring abandonment or demolition, but it often brings a change of occupants and can lead to a shortened physical life span. A good example is provided by the first-tier and second-tier suburbs of many cities, where housing built before 1910 suddenly became obsolescent for many potential occupants because of the lack of off-street parking and garage space for automobiles. Meanwhile, the trend away from large families, combined with a sharp rise in the cost of domestic servants, made large, free-standing town mansions something of an anachronism by the 1920s, with the result that many mansions were converted into nonresidential uses. Technological obsolescence occurs when the functional design and equipment of housing and the neighborhood infrastructure become superseded. Advances in kitchen technology and heating and cooling systems and the provision of swimming pools, bike trails, and community centers are among the more important triggers of such change.

Given that most tracts of housing are developed for a specific and relatively homogeneous group, the principal point about social demographic *changes in place* (i.e., changes experienced by a given population in a particular locality) concerns the gradual aging of the original "colonizing" cohort (i.e., the initial group of residents). As we have seen, however (p. 302), certain stages of the household life cycle tend to trigger moves to different kinds of accommodation and different parts of the city. Although those households who are unwilling or unable to move may have a life span that approximates that of the housing itself (say 60 years: formed around the age of 25 and dissolved by death at about 85), the tendency for family life cycle changes to trigger a change of residence means that the periodicity of demographic change tends to be much shorter than that of physical deterioration and obsolescence.

Consequently, successive rounds of **filtering** will bring about a *changing composition* of inhabitants, generally resulting in an influx of younger and slightly less affluent households to the neighborhood. In addition, of course, physical deterioration, structural and technological obsolescence, and the aging of the original "colonizing" cohort will collectively have changed the neighborhood to the point where further changes are induced, eventually to the point where the neighborhood is "invaded" by representatives of a significantly different socioeconomic, demographic, ethnic, or lifestyle group.

REDEVELOPMENT AND REINVESTMENT

All these changes, in turn, translate into changing opportunities for *investment*. Other things being equal, most neighborhoods attract a certain amount of investment over the course of their lifetimes in terms of renovations and improvements made by homeowners and landlords. Where physical deterioration, obsolescence, and social change conspire to discourage such activity, *disinvestment* may take place—deliberately neglecting routine maintenance, putting homes, apartment buildings, and vacant land on the market, or abandoning them altogether. Other things are not always equal, however.

First, differences usually exist between neighborhoods in the rate and nature of change, affecting the landscape of investment opportunities. For example, a neighborhood may be physically quite sound and socially and demographically very stable—characteristics that ordinarily mean stability without major reinvestment. Such a neighborhood may nevertheless be considered ripe for *redevelopment* or *reinvestment* because of the differences between current rates of return on property in the area and the rates of return anticipated from investment in a change in neighborhood character or land use. Put in crude terms, the anticipated profits from established low-income housing may be significantly lower than the profits anticipated from a new shopping complex in the same location.

Second, neighborhood change also has to be seen in relation to changes in market demand. For example, a neighborhood may have changed to the point where modest investments are able to capture new or expanding markets through conversions (from rental apartments to upscale condominiums, from industrial loft space to residential lofts, from town mansions to funeral homes, and so on).

Finally, differentials exist between investment opportunities and returns in real estate and those in other sectors of the economy. These differentials are tied into the family of economic rhythms described in Chapter 1: **Kuznets cycles**, in particular, which are echoed in building cycles of the sort depicted in Figure 6.9.

NEIGHBORHOOD LIFE CYCLES

Drawing on these observations about neighborhood change, we can make some tentative generalizations about the typical sequence and interdependent relationships that constitute the neighborhood life cycle. Table 13.1 illustrates these relationships for a homogeneous suburban neighborhood, showing how the interactions between physical deterioration, obsolescence, sociodemographic change in place, and sociodemographic change induced by filtering result in a five-stage life cycle.

1. *Suburbanization.* The beginning of the life cycle, characterized by low-density, single-family housing occupied by young families of relatively high socioeconomic background.

2. *In-filling.* Multifamily and rental dwellings are added on vacant lots, increasing the density and decreasing the social and demographic homogeneity of the neighborhood.

3. *Downgrading.* The longest phase of the life cycle. A period of slow but steady deterioration and depreciation in the housing stock, of aging in place, and of increasing population turnover.

4. *Thinning Out.* The beginning of the end: high population turnover bringing social and demographic change; conversion and demolition of some residential units.

TABLE 13.1

Summary of Neighborhood and Housing Life Cycles

	Investment and construction		Social attributes				
Sequence	Dwelling type (predominant additions) and tenure	Levels of investment and	Population density	Household and family structure	Socioeconomic background, income	Turnover, migration mobility	Other characteristics
Suburbanization (new growth) "homogeneity"	Single-family (low density; multiple); owner-occupied	High	Low (but increasing)	Young families; small, large households	High (increasing)	High net immigration; high mobility; turnover	Initial development stage; cluster development; large-scale projects
In-filling (on vacant land)	Multifamily; rental	Low; decreasing	Medium (increasing slowly or stable)	Aging families; older children; more mixing	High (stable)	Low net immigration; low mobility; turnover	First transition stage—less homogeneity in age, class, housing; first apts.
Downgrading (stability and decline)	Conversions of existing dwellings to multifamily; rental	Very low	Medium (increasing slowly); population total down	Older families; fewer children	Medium (declining)	Low net out-migration; high turnover	Long period of depreciation and stagnation; some nonresidential "succession"
Thinning out	Nonresidential; construction–demolition of existing units	Low; increasing	Declining (net densities may be increasing)	Older families; few children; nonfamily households	Declining	Higher net out-migration; high turnover	Selective nonresidential "succession"
Renewal	(a) Public housing; rental	High	Increasing (net)	Young families; many children	Declining	High net immigration; high turnover	The second transition stage; may take either of two forms depending on conditions
	(b) Luxury high-rise apts. and townhouses	High	Increasing	Mixed	Increasing	Medium	
Rehabilitation and gentrification	Conversions	Medium	Decreasing (net)	Few children	Increasing	Low	

Source: Based on *L. Bourne,* The Geography of Housing, *London: Edward Arnold, 1981, p. 24.*

Box 13.1   Stability and Change: A Typology of Neighborhoods

Seeing the residential kaleidoscope in terms of neighborhood life cycles helps to draw attention to the fact that each city is divided not just into neighborhoods of different composition but also of different rates of *population turnover* and different rates of *sociodemographic change*. Geographer Eric Moore has captured these important dimensions of urbanization in a four-fold typology (Figure 13.3) that is based on simple contrasts between high and low levels of residential mobility (i.e., population turnover) and between sociodemographic stability and rapid change in the sociodemographic profile of a neighborhood.

Type I situations, characterized by high mobility and rapid change, include the classic "invasion-succession" sequence that is most frequently encountered these days where a low-income ethnic minority is "invading" the territory of another group (Type Ia). Other Type I situations include the gentrification of inner-urban neighborhoods and black suburbanization (both Type Ib), and the high levels of residential mobility and rapid social change associated with the intrusion into established neighborhoods of locally undesirable land uses (LULUs) such as freeway construction, refuse incinerators, and so on (Type Ic).

Type II situations, with stable characteristics yet high levels of mobility, are more common than Type I situations. They are typified by newer suburbs of starter homes for newly established middle-income household or neighborhoods that receive wave after wave of upwardly mobile households with more-established families and higher incomes (Type IIa). A very different case is represented by neighborhoods that specifically cater to transient populations, such as the districts of small rented apartments close to the CBD occupied by young single people. Type IIb situations are best illustrated by slum districts that serve as ports of entry for low-income immigrants and in-migrants.

Type III situations, where neighborhood change is associated with relatively low levels of population turnover, is typified by neighborhoods that experience a gradual demographic change as a cohort of households' ages in place (Type IIIa). They also encompass a large spectrum of middle-income neighborhoods whose socioeconomic composition steadily changes with the filtering of the housing stock down the socioeconomic ladder as a result of slow but steady and selective out-migration and in-migration (Type IIIb).

Finally, *Type IV* situations, where low rates of mobility are accompanied by stable neighborhood profiles, find their strongest expression in the tightly knit ethnic neighborhoods of inner-urban areas and in some of the most affluent and well-established white suburbs. All such neighborhoods derive their stability from the lack of incentives for their residents to move elsewhere in the city.

	Neighborhoods experiencing significant/rapid socio-demographic change	Neighborhoods experiencing little/slow socio-demographic change
	I	**II**
Neighborhoods experiencing high mobility	(a) Rapid change resulting from ethnic, social, or racial conflict within area. (b) Change resulting from area being assigned high social value by specific group. (c) Change resulting from rapid deterioration of physical environment (particularly due to location of public facilities).	(a) Inflexible housing catering to small range of household types. (b) Neighborhood is a transit point for in-migrants from rural areas and other urban areas.
	III	**IV**
Neighborhoods experiencing low mobility	(a) Flexible housing catering to many household types. Slow aging of population and selective out-migration by age. (b) Deteriorating housing with selective in-migration by socioeconomic background.	(a) Tightly structured social networks, particularly for ethnic communities, tie individuals to their neighborhood.

FIGURE 13.3   A typology of neighborhood change.

5. *Renewal or Rehabilitation and Gentrification.* Renewal ends the neighborhood life cycle abruptly and begins a new one in the form of new tracts of housing, usually in some high-density format that reflects the neighborhood's (now) relatively central location. In certain cases, rehabilitation and gentrification can extend the neighborhood life cycle through conversion and reinvestment.

This example shows very clearly how housing "acts both as a determinant and a consequence of neighborhood change."[1] It also helps us to see the residential kaleidoscope as a result in part of the juxtaposition of neighborhoods at different stages of successive life cycles. New townhouses, condominiums, and apartment buildings in central city locations correspond to neighborhoods that have already experienced at least one cycle of structural obsolescence and that have now been redeveloped. Nearby are areas of older housing that have become technologically obsolescent for most households because of their lack of amenities. As a result, successive neighborhoods of this sort have been "invaded" by the very lowest socioeconomic groups.

Where the cycle of physical deterioration has been completed, some neighborhoods stand scarred and partially abandoned; in others, **urban renewal** has made room for public housing projects that have begun a new life cycle of their own. Where the original central city housing was built to a higher standard, the cycle of physical deterioration is still incomplete. Neighborhoods of larger dwellings—built to house prosperous nineteenth-century merchants and industrialists—remain as the basis of residual neighborhoods that have filtered down the socioeconomic ladder to become rooming house areas, with the structural and technical obsolescence of large units being resolved by subdivision and multiple occupancy. Pockets of well-built but smaller dwellings remain as the basis of gentrifying neighborhoods: areas that had slowly filtered down the socioeconomic ladder but that have been "invaded" by younger, middle-income households who have invested real capital or sweat equity, or both,[2] in renovating the housing.

Farther out are the city's middle-income suburbs which, because they date mostly from the 1960s onward, have suffered only moderately from physical deterioration and structural or technological obsolescence. Most still contain a significant proportion of their original "colonizing" population, topped up by households of similar types. The chief exceptions are the older, innermost suburbs, which have reached the thinning-out stage, with a large number of younger families moving in and the conversion and demolition of some homes to make way for commercial developments and highway improvements. Finally, the newest suburbs, in the outermost reaches of **urban realms** and in occasional pockets of development, represent neighborhoods at the beginning of their first life cycle.

HOUSING MARKETS

Any consideration of just how these patterns of residential mobility and neighborhood change come about must take place with reference to the operation of housing markets and the special nature of housing as a commodity. Housing, as geographer David Harvey has observed, "is fixed in geographic space, it changes hands infrequently, it is a commodity which we cannot do without, and it is a form of stored wealth which is subject to speculative activities in the market. . . . In addition, the house has various forms of value to the user and above all it is the point from which the user relates to every other aspect of the urban scene."[3] In market terms, therefore, housing extends well beyond the shelter provided by the dwelling itself to include a complex package that is often referred to in terms of housing services. We can identify four main aspects of housing services:[4]

1. Shelter and privacy.
2. Satisfaction and status associated with the size and quality of the dwelling and the prestige of the address.
3. Environmental quality, including both the quality of the physical environment (trees, vistas, parks, sidewalks, bike trails) and the quality of the social environment.
4. Accessibility to places of work, schools, shopping, friends, sport and recreational facilities, and other services and amenities.

The net utility of these services is generally referred to as the **use value** of housing. Because it depends a great deal on the needs and preferences of particular households, the use value attributed to a particular dwelling will tend to vary according to socioeconomic background, household type, lifestyle, and so on.

The role of housing as a form of stored wealth adds a fifth aspect to housing services:

5. Equity (for owners)—the financial return on an investment in housing (specifically, the difference between the market value of a dwelling and the amount of any outstanding mortgage debt on the property) that is (for owner-occupiers) tax-free. In this context we should note that the equity value of housing, along with that of other real estate investments, ebbs and flows with economic long waves. During the **Kondratiev** upswing, equity tends to increase rapidly; in the downswing, its growth is slowed or even reversed, so that owner-occupier housing can be more of a liability than an asset. The potential for gaining unearned income through equity increases, together with the use value of a dwelling, will determine its **exchange value** in the marketplace.

At this point we are confronted by the fact that by no means all of the housing market is for owner-occupier dwellings. Other forms of tenure, particularly private rentals, account for a large proportion of the stock of available housing in every city; so it is less useful to think in terms of overall urban housing markets than to think in terms of a series of *submarkets*. The changing composition of these housing submarkets is both a product of urbanization processes and a determinant of sociospatial differentiation. **Housing submarkets** can be delimited in terms of dwelling type, price range, and location, but *tenure* represents the single most important factor.

URBANIZATION AND THE TENURE TRANSFORMATION

The rise of homeownership has been not only the greatest single aspect of the transformation of urban housing but also one of the most important elements in the social and cultural evolution of American urban life. Although homeownership and the primacy of private property rights have been important from the start, it was only after World War II that homeownership became the dominant form of tenure in the United States. In mercantile settlements only a small proportion of all households owned their dwellings:

Many employees, especially the unmarried ones, lived in their employers' homes; some rented their own,

from landlords who were either mercantile capitalists or rural landholders and who had invested in providing housing for urban dwellers. Payment of rent required a regular income, and many families ensured that they could afford this by taking in lodgers.[5]

In the early industrial city, swollen with migrants, the evolution of housing markets was a critical element in the articulation of industrial society. At first some employers found that they needed to provide housing for their workers, either for rent or rent-free, but with lower wages. These initiatives ranged from temporary, barracks-like accommodations to substantial cottages, row houses, and tenements that were to become important landmarks in the history of urban planning. They could not cope, however, with the phenomenal flood of immigrants. More important, employers soon realized that factory housing tied up excessively large amounts of capital that might otherwise be invested in profit-making activities. Workers, for their part, found that factory housing was a mixed blessing, because it made them dependent, restricting their mobility in the labor market. Consequently:

As towns grew and developed into competitive labor markets rather than locales monopolized by a single employer, so the provision of housing was separated from the provision of work. Entrepreneurs perceived that investment in the former might be a profitable use of capital.[6]

Because the bulk of the market for housing consisted of unskilled and semiskilled industrial workers, the majority of this housing investment took the form of cheap rental units. Because status and power were closely linked to the ownership and control of property, landlordism became an important means of upward mobility for entrepreneurial groups of shopkeepers, small merchants, and people in professional occupations—even though the income they received from rents often amounted to a relatively modest level of profit. The net effect was the evolution of a highly competitive, **generalized housing market** that historical geographer James Vance, Jr. has identified as one of the single most important preconditions for the transformation of urban form and residential structure.[7]

The parallel emergence of a generalized submarket for owner-occupied housing was slower and less dramatic at first because of the smaller numbers and slower growth of white-collar workers. It was not until the appearance of the streetcar and the development of **streetcar suburbs** at the end of the nineteenth century (Chapter 5) that the owner-occupier submarket became an important element in the dynamics of American urbanization. From that point on, however, it rapidly became a key element that has been reflected, as we have seen, in the development of extensive tracts of low-density urban sprawl and an increasing degree of residential differentiation according to each household's ability to pay for different grades of housing and different packages of housing services.

The overall proportion of owner-occupied dwellings in U.S. cities rose from 20 percent in 1920 to 44 percent in 1940, 60 percent in 1960, 66 percent in 1980, and 68 percent in 2000 (Figure 13.4). There are four principal reasons for this trend:

1. *Increasing affluence among a progressively wider section of society*, which made homeownership possible for a greater proportion of households. Meanwhile, **economies of scale** and other innovations in the development and construction industries kept down the costs of suburban single-family dwellings, reinforcing the basic affordability of home ownership.

2. *Increasing recognition of the benefits of homeownership* in terms of
   - establishing social status and realizing an important part of the American Dream of independence based on property rights.
   - achieving residential segregation and pursuing exclusionary social and political strategies.
   - acquiring financial benefit through equity gains that in turn help to finance upward social mobility on the housing "ladder."
   - fulfilling, in many cases, familistic lifestyles.

3. *Increasing recognition of the economic and political significance of homeownership*. There were several aspects to this factor.
   - The role of homeownership in the circulation of capital: Traditionally, saving for the substantial down payment needed for entry into the homeownership submarket created a pool of capital that financial institutions could use to lend to entrepreneurs to invest in industrial projects; interest payments on existing mortgages, meanwhile, ensure a steady additional flow of (potential) investment capital; and financial institutions can buy and sell the mortgages themselves and use them to leverage further sources of investment capital. Finally (but by no means least), homeownership promotes the circulation of capital by creating a platform for all kinds of consumption, from furniture and furnishings to yard care equipment, that stimulate cycles of investment/production/consumption/profitability in a variety of industries.
   - The role of homeownership as a regulatory economic mechanism: As we saw in Chapter 6, the size of the housebuilding industry and its **multiplier effects** in relation to the market for consumer goods have made it a key element in **Keynesian** economic management in the past: one of the principal levers that governments have been able to manipulate in boosting or damping down the economy.
   - The role of homeownership in promoting social and political stability: As suggested in Chapter 6, the more people there are who have a stake in the private property market, the less likely they are to behave in ways that threaten economic and political stability. As the U.S. National Committee on Urban Problems observed in 1968 (in the wake of widespread urban unrest and rioting), "Home ownership encourages social stability and financial responsibility. It gives the homeowner a financial stake in society. . . . It helps eliminate the 'alienated tenant'

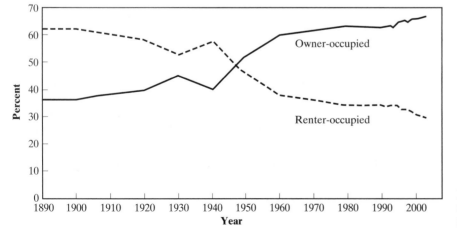

FIGURE 13.4   Tenure of occupied nonfarm dwelling units in the United States, 1890–2003.

psychology,"[8] The more people there are who carry the significant encumbrance of debt that comes with the award of a mortgage, the more individual and social incentives there are to protect and enhance the exchange value of property, to support employment stability (even at the cost of foregone wage raises), and to support the institutionalization of economic and political affairs that constitutes **organized capitalism** (Chapter 1).

As a result of the increasing recognition of these economic and political aspects of homeownership, *successive governments have actively fostered the growth of homeownership*. In broad terms in the United States, they have done so through two main sets of policies, those aimed at:

- Cultivating and protecting savings and loan institutions involved in financing homeownership. In addition to insuring mortgage loans made by private institutions for home construction or purchase (Chapter 6), the institutions themselves received privileges and concessions in terms of corporate law and taxation.

- Encouraging households to purchase rather than rent (Figure 13.5). Two particular aspects of income tax policy are important here: the inclusion of interest payments on mortgage loans for residential property as allowable deductions before tax rates are applied (interest payments on only one property are allowable, however) and the exemption of equity gains from capital gains tax.

4. *A sharp decline in the profitability of rental units.* This decline is also the result of a combination of several factors:

- The introduction and enforcement of more rigorous building standards and housing codes, which eliminated the supply of the cheapest rental units.

- The relatively slow rise in the incomes of tenants in privately rented accommodation, which has made it difficult for landlords to raise rents.

- Rent control legislation, initially introduced in order to divert investment capital and civilian labor away from housing and into defense industries during World War II.

- The increasing physical deterioration of the rental stock, much of which had been constructed in the nineteenth century.

FIGURE 13.5   Moving in. A couple putting up wallpaper in a room of their new home. A scene from the 1950s, replicated across the United States as more and more households became able to afford to purchase their own homes.

- Taxation policies that have made it unattractive for landlords to maintain or improve their rental properties.

- The decrease in demand for rental accommodation as a result of the government-supported financial advantages associated with homeownership.

As a result, progressively fewer rental units were built, and increasing numbers of rental units were demolished, abandoned, sold into owner-occupancy, or converted to nonresidential uses. Over the long haul, then, American cities experienced an almost complete reversal of the relative importance of the owner-occupier and rental submarkets.

HOUSING AFFORDABILITY

There was a perceptible decline in the overall level of homeownership between the early 1980s and the early 1990s (Figure 13.4). Once again we must invoke the economic, social, and demographic changes since the early 1970s as an explanatory context. The **stagflation** of the economy caused prices to rise sharply while, for many, incomes were stagnant. Moreover, after 1973 house prices tended to rise faster than the overall rate of economic inflation, mainly because of rapidly rising construction and finance costs. By 1983 homeowners had to use, on average, nearly 20 percent more of their incomes to pay for their homes than in 1973. This increase reflects the onset of an *affordability crisis* for homeownership. Subsequent decreases in mortgage interest rates alleviated the crisis somewhat, but the affordability of housing has become a chronic issue and a constant feature of the urban scene.

The decreased affordability of homeownership did not affect demand quite as dramatically as might be expected. For some households, it underscored the urgency of participating in the long-term benefits of homeownership; for many, the possibility of capturing large windfall gains in a rapidly inflating market represented an enticing prospect. If households could only get a foothold on the homeownership ladder, it seemed, their financial security would be assured. Houses that had traded for $35,000 at the beginning of the 1970s were being resold for $60,000 just a year or two later and for over $100,000 by 1985. Although median single-family house values more than doubled between 1950 and 2000, they rose fastest during the 1970s (43%) (Figure 13.6). Many households, therefore, opted to become "house poor," meeting mortgage payments at the expense of consumption in other areas, by deferring or abandoning plans to have children, by homemakers reentering the labor market much sooner after having children, and by householders taking second jobs. Developers and builders responded, meanwhile, by building smaller homes at higher densities: townhouses and condominiums that were also better suited in layout and design for smaller, two-income households (Figure 13.7). Overall, the number of multiunit dwellings completed by the private sector increased from less than 270,000 in 1975 to more than 440,000 in 1980 and to nearly 700,000 by 1985.

FIGURE 13.7   Garage townhouses. One of a number of recent responses by developers to the need for affordable housing. Smaller homes at higher densities are also better suited in layout and design for smaller, two-income households.

Nevertheless, the economic trends of the 1970s and 1980s put homeownership out of reach for an even greater number of households. By the early 1980s the costs of homeownership had risen to the equivalent of 36 percent of median family income. For the first time since the Great Depression of the 1930s, between 1980 and 1986 the overall proportion of U.S. households buying homes actually decreased. This trend was accompanied, however, by a resurgence in the development of housing units for rent (from 196,100 in 1980 to 365,200 by 1985). In part, of course, this resurgence was simply a response to the affordability problems of homeownership. But it was also a product of some very different trends, including the recentralization of high-income jobs in finance and business services in downtown areas (resulting in the increased attractiveness of in-town apartments) and the eclipse of familistic lifestyles by a variety of materialistic, "active," and less traditional lifestyles (resulting in homeownership being traded for tenant status in highly packaged settings and higher disposable incomes).

During the 1990s the overall level of homeownership and housing starts rebounded (see Figure 13.4 and Figure 6.9). Housing prices climbed faster than both household incomes and overall prices, however, causing the ratio of house prices to incomes to increase among all owners and purchasers of homes. Nevertheless, housing affordability—the percentage of the median household income needed to meet the mortgage payment on the median priced home—improved due to reductions in mortgage interest rates (see Table 4.3). During the 1990s homeowners were spending on average about 20 percent of their incomes to pay for their

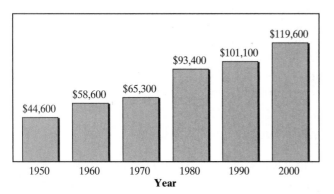

FIGURE 13.6   Median values of owner-occupied single-family homes in the United States, 1950–2000, in 2000 dollars.

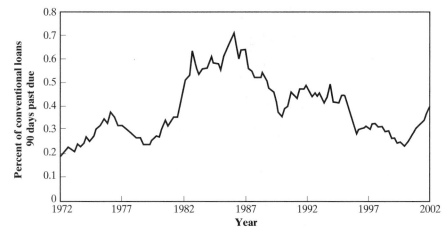

FIGURE 13.8   Moving out. Serious mortgage delinquencies were all too common in the early 1980s, as families who had stretched their resources to purchase homes found they could not meet mortgage payments during a phase of economic recession in the United States.

homes (Table 4.3). Perhaps the strongest indicator of improved homeownership affordability was the rising homeownership rate among households in the bottom 20 percent of the income distribution. The evidence suggests, however, that the low-income renters who shifted to homeownership were not the poorest of the poor.[9]

Although minorities have accounted for more than 40 percent of the net growth in owners since the early 1990s, the homeownership gap between whites and minorities has narrowed only slightly. Black and Hispanic homeownership rates in 2001 lagged behind that of whites by 25 percent. In addition, the growth in the subprime lender share of loans to low-income minority homeowners rose dramatically—the subprime lender share of home purchase loans in low-income, predominantly minority communities increased from 2.4 to 13.4 percent, while the share of subprime refinance loans rose from 6.8 to 27.5 percent. Although reaching traditionally underserved markets, subprime loans—loans to borrowers with weak credit histories—come with higher interest rates and fees than standard loans, as well as higher default rates (Figure 13.8). Additionally, it is estimated that between 10 and 35 percent of borrowers who obtained mortgages in the subprime market could have qualified for a conventional loan.[10] This situation raises concerns because of the dramatic increase in the number of homeowners who spent more than half of their income on housing in the late 1990s (from 5.8 million in 1997 to 7.3 million in 2001). Meanwhile, improvements in homeownership affordability as a result of reductions in mortgage rates depend on falling interest rates, which can go only so low.

Housing affordability among renters improved slightly during the 1990s, as median gross rent as a percentage of household income fell for the first time since 1950 (Figure 13.9). Utilities (gas and electricity) expenses rose less rapidly than contract rent over this period, holding down the ratio. Nationally, renter households spent 25.5 percent of their pretax income on rent in 2000, down from 26.4 percent in 1990. This overall trend in rental housing affordability, however, masks an increase in the cost-to-income ratio among the poorest.[11] Making matters worse, the shortage of affordable market-rate rentals is dire. In 2001 the 9.9 million renters in the lowest 20 percent of the renter income distribution outnumbered the supply of affordable lowest-cost rental units by fully 2 million.[12] A contributing factor was the decline in the number of multiunit dwellings completed by the private sector, from nearly 700,000 in 1985 to less than 300,000 in 1990 and 1995 and to just under 340,000 in 2000. A continuation of the shift toward the production of more expensive apartments that began in the 1980s further thinned the supply of affordable rental housing. The supply of federally subsidized housing has also fallen. Already, 90,000 fewer units are available because private owners either have opted out of federal programs or have prepaid their federally subsidized mortgages in order to capitalize on higher prevailing market rents. Meanwhile, the demolition of public housing units has eliminated many badly deteriorated units without providing one-for-one replacement.[13]

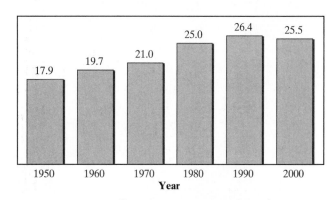

FIGURE 13.9   Gross rent as a percentage of household income, 1950–2000, in 2000 dollars.

At the opening of the new millennium, the nation faces a widening gap between the demand for affordable housing and the supply of it. The causes are varied—rising housing production costs in relation to family incomes, inadequate public subsidies, restrictive zoning practices, adoption of local regulations that discourage housing development, and loss of units from the supply of previously subsidized federal housing . . . . And despite civil-rights and fair housing guarantees, the housing shortage hits minorities hardest of all.[14]

PUBLIC HOUSING

In U.S. cities public housing is conspicuous by its absence. Cities in most other developed countries contain a substantial sector of public housing that has been provided, through one mechanism or another, to address acute problems of housing need that have arisen because the private market has been unable to provide decent housing for low-income households and still make a profit. In Britain, for example, public housing typically accounts for about 20 percent of the housing stock in cities. In many other developed countries, public housing accounts for between 10 and 20 percent of the urban housing stock; but in the United States, the figure is closer to 1.5 percent. Another feature that makes U.S. public housing distinctive is that much of it is provided through rent subsidies to private landlords (instead of being provided through public housing stock by government housing agencies).

Yet U.S. cities have just as much need for public housing. Although they were fortunate (unlike many of their counterparts in Europe and Japan) in escaping the devastation of wartime bombing, U.S. cities have certainly experienced acute problems of housing need. Slums have persisted within every large city for decade after decade. A succession of studies has shown that about one-third of all families in central cities are in need of basic housing, which they cannot afford,[15] and homelessness has risen alarmingly (see Chapter 15). It is, therefore, instructive to examine how and why American urbanization has not produced a substantial sector of public housing.

In fact, there was a time when public housing might have developed into a substantial sector. The initial legislation mandating public housing was passed in 1937 in the political climate of the post-Depression New Deal (see p. 507). Although public housing was widely seen as ideologically distasteful (as it had been in most other countries), the depth of the Depression was seen as having created exceptional needs, and the program was able to be "sold" politically in terms of employment creation rather than housing subsidy. The program established a federal government presence in urban housing, but it was not until after World War II that a more extensive and ambitious national program was launched. The Housing Act of 1949 linked public housing construction to slum clearance programs and represented a significant commitment to subsidized housing for the poor (Figure 13.10). Strong opposition to the act was

FIGURE 13.10
Public housing built under the provisions of the 1949 Housing Act. The relaxed atmosphere depicted in this photograph soon deteriorated to a more oppressive one as public housing in the United States became widely regarded as a receptacle for "problem" families.

deflected by a coalition of different interest groups, all with different agendas and different assumptions about the main purpose of the legislation. Political liberals (thinking of both the slum clearance and the provision of public housing) saw it as a means of eliminating slums and rehousing the poor, but business interests (thinking mainly of the slum clearance) saw it as a means of bolstering central city property values, while many local politicians (equating slum clearance with urban renewal) saw it as a means of bolstering their tax base and luring back some of the more affluent consumers and taxpayers who had moved to the suburbs.

The 1949 Housing Act authorized the construction of 810,000 low-rent units over six years. The overall contribution of this program to the housing stock of U.S. cities has been minimal, however. The 810,000 units called for under the 1949 legislation took 25 years to materialize. It did so, moreover, in the form of very poor quality housing that was localized in some of the worst and most unpopular districts within cities. Meanwhile, the rate of construction of public housing was swamped by the scale of urban poverty and the rate of physical deterioration of the pool of inexpensive private rental units. By the mid-1970s public housing had become synonymous with failure, and the prospect of establishing a substantial program of public housing had vanished, at least for the foreseeable future. Looking back, we can see several reasons why public housing did not take root in the United States:

- An exceptionally strong free-enterprise ethic (Figure 13.11) in which public housing aroused suspicions of "creeping socialism." In this context it is not insignificant that the crucial period of debate about public housing took place during the notorious McCarthy witch-hunt for communist sympathizers.

FIGURE 13.11 Free-market propaganda, cleverly wrapped in patriotism and democratic ideals, was organized by industrial, financial, and real estate organizations to minimize the "creeping socialism" of public housing legislation and other liberal urban policies in the United States.

- Racial bias and discrimination that caused the general population to be reluctant to support a program whose beneficiaries were likely to be mostly African American or Hispanic.

- Highly organized and well-funded opposition from the National Association of Real Estate Boards, the National Association of Home Builders, the United States Savings and Loan League, the U.S. Chamber of Commerce, the Mortgage Bankers Association of America, and the American Bankers Association.

- Overly restrictive terms and conditions governing the construction and management of public housing. Because of a dollar limit on construction costs (pegged at $2,400 between 1949 and 1965) some tenants were given homes that had no bathroom doors, no toilet seats, and no baths or showers. Strict means-testing for potential tenants was commonly coupled with a rule that the household had to vacate the property as soon as its income exceeded the qualifying minimum by 25 percent or more. As such, public housing was programmed to be no more than a jerry-built haven for the chronically poor. Public housing projects soon came to be stigmatized as "instant slums," their residents stigmatized as economically incompetent and socially undesirable. Many of the residents themselves, feeling **labeled** and with no security of tenure, found no reason to care for their homes or their neighborhood, the net result helping to fuel the arguments of opponents of public housing, who argued that it would be a breeding ground for social problems (Figure 13.12).

HOUSING SUBMARKETS

Although tenure is the single most important delimiter of housing submarkets, it is important to recognize that urban change is mediated in practice through submarkets that are also delimited as a result of other attributes. Six kinds of constraints collectively shape urban housing submarkets:[17]

1. *Supply restrictions:* There is limited availability of certain kinds of housing that are in demand but that cannot easily be reproduced (e.g., nineteenth-century brownstones or low-cost older units).

2. *Accessibility restrictions:* Some houses may have a unique location, which conveys an additional benefit (or liability) in terms of accessibility (e.g., single-family housing within walking distance of a neighborhood shopping district).

3. *Neighborhood restrictions:* For various reasons, particular small areas can and do become especially attractive (or unattractive), for which entry is limited and people will pay a premium (or discount).

FIGURE 13.12
A self-fulfilling prophecy: underfunding and overly restrictive tenancy regulations led to the failure of many public housing projects, including this one—the Robert Taylor Homes—in the process of demolition in Chicago.

4. *Institutional restrictions:* Perhaps the most obvious has been the practice of **redlining** (see p. 367), in which mortgage-lending agencies refused to lend in certain areas. Other examples include the effects of building codes, **land use zoning**, and planning regulations.

5. *Racial, ethnic, and class discrimination:* Certain families are limited in their search for and choice of housing because of direct exclusion.

6. *Information restrictions:* Households have different access to information on housing opportunities and on how the market works.

BOX 13.2   PUBLIC AND PRIVATE HOUSING IN EUROPEAN CITIES[16]

Apartment living is common in Europe—residents of all income groups own or rent apartments. Apartments are a good land use choice in high-density and compact European cities where space is at a premium and land values are high. Traditionally, instead of growing outwards, cities grew upwards to the limit of a city's height regulations.

The multistory apartment house originated in northern Italy to accommodate the wealthy during the Renaissance. By the early eighteenth century, the apartment house had spread to the larger cities in continental Europe and Scotland. Until the invention of the elevator, social stratification within individual buildings was vertical: Wealthier families occupied the lower floors, while poorer residents lived in smaller apartments above. Horizontal social stratification also developed within apartment blocks. The large expensive units were located in the front of buildings, with small low-rent units facing the rear. By the late eighteenth century, as the Industrial Revolution spurred increasing urbanization, apartment blocks had spread

to medium-size cities. Speculators built large-scale standardized tenements for middle-income occupants and barracks for low-income residents.

The two-story terrace of single-family houses is distinctive to England, Wales, and Ireland. This row house tradition can be traced back to efforts to restrict congestion in London in the late 1500s that made it illegal for more than one family to rent a new building. The narrow multistory house with an apartment on each floor, a variation of the row house, is found along the North Sea coast from cities like Rouen and Lille in northern France to Bremen and Hamburg in northern Germany.

The serious housing shortage that started with the Great Depression of the 1930s was exacerbated by the lack of construction and significant destruction during both world wars. The public housing programs that began in Vienna in the early 1920s were stepped up across Western Europe after World War II. Modern architecture and urban design principles were combined with low-cost factory production

(Continued)

BOX 13.2   PUBLIC AND PRIVATE HOUSING IN EUROPEAN CITIES   *(Continued)*

methods. This resulted in the replacement of the war-damaged historic houses and dilapidated nineteenth-century tenements in the central parts of cities by monotonous high-rise apartment buildings.

In the 1950s and 1960s most governments adopted a policy of metropolitan decentralization of public housing to inexpensive open land at the edge of cities. Modern high-rise apartment blocks were concentrated in large peripheral estates known by their French name—*grands ensembles*. Developments like Park Hill in Sheffield, Sarcelles in northern Paris, and Chorweiler in Cologne contained 1,000 or more apartments at densities as high as 120 residents per acre.

Traditionally, West European governments spent more on public housing construction than on rent subsidies. Because the poorest segment of society cannot afford public housing, less stigma is attached to public housing in Europe compared to North America. The very poorest residents are distributed instead throughout European cities in the cheapest private rental units. These poor quality dwellings are found in damp basements, attic spaces, and run-down rear apartments. Where a stigma is associated with public housing, it applies only to specific kinds of government-subsidized developments. These include the purpose-built housing projects that replaced the peripheral shantytowns of poor rural immigrants and postwar refugees in Spain, Italy, France, and Germany.

Historically, the severity of need and the political leanings of governments determined the amount of public housing provided within each city and country. The amount was highest in cities with serious housing shortages and liberal municipal governments such as Edinburgh and Glasgow in Scotland where the number of public units grew to well over half the total housing stock. Until the 1970s, public housing comprised 20 to 30 percent of the total in England, France, and Germany, and 10 percent in Italy. Public housing represented only 5 percent or less of the housing in the more affluent and conservative Swiss cities. Since the 1970s, however, the amount and percentage of public housing have declined due to government cost-cutting privatization programs.

In general, the cities that developed under socialism throughout Eastern Europe were less spatially segregated than those that evolved under capitalism. Certainly, mansions—the prewar residences of the social elites—were used for political purposes to house party officials, foreign delegations, or institutes. But housing was viewed as a right, not a commodity; each family was entitled to its own home at reasonable cost.

In the face of the tremendous housing shortfalls following World War II, from the 1950s to the 1970s the Communist governments began massive building programs of *housing estates:* typically 11-story, prefabricated, multifamily apartment blocks. These apartments were small (about 460 to 650 square feet), poorly constructed, and almost universally disliked by residents. Often, housing estates were built in large clusters, sometimes forming massive concrete curtains, usually on land near the edge of cities. Because of this, urban population densities, which are typically higher in Eastern Europe than in North America, could actually *increase* near the urban periphery (Figure 13.13).

The housing estates were often constructed in groups to form a **neighborhood unit** consisting of three to four apartment blocks arranged into a quadrangle surrounding shops, green space, and play areas for children (see Chapter 17). The purpose of these units, in keeping with socialist planning philosophy, was to promote greater social interaction; however, the neighborhood unit concept, along with the construction of housing estates, was abandoned by the mid-1980s after governments recognized the futility of trying to achieve social engineering through physical planning.

FIGURE 13.13   Former socialist suburban housing estate in East Berlin, Germany.

The overall result is a series of housing submarkets. A study of housing submarkets in Minneapolis–St. Paul found 14 distinctive spatial submarkets (Figure 13.14);[18] an analysis of central Baltimore's housing market found 13.[19] The Baltimore study revealed the importance of financing (an institutional restriction), emphasizing the highly structured relationship between household characteristics (particularly income) and the

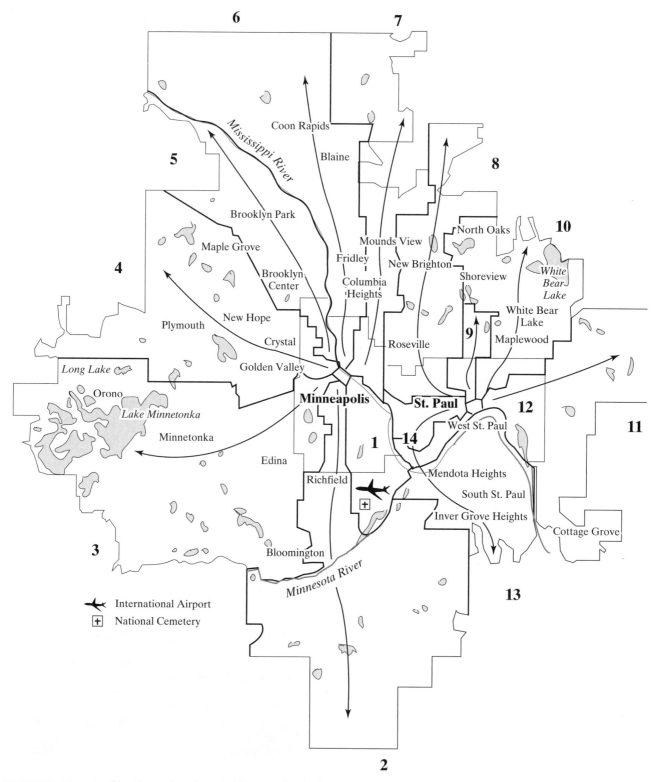

FIGURE 13.14   Housing submarkets in Minneapolis–St. Paul.

availability of mortgage funds from different sources. This geographic structure of submarkets forms a "decision environment" within which individual households make housing choices. Most of these choices, however, are likely to reinforce this structure and contribute to neighborhood stability rather than neighborhood change.

But, as we have seen, neighborhoods do change as physical deterioration, obsolescence, and sociodemographic change in place induce changes in levels of investment and residential mobility. Thus the ebbs and flows of households and housing finance gradually transform the geographic structure of housing markets—the basic framework of the city's social geography. These ebbs and flows, however, are mediated by key decision makers who operate within the context of particular professional and institutional settings. In the rest of this chapter we explore in detail the linkages between neighborhood change or stability and residential mobility, paying particular attention to the decision-making roles of both households and housing professionals.

BOX 13.3 NEIGHBORHOOD STABILITY IN WEST EUROPEAN CITIES[20]

West European cities enjoy remarkable neighborhood stability. Europeans move much less frequently than North Americans, and homeowners maintain their characteristic solidly constructed concrete block, brick, and stone housing. As a result, older neighborhoods at or near the center of large cities enjoy remarkably long lives despite suburbanization.

The districts of handsome mansions built by speculative developers for wealthy families in the seventeenth, eighteenth, and early nineteenth centuries remain stable high-income neighborhoods, such as Belgravia, Bloomsbury, and Mayfair in central London (Figure 13.15). High-income suburban neighborhoods developed typically in the western part of older industrial cities, upwind of industrial smokestacks and residential chimneys.

Wealthy residents, in fact, have remained at or near the city center in Western Europe since before the Industrial Revolution. Higher taxes on city land until the late nineteenth century kept the poorest residents and immigrants outside the city walls. Beginning with Paris in the second half of the nineteenth century, this tradition was strengthened by the replacement of areas of slums and former city walls with wide boulevards and imposing apartments (see Figure 12.19).

Since the eighteenth century in Western Europe, however, urban growth has spread to suburban zones and has even enveloped free-standing villages and towns. Yet these separate urban centers became distinct quarters within the expanding city as they maintained their long-established social and economic characteristics, major landmarks, and shopping streets. During the second half of the nineteenth century, **annexations** of groups of these suburban quarters produced distinctive city districts with their own shopping areas and government institutions.

In the last few decades governments in older industrial cities have funded urban renewal projects designed to attract higher-income residents to the revitalized parts of central areas. The success of these large-scale city center redevelopments has given rise to gentrification in the surrounding area. Demand for housing that has the potential to be renovated for higher-income occupants, however, has raised property values in certain areas and pushed out lower-income residents.

FIGURE 13.15   Belgrave Square in Belgravia, central London. Districts of handsome mansions such as this, which were built by speculative developers for wealthy families in the seventeenth, eighteenth, and early nineteenth centuries, remain stable high-income neighborhoods.

RESIDENTIAL MOBILITY AND NEIGHBORHOOD CHANGE

Residential mobility is a central facet of urban social geography, for it provides a spatial expression of the link between the individual household and the social structure, between the household's life-world and its biographical situation, between internal culture-building processes and the spatial template of the city. The residential choices of individual households in aggregate define the social areas of the city. The rapid transformation of American central cities and the plight of their dwindling fiscal base was caused in part by the massive migration of middle-class families to the suburbs in the 1960s. Similarly, the anticipated revitalization of the downtown areas of some cities in the 1970s . . . was brought about by the reverse migration of childless professionals to "gentrified" central neighborhoods. But there is a two-way relationship between individual and aggregate levels, for at the same time the individual's pattern of choices is constrained by the preexisting set of spatial opportunities in the city and the household's own biography—those characteristics of income, stage in the life cycle, ethnic status and lifestyle which will close off certain housing options to it and substantially reduce its range of choice.[21]

This quotation from the work of social geographer David Ley provides a good description of the overall relationship between residential mobility and urban residential structure. Although migration actively creates and remodels the social and demographic profile of neighborhoods, it is also conditioned by the existing pattern of neighborhoods. (In addition, households moving into neighborhoods are likely to find that they in turn are subsequently affected by the attributes and behavior of their new neighbors—as we will see in Chapter 14.) This relationship is outlined in Figure 13.16, which emphasizes the effects of household mobility and residential structure on each other. The residential mosaic is seen as a cumulative product of residential mobility, which in turn is seen as a product of housing opportunities and household needs and expectations. These needs and expectations, in their turn, are influenced by what Ley calls the household's biography (its income, lifestyle, household type, ethnicity, and so on) and by its knowledge and perceptions of housing opportunities.

RATES OF TURNOVER: MOVERS AND STAYERS

We can take these relationships seriously only if it can be shown that rates of residential mobility in cities run consistently at reasonably significant levels. In fact, the turnover of addresses in most U.S. cities is strikingly high: About 1 in every 12 metropolitan households moves to a new home each year. Naturally, this figure masks some important differences between metropolitan areas, between central city and suburban areas, between neighborhoods, and between tenure sectors. Table 13.2 illustrates the extent of these differences between owners and renters at the intrametropolitan scale and by family type, age, and ethnicity.

These differentials suggest that residential mobility is a selective process, with some neighborhoods being dominated by households with little propensity ever to move. These households have been called *stayers*. Among "stayers," owners, middle-aged and elderly

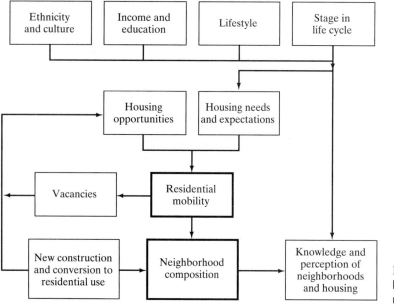

FIGURE 13.16   Relationships between housing demand, residential mobility, and neighborhood change.

TABLE 13.2

Tenure Choices of U.S. Households that Changed Residence in 2001 (in Thousands)

Previous tenure	Own			Rent		
Current tenure	Own	Rent	All	Own	Rent	All
Total	4,783	2,315	7,098	5,028	11,172	16,200
Metropolitan status						
Central city	856	858	1,714	1,347	5,066	6,413
Suburban	2,727	982	3,709	2,684	4,311	6,995
Non-metropolitan	1,200	475	1,675	997	1,795	2,792
Family type						
Married without children	1,551	292	1,843	1,236	1,232	2,468
Married with children	1,751	327	2,078	1,624	1,931	3,555
Single parent	260	402	662	555	2,188	2,743
Other family	243	135	378	252	631	883
Single person	786	972	1,758	966	3,520	4,486
Other nonfamily	191	186	377	395	1,669	2,064
Age of household head						
Under 35	1,068	808	1,876	2,554	6,625	9,179
35–44	1,448	606	2,054	1,342	2,312	3,654
45–54	982	442	1,424	660	1,230	1,890
55–64	705	202	907	308	518	826
65+	579	256	835	163	486	649
Ethnicity of household head						
White	4,152	1,634	5,786	3,605	6,579	10,184
Hispanic	269	238	507	556	1,735	2,291
Black	213	325	538	534	2,118	2,652
Asian/Other	149	117	266	333	740	1,073
Selected stated reasons for moving						
Needed a larger house or apartment	866	57	923	433	1,359	1,792
Wanted a better quality house (apartment)	633	58	691	309	1,065	1,374
To establish own household	135	156	291	833	717	1,550
To be closer to work/school/other	269	170	439	178	1,120	1,298
Change from owner to renter OR renter to owner	68	92	160	1,330	46	1,376
Wanted lower rent or less expensive house to maintain	100	72	172	79	763	842
New job or job transfer	497	303	800	292	1,313	1,605
Married, widowed, divorced, or separated	296	370	666	113	316	429
Disaster loss (fire, flood, etc.)	15	19	34	16	57	73
Forced to leave by the government (eminent domain)	18	0	18	7	40	47

Source: *Joint Center for Housing Studies of Harvard University,* The State of the Nation's Housing: 2003, *Cambridge, M.A.: Harvard University, 2003, Table A-8, p. 38.*

households, and households with low-to-middle socioeconomic background tend to be overrepresented. A major contribution of such households to the city's residential mosaic is to impose a relatively high degree of stability on certain neighborhoods. *Movers,* whose impact, as we have seen, may either reinforce neighborhood composition or initiate change, tend to be younger, to be renters, and to be at one extreme or another of the socioeconomic ladder.

Given that people age, adopt new lifestyles, experience changes in income levels, and so on, it follows that stayers can become movers, and vice versa. In addition, research has shown that there is an independent duration-of-residence effect, whereby the longer a household remains in a dwelling the less likely it is to move, mainly because of the emotional attachments that develop toward the home and the immediate neighborhood and because of the social networks developed there. These attachments are explored in Chapter 14, but it is worth noting here that research has also supported the idea that people can be thought of as *locals* or *cosmopolitans*, depending on their natural propensity to develop such attachments.

THE IMPACT OF NEW ARRIVALS TO THE CITY

A second important distinction to make in relation to patterns of residential mobility concerns the difference between intraurban moves (i.e., *within* the metropolis) and other kinds of mobility—intermetropolitan migration, inmigration from nonmetropolitan areas, and immigration directly to metropolitan areas. In the typical American metropolis, intermetropolitan migration, inmigration, and immigration account for up to one-third of all moves—substantially less than in the past, when migration and immigration not only fueled urban growth to a much greater extent but also dominated neighborhood dynamics.

The impact of these long-distance moves today can best be divided into two categories: the arrival of low-income migrants and the arrival of middle-income and high-income migrants. As in Chicago of the 1920s and 1930s (see p. 311), flows of low-income migrants and immigrants are mainly focused on inner-city neighborhoods with inexpensive housing, with a significant degree of localized differentiation according to ethnicity and place of origin. In addition, some of these flows are now directed to first-tier and second-tier suburbs, where established ethnic communities provide a port of entry for new arrivals. A good example is the "Little Saigon" area in the Clarendon neighborhood of Arlington, Va., a 60-year-old suburb of Washington, D.C. Such flows reflect the continuing importance of **chain migration**, whereby migrants who were encouraged and assisted in their move by friends and relatives from their place of origin encourage other friends or relatives to join them, helping them to find accommodation and jobs when they arrive.

The impact of middle-income and high-income migrants, in contrast, is much less focused in spatial terms. They do, however, exhibit certain broad regularities in spatial behavior. An occupationally and geographically mobile group of cosmopolitan movers, they are often constrained by limited amounts of time available to search for a home. In addition, because they are willing and able to move in search of better jobs or career advancement, if they do purchase a home rather than rent, they tend to opt for homes that are likely to be relatively easy to resell. Finally, experience may have alerted them to the possibilities of moving into a neighborhood only to find that their own values and lifestyle are out of tune with the established character and social networks of the neighborhood. The net result is a tendency to select newly built homes in suburban tracts or packaged developments, moving on after a year or two, when they have developed a better sense of the geography of their new city.

INTRAURBAN MOVES

Moves within metropolitan areas account for at least two-thirds of the total turnover. In addition to the follow-up moves of newly arrived migrants, they include a great variety of adjustments as new households are formed, old ones split up, and existing ones respond to changed circumstances. Observations of the overall patterns of movement have revealed some important generalizations about these moves.

- Most moves take place over relatively short geographical distances—about one-sixth to one-quarter of the diameter of the city. This fact is generally taken to reflect the importance of people's desire to preserve local ties and, in particular, children's ties to friends and schools. It is also related to a second important tendency:
- Most moves are to dwellings and neighborhoods of status and character similar to those left behind. Household changes in housing and neighborhood status are, in other words, most often marginal.
- There is a marked directional bias to patterns of movement, with many households staying within the same sector or quadrant of the city. This bias is generally taken to be a product of the first two tendencies, combined with the effects of familiarity on people's search behavior. Most metropolitan areas are so large that many people prefer to consider first the housing vacancies that occur within the part of the metropolis that they know best. This preference ties in conveniently with the concept of urban realms (p. 161).
- There is a strong tendency for the timing of intraurban moves to be tied to the rhythms of business cycles. Economic upswings, increased employment opportunities, higher wages, and higher consumer confidence lead to an increase in the effective demand for new housing and to increased demand for trading up to bigger and better homes, thus initiating a flurry of relocations.

Analyses of the geographical interrelationships between patterns of residential mobility and neighborhood housing attributes and socioeconomic attributes have also shown that, in addition to these broad generalizations, housing *type* (as measured by owner-occupancy rates, the incidence of single-family dwellings, and the number of rooms per dwelling) rather than housing *quality* (as measured by housing values, housing age, and the number of bathrooms per dwelling) is the chief determinant of intraurban mobility.[22] The same studies have shown that it is family status rather than socioeconomic status that is a function of housing type (and therefore of mobility), thus underscoring the importance of changes in the household life cycle in triggering residential mobility (see p. 302). As we might expect from our knowledge of changing metropolitan form and changing patterns of residential differentiation, however, these relationships have become progressively weaker since 1950, thus pointing to an increasing complexity in the pattern and logic of residential mobility.

REASONS FOR MOVING

We can begin to understand this complexity a little better by investigating the behavior of individual households, seeking answers to the question of why families move by distinguishing between voluntary and involuntary moves and between "push" and "pull" factors (Figure 13.17).[23]

Involuntary moves can account for a significant and surprisingly high proportion of intraurban mobility—between 15 and 25 percent. They are necessitated by events totally beyond the household's control: property demolitions, **eminent domain** proceedings, evictions for rent arrears or defaulting on mortgage repayments, and disasters such as fire or flood. In addition to these purely involuntary moves are moves that are *induced* by unwanted or unforeseen circumstances, such as divorce, ill health, a death in the family, or

corporate relocation (a change in the location of the place of work—because of office decentralization, for example) (Table 13.2). On average, such moves account for a further 15 percent or so of all intraurban mobility.

People give a wide variety of reasons for purely *voluntary* moves. Most are adjustment moves that are intended to alter the type and quantity of housing. Classic migration theory groups these into **push factors** and **pull factors**. The most frequently cited push factor by far is the lack of sufficient space—or, to be precise, the *feeling* of insufficient space for household needs. Once again, this factor underscores the relevance of changes in the household life cycle. Other frequently cited push factors include the costs of maintenance and repair, structural or technological obsolescence in the dwelling, aspects of the physical environment of the neighborhood (e.g., poor upkeep of neighboring properties, heavy traffic), and characteristics of the social environment (e.g., too many or too two few children, too much or too little street life). The most frequently cited pull factors by far are those associated with a change of employment. Other pull factors include accessibility to shops, amenities, and friends, the attractions of better-quality schools and public services, the chance to switch housing submarkets (usually from renting to owning), and the attractions of particular settings in pursuing particular lifestyles (Table 13.2).

UNDERSTANDING HOUSEHOLD BEHAVIOR: THE DECISION TO MOVE

In practice, of course, households are "pushed" and "pulled" by a variety of factors at once. In addition, the net effect of push and pull factors may not be enough to cause a household to look for alternatives. Even if a

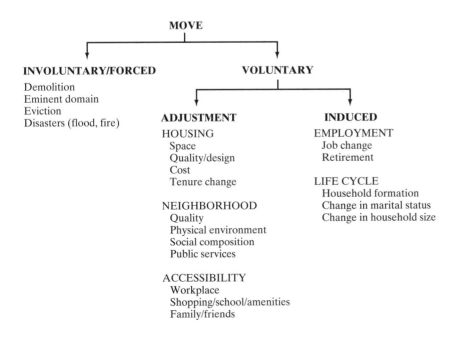

FIGURE 13.17 Reasons for household relocation.

search among potential alternatives is begun, a household may not necessarily eventually move: Adjustments can be made to perceived wants and needs, to the dwelling itself, and even to the neighborhood. These possibilities are accommodated in Figure 13.18, which is based on a conceptual model of the household relocation and search process first developed by urban geographers Larry Brown and Eric Moore.[24]

This model recognizes that a household will derive some degree of "utility" from its current dwelling: a product of its own biography and of the qualities and attributes of the dwelling itself and the neighborhood. This utility may not be positive, however. "Stressors," arising from the interaction of a variety of push and pull factors, may reduce the utility of the dwelling so that the household feels forced to do something about the situation (point A on Figure 13.18).

One course of action is to address directly the shortcomings of the dwelling or neighborhood. Dwellings can be remodeled, extra space can be added, kitchens can be refitted, and so on. Neighborhood shortcomings can be addressed through joining or forming residents' associations or action committees, by confronting troublesome neighbors, by political action, and so on.

If such action is not feasible, or if it is unsuccessful, a second course of action may be pursued: modifying the household itself as a means of coming to terms with existing conditions. Plans to have children may be deferred or abandoned, household lifestyle may be modified or constrained, or, more likely, household aspirations and expectations may be lowered. The older people get, the more adept they become at lowering their aspirations to fit their

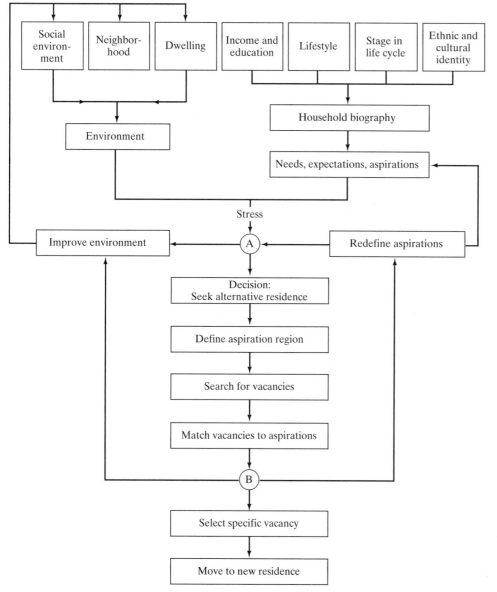

FIGURE 13.18 A model of the residential location decision process.

experience of reality. This, then, is an important explanation for the tendency for older households to become stayers.

UNDERSTANDING HOUSEHOLD BEHAVIOR: THE SEARCH FOR ALTERNATIVE PLACES TO LIVE

The third course of action is to seek alternative places to live. Figure 13.18 shows that there are three stages to this course of action: (1) defining the "aspiration region" (i.e., deciding on the attributes to look for in a new home), (2) searching for dwellings that fall within the aspiration region, and (3) comparing possible alternatives. The importance of these stages is that *different kinds of households behave differently at each stage, thus reflecting and sustaining the processes of social and residential differentiation.* Households with different biographies and income constraints will set out to look for vacancies with quite different housing goals in mind. In addition, some households will develop a much clearer idea of what they are looking for, depending in part on the personality of household members but mostly on their ability to organize their thoughts (partly a function of education) and on their overall knowledge and experience of housing and neighborhood types.

The search process itself is much more closely linked to household socioeconomic attributes. Typically the search does not cover the whole city, or even whole housing submarkets, but takes place within the households' *awareness space*, which is a product of its *action space* and its *information space*, both of which are closely linked to income, education, and occupational status. Consequently, while some households rely on word-of-mouth within a very localized search space, some systematically pore over newspaper advertisements, while others base their search on wide-ranging neighborhood visits and video presentations, guided by real estate agents with computerized multiple listings.

The search is not simply a function of the range and quality of information, however. Households differ in the extent to which they face barriers and constraints during the search process. Lack of personal transportation is an obvious barrier that is directly related to socioeconomic status. More important, but less clearly a function of socioeconomic status, is the constraint of time. Each household has to balance (1) the probability of making a better decision after an exhaustive search against (2) the costs—both real and psychological—of doing so. Where time is limited, the anxiety produced by lack of success may result in a modification of the household's aspiration region, a restriction of its search

space, or a shift in its use of information sources. The pressure of time may also simply lead to confused or irrational decision making. Then again, the longer the search goes on, the greater the household's knowledge of the housing market, although there comes a point, as in shopping for anything else, where the cost of missed opportunities outweighs the value of additional knowledge.

Having identified possible alternatives, households must make a choice. The better defined the household's aspiration region, the more clear-cut this decision should be, though research has suggested that, having gone through the search process, most households do not refer rigidly to their original formulations. Rather, they are likely to compromise a good deal. In the final decision, neighborhood quality (and in particular the quality of neighborhood schools, for households with children of school age) seems to weigh more heavily than housing quality or accessibility, and interior style and appearance weigh more heavily than exterior style and appearance.[25] Households that are unable to find vacancies that fall within their aspiration region at this point (B in Figure 13.18) must revert to one of the two other possible courses of action—addressing the shortcomings of their existing dwelling or neighborhood, or redefining their aspirations and needs.

Empirical analyses of people's behavior in large and complex housing markets show that their search patterns tend to be quite restricted. Most households do not, for example, search anywhere near all of the housing submarkets within their affordable price range. Rather, they are likely to attempt to reduce search costs and uncertainty by searching within a single housing submarket. Searches tend to be anchored around familiar community areas and key nodes such as workplace locations, and most households are very persistent in restricting themselves to one or two preferred local areas.[26]

One important factor that is not explicit in Figure 13.18, is discrimination and bias within housing markets. Most scholars agree that this is an important factor in shaping patterns of residential location, along with affordability, job location, household preferences, and household information. There is, however, considerable debate as to just *how* important discrimination is within the overall context of the residential kaleidoscope.[27] The various *ways* in which bias and discrimination take place are discussed below (pp. 364–374).

CONCEPTS THAT LINK MOBILITY WITH NEIGHBORHOOD CHANGE

Several concepts help to make the connections between this model of individual household behavior

FIGURE 13.19   A neighborhood that filtered down the socioeconomic ladder. A neighborhood can start as a middle-income community and subsequently be "invaded" by a series of successively lower-income groups.

and the dynamics of neighborhood change. We have already referred to several of them. The concept of the **household life cycle**, for example, involves explicit links between residential mobility and housing and neighborhoods. In their classic form, these links tend to contribute to a zonal pattern of neighborhood differentiation based on sociodemographic cleavages (see p. 302). Ecological ideas concerning *invasion* and *succession* (p. 312) provide an additional conceptual basis that helps to account for links between residential mobility and the dynamics of ethnic clusters. Hoyt's emphasis on the roles of filtering and **vacancy chains** (p. 136), meanwhile, helps us both to understand the less dramatic changes to urban structure as households make marginal adjustments to their circumstances and to understand how individual household relocation decisions are inextricably related not only to neighborhood change (Figure 13.19) but also to the opportunities available to other households.

Another useful conceptual framework is that of *loyalty*, *voice*, and *exit* as alternatives to unwanted neighborhood change.[28]

People who don't like the way things are going in their neighborhood have three broad alternatives. First, they can resign themselves to the situation and, doing nothing, just tolerate it. . . . When problems are sufficiently bad or long-standing, however, they can (second) leave the neighborhood altogether, or (third) they can stay and make some attempt to change things.[29]

The first of these options is "loyalty," the second is "exit," and the third is "voice." They represent *collective* strategies that correspond to the three courses of action included in the household decision-making model discussed above (redefining aspirations, addressing housing deficiencies, and moving, respectively). Research suggests that it is the voice option that is generally preferred by both higher-income and lower-income residents who feel threatened by unwanted neighborhood change. This preference clearly has important implications for the relationship between neighborhood change and local politics, a topic that we pick up again in Chapter 16. For the moment, however, the conceptual framework provided by loyalty/voice/exit as collective options is important for two other reasons: It helps to emphasize the *interdependence* of household decision making, and it helps us to understand how neighborhood *stability* can occur in the face of powerful processes of urban change.

HOUSING MARKET GATEKEEPERS, BIAS, AND DISCRIMINATION

Decisions about where to live and when to move are constrained not only by income, barriers to the search process, and the interdependence of people's actions, but also by the decisions and behavior of key groups who influence the supply of housing and housing finance. We saw in Chapter 11 how speculators, developers, builders, realtors, and public officials, as well as individual households, can be regarded as "city makers."

Yet city making does not end once housing has been built and sold. The continual restlessness of people and capital constantly remakes the sociodemographic landscape in response to new needs and opportunities (and in so doing inevitably modifies the physical landscape). At the center of this process are the "exchange professionals" who facilitate residential mobility: realtors, mortgage financiers, insurance agents, appraisers, landlords, and so on.

Although these exchange professionals project themselves as being socially neutral, in certain circumstances they may act as "social gatekeepers," facilitating the residential mobility of certain groups and the financing of housing in certain neighborhoods but limiting the mobility of other groups and suppressing the capital available to other neighborhoods. This gatekeeping is not necessarily the result of conscious attempts to influence the pattern of flows of people and capital. In some circumstances exchange professionals may not even be aware of the gatekeeping consequences of their actions.[30] Given the importance of their roles in the marketplace, however, the dynamics and constraints of their professional environments make a certain amount of bias almost inevitable.

Every profession tends to develop its own distinctive professional ideology or view of the world, and individual professionals are influenced and guided by these perspectives in making their day-to-day decisions.[31] These professional ideologies are the product of a variety of factors, including

- selectivity in recruitment (including the self-selection that draws people with certain backgrounds, interests, and dispositions to certain careers).
- values, attitudes, and priorities imbued through education and training.
- the criteria of professional success sponsored by professional journals, magazines, and websites.
- the reward system and career structure.

Within every profession, however, there is also scope for deliberate discrimination: consciously wielding professional power or influence in order to deny members of particular groups access to housing in certain neighborhoods or to restrict the flow of finance to certain properties or neighborhoods. This discrimination may be personally motivated, but more likely it is the result of the exploitation of social prejudices for personal (or corporate) financial gain. Civil rights legislation has made such activity illegal in the United States, but most cities still bear the imprint of housing discrimination of the pre-civil rights era. In addition, discrimination remains endemic to American housing markets because it is very difficult to police.

REAL ESTATE AGENTS AS SOCIAL GATEKEEPERS

The term "social gatekeeper" was first used in a study of real estate agents in New Haven, Conn., in 1955.[32] The key to understanding the professional behavior of real estate agents is in their reward system, which is normally a commission based on a percentage of the sale price. Ideally, therefore, real estate agents would like to see a situation where house prices are high *and* there is a high turnover rate. This is by no means the usual situation, however, and in many circumstances high prices work against a high turnover or vice versa. In addition, real estate agents can find themselves working on behalf of both buyers and sellers in the same market, thereby not only dividing their loyalties but also complicating their attitudes toward asking prices. As a result, real estate agents develop a keen sense of the residential kaleidoscope as an opportunity structure that has to be managed and exploited carefully. Neighborhoods with high prices (and therefore high potential commissions) must be monitored for changes that might depress exchange values, neighborhoods with low turnover rates must be carefully monitored for signs of movement, and lower-priced neighborhoods must be monitored for signs of gentrification. It is a short step for some unscrupulous real estate agents (a small minority within their profession, it must be emphasized) to encourage or contrive such changes.

The principal activity in this respect has always been **steering**: keeping like with like, deterring households from moving into neighborhoods occupied by households of a different socioeconomic background or ethnicity in order not to jeopardize local prices. In 1955 real estate agents could openly explain steering in this way:

> People often try to get in higher class areas than they'll be accepted in. We just don't show them any houses in those areas. If they insist, we try to talk them out of it one way or another. I've purposely lost many a sale doing just that. It pays in the long run. People in the community respect you for it and they put business your way.[33]

The result was "slammed door" discrimination, mainly against African Americans. In 1968, Title VIII of the Civil Rights Act made such activity illegal. Discrimination continued, however. A 1989 report by the U.S. Congressional Research Service described how "revolving doors" have replaced "slammed doors":

> The new technique—apparently developed to evade statutory prohibitions against discrimination—consists essentially in deceiving a minority applicant into believing that he or she would be welcome as a tenant or

owner, but in withholding information about available housing. . . . In a sales situation, the applicant may be told the seller has already accepted another sales contract or told the owner had changed his mind and taken the dwelling off the market. Everything is done in such a way that the applicant will go away believing that nothing is available and without the slightest awareness that he or she has been discriminated against. The black or Hispanic applicant has been led in and then conducted out, so to speak.[34]

In 2000 the Urban Institute undertook a "paired testing" study on behalf of the U.S. Department of Housing and Urban Development (HUD), in which 4,600 pairs of testers—one minority and the other white—posed as otherwise identical home seekers and visited real estate or rental agents to inquire about the availability of advertised housing units that were for sale or rent in 23 metropolitan areas across the United States.[35] This methodology is designed to provide direct evidence of differences in the treatment experienced by minorities and whites in their search for housing. Because the Urban Institute conducted a previous study for HUD in 1989, we can measure the extent to which housing discrimination has changed in just over a decade.

The results of the 2000 study of metropolitan home sales markets showed that there had been a decline in the level of discrimination experienced by blacks and Hispanics seeking to buy a home between 1989 and 2000 (Table 13.3). During that period, however, steering on the basis of neighborhood

racial composition increased for black homebuyers—by 10 percent in the case of homes recommended and by almost 8 percent in the case of the homes that were inspected. So although black homebuyers faced less discrimination overall in 2000, blacks and whites were apt to be recommended and shown homes in different neighborhoods. Blacks were more likely to be steered to neighborhoods that were predominantly black compared to the neighborhoods recommended to comparable white homebuyers. Hispanic homebuyers, in particular, faced significant discrimination from real estate agents when it came to help in obtaining financing. Between 1989 and 2000, the overall incidence of white-favored treatment on financing assistance (offers of help with financing, lender recommendations, and discussion of down payment requirements) rose from 33.3 percent to 38.6 percent.

The results of the 2000 study of metropolitan rental markets showed that there had been a modest decrease in discrimination toward blacks seeking to rent a housing unit. Hispanic renters, however, did not experience this downward trend (Table 13.3). Hispanics, for example, were more likely to be quoted a higher rent for the same housing unit than their white counterparts in 2000 compared to 1989. In addition, Hispanic renters were more likely to experience discrimination in their housing search than African-American renters in 2000 (Table 13.3).

Steering and "revolving door" tactics are not the only kinds of intervention pursued by unscrupulous

TABLE 13.3

Discrimination by Real Estate and Rental Agents Against Minorities in U.S. Metropolitan Housing Markets, 1989 and 2000

	1989	*2000*	*Examples of discrimination*
Homebuyers			
Black	29.0%	17.0%	White homebuyers were more likely to be able to inspect available homes, to be shown homes in predominantly white neighborhoods, and to receive more information/assistance in financing as well as more encouragement.
Hispanic	26.8%	19.7%	White homebuyers were more likely to receive information/assistance in financing and to be shown homes in non-Hispanic neighborhoods.
Renters			
Black	26.4%	21.6%	Whites were more likely to receive information about available housing units, had more opportunities to inspect available units, and were more likely to be offered rental incentives.
Hispanic	23.7%	25.7%	Whites were more likely to receive information about available housing units, had more opportunities to inspect available units, and were more often quoted a lower rent for the advertised unit than similarly qualified Hispanic renters.

Source: *The Urban Land Institute,* Discrimination in Metropolitan Housing Markets: National Results from Phase 1 HDS 2000, *Washington, D.C.: U.S. Department of Housing and Urban Development, 2002.*

real estate agents. Local submarkets can be manipulated by **block busting**, whereby prices are deliberately driven down, temporarily, allowing real estate agents or their associates to buy up as many properties as possible before restoring equanimity to the market and then selling to a new group of purchasers. Prices can be driven down in various ways: by steering minority households *toward* a lower-income white neighborhood, for example, supplemented by scare tactics designed to hasten "white flight"—posting bogus FOR SALE signs, or hustling for listings. In extreme cases outsiders may be hired to commit petty acts of vandalism that are intended to signal the social deterioration of the neighborhood. Alternatively, neighborhood price decline can be "seeded" by purchasing houses only to leave them vacant and neglected. As residents see the neighborhood beginning to slip, more will put their homes on the market. When a sufficient number of dwellings has been acquired, the real estate agents or their associates will hope to derive a large profit, either by selling them one by one to members of an incoming group or by selling them all to a single developer seeking a large lot for a big project.

There is also evidence of more subtle forms of bias through advertising. An analysis of the print media advertising practices of a major real estate company in Milwaukee found that, compared with homes listed in white neighborhoods, those of comparable value (and with comparable levels of demand) listed in African-American neighborhoods were advertised less frequently in the Sunday metropolitan newspaper, less frequently selected as open houses, more often advertised with "one-line" entries, and less frequently described in favorable terms.[36]

MORTGAGE FINANCE MANAGERS AS SOCIAL GATEKEEPERS

Discrimination against minorities in the metropolitan home sales market may also arise in subsequent contacts involving applications for loans. Bank officers and savings and loan association managers tend to operate within decision-making frameworks that are fairly closely circumscribed by company policy, nationally established money market rates, and federal law. Nevertheless, these lenders still exercise a considerable amount of discretion in allocating mortgages, so that they are able to play a pivotal role in deciding "who lives where, how much new housing gets built, and whether neighborhoods survive."[37]

The chief allegiance of commercial lending institutions is not to borrowers but to investors. Funds are therefore not to be risked on loans to vulnerable households, for unconventional properties (since they may be difficult to sell if the borrower defaults),

or for properties in unpromising neighborhoods (also difficult to resell in the event of a default). Risk minimization means attaching a good deal of weight to creditworthiness, with special emphasis on income stability. This practice tends to favor salaried middle-income households while making it more difficult for wage earners, the self-employed, and female-headed households to secure a mortgage. Risk minimization also involves categorizing applicants and properties in terms of stereotypes. Since mortgage finance officers traditionally have tended to be predominantly white, male, middle-income and married with children, such stereotyping tends to work to the disadvantage of minority groups, female-headed households, other nontraditional household types, and people with unconventional lifestyles.

The easiest means of stereotyping properties is on the basis of neighborhood quality, drawing a line around "high-risk" neighborhoods and using this as a basis for determining loans. This practice, known as **redlining**, also results in a bias against minorities, female-headed households, and other vulnerable groups, since they tend to be localized in high-risk neighborhoods. A further problem with redlining is that it becomes a self-fulfilling prophecy, as neighborhoods starved of mortgage funds become progressively more run-down and so increasingly risky. We can summarize the consequences of redlining in terms of seven stages:[38]

1. The cheaper housing typical of redlined areas is effectively placed beyond the means of many lower-income households, since with regular mortgages difficult or impossible to obtain, loans can be secured only by means of higher down payments, higher interest rates, and shorter loan maturities.

2. The worst dwellings, denied mortgage loans even from more expensive sources, come to be left vacant and neglected, thus initiating a spiral of neighborhood decline that is difficult to stop.

3. The neighborhood declines to the point where even home improvement loans are difficult to obtain, leading to rapid physical deterioration and, in some cases, abandonment.

4. As conditions worsen, property insurance becomes expensive and difficult to get. Neighborhood businesses begin to fail, taking essential services and important sources of cash flow out of the neighborhood.

5. In the face of accelerating decline in property values, those who are able to do so begin to move away, leaving behind a residual population dominated by elderly and disadvantaged households.

6. The city treasury feels the results of neighborhood decline as property tax revenues decline while the demand for public services increases.

7. The entire city is drained of so much revenue that public services have to be cut back across the whole city or tax rates have to be increased, or both.

The severity of these problems, coupled with the fact that it was most often minority neighborhoods that were the worst casualties, led to a series of laws designed to eradicate redlining in the United States. These laws included the Civil Rights Act (federal Fair Housing Act) of 1968, the Home Mortgage Disclosure Act of 1975, the Equal Credit Opportunity Act of 1976, the Community Reinvestment Act of 1977, and the Fair Housing and Equal Opportunity Act of 1989. Nevertheless, many cities still carry the blight of neighborhood decline initiated (or at least reinforced) by redlining.

More importantly, it is clear that, regardless of the laws that prohibit discrimination against either people or places, bias and discrimination are a continuing element within U.S. housing markets. A study of mortgage applications in a sample of 19 cities in 1990 by the Federal Reserve Board showed that whites were approved for mortgages more often than African Americans or Hispanics *in every city and in every income group, in predominantly minority and in white neighborhoods, in rich communities and in poor.*[39] Black applicants for conventional purchase loans were nearly two and a half times more likely to be turned down

for a mortgage than white applicants; Hispanics were one and a half times more likely to be turned down (Table 13.4). The disparities in rejection rates persisted even when income was held constant. In fact, white borrowers in the lowest income category (20 percent or more below the median income of the census tract) were approved for home mortgages more often than black borrowers in the highest income group (20 percent or more above the median income of the census tract) (Table 13.4).

An analysis in 2001 of mortgage applications in a sample of 68 metropolitan areas by ACORN (Association of Community Organizations for Reform Now) produced results surprisingly similar to the 1990 Federal Reserve Board study (Table 13.4).[40] Although overall rejection rates were higher in 2001, the relative disparities were largely unchanged. In addition, as in 1990, the disparities in rejection rates in 2001 remained when income was held constant. In 2001, however, the disparities were higher among the high-income applicant groups—high-income black applicants for conventional purchase loans were nearly three times more likely to be turned down for a mortgage than their white counterparts; Hispanics were more than twice as likely to be turned down. This situation has implications for minority suburbanization.

The overall figures, of course, obscure local variations: Some banks rejected applications from African Americans as many as four times more frequently than those from whites.[41] There was also significant variation among metropolitan areas. In Milwaukee, for example, blacks were more than five times more

TABLE 13.4

Rejection Rates for Conventional Mortgage Applications for Home Purchase in the United States, 1990 and 2001

	Rejection rates 1990	White-Minority disparity 1990	Rejection rates 2001	White-Minority disparity 2001
(a) By ethnicity				
White	14.4%		17.2%	
Hispanic	21.4%	1.5	26.2%	1.5
Black	33.9%	2.4	39.7%	2.3
(b) By income and ethnicity				
Upper-income white	8.5%		7.4%	
Upper-income Hispanic	15.8%	1.9	15.2%	2.1
Upper-income black	21.4%	2.5	21.6%	2.9
Lower-income white	23.1%		24.7%	
Lower-income Hispanic	31.1%	1.3	32.1%	1.3
Lower-income black	40.1%	1.7	40.9%	1.7

Source: G. B. Canner and D. S. Smith, Home Mortgage Disclosure Act: Expanded Data on Residential Lending, *Federal Reserve Bulletin 77(11), 1991, Fig. 5, p. 870, Fig. 6, p. 872 (for 1990 figures); Association of Community Organizations for Reform Now (ACORN),* The Great Divide: Home Purchase Mortgage Lending Nationally and in 68 Metropolitan Areas, *Washington, D.C.: ACORN, 2002, pp. 17 and 21 (for 2001 figures).*

TABLE 13.5

Disparities Between Rejection Rates for Conventional Mortgage Applications by Ethnicity and Income
in Selected Metropolitan Areas, 2001

	White rejection rates	*Black rejection rates*	*White-Black disparity*	*Hispanic rejection rates*	*White-Hispanic disparity*
(a) All Incomes					
Milwaukee, Wisc.	5.7%	32.0%	5.6	16.4%	2.9
Memphis, Tenn.	10.5%	26.9%	3.5	30.8%	2.9
Hartford, Conn.	7.0%	24.1%	3.5	18.2%	2.6
Waterbury, Conn.	8.2%	23.5%	2.9	21.6%	2.6
Stockton-Lodi, Calif.	11.3%	20.4%	1.8	16.8%	1.5
Ft. Lauderdale, Fla.	11.9%	22.9%	1.9	15.9%	1.3
Ft. Worth-Arlington, Tex.	18.8%	40.3%	2.1	29.6%	1.6
(b) Upper-income					
Waterbury, Conn.	4.7%	28.6%	6.1	17.4%	3.7
Cleveland, Ohio	4.4%	26.2%	5.9	21.6%	4.9
Milwaukee, Wisc.	3.4%	16.1%	4.7	15.3%	4.4
Memphis, Tenn.	6.4%	26.9%	4.2	24.5%	3.8
Orange County, Ca.	10.0%	19.6%	1.9	15.3%	1.5
San Antonio, Tex.	10.1%	20.9%	2.0	16.7%	1.7
(c) Lower-income					
Chicago, Ill.	12.4%	40.6%	3.3	23.3%	1.9
Oakland, Calif.	18.4%	49.6%	2.7	32.7%	1.8
San Francisco, Calif.	20.9%	50.0%	2.4	61.4%	2.9

Source: *Association of Community Organizations for Reform Now (ACORN),* The Great Divide: Home Purchase Mortgage Lending Nationally and in
68 Metropolitan Areas, *Washington, D.C.: ACORN, 2002, pp. 13–15.*

likely than whites to be denied a conventional purchase loan in 2001; Hispanics were nearly three times
more likely to be denied (Table 13.5). In contrast, in
Ft. Lauderdale, Fla., blacks were slightly less than
twice as likely to be turned down as their white counterparts, while Hispanics were less than one and a half
times as likely to be turned down.

Although slightly lower overall, the rejection rates
for federally backed (**Federal Housing Administration [FHA]**, Veterans Administration [VA], and the
Farmers Home Administration [FmHA]) mortgages
showed similar patterns of disparities (Table 13.6).
This situation raises concerns because FHA loans
make up a disproportionate share of the financing
used by black and Hispanic homebuyers. In 2001 federally backed loans accounted for 42 percent of all
home purchase loans to black applicants and 38 percent of home purchase loans to Hispanics (compared
to 19 percent for white applicants). In addition, in recent years, HUD has identified abuse of the FHA program by lenders and sellers who have tried to take
advantage of first-time homebuyers and have used
FHA loans to carry out "property flipping" scams.
Property flipping involves purchasing distressed properties at a negligible price, and then, after minimal

cosmetic or even no work being done to the property, selling it at a price far above what it is worth. Victims of property flipping are often unsuspecting
low-income, minority first-time homebuyers.[42]

INSURANCE AGENTS
AS SOCIAL GATEKEEPERS

Discrimination against minorities in the metropolitan
home sales market may also arise in contacts involving
applications for home insurance. Discrimination in underwriting may seem less noteworthy than discrimination in lending, but its impact is no less profound. A
lender will not provide a mortgage unless a potential
homebuyer first obtains property insurance. The insurance policy is required because it minimizes the risk of
financial loss to the lender in the event that the house
is damaged or destroyed.

Households who experience problems of insurance availability tend to be located within central city
neighborhoods, often with high concentrations of
minority residents. Certain risk factors—such as
older wood frame homes, electrical and heating systems that have not been updated, and higher theft

TABLE 13.6

Rejection Rates for Federally Backed Mortgage Applications for Home Purchase in the United States, 1990 and 2001

	Rejection rates 1990	*White-Minority disparity 1990*	*Rejection rates 2001*	*White-Minority disparity 2001*
White	12.1%		7.0%	
Hispanic	18.4%	1.5	11.0%	1.5
Black	26.3%	2.2	12.8%	2.3

Source: *G. B. Canner and D. S. Smith*, Home Mortgage Disclosure Act: Expanded Data on Residential Lending, *Federal Reserve Bulletin 77(11), 1991, Fig. 5, p. 870, Fig. 6, p. 872 (for 1990 FHA, VA, and FHA (now FmHA (Farmers Home Administration)) figures); Association of Community Organizations for Reform Now (ACORN)*, The Great Divide: Home Purchase Mortgage Lending Nationally and in 68 Metropolitan Areas, *Washington, D.C.: ACORN, 2002, pp. 17 and 21 (for 2001 FHA, VA, FmHA figures).*

rates—are certainly greater for central cities than for suburbs.[43] A recent study of loss costs in eight major metropolitan areas in the United States published by the Insurance Research Council found that the frequency of claims, the size of the claims, and consequently industry costs per insured home were higher for urban than suburban policy holders.[44] At the same time, anecdotal and quantitative evidence points to the existence of discrimination on the part of property insurers based on racial stereotyping.

Systematically ascertaining the extent of discrimination is difficult due to the limitations of current data availability—the insurance industry is not subject to the federal disclosure requirements that apply to home lenders. Nevertheless, a recent statistical analysis of 33 U.S. metropolitan areas by the National Association of Insurance Commissioners, a trade association of state law enforcement officials who regulate the insurance industry, found that the racial composition of the neighborhood remained significant in the number and cost of policies after controlling for risk factors

covering loss experience and other demographic factors.[45] Paired testing of major insurers in nine cities by the National Fair Housing Alliance found evidence of illegal discrimination in the shares of tests: Chicago (83 percent), Atlanta (67 percent), Toledo (62 percent), Milwaukee (58 percent), Louisville (56 percent), Cincinnati (44 percent), Los Angeles (44 percent), Akron (37 percent), and Memphis (32 percent).[46] Fair housing organizations have filed a series of lawsuits and administrative complaints against some of the country's largest insurers (and lenders) that have resulted in settlements in the millions of dollars. Recent progress in addressing discrimination in underwriting include voluntary educational, mentoring, and outreach initiatives on the part of insurance companies and the establishment in 1994 of the National Insurance Task Force (NITF), comprising insurance companies, government regulators, and community groups, which has the mission of developing partnerships between the insurance industry and community groups.[47]

PUTTING IT ALL TOGETHER: THE EXAMPLE OF GENTRIFICATION

It should not be surprising—given what we have seen in previous sections about the differential physical deterioration of neighborhoods, fluctuations in the flow of investment capital, variability in household behavior, changing institutional constraints, and the influence of key gatekeepers—that neighborhood change is not easy to predict. For the same reasons, explanations of changes that *have* taken place are difficult to establish and are often hotly contested among academics, city planners, policy makers, and politicians.

Figure 13.20 brings together the major elements influencing neighborhood change in general terms.

Here, households are seen as one of a number of interdependent principal actors whose actions are set within the context of the financial climate, legal frameworks, public policies, institutional practice, and professional ideology. What Figure 13.20 cannot capture, however, are the relative importance and specific interactions of particular elements in relation to any given type or case of neighborhood change. Consequently, it is possible to "explain" neighborhood change in different ways, depending upon the emphasis given to different elements. In this section we examine the case of gentrification (Figure 13.21) as a particular type of neighborhood change, showing how various

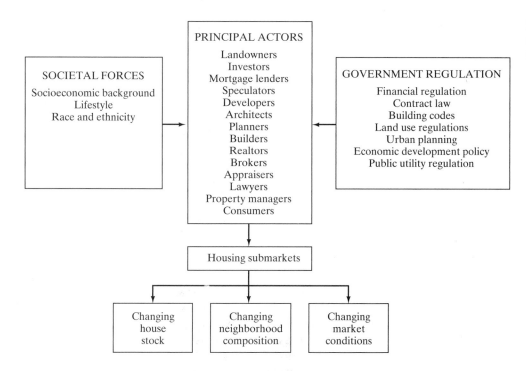

```
┌─────────────────────────────┐
│       PRINCIPAL ACTORS       │
│         Landowners           │
│          Investors           │
│       Mortgage lenders       │
│         Speculators          │
│          Developers          │
│          Architects          │
│           Planners           │
│           Builders           │
│           Realtors           │
│            Brokers           │
│          Appraisers          │
│           Lawyers            │
│       Property managers      │
│           Consumers          │
└─────────────────────────────┘

┌─────────────────────────┐        ┌──────────────────────────────────┐
│    SOCIETAL FORCES      │        │      GOVERNMENT REGULATION         │
│ Socioeconomic background│  →   ← │        Financial regulation        │
│        Lifestyle        │        │          Contract law              │
│    Race and ethnicity   │        │          Building codes            │
└─────────────────────────┘        │       Land use regulations         │
                                   │          Urban planning            │
                                   │   Economic development policy       │
                                   │      Public utility regulation      │
                                   └──────────────────────────────────┘

              ┌───────────────────────┐
              │   Housing submarkets   │
              └───────────────────────┘

   ┌──────────────┐   ┌──────────────────┐   ┌──────────────┐
   │  Changing    │   │   Changing       │   │  Changing    │
   │   house      │   │  neighborhood    │   │   market     │
   │   stock      │   │  composition     │   │ conditions   │
   └──────────────┘   └──────────────────┘   └──────────────┘
```

FIGURE 13.20   Actors and institutions affecting neighborhood change.

elements depicted in Figure 13.20 can be interpreted and how they in turn are connected to broader shifts in the trajectory of economic, social, cultural, political, and urban change.

Gentrification is simultaneously

> . . . a physical, economic, social and cultural phenomenon. [It] commonly involves the invasion by middle-class or higher-income groups of previously working-class neighborhoods or multi-occupied "twilight areas" and the replacement or displacement of many of the original occupants. It involves the physical renovation or rehabilitation of what was frequently a highly deteriorated housing stock and its upgrading to meet the requirements of its new owners. In the process, housing in the areas affected, both renovated and unrenovated, undergoes a significant price appreciation. Such a process of neighborhood transition commonly involves a degree of tenure transformation from renting to owning.[48]

Although gentrification has been most pronounced in **world cities** like New York and London and in regional **nodal centers** that have evolved

FIGURE 13.21   Houses in a gentrified neighborhood of the Richmond district of San Francisco—part of the streetcar suburb illustrated in Fig. 5.11.

from older urban cores, such as Minneapolis and Philadelphia in the United States or Manchester and Glasgow in the United Kingdom, by the early 21st century it has become virtually global in its incidence.[49] Research on gentrification in the United States suggests that it is increasing and involves between 1 and 5 percent of urban households and the displacement of an estimated 900,000 households each year.[50]

> Displacement imposes substantial hardships on some classes of displacees, particularly lower-income households and the elderly. Although some displacees report finding similar or improved dwelling units and neighborhoods, a substantial number report a deterioration in post-move dwelling units and/or neighborhood quality. Rents almost always increase, modestly for some households, substantially for others. Lower-income outmovers are particularly hit, finding the least satisfactory alternative units and neighborhoods and facing the highest proportional shelter-cost increases. For elderly displacees, the neighborhood studies show particular hardships.[51]

The significance of gentrification really lies in its qualitative, symbolic, and ideological implications for urban change. As an example of "Type I" neighborhood change (p. 345), it involves dramatic changes in neighborhood character, with a good chance for social conflict. Because it creates improvements to the built environment, encourages new retail activity, and results in the expansion of the local tax base without necessarily drawing heavily on public funds, it has become an important symbol and prospect for urban change for ideological conservatives. Because it fosters capital accumulation, caters to the consumption patterns of higher-income groups, and results in the displacement of vulnerable and disadvantaged households, it has become emblematic of urban restructuring and a portent of urban change for ideological liberals. Because it can be seen as the product of a wide variety of factors, it has become the focus of theoretical debates contrasting the effects of production and consumption, supply and demand, capital and culture, and gender and socioeconomic background in accounting for neighborhood change.

The scope of these debates is important, partly because it shows how the same causal factors can be seen very differently in relation to one another, but most importantly because it shows how the elements depicted in Figure 13.20 are in turn connected to broader contextual changes in the trajectory of economic, social, cultural, and political life—as depicted in Figure 1.4. The scope of the debates has been framed around two main sets of ideas. The first, associated in particular with the work of David Ley,[52] emphasizes the importance of occupational, social, and cultural shifts in influencing patterns of demand. This *humanistic* argument gives priority to human agency and consumer

preferences and can be paraphrased as follows. The increased pool of professional, administrative, managerial, and technical workers, together with the politicization of middle-income interest groups and the emergence of postmodern cultural sensibilities, has generated an emerging group of potential gentrifiers. Because many of these potential gentrifiers are employed in central city settings, and because their postmodern sensibilities lead them to reject modern homes in downtown apartments or suburban tracts in favor of settings with some history, human scale, and ethnic and architectural diversity, they are attracted to older inner-city neighborhoods. Once established in sufficient numbers, gentrifiers are able to consolidate their lifestyle through local political influence, enhancing their neighborhood and attracting still more gentrifiers by electing representatives who "deliver" better security, environmental improvements, and the preservation of historic buildings. Real estate agents and mortgage finance managers further reinforce the process by exploiting the trend while commercial investors reinforce the process through investments such as in upscale clothing boutiques, chic wine bars, and bookstores with coffee shops. The chief actors in this perspective, however, are the individual households who initiate and sustain the demand for "gentrified" settings.

In contrast, the alternative interpretation of gentrification, associated mainly with the work of Neil Smith,[53] sees the chief actors and pioneers as real estate agents and developers. This *structuralist* argument gives priority to the process of capitalist economic development and, in particular, moves by capital to arrest declining rates of profit and can be paraphrased as follows. The suburbanization of economic activity and households has steadily distorted the classical land value gradient (that fell steadily from the **CBD,** to the periphery of the city: see p. 133). In particular, land values in inner-city neighborhoods have fallen relative to the CBD and suburban nodes, creating a "valley" in the land value gradient. This valley intensified during the "Freeway and Metropolitan Sprawl" era (see pp. 150–157), resulting in the "devalorization" of inner-city neighborhoods, a situation in which the rent from land uses allocated under the market conditions of earlier (preautomobile) times was significantly less than the **ground rent** that could be obtained under new uses. This situation is the **rent gap** (see p. 276), the fundamental precondition for gentrification, which is then initiated by three types of developers: (1) professional developers who purchase property, redevelop it, and resell for profit; (2) occupier-developers, who buy and remodel property and inhabit it after completion; and (3) landlord developers, who rent property to tenants after rehabilitating it. The role of gentrifying households, meanwhile, is interpreted less in terms of individual consumer preferences and more in terms of socioeconomic relationships, intertwined with the dynamics of culture and politics. To varying degrees in different cities, gentrification has been incorporated into the **neoliberal policies** of city governments (see Chapter 17, pp. 514–518). Gentrification is therefore a back-to-the-city move by private-sector capital that is directly facilitated by public-sector policies. Smith sees this move—together with deregulation, **privatization**, and the other neoliberal reforms of the 1980s and 1990s—as a new form of revenge by the powerful in society for the moral and economic decline of city life following the social reforms of the 1960s (hence the term **revanchist city**—the French word *revanche* meaning revenge).

Drawing on both sets of ideas, three conditions can be posited as being necessary to explain the occurrence of gentrification:[54]

1. A pool of potential gentrifiers. This pool can be traced to the production and concentration of key fractions of professional, administrative, managerial, and technical workers in major cities around the world. It is argued that it is a product of the restructured social and spatial **division of labor** associated with the onset of **advanced capitalism** (see Chapter 4 for a review of this restructuring in relation to the U.S. **urban system**).

2. A supply of potentially gentrifiable inner-city property (the rent gap theory). But the existence of a rent gap does not necessarily lead to gentrification:

   > Without the existence of a pool of potential gentrifiers and available mortgage finance, gentrification will not occur however great the rent gap and however great the desire of developers to make it happen. And where appropriate housing stock does not exist in sufficient quantity, as for example in cities such as Dallas, Phoenix and other new southern and western U.S. cities, gentrification may be very limited.[55]

3. A degree of effective demand for inner-city property from potential gentrifiers. Preferences for inner-city settings

   > depend on both the growth of service class job opportunities downtown, and on demographic and lifestyle changes which have seen large numbers of women enter the labor force and growing numbers of both single households and dual career childless couples. For these groups, with a high disposable income, inner-city locations offer proximity to employment and to restaurants, arts and other facilities.[56]

Finally, it should be noted that other theorists put the emphasis elsewhere. Liz Bondi and Alan Warde, for example, both emphasize the *interaction of class and gender* in understanding gentrification.[57] From this perspective, the location of dual-earner households in the inner city is a solution to problems of access to work and home and of combining paid and unpaid labor for married middle-income women and men in well-paid career jobs. This development, of course, is related to other broad sociodemographic shifts, including the restructuring of family life that has been reflected in the postponement of child bearing, decreased completed family size, and closer spacing of children. These trends result in small, affluent households prepared to pay high prices for sought-after housing because they benefit more from the reduction in commuting costs associated with inner-urban residential locations than do those with only one adult working in the city center."

Based on their work on gay involvement in gentrification, writers such as Larry Knopp and Mickey Lauria have introduced sexuality to the equation.[58] They argue that the decline in traditional manufacturing employment in many North American cities, combined with the increase in administrative, managerial, and **producer services** jobs in the CBD, drew many first-generation and second-generation openly gay and lesbian people to inner-city neighborhoods. For a variety of reasons, gay people (especially gay men) have traditionally been disproportionately represented in these types of jobs. The economic draw of these jobs, combined with the heterosexism and homophobia experienced in suburban "family" neighborhoods, has made central cities very attractive to gay and lesbian people. In addition, the inexpensive and renovatable housing stock in depressed neighborhoods gave gay people (and again, particularly gay men, because of their higher earning power as male wage earners) the opportunity to develop a territorial and economic base for establishing a political voice and for developing community resources.

Writers such as Sharon Zukin and Rosalyn Deutsche, meanwhile, have emphasized the role of the avant-garde and the contextual shift toward a "society of the spectacle" (see p. 326) in which stylish materialism is increasingly important.[59] In larger metropolitan areas after the 1960s, they point out, postmodernism made the avant-garde less high-culture and elitist and more important as cultural intermediaries. Politicians, speculators, and developers came to see avant-garde art and culture as a crucial element in any new project. Meanwhile the avant-garde themselves, with a lifestyle that focuses very much on identity, appearance, presentation of self, fashion design, decor, and symbolism, have been critically important not only in pioneering the "resettlement" of rundown, low-rent areas but also in simultaneously providing these areas with the "designer" touch of radical chic necessary for them to be seen in a new light by the newly affluent and aestheticized professional service workers.

All writers on gentrification acknowledge that both economic and cultural processes are at work, so the crucial issue is *which factor is most important.* This might seem like some esoteric academic debate, but it has important implications, such as for neighborhood planning by city governments and for political action by community groups and individuals adversely affected by gentrification. If the forces of capital are seen as overwhelmingly dominant, as in the structuralist explanations, then human agency can achieve relatively little without wholesale reforms of the operation of capital markets. If, however, more importance is given to the autonomous role of cultural movements, as in the humanistic arguments, then these can influence the nature of capitalist development itself. What is clear from the research, however, is that the relative importance of economic and cultural factors varies in different cities. In Mexico City, for example, gentrification is not nearly as highly capitalized or widespread as in New York City. In the Caribbean, the increasing interrelationships between gentrification and global capital investment tend to filter through the tourist industry, giving gentrification there its own distinctive flavor.[60]

FOLLOW UP

1. Be sure that you understand the following key terms:

    action space
    awareness space
    blockbusting
    building cycles
    chain migration
    depreciation curve
    eminent domain
    exchange value
    generalized housing markets
    gentrification
    information space
    neighborhood life cycle
    redlining
    revanchist city
    spatial submarkets
    steering
    use value

2. Consider your own neighborhood or one that you know. How does it fit within the conceptualization of neighborhood life cycles described on pages 344–346? What signs of change can you see in the neighborhood? How is this change related to changes in nearby neighborhoods and to changes in the rest of the city?

3. List the residential moves that your family (or one that you know) has made within a city. How can these moves be explained (1) individually and (2) collectively?

4. Journals sometimes contain a series of articles that represent a back-and-forth academic debate about some important topic—such as one about gentrification between Chris Hamnett and Neil Smith. In this case Hamnett's initial paper reviewed the major theories of gentrification, Smith's commentary took issue with some of Hamnett's arguments (and to being characterized as a "blind man" whose theory only perceived part of the "elephant" of gentrification), and Hamnett then responded to Smith. Try to get a copy of these three articles and see what you think—do you tend to agree more with Hamnett's arguments or Smith's or do you find yourself somewhere in between? The articles are in the *Transactions of the Institute of British Geographers.* You should read them in chronological order: (1) Hamnett, "The blind men and the elephant: the explanation of gentrification" 16(1991): 173–89; (2) Smith, "Blind man's buff, or Hamnett's philosophical individualism in search of gentrification" 17(1992): 110–15; and (3) Hamnett, "Gentrifiers or lemmings? A response to Neil Smith" 17(1992): 116–19.

5. Update your *portfolio.* Remember to keep a note of any questions that are raised by your reading and to reflect from time to time on the overall framework provided by Figure 1.4. Now that you have covered more material, you may also find it useful to review earlier sections of your portfolio and perhaps add new material or observations.

KEY SOURCES AND SUGGESTED READING

Bier, T. 2001. *Moving Up, Filtering Down: Metropolitan Housing Dynamics and Public Policy.*Washington, D.C.: The Brookings Institution. (*http://www.brook.edu/es/urban/publications/bierexsum.htm*).

Bondi, L. 1991. Gender Divisions and Gentrification: A Critique. *Transactions of the Institute of British Geographers* 16: 190–98.

Burchell, R. W. and D. Listokin. 1995. Influences on United States housing policy. *Housing Policy Debate* 6: 559–617.

Cadwallader, M. T. 1992. *Migration and Residential Mobility.* Madison, Wis.: University of Wisconsin Press.

Clark, W. A. V. and S. D. Withers. 1999. Changing Jobs and Changing Houses: Mobility Outcomes of Employment Transitions. *Journal of Regional Science* 39: 653–73.

Cross, M. and M. Keith, eds. 1993. *Racism, the City and the State.* London: Routledge.

Doucet, M. and J. Weaver. 1991. *Housing the North American City.* Montreal, Canada: McGill-Queens University Press.

Feagin, J. and R. Parker. 1990. *Building American Cities.* Englewood Cliffs, N.J.: Prentice Hall. See especially Chapter 5 on gentrification.

Madanipour, A., G. Cars, and J. Allen. 1998. *Social Exclusion in European Cities: Processes, Experiences and Responses.* London: Jessica Kingsley Publishers.

Rusk, D. 2001. *The "Segregation Tax": The Cost of Racial Segregation to Black Homeowners.* Washington, D.C.: The Brookings Institution. (*http://www.brook.edu/es/urban/publications/ruskexsum.htm*).

Smith, N. 1996. *The New Urban Frontier: Gentrification and the Revanchist City.* London: Routledge.

Squires, G, ed. 1997. *Insurance Redlining: Disinvestment, Reinvestment, and the Evolving Role of Financial Institutions.* Washington, D.C.: The Urban Institute Press.

Squires, G, ed. 2003. *Organizing Access to Capital: Advocacy and the Democratization of Financial Institutions.* Philadelphia, Pa.: Temple University Press.

Farnsworth Riche, M. 2001. *The Implications of Changing U.S. Demographics for Housing Choice and Location in Cities.* Washington, D.C.: The Brookings Institution. (*http://www.brook.edu/es/urban/riche/richeexsum.htm*).

Urban Institute and Brookings Institution. 2003. *Rethinking Local Affordable Housing Strategies: Lessons from 70 Years of Policy and Practice.* Washington, D.C.: The Urban Institute and The Brookings Institution. (*http://www.brook.edu/es/urban/housingreview.htm*).

van der Vlist, A. J., C. Gorter, P. Nijkamp, and P. Rietveld. 2002. Residential Mobility and Local Housing-Market Differences. *Environment and Planning A* 34: 1147–64.

RELATED WEBSITES

U.S. Department of Housing and Urban Development (HUD): *http://www.hud.gov/*
HUD's website contains information and publications about housing and community development in the United States, including material on public housing, affordable housing, homeownership, the U.S. Fair Housing Act, and the U.S. Fair Lending Act.

National Housing Institute: *http://www.nhi.org/*
The National Housing Institute is an independent nonprofit organization that is concerned about issues related to affordable housing and community development in America. Its website includes current and back issues of *Shelterforce, the Journal of Affordable Housing and Community Building*, that contain articles on issues that affect affordable housing, such as jobs, education, poverty, unemployment, and disinvestment.

National Community Reinvestment Coalition: *http://www.ncrc.org/*
The National Community Reinvestment Coalition (NCRC) is a national nonprofit organization that seeks to unite the grassroots efforts of nonprofit organizations around the United States to increase the availability of private capital in traditionally underserved communities. The website offers free online access to its newsletter, *Reinvestment Works*, and to research and other information on national policy and legislation, including the U.S. Community Reinvestment Act, which outlaws the practice of redlining by banks and other financial institutions.

Public Broadcasting Stations (PBS): *http://www.pbs.org/pov/pov2003/flagwars/*
This PBS website was set up to accompany *Flag Wars*, a documentary about gentrification. The site contains information about gentrification, and interactive maps of the three neighborhoods in the video (New York's West Harlem, San Francisco's Mission District, and Olde Towne East in Columbus, Ohio) that feature interviews with residents as well as U.S. Census data and historical information about each neighborhood. There is also a web page offering insights from a group of community development professionals and activists concerning urban development and the displacement of long-time residents.

14

URBANIZATION, URBAN LIFE, AND URBAN SPACES

Urban spaces are created by people, and they draw their character from the people that inhabit them. As people live and work in urban spaces, they gradually impose themselves on their environment, modifying and adjusting it as best they can, to suit their needs and express their values. Yet at the same time people themselves gradually accommodate both to their physical environment and to the people around them. There is thus a continuous two-way process, a sociospatial dialectic, in which people create and modify urban spaces while at the same time being conditioned by the spaces in which they live and work. Neighborhoods and communities are created, maintained, and modified; the values, attitudes, and behavior of their inhabitants, meanwhile, cannot help but be influenced by their surroundings and by the values, attitudes, and behavior of the people around them. In addition, the ongoing processes of urbanization contribute economic, demographic, social, and cultural forces that must be accommodated within the sociospatial dialectic.

CHAPTER PREVIEW

In this chapter we explore the complex relationships between urban settings and individual and social behavior. The objective is to establish the main lineaments of the sociospatial dialectic in which people create and shape their environment while at the same time being conditioned by it. An additional objective is to show how this two-way relationship has changed as cities themselves have changed.

We will see that there is a long history of ideas concerning the effects of urbanization on people's behavior. We review the most important of these ideas, focusing in particular on the concept of the "moral order" of urban society and on the very influential theory that urban settings precipitate distinctive modes of social interaction and ways of life. This review leads to a closer examination of the issues in the context of different kinds of urban settings. We will see, for example, why we should expect quite different outcomes in low-income, inner-city neighborhoods than in affluent suburbs.

The relationships between society and space are further explored in relation to the phenomenon of "community." This exploration, in turn, leads to an examination of territoriality and of the implications of the ways in which people perceive various elements of their urban environments. All this background, in turn, helps in understanding the importance of people's everyday "lifeworlds." We will see how these lifeworlds are established through fundamental geographical parameters of individual "time-space" routines that help to structure the very fabric of society.

We conclude by examining one particular dimension of the sociospatial dialectic that tends to be obscured by emphases on specific social groups and specific kinds of settings: women's spaces and women's places in cities. With this theme we will also see how long-term changes in the political economy of urbanization have conditioned sociospatial relationships.

SOCIAL LIFE IN CITIES

Rural life is usually portrayed as "natural"—that is, close to nature. City life and city spaces are portrayed as "artificial." Rural society is portrayed as stable and neighborly; urban society is portrayed as volatile and individualistic. Rurality is associated with tradition and familiarity; cities are associated with novelty and variety. Overall, the experience of living in cities is characterized in rather negative terms. Art, literature, public opinion, and social theory have tended to portray cities as "necessary evils," places that offer economic opportunity and accessibility to a broad spectrum of amenities but that are somehow "unnatural"—their sheer size and density leading to personal and social stress. We have already seen how Jeffersonian notions of the virtues of pastoral life, Thoreau's emphasis on the importance of nature, and Turner's celebration of the frontier experience have exerted a powerful influence on American culture (see p. 248). Surveys of American literature and intellectual thought have shown how city spaces are most often depicted as arenas of conflict and incubators of loneliness, isolation, and deviance[1]—themes that continue to be reflected in movies, television series, and popular music.

At the same time, the catalytic qualities of urban spaces in generating and accommodating many of the innovative and expressive qualities of individual and social life are widely recognized. Much depends on the eye of the beholder. For some people rural settings provide human scale and a comfortable pace and cities present garish sights and overwhelming

and frenetic experiences. For others, rural settings are dull and suffocating, while cities offer exciting sights and experiences.

Making sense of this ambiguity and ambivalence has been a preoccupation of theorists ever since the urban transformation that produced the early industrial city (see p. 117). Amid slowly changing patterns of land use and residential segregation, each district and neighborhood provides a setting in which particular social and cultural groups act out their daily lives; there is an intimate relationship between space and society. People create and change urban spaces according to their needs, but at the same time the physical and sociocultural attributes of particular settings shape and constrain their behavior—both individually and collectively. The relationship between urban space and urban society can thus be characterized as a continuous **sociospatial dialectic**[2] in which each shapes and reshapes the other.

We must also recognize that, within the sociospatial dialectic of overall relationships between space and society, people create individual *lifeworlds* for themselves and for one another: ways of being, ways of seeing, and ways of knowing that are shaped by personal experience through daily routines in time and place.[3] Because of the general tendency for sociospatial segregation, urban spaces—districts and neighborhoods—often generate widely shared ("intersubjective"), taken-for-granted worlds with common horizons. But sociospatial segregation is never absolute, and some individuals will inevitably construe their lifeworlds in uniquely personal ways. As a result, the relationships between space, society, and ways of life are complex and contingent. As we will see, however, early and influential theorists emphasized the effects of the strangeness, artificiality, and individualism of urban settings on people's behavior: a deterministic perspective that has had a profound effect on ideas in urban geography, urban sociology, and all the related disciplines.

THEORETICAL INTERPRETATIONS OF URBAN LIFE

Our stock of theories concerning urban life stems from the work of European social philosophers such as Ferdinand Tönnies (1855–1935), Emile Durkheim (1858–1917), and Georg Simmel (1858–1918), who sought to understand the social and psychological implications of the shocking changes brought about by the unprecedented urbanization associated with the Industrial Revolution. At the heart of their analyses was an association between the scale of society and the nature of social organization. Their argument ran as follows. In preindustrial society, small, fairly homogeneous populations contain people who know one another, perform the same kind of work, and have the same kind of interests. They thus tend to look, think, and behave alike, reflecting a consensus of values and norms of behavior (the rules and conventions of proper and permissible behavior). In contrast, the inhabitants of large cities find themselves living at unprecedented densities and organized along new lines as a result of economic specialization.

In such a setting there is, inevitably, contact not only with more people but also with more *kinds* of people. But close relationships with family and friends are less easily sustained because of the fragmentation of life: the factory whistle and the office clock impose rigid new constraints on people's use of time, while the functional separation of land uses results in the spatial separation of home, workplace, and leisure activities. Family life becomes more circumscribed, and extended family systems have to give way to compact, nuclear families. At the same time, social differentiation brings about a divergence of lifestyles, values, and aspirations, thus weakening social consensus and cohesion and threatening to disrupt social order. This divergence, in turn, leads to attempts to adopt "rational" approaches to social organization, to the proliferation of formal social controls, and, where these attempts are unsuccessful, to an increase in social disorganization and deviant behavior.

THE "MORAL ORDER" OF CITY LIFE

The earliest theorists of urban life were, like all of us, products of their time. In wrestling with the implications of the emergence of huge concentrations of population, one of their central concerns was with the "moral order" of urban populations: the norms of behavior and patterns of social interaction that were clearly different (and, for many, disturbingly different) from the well-worn conventions, rigid social hierarchies, and close-knit groups of rural and small-town life. It was in this context that Ferdinand Tönnies, one of the founding fathers of sociology, established the framework for theories of urban life by conceptualizing patterns of human association in terms of a continuum marked by polar extremes.[4] At one end was the concept of *Gemeinschaft* (which can be translated as "community"), in which

- the basic unit of organization is the family or kin-group.
- social relationships are characterized by depth, continuity, cohesion, and fulfillment.
- people are bound together in caring, familial ways.
- controls over individual behavior are exerted through the informal discipline of family and neighbors.

At the other extreme was the concept of *Gesellschaft* (which can be translated as "society"), in which

- social relationships are founded on the rationality, efficiency, and contractual obligations stemming from patterns of economic organization.
- a large proportion of social interaction tends to be short-lived and superficial.
- people are bound together by formal ties to institutions and organizations.
- controls over individual behavior are exerted through impersonal, institutionalized codes.

Although Tönnies did not equate *Gemeinschaft* with rural life and *Gesellschaft* with urban life in simple or rigid fashion, the inference was clear: industrial urbanization brings a transition in the very fabric of social life. The same assumptions were elaborated by Emile Durkheim (Figure 14.1), another of the founding fathers of sociology.[5] His observations on the changing basis of the bonds sustaining human groups led to the idea of two forms of social "solidarity." The first, based on similarities between people, he called **mechanical solidarity**. The second, based on the existence of differences stemming from specialized economic roles, he

BOX 14.1 "SEX AND THE CITY": PROSTITUTION[6]

Cities have often provided opportunities to transgress moral codes, and one important manifestation of this has been prostitution—the granting of sexual favors in exchange for monetary reward (usually, but not exclusively, by women for men). This is often called the world's "oldest profession," but the actual term "prostitute" came into prominence only in the eighteenth century. In older societies sexual favors outside marriage were often granted by women who were courtesans, mistresses, or slaves. The crucial point is that these women were often known to those procuring the sexual favors.

With the development of large cities, prostitution changed in character in that the women and their clients frequently did not know each other. The reason for this change is fairly obvious; in small-scale agrarian societies, people were much more likely to be familiar with each other, whereas in cities there was a much greater chance of anonymity. In addition, economic destitution in the early industrial cities meant that prostitution was often the only effective means for some women to earn an income.

Prostitution was rife in many nineteenth-century cities. In London in the mid-nineteenth century, it has been estimated that there were 8,000 or more women making a living through prostitution.[7] Because it was against the law for women to stand in one place to attract clients, most solicited trade by walking the streets—as when Jack the Ripper was targeting prostitutes in London in 1888. Men of all social backgrounds used prostitutes, and shops in the Strand and Haymarket areas of London advertised "beds to let," often by the hour.

Despite the continual danger of physical assault from their clients and the very high risk of contracting sexually transmitted diseases, prostitutes could earn far more on the streets than they could through work in the poorly paid industries of the time. But a double standard operated—although prostitutes were frequently persecuted, the men who used their services were not. Prostitution, therefore, reflects patriarchal gender relations (the inequalities in power relations between men and women).[8]

Prostitution has continued to be a source of much conflict in the twentieth and twenty-first centuries. In the United States and the United Kingdom, for example, prostitution is found in distressed inner-city neighborhoods. Many of the residents object to the women seeking "custom" on their streets. This has led to community campaigns to expose "curb crawlers" (those who drive through these residential areas looking for prostitutes). Such campaigns can involve cooperation between the police and local community groups, although the result is often that prostitution is merely displaced elsewhere in the city.

Prostitution illustrates well the importance of recognizing the different "voices" in the city. It seems clear, for example, that the majority of women are coerced into prostitution through economic disadvantage and often experience considerable physical and psychological harm from both their pimps and clients. We should recognize at this point that the legal and social restrictions surrounding prostitution are full of hypocrisy and contradiction. In the United Kingdom, although prostitution is legal in principle, there are prohibitions on soliciting on the streets or living off "immoral earnings." As Duncan notes, widespread condemnations of prostitutes as being exploited by men or suffering from "false consciousness" exclude prostitutes from the freedom to control their own bodies in safe conditions free from police harassment.[9] It is little wonder then that the discourses surrounding prostitution also display ambiguity and contradiction: portraying women as either victims of sex-hungry predatory males or autonomous providers of social services.

---

called **organic solidarity**. This term was based on a biotic analogy, the idea being that in a complex organism (modern, urban society), every single "organ" (social group) is mutually dependent on the rest.

Anomie and Deviant Behavior

Durkheim believed that the transition from mechanical to organic solidarity as the dominant form of social organization was a direct result of three factors: the sheer number of people involved in urban settings, the density of urban residential settings (remember that he was writing only just after the introduction of the streetcar), and the dynamism of cities, sustained by improved transportation and communication systems, which put larger and larger numbers of different sorts of people in contact with one another. It is important to note that Durkheim did not see this idea of the "dynamic density" of urban life as inherently bad. The loss of the mechanical solidarity provided by family and neighborhood would, he believed, be compensated for by the growth of organic solidarity provided by the workplace. He did, however, recognize that there may be circumstances in which the regulation and peaceful

FIGURE 14.1　Emile Durkheim (1858–1917), one of the founding fathers of social science. His work provided important insights about urban life and its effect on individual and group behavior.

coexistence of organic solidarity might break down, leading to a social condition of **anomie**. The term literally means "normlessness," but it is more accurate to think of situations in which the norms of personal and social behavior are so weak or muddled that some people become socially isolated, confused, or uncertain about how to behave, while others are easily able to challenge or ignore social conventions. One of the consequences of such circumstances is an increase in **deviant behavior**, one aspect of which Durkheim himself wrote about in his 1897 book *Suicide*.[10]

In contrast to the emphasis given by Tönnies and Durkheim to the contextual effects of urban life on social organization, Georg Simmel believed that all social relationships and all forms of social organization stem from basic human drives, interests, and psychic states. As a result, according to Simmel, people enter into relationships in pursuit of a variety of objectives: survival, acquisitiveness, status, power, erotic and spiritual instincts, and so on. People's lives, he concluded, are acted out amid tensions and conflicts between their individual creativity and the various institutionalized forms of social interaction and social organization that have evolved in order to fulfill generalized human needs. In this context, Simmel saw urbanization as both constraining and liberating. The constraints to self-fulfillment and creativity, he argued, had two main dimensions.

One dimension was the intensification of objectivity and rationalization that accompanied the evolution of an industrialized, urbanized society. In particular, the emergence of a highly developed money economy creates conditions that foster changes in the way people are able to relate to one another, their associational life, their work, and their culture. This, he argued, rewards intellectual over emotional development and threatens to dry up the wellsprings of the mind and self.

The other dimension of modern urban society that threatened to constrain self-fulfillment and creativity concerns the intensification of external stimuli. These include the bombardment of the psyche by the sheer numbers of people and the fast pace of city life (dynamic density again) and the bewildering plurality of styles accommodated by city life—compounded by their transitory and illusory nature.

The overall effect of these two constraining dimensions of urbanization, Simmel suggested, was the emergence of a metropolitan "way of life." In his 1905 essay on "The Metropolis and Mental Life"[11] he pointed to several key attributes of this way of life. Blasé attitudes and antipathy toward others emerge in response to the *psychic overload* generated by the unexpected, discontinuous, and rapidly changing stimuli and situations encountered in the city. At the same time, reserve and privacy emerge in response to the need for contemplation and inner reflection. But these defenses against overstimulation make it difficult for others to express their subjective personalities, so that a third distinctive component of the metropolitan way of life is fostered: heightened and exaggerated forms of display, from dress to architecture.

Liberating Aspects of Urban Life

It is here that we can see the liberating side of urbanization: The "lonely crowd" created by dynamic density of city life *allows* reserve and privacy to flourish; the plurality of styles accommodated by city life gives full scope to the variety of human creativity. The aestheticization of everyday life had been in fact a central preoccupation of some nineteenth-century theorists and commentators. Following the novelist Honoré de Balzac (1799–1850), they pointed to the emergence of artistic countercultures, the *bohème* and *avant-gardes* of metropolitan Paris, Berlin, and Vienna as intermediaries in stimulating, formulating, and disseminating a new sense of aesthetics and consumption. According to this perspective, urban life is infused with an intoxicating and constantly changing flow of commodities and sensations.[12] Department stores and shopping arcades become elements in a dream world of consumption; city streets become stages upon which people can act out multiple roles; crowds become audiences for the *flâneur*—the stroller wanting to see and be seen, the dandy "who makes of his body, his behavior, his feelings and passions, his very existence, a work of art."[13]

When theorists turned their attention to the urban experience in the mushrooming cities of the United States, however, they paid little attention to the liberating dimensions of urban life and the individuality of human creativity. Rather, it was the negative aspects of urban life that came to be writ large. This emphasis was in large measure due to the influence of Louis Wirth (1897–1952), a German immigrant who had studied under Simmel and subsequently came to study sociology at the University of Chicago. His particular distillation of ideas in an essay entitled "Urbanism as a Way of Life," published in 1938,[14] was to become one of the single most influential treatises in urban sociology, environmental psychology, and urban geography—even to the point of straitjacketing many people's ideas about urban life.

URBANISM AS A WAY OF LIFE

Writing from the same perspective as Tönnies and Durkheim, drawing heavily but selectively on the ideas of Simmel, and strongly influenced by colleagues in the Chicago School of Human Ecology (see p. 311), Wirth developed a deductive theory that was based on what he saw as the three fundamental attributes of urbanization:

1. The increased number of people.
2. The increased (physical) density of living.
3. The increased heterogeneity of populations.

Each of these attributes, he reasoned, gives rise to a slightly different cluster of outcomes. Together, the outcomes are mutually reinforcing, generating a distinctive web of attitudes, behavior, and social organization that characterize **urbanism**: a way of life that is an inevitable consequence of urbanization. Although Wirth's theory involved a complex interweaving of hypothesized relationships, the outcomes can be summarized in terms of changes that affect individuals and changes that affect society at large.

Wirth's deductions led him to the conclusion that the combined influence of increased size, density, and heterogeneity of urban populations would affect individual behavior because people had somehow to cope with a great number and variety of physical and social stimuli. The overall result, he argued, was that city life leads people to become more withdrawn, emotionally buffered to the point where they are naturally aloof, brusque, and impersonal in their dealings with others. Furthermore, the general loosening of interpersonal bonds through this adaptive behavior tends to leave people *unrestrained*, so that ego-centered and unconventional behaviors are fostered. They are also left *unsupported* in times of crisis, leaving them vulnerable to neurosis, alcoholism, suicide, or some form of deviant behavior.

At the same time, Wirth draws a parallel picture of social change associated with the increased size, density, and heterogeneity of urban populations. Economic competition and the division of labor provide the basis of spatial differentiation that is reinforced by impulses of social attraction and social avoidance, creating a mosaic of social worlds and resulting not only in the segregation of different income groups, ethnic groups, and household types but also the *fragmentation of social life* between home, school, workplace, friends, and family. Wirth allowed that this segregation and fragmentation tend to foster a general increase in tolerance. At the same time, though, the fact that people have to divide their time and attention among highly differentiated people and places weakens the support and control provided by primary groups such as family, friends, and neighbors. Moreover, this trend is reinforced by the weakening of social norms that is a consequence of the divergent interests, lifestyles, and cultures associated with the increasing heterogeneity of urban populations.

The overall societal response is to replace the support and controls formerly provided by primary social groups with "rational" and impersonal procedures and institutions (welfare agencies, criminal codes supported by police forces, etc.). According to Wirth, however, such an order can never adequately replace a communal order based on consensus and the moral strength of small primary groups. As a result, urbanism brings a loosening of the social fabric, allowing ego-centered and unconventional behavior to flourish and allowing personal crises to snowball into social problems.

Wirthian theory thus points to the inevitability of situations and settings in which city life—or, to be precise, life for certain people in certain parts of the city—breaks the bounds of convention and normality, often to the disadvantage not only of the people and neighborhoods involved but also of urban society as a whole. This aspect of urban life can be developed in several ways. The idea of psychological overload resulting from complex or unfamiliar environments has been investigated by environmental psychologists and was popularized by Alvin Toffler in his best-selling book *Future Shock*.[15] Toffler used the notion of "future shock" to describe the outcome of the need for people in modern (urban) settings to "scoop up and process" more and more information. Scientific studies of people's response to such overstimulation have shown that several different strategies are common. One involves the elimination from perception of the most unwelcome aspects of reality. In its most extreme form, it can result in the construction of a mythological world that becomes a substitute for the real world.

In such a world the person concerned may consider odd behavior to be "normal." Others, of course, all too easily view that person as just another of the city's eccentrics.

This analogy points to another, more common strategy: managing several distinct roles or identities at once. According to Wirthian theory, this strategy is characteristic of urbanism because of the geographical and functional separation of the "audiences" to which different roles are addressed: family, neighbors, coworkers, and so on. People are consequently able to present very different "selves" in different social contexts. The city, with its wide choice of different settings, roles, and identities, becomes a magic theater (Figure 14.2). Unfortunately, the anonymity afforded by the ease of slipping from one role to another not only helps to relieve the stresses and strains of urban life but also facilitates deviant and criminal behavior. Furthermore, the strain of having to sustain different and perhaps conflicting roles or "selves" over a prolonged period may, for some people, contribute to mental illness or to deviant behavior.

The most widespread strategy for many people, however, is simply to withdraw, to become aloof, and to mind their own business. The social psychologist Stanley Milgram documented various manifestations of this strategy,[16] the most striking of which is the collective paralysis of moral order reflected in the lack of bystander intervention when something clearly antisocial, deviant, or unlawful occurs. The most widespread examples are those involving rudeness, petty theft, or vandalism. The most vivid examples are those involving violence. Milgram cites the highly publicized example of the murder of Catherine Genovese, who was stabbed to death in a respectable district of Queens, New York. Her death was reputedly witnessed by more than 40 people, none of whom attempted even to call the police. The behavior of these witnesses was itself deviant in a way; but the significance of such behavior from the point of view of Wirthian theory is in the way that it can foster the spread of more serious forms of deviancy by eroding social responsibility and social control.

What has been of special interest to many geographers and sociologists is the clear variability in the amount and type of deviant behavior from one neighborhood to another. Deploying Wirthian ideas, an explanation can be traced to the much greater size, density, and heterogeneity of the populations of certain areas, particularly those that are low-rent neighborhoods receiving the brunt of low-income migrants and immigrants. Here, the effects of urbanization go far beyond the generalized notion of a way of life characterized by aloofness, withdrawal, ego-centered behaviors, and the substitution of impersonal and institutionalized controls for self-control and social constraints. In these neighborhoods, the argument runs, the combined stresses of size, density, and heterogeneity lead not only to increases in anomie and deviant behavior but also to a more widespread breakdown of moral order, social solidarity, and social control: a condition known as **social disorganization**.

THE PUBLIC AND PRIVATE WORLDS OF CITY LIFE

Although such interpretations of urban life have proven popular and remarkably durable, they are only partially sustained by the evidence of scientific analysis. They are also vulnerable to criticism on the grounds that human beings are in fact more flexible and inventive than allowed for within the deterministic cause-and-effect logic of Wirthian theory. Experiments designed to measure people's willingness to be

FIGURE 14.2   Actors in the "magic theater" of the city. Two punkers argue with an evangelist on a sidewalk in Camden Town, London.

helpful and cooperative, together with studies of certain categories of crime and interpersonal conflict, tend to support the Wirthian scenario, with successively larger towns and cities exhibiting less helpful and more conflictual and deviant behavior.[17] At the same time, studies that have compared the number or quality of friendships and personal relations have found no difference between different-sized settlements, while experiments focusing on people's psychological states have found that stress and alienation are just as prevalent, if not more so, in rural and small-town communities.

One way of reconciling the apparent inconsistency of these findings is to reexamine the notion of urban settings, recognizing a distinction between the public and the private spheres of life in urban settings (Figure 14.3). The public spheres are characterized by situations in which people are strangers, situations that require a special etiquette: reserved, careful, nonintrusive. People must be—or at least appear to be—indifferent to others. But this is *situational behavior, not a psychological state*. From this perspective the city dweller "did not lose the capacity for the deep, long-lasting, multi-faceted relationship. But he (sic) *gained* the capacity for the surface, fleeting, restricted relationship."[18] Sociologist Claude Fischer has drawn on this distinction, suggesting that urbanism can be regarded as involving distrust, estrangement, and alienation from "other people" in the wider community but not from neighbors or associates at work or school. In his words, "Urbanism produces fear and distrust of 'foreign' groups in the public sphere, but does not affect private social worlds."[19] In Wirthian terminology this view means that urbanism can accommodate both moral order and social disorganization.

CHANGING METROPOLITAN FORM AND NEW FORMS OF URBANISM

It should be remembered that Wirth's observations and theories were based on a very specific phase of urbanization: one fueled by European immigration, driven by industrialization, and only just beginning to feel the effects of low-density decentralization made possible by buses, trucks, and automobiles. The emergence of a new metropolitan form (described in Chapter 6), characterized by polarized central city cores surrounded by sprawling, low-density **urban realms** of **neo-Fordist** development, raises the possibility of a new form of urbanism based on new ways of life—urban and suburban certainly, but also those associated with **splintering urbanism**.

The changes brought about by suburbanization, by the intensification of economic differentiation, and by the proliferation of lifestyles have made it increasingly difficult and unrealistic to think in terms of entire urban or metropolitan populations being shaped in their behavior by variations in the size, density, and heterogeneity of their surroundings. Meanwhile, it has become increasingly apparent that the ecological parameters of socioeconomic background, occupational role, family structure, ethnicity, and lifestyle (described in Chapter 12) operate as powerful determinants not only of patterns of marriage and relationships, friendship,

FIGURE 14.3   The sidewalk is usually a neutral zone in urban space, part of the "public world" shared by people of different backgrounds.

Box 14.2 Homosexuality and the City[20]

Homosexuality involves emotional and sexual attraction between people of the same sex. Homosexuality has existed in all cultures, with sodomy long prohibited by religious teaching, although not marked out as a particular homosexual act. It was only in the mid-nineteenth century that the term homosexuality was devised, thereby denoting homosexuals as a distinct and separate group. The association of the term with persecution in the twentieth century led many to prefer alternative terms, such as "gay" or "queer" (the usually pejorative word "queer" being deliberately used in a somewhat ironic way to acknowledge the repressive character of social constructions of homosexuality).

Cities have had a profound impact on the development of homosexuality. In general, cities have provided greater anonymity and tolerance of alternative lifestyles compared with the hostility toward gays and lesbians manifest in rural communities. Since the 1960s and the civil rights movement, there have been profound changes in the nature of homosexual spaces within cities that reflect broader social and political changes in society. Together with the race riots, student protests, and anti-Vietnam War demonstrations of the 1960s in the United States, there were a number of gay protests. The most famous of these took place in New York in 1969 when gays and lesbians rioted after a police raid on the Stonewall Inn. Following this, the Gay Liberation Front (GLF) was formed in New York in 1971 to make vocal gay demands for equality. Arguably, there is now greater tolerance of homosexual activity by "straight" segments of society, although the extent of this tolerance should not be exaggerated—the ongoing political struggles to make gay marriages and civil unions legal in the United States being a case in point. Nevertheless, the hidden and covert character of gay activity in some areas has been replaced by distinctive residential districts composed of substantial proportions of gay people in which gay lifestyles are explicitly displayed.

Weightman was among the first to draw attention to areas in U.S. cities with distinctive gay lifestyles.[21] He noted that gays were playing a leading role in the process of **gentrification** in some inner-city neighborhoods, often displacing the poorer residents. Without doubt, the most famous of these residential districts is the Castro district of San Francisco. The origins of this district can be traced back to World War II. Gays and lesbians serving in the armed forces were often discharged in San Francisco and preferred to establish homes in the city rather than returning to face the prejudice of their home communities. Another well known gay space is the Marigny district of New Orleans. As in San Francisco, this district is located in a culturally mixed, relatively tolerant area. Knopp[22] describes how property developers and speculators rapidly took advantage of the gay population's demand for property in this area. In particular, they exploited the **rent gap** (p. 276) by artificially inflating the values of properties through bribing private appraisers. Paradoxically, the resultant influx of middle-income gays caused many of the existing middle-income gays in the area to become concerned over the preservation of historically important areas in the city rather than any broader issues affecting the gay community.

Male homosexuality has received much more analysis in urban geography than lesbianism, partly because gay residential areas are generally more visible and easily mapped and studied. Women in general have fewer financial resources than men, and they also face the threat of male violence. Lesbians therefore share a desire for relatively inexpensive housing as well as a concern with personal safety. These factors, combined with the pressures of a predominantly heterosexual society, mean that lesbian residential areas are less overt than gay spaces. Nevertheless, lesbian spaces in cities have been mapped and analyzed, including housing cooperatives and bars, restaurants, nightclubs, cinemas, and bookshops with provision for lesbians.[23]

Such mapping of gay spaces in cities has extended our knowledge, but these studies have also been criticized in recent years. In particular, it has been argued that the work on *gay ghettos* and *enclaves* have been limited in their contribution to an understanding of the processes that lead to such spatial clustering.[24] To begin with, most of these areas cannot be thought of as exclusively gay. In addition, many gays and lesbians live outside these parts of the city. The crucial point is that such a focus on gay spaces tends to conceptualize them as different when, in reality, *all* spaces in cities are socially constructed in a sexualized manner. For example, research has focused on the processes through which certain public spaces, such as hotels and restaurants, become constructed as heterosexual.[25] Even the domestic environment in the home is constructed around a

(Continued)

BOX 14.2   HOMOSEXUALITY AND THE CITY   *(Continued)*

heterosexual-based notion of sexual reproduction (see the Box entitled "Discrimination by Design: Domestic Architecture and Gender Differences" later in this chapter).[26]

It has been suggested that the spatial concentration of gay people would, as in the case of ethnic minorities, provide a base for political mobilization against repression and discrimination. However, many have questioned the efficacy of such a separatist strategy. It has been argued that such gay ghettos and enclaves have helped to maintain the notion of gay lifestyles as separate, different, deviant, and sinful. In the wake of HIV and AIDS, however, such areas have assumed a new role as

focal points for support networks and healthcare services.

Yet another change in gay spaces in recent years has been the commercialization of these areas as entrepreneurs have sought to exploit the high incomes of some gay households. For example, the Soho area of London has developed into a thriving gay commercial scene in what has been described as the world's Pink Capital. Clubs, bars, and shops catering explicitly to gays have heightened the visibility of gay lifestyles but have tended to cater to specific types of gays—the young and wealthy—with older and poorer gays more likely to be excluded. So a new form of economic division has been imposed on divisions based on sexuality.

and social interaction but also of personality development, sociopolitical attitudes, and cultural values. From this perspective the urban mosaic thus sustains many different "ways of life," while the significance of "urbanism" is relegated to the notion of the overall variety implied by these relationships.

Urban Villages

One of the first writers to recognize the links between ecological subareas and ways of life was sociologist Herbert Gans. Gans emphasized the cohesion and intimacy of distinctive social worlds based on ethnicity, kinship, neighborhood, occupation, and lifestyle, strongly rejecting the idea that urban life in any way diminishes these social worlds. His conclusions were based on an intensive study of the Italian quarter in the West End of Boston. So intense was the cohesion there (and so different from the prognoses of Wirthian theory) that Gans titled his book *The Urban Villagers*.[27]

Studies of "urban villages" in other cities have found the dominant "substructures of social interaction" to be cast in terms of socioeconomic status as often as race or ethnicity; but the common denominators are those of older, central city neighborhoods with a long history of immobility. These attributes give us a strong clue as to the necessary conditions for the emergence of a "village" type of social world. It is the relative immobility—personal, occupational, and residential—of lower socioeconomic groups (of all ethnic backgrounds) that is the key. Immobility results in a strengthening of "vertical bonds" of kinship and "horizontal bonds" of friendship.

The typically high degree of residential proximity (see p. 297) between family members traditionally in low-income areas like Boston's Italian quarter not only made for a greater intensity of interaction between

kinfolk but also facilitated the important role of the matriarch in providing practical support for family life—looking after grandchildren, for example—and in passing on attitudes, information, beliefs, and norms of behavior. Primary social interaction between friends is also reinforced by the residential proximity that results from immobility. Relationships formed among a cohort of children at school are carried over into street life, relationships, and, later on, leisure (Figures 14.4 and 14.5). Finally, the shared and repeated experience of hard times can foster the development of close-knit

FIGURE 14.4   "Village" life in the city. Shared spaces and shared experiences with familiar faces can promote a villagelike community atmosphere.

FIGURE 14.5   Street play can be an important dimension of neighborhood social life.

and overlapping social networks in low-income areas. With everyone "in the same boat" and little population turnover, there develops a mutuality of feeling and purpose that can become the mainspring of the social institutions, way of life, and "community spirit" associated with urban villages.

This community spirit can be a brittle phenomenon, however. The stresses and tensions that follow from economic insecurity and deprivation underlie the mutuality of urban villages. Studies of life in urban village-type neighborhoods have described a good deal of conflict and disorder, and the chief single cause seems to be the absolute lack of space. Crowding leads to noise problems, inadequate play space, and inadequate laundry facilities and is associated with personal stress and fatigue. Children especially are likely to suffer from the psychological effects of the lack of privacy that inadequate space imposes. The links between crowding, stress, and deviant behavior are difficult to establish scientifically, however, particularly since so many other potential causal factors are present.[28]

Such stresses and tensions would seem to point to the operation of Wirthian processes. But Herbert Gans has suggested that urban villagers develop a *subculture* that short-circuits Wirthian determinism. Gans defines a subculture as

> an organized set of related responses that had developed out of people's efforts to cope with the opportunities, incentives and rewards, as well as the deprivations, prohibitions and pressures which the natural environment and society—that complex of co-existing and competing subcultures—offer to them.[29]

It follows that, rather than thinking of urbanism as a way of life, we should think of urbanism as fostering distinctive ways of life for different social/demographic/economic/spatial groups. Claude Fischer has taken this argument a step further, suggesting that these ways of life, or subcultures, will be intensified by the conflict and competition inherent to urbanization and that new subcultures will spin off as new social groups (generated by the arrival of new migrants and immigrants and by the evolution of new **class fractions**) reach the "critical mass" required to sustain cohesive social networks.[30] Although Fisher did not address the spatial framework associated with these networks, others have pointed out that their continued existence depends to a large extent on avoiding conflict with other subcultures. Conflict can be avoided by the development of implicit behavioral limits: a sort of "social contract" between subcultures. But it seems to be best avoided by means of spatial segregation. Thus conflict avoidance provides a further dynamic that contributes to the development of urban villages.

The idea of subcultures and ways of life fits conveniently with the long-standing concept of **cultural transmission**, whereby distinctive norms of behavior, sometimes norms that would be seen as different by society at large, are passed from one generation to another within a local environment. This concept was given prominence by sociologists of the Chicago School in the 1930s and 1940s, who were particularly interested in those "natural areas" of the city in which the intergenerational transmission of cultural values and attitudes seemed to foster criminal and delinquent behavior.[31]

Also relevant here is the concept of **neighborhood effects**, whereby people tend to conform to what they perceive as local norms in order to gain or maintain the respect of their local peer group. Evidence from studies of voting behavior and attitudes toward education, as well as studies of criminality and delinquency, tends to support the idea that people act and think in ways that are not always in their best interests objectively but that, rather, conform to the opinions and images of the neighborhood population with whom they interact. One example of such neighborhood effects is the paradoxical syndrome of "suburban poverty." This is a term used to describe the result of neighborhood effects that impose middle-income consumption patterns on some incoming households whose incomes are not quite sufficient to service a new mortgage *and* "keep up with the Joneses" but who nevertheless feel obliged to conform with their neighbors' levels and patterns of consumption.

Suburban Communality, "Habitus," and Postmodern Lifestyles

The first interpretations of suburban life, following the traditions of Tönnies, Durkheim, Simmel, and Wirth, saw suburbia as the manifestation of one particular side of urbanism: the adaptive withdrawal from the "psychic overload" of the city. Lewis Mumford, for example, wrote that American suburbs represented "a collective attempt to lead a private life."[32] Early sociological studies of suburban life endorsed this view.[33]

But subsequent investigations and, indeed, the evolution of suburban and splintering urbanism settings, have shown this perspective to be incomplete. Once again it was Herbert Gans who took the lead, demonstrating in his study of Levittown that suburban neighborhoods did in fact exhibit localized social networks with a considerable degree of cohesion.[34] Several factors can be identified as contributing to this cohesion, including the social and demographic homogeneity of suburban neighborhoods, the physical isolation of outlying neighborhoods, and a "pioneer eagerness" to make friends in newly developed tracts. The cohesiveness of suburban communities is further strengthened by social networks related to local voluntary associations of various kinds (such as parent-teacher associations or charitable organizations) and to the interaction between parents before, during, and after organized activities for children (team sports, dance classes, cheerleading practice, band and orchestra recitals). This purposive kind of cohesiveness has been described as producing *communities of limited liability*. According to this perspective, suburban neighborhoods represent just one version of a series of different kinds of communities of limited liability, others being based around workplace, political, lifestyle, or other interests. So instead of suburbanization leading to the breakup of communities, it can be thought of as breaking down communality into an increasing number of semi-independent groups, only some of which are locality-based.

This interpretation allows us to accommodate the increasing importance of lifestyle as a basis for residential segregation, as discussed in Chapter 12. French sociologist Pierre Bourdieu has pointed to the close relationships between class and lifestyle thorough his concept of **habitus**. A social group has a habitus if it has a distinctive set of values, ideas, and practices: a collective perceptual and evaluative schema *that derives from its members' everyday experience* and operates at a subconscious level, through commonplace daily practices, dress codes, use of language, and patterns of consumption. The result is a distinctive pattern in which "each dimension of lifestyle symbolizes with the others."[35] This schema, suggests Bourdieu, evolves in response to certain objective conditions (including class and lifestyle) but once established serves to reproduce and sustain those conditions. It is the subtlety of the habitus of different groups that provides important clues to city dwellers about social distance and group membership; and it is their richness and variety that lend vitality to city life and give color and texture to the residential kaleidoscope.

As we saw in Chapter 12, however, one of the most significant trends in terms of lifestyles was the influence of the superficial, transitory, and eclectic tendencies of postmodern culture during the 1970s and 1980s, in particular. In this new kind of cultural context, it is argued, people have a greater capacity to explore a much wider spectrum of sensations and experiences. Through settings such as shopping malls, museums, galleries, movie theaters, and theme parks and, above all, through media such as glossy "lifestyle" magazines, television, and the Internet, an aesthetic sensibility is created in which masses of people come together in temporary emotional communities.[36] These are rooted in the materialism and hedonism of the "society of the spectacle" (see p. 326) and are centered on intermittent periods of empathy and shared self-imagery that are attached to the symbolic properties of particular urban settings, collective experiences, or material possessions. In this context the sense of belonging to a lifestyle community was often exploited (and sustained)—in the 1980s especially—by commercial advertising: the "Pepsi Generation," the "United Colors of Benetton," and the "New Generation of Olds," for example. The result was a blur of ephemeral, affective communities, fluid *postmodern lifestyles* brought together around the icons of materialistic lifestyles, the inconspicuous consumption and more casual attitudes toward fashion in the 1990s notwithstanding (see p. 327).

Implicit in the notions of habitus, lifestyle communities, and communities of limited liability is the idea that the phenomena they are describing somehow lack the mutuality, the permanent but intangible community spirit that is characteristic of the urban village. It should not be assumed, however, that community spirit cannot develop from such settings. When territorial exclusivity, amenities, or property values are threatened, the low-key, purposive communality of suburban America can flare into an intense mutuality and a tangible sense of community spirit. This reaction has been described as *status panic* or *crisis communality*. Rarely, however, does it survive the resolution of the perceived threat. This phenomenon emphasizes yet again the complexity of relationships between society and space: It is not at all easy to say which situations reflect the existence of "community," let alone which of those are also congruent with a particular territory.

COMMUNITY AND TERRITORY

As we have seen, the classical sociological theories of Tönnies, Durkheim, Simmel, and Wirth hold that communities should not exist at all in contemporary cites; or, at the most, that they exist only in a weakened form. This theoretical perspective can be summed up as *community lost*. The view that urbanization actually fosters a variety of cultures, leading to the development of urban villages, can be summed up as *community saved*. The view that urbanization creates the possibility for people to participate in one or more of a number of communities of limited liability, only some of which are place-based, can be summed up as *community transformed*. The view that contemporary urbanization fosters ephemeral, emotional communities based on the icons of materialistic lifestyles can be summed up in terms of *community commodified*.

Apart from anything else, these different views make it necessary for us to be very careful in using the terms community and neighborhood. They are probably best used simply as general terms that cover a range of situations relating to place, space, and social interaction. It is useful to think in terms of a loose hierarchy in which the base unit is the *neighborhood*—a territory that contains people of broadly similar demographic, economic, and social characteristics, but that is not necessarily very significant as a basis for social interaction. We can perhaps think of some lifestyle clusters, such as "Pools and Patios" and "Towns and Gowns" described in Chapter 12, as building blocks in which the sociospatial dialectic may or may not foster the kind of social interaction that is associated with *community*, the next level in the hierarchy. Communities (in this sense) exist where a degree of social coherence develops on the basis of interdependence, which in turn produces a degree of uniformity of custom, taste, and modes of thought and speech. They are thus taken-for-granted worlds defined by reference groups that *may* be territorially based but may equally be school-based, work-based, or lifestyle-based.

The third tier of the hierarchy is represented by *communality*, a form of community *at the level of consciousness*. This is where a degree of community spirit is apparent, where a sense of "us" and "them" is prevalent. Like community, however, communality does not necessarily have to be based on territorial cohesiveness. So, as the U.S. National Commission on Neighborhoods pointed out, each neighborhood (or community) "is what its inhabitants think it is."[37] This assertion means, in turn, that definitions and classifications of neighborhoods and communities must depend on people's conditions and perceptions of social and physical space and on the social and geographical scales of reference they use—to which we now turn.

Cognition, Perception, and Mental Maps of the City

People use the filters of their own personal experience, knowledge, and values to deal with the stream of stimuli that they encounter in urban settings. The result is that they modify their real-world, objective experiences, creating in their minds a series of partial, simplified (often distorted), and flexible inner representations, or "mental maps." These inner representations form the basis not only for various aspects of people's behavior, but also for the experiential sense of place and sense of being that are central to community and communality.

Because mental maps are the product of personal knowledge, experience, and values, the same social and physical stimuli may evoke different responses from different individuals, each person effectively living in his or her "own world." Figures 14.6 and 14.7, for example, illustrate the two very different "worlds" of two residents of Philadelphia and Washington, D.C., the organization and content of their sketches suggesting differences in lifestyle and familiarity with different parts of the city. Analyses of sketches of this sort have shown that, in general, they vary in two ways: (1) in terms of objective accuracy and (2) in terms of the type of element emphasized, the main difference being between those who emphasize *sequential* elements (such as roads and pathways) and those who emphasize *spatial* elements (such as individual buildings, landmarks, and districts). Like the creators of Figures 14.6 and 14.7, most people orient their mental maps around their own neighborhood or

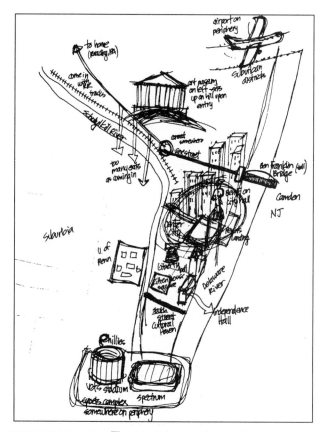

FIGURE 14.6   The mental map of a resident of Philadelphia.

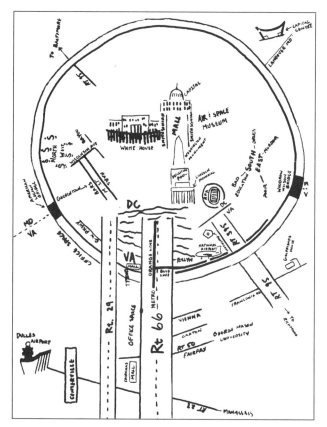

FIGURE 14.7   The mental map of a resident of Washington, D.C.

around the **central business district** (**CBD**), and most tend to tidy up and simplify reality.

One of the key elements in shaping people's mental maps is the **cognitive distance** between the elements that they include in their maps. Cognitive distance is the basis for the arrangement of the spatial information in mental maps, while **social distance** (see p. 297) is the basis for the arrangement of much of the rest of the information. Cognitive distance is generated from the brain's perception of the distance between visible objects, a perception that is influenced by land use patterns, the distinctiveness of objects, and the impact of symbolic representations of the environment such as maps and road signs.

In general, cognitive distance seems to be a function of the number, variety, and familiarity of *cues* or stimuli that are encountered along commonly used pathways through the city. These pathways constitute the framework of a person's *action space*. A simple illustration of the effect of familiarity on spatial cognition is the way that the return leg of a new trip usually seems shorter. Elasticity is also an important attribute of cognitive distance, something that is illustrated by the fact that many people judge the same physical distance in different ways, depending on orientation and context. Thus, for example, the cognitive distance

from a suburb to a downtown shopping area is often less than the cognitive distance from the downtown to the same suburb. Similarly, cognitive distance to attractive features such as parks and amenities tends to underestimate the actual physical distance, while cognitive distance to unattractive features such as parking lots and run-down neighborhoods tends to overestimate actual distance.

Although cognitive distance is the basis for the arrangement of the elements that people include in their maps, the elements themselves are what give mental maps their distinctiveness. People do not have a single image or mental map that can be consulted or recalled at will. Rather, each person appears to be able to draw upon a series of latent images that are unconsciously operationalized in response to specific spatial tasks: navigating across town, walking around a shopping district, or searching listings for new homes, for example. These latent images, like the cognitive distances between the elements they contain, are largely a function of people's action space, but they are also derived in part from a broader *information space* or *awareness space* that is built up from a variety of sources and experiences: TV and radio coverage, the Internet, newspaper reports, books, conversations, and so on. As these sources of information change, so the latent images change. Meanwhile, as people's tasks change, so different images are called upon. In this context, a useful distinction can be made between the *designative images* that relate to the cognitive organization of space and the *appraisive images* that represent various feelings and judgments about particular components of the city.

Following the work of urban designer Kevin Lynch, we can recognize five principal elements in people's designative mental maps of cities: paths, edges, districts, nodes, and landmarks.[38]

1. Paths are the channels along which the observer customarily, occasionally, or potentially moves. They may be streets, walkways, transit lines, railroads. For many people, these are the predominant elements in their image. People observe the city while moving through it, and along these paths the other environmental elements are arranged and related.

2. Edges are the linear elements not used or considered as paths by the observer. They are the boundaries between two phases, linear breaks in continuity: shores, railroad cuts, edges of development, walls. They may be barriers more or less penetrable, which close one region off from another; or they may be seams, lines along which two regions are related and joined together.

3. Districts are the medium-to-large sections of the city, which the observer mentally enters, and which are recognizable as having some common, identifying character.

4. Nodes are the strategic spots in a city into which observers can enter, and each is an intensive focus to and from which they are traveling. Nodes may be primarily junctions, places of a break in transportation, a crossing or convergence of paths, or they may simply be concentrations which gain their importance from being the focus of some use or physical character, such as a street-corner hangout or an enclosed square. Some of these nodes are the focus and embodiment of a district, over which their influence radiates and for which they stand as a symbol (Figure 14.8).

5. Landmarks are another type of point-reference, but in this case the observer does not enter them. They are usually a rather simply defined physical object: a sign, a clocktower, or a tall or otherwise distinctive building.

Lynch derived these five elements from a series of interviews and an examination of people's sketch maps. He then pioneered the construction of maps of *aggregate* perceptions of urban space, using lighter or darker symbols to indicate the proportion of people who had mentioned or sketched a particular element in recalling their own individual mental map. Figure 14.9, an example of this kind of map, shows the collective designative image of Los Angeles held by the residents of Westwood, an affluent neighborhood located between Beverly Hills and Santa Monica. This map illustrates very clearly the tendency for people to simplify and exaggerate, but it also shows that these tendencies do not result in neat and tidy representations. Districts are structured with nodes, defined by edges, penetrated by paths, and sprinkled with landmarks.

Lynch used his work to focus on the *legibility*, or imageability, of cities, believing that people would prefer to live in more legible environments. These, he argued,

FIGURE 14.8   Pioneer Courthouse Square, a "node" that has been planned into the structure of downtown Portland, Ore.

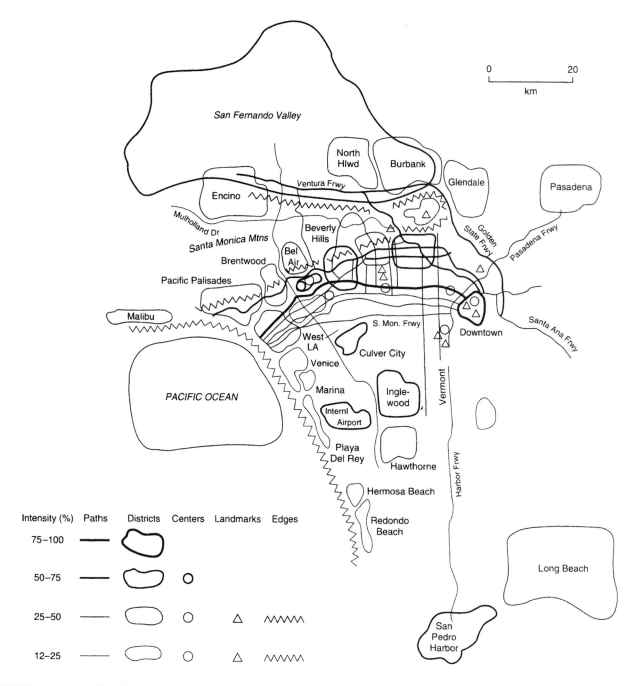

FIGURE 14.9   The designative image of Los Angeles, as derived from a sample of residents from Westwood, an affluent middle-income community.

are those settings where built form is clear and straightforward, with continuity and rhythm in paths and edges, where districts are clearly separated, and where there are dominant nodes and landmarks. So, for example, Lynch suggested that a city like Los Angeles, with few of these qualities, would be less legible (and therefore less livable) than a city like Boston, with its highly structured elements.

Subsequent research has provided broad support for such an interpretation. It has also suggested, however, that it is unwise to generalize too much about mental maps. Urban environments can be legible in different

ways to different people. Some cities are legible merely because of the dominance of a clearly connected set of paths, others because of the clarity of a series of landmarks. Some people prefer *illegible* environments, possibly because of abstract qualities such as "quaintness" or "intimacy." Most important of all from the point of view of relating people's mental maps to their behavior (and thus being able to shed more light on the processes involved in the sociospatial dialectic) is that different groups—by age, gender, socioeconomic background, and neighborhood—tend to have very different mental maps of the same environment. This

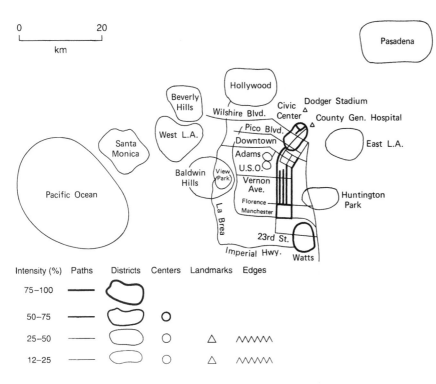

FIGURE 14.10   The designative image of Los Angeles, as derived from a sample of black inner-city residents.

difference is the result, of course, of differences in their respective action and awareness spaces. Figure 14.10 illustrates the kind of difference that can result; the collective designative image of Los Angeles held by low-income black inner-city residents contrasts noticeably with that of Westwood residents depicted in Figure 14.9. The world of the black residents was much more limited geographically—restricted by a lack of disposable income, dependence on a public transportation system that is focused on the urban core, and discrimination. Note the prominence of the corridor between the downtown and the Watts neighborhood.

The collective designative image of the low-income Hispanic residents of Boyle Heights (Figure 14.11) contrasts strikingly even with that of the black residents. The residents of Boyle Heights, a low-income, Spanish-speaking neighborhood, were isolated from the rest of the city both by language barriers and by geographical immobility. Their world barely extended to the city as a whole. Less than half of a sample of residents from the neighborhood had designative images that included elements from beyond the immediate neighborhood, and those elements included only one district (the downtown) and three landmarks: City Hall, Union Station, and the bus depot.

Appraisive Images

In many circumstances it is not the sophistication of these designative images that is important so much as the feelings that people attach to specific elements in their mental maps. A specific node or district, for example, may be regarded as attractive or unattractive, exciting or dull; or, more likely, it may evoke a combination of feelings. These reflect the *appraisive images* that people draw upon in order to steer their way through the city, that they use in making decisions about where to live, shop, or play, and that they refer to in making subjective judgments about neighborhoods and their residents. One very vivid and specific aspect of appraisive imagery concerns people's perceptions of threats to personal safety. In larger cities certain districts are generally perceived to be dangerous. At a more detailed scale, residents familiar with such neighborhoods develop "fear maps" (Figures 14.12 and 14.13) that highlight danger points near gang hangouts, abandoned buildings, crack houses, drug markets, and so on.

In overall terms, appraisive imagery can be thought of as the product of three main evaluative dimensions:[39]

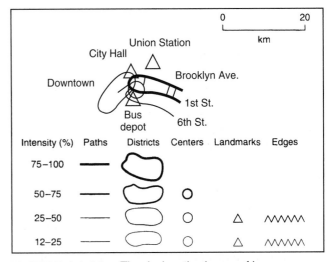

FIGURE 14.11   The designative image of Los Angeles, as derived from a sample of residents from Boyle Heights, a poor Hispanic neighborhood.

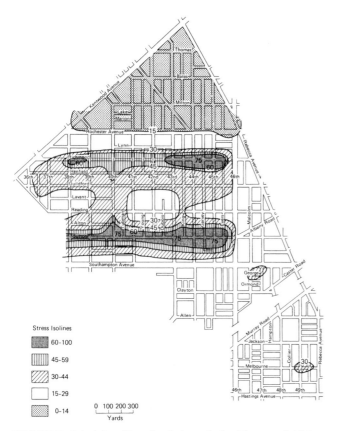

FIGURE 14.12   Perceived stress in the Monroe district of Philadelphia (the darker the shading, the higher the stress).

1. The *interpersonal environment* that is based on the social attributes of different parts of the city and that is particularly influenced by perceived social distance, by perceptions of the friendliness, helpfulness, snobbishness, pride, power, contentment, and so on of different groups, and by perceptions of the tranquility and safety of different settings.

2. The *impersonal environment* of streets and buildings that is based on the aesthetics of particular settings, their tidiness, level of maintenance, amenity value, and traffic noise.

3. The *locational* attributes of different parts of the city—accessibility to freeways and transportation systems and proximity to amenities, attractive settings, or potential nuisances.

Perhaps the best-developed aspects of appraisive imagery concern people's images of their *home area*. They are also the most important aspects of appraisive imagery in the present context since they relate directly to the issues of community and territory. As we might expect from a reading of Gans and Fischer, plenty of evidence points to the existence of some kind of neighborhood attachment among people who have lived in a particular area for any length of time. Research has shown that some 80 percent of people are able to draw upon a mental model, or schema, of the area in which their daily lives are played out and that

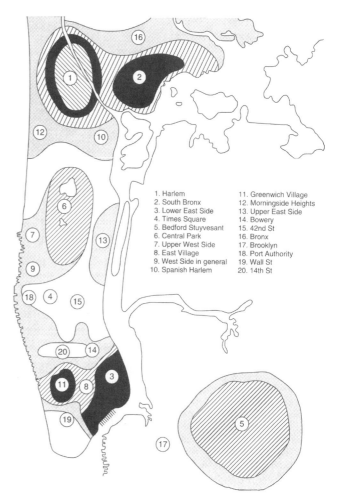

1. Harlem
2. South Bronx
3. Lower East Side
4. Times Square
5. Bedford Stuyvesant
6. Central Park
7. Upper West Side
8. East Village
9. West Side in general
10. Spanish Harlem
11. Greenwich Village
12. Morningside Heights
13. Upper East Side
14. Bowery
15. 42nd St
16. Bronx
17. Brooklyn
18. Port Authority
19. Wall St
20. 14th St

FIGURE 14.13   The geography of fear in New York. The darker the shading, the more widespread the fear among New Yorkers of that particular part of the city. The numbered locations list, in rank order, the 20 "most fearsome" neighborhoods.

they identify with as their home area. Although these schemata are highly personal and idiosyncratic (with neighbors varying quite a great deal in the boundaries they place on their particular home area), they exhibit some interesting regularities that provide important clues about the relationships between community and **territoriality**. First, the area covered by people's schemata is broadly consistent at about 100 acres. Second, this territorial extent is not influenced by variations in population density: the home areas of suburban residents cover the same amount of "turf" as central city residents. Social psychologist Terence Lee has developed a hierarchical topology, based on analyses of people's responses to questions about their home area:[40]

- *The social acquaintance neighborhood*—a small area within which, apart from a few close ties of kinship or friendship or both, people keep to themselves.

- *The homogeneous neighborhood*—a schema that extends to a greater awareness of the physical aspects of the setting as well as the social aspects of neighbors. The homogeneous neighborhood comprises not only "people like us" but also "people who live in homes like ours."

- *The unit neighborhood*—a larger territory, a "named" district: "our neighborhood as others see it." Unit neighborhoods usually contain a variety of shops and amenities and some variability in socioeconomic background and household types. They approximate planners' and politicians' conceptions of urban neighborhoods.

This topology not only reflects the importance of *spatial scale* in relationships between community and territory but also points to the multidimensionality of people's sense of place. We now turn, therefore, to a closer examination of the affective bonds between people and places, looking in particular at the importance of localized lifeworlds and the significance of the symbolism associated with particular sociospatial settings.

LIFEWORLDS AND THE "STRUCTURATION" OF SOCIAL LIFE

The home-area schemata described above provide an important framework for many aspects of daily life in cities. This dimension of urbanization, however, can be understood only through insights that are derived from the insiders' point of view. It requires an appreciation of how meanings are assigned to people and places, how people obtain and process information about people and places, how they "learn" sociospatial settings, and how they react to changes in them. Put another way, it requires "the contextual interpretation of subjectively meaningful social action," a "form of empathetic understanding gained from the adoption of the subject's own perspective."[41] Using this approach, it becomes apparent that every place, every locality within the city is socially constructed. That is, they are given meanings that are contingent on the people who use them and live in them. Cultural geographer Yi-Fu Tuan puts it this way: "Place is a center of meaning constructed by experience."[42] Thus it is possible for localities to have "multiple realities," as different groups of people construct different meanings around them. As social geographer David Ley emphasizes, "Place is relational to a purpose and an intent. As such, any place has a *multiple reality* for the plurality of social groups that encounter it."[43]

A good starting point in understanding these insiders' views is the concept of the *lifeworld*, the setting (or settings) for everyday life, lived reflexively, with natural, unreflective attitudes and behaviors constituting a taken-for-granted social world (Figure 14.14) and a meaningful personal context in which every object has its purpose and every place its significance.[44] It follows that lifeworlds form as a result of established routines—in the workplace, around the home, around a set of park benches, in a bar, at a club. But in order for individual routine to become a collective lifeworld, a certain amount of *intersubjectivity* is required. People need to have access to one another's feelings and meanings in order for the experience of everyday life to be a shared experience. This intersubjectivity is constructed and maintained in specialized systems of communication and symbolization. Conformity to a given code of communication and symbolization—through styles of speech, specialized vocabulary, dress codes, use of personal space and body language, use of humor, unspoken rules of interaction, etiquette, and rituals—defines an in-group (members of the lifeworld) and an out-group (the rest). Such conformity requires the conscious suppression of individualistic traits, and it brings about the *un*conscious suppression of individualistic and idiosyncratic traits and the development of cognitive and attitudinal consensus.

A useful concept in this context is the notion of a *structure of feeling* developed by Raymond Williams.[45] It is a concept that incorporates both the sense of

FIGURE 14.14   The taken-for-granted lifeworld. This photograph captures something of the distinctive lifeworld of part of Harlem in New York City.

Box 14.3   Disability and the City[46]

In most studies of cities the able-bodied character of people is taken for granted, leaving little room for the analysis of disability.[47] In recent years, however, geographical and other social scientific research has begun to address issues of disability.[48] A key issue is what is meant by disability. Rather than being a physical defect affecting particular individuals, some argue that disability is above all a **social construction**. So the social construction of "disability" is bound up with the attitudes and structures of oppression in an able-bodied society, rather than the failing of a particular individual.

In societies that view the role of medicine as primarily one of making sick people well, disability is often seen as something unhealthy. But if sufficient facilities were provided, being disabled could be regarded as something akin to being shortsighted— wearing glasses or contact lenses not generally being regarded as a disability. This is a good example of how social scientific research can challenge taken-for-granted assumptions. What seems like common sense can often be just a highly mistaken function of the way we have been taught to think about the world.

The socially constructed character of disability was highlighted in a classic study of attitudes toward blindness. In the United States blindness was regarded as an experience of loss requiring counseling; in the United Kingdom as a technical issue requiring aids and equipment; and in Italy as a need to seek consolation and salvation through the church and religion.[49] Different societies, therefore, produce differing definitions of disability. In the United States, for example, following the politicization of disability rights, a multimillion-dollar disability industry has emerged. Disability has, therefore, become a major source of income for doctors, lawyers, rehabilitation professionals, and disability activists.

Cities frequently contain numerous barriers to mobility and access for disabled people. Typical problems include high curbs, steep steps, no ramps for wheelchairs, narrow doors, and the absence of information in Braille. In addition, elevators for those who are disabled are often badly signposted, with inaccessible buttons, and in unattractive locations (such as when they are the service elevators next to kitchens). Public transportation systems can also pose problems for people with disabilities. Often the problem boils down to one of cost, with inadequate funds being made available for adequate conversion of premises for disabled access.

The problem is also one of dominant attitudes in society.[50] All too often, architects, planners, and the public at large have assumed that disability leads to immobility. Consequently, the needs of disabled people are ignored. The barriers are therefore social and psychological as much as physical. Cost constraints on the provision of disabled access therefore reflect wider sets of social values toward disability. These barriers prevent disabled people from fully participating in social life and in paid employment. A study in Ontario, Canada, found that 80 percent of disabled people lived in relative poverty because of their exclusion from the job market and limited support programs from both the public and private sectors.[51] Census figures indicate that people of working age (16 to 64) in the United States were less likely to be employed if they were disabled. Although in 2000 about 80 percent of working-age men without a disability were employed, only about 60 percent of those with a disability worked. Among women of working age, the respective employment rates were just over 67 and 51 percent.[52]

Most recently, in many cities the issue of disability has been taken more seriously. In the United States, the Americans with Disabilities Act of 1992 requires businesses to provide wheelchair access. In the United Kingdom, although progress is patchy, many local governments now have Disability Officers whose task it is to improve disabled access in city centers. These improvements are in part a response to the increased activities of various disability rights movements. These groups have begun to campaign on issues of income, employment, civil rights, and community living rather than the older issues of institutional care.

As with all **urban social movements**, there have been setbacks as well as advances. For example, in the United States corporate interests have lobbied before the judiciary that the Americans with Disabilities Act is an unnecessary restriction on private property rights and therefore an infringement of the Fifth Amendment of the Constitution.[53] Similarly, in the United Kingdom the Thatcher governments of the 1980s progressively relaxed accessibility standards for disabled people.[54] In the late 1990s the Labor government set about reforming the benefits system to limit entitlements for disabled people in an effort to encourage disabled people to take up paid employment, given that an estimated 75 percent of disabled adults relied on some form of government support.

place and the views of the world that stem from the intersubjectivity of the lifeworld, but it extends to conscious attitudes and feelings about the future as well as the past and present. These feelings add up to a "structure" in the sense that they comprise an interlocking set of cognitive elements. As cultural geographer Peter Jackson observes, the idea of a structure of feeling has more than a passing similarity with Pierre Bourdieu's concept of habitus (see p. 388).[55] In practice, however, both depend on intersubjectivity that is maintained through shared codes of symbolism and communication, and these in turn depend heavily on the existence of established routines in time and space.

TIME-SPACE ROUTINES

The importance of established routines in facilitating the development of codes, intersubjectivity, and conformity has been emphasized by geographer David Seamon, who has used the concept of *time-space routines* to describe the existential importance of repetitive situations for individuals and the concepts of *place ballets* to capture the unfolding of multiple time-space routines within a single setting.[56] The most useful approach to time-space routines, however, comes from the work of Torsten Hägerstrand and the Lund School of Geography in Sweden. In simple but effective fashion, Hägerstrand has captured the constraints of both space and time on individuals in their everyday lives. His model (Figure 14.15) details the way that people trace out "paths" in time and space, moving from one place (or "station," in Hägerstrand's terminology) to another in order to fulfill particular purposes ("projects").

According to Hägerstrand, the combined effects of three kinds of constraints define any individual's capacity to move from one station or project to another within the broader environment:

- *Capability constraints*—which include the need for a minimum amount of sleep (thus limiting the time available for traveling) and the means of transportation that are available (thus determining

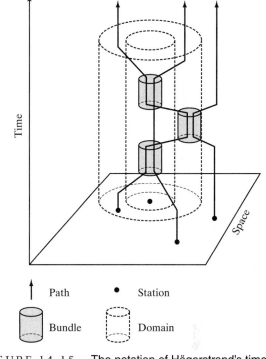

FIGURE 14.15   The notation of Hägerstrand's time geography.

the radius of the territory that can be included in the daily life path).

- *Authority constraints*—which include laws, rules, and customs that limit accessibility to certain places and curtail participation in certain activities. Examples include highway speed limits and laws governing alcohol consumption.

- *Coupling constraints*—which stem from the limited periods during which particular projects are accessible. These are determined by the opening hours of businesses, the scheduling of public transport, and the timing and duration of rush hours.[57]

The *time budget* of any individual within his or her environment can be traced graphically, as in Figure 14.16, the effective range of that individual being described by a prism (or series of prisms) whose

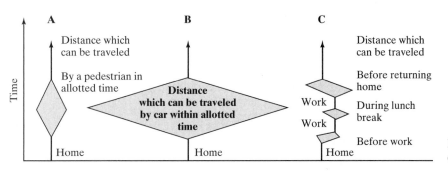

FIGURE 14.16   Time budgets and daily prisms.

BOX 14.4 STRUCTURATION: TIME AND SPACE IN EVERYDAY LIFE

"Bundles" of activity in time and space are central to the important concept of the "*structuration*" of social life advanced by social theorist Anthony Giddens.[58] Giddens emphasizes the fundamental importance of the "binding" of time and space on the very fabric of social organization and the conduct of social life. The concept of structuration is based on the interdependence between (1) social systems and structures and (2) purposeful individual behavior. It is often couched in terms of (social) *structure* and (human) *agency* respectively. According to Giddens, the continuity of day-to-day life and the existence of bundles of routinized activity foster *social integration* as individuals respond reflexively and purposely to their shared experiences in time and place. In so doing, they at once shape and are influenced by the structural properties of social systems—the frameworks of class, gender roles, educational stratification, legal codes, and so on.

Insofar as routinized social practices are consistent over particular spans of time and space, these practices flow from and channel back into structural relations that reach beyond localized bundles to influence interactions with and between others who are *absent* in time or space. This influence helps to sustain *system integration*, the extension of social relations across both time and space, which Giddens calls *time-space distanciation* (Figure 14.17). At an intermediate scale, and connecting the *time-space routinization* of the lifeworld with the time-space distanciation of society, is a hierarchy of locales, similar to Hägerstrand's domains. This hierarchy represents a "regionalization" that reflects multiple sociospatial networks based on the differential distribution of resources (economic power) and authority (political power).

This idea gives us a very useful framework for conceptualizing urban life and the way that human landscapes are created and sustained. Geographers Michael Dear and Jennifer Wolch use a similar approach in showing, as they put it, "how territory shapes social life":

> Human landscapes are created by knowledgeable actors (or agents) operating within a specific social context (or structure). The structure-agency relationship is mediated by a series of institutional arrangements which both enable and constrain action. Hence three "levels of analysis" can be identified: structures, institutions and agents. Structures include the long-term, deep-seated social practices which govern daily life, such as law and the family. Institutions represent the phenomenal forms of structures, including, for example, the state apparatus. And agents are those influential individual human actors who determine the precise, observable outcomes of any social interaction.[59]

The concept of structuration, together with the idea of a hierarchy of locales each containing various lifeworlds, provides a strong but flexible framework for dealing with the social construction of reality. All physical space is open to the social construction of meaningful reality. Some spaces are even the object of multiple realities, being socially constructed in different ways by different groups. Space is thus invested with an ideological content that reflects the values, attitudes, and cognitive structures of the groups involved. This ideological content has important implications for the overall trajectory of urbanization because of the way that it serves to reproduce and sustain the structure and dynamics of the relations between social groups.

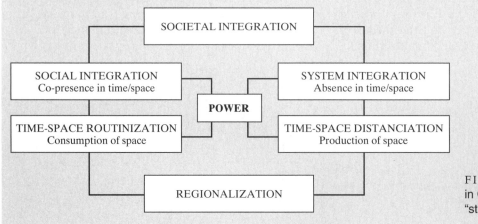

FIGURE 14.17 Elements in Giddens's concept of "structuration."

shape depends on capability constraints. Every stop, for whatever project, causes the range of the prism (or subprisms) to shrink in proportion to the length of stay. In addition, both people's time budgets and their activity patterns are recognized as being shaped by—and contributing to (the sociospatial dialectic again)—wider structural features of the social systems in which they operate. These features include the various institutions that control the authority constraints and coupling constraints relevant to particular constellations of stations ("domains"). Shared domains and routines create coherent time geographies, the intersections of individual paths forming "bundles" of social activity in time and space.

GENDERED SPACES: WOMEN, HOME, AND COMMUNITY

One important dimension of the sociospatial dialectic that tends to be concealed by emphases on particular social groups or particular localities is that of gender. When the existence of women is acknowledged, neighborhoods and lifeworlds look somewhat less consistent. "Middle-income" neighborhoods, for example, may well be fairly homogeneous in terms of the occupation and socioeconomic background of the male heads of households that society and most social scientists use in categorizing neighborhoods, but, among others, they also include those women whose labor-market status as unskilled, semiskilled, and low-paid workers would categorize them as low-income. Conversely, "low-income" neighborhoods are often home to white-collar female workers.

Geraldine Pratt and Susan Hanson have used this point in arguing that social status is reproduced in different ways for men and women. Men often experience class consistently, with their family-class position being similar to their sense of class developed through their work. Some married women experience class less consistently. Their family-class position, if defined by their husband's occupation, may not coincide with their own class position as experienced through their own paid employment.[60] In addition, traditionally many women's experience of social status has varied significantly through time as a result of childrearing roles that have taken them out of the paid labor force for long periods, often forcing them to return in inferior positions or in totally different occupations. For the same reason, some women experience locality differentially, "homemakers," for example, spending a much greater proportion of their time in localized lifeworlds centered on the home and the social acquaintance neighborhood.

THE CREATION OF WOMEN'S SPACES

We have to go back to the early industrial city in order to understand the evolution of women's roles and women's spaces. For households caught up in the unprecedented vortex of industrialization and urbanization, the demands of daily routine required a rationalization of family life. The new rhythm of life dictated by office clocks and factory whistles made it difficult for the family to fulfill some of its traditional roles, while increased residential mobility tended to isolate the families from their larger network of kin. With fewer relatives available to perform informal service functions, families found themselves having to give up not only their economic function (as a unit of production) but also having to shed many traditional responsibilities: educating children, supporting the elderly, and caring for the sick, for example. The corollary of these changes was the emergence of a whole range of new service professions and establishments such as tailors, barbers, doctors, restaurants, and laundries. In many ways the major distinction was between workplaces and home; but this distinction had very different consequences for men and women.

Although women were drawn increasingly into the formal economy, their participation was often transitory, depending on the stage in the **household life cycle** and on the cyclical fluctuations of the industrial economy. For less-skilled and less-well-educated women in working-class households, opportunities for paid work outside the home were further constrained by perceptions that cast women as both physically weaker and emotionally less stable than men. As a result, many women could obtain money to supplement family income only in domestic settings (theirs or other people's), undertaking piecework such as sewing or providing surrogate family functions such as casual cleaning, childcare, taking in washing, or providing houseroom and domestic services to lodgers. The domestic setting also became the basis for the lifeworld of middle-income women, though for rather different reasons.

For middle-income women, it was the *suburban* domestic setting that framed the routine of everyday life. In this context it is important to appreciate the suburbs as phenomena that were socially constructed. According to historian Margaret Marsh, this construction originally had two components. Although the domesticity now associated with suburban settings was initially a female ideal, centered around the cultural institution of family, it was seen at the time as something that could best flourish amid the markets, bakeries, and neighborly support provided by cities, not suburbs. Women domestic reformers did not urge their readers to leave the city but to develop domestic ideals within it. The initial suburban ideal, Marsh points out, was a male construct, derived (as we saw in Chapter 10) from Jeffersonian political philosophy and Emersonian social philosophy, serving the Romantic ideals of Nature and the Frontier and recognizing the imperatives of property ownership.[61] It was the success of that ideal that brought the middle-class female ideal of domesticity to the suburbs, where, by the late nineteenth century, the two ideals fused to form the basis for the **social construction** of suburban life. Thus:

> The clear separation of work from home, the insistence on social distancing, the treatment of the home as a feminine domain, the importance attached to domestic privacy and the exclusion of the vulgar prying multitude can all be seen as parts of a code of individual responsibility, male economic dominance, female domestic subordination, and family-nurtured morality *which served to give the bourgeoisie a social identity and marked them off from the upper class and lower orders.*[62]

None of this could happen simply through the force of ideals, however. As well as being socially constructed, suburban middle-class life had to be physically designed and built, and models of appropriate behavior had to be propagated. An important influence in this context was Andrew Jackson Downing's book *The Architecture of Country Houses,*[63] which went through nine printings between its publication in 1850 and the end of the Civil War. Downing took the Jeffersonian and Emersonian equation between Nature/Rurality and Morality/Democracy to the level of individual dwellings. The home, he argued, must express in size, style, layout, and ornamentation the morality of its inhabitants. He presented a series of illustrative designs (though they were in fact taken from the work of English architect J. C. Loudon) as illustrative of this American ideal. The market was soon awash with pattern books "whose authors considered themselves a critical component of the creation of a democratic republic."[64]

The female ideal of domesticity was similarly advanced through print media. Here, articles and advertisements in popular magazines played the most important roles. One of the most important was the *Ladies Home Journal*, which, under the editorship of Edward Bok between 1889 and 1912, spelled out for its readers the canons of Good Taste in everything from domestic architecture and interior design to domestic manners and social comportment. Bok's evangelical mission was:

> to reform and simplify the American home and to keep women in it. He enlisted architects, suffrage leaders, reformers, novelists, statesmen, and even presidents to assure his readers that the right kind of home environment could preserve the family, strengthen the nation, and thereby give women more than enough meaningful work to do. For him, women belonged in these homes, and each one was to be separate from the others.[65]

Meanwhile, advertisements in popular journals further advanced the twin ideal of suburban domesticity in selling consumer goods. A detailed analysis of advertisements in the *Saturday Evening Post*, for example, shows that women were given responsibility through the earliest advertising campaigns for steering the household's adaptation to its new suburban surroundings. This program not only helped to reaffirm the ideal of suburban domesticity but also helped to crystallize gender roles and gender relations.[66]

Referring for a moment to the concept of **social reproduction**, we can see how these changes created a sociospatial setting that ensured not only the social reproduction of the middle class but also the social reproduction of gender roles (Figure 14.18). Men set forth into the world of work while women, left to make what they could of day-long isolation, busied themselves with creating a comforting and uplifting environment.

Over time, as more and more Americans aspired to replicate the lifestyle of suburban affluence and as levels of living steadily rose, the social construction of suburban domesticity, together with the associated gender roles, came to incorporate a broad band of socioeconomic groups. The process was consolidated in various ways by the federal government and private corporations, who "made clear that the nuclear family, with a male breadwinner and a female homemaker, ought to be considered the 'normal' type of family. All three **greenbelt towns**, for example, excluded all families with wives employed outside the home; banks, schools and insurance companies all across the country began to bar the employment of married women, and the women's magazines urged married women with jobs to quit so that men could take their places."[67] By the 1960s feminist Betty Friedan could justifiably claim that there had emerged a "mystique" of feminine fulfillment—the idea that the suburban housewife could find true fulfillment through her husband, children, and home—that had become "the cherished and self-perpetuating core of contemporary American culture."[68]

FIGURE 14.18   The "ideal of suburban domesticity."

CHANGING ROLES, CHANGING SPACES

Yet economic and social change was already overtaking the ideal of suburban family life. Some women had been drawn into the labor force during World War I to fill civilian jobs vacated by enlisted men; some were drawn into "women's work" as industrialization sought out more and more product lines for mass production to exploit (Figure 14.19). World War II had increased the participation of married women in the labor force to an unprecedented 25 percent (Figure 14.20); divorce rates, having increased slowly throughout the twentieth century (with the exception of the sudden peak around the time of World War II), suddenly accelerated in the late 1960s (Figure 14.21) in response to changing social attitudes (see p. 319). By 2000 only 25 percent of all households in the United States consisted of a married couple with their own children (down from 40 percent in 1970), while nearly 30 percent were headed by women (both with and without children). Meanwhile, the labor force participation rate of women had risen to 60 percent. The context of these demographic and social changes of the 1960s and 1970s was described in Chapter 4 in relation to shifts in the dynamics of the **urban system**. But the same changes were implicated in the restructuring of urban life and intraurban space. The increase in couples' average age at marriage, the increases in the number of single and childless women, and the dramatic rise in the

FIGURE 14.19   Women at work in 1912. Assembling dolls at the Shrenhat Toy Company, Philadelphia.

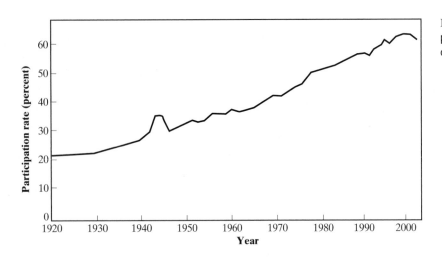

FIGURE 14.20   Labor force participation rate of females aged 16 and over, 1920–2002.

number of female-headed households all contributed to a huge pool of women whose work and residential choices were being exercised without reference to male preferences and without financial assistance from (or dependence upon) men.

Social geographers Robin Law and Jennifer Wolch have noted three principal ways in which these trends have affected urban space:

1. The transformation of suburbia. The participation of women in the paid labor force has undermined the communality of suburban neighborhoods by dramatically reducing the number of full-time housewives available to perform the unpaid volunteer community services and sustain the spatially bound friendships and acquaintanceships necessary to a sense of community.

2. The creation of distinctive new urban spaces through the residential behavior of two-worker households without children (DINKs), reflecting their generally higher disposable income, a different set of consumption patterns (such as priority on restaurants rather than schools), and

the need to balance the commuting trips of the two workers.

3. The creation of distinctive new urban spaces through the residential constraints experienced by female-headed families, who tend to become localized in central city neighborhoods because of low incomes and the need for urban services such as public transport and accessibility to services such as child care.[69]

They also pointed out several ways in which these changes have affected urban life, time-space routines, and (therefore) lifeworlds. The increased labor market participation among women, for example, is clearly bound up in the "new materialism" and materialistic lifestyles described in Chapter 12. But access to market goods and services is closely circumscribed by income, so differences have emerged in the *form* in which materialism is inscribed onto urban life. There is an increasing difference between the mass-produced goods and standardized services consumed by lower-income groups and the customized, personalized consumption of upper-income groups. One effect of this difference, as Law and Wolch pointed out,

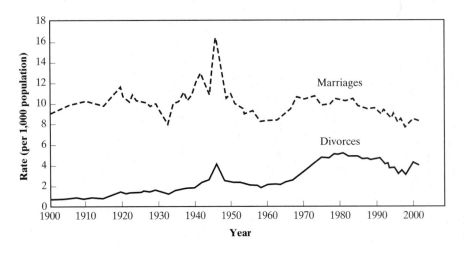

FIGURE 14.21   Marriage and divorce rates, 1900–2001.

is that the spaces and times of personal fulfillment, entertainment and leisure for the affluent are also spaces of work (exploitation) for those in labor-intensive service jobs. Evenings, weekends, public holidays are times of work for more and more people, as are the public sites of entertainment, restaurants, shopping malls and resorts. And more and more, the requirement of that labor is that the act of labor (work-for-wage) be presented as an act of care (work-for-love). The requirement is not only that workers perform the task but that they accompany it with emotion: the sincere smile, the warm greeting . . . . Thus personal life and "home" (true affection, sincerity, personal commitment) become colonized and blurred with "work."[70]

The overall increase in female participation in the labor market has also, they pointed out, accelerated the time-space routine of both middle-income and low-income households. Two-worker families with children, together with single-parent families, have had increasingly to rush themselves and their children through the day, until (perhaps) an extended bedtime allows for a few moments of "quality time." Since women traditionally tended to remain burdened with an unequal domestic division of labor as a result of the legacy of the ideal of suburban domesticity and the "mystique" of feminine fulfillment, they tended to suffer most from pressures on the use of time. For them the "prism" of opportunities in time and space was too often replaced by a "prison" of coupling constraints. For this reason, many women now perceive central city environments as being most supportive and appropriate to their needs, just as they did before the advent of the male suburban ideal. In these settings, it is suggested, higher densities make it easier to develop social support networks and to organize collectively to assist with domestic responsibilities such as childcare. The domestic ideal, fused for so long with the suburban ideal, thus promises to break free, bringing a further round of change in the sociospatial dialectic.

BOX 14.5   DISCRIMINATION BY DESIGN: DOMESTIC ARCHITECTURE AND GENDER DIFFERENCES[71]

One well-worn theme in architectural theory has been the manifestation of "masculine" and "feminine" elements of design. For the most part this has involved crude anatomical referencing. Skyscrapers, as phallic towers for example, can be seen to embody the masculine character of capital. As some feminist interpretations of architectural history have shown, however, the silences of architecture can be more revealing than crude anatomical metaphors. Elizabeth Wilson highlighted the way that Modernist architecture, although being self-consciously progressive, had nothing to say about the relations between the sexes.[72] It changed the shape of housing units without challenging the functions of the domestic unit. Indeed, the **Bauhaus School**, vanguard of the Modern Movement, helped reinforce the gender division of labor within households through Breuer's functional Modern kitchen.

The internal structure of buildings embodies the taken-for-granted rules that govern the relations of individuals to each other and to society just as much as the overall plan and internal structure of cities. The floor plans, decor, and use of domestic architecture have in fact represented some of the most important encodings of patriarchal values (involving the broad system of social arrangements and institutional structures that enable men to dominate women). As architects themselves have so often emphasized, houses cannot be regarded simply as utilitarian structures but as "designs for living." The strong gender coding built into domestic architecture has been demonstrated in analyses ranging from Victorian country houses to bungalows and tenements.[73]

Today, the conventional interpretation of suburban domestic architecture recognizes the way that the ideals of domesticity and the wholesomeness of nuclear family living are embodied in the feminine coding given to the "nurturing" environments afforded by single-family homes that center on functional kitchens and a series of gendered domestic spaces: "her" utility room, bathroom, bedroom; "his" garage, workshop, study. The importance of these codings rests in the way that they present gender differences as "natural" and thereby universalize and legitimize a particular form of gender differentiation and domestic division of labor.[74]

We must recognize, however, that even as buildings and domestic spaces are designed to symbolize and codify gender roles, their meanings become contested and unstable, particularly in relation to the complex conflicts and compromises between class and gender interests that characterize the sociospatial dialectic. Meanwhile, changes in the dynamics of household formation, in employment patterns, and in the design professions themselves all conspire to continuously modify the received meaning of domestic design.[75]

FOLLOW UP

1. Be sure that you understand the following key terms:

   anomie
   appraisive imagery
   cognitive distance
   communities of limited liability
   crisis communality
   cultural transmission
   designative images
   Gemeinschaft
   Gesellschaft
   habitus
   lifeworld
   mechanical solidarity
   neighborhood effects
   organic solidarity
   place ballets
   psychic overload
   social disorganization
   sociospatial dialectic
   structuration
   time-space distanciation
   urbanism
   urban legibility
   urban villages

2. At the beginning of the chapter (p. 378) it is asserted that the idea of the city as an arena of conflict and as an incubator of loneliness, isolation, and deviance is one that is reflected in movies, television series, and popular music. Select one of those media and compile a list of examples that illustrate the point.

3. Construct a time-space path of your typical day and annotate it to show when, where, and how your path coincides with those of others in ways that contribute toward promoting intersubjectivity and distinctive lifeworlds.

4. Watch a movie: *Boyz N The Hood* (Columbia TriStar, 1991; Director John Singleton) provides a vivid portrayal of neighborhood life in south-central Los Angeles. Critically review this portrayal in the context of concepts such as lifeworld, community, neighborhood, structuration, and the sociospatial dialectic.

5. Add to your *portfolio.* One thing you can do to personalize this section is to create portrayals of mental maps (your own or those of a friend).

KEY SOURCES AND SUGGESTED READING

Bell, D. and G. Valentine, eds. 1994. *Mapping Desire: Geographies of Sexuality.* London: Routledge.

Chouinard, V. and V. A. Crooks. 2003. Challenging Geographies of Ableness: Celebrating How Far We've Come and What's Left to be Done. *The Canadian Geographer* 47: 383–85.

Clark, C. E., Jr. 1986. *The American Family Home.* Chapel Hill, N.C.: University of North Carolina Press.

Cullen, J. D. and P. L. Knox. 1982. The City, the Self and Urban Society. *Transactions, Institute of British Geographers* 7: 276–91.

Davies, W. and D. Herbert. 1993. *Communities within Cities: An Urban Social Geography.* London: Pinter.

England, K. 1991. Gender Relations and the Spatial Structure of the City. *Geoforum* 22: 135–47.

Entrikin, N. 1991. *Betweenness of Place: Towards a Geography of Modernity.* Baltimore, Md.: Johns Hopkins University Press, 1991.

Eyles, J. 1989. The Geography of Everyday Life. pp. 102–17 in *Horizons in Human Geography,* ed. D. Gregory and R. Walford. Totowa, N.J.: Barnes and Noble.

Gottdiener, M. 2000. *New Forms of Consumption: Consumers, Culture, and Commodification.* Lanham, Md.: Rowman & Littlefield.

Hanson, S. and G. Pratt. 1995. *Gender, Work and Space.* London: Routledge.

Imrie, R. 1996. *Disability and the City: International Perspectives.* London: Paul Chapman.

Jackson, P. 1989. *Maps of Meaning.* Boston: Unwin Hyman.

Law, R. M. and J. R. Wolch. 1993. Social Reproduction in the City: Restructuring in Time and Space. pp. 165–206 in *The Restless Urban Landscape,* ed. P. L. Knox. Englewood Cliffs, N.J.: Prentice Hall.

Ley, D. 1983. *A Social Geography of the City.* New York: Harper and Row. See especially Chapter 4, "The city of mind and action," Chapter 5, "The urban sense of place," and Chapter 6, "The geography of informal social groups."

Marsh, M. 1990. *Suburban Lives.* New Brunswick, N.J.: Rutgers University Press.

Pocock, D. and R. Hudson. 1978. *Images of the Urban Environment.* London: Macmillan.

Smith, M. P. 1980. *The City and Social Theory.* Oxford, UK: Blackwell.

Spain, D. 1992. *Gendered Spaces.* Chapel Hill, N.C.: University of North Carolina Press.

Strong-Boag, V., I. Dyck, K. England, and L. Johnson. 1999. What Women's Spaces? Women in Australian, British, Canadian and U.S. Suburbs. pp. 168–86 in *Changing Suburbs: Foundation, Form and Function,* eds. R. Harris and P. J. Larkham. London: E & FN Spon.

Tuan, Y-F. 1977. *Space and Place: The Perspective of Experience.* Minneapolis, Minn.: University of Minnesota Press.

Weisman, L. K. 1992. *Discrimination by Design. A Feminist Critique of the Man-Made Environment.* Urbana, Ill.: University of Illinois Press.

Wilson, E. 1991. *The Sphinx in the City. Urban Life, the Control of Disorder, and Women.* Berkeley, Calif.: University of California Press.

RELATED WEBSITES

CSISS Classics: *http://www.csiss.org/classics/*
The CSISS Classics website provides summaries and illustrations of men and women who have made major contributions to spatial thinking in the social sciences during the last two centuries. These "spatial innovators" include Charles Booth, Ernest Burgess, Walter Christaller, Torsten Hägerstrand, Kevin Lynch, Robert Park, Georg Simmel, and Sam Bass Warner.

Smith College, Women's Studies Program:
http://www.smith.edu/wst/gradlinks.html
This website contains numerous links to graduate programs in women's studies in the United States, Europe, and elsewhere. It also offers links to other websites that contain links to undergraduate programs in women's studies, including the comprehensive Artemis Guide to Women's Studies Programs (*http://www.artemisguide.com/*).

University of Kansas:
http://www.people.ku.edu/~jyounger/lgbtqprogs.html
This website contains numerous links to undergraduate and graduate programs in the United States and Canada in lesbian, gay, bisexual, and transgender programs.

Americans with Disabilities Act:
http://www.usdoj.gov/crt/ada/adahom1.htm
The Americans with Disabilities Act (ADA) website provides information and technical assistance on disability related to legal, employment, transportation, education, healthcare, and housing and building design issues. The *DisabilityInfo.gov* link connects this website to U.S. government and other websites with disability-related information.

15

PROBLEMS
OF URBANIZATION

Cities have always been seen as problematic: so much so, as we saw in Chapter 14, that American culture is distinctively antiurban. Cities have been widely regarded as necessary evils, essential for economic development but somehow unnatural, dangerous, and corrupt. It should not be surprising, of course, that cities are so often the site of problems. With so much economic, demographic, social, political, and cultural change taking place in cities, it is inevitable that there are unanticipated and unintended consequences and that tensions, contradictions, and conflicts occur between different groups. Some of these problems stem from society as a whole; they are particularly stark in cities simply because metropolitan areas contain the bulk of the population and are the crucibles of change. Some of them, however, are caused or at least intensified by urbanization and the nature of urban settings.

CHAPTER PREVIEW

The purpose of this chapter is not to provide an exhaustive review and inventory of urban problems. Rather, the objective is to show how the processes involved in urbanization are inherently problematic, with space and place often playing key roles. We focus on four major sets of problems—poverty, criminal violence, homelessness, and infrastructure problems. First, however, we will discuss how society came to perceive and define certain issues as "problems" and how certain perceptions have changed, persisted, or been modified as urbanization has moved from the early industrial city to today's **neo-Fordist** metropolis.

One of the most persistent themes to emerge from this review is that of poverty, slum neighborhoods, and the behavior that is associated with slum settings. These issues are examined in detail, providing an opportunity to consider theories and concepts related to the causes and effects of poverty, slums, and localized social problems. Discussion of these issues is followed by an examination of the geography of criminal violence and a review of the effects of crime on urbanization and urban life. Once again, the objective here is to demonstrate the two-way causality inherent to urbanization: Certain settings give rise to high levels of criminal violence, which, in turn, affect the whole dynamic of spatial organization. Homelessness is likewise presented in the context of its wider causes and effects, with particular attention being given to the ways in which recent changes in the political economy of urbanization have led to new forms of homelessness.

In the final section of the chapter the focus shifts away from social problems toward the physical problems associated with environmental degradation and decaying infrastructure. As we will see, these problems are intimately linked to the economic, political, and fiscal issues that dominate politics, policymaking, and planning.

PROBLEM? WHAT PROBLEM?

It is important to make a distinction at the beginning of this chapter between problems *in* cities and problems *of* cities.[1] Problems *of* cities are those that are brought about or amplified in some way by the attributes (environmental, demographic, social, economic, political, cultural, and so on) of urban settings. The anomie, social disorganization, and **deviant behavior** of certain neighborhoods, as interpreted by Durkheim and the Chicago School (see Chapter 14), provide classic examples. Another example stems from the **neighborhood effects** that we have already encountered (p. 301): local value systems that socialize residents into distinctive (and sometimes problematic) attitudinal and behavioral patterns. Problems *in* cities are problems associated with the broader sweep of socioeconomic change. They are "urban" problems only to the extent that people and jobs are concentrated in urban settings. The settings themselves are not seen as causal factors. Rather, the roots of such problems are to be found (1) in the tensions and contradictions arising from the economic and social relations that underpin particular forms of economic and social organization and (2) in the conflicts and frictions arising from the economic and social restructuring involved in moving from one stage of development to another.

In practice it is not always easy to make this distinction between problems in and problems of the city. A good example is provided by one of the biggest problems of all: poverty and deprivation. Poverty is a problem *in* cities in that it is fundamentally the product of an income distribution that reflects the workings of a competitive social and economic system based on unequal distributions of resources, skills, and power. It is also a problem *of* cities in that it is compounded and intensified by the settings in which the majority of the poor find themselves: physical and social environments that inhibit an individual's life chances and contribute to cumulative, negative spirals of community well-being.

It will also be helpful if we recognize at the outset that some issues are widely regarded as problems while some are tolerated (or overlooked) as part of life, something that "comes with the territory." In short, situations are problematic only if they are, first, perceived as such and, second, generally regarded as being of substantive importance. The first question, therefore, is: "Problem? What problem?" Environmental pollution provides a good example of the kind of situation that can represent a serious threat long before it is acknowledged as a problem. Some problems, in contrast, become recognized only as a result of their **social construction**. Thus, for example, the Chinese population of towns and cities up and down the U.S. West Coast came to be seen as a problem in the mid-nineteenth century as a result of a mixture of beliefs, misconceptions, and prejudices that characterized Chinese customs and behavior as a threat to public health. "All great cities have their slums and localities where filth, disease, crime and misery abound," conceded a Special Committee of San Francisco's Board of Supervisors in

1885, "but," they continued, "in the very best aspect which Chinatown can be made to present, it must stand apart, conspicuous and beyond them all in the extreme degree of all these horrible attributes, the rankest outgrowth of human degradation that can be found upon this continent."[2] In fact, subsequent scientific research showed that Chinese quarters were no worse than other slums: The Chinese had been made scapegoats for a more general breakdown of social order and the failure of public health systems.[3]

The second key question is: "Problem *for whom*?" Homelessness provides a good example of a long-standing issue that for many years was considered not to be a serious social problem. Even in the early 1980s, when American cities were estimated to contain between 250,000 and 300,000 homeless people, the "problem" was invisible to many Americans and denied by the country's leadership. Only after intensive publicity campaigns has homelessness come to be widely recognized as a problem. This example also points to another source of ambiguity. Because problems are very much a function of the

eye of the beholder, they are often *relative*. How much homelessness must there be before it constitutes a problem? The answer often depends on how much of a threat homelessness poses: Do other city residents believe that homeless people threaten their lifestyle or their property values? Do they offend people's sense of equity or intrude on their sense of aesthetics? Such differences in perception are the stuff of politics, which we consider systematically in Chapter 16.

Perceptions also change over time, however, as economic and social restructuring alters social relations. In this chapter we begin by reviewing very briefly some of the changes that have occurred in the major problems associated with urbanization (i.e., problems *of* cities) in the more developed countries of the world. This is an important exercise, because it gives us some perspective on contemporary problems. Against this background, four of the major problems of contemporary cities—localized poverty, criminal violence, homelessness, and environmental and infrastructure problems—are examined in closer detail.

FROM HAUNTS OF VICE TO GANG WASTELANDS—AND BACK

The central paradox of urbanization has always been that although cities have been essential as crucibles of economic development and production, they have brought with them a variety of negative and unprecedented side effects.

PROBLEMS OF THE EARLY INDUSTRIAL CITY

The industrial logic that underpinned urbanization in the early industrial cities of Europe and North America was based on a division of labor and on **external economies**, **agglomeration economies**, and **localization economies** (see p. 65) that required vast numbers of workers to be in close proximity in order to staff the complex and interdependent new systems of production, storage, and distribution. But this same concentration of workers also resulted in acute problems of crowding, pressure on the infrastructure of roads, bridges, and waste disposal systems, and greatly increased risk from fire and infectious diseases.

Within this context, the **division of labor** was starkly reflected in economic inequality. Social interaction was also affected: The heterogeneity of city populations, as well as their sheer size and density, apparently led to anomie, social disorganization, and deviant behavior, as we saw in Chapter 14. Meanwhile, social conflict and political volatility increased partly in response to all these pressures and partly as a result of increased political consciousness and sensitivity

(propagated by the widespread emergence of pamphlets and popular newspapers but underpinned by the increased literacy and numeracy required within the new division of labor).

It was the physical dimensions of these side effects that were most clearly perceived as problems. They were also broadly seen as *urban* problems rather than economic, social, or political problems: problems *of* the city rather than problems *in* the city. Adna Weber, summarizing the literature on nineteenth-century urbanization in 1899, wrote in terms of a "theory of city degeneracy," pointing to what he saw as incontrovertible evidence that "cities are the site, and city life the cause, of the deterioration of the race."[4] Among the evidence available to Weber was the conclusion of British statistician G. B. Longstaff:

> That the town life is not as healthy as the country is a proposition that cannot be contradicted. . . . The narrow chest, the pale face, the weak eyes, the bad teeth, of the town-bred child are but too often apparent. It is easy to take an exaggerated view either way, but the broad facts are evident enough; long life in towns is accompanied by more or less degeneration of race.[5]

Another of Weber's sources went further, emphasizing mental "degeneration":

> The inhabitant of a large town, even the richest who is surrounded with the greatest luxury, is continually exposed to the unfavorable influences which diminish his

vital powers. . . . He breathes an atmosphere charged with organic detritus; he eats stale, contaminated, adulterated food; he feels himself in a state of constant nervous excitement, and one can compare him without exaggeration to the inhabitant of a marshy district. The children of large towns who are not carried off at an early age . . . develop more or less normally until they are 14 or 15 years of age, are up to that time alert, sometimes brilliantly endowed, and give highest promise. Then suddenly there is a standstill. The mind loses its facility of comprehension and the boy, who only yesterday was a model scholar, becomes an obtuse, clumsy dunce, who can only be steered with the greatest difficulty through his examinations. With these mental changes, bodily modifications go hand in hand.[6]

Although cities in general were seen as fostering physical and intellectual degeneration, the slums that were discovered in every large city between 1840 and 1875 were seen as "haunts of vice" that harbored *moral* degeneration, which was in turn seen as a fundamental cause of both poverty and disease. In the words of a leading New York City housing reformer:

> Vice, crime, drunkenness, lust, disease and death here hold sway, in spite of the most powerful moral and religious influences. . . . Their intellects are so blunted and their perceptions so perverted by the noxious atmosphere which they breathe, and the all-pervading filth in which they live, move, and have their being, that they are not susceptible to moral or religious influences. . . . [This is] a depraved physical condition which explains the moral deterioration of these people and which can never be overcome until we surround them with the conditions of sound health.[7]

It was this focus on the degeneracy inherent to urban life (Figure 15.1) that fostered the antiurbanism of the Romantics, encouraged the pastoral ideals of the **American Renaissance**, and gave impetus to the European Garden City movement (see p. 248), to the first U.S. planned suburbs (see p. 143), and to Arcadian Classicism and the City Beautiful movement in the United States (see p. 250). As we will see in Chapter 17, it also gave direction to the first crucial efforts at urban reform and planning in Europe and North America.

PROBLEMS OF THE INDUSTRIAL CITY

By the late nineteenth century in North America, perceptions of the nature of urban problems had changed, along with changes in the overall dynamics of urbanization. The poverty and disease of the slums were still perceived as the main problems, but they came to be associated more with concentrations of particular ethnic groups. Instead of city life in general and the slums in particular "breeding" the physical and moral degeneration that seemed to lead to depravity, drunkenness,

FIGURE 15.1   Haunt of vice. The Short Tail Gang in its hideaway beneath a pier at Jackson Street in New York, photographed by Jacob Riis in 1888.

pauperism, prostitution, and religious infidelity, it was now concentrations of immigrants that represented the root of perceived urban ills.

The slum problem became the ghetto problem. Immigrants (Figure 15.2), previously seen as the backbone of the country, were now recast in the role of public menace. As we have seen, the Chinese population (Figure 15.3) bore the brunt of this prejudice in West Coast towns and cities. Elsewhere, it was the newcomers from southern and eastern Europe (who by now were responsible for much of the demographic growth and increased labor power of cities) who found themselves held accountable for many of the shortcomings of urban life. Joseph Kirkland, describing life among the poor in Chicago in 1895, asserted that:

> For depth of shadow in Chicago low life one must look to the foreign elements, the persons who are not only of alien birth but of unrelated blood—the Mongolian, the African, the Sclav, the semitropic Latin. Among them may be found a certain degree of isolation, and therefore of clannish crowding; also of contented squalor, jealous of inspection and interference. It is in the quarters inhabited by these that there are to be found the worst parts of Chicago, the most unsavory spots in their moral and material aspects.[8]

Within just a couple of decades the upward socioeconomic mobility and outward geographical mobility of many of the immigrants and their children

FIGURE 15.2
Immigrants, formerly portrayed as the cornerstone of a new country, came to be portrayed in the late nineteenth century as a problem and a menace.

had undermined this perspective on urban problems. Immigrant **ghettos** consequently came to be seen as short-term havens that, over the longer term, became "ramps" for economic advancement. It was organized crime, rather than endemic criminality, that instead appeared to be the main obstacle to civic order. Meanwhile, African-American ghettos, newly formed by the migration of rural African Americans from the South after 1910, replaced immigrant ghettos (in the social construction of urban problems) as the chief nodes of social and economic problems.

By this time cities had also changed in other ways, bringing new issues and new perceptions of the nature of urban problems. Cities were not only much bigger, but they were also more "modern," faster-paced, and with much more specialization and

FIGURE 15.3   Chinatown, San Francisco, about 1900. An ethnic community that was unjustly blamed, like its counterparts in other cities, for citywide public health problems.

compartmentalization in economic affairs and social life. Alienation, along with mental and physical stress, came to be more widely perceived as major problems. As the ideas of Louis Wirth and the Chicago School (see Chapter 14) gained currency, these problems came to be seen increasingly as a product of the breakdown of "normal" patterns of social interaction and social organization. Once again, though, it was the slum that came to be perceived as the location of the problem: the place where stresses were intensified to the point where deviant behavior of all kinds was generated.

PROBLEMS OF THE FORDIST ERA

After World War II, a combination of

1. **Keynesian** economic management (see p. 76);
2. improved public health based on scientific medicine, improved diets, and environmental regulation;
3. an unusually long period of economic growth and prosperity; and
4. an unprecedented array of welfare programs

made impressive inroads into the previously intractable problem of slums and poverty. In this era of optimism and can-do attitudes, one of the persistent problems was seen as social and institutional discrimination against black people, not least because it was trapping black households in central city neighborhoods.

The central cities themselves also came to be seen as problematic. Their aging neighborhoods, suddenly reaching structural and technological obsolescence (Chapter 13) just as cities were decentralizing, now contained blighted tracts (Figure 15.4) that became emblematic of urban problems. Nevertheless, the overall affluence of the period led to a feeling of optimism. Urban problems, it was felt, were a residual legacy that would be tidied up in short order: Slums and poverty would soon be a thing of the past. The maturing Baby Boom generation, taking prosperity for granted, soon began to look beyond economic and social inequality, uncovering environmental issues as a much greater cause for concern.

PROBLEMS OF THE "NEO-FORDIST" ERA

In the mid-1960s the United States was rudely awakened from the Dream of the Affluent Society. Widespread civil unrest, including riots, looting, and arson, served notice that racial tension was still a major problem. With attention focused once more on inner-city slums, poverty was rediscovered in the midst of

FIGURE 15.4   Urban decay in the South Bronx.

affluence. But before the magnitude of the problem could be properly grasped, the overall dynamics of urbanization were jarred by the economic recession that was triggered by the 1973 oil price increases (see p. 108). **Deindustrialization** and its problems of unemployment and **underemployment** now came into focus, and price inflation became problematic, bringing with it the issue of housing affordability. City budgets were suddenly unable to sustain the services and welfare programs that had been built up during more affluent times, resulting in acute fiscal crises (see p. 475) that were soon followed by drastic changes in the role and orientation of city governments (see pp. 478–480).

As the economic boom of the 1980s succeeded these crises, the nature of urban problems came to be redefined once more. Prosperity brought renewed concern, in many cities, for the costs of urban growth: traffic congestion, air pollution, escalating house prices, and backlogs in public service provision. For some observers, though, the sociocultural changes that began in the 1980s gave new cause for concern. New kinds of stresses and strains emerged from the combination of intensified materialism and labor market changes (see p. 100) that drew more and more women into the labor force and contributed to the restructuring of family and community life. In this context it was not only the slums but also the suburbs that seemed to be the location of urban problems. Teenybopper mall rats and "Valley Girls" came to be emblematic of the shallow materialism endemic to the outer fringes of the Galactic Metropolis. Much worse, however, was the darker side of social and economic change in the suburbs: a landscape of latchkey kids and teenagers with socialization problems; a place where economic, social, and family pressures contributed to unprecedented rates of teen suicide and behavioral problems.[9]

Yet the more things change, the more they can seem to stay the same. Poverty and slums were rediscovered once again beginning in the 1980s, with the relationships between them reformulated this time around to reflect the latest impulses in the dynamics of urban change. Economic restructuring and the retrenchment of welfare programs had spread and intensified poverty and contributed to homelessness on an unprecedented scale. Changing social attitudes, meanwhile, had contributed to an increase in divorce and single-parent households. One result was the **feminization of poverty** and homelessness. This in turn was compounded by spatial restructuring, which resulted not only in the decentralization of many better-paid jobs but also in the departure of many social role models from low-income neighborhoods.

The result was that the problem came to be reformulated in terms of a geographically, socially, and economically isolated group of truly disadvantaged poor people who were seen at once as victims and as incubators (see p. 424). Here were residual populations of vulnerable and disadvantaged households, casualties of structural economic change. But here also were the inheritors of urban degeneracy, within a whole spectrum of new pathologies that included senseless and unprovoked violence, premeditated, predatory violence, domestic violence, the organized but mostly latent violence of street gangs, and high levels of HIV/AIDS—all closely associated with drug taking and drug dealing. With this interpretation, then, we have returned to the theme of cities as haunts of vice.

What can we learn from this brief sketch of the history of urban problems? First, we should recognize that problems have to be seen not only in the context of the broader sweep of urban change but also through the eyes of beholders whose own position in this same history colors their perceptions. Second, we should recognize the complexity of the cause-and-effect web inherent to all urban problems: Each strand of the web is part of the sociospatial dialectic, not simply a series of irritations and unwanted by-products of urbanization. The remainder of this chapter is designed to illustrate this complexity in relation to some of the most critical problems of contemporary urban change in developed countries. Third, given this perspective, we can begin to appreciate the changing imperatives of urban governance, policy making, and planning that are the subject of Chapters 16 and 17.

SLUMS AND POVERTY AREAS

Poverty is arguably the most compelling problem both *in* and *of* cities. In reality, poverty means not only low incomes, few material assets, and a low quality of life, but also *persistent* low income and little prospect of acquiring material assets or an improved quality of life. More important, poverty comes as a package: low income is at the core, but it is inextricably bound up with poor diet, poor environment, poor physical health, the psychological stress of continuously having to make ends meet, and the economic, social, and political disadvantages of being often unfairly stigmatized as a "loser" in an increasingly competitive society.

In general terms, the core of poverty—low income—can be understood as a product of:

1. the income distribution inherent to an economic system based on relatively free competition and private profit;
2. uneven distributions of resources and opportunities between places and between socioeconomic and ethnic groups; and/or

3. institutional shortcomings, specifically involving:
   - inadequate participation and representation within the political process, and/or
   - inappropriate structure or malfunctioning of the welfare system.

To the extent that these factors may well be more intense in some cities or neighborhoods, the resultant poverty can be considered to be a problem *of* urbanization. Cities experiencing deindustrialization, for example, are likely to have high levels of unemployment and depressed wage rates within their local labor markets.

But poverty also needs to be understood in terms of the full package of disadvantages that come with low incomes, and it is in this context that poverty can be seen more clearly as a problem of urbanization. As we saw in Chapter 5, the logic of urban land use dictates that low-income households are left to live at high densities in the cheapest housing. Such households live in crowded conditions in run-down housing in older neighborhoods: classically in the **zone in transition** around the **CBD** or adjacent to older industrial districts and corridors. Filled with low-income households and nearing the end of their physical lifecycle (see Chapter 13), such neighborhoods are subject to a **spiral of decay** that can result in their becoming *slums*, where maintenance and repair are, at best, makeshift and where physical deterioration is accompanied by an accumulation of garbage and graffiti.

The spiral of decay begins with substandard housing (a product of physical deterioration and structural and technological obsolescence: see Chapter 13) occupied by low-income households who can afford to rent only a minimal amount of space (Figure 15.5). The resultant overcrowding not only causes greater wear and tear on the housing itself but also puts intense pressure on the neighborhood infrastructure of streets, parks, schools, and so on. The need for maintenance and repair increases very quickly, but it is not met. Individual households cannot afford to pay for dwelling maintenance or repair on any significant scale. Landlords have no incentive to do so, since they have a "captive" market. Public authorities are often indifferent to the needs of such neighborhoods because of their relative lack of political power, or they simply feel overwhelmed by the costs of rehabilitating and maintaining

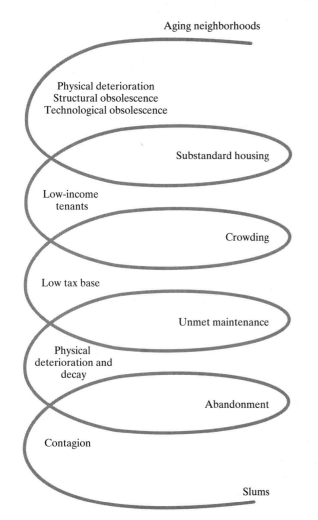

the neighborhood infrastructure. The same spiral of decay afflicts shops and privately run services such as restaurants and hairdressers. With a low-income clientele, profit margins must be kept low, leaving little to spare for upkeep or improvement. Many small businesses fail or relocate to more profitable settings, leaving commercial property vacant for long periods. In extreme cases such property becomes abandoned, the owners unable to find either renters or buyers.

Some residential buildings may also be abandoned, though abandonment of residential property usually requires the loss of a significant proportion of landlords' captive market of low-income households. In the 1960s residences began to be abandoned in some of the older industrial centers as a result of the decentralization of manufacturing jobs. Faced with escalating maintenance costs and rising property taxes and unable to increase revenue because of rent controls and the depressed state of the inner-city housing market, some landlords simply wrote off their property by abandoning it to long-term vacancy. In desperation, some property owners (both commercial and residential) resorted to arson in an attempt to at least salvage insurance monies.

Abandonment in any case increases the risk of accidental fire caused by the homeless or others. Fire-blackened ruins or just boarded-up, abandoned property can be an important element in the spiral of neighborhood decay because of the **contagion effect** that they introduce (Figure 15.6). A detailed study of abandoned buildings in Philadelphia by geographer Michael Dear suggests that abandonment can be viewed as a contagious process with two broad stages.[10] Initially there emerges a broad scatter of loosely defined clusters of abandoned structures, each consisting of several "microgroups" of two or three abandoned units. Over several years, such patterns are intensified, the broad scatter being maintained but the abandoned units increasing in number.

The reasons for this localized contagion effect are closely related to the depressing effects of long-term vacancy on the value and desirability of adjacent property. The fact that abandoned buildings make good sites for crime (especially the sale and use of illegal drugs) and vandalism provides a further impetus for the abandonment of nearby property. The downward spiral is further reinforced by what Dear calls the "psychological abandonment" of the wider area by realtors, financiers, and landlords (who become more systematic in their disinvestments) and by public agencies (which begin to cut back on maintenance and service delivery).

THE CYCLE OF POVERTY

There is, meanwhile, a dismal cycle that intersects with these localized spirals of decay. The **cycle of poverty** (Figure 15.7) also begins with low incomes, poor housing, and overcrowded conditions. Such conditions are

FIGURE 15.5   The spiral of neighborhood decay.

FIGURE 15.6   Abandoned property such as this represents a risk to public health and safety, provides a niche for illegal activity, and has a negative effect on the value of neighboring property because of the contagion effect.

unhealthy, extreme overcrowding contributing to susceptibility to physical ill-health, which is compounded by poor diets that are also a result of low incomes. Ill-health causes absenteeism from work, which results in decreased income. Absenteeism from school through illness may also contribute to the cycle of poverty by constraining educational achievement, limiting occupational skills and so leading to low wages. More significantly, crowding also produces psychological stress that contributes not only to ill-health (and therefore reinforces the cycle of poverty) but also to behavioral responses.

This cycle, of course, is related to the ecological ideas first put forward by Louis Wirth and the Chicago School of sociologists (see pp. 311–314): High-density slums represent distinctive "natural areas" in which "normal" adaptive behavior is often insufficient to cope with the pressures not only of crowding but also of a heterogeneous population living under chronic economic stress (Figure 15.8). As a result, such areas tend to foster **anomie**, **social disorganization**, and a variety of pathological behaviors, including crime and vandalism. Such conditions not only affect individuals' educational achievement and employment opportunities but can also lead to **labeling**, through which all residents may find that their neighborhood's poor image affects their employment opportunities. The

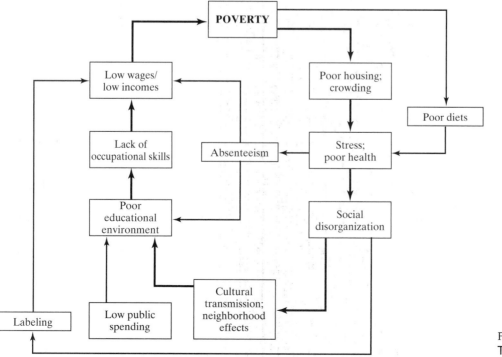

FIGURE 15.7
The cycle of poverty.

FIGURE 15.8   The disadvantages are not insurmountable, but growing up in this kind of setting brings a high probability of staying trapped in a localized cycle of poverty.

lifeworld of slum neighborhoods, meanwhile, is often seen as a critical element in the cycle of poverty, a combination of **cultural transmission** and neighborhood effects (see p. 301) leading to the transmission of modes of speech, codes of behavior, and attitudes to schooling, work, and society that operate as handicaps in wider educational and labor markets, thus contributing to the "trap" of low incomes and slum settings.

Probably the most important element in this cycle of poverty, however, is the educational setting of slum neighborhoods. Schools tend to be not only obsolete and physically deteriorated, like their surroundings, but also unattractive to teachers—partly because of the physical environment and partly because of the social and disciplinary environment. Because of the traditional indifference of public agencies to the political and social significance of their surrounding communities, often the schools are also resource-poor with relatively small budgets for staff, equipment, and materials. Over the long term, poor educational resources translate into poor education, however positive the values of the students and their parents. Poor education limits occupational choice and, ultimately, results in lower incomes. Faced with the evidence all around them of unemployment and low-wage jobs at the end of school

careers, students and parents inevitably find it difficult to sustain positive values about education. The result then becomes a self-fulfilling prophecy.

Some observers have given particular emphasis to the role of values and attitudes in perpetuating the cycle of poverty, arguing that there eventually emerges a culture of poverty.[11] In essence, it points to a vicious cycle that links lack of opportunity and lack of aspiration. The poor, according to this perspective, see the improbability of achieving any kind of material or social success and so adapt their expectations and behavior, becoming so accustomed to deprivation, so defeated, and so withdrawn that they become both unable and unwilling to seize opportunities for educational or occupational advancement when they do occur. Indolence, distrust, suspicion, and introversion become a way of life, defining a distinctive **culture of poverty**. This scenario has proven particularly popular with observers whose political inclinations are to "blame the victim." There is, however, considerable room for debate as to the actual extent of such phenomena. Indeed, there is no empirical evidence to suggest that the urban poor have values and aspirations—let alone a culture—significantly different from the rest of society.

Finally, we should note that some slums are also ghettos (though not all ghettos are slums). As we have seen, ethnic and racial segregation fulfills several functions, including cultural preservation, defensive functions, support functions, and "attack" functions (p. 306). Although these may help to mitigate the cycle of poverty, we should also note that ghettos are, by definition, ethnic or racial concentrations that stay in existence over several generations mainly because of charter group attitudes and discrimination (p. 364). The point is that this discrimination is not limited to principles of **social distance** and spatial segregation. It is part of a more general pattern of discrimination that affects education and labor markets as well as housing markets. In the case of the ghetto poverty, all three come together, reinforcing the cycle of poverty and intensifying the disadvantages of the minority poor.

Poverty Areas

We can put these conceptual frameworks into better perspective by reviewing the actual pattern and extent of poverty in U.S. cities. In the United States poverty is officially defined in terms of a set of money income thresholds that vary by family size and composition, indexed against the Consumer Price Index. In 2002 the poverty level for a typical family of four was $18,400.[12] By this criterion, there were nearly 35 million people living in poverty in the United States, just over 78 percent of them living in metropolitan areas. As Figure 15.10 shows, this incidence has fluctuated at about the same level (between 12 and 15 percent of the population) since falling dramatically in

BOX 15.1   POVERTY, STRESS, AND CIVIL DISORDER

Slum settings inevitably create stress and tensions, and it is perhaps not surprising to find that rioting and civil disorder break out from time to time. There has been a long but intermittent history of such civil disorder in the United States. Many of the early incidents involved minority migrants and immigrants being attacked by mobs of established residents who had been incited to anger at the perceived threat to their health, their wages, and their jobs. Significant examples occurred in East St. Louis in 1917, in Chicago in 1919, and in Detroit in 1943. Rioting *within* ghettos, more a form of mass protest and an expression of anger and desperation, first occurred in Harlem, New York City, in 1935 and 1943, the riots in both cases consisting of African Americans massing in the streets, breaking store windows, looting, and starting fires. The incidents were confined entirely to Harlem, the only whites victimized being those who were encountered on local streets.

Riots like these occurred in every major American city during the mid- to late 1960s. Although their immediate effects were quite localized, their impact on American politics and urban affairs was profound. Poverty was rediscovered, racism was confronted, and specifically targeted urban policies were developed.[13]

As the "long hot summers" of the late 1960s progressed, a clear pattern emerged. The riots were typically triggered by a police incident. Many African Americans saw the prejudice and brutality of white police officers who typically patrolled their neighborhoods as symbolic of the broader discrimination they experienced. Specific incidents served as catalysts for chain reactions of frustration and resentment. Once under way, the typical riot lasted for several days and nights, with massive crowds in the streets looting and setting fire to stores and preventing police and firefighters from entering the area. Large proportions of local residents were involved— 30 to 50 percent was common—but few attempts were made to spread riot activities into other areas of the city.

The scale and intensity of the riots can be conveyed by a few examples. In two of the earliest incidents, riots in Harlem and Bedford-Stuyvesant (in Brooklyn) in 1964 lasted for six nights, resulting in one death, 118 injuries, and 465 arrests. The following year saw many more incidents, including the notorious Watts riots in Los Angeles, in which 30,000–35,000 adults participated, burning, looting,

or damaging almost 1,000 buildings, creating property damage estimated at $40 million, and resulting in 34 deaths, over 1,000 injuries, and nearly 4,000 arrests. The peak year of rioting was 1967, with major incidents in Cleveland, Newark, and New Haven. The most serious incident, however, was in Detroit, where more than 75,000 were on the streets for three days and nights. This time 43 were left dead, over 2,000 were injured, 5,642 were arrested, and the bill for property damage was $200 million. In that fourth year of rioting, President Lyndon Johnson appointed the Kerner Commission to investigate the causes of what had become a major national issue. *The commission's report pointed to the underlying racism of American society, acknowledged bad relations between black communities and police and public agencies as the immediate cause of most incidents, and stressed the economic and social distress of African Americans in inner-city neighborhoods.*

A generation later the country's worst-ever incident provided a depressing echo of the 1960s (Figure 15.9). After the acquittal in May 1992 of police officers who had beaten Rodney King, an African-American motorist, rioting in Los Angeles (including in the Watts neighborhood) resulted in more than 1,000 fires, left more than 5,000 buildings damaged, and produced a death toll of 52. More than 8,000 people were arrested, the bill for property damage was in excess of $5 billion, and thousands of workers were temporarily unemployed because of riot damage.

In the early twenty-first century the underlying causes of these incidents have not yet been adequately addressed, as evidenced by the riots that continue to be sparked in black inner-city neighborhoods. In June 2003 rioting broke out in Benton Harbor, Mich., after the death of Terrance Shurn, a black Benton Harbor resident, whose speeding motorcycle crashed into a building as police were chasing him into the city. Benton Harbor is a city of 12,000 people about 100 miles northeast of Chicago that is 92 percent black. Across the river, St. Joseph, a city of 8,800, is 90 percent white. Hundreds of people rioted in the streets near where the crash occurred, setting fires and attacking passers-by. At least 11 people were hurt. Rioters fired shots at emergency personnel, and eight police and fire vehicles and two private vehicles were damaged. Property damage was estimated at $500,000—the fires completely burned 21 homes, including the one that Shurn struck, while seven more were damaged.

(Continued)

BOX 15.1   POVERTY, STRESS, AND CIVIL DISORDER   *(Continued)*

FIGURE 15.9
A neighborhood in South Central Los Angeles following the 1992 riots that swept the city for days after the acquittal of police officers who had beaten Rodney King, an African-American motorist.

the early 1960s as a result of economic prosperity and the introduction of a welfare safety net.

Within metropolitan areas, the Census Bureau has defined **poverty areas** as contiguous census tracts in which at least 20 percent of the households have an income below the official poverty level; census tracts with 40 percent or more below the poverty level have been defined as "extreme poverty areas." Other researchers have also identified "high poverty areas"—those in which at least 30 percent of households have an income below the poverty level.[14] These areas—typically clustered in central cities—provide a good indicator of the nature and extent of poverty as a problem of urban change. Data from the period since 1970 are especially

illustrative, since they cover a period of particularly intense restructuring in the 1970s and 1980s across the urban system and within metropolitan areas.

Figure 15.11 shows the spatial pattern of poverty areas in six very different cities in 1970 and 1980. What the maps reveal is typical: a concentration in inner-city neighborhoods that spreads outward during the 1970s. What was happening was that people both above and below the poverty line were moving out of poverty-area neighborhoods in the 1970s, partly in response to the decentralization of jobs and partly in response to increasing levels of violent crime and a deteriorating quality of life. Those above the poverty line moved out faster and in greater numbers than the

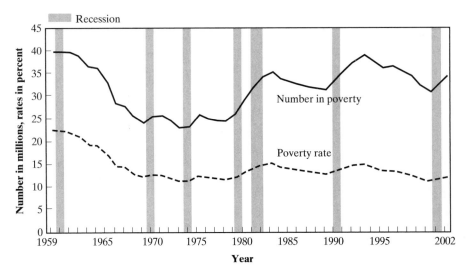

FIGURE 15.10   Poverty in the United States, 1959–2002.

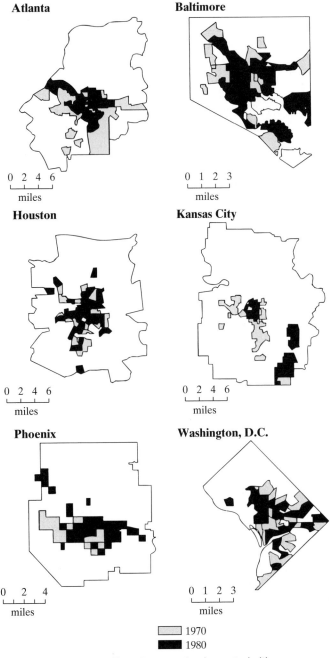

Atlanta

0 2 4 6
miles

Baltimore

0 1 2 3
miles

Houston

0 2 4 6
miles

Kansas City

0 2 4 6
miles

Phoenix

0 2 4
miles

Washington, D.C.

0 1 2 3
miles

☐ 1970
■ 1980

FIGURE 15.11   Poverty areas in six central cities, 1970 and 1980.

poor, however, while few of the extremely poor were able to move out at all.

Between 1970 and 1990 the spatial concentration of the poor increased dramatically in many U.S. metropolitan areas. The number of people living in extreme poverty areas doubled; the probability that a poor black child was living in such an area increased from about one in four to one in three; and the areal extent of the blighted sections of many central cities increased even more dramatically. In contrast, poverty—measured at the family level—did

not increase. As a result, the poor became more spatially isolated from the social and economic mainstream of society.[15]

During the economic expansion in the 1990s, poverty became markedly less concentrated. The share of the metropolitan poor who lived in extreme poverty neighborhoods dropped to 12 percent in 2000 (having increased from 13 to 17 percent in the 1980s) as the number of people living in extreme poverty areas declined significantly, by 2.5 million or 24 percent (Figure 15.12). The change in concentrated poverty was associated with a corresponding increase in high poverty areas in the suburbs—especially the older inner-ring suburbs—of the 100 largest U.S. metropolitan areas (to 15 percent in 2000, up from 11 percent in 1980). During that time, however, the central cities in these metropolitan areas continued to retain a dominant, if decreasing, proportion of high poverty areas—at 62 percent in 2000, down from 67 percent in 1980 (Figure 15.13).[16]

The composition and share of the poor living in poverty areas in cities also changed a good deal. In 1980, for example, blacks were the predominant racial group (comprising more than 60 percent of total population) in nearly half (48 percent) of all high poverty areas, and these areas accounted for more than half (54 percent) of the poor population in high poverty areas. By 2000 the share of all high poverty areas that was predominantly black had declined markedly to 39 percent of the poor population in these areas (down from 48 percent in 1980) and to only one-third of the total, while those that were predominantly white fell from 18 percent to 14 percent. In contrast, high poverty areas that were predominantly Hispanic increased from 13 to 20 percent (Figure 15.14).

Comparisons across the **urban system** show that these overall figures mask a great deal of variation, reflecting the changes that were being experienced in the functional organization of the urban system (Chapter 4). Industrial cities such as Baltimore, Buffalo, Chicago, Detroit, and Rochester, for example, were distinctive for particularly large increases in the incidence of poverty and in the proportion of the poor (African American, Hispanic, and white) living within their poverty areas during the 1970s and 1980s. A few cities (Birmingham, Jacksonville, Louisville, and Omaha) were distinctive because of very high increases (around 150 percent) in the proportion of their poverty-area poor who were unemployed and for large increases (about 30 percent) in the proportion of poverty-area families headed by women. The Midwest and the South experienced comparable significant decreases in the percentage of the poor population living in high poverty areas between 1990 and 2000; the Northeast experienced a slight decrease; while there was a slight increase in the fairly low percentage in the West (Table 15.1).[17]

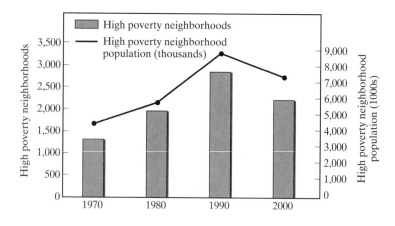

FIGURE 15.12   Extreme poverty areas and population in U.S. metropolitan areas, 1970–2000.

Table 15.2 shows the 15 metropolitan areas with the greatest decreases in extreme poverty area population during the 1990s—all of which were located in the Midwest and South. Detroit's improvement was substantially better than that of any other metropolitan area. Its experience reflected that of many of the other 199 metropolitan areas (out of 331) that saw declines in the number of people living in extreme poverty areas in the 1990s. During the 1970s and 1980s, they experienced population loss at the same time as they increased in spatial extent—the depopulating and impoverished urban cores became surrounded by growing and prosperous suburbs. But just as central cities had borne the brunt of the fiscal, social, and economic burden of concentrating poverty in the 1970s and 1980s, so they became the beneficiaries of its reduction during the economic expansion in the 1990s.[18]

Figure 15.15 shows the extreme poverty areas in Detroit. Between 1970 and 1990 the number of extreme poverty areas increased significantly so that by 1990, nearly half of the land area of the central city of Detroit comprised these areas. In the 1990s this trend reversed. Figure 15.16 shows the growth of extreme poverty areas during the 1970s and 1980s in Dallas. Their growth was associated with sharply lower oil prices following the collapse of the OPEC oil cartel and in the wake of rapid suburban development. Plano, a **boomburb** just north of Dallas, was

for years the fastest growing city in the country. In the 1990s, however, Dallas experienced substantial redevelopment, including condominium and apartment developments, just north of the downtown and along Interstate 45.[19]

Table 15.3 shows the 15 metropolitan areas with the greatest increases in extreme poverty area population during the 1990s—seven of which are located in California. A total of 91 of 331 metropolitan areas experienced some increase in persons living in extreme poverty areas. Figure 15.17 shows the expansion of extreme poverty areas in Los Angeles during the 1970s and 1980s. Several factors may account for the divergence of Los Angeles from the national trend in the 1990s.

First, the city experienced a deadly and destructive riot after the Rodney King verdict in 1992, and further heightening of racial tension due to the trial of O. J. Simpson in 1995. The riot and its aftermath almost certainly accelerated middle-class flight from the central city area, and the trial emphasized racial divisions in the region. Second, the Los Angeles region experienced tremendous immigration from Mexico and other central and South American countries . . . . Third, the recession of the early 1990s was particularly severe in Southern California, and the economic recovery there was not as rapid as in other parts of California (such as the San Francisco/Silicon Valley area) that benefited from the Internet boom.[20]

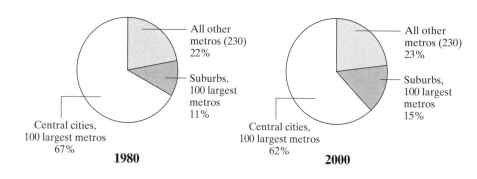

FIGURE 15.13   High poverty areas in the United States by metropolitan location, 1980 and 2000.

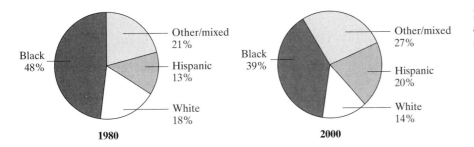

1980   **2000**

FIGURE 15.14   High poverty areas in the United States by race/ethnicity, 1980 and 2000.

TABLE 15.1

Changes in Concentrated Poverty by Region and Type of Metropolitan Area, United States

Region	Number of metropolitan areas	Population (millions) 2000	Percent population growth 1990-00	Poor population in high poverty areas (thousands)			Percent of poor population in high poverty areas		
				1980	1990	2000	1980	1990	2000
Northeast	59	47.6	6	1,377	1,419	1,605	29	31	30
Midwest	78	47.0	9	1,029	1,710	1,165	26	36	26
South	128	73.9	19	1,943	2,792	2,319	29	34	25
West	65	54.6	19	533	1,182	1,613	14	22	24
Total U.S.	330	223.1	14	4,883	7,103	6,701	25	31	26

Source: *G. T. Kingsley and K. L. S. Pettit,* Concentrated Poverty: A Change in Course, *Washington, D.C.: Urban Institute, 2003, Table 3, p. 6 (http://www.urban.org/UploadedPDF/310790_NCUA2.pdf).*

FIGURE 15.15   Extreme poverty areas in Detroit, 1970–2000.

TABLE 15.2

Top 15 U.S. Metropolitan Areas by Decline in Population of Extreme Poverty Areas (Census Tracts), 1990–2000

	Decline in population	*% Decline in population*	*Decline in census tracts*
Detroit, Mich.	313,217	74.4	97
Chicago, Ill.	177,908	43.1	73
San Antonio, Tex.	107,272	70.1	18
Houston, Tex.	77,662	47.8	27
Milwaukee, Wis.	63,357	45.0	16
Memphis, Tenn.	61,924	43.6	11
New Orleans, La.	57,332	34.6	18
Brownsville, Tex.	50,559	37.1	4
Columbus, Ohio	48,020	55.4	11
El Paso, Tex.	44,489	40.2	4
Dallas, Tex.	41,805	45.3	19
St. Louis, Mo.	38,866	35.5	13
Lafayette, La.	33,978	54.8	10
Minneapolis–St. Paul, Minn.	32,005	40.5	18
Flint, Mich.	31,631	61.2	6

Source: *P. A. Jargowsky,* Stunning Progress, Hidden Problems: The Dramatic Decline of Concentrated Poverty in the 1990s, *Washington, D.C.: The Brookings Institution, 2003, Table 1, p. 6 (http://www.brookings.edu/es/urban/publications/jargowskypoverty.pdf).*

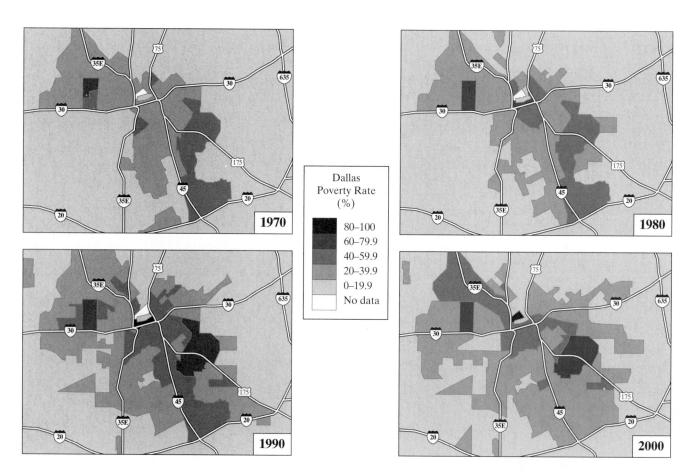

FIGURE 15.16   Extreme poverty areas in Dallas, 1970–2000.

TABLE 15.3

Top 15 U.S. Metropolitan Areas by Increase in Population of Extreme Poverty Areas (Census Tracts), 1990–2000

	Increase in population	*% Increase in population*	*Increase in census tracts*
Los Angeles, Calif.	292,359	109.2	81
Fresno, Calif.	60,005	68.7	12
Riverside–San Bernardino, Calif.	58,669	260.5	12
Washington, D.C.	56,954	276.4	14
Bakersfield, Calif.	42,622	190.8	9
San Diego, Calif.	33,274	86.1	10
McAllen–Edinburg–Mission, Tex.	28,117	12.0	(1)
Providence, R.I.	22,186	235.4	3
Chico–Paradise, Calif.	16,675	103.3	4
Middlesex–Somerset–Hunterdon, N.J.	14,020	n.a.*	3
Wilmington–Newark, Del.	12,349	276.3	2
Bryan–College Station, Tex.	11,746	29.4	4
Visalia–Tulare–Porterville, Calif.	11,176	60.1	2
Rochester, Tex.	9,989	29.8	0
Monmouth–Ocean, N.J.	9,114	318.3	(1)

* Middlesex–Somerset–Hunterdon, NJ, had no census tracts with poverty rates of 40 percent or higher in 1990.

Source: *P. A. Jargowsky,* Stunning Progress, Hidden Problems: The Dramatic Decline of Concentrated Poverty in the 1990s, *Washington, D.C.: The Brookings Institution, 2003, Table 2, p. 9 (http://www.brookings.edu/es/urban/publications/jargowskypoverty.pdf).*

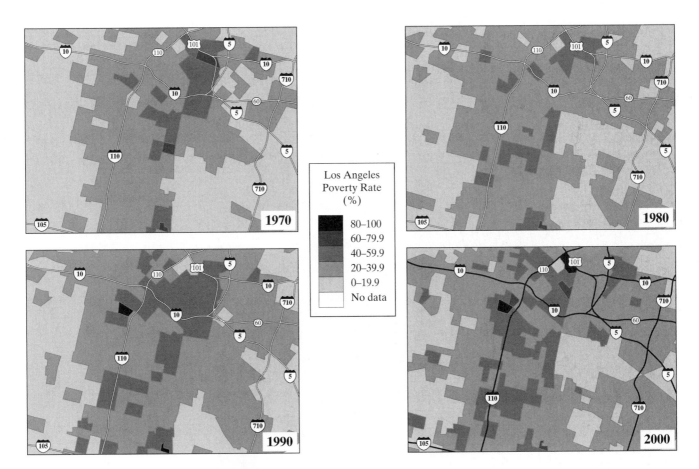

FIGURE 15.17   Extreme poverty areas in Los Angeles, 1970–2000.

Certainly, conditions in high poverty areas generally improved in the 1990s. For example, the proportion of adults without a high school degree fell from 48 to 43 percent, the proportion of female-headed families with children fell from 53 to 49 percent, and the proportion of women over 16 who were working increased from 40 to 42 percent. At the same time, however, conditions in other parts of most metropolitan areas also improved—so disparities in conditions did not diminish much, if at all.[21] In addition, economic downturns and the increasing concentration of poverty in many older inner-ring suburbs may lead to a further reversal of the improving national trend of the 1990s.[22]

Dual Cities?

Some evidence supports the idea that the extreme poverty areas are the expression of localized cycles of poverty *in which disadvantages are accumulated and deprivation is intensified*. Poor people in these areas experienced significantly higher rates of unemployment than the poor who lived in areas of less-concentrated poverty, and they were more dependent on welfare (Figure 15.18), more likely to live in single-parent households, and more likely to have dropped out of high school. They were also *much* more likely to be African American.

These are the urban poor who have become relatively isolated not only from the economic mainstream but also from the social values and behavioral patterns of the rest of society. Sociologist William Julius Wilson defined the truly disadvantaged as:

> . . . that *heterogeneous grouping of families and individuals who are outside the mainstream of the American occupational system*. Included . . . are individuals who lack training and

skills and either experience long-term unemployment or are not members of the labor force, individuals who are engaged in street crime and other forms of aberrant behavior, and families that experience long-term spells of poverty and/or welfare dependency.[23]

A more concise conceptual definition has been suggested by Martha Van Reitsma:

> . . . those who are weakly connected to the formal labor force and whose social context tends to maintain or further weaken this attachment.[24]

Van Rietsma, like Wilson, pointed to macroeconomic restructuring as the fundamental reason for this situation. The existence of large numbers of people with only weak connections to the formal labor force is attributed to the **spatial mismatch** between people and jobs that intensified as many of the low-skill jobs traditionally found in inner-city areas have been relocated, to be replaced mainly by jobs requiring higher skills.[25] Figure 15.19 shows the average values for a spatial mismatch measure in 2000—the dissimilarity index for total employment and that for retail employment—by race and ethnicity, weighted by population. What is most striking are the clear racial/ethnic differences that persist in the extent of the mismatch between people and jobs. The greatest differences are between whites and blacks, with Asians and Hispanics lying between these values. In 2000 more than half the black population would have had to relocate to achieve an even distribution of blacks relative to jobs, compared to only one-third of whites.

Figure 15.20 shows the association between the spatial mismatch and housing segregation in U.S.

FIGURE 15.18   Unemployed people waiting to collect welfare benefits at an unemployment office in New York.

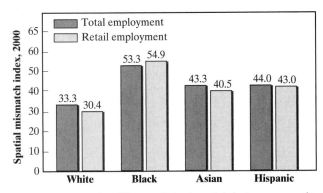

FIGURE 15.19 The spatial mismatch between people and employment by race/ethnicity, U.S. metropolitan areas, 2000.

metropolitan areas in 2000. Metropolitan areas with higher levels of black-white residential segregation show a greater degree of spatial mismatch between blacks and jobs. As residential segregation between blacks and whites increases, the physical separation of blacks from total employment also increases. At one extreme—Portland, for example—the mismatch index between blacks and jobs is just 19 percent, and only 26 percent of blacks would have needed to relocate in 2000 to achieve an even distribution of blacks and whites across all neighborhoods. At the other extreme, the mismatch index for blacks and total employment was 71 percent in Detroit, a metropolitan area where 85 percent of the black population would have had to relocate to achieve an even black-white residential distribution.[26] These figures reflect the fact that a large proportion of the black inner-city population remains "trapped," unable to afford housing or transportation that would make suburban jobs accessible to them, and dependent on welfare or various "informal" sources of income, including illegal activities such as prostitution and drug dealing.

The "social context" referred to by Van Rietsma is partly a reference to the role of local attitudes and behaviors as outlined above in relation to the idea of a cycle of poverty—the tendency to drop out of school,

for example. But it also refers to the outcomes of some of the social, cultural, political, and demographic trends that have been associated with the economic and metropolitan restructuring since the 1960s. One of the most important of these trends is the increase in the number of black teenage unwed mothers (see p. 324) that has led to dramatic increases in black single-parent households and to the feminization of poverty. Another is the selective migration of younger and higher-skilled blacks from central cities to the suburbs. This migration has left a residual population of unskilled and elderly persons along with female-headed households in inner-city neighborhoods. It has left them, additionally, with significantly fewer role models. As a result, there has been an erosion of mainstream social norms and an increase in social disorganization and deviant behavior.

But these linkages between labor markets, poverty, migration, household structure, race, gender, attitudes, and behavior have proven very difficult to measure scientifically. Even so, the economic, social, and metropolitan restructuring since the 1960s has resulted in both qualitative and quantitative changes in American slum and poverty-area settings. In this context it is important to see these changes as part of a broader shift in the dynamics of urbanization—one that has introduced new patterns of vulnerability, fragmentation, and disadvantage while intensifying the occupational and spatial polarization of the city. Sociologist Manuel Castells has argued that the overall result is a **Dual City** in which downsized workers, the poor, and the homeless share excluded spaces, fragmented along racial and ethnic lines. Just as the slums of nineteenth-century U.S. cities were "haunts of vice," so these "excluded" spaces are today the places that are most often associated with the most intense concentrations of social pathologies. As Castells observes, "Downgraded areas of the city serve as refuges for the criminal element of the informal economy, as well as reservations for displaced labor, barely maintained on welfare."[27] In the next section we examine one particular aspect of pathological behavior—criminal violence—that is widely identified as one of the salient problems of contemporary urbanization.

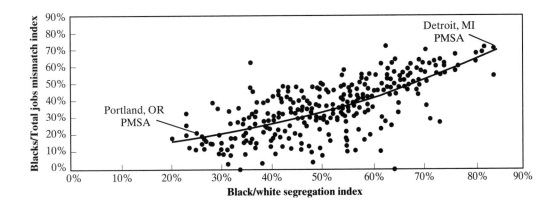

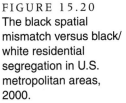

FIGURE 15.20 The black spatial mismatch versus black/white residential segregation in U.S. metropolitan areas, 2000.

CRIMINAL VIOLENCE

The United States is "the most violent and self-destructive nation on earth" was the conclusion of a 1991 report of the Senate Judiciary Committee on violent crime (homicide, rape, robbery, assault). Since the early 1990s, however, violent crime rates have declined to the lowest levels recorded recently in the United States (from about 48 incidents per thousand people in 1973 to about 23 per thousand in 2002) (Figure 15.21).[28] Like poverty, social pathologies such as criminal violence are the product of the interplay of several causal processes; like poverty, they are at once problems *in*, as well as problems *of*, the city. In addition, poverty and social pathology have for a long time been seen as causally linked, poverty and its associated environments seeming to foster deviant attitudes and to create stress and resentment that trigger deviant behavior.

We can recognize four main explanatory linkages between urban context and violent crime.[29]

1. *Economic Effects.* Among the reasons for hypothesizing direct causal linkages between poverty and violent crime are ideas that:
   - the poor as a group develop a culture that embraces violence as an expression of toughness or a measure of status and identity;
   - feelings of relative deprivation lead to anger and resentment that are channeled into violence;
   - psychological reactions to feelings of relative deprivation, together with psychological stresses resulting from poverty, serve to limit intergroup and intra-group interaction, thus undermining both formal and informal mechanisms of social control and, consequently, removing many of the restraints that would otherwise suppress violence.

Research has confirmed clear links between poverty, inequality, and violent crime—particularly homicide (Figure 15.22). But researchers have found it difficult to isolate these effects from the mediating and interactive effects of other factors, such as racial composition, racial inequality, social disorganization, and family composition.

2. *Social Disorganization.* Even without the presence of significant levels of poverty or inequality, high rates of population turnover and high levels of diversity in the social, demographic, cultural, and ethnic attributes of community populations can sometimes inhibit the formation of local ties, weaken the effectiveness of local social institutions, and leave communities ill-equipped to maintain social order. Research has shown clear links between criminal violence (particularly assaults) and rapid ecological change, even when the change is positive, involving neighborhood upgrading.

3. *Demographic Effects.* Since the age distribution of offense rates shows a clear peak in the late teens and early twenties, the argument here is that sociodemographic changes in neighborhood composition are related to changes in urban crime rates. Research has provided some evidence of such a linkage, though it was underpinned by differential effects of economic restructuring on specific sociodemographic groups.

4. *Effects of Lifestyle and Routine Activities.* The linkage between crime and people's lifestyle or routine patterns of activity is based on the observation that for a crime to occur there are three preconditions: (a) the presence of a motivated offender, (b) the absence of capable guardians or protectors, and (c) the presence of a suitable victim.

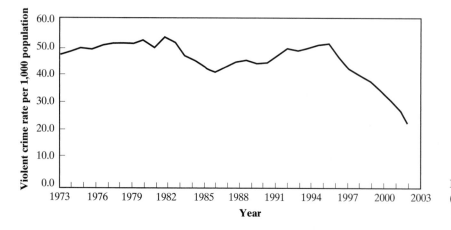

FIGURE 15.21 Violent crime rate (assault, rape, robbery, and homicide) per 1,000 people, 1973–2002.

FIGURE 15.22
Violent crime has become endemic to many poor inner-city neighborhoods.

It follows that victimization rates vary from one community to another because the time-space routines of different socioeconomic, demographic, ethnic, and lifestyle groups result in different degrees of proximity, *in space and time*, between potential offenders and potential victims. Research on such linkages, however, has shown that while they help to explain patterns of property-related crimes they do not explain patterns of criminal violence.

SPATIAL PATTERNS

Given this discussion, the spatial patterns associated with violent crime hold few surprises, whether the focus is on the geography of offenders' residences or the geography of crime locations. The classic pattern of *offenders' residences* was established long ago in the ecological studies of the Chicago School of the 1930s: a steady gradient, with low rates in the suburbs and a peak in the zone in transition that contains the worst slums and poverty areas. This basic pattern varies according to the nature of the offense and has been modified by the changes that have occurred in metropolitan structure since the 1930s. Nevertheless, the conclusions of a landmark study in 1960 of the geography of crime in Seattle hold true: High crime areas "are generally characterized by all or most of the following factors: low social cohesion, weak family life, low socio-economic status, physical deterioration, high rates of population mobility and personal disorganization."[30]

In broad terms, classic *occurrence patterns* are very similar. A large literature on the ecology of violent crime has not only established a strong correlation between low-income neighborhoods and crimes of violence but also points to the tendency for victims to be drawn overwhelmingly from the family, friends, neighbors, and local associates of criminals.[31] This literature also points to the importance of the microscale physical environment because of its influence on *opportunities* for crimes of violence. The U.S. National Commission on the Causes and Prevention of Violence, which reported in 1969, concluded that accessibility, visibility, control of property, residential density, and state of physical repair were the most significant aspects of the microenvironment of violent crime.

It was factors such as these that led architect Oscar Newman to stress the importance of **defensible space**. As we saw in Chapter 10, Newman was the author of a damaging attack on Modernist housing that argued that much of the vandalism, burglary, and criminal violence of inner-city neighborhoods was related to the "designing out" of territorial definition and delineation in the Modernist apartment blocks, duplexes, and triplexes that had come to dominate many inner-city neighborhoods. Once the space immediately outside the dwelling becomes public, Newman suggested, nobody feels obliged to supervise it or defend it against intruders.[32]

In recent years the patterns and intensity of offenses have seemed to many observers to have changed, generating increased concern about criminal violence as a key element in an increasingly dystopian scenario

BOX 15.2   A PROFILE OF DRUG INDICATORS: WASHINGTON D.C.

The White House Office of National Drug Control Policy's Drug Policy Clearinghouse produced a profile of Washington, D.C. in 2003 that contains information on drug use, drug trafficking, crime, enforcement, and drug programs.[33] About 60 open-air markets along major commuting routes and within public housing projects provide dealers with a constant supply of customers. The data on drugs and drug use reveal marked variations within and among the different racial, socioeconomic, and age groups across the city.

Cocaine:	Violence associated with the crack cocaine trade remains high. Crack is primarily abused in low-income, inner-city housing projects. Although the majority of powder cocaine users are black people over age 30 who inject the drug, an emerging group has been reported: young adult white males (aged 18–30) who snort the drug.
Heroin:	The heroin open-air markets located near methadone clinics sell lower-purity heroin to long-term heroin addicts. The markets along commuter routes offer high-purity heroin that is purchased and snorted by predominantly younger and more suburban users from Maryland and Virginia.
Marijuana:	Marijuana is the most readily available, least expensive, and widely abused illegal drug in D.C. About 20 percent of Washington, D.C. high school students surveyed in 2001 reported being current users of marijuana. Over 43 percent of 12th graders reported using marijuana at least once during their lifetime.
Methamphetamine:	Teens and young adults abuse methamphetamine, particularly crystal methamphetamine, in combination with other drugs at nightclubs and raves.
Club Drugs:	MDMA, Ketamine, GHB, crystal methamphetamine, and other hallucinogenic and stimulant drugs have been readily available in D.C. for about a decade. Counterfeit MDMA tablets are sometimes sold containing substances such as ketamine, methamphetamine, cocaine, or PCP. GHB is sold at nightclubs, raves, bars, and universities.
Other Drugs:	Prescription painkillers and methadone are in high demand and readily available. Hydromorphone and OxyContin are sometimes used as heroin substitutes.

of inner-city life. Criminal violence seemed to be on the increase because it had not only spilled out onto the streets but also had become a daytime as well as a nighttime activity. At the same time, the menace of violence had increasingly extended beyond the slums. Delinquency had escalated from vandalism and larceny to senseless violence: beatings and shootings triggered because an innocent passer-by has the wrong look or is wearing the wrong colors or because a driver in the next lane at traffic lights is playing the wrong music. Fists, knives, and handguns have been augmented by semiautomatic weapons and sniper rifles. The city's most fearsome threat is now that of becoming embroiled in "random" violence: muggings, drive-by shootings, gang rapes. As a result, the city's slums and poverty areas are not just excluded from the mainstream life of the city, they are carefully avoided by all outsiders except those whose work takes them there. In fact, the murder, aggravated assault, forcible rape, and robbery rates in U.S. cities have all decreased since the early 1990s (see Figures 15.21 and 15.23).

As we would expect, the experience of individual cities differs considerably (Figure 15.23). In 2001, for example, the incidence of aggravated assaults was nearly two and a half times as high in Baltimore and Atlanta as it was in San Antonio and San Diego; the incidence of rapes was more than twice as high in Detroit as in Washington, D.C. The murder rate in Dallas in 2002 was more than twice as high as that in San Francisco. In 2002, Washington, D.C. was the country's Murder Capital yet again. With 262 homicides in 2002 (down from 487 in 1991), the city had a homicide rate of 46 per 100,000 residents, compared to 42 and 38 per 100,000, respectively, in Detroit and nearby Baltimore.

Within cities, the basic spatial pattern of criminal violence shows little change, despite the newspaper reports of random violence spilling out of the slums or the recently declining rates of violence. In Washington, D.C., for example, there is a very close association between patterns of homicide, rape, and assault and neighborhoods of poor blacks and of individuals with low levels of education.[34] The vast majority of the homicide victims in the District are, like their assailants, black.[35]

In part, the criminal violence described in the preceding paragraphs can be attributed to the same shifts in the dynamics of urbanization that have intensified occupational and spatial polarization, introduced new patterns of vulnerability and disadvantage, and created the "dual city." In part, it is attributable to the escalation of drug abuse. By the 1980s heroin addiction had become an endemic problem in the central parts of American cities. The cost of drug addiction—estimated at about $3,600 a month[36]—leads directly to an escalation in crime, particularly burglary and armed street robbery, and to a further degeneration in social order. When crack cocaine appeared in the mid-1980s, the problems multiplied and intensified. The link between homicide and drugs, in particular, became direct. Two-thirds of the homicide victims in the District of Columbia in recent years have in fact been found to have drugs in their systems, and a similar

---

BOX 15.3   CRIME AND CORRUPTION IN THE CITIES OF THE RUSSIAN FEDERATION[37]

As the Soviet Union began to break up, crime and corruption began to appear in cities on an unprecedented scale. *Perestroika* and *glasnost* allowed new criminal networks to emerge amid the rapidly changing political and social environment in cities. There was a proliferation of ethnic *mafiyas*—Chechens, Azeris, Georgians, and so on—and a rapid spread of corruption that has since become ingrained in the business world and the political system. Organized crime and widespread corruption have subverted democratic governance, affecting the nature of economic and social development in many cities.

Privatization and market reform created unprecedented opportunities for corruption and organized crime. In the institutional vacuum that followed the breakup of the Soviet Union, there was no accepted code of business behavior, no civil code, no effective banking or accounting systems, and no procedures for declaring bankruptcy. Security agencies were disorganized and bureaucratic lines of command were blurred. Before long almost all small private businesses were paying protection money to criminal groups, while an estimated 70 to 80 percent of larger companies and commercial banks had criminal connections. The U.S. Drug Enforcement Agency has estimated that in 2000 organized crime controlled up to 25 percent of the commercial banks in Moscow.

Organized crime did not consist just of the traditional activities of prostitution, drugs, auto theft, and protection rackets; it also included the more lucrative business of illegal traffic in weapons, nuclear materials, oil, rare metals, other natural resources, and currency. Levels of criminal violence escalated in parallel with rising levels of crime and corruption, so that by the mid-1990s contract killings had become a way of life in the business world of Russian cities.

In 1999, U.S. investigators discovered that billions of dollars were being laundered through Bank of New York accounts, apparently shunted through a maze of companies traced to Russia. This was done largely in the name of Benex Worldwide Ltd., a company that was said to be controlled by Semyon Mogilevich, allegedly a vicious individual known as the "brainy don." It is not clear exactly where the money came from: It may simply have been capital that investors were taking out of the shaky Russian economy; it may have been looted from **International Monetary Fund** loans; it may have come from revenues from state assets such as oil or aluminum; or it may have come from organized crime.

One result of this has been that crime and corruption have taken away a significant proportion of resources that would otherwise have gone into restructuring the urban economies in Russia. This loss of investment for cities, combined with the uncertainties created by crime and corruption, has restricted the role of cities as engines of economic growth. Crime and corruption have also stifled the emergence of a civil society with a democratic base in cities. The scale of crime and corruption in the Russian Federation has also become an issue of geopolitical importance. Russian *mafiyas* have extensive connections to international organized crime; corrupt politicians and businesses have laundered so much money through the international financial system that the ramifications are truly global in scale. Global organized crime, particularly the Sicilian Mafia and the Colombian drug cartels, seized the chance to link up with Russian *mafiyas* to launder huge sums of money, to circulate counterfeit dollars by the millions, and to establish smuggling networks. The Russian *mafiyas*, in turn, seized the chance to extend their operations to the rest of the world. Today the scope of Russian *mafiya* organizations ranges from prostitution rings in major tourist hubs outside Bangkok, Thailand, to heroin trafficking between Afghanistan and major cities in Europe.

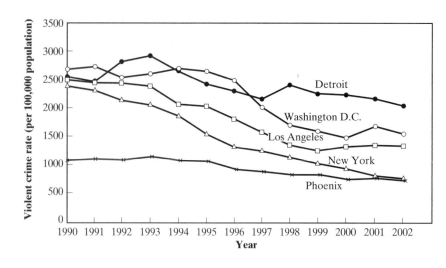

FIGURE 15.23   Violent crime rate (assault, rape, robbery, and homicide) per 100,000 people, selected cities, 1990–2002.

proportion of defendants in the District's Superior Court have been found to be drug users. Some of the homicides are the result of execution-style killings, casualties of the turf wars that have proliferated in the absence of organized crime and the presence of an oversupply of aggressive people seeking to escape poverty or pay for their own habit by dealing drugs.[38]

THE EFFECTS OF CRIME ON URBANIZATION AND URBAN LIFE

Just as criminal violence can be seen partly to be produced and substantially mediated by processes of urbanization, it can also be seen to influence these processes. As such, it is a good example of the **sociospatial dialectic** (see p. 378), the continuous relationship in which urban space and urban society shape and reshape each other.

One of the best-documented effects of crime on urbanization concerns house values and rents.[39] A relatively small change in the incidence of crime—a shift of 4 to 5 percent—can affect aggregate house values in a large city by hundreds of millions of dollars (and affect property tax revenues by tens of millions of dollars, thereby affecting the city's ability to deliver services and maintain infrastructure). In fact, of course, these changes are highly localized, so that one of the main effects, in terms of the sociospatial dialectic of urbanization processes, is on the dynamics of neighborhood change (Chapter 13).

Closely related to changes in house values are changes in people's housing and neighborhood satisfaction. Yet although high crime rates have been shown to produce significant increases in local residents' desire to move out, research has shown that actual residential mobility is not affected. Quite simply,

the residents of high-crime neighborhoods cannot afford to move out—everywhere else is more expensive. Being trapped amid the prospect of crime and criminal violence reduces community quality of life and induces strong feelings of fear and powerlessness

FIGURE 15.24   Crime and fear of crime can have important effects on urban life. The large number of Crime Watch signs in residential neighborhoods of all kinds attests to the extent of the problem.

that become contagious. Sensationalist reporting and repressive styles of policing often amplify these feelings. Fear of crime consequently becomes a significant element in the politics and behavior of other neighborhoods (Figure 15.24), even of the whole city. Fear also leads to changes in the built environment, as targets are "hardened" (Figure 15.25), and in methods of policing and social control, as electronic surveillance becomes the norm (see p. 269).

Meanwhile, back in the high-crime areas, there often emerges a social cleavage between "roughs"—those involved with crime or drugs, or both—and "respectables"—usually a majority. The latter have to adapt their lifeworld to the presence of the former. In practice, they have to watch cars carefully to avoid being caught in a drive-by shooting, steer clear of the neighborhood whenever possible, lock themselves into their apartments when they are home, draw shades and blinds and watch television from positions well away from windows, and stuff towels around doors to keep out crack fumes from the stairwells.[40] An additional form of social cleavage can emerge within the "roughs," if there are

sufficient numbers. In addition to the hierarchies within and between gangs and the divisions between offenders whose activities are accorded different degrees of status, the type of drug preferred by users can provide the basis for microlevel patterns of social interaction. "[Heroin] junkies despise the 'pipeheads' whose entire daily existence is often spent getting high or looking for rocks. Crack smokers, meanwhile, consider themselves superior to heroin junkies, who litter the complex with needles"—and will shoot up anywhere, cooking their dope with rainwater from the ground and using the mirrors of parked cars to locate veins in necks or groins.[41]

At the macroscale the effect of high crime rates in inner-city areas is to inhibit in-migration by the upwardly mobile young professionals who might otherwise be likely to act as a catalyst for neighborhood upgrading. Similarly, investment in retail and service businesses is deterred and disinvestment accelerated, while local government services are cut back or eliminated. In short, high crime rates accelerate and intensify the spiral of decay and reinforce the cycle of poverty.

(a)

(b)

FIGURE 15.25   Urban spaces have become "hardened" in response to increasing crime rates.

Box 15.4 Terrorism and Cities[42]

Not only has terrorism taken on a global reach, it is also apparent that cities have become a preferred location for large-scale terrorist attacks. Between 1993 and 2000 there were over 500 terrorist incidents in more than 250 cities around the world. Even before the terrorist attacks on the World Trade Center and the Pentagon in September 2001, cities had become the central venues of terrorist attacks.

There are several reasons for this. First, cities—especially **world cities**—have considerable symbolic value. They are not only dense agglomerations of people and buildings but also symbols of national prestige and military, political, and financial power. A bomb in London's Underground or a poison gas release in a Tokyo metro arouses international alarm. This kind of event will be communicated instantly to a world audience. Second, the assets of cities—densely packed and with a large mix of industrial and commercial infrastructure—make them rich targets for terrorists. Third, cities have become nodes in vast international networks of communications. This is a reflection not only of their power, but also of their vulnerability. A well-placed explosion can produce enormous reverberations, paralyze a city, and spread fear and economic dislocation. Finally, word gets around more quickly and socialization proceeds more rapidly in high-density localities. These kinds of environments can be an abundant source of recruits for terrorist organizations.

Terrorism takes a toll on cities in a variety of ways. In Belfast, for example, security forces have used "fortress architecture" and principles of defensible space to territorially control and protect designated areas of the city from terrorist attacks linked to the Irish Republican cause. This has been most visible in the central shopping area where access was barred, first by concrete blockers and barbed wire, and later by a series of high metal gates that became known as "the ring of steel."[43]

The impacts of the 2001 terrorist attacks on New York City have been significant—and not limited to the death and destruction that was targeted on lower Manhattan on September 11. The cost of doing business in New York has gone up, even for workers and companies quite distant from the attack site. Some of these costs include[44]

- Slower and less efficient transportation, because of the physical disruption of transport links and the delays caused by increased security.
- Greater security and counterterrorism spending.
- Higher insurance premiums.
- Corporate decisions dictated by concerns about terrorism rather than business efficiency.
- The emotional toll on workers caused by concerns about future attacks.

Central London has sought to reduce the real and perceived threat of terrorist attacks that were until recently almost exclusively associated with the Irish Republican cause. Physical and increasingly technological approaches to security have been adopted at increasingly expanded scales. In 1989 the prime minister, Margaret Thatcher, ordered iron security gates to be installed at the entrance to Downing Street as a means of controlling public access (Figure 15.26).

In 1993 what was referred to in the media as a Belfast-style "ring of steel" was put into place, securing all entrances to the central financial zone of the City of London (the Square Mile). The 30 entrances to the City were reduced to 7, with roadchecks manned by armed police. Locally, the "ring of steel" was referred to as the "ring of plastic" because restricted access was based primarily on funneling traffic through rows of plastic traffic cones. This was viewed as a more symbolic and technologically advanced approach to security that avoided the "barrier mentality" in Belfast. Over time the spatial scale of the security cordon was increased to cover 75 percent of the Square Mile (Figure 15.27).[45]

The security cordon, as a territorial approach to security, was augmented by retrofitting the closed circuit TV (CCTV) system. The police, through its "Camera-Watch" partnership effort, encouraged private companies to install CCTV. At the seven entrances to the security cordon, 24-hour Automated Number Plate Recording (ANPR) cameras, linked to police databases, were installed. Within a decade the City of London had been transformed into the most surveilled space in the United Kingdom, and perhaps in the world, with more than 1,500 surveillance cameras in operation, many of which are linked to the ANPR system.[46]

(Continued)

Box 15.4   Terrorism and Cities   (*Continued*)

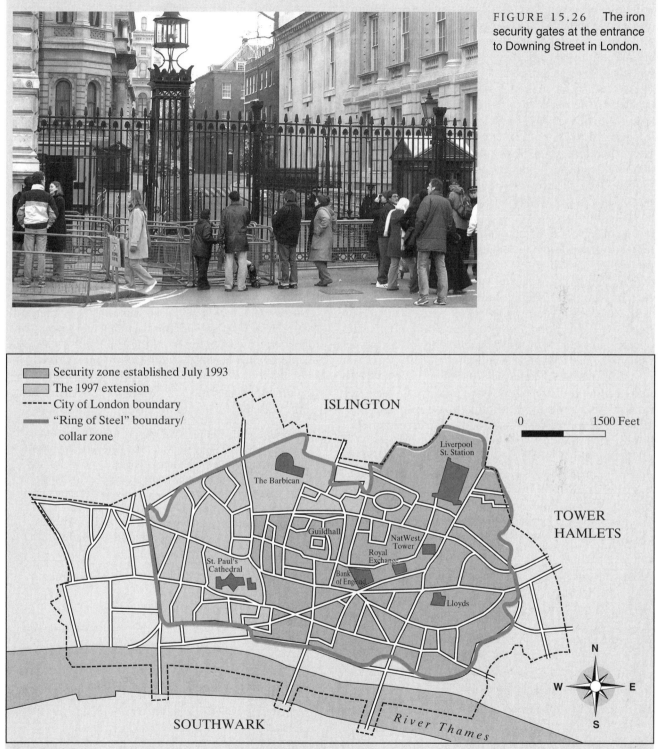

FIGURE 15.26   The iron security gates at the entrance to Downing Street in London.

Security zone established July 1993
The 1997 extension
City of London boundary
"Ring of Steel" boundary/ collar zone

ISLINGTON

1500 Feet
0

Liverpool St. Station

The Barbican

TOWER HAMLETS

Guildhall
NatWest Tower
Royal Exchange
St. Paul's Cathedral
Bank of England
Lloyds

N
W   E
S

SOUTHWARK

River Thames

FIGURE 15.27   The security cordon covering the City of London.

HOMELESSNESS

"Homelessness" wrote sociologist Peter Rossi, "is most properly viewed as the most aggravated state of a more prevalent problem, *extreme poverty*."[47] There is a very fine line between being precariously housed and being literally homeless; and between short-term (episodic) homelessness and long-term (chronic) homelessness. These distinctions make for some slippery definitional problems. The most widely used definitions of homelessness involve chronic homelessness—not having customary and regular access to a conventional dwelling. They therefore include people sleeping in shelters, flophouse cubicles, emergency dormitories, and missions as well as those sleeping in doorways, bus stations, cars, tents, temporary shacks, and cardboard boxes and on park benches and steam grates. They do not include people living in **single-room occupancy (SRO) hotel** rooms, in lodgings, rooming houses, or the homes of relatives or friends. Nor do they include people living in backyard structures and taped-up garages without kitchen or toilet facilities.[48] Homelessness, then, is not simply the lack of a regular place to sleep and receive mail but also a condition of extreme poverty and a state of *disaffiliation*, without lasting or supporting ties to family, friends, and neighbors.

Like poverty, homelessness has always been a feature of American cities. It was first perceived to be a serious problem during the Great Depression of the 1930s, when economic dislocation led to an accumulation of an estimated 200,000 hobos and tramps. The problem receded quickly during World War II and virtually disappeared in the affluence of the postwar era. But the number of homeless people began to rise sharply again with the economic and social polarization and dislocation resulting from the restructuring of the 1970s. By the mid-1980s, homelessness was once more beginning to be perceived as a serious problem. Estimates of the number of homeless people vary a great deal. The Urban Institute estimated in 1999 that between 2.3 and 3.5 million people are likely to experience a spell of homelessness at least once during a year.[49] The U.S. Department of Housing and Urban Development, drawing on different measurement approaches, estimated the total number of homeless at any one time to be between 250,000 and 300,000. The U.S. Census Bureau did not release estimates of the total number of homeless people in 2000,[50] but the census counted 170,706 people alone in emergency and transitional shelters in April. Estimates of the number of homeless people can be criticized in terms of their methodology and, therefore, their accuracy. What is not disputed is that homelessness increased significantly from the mid-1970s onward.

What made the homelessness of the 1980s and 1990s such a striking problem was not just the scale of the phenomenon but also its nature. Whereas homelessness had previously involved mostly white adult males, relatively few of whom actually had to sleep outdoors, the "new" homeless of the 1980s and 1990s were of mixed race, included significant numbers of women and children, and were less likely to be able to find shelter indoors (Figures 15.28 and 15.29). Martha Rosler summarized the consequent change in perceptions:[51]

> Until recently, people who lived on the streets were labeled tramps, bums, vagrants and derelicts. Depression imagery prevailed. The stereotypical "Bowery Bum" was perceived as an alcoholic male transient of no particular race (though in fact such a person was overwhelmingly likely to be Caucasian). By the turn of the 1980s, the stereotypical street denizen had become a deranged hebephrenic bag person, smelly and threatening, a person

FIGURE 15.28   Homeless men in Washington, D.C.

FIGURE 15.29 The new homelessness. One of thousands of homeless women in central Los Angeles.

evicted from a state-run mental institution. Lone homeless women, or "bag ladies," became a familiar mass-culture image. Lucille Ball played one on television.

As the decade progressed and homelessness became endemic, the image of homelessness, instead of gaining depth, was broadened to cover a more varied population, including displaced, primarily black, inner-city down-and-outers and vets; then inner-city mothers and children; then refugees from the rust belt and the foreclosed family farm—including family groups now perceived as possibly white.

When the media discovered homelessness, Rosler points out, the actual dimensions of homelessness became lost amid stereotypes that ranged from cast-up Middle Americans to deranged, crack-addicted, and HIV-infected welfare mothers. Nevertheless, social science research has captured the actual dimensions. There are several characteristics of the "new" homeless:[52]

1. They are worse off than the homeless in the past in terms of employment and income. Compared with the homeless in a benchmark study of 1958, the homeless of the 1980s and 1990s were less likely to have a steady job (20 percent in 2001 compared with 28 percent in 1958); in addition, the real (inflation-adjusted) value of the minimum wage is now far less than it was when it was introduced in the 1970s (see Table 12.6).

2. They are worse off in terms of shelter. Whereas, for example, only about 100 people slept out on the streets of Chicago in 1958, surveys in the mid-1980s found between 500 and 1,400;[53] in 2002, 18,065 people were served in city funded shelters.[54]

3. More of them are families. Families with children comprise between 34 and 40 percent of the homeless population and are the fastest growing segment. It has been estimated that between 900,000 and 1.35 million children may experience homelessness at least once during a year.

4. More of them are women. Using the 1958 benchmark again, the proportion of women among the homeless population has gone up from 2 or 3 percent to more than 20 percent. Most homeless families, 84 percent, are now headed by single women.[55]

5. They are younger, the median age being the early years of adulthood rather than middle age. In addition, there are proportionally fewer elderly among the homeless now (although their absolute number has grown during the past three decades).

6. More of them are African American. Although the figures vary from city to city, African Americans now account for about 40 percent of the homeless, compared with less than 25 percent in the 1950s.

Another important change is that homelessness in the 1980s and 1990s spilled out from its traditional "Skid Row" settings. The geography of homelessness is in fact complex, reflecting several aspects of change in metropolitan structure and dynamics. One of the most significant of these was the emergence of **service-dependent ghettos** in the inner city as welfare dependency came to be increasingly prevalent in slum and poverty-area populations and the zone in transition increasingly became the location for the services offered by an expanding array of public and voluntary welfare agencies—institutions that in turn attracted the service-dependent populations of the poor, the disadvantaged, and the homeless.[56] At the same time, new, flexible **land use zoning** schemes made it possible for missions, shelters, and other services for homeless and impoverished people to be created in a wider range of inner-city locations.

Meanwhile, urban renewal projects were systematically targeting traditional skid row settings. Another important factor was the decriminalization of public drunkenness and vagrancy. Swamped by increasing numbers of cases and with more pressing crimes to deal with, police no longer warned the homeless away from public places or picked them up for sleeping on the streets. As a result, homelessness became more widely suffused through urban space and, consequently, much more visible. This visibility, in turn, led to more repressive reactions: the use of urban design to discourage the homeless from entering and (where such subtlety might not be sufficient) the installation of convex "bum-proof" benches and of sprinkler systems programmed to switch on at random times through

the night and the deployment of private security forces. Meanwhile, the dynamics of central city revitalization began to encroach on the inner-city spaces that had previously been the preserve of welfare-dependent populations and welfare agencies. More of the homeless were pushed out of the central city altogether into shelters that were opened in converted properties in the first-tier and second-tier suburbs and into the parks and public spaces of edge cities.

THE CAUSES OF HOMELESSNESS

The immediate cause of homelessness is an excess demand for the very cheapest accommodations. People who do not have enough money to compete effectively at the bottom end of **housing submarkets** and who do not have friends or relatives to give them accommodation become homeless. There are, however, many

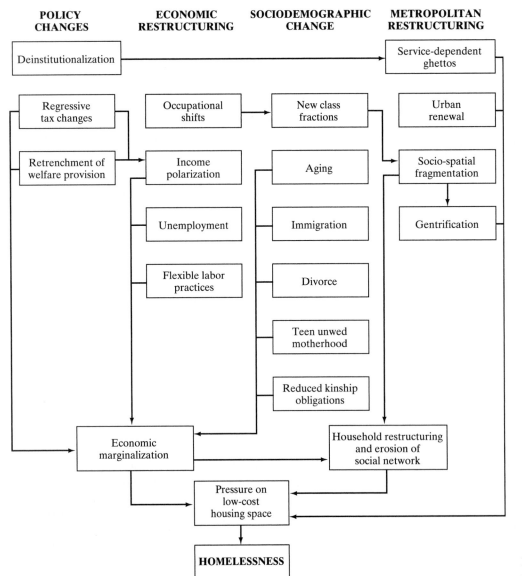

FIGURE 15.30 The causes of homelessness.

causes of pressure on low-cost housing space, as Figure 15.30 shows. In broad terms, the excess demand for cheap accommodation that led to the dramatic increases in homelessness beginning in the 1970s was the product of interaction between four sets of trends: economic restructuring, sociodemographic change, metropolitan restructuring, and policy changes. The nature of these trends is already familiar: They are central to the evolutionary adaptation, restlessness, and cyclical changes that have been recurring themes of this book. Some of the major factors that have contributed to the increase in homelessness are surveyed here.

The occupational shifts and new **class fractions** generated by the evolution of the economy toward an **advanced capitalism** characterized by **informational cities** (p. 86) have contributed to a *sociospatial fragmentation* (pp. 318–331) that has resulted in, among other things, **gentrification** (p. 370). Gentrification, in turn, has removed substantial amounts of moderately low-cost housing from the market, displacing many households into lower-cost segments of the market, ultimately putting pressure on the lowest-cost housing.

The *income polarization* associated with economic restructuring and changing labor markets, as we have seen in this chapter, has contributed to increases in poverty and economic marginalization, leading directly to increased competition for cheap accommodation. By 1985 the earnings for young workers (aged under 35) had declined to about 80 percent of the 1968 level, in constant dollars, and their earnings rose by only 1.2 percent between 1985 and 1995 (but by nearly 16 percent between 1995 and 2001). The slow income growth trend into the 1990s was compounded by the effects of policy changes that were themselves a product of a political climate associated with the same economic shifts. The **new conservatism** of the Reagan and first Bush presidencies featured two policy shifts that were particularly important in this respect:

1. *Tax Changes.* Designed to cater to middle-income voters and to stimulate economic growth through middle-income materialism, tax changes resulted in a highly regressive shift in the burden of taxation. As we saw in Chapter 12, between 1979 and 2000 the average after-tax income of the poorest 20 percent of households increased by only about 9 percent (compared to 68 percent for the richest 20 percent) (see Table 12.5). During that time, the share of after-tax income of the poorest 20 percent of households fell from about 7 to 5 percent (while that of the richest 20 percent rose from about 42 to 51 percent) (see Figure 12.35).

2. *Cutbacks on Public Spending for Welfare Services* (see also Chapter 17). When this retrenchment began in the late 1970s, discretionary programs (including most housing programs) bore the brunt of the cutbacks. In 1982 $1 billion was cut from child nutrition programs and $1.7 billion was removed from the food stamp program. Public housing programs came to a virtual standstill. More than 150,000 people (many of them with psychiatric disabilities) were removed from Social Security Disability Insurance rolls as a result of more stringent interpretations of eligibility. As retrenchment gathered momentum during the 1980s, the erosion of income support and housing programs for the poor became so pronounced that the period came to be known in policy circles as the War on Welfare. The average monthly value of Aid to Families with Dependent Children (AFDC) and Temporary Assistance for Needy Families (TANF) benefits declined by almost one-quarter, in constant dollars, between 1977 and 2000 (Figure 15.31).

Structural unemployment, which rose significantly in the 1970s as a result of economic recession and restructuring (p. 82), had a differential impact, affecting young males more than others. Whereas the unemployment rate in 1968 for men under 35 years of age was less than 5 percent, it rose to a high of 15 percent in 1980, declining to 13 percent in 1984 and fluctuating between 12 and 14 percent through 1990 (declining to just under 9 percent by 2003). For younger black males, unemployment reached disastrous proportions: 40 percent among black males aged under

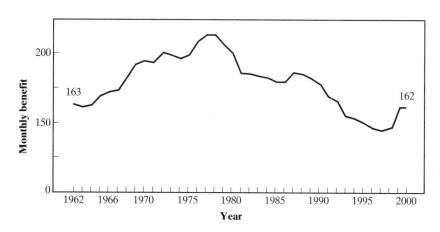

FIGURE 15.31 Average monthly Aid to Families with Dependent Children (AFDC) and Temporary Assistance for Needy Families (TANF) benefits per recipient in constant dollars.

25 in 1985 (declining to about 25 percent by 2003). These figures, of course, do not reflect potential workers who had given up looking for work and had become disengaged from the labor market.

Flexible labor strategies (involving temporary, part-time, and short-term contract jobs) that emerged with the shift away from **Fordist** approaches (p. 319) segmented the labor force in new ways, creating a large "secondary" labor market in which many workers had low wages, little or no security, and no fringe benefits. This secondary labor market added to the number of economically marginal households, creating still further pressure on the limited amount of low-cost housing space.

A major policy shift in the care of mentally disabled and impoverished people resulted in the **deinstitutionalization** of many individuals. The traditional settings of care, such as asylums, hospitals, and orphanages, had come into question by the 1950s because of the shortcomings of crowded and antiquated institutions, an increasing conviction that prolonged institutionalization often did more harm than good, and an increasing sensitivity to the civil rights of impoverished and mentally disabled people. With the expansion of spending on welfare in the 1960s, federal funds were made available to encourage the transition from institutionalized care to grassroots care that typically involved a combination of independent living and local service delivery points. More recently, some people with HIV/AIDS have lost their jobs due to discrimination, fatigue, or periodic hospitalization caused by HIV-related illnesses; they have also found their incomes drained by healthcare costs. They and others can be denied services or be prematurely discharged from healthcare facilities due to managed care arrangements. Large numbers of vulnerable, disabled, and low-income individuals thus gravitated to the service-dependent ghettos that had developed in inner-city areas, adding more pressure to the localized submarkets of low-cost housing.

Urban renewal has eroded the stock of older, low-cost housing, partly in response to public policies (p. 509) and partly in response to changing patterns of land values and **ground rents** (p. 274). Almost all the cubicle hotels that had characterized the skid rows of the 1950s and 1960s had been demolished by 1980. Many SRO hotels also disappeared—about half the total stock of SRO units was destroyed in the 1970s. Between 1970 and the mid-1980s, an estimated 1 million SRO units were demolished.[57] Of New York City's 129,000 $200-per-month or less SRO units in 1960, 113,000 units—87 percent—were gone by 1983 (and SROs continued to disappear in the 1980s and 1990s); in Chicago, 18,000 SRO units—19 percent of the stock—disappeared between 1973 and 1984 (12,500 remained in 2003); and in Los Angeles, more than half of the downtown SRO units disappeared between

1970 and 1985.[58] As a direct result, the price of SRO housing rose significantly.

Sociodemographic changes contributed to pressure on this diminishing pool of low-cost housing through the increased numbers of vulnerable households seeking shelter as a result of several interconnected trends. Dramatic increases in divorce rates (p. 402) and the number of teenage unwed mothers (p. 324) led to the feminization of poverty. Meanwhile, the combination of economic hardship and an increasing intergenerational gap in lifestyles and norms of behavior led to the evolution of new limits on kinship obligations. In short, fewer families were both able and willing to shelter and care for children and grandchildren who had fallen on hard times. The resulting social dislocation and economic marginalization was compounded by the *sociospatial fragmentation* that, in the process of restructuring urban space (pp. 318–331), inevitably eroded some of the long-established social networks of traditional low-income neighborhoods. High rates of domestic violence have also been identified as an important factor in homelessness among women.[59]

Finally, two other sociodemographic trends added to the number of low-income households and, therefore, intensified the pressure on low-cost housing. One is the *aging* of the population (p. 302), which has also contributed to an increase in the number of single-person households. The effects of this trend have been mitigated, however, by the fact that welfare retrenchment did not as adversely affect the elderly (largely because of their weight of influence at the ballot box). In fact, the constant-dollar value of Social Security and Medicare benefits increased by 280 percent between 1960 and 2000, thanks to increases in benefit levels and the indexing of benefits to the Consumer Price Index (Figure 15.32). In contrast, increased *immigration* (p. 318) brought large numbers of households, many of whom were unaware of their welfare rights and who, as a group, often faced discrimination in both the welfare system and the labor market.

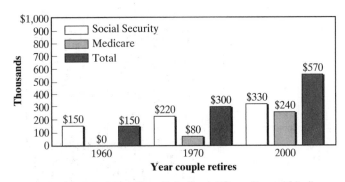

FIGURE 15.32   Growth in Social Security and Medicare lifetime benefits for an average-wage/low-wage two-earner couple, 1960–2000 (discounted to present value at age 65).

INFRASTRUCTURE AND ENVIRONMENTAL PROBLEMS

For many people the stereotypical image of urbanization in the older industrialized national economies includes congested traffic, smoke-belching factories, and a striking lack of vegetation and wildlife. Implicit in such images are an overloaded infrastructure and an abused and degraded environment, all part of the "necessary evil" that comes with cities' primary roles as economic production and distribution points. Thanks to improved public health, urban planning, more rigorous environmental regulation, and sustained investment in public works, significant advances have been made in managing urban environmental quality. Nevertheless, cities are more than ever like monstrous organisms that consume vast quantities of oxygen, water and organic matter, fossil fuels, and foodstuffs and that release huge quantities of waste products into the air, into rivers, lakes, and oceans, and onto the land.

In 2001 in the United States alone, more than 6 billion pounds of toxic chemicals were released into the air, water, and land.[60] The spatial pattern of these incidents reflects the division of labor within the urban system, with some of the heaviest releases occurring in counties in the older, well-established industrial centers of the eastern United States and with concentrations in the "chemical corridor" of the Northeast and in the petrochemical complexes surrounding the Gulf Coast (Figure 15.33).

Although urban environments are now highly regulated and although sophisticated transportation and waste disposal systems have been developed to allow them to continue to function, they are, as physical geographer Ian Douglas notes, *dependent systems*. They are dependent on (1) continuous expenditure and capital investment; (2) efficient governance; and (3) planning, management, and monitoring. Ultimately, therefore, they are dependent on finance, both public and private, in order to sustain their life-support functions.[61] In this context many of the infrastructural

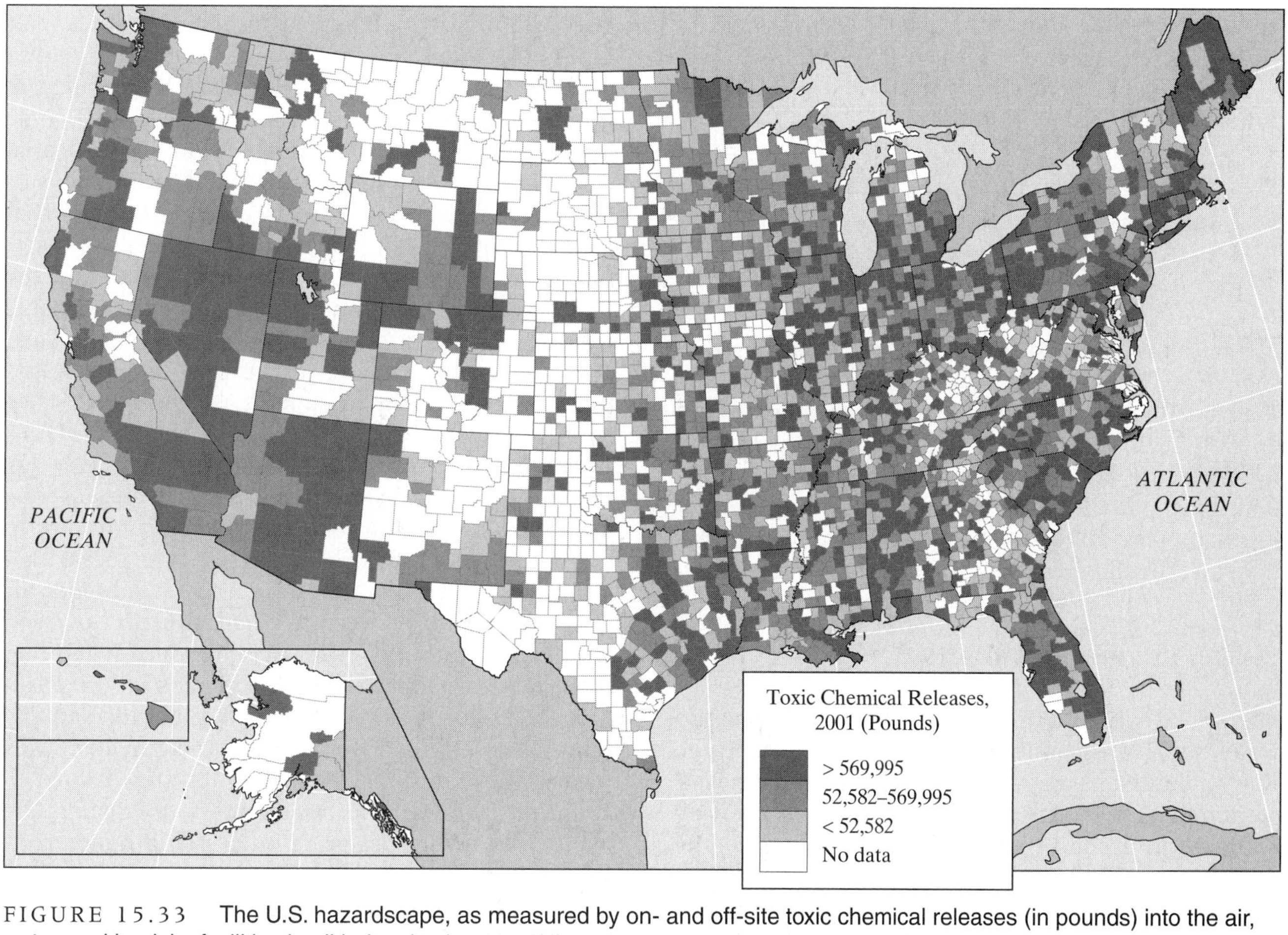

FIGURE 15.33 The U.S. hazardscape, as measured by on- and off-site toxic chemical releases (in pounds) into the air, water, and land, by facilities in all industries in 2001. "High" = the 33 percent of counties with releases greater than 569,995 pounds; "medium" = the 34 percent of counties with releases between 52,582 and 569,995 pounds; "low" = the 33 percent of counties with releases less than 52,582 pounds.

and environmental problems of cities can be seen as the result of investment and operating expenditures that:

- are persistently insufficient.
- significantly lag behind urban growth.
- decrease faster than city population.

It is also important to recognize the functional interdependence between the economic and environmental dynamics of cities and their relationships to the need for services and infrastructure:

> The material demands of the urban population are the cause of profound changes in the physical character of the city and its surroundings. Modification of the natural flows of energy, water and materials goes hand in hand with the creation and maintenance of artificial flows of the same commodities. The character of the land surface changes, affecting the radiation balance, the rainfall-runoff relationship, sources and supplies of sediments and solutes, infiltration and groundwater levels, soil chemistry and plant and animal habitats. Such changes are both continuous and episodic. The impacts become apparent irregularly, often in extreme or unusual climatic conditions such as floods caused by exceptional rains whose height and duration is affected by the large proportion of impermeable roofed and paved surfaces in the urban catchment area. Similarly, temperature inversions can trap the gaseous pollutants and particulates emitted by urban chimneys and exhausts, causing smog. Both floods and smog lead to accidents, damage, injury and even loss of life, which in turn place demands on urban services.[62]

It is not possible to do full justice here to this kind of complexity. It is possible, however, to illustrate the functional interdependence between economic and environmental change and the relationship of both to infrastructure and service needs. The following sections attempt to do this by pointing to three important issues: problems of water supply, problems of air pollution, and the threatening crisis in urban infrastructure.

WATER SUPPLY PROBLEMS

There are two main dimensions to the problem of urban water supplies: quantity and quality. Cities consume vast amounts of water. More important, urban water consumption has increased dramatically. Public-supply withdrawals in the United States (that are supplied by public and private freshwater suppliers) were more than 43 billion gallons in 2000, compared to only 14 billion gallons in 1950. Public-supply water is supplied for domestic, commercial, industrial, or thermoelectric power

uses. In 2000 about 242 million people depended on water from public suppliers (another 43.5 million people were self-supplied—usually from a well).

Not only do America's metropolitan areas accommodate more people than ever, but people's consumption of water has risen in parallel with increased consumerism. The increased prevalence of washing machines, dishwashers, showers and whirlpool bathtubs, car washes, and suburban lawns has helped to push consumption to unprecedented levels. A typical family of four uses about 340 gallons of water *each* day. Up to 40 percent of indoor water consumption is used to flush toilets, 30 percent is used in showers and bathtubs, 20 percent is used by washing machines, 5 percent by dishwashers, and only about 3 percent is used for cooking and drinking. Despite the Energy Policy Act of 1992 requirement that all new home-use toilets be "low-flow" (using 1.6 gallons per flush instead of the 3.5 to 5 gallons in traditional toilets), the United States still uses about three times more fresh water per person than the rest of the world.[63]

Consumption on this scale has become a problem in many cities. Inland cities that depend heavily on groundwater have suffered from serious *subsidence*. The withdrawal of large quantities of groundwater lowers the water table and reduces the water held in voids between mineral particles. The sheer weight of urban development causes these particles to become more closely compacted, leading to subsidence. Coastal cities that depend heavily on groundwater have suffered from a different problem: *seawater intrusion* that ruins the quality of groundwater supplies. Seawater intrusion has been a problem for many cities along the East and Gulf coasts and a severe problem for some cities in Florida.

In some other regions, because groundwater supplies are insufficient to support urbanization, inter-basin transfers of water have had to be arranged. Los Angeles, for example, had to tap the Owens River watershed, east of the Sierra Nevada Mountains, as long ago as 1913. Less than 20 years later this source had to be supplemented by an aqueduct (Figure 15.34) that tapped into the Colorado River. Federal subsidies for such transfers have recently been reduced, however, while states such as Nevada and Arizona, aware that their own economic growth will depend on urbanization that will require water supplies, are increasingly reluctant to sell their water rights. The growth of municipal and industrial demands for water has led to conflicts over the distribution of water rights. City planners now see water resources as a major constraint to growth and increased economic development, especially in the arid regions of the West and Southwest.[64] Finally, these fundamental problems of supply are compounded by infrastructure problems;

FIGURE 15.34 The California Aqueduct, a federally subsidized lifeline for Los Angeles.

leaky water mains, for example, afflict most cities and can account for losses of 10–30 percent of total consumption in the United States and Canada. Detroit, for example, loses an estimated 35 billion gallons—17 percent—of the fresh, clean water it pumps through the system every year—at a cost in 2000 of $23 million for water that never reached homes and businesses.[65]

The Congressional Budget Office (CBO) has estimated that for the years 2000 to 2019 annual investment in the drinking water systems in the United States would need to average about $16 billion in order to maintain desired levels of service for customers, meet standards for water quality, and maintain and replace their assets cost-effectively. The CBO's estimate of actual investment spending in 1999 was only $11.8 billion.[66]

Concerns over water *quality* are important not only because of potential public health hazards, potential employment losses in industries that require sound water, and potential hazards to wildlife and vegetation, but also because of potential reductions in the quantity of available supplies. Beginning in the 1970s, every U.S. state has had reported cases of groundwater contamination. According to the National Research Council, initial remediation of the 300,000 contaminated groundwater sites in the United States will cost up to $1 trillion over the next 30 years.[67]

A 2001 industry study found pathogens and "fecal indicator" bacteria at significant levels in soil and trench water at water main break repair sites across the United States. Of the 619 waterborne disease outbreaks that the Centers for Disease Control and Prevention tracked between 1971 and 1998, 18 percent were due to germs in the distribution system. The worst outbreak was in Milwaukee in April 1993, when 403,000 people, about half the city's population, became ill from an outbreak of the intestinal parasite *Cryptosporidium*. In healthy humans this parasite—which is excreted by cattle or other animals and washed into lakes and reservoirs, easily finding its way through obsolescent water purification systems—causes stomach upsets and diarrhea. It can be dangerous for people with weak immune systems: 104 died in Milwaukee's outbreak.[68]

In addition to coliform bacteria from septic tanks and sewage, chlorides, oil, phosphates, and nondegradable toxic chemicals from industrial wastewater, dissolved salts and chemicals from highway deicing, and nitrates and ammonia from fertilizers and sewage are all regular threats to city water supplies. One reason for the increasing scale of the problem is that many older local water treatment systems were designed to cope with bacteria rather than chemicals and toxic substances. Another is that in many cities the sewer system was engineered in combination with the storm drainage system. Treatment plants attached to such systems are easily overloaded after heavy rain, with the result that plant managers sometimes have to open floodgates, releasing a yearly 1.2 trillion gallons of storm water overflows from combined sewer systems that carry untreated sewage into rivers and lakes.[69] A third reason is the breakdown of treatment plants that has become increasingly common as urban growth has outstripped investment in maintenance and construction. The CBO estimated that annual investment in the country's wastewater systems would need to average about $17 billion between 2000 and 2019 in order to maintain the levels of service desired by customers, meet water quality standards, and maintain and cost-effectively replace their assets. The CBO's estimate of actual investment spending in 1999 was only $9.8 billion.[70]

AIR POLLUTION

Cities not only generate a great deal of air pollution but also create microclimates and change atmospheric conditions in ways that intensify the problem. The built environment, overall, has a lower *albedo* than natural earth surfaces. That is, it is less able to reflect incoming solar radiation and so absorbs more heat. In addition, the atmosphere in cities is warmed by the release of heat from fossil fuels and, in summer, from air conditioning units. The aggregate effect is for cities to become **heat islands**, with average temperatures one or two degrees Fahrenheit above those of the surrounding countryside. During heat waves, this heat island effect is particularly pronounced and can lead to average temperatures that are 10 to 15 degrees Fahrenheit above those of the surrounding countryside. Several successive days of such intensified heat can increase both physical and mental stress. On such days mortality increases among the sick and the elderly, tempers fray, and tensions rise.

The heat island effect also leads to a distinctive pattern of local air circulation on days when there is no regional air movement. Light surface winds are drawn toward the city center, then rise above the city to descend slowly to the ground at the edge of the built-up area. This pattern is associated with the development of an **urban dust dome** in which fumes, soot, and chemicals are trapped in the air over the city. This kind of phenomenon becomes particularly pronounced in circumstances where a temperature inversion develops and a lid of relatively warm air effectively flattens the dust dome, keeping pollution low over the city. When exposed to strong sunshine, such concentrations of pollution can be transformed through photochemical reactions into an orange-brown *smog*.

About 160 million tons of pollution are released into the air in the United States each year.[71] Of this, about half is generated by transportation, just under a fifth is generated by stationary fuel combustion (in power plants, etc.), and just under one-tenth is generated by industrial processes (Figure 15.35). Among the main pollutants are

- *Sulfur oxides* that are chiefly the product of burning fossil fuels (mainly coal and oil). In addition to aggravating respiratory conditions, they combine with atmospheric water to form dilute sulfuric acid that is a major component of **acid rain**, which can kill trees and erode the surfaces of buildings, monuments, and statues.

- *Nitrogen oxides* that are the product of burning fossil fuels at high temperature (as in power plants and automobiles). Like sulfur oxides, these emissions aggravate respiratory tract problems and combine with atmospheric water to produce acid rain. In addition, nitrogen oxide reacts with oxygen to form nitrogen dioxide, a reddish-brown gas that reduces visibility.

- *Carbon monoxide.* Chiefly derived from automobile emissions (when carbon in fuel is not burned completely), this colorless and odorless gas is

FIGURE 15.35   Urban air pollution: downtown Los Angeles skyline shrouded in smog.

poisonous in high concentrations and even at lower concentrations can impair psychomotor functions and the cardiovascular system. Motor vehicle exhaust contributes about 60 percent of all U.S. carbon monoxide emissions.

- *Carbon dioxide.* Another by-product of burning fossil fuels, it has no directly harmful effects on plant or animal life but is now recognized as a serious threat because of its role in creating the **greenhouse effect**. The possibility of global warming poses the specific problem for coastal cities of rising sea levels, which would increase the probability of flooding; it might also cause an intensification of the hydrological cycle, which would in turn exacerbate problems of water supply.

- *Photochemical oxidants,* such as ground-level ozone (the primary constituent of smog), that are produced by the effects of heat and sunlight on nitrogen oxides and volatile organic compounds (VOCs). The resultant smog is an irritant to eyes and the respiratory system and can kill some trees and ornamental plants. The 1990 U.S. Clean Air Act amendments require that reformulated gasoline containing oxygen to reduce harmful emissions of ozone be used in cities with the worst smog pollution. Reformulated gasoline is gasoline that is blended to burn cleaner and reduce smog-forming and toxic pollutants in the air.

- *Particulate matter,* an airborne mixture of solid particles and liquid droplets, that is formed by sulfur dioxide, nitrogen oxides, ammonia, VOCs, and direct particle emissions. The resultant smog (containing dust from roads or soot [black carbon] from combustion sources) can reduce visibility and be an irritant to the respiratory system. Airborne particles can also damage vegetation and cause damage to paints and building materials.

- *Fluorides,* mainly the product of industrial processes, cause eye irritation, respiratory inflammation, nosebleeds, and damage to trees and flowering plants.

- *Lead,* a heavy metal pollutant produced mainly by industrial processes, primarily metals processing, with the highest air concentrations usually found in the vicinity of smelters and battery manufacturers. It is a cumulative poison that can be inhaled, absorbed through the skin, or ingested

---

BOX 15.5　LOWER MANHATTAN AIR QUALITY FOLLOWING THE COLLAPSE OF THE TWIN TOWERS[72]

The impacts of the burning and collapse of the twin towers of the World Trade Center on September 11, 2001, have been significant—and not limited to the catastrophes of the lives lost and economic disruptions. The environmental impacts of the disaster have also alarmed air pollution experts and atmospheric scientists.

The U.S. Environmental Protection Agency (EPA) and other federal, state, and local agencies collected extensive environmental monitoring data from the World Trade Center site and nearby areas in Manhattan, Brooklyn, and New Jersey. For several months the EPA took samples of air, dust, water, river sediments, and drinking water and analyzed them for the presence of pollutants that might pose a health risk to response workers at the World Trade Center site and to surrounding residents (see *http://www.epa.gov/wtc/monitoring.html* for the monitoring summaries). The U.S. Department of Energy also commissioned independent research scientists from several California universities to conduct a wide range of tests on air quality in and around the World Trade Center site for three months following the disaster.

This monitoring revealed that smoke and dust from the destroyed buildings exposed the response workers and residents of lower Manhattan to some of the highest levels of air pollution ever detected anywhere in the world. The dreadful fires that raged and then smoldered in the rubble of the collapsed buildings turned glass, concrete, computer equipment, carpeting, plastic, and a wide range of other materials into an intense and persistent aerosol fallout that people could see and smell for months after the disaster (Figure 15.36). Although only very low levels of asbestos and lead were detected, very high levels of sulfur and sulfuric acid, titanium, nickel, and silicon were present. For many of the pollutants, the exact health effects are unknown. Recent students have shown, however, that these tiny particles can penetrate lungs and be absorbed into the bloodstream, possibly leading to elevated rates of cancer, heart attacks, and respiratory disease.

(Continued)

BOX 15.5   LOWER MANHATTAN AIR QUALITY FOLLOWING THE COLLAPSE OF THE TWIN TOWERS   *(Continued)*

FIGURE 15.36   World Trade Center disaster site, September 13, 2001, as urban search and rescue specialists continued to search for survivors among the wreckage and airborne pollution.

Although the high levels of outdoor pollution had largely dissipated by the spring of 2002, pollution inside the homes and businesses near the disaster site remained a concern. Despite the thorough cleaning of the interior of nearby buildings, the discarding of items such as towels and sheets, and government assurances that the air was safe, many residents and office workers said that the dust and toxins released by the collapsing towers had made them sick, disrupted their lives, and aroused fears about long-term health problems. They feared that toxic particles, many of them more than 100 times smaller than the diameter of a human hair, could easily slip past windows and

doors and collect on window furnishings, carpets, and the cracks in the floorboards.

Many firefighters and other responders who worked without respirators at Ground Zero reported an increase in respiratory ailments, such as the persistent "World Trade Center cough." Other reported symptoms ranged from nosebleeds and shortness of breath to dizziness, headaches, and asthma. Some of the most severely affected firefighters, who were among the 24-hour search teams in the first weeks after the disaster, have applied for early retirement or disability benefits because of persistent respiratory problems.

from vegetables that have absorbed lead via air pollution. It has adverse effects on metabolism, blood, and kidney functions. Even at low doses, lead exposure is associated with damage to the nervous systems of fetuses and young children, resulting in learning deficits and lowered IQ. Thanks to the phaseout of leaded gasoline by the U.S. Environmental Protection Agency since the 1970s, culminating in the banning of lead in motor vehicle gasoline after 1995, lead pollution has decreased significantly. The 2002 average air quality concentration for lead was 94 percent lower than in 1983. Emissions of lead decreased 93 percent over the 21-year period from 1982 to 2002.

Finally, it is worth noting that air pollution is associated with definite spatial distributions, with significant gradients between one part of a city and another. In general, the highest pollution occurs near the largest concentrations of traffic and industry. What is less obvious but of much greater significance is that the residential settings with the highest levels of air pollution tend to be low-income neighborhoods and, in particular, poverty areas and slums. Children are at greater risk because they are generally more active outdoors and their lungs are still developing. The elderly and people with heart or lung diseases are also more sensitive to some types of air pollution.[73]

INFRASTRUCTURE CRISIS

The urban infrastructure of roads, bridges, parking spaces, transit systems, communications systems, power lines, gas supplies, street lighting, water mains, sewers, and storm drains is crucial not only to economic efficiency and productivity but also to public health, safety, and the quality of life. In the early 1980s a series of highly publicized infrastructure failures (including the collapse in 1983 of the Mianus River Bridge on the Connecticut Turnpike that killed three motorists and the broken water main in the same year that interrupted power to businesses in New York's garment district for a week) underlined the importance of urban infrastructure and drew attention to some important shortcomings. In 1983, Pat Choate and Susan Walter published a systematic survey whose title—*America in Ruins*—conveyed the sense of crisis that had developed.[74]

The federal government, meanwhile, had created a National Infrastructure Advisory Committee, which submitted its final report[75] to the Joint Economic Committee of Congress in 1984. The report suggested that $1.15 trillion would have to be spent by the year 2000 to repair, maintain, and develop basic infrastructure (transportation, water, and sewer systems) simply to avoid major quality of life and productivity problems. The committee estimated that $714 billion would be available from federal and state revenues. This sum left a shortfall of $436 billion that seemed unlikely to be met because of the inherent nature of infrastructure provision: unglamorous and easily postponed because it rarely enjoys priority support from any politically significant group. Congress immediately established the National Council on Public Works Improvement to study the problems in more detail; its 1988 report, *Fragile Foundations: A Report on America's Public Works*, which concluded that the quality of

BOX 15.6 HIGH-SPEED RAIL IN AN INTEGRATING EUROPE[76]

The increasing integration of Europe, following fundamental geographic principles, relies heavily on policies that increase accessibility and spatial interaction among cities. One goal was to link peripheral cities like Lisbon, Seville, and Rome with cities like Paris and Brussels that are located at the core of the EU. This would boost economic efficiency, reinforce the social and political cohesiveness of the EU, and in the process could create up to 1 million new jobs.

In 1994 the European Union (EU) approved a far-reaching plan for a series of trans-European networks that would weld together Europe's patchwork of national road, rail, air, and water transportation systems. The $400-billion budget included $250 billion for investment in 20,000 miles of high-speed track (Figure 15.37).

The relatively short distances between major cities in Europe makes it ideally suited for rail travel; Europe is less suited to air travel because of the high population densities and traffic congestion around airports. Even before the terrorist attacks of September 11, 2001, the delays caused by check-in times and accessibility to airport terminals meant that it was quicker to travel between many major European cities by rail than by air. Europe's high-speed rail network—in direct competition with the

airlines—already connects a number of cities at the core of Europe. The "PBKAL web" connects Paris, Brussels, Köln (Cologne), Amsterdam, and London (via the so called Chunnel—Channel Tunnel).

Improved locomotive technologies and specially engineered tracks and rolling stock make it possible to achieve speeds of 180 to 250 miles per hour. New tilt-technology railway cars—designed to negotiate tight curves by tilting the train body into turns in order to counteract the effects of centrifugal force—are being introduced in many parts of Europe in order to maximize speeds on conventional rail tracks. German Railways, for example, introduced third-generation inter-city express trains with a maximum speed of 205 miles per hour in 2000.

Europe's high-speed rail network is already leading to a restructuring of the urban hierarchy. Because the high-speed rail routes have only a few scheduled stops—to avoid the time penalties incurred by decelerating, stopping, and accelerating—the system has created winners and losers across the urban system. In relative terms, the cities with scheduled stops are now more accessible and so more attractive for economic development than those cities without scheduled stops.

(Continued)

Box 15.6   High-Speed Rail in an Integrating Europe   *(Continued)*

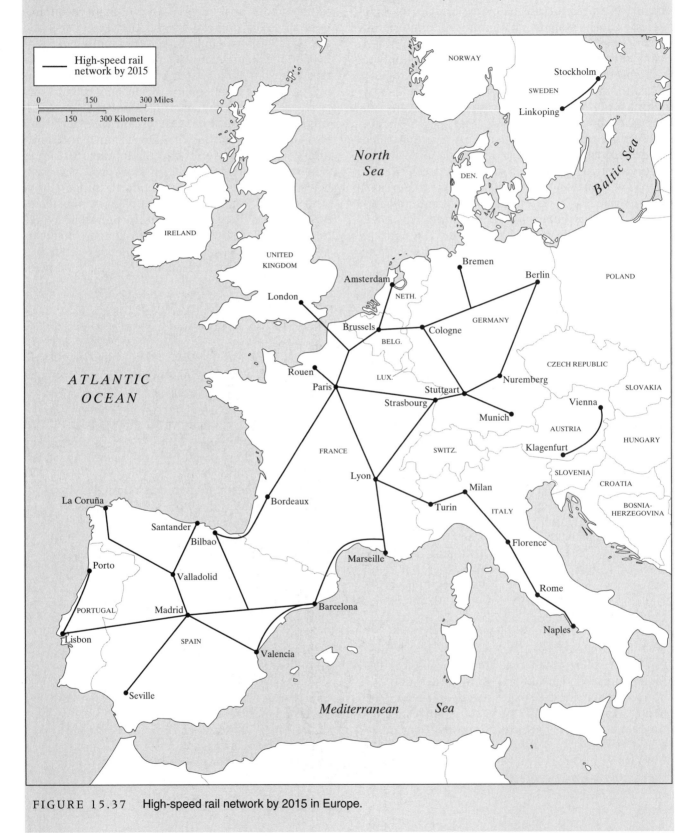

FIGURE 15.37   High-speed rail network by 2015 in Europe.

America's infrastructure was barely adequate to meet current requirements and inadequate to meet the demands of future economic development, stressed the importance of creating an agenda upon which Congress could act. By 2001, however, the American Society of Civil Engineers (ASCE) estimated that it would cost $1.3 trillion over five years to bring the country's infrastructure up to acceptable standards; its 2003 report found little improvement and upped this estimate to $1.6 trillion.[77]

It is now very clear that the root of the infrastructure crisis is a combination of aging central cities and declining capital investment in the face of public budgetary crises, voter opposition to infrastructure projects, and—given the threat of possible terrorist attacks on critical infrastructure—the diversion of available public funding away from infrastructure maintenance and toward infrastructure security measures. The key infrastructural elements of most central cities were put in place more than a century ago and have long outlasted their designed life span (Figure 15.38). Most investment in new infrastructure, meanwhile, has taken place in new suburbs, boomburbs, and **edge cities**. Federally funded infrastructure investment increased significantly in the 1930s as a result of New Deal make-work programs (see p. 507), and federal, state, and local outlays on infrastructure increased in the affluent postwar period. During the period between the early 1960s and late 1990s, however, infrastructure was increasingly neglected compared to other expenditures (Figure 15.39). The impact of this trend on infrastructure provision has been compounded by inflation in the costs of building construction, which has consistently been higher than general economic inflation.

As a result, many cities, if not imminently facing the irretrievable decay and ruin of their infrastructure, face serious problems. For example, according to a 1998 estimate by the New York City Office of the Comptroller, the city would need to spend almost $92 billion over the next 10 years to get its crumbling infrastructure into a good state of repair. Meanwhile, the Comptroller's Office calculated that the 10-year funding projection was only $52 billion. The city is responsible for more than 6,000 miles of water mains, 600 miles of subway tracks, 5,700 linear miles of streets, 78 police precincts, 223 firehouses, 6 community colleges, 23 health facilities, and more than 1,000 school buildings (half of which are more than 50 years old).[78]

In New York City alone, up to 3,000 miles of water mains and 1,000 miles of sewers are at or near the end of their expected life spans. Across the United States, chronic leaks in pipes already consume 20 percent of

the water carried by many aging city supply systems, and more water mains crack or fail each year. By 2016 the U.S. Environmental Protection Agency predicts that more than 50 percent of the country's sewer pipes will be in poor or very poor condition or broken, up from only 8 percent of the estimated 600,000 miles of sewer lines that were in poor shape or worse in 2000.[79]

In many cities the most visible and acute infrastructure problems are those associated with the shortcomings of roads and transit systems. Suburban gridlock became an increasingly common phenomenon during the 1970s as new automobile and truck registrations overtook highway capacity and outstripped highway construction (Figure 15.40). Freeway rush-hour traffic that was officially "congested"[80] was virtually unheard of during the 1960s. By 1975 it

FIGURE 15.38   Infrastructure collapse. In April 1992 the collapse of part of a wall in disused underground freight tunnels beneath Chicago's business district caused over 250 million gallons of river water to flood a large section of the downtown, causing $500 million in damage—all for want of a $30,000 repair.

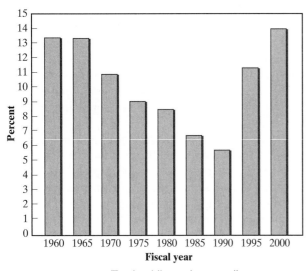

FIGURE 15.39   Total public works spending as a proportion of total public expenditures, 1960–2000. Public works include highways, airports, water transportation terminals, sewerage and solid waste management, water supply, and mass transit systems.

FIGURE 15.40   Gridlocked traffic.

had reached 41 percent of freeway rush-hour traffic, and by 1985 it had reached 56 percent. By the beginning of the 1990s, rush-hour traffic on the likes of Interstate 25 southeast of Denver, the Dan Ryan Expressway in Chicago, the Ventura Freeway in Los Angeles, Route 101 south of San Francisco, and the Beltway (I-495) around Washington, D.C., was routinely stop-and-go, averaging about 12 mph. Between rush hours, traffic was merely "congested."

Based on analysis of the performance of the transportation systems in a group of 75 small to very large U.S. urban areas—ranging in size from Boulder, Colo. and Laredo, Tex., to Los Angeles and New York—traffic congestion continues to increase. The average annual delay per person in these 75 urban areas in 2001 was 26 hours. There were 13 urban areas with delays per person in excess of 30 hours, reflecting the very large delays in urban areas with populations larger than 3 million (Figure 15.41). In 2001 congestion—based on wasted time and fuel—cost nearly $70 billion in the 75 urban areas, compared to less than $10 billion in 1982. The delays in these urban areas represented 3.5 billion vehicle hours and 5.6 billion gallons of fuel in 2001; meanwhile, the average cost per person was $520.[81]

Part of the reason for this congestion was of course the continual increase in automobile and truck registration (Figure 6.1). This factor was compounded by the crosstown and intersuburban traffic resulting from the decentralization of both jobs and homes since the 1950s. Since 1960 the share of work trips that began and ended in suburban locations has increased from less than 30 percent to more than 60 percent. It was also compounded by a slowdown in capital investment in highway infrastructure and by delays caused by maintenance and repairs to an increasingly degraded infrastructure. By the early 1980s more than 70 percent of the 125,000 miles of interstate highways within the boundaries of metropolitan areas were classified as being in only "fair" or "poor" condition. By 1990 almost all these highways were approaching the end of their design lives, with the resulting rehabilitation and improvements causing additional traffic delays and escalating expenditures.

Meanwhile, after the General Accounting Office criticized the state of the interstate highway system in 1991, Congress focused its funding efforts on this

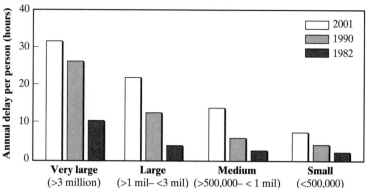

FIGURE 15.41   Trends in traffic delays in a selection of 75 U.S. urban areas, 1982–2001.

aging infrastructure. Capital outlays—expenditures of federal, state, and other funds—for projects on interstate highways have generally increased since 1992. State spending of federal, state, and local funds for interstates grew from $13 billion in 1992 to over $16 billion by 2000. Besides capital outlays, states also had routine interstate maintenance outlays that averaged $1.6 billion each year. These outlays, which generally consist of nonfederal funds, are for routine tasks such as sealing cracks and patching potholes that help keep the road surfaces in good condition but do not cover capital improvements like resurfacing, rehabilitation, or reconstruction.[82]

As a result, the interstate highway system's physical condition (road surface and bridges) has improved (with 8.6 percent or 3,897 miles of road surface in poor condition in 1990 compared to 3.4 percent or 1,560 respectively in 2000, while the number of structurally deficient bridges declined by over 22 percent between 1990 and 2000). Meanwhile, traffic

congestion on interstate highways has grown because the pressures from the many factors that contribute to congestion—population growth, the number of licensed drivers, vehicle miles traveled, and freight movement by truck—have continued to outpace the increase in capacity in terms of new lane miles (Figure 15.42).

Although the transit systems fostered by federal policies helped to prevent highway congestion from becoming even worse—with transit ridership increasing by 15 percent between 1995 and 2001 alone—they could not compete with Americans' love for—and dependency on—automobiles. Meanwhile, many people cannot take advantage of public transportation. The Federal Transit Authority has estimated that 25 percent of America's urban population does not live within walking distance of mass transit. In many instances suburb-to-suburb commuters are not served, and 30 percent of the country's suburbs have no mass transit service at all.

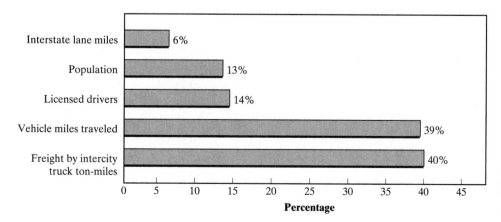

FIGURE 15.42   Variables contributing to traffic congestion, percent change, 1990–2000.

BOX 15.7   LONDON'S TRAFFIC CONGESTION CHARGE[83]

The number of vehicles worldwide has doubled since 1980 and is set to double again during the next two decades. Highways in many cities are heading toward gridlock. The record for the world's longest-ever traffic jam (109 miles) is held by the road between Paris and Lyon. The United Kingdom is Europe's most congested country (Figure 15.43). Governments are struggling to address the problem.

Building more roads as part of the solution is incredibly expensive, cannot keep up with increasing demand, and often encounters voter objection. In the central parts of cities like London and Paris, there is just no room to build new roads. Some cities have tried regulation to reduce congestion. Many European cities have created extensive pedestrian shopping areas along the narrowest streets in the oldest parts of the urban core. Rationing has been tried—with varying degrees of success (see, for example, the Box in Chapter 9 entitled "How Rationing Can Backfire: The 'Day Without a Car' Regulation in Mexico City"). Traffic congestion charges are another option. Singapore's scheme has been operating since the 1970s and is now electronic. Melbourne and Oslo have toll charges for specific roads.

With average traffic speed having fallen to less than 10 miles per hour during the day—the lowest average speed since cars were invented—and air pollution and traffic noise a constant problem, London in 2003 introduced a £5 per day congestion charge for motorists driving into an eight-square-mile zone within the Inner Ring Road—the most central and heavily congested part of the city (Figure 15.44). Motorists pay in advance or on the day of travel using a variety of options (online, by telephone, by mobile phone text message, in person at selected retail outlets, post offices, gasoline stations, at self-service machines in some car parks, or by mail), Between 7:00 a.m. and 6:30 p.m., 800 cameras at 400 points scan the license plates of the 250,000 or so motorists that typically enter the zone during the work week. This information is matched each night against a database of drivers who have paid the charge, and anyone who has failed to pay by midnight is liable for fines of up to £120. Persistent nonpayers can find their vehicles "clamped" or towed. Exceptions include taxis, emergency services, moped riders, vehicles powered by alternative fuels, and disabled people. Residents of the zone are given a 90 percent discount. The central London scheme could lead to a wider plan for the country—vehicles

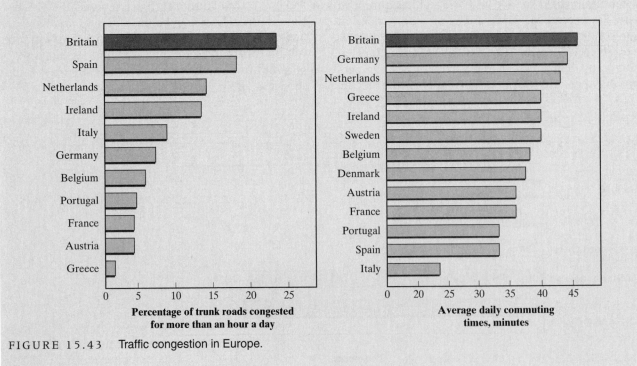

Percentage of trunk roads congested for more than an hour a day

Average daily commuting times, minutes

FIGURE 15.43   Traffic congestion in Europe.

(Continued)

Box 15.7   London's Traffic Congestion Charge   (*Continued*)

would be fitted with a transponder and their movements monitored by satellite.

While reducing traffic levels and easing severely congested roads, the congestion charge, unlike regulation, can also raise revenue to reinvest in the city's transportation system. In London the estimated benefits are a 15 percent reduction in traffic and a 25 percent reduction in traffic delays. The estimated £125 million net annual revenues are earmarked for improving the city's bus service. Of course, many motorists who drive into central London prefer not to take the bus.

FIGURE 15.44
Motorists who drive into the eight square mile congestion charging zone in London—the most central and heavily congested part of the city—must pay a fee to enter.

FOLLOW UP

1. Be sure that you understand the following key terms and concepts:

   culture of poverty

   cycle of poverty

   defensible space

   Dual City

   feminization of poverty

   labeling

   poverty areas

   service-dependent ghettos

   social construction of problems

   spatial mismatch

   spiral of decay

   urban dust domes

2. If you have access to a video library, you should arrange to watch an educational video about inner-city problems and then consider what you think about the arguments and explanations put forward. Two classics are *Race, Place, and Risk: Black Homicide in Urban America*, in which Dr. Paula McClain discusses the issue of black urban homicide (Charlottesville, Va.: Medical Television Services, 1993), and *Solving Black Inner-City Poverty*, in which Bill Moyers interviews Dr. William Julius Wilson about inner-city black poverty (Princeton, N.J.: Films for the Humanities and Sciences, 2004).

3. Poverty, it is noted on page 413, is arguably the most compelling problem both *in* and *of* cities. Find out how many people in your town or city are below the official poverty line (you will need to

look in your library or on the Web for census materials and publications). What can you find out about the attributes of these people—their age, gender, race, and place of residence? To what extent do they seem to be localized in particular neighborhoods?

4. Add the results of your research to your *portfolio*. You may also find that a systematic review of your city newspaper is revealing; look for newspaper stories in your library or online that deal with urban problems and classify them according to the type of problem (crime, environment, infrastructure). Review your findings in relation to the themes discussed in this chapter, paying special attention (1) to the issues of problems of and in cities and of the social construction of problems and (2) to any consistent spatial patterns that show up in news stories about particular kinds of problems. How do these findings relate to what you were able to find out about poverty in the city?

KEY SOURCES AND SUGGESTED READING

Belmont, S. 2002. *Cities in Full: Recognizing and Realizing the Great Potential of Urban America.* Chicago, Ill.: Planners Press, American Planning Association.

Burchell, R. 2002. *Costs of Sprawl: 2000.* Washington, D.C.: National Academy Press.

Downs, A. 2004. *Still Stuck in Traffic: Coping with Peak-Hour Traffic Congestion.* Washington, D.C.: Brookings Institution.

Downs, A. 2004. *Traffic: Why It's Getting Worse, What Government Can Do.* Washington, D.C.: Brookings Institution. (*http://www.brookings.edu/comm/policybriefs/pb128.htm*).

Fasenfest, D., J. Booza, and K. Metzer. 2004. *Living Together: A New Look at Racial and Ethnic Integration in Metropolitan Neighborhoods.* Washington, D.C.: Brookings Institution. (*http://www.brookings.edu/urban/publications/ 20040428_fasenfest.htm*).

Fowler, E. P., and D. Siegel, eds. 2002. *Urban Policy Issues: Canadian Perspectives,* Toronto, Canada: Oxford University Press, 2nd ed.

Harrigan, J. and P. Martin. 2002. Terrorism and the Resilience of Cities. *Economic Policy Review* 8: 97–116.

Hughes, M. A. 1990. Formation of the Impacted Ghetto: Evidence from Large Metropolitan Areas, 1970–1980. *Urban Geography* 11: 265–84.

Kreimer, A., M. Arnold, and A. Carlin, eds. 2003. *Building Safer Cities: The Future of Disaster Risk.* Washington, D.C.: World Bank.

Law, R. and J. Wolch. 1991. Homelessness and Economic Restructuring. *Economic Geography* 12: 105–36.

Office of Juvenile Justice and Delinquency Prevention. 2001. *The Growth of Youth Gang Problems in the United States, 1970–98.* Washington, D.C.: U.S. Department of Justice, Office of Justice Programs.

Orlebeke, C. J. 2002. The Evolution of Low-Income Housing Policy, 1949–1999. *Housing Policy Debate* 11: 489–520.

Palmer, J. L., ed. 1990. *The Vulnerable.* Washington, D.C.: The Urban Institute Press.

Polakow, V., and C. Guillean, eds. 2001. *International Perspectives on Homelessness.* Westport, Conn.: Greenwood Press.

Rose, H. and P. McCain. 1990. *Race, Place and Risk.* Albany, N.Y.: State University of New York Press.

Rusk, D. 1999. *Inside Game/Outside Game: Winning Strategies for Saving Urban America.* Washington, D.C.: Brookings Institution.

Sanders, R., and M. T. Mattson. 1998. *Growing Up in America: An Atlas of Youth in the U.S.A.* New York: Macmillan Library Reference USA; Prentice Hall International.

Sommer, H. 2001. *Homelessness in Urban America: A Review of the Literature.* Berkeley, Calif.: Institute of Governmental Studies Press, University of California. (*http://urbanpolicy.berkeley.edu/pdf/briefbook.pdf*).

Squires, G. D., and S. O'Connor. 2001. *Color and Money: Politics and Prospects for Community Reinvestment in Urban America.* Albany, N.Y.: State University of New York Press.

Taylor, R. B. 1991. Urban Communities and Crime. pp. 106–34 in *Urban Life in Transition,* eds. M. Gottdiener and C. G. Pickvance. Newbury Park, Calif.: Sage.

Turner, S. M. 2003. *Homeless in America: How Could It Happen Here?* Detroit, Mich.: Gale Group.

Wiewel, W., and J. Persky, eds. 2002. *Suburban Sprawl: Private Decisions and Public Policy.* Armonk, N.Y.: M. E. Sharpe.

U.S. Conference of Mayors. 2003. *Hunger and Homelessness Survey: A Status Report on Hunger and Homelessness in America's Cities: A 25-City Survey.* Washington, D.C.: U.S. Conference of Mayors (*http://www.usmayors.org/ uscm/hungersurvey/2003/onlinereport/ HungerAndHomelessnessReport2003.pdf*).

Wright, J. D., B. A. Rubin, and J. A. Devine. 1998. *Beside the Golden Door: Policy, Politics, and the Homeless.* New York: Aldine de Gruyter.

RELATED WEBSITES

Urban Institute: *http://www.urban.org/*
    The Urban Institute, a nonpartisan economic and social policy research organization, provides useful information and publications on urban issues. Its Center on Metropolitan Housing and Communities, for example, focuses on quality of life issues in urban communities.

U.S. Conference of Mayors: *http://www.usmayors.org/*
    The U.S. Conference of Mayors is the official nonpartisan organization of the country's 1,183 U.S. cities with populations of 30,000 or more. The website contains a wealth of reports and data—such as on homelessness, poverty, infrastructure, brownfields, and security issues—that reflect the goals of the organization: to promote the development of effective national urban and suburban policy, strengthen federal-city relationships, ensure that federal policy meets urban needs, provide mayors with leadership and management tools, and create a forum in which mayors can share ideas and information.

U. S. Department of Homeland Security: *http://www.dhs.gov*
    Since being established following the terrorist attacks of September 11, 2001, the Department of Homeland Security has generated useful reports and data sets related to urban terrorism that are available on its website. Topics include emergencies and disasters, travel and transportation, immigration and borders, and research and technology.

Recycle City: *http://www.epa.gov/recyclecity/*
    The U.S. Environmental Protection Agency's Recycle City website was launched on Earth Day 1997 and provides useful information about household waste recycling in cities.

16

THE POLITICS OF CHANGE: URBANIZATION AND URBAN GOVERNANCE

In this chapter we examine urban governance and the politics of changing urban geographies. In keeping with the overall approach of this book, the emphasis is on the *relationships between urbanization and the changing roles of urban government, and on the changing politics of urban development.* Our concern is with the generalizations that can be made about the influence of urban change on urban governance and vice versa rather than with the details of governmental organization, the structures of political participation, or the political careers of influential urban leaders. Similarly, we look beyond "symbolic politics" (the "hot" issues that provide easy copy for parochial newspapers while satisfying the needs for political figures to stay in the public eye)[1] to focus on long-term politics that determine "who gets what, when, and how."[2]

CHAPTER PREVIEW

The purposes of this chapter are (1) to show how, as the economic base of cities evolved, the fortunes of different socioeconomic groups changed and cities themselves generated new problems and challenges; and (2) how, as cities became larger and more complex, the scope of urban government broadened to include the regulation and provision of an increasingly wide range of infrastructural elements, goods, and services—all of which had a direct and sometimes profound effect on the social geography as well as the built environment of cities.

We will also see that the combination of sociospatial change, changing urban problems, and the changing scope of urban government resulted in a succession of different types of power holders; different, that is, in terms of their motivations and objectives. The ethos and orientations of urban government, reflecting these changes, in turn fostered further changes in the nature and direction of urban development. In the broadest terms, it is possible to recognize six distinctive phases in the evolution of urban governance and urban politics:

1. Laissez-faire in the mercantile city (1790–1840)
2. Municipal socialism and the rise of machine politics (1840–1875)
3. Boosterism and the politics of reform (1875–1920)
4. Metropolitan fragmentation and formation of progrowth coalitions (1920–1945)
5. Cities as growth machines and service providers (1945–1973)
6. Fiscal crisis and entrepreneurial politics (1973–present).

The following sections provide brief summaries of each of these, followed by outlines of the most important aspects of the relationships between urbanization, the ethos and organization of urban governance, and the politics of urban development, particularly insofar as they help us to understand contemporary urbanization. In the concluding section of the chapter, we turn to conceptual and theoretical perspectives on urban governance and politics to answer three broad questions raised by the changes we have reviewed: (1) How is power structured within cities and how does its structure change with the changing dynamics of urbanization? (2) How can we interpret the role of central governments in relation to urban development? and (3) How can we interpret local conflicts in relation to patterns and processes of urban change?

BOX 16.1   URBAN GOVERN*ANCE*

Urban governance is a multifaceted and at times a contentious concept. Some aspects of urban governance, however, are commonly agreed upon.

- Urban govern*ance* is not the same as urban govern*ment*. Governance is a concept that recognizes that power exists outside as well as inside the formal authority and institutions of government. Most definitions of governance include three main groups: government, the private sector, and civil society.

- Urban governance emphasizes "process." It recognizes that any decision-making that affects cities is based upon the complex interrelationships among many individuals and

groups who have different goals and priorities. It is the reconciliation of these competing goals and priorities that lies at the heart of the concept of urban governance.

The United Nations Human Settlements Program (UN-HABITAT) has proposed the following definition:

> Urban governance is the sum of the many ways individuals and institutions, public and private, plan and manage the common affairs of the city. It is a continuing process through which conflicting or diverse interests may be accommodated and cooperative action can be taken. It includes formal institutions as well as informal arrangements and the social capital of citizens.[3]

LAISSEZ-FAIRE: GOVERNMENT AND POLITICS IN THE MERCANTILE CITY (1790 TO 1840)

The earliest phase of the evolution of urban governance and politics, roughly coincident with the era of the mercantile city (i.e., to about 1840), was dominated by the doctrine of "utilitarianism," a laissez-faire philosophy that rested on the assumption that the maximum public benefit would result from unfettered market forces. This phase was also the "Age of the Common Man," with rapid demographic growth and an expanded electoral franchise combining to wrest urban political power from the exclusive control of the literate and the wealthy. This meant that an oligarchy of local merchants and wealthy landowners came to preside over urban affairs. But because of the pervasive ideology of free enterprise, urban government was weak and disorganized. And because of the pervasive opportunism in political life, it was also endemically corrupt, with victorious politicians routinely handing out jobs and rewards to friends and supporters. The result was that urban governance or politics rarely affected urban development in a direct or beneficial way. The indirect consequences were that cities became increasingly perplexing, their overall economic growth being accompanied by social polarization and mounting disorder, disease, and congestion.

The basic system of local government in most of the United States dates from the early nineteenth century, when counties, municipalities, and school districts were set up in order to assist state governments in carrying out their responsibilities. Municipalities were established in order to provide essential local services.

School districts were established as *independent* units because of a strong conviction that public education was important enough to society to warrant its economic and political freedom from other governments.

A notable exception to this pattern of local government was the township, which originated during the colonial period and is still prevalent in New England. In addition to the principle of elected representatives guiding local affairs according to their interpretation of the "public good," an important feature of the New England township system is the element of participatory democracy provided by town meetings. Town meetings involve qualified voters who come together to approve town budgets, to resolve controversial policy issues, and to formulate basic policies; they elect a town board to implement the budget and carry out policy. A few municipalities in the New England township system, however, have *representative* town meetings, involving a system in which voters can attend and participate in the town meetings but at which only their elected representatives can vote (Table 16.1).

TABLE 16.1

A Survey of the Forms of Local Government in the United States by Size of Jurisdiction and Location, 2001

	Mayor-council		City manager-council		Commission		Town meeting		Representative town meeting	
	No.	%	No.	%	No.	%	No.	%	No.	%
Total reporting (4,244)	1,611	38.00	2,248	53.00	49	1.20	236	5.60	38	0.90
Population size										
Over 1,000,000	2	0.05	2	0.05	0	0.00	0	0.00	0	0.00
500,000–1,000,000	7	0.16	1	0.02	0	0.00	0	0.00	0	0.00
250,000–499,999	12	0.28	11	0.30	0	0.00	0	0.00	0	0.00
100,000–249,999	29	0.68	75	1.77	0	0.00	0	0.00	0	0.00
50,000–99,999	60	1.41	161	3.79	2	0.05	0	0.00	3	0.07
25,000–49,999	164	3.86	302	7.12	1	0.02	5	0.12	11	0.30
10,000–24,999	326	7.68	562	13.24	18	0.42	59	1.39	11	0.30
5,000–9,999	353	8.32	503	11.85	13	0.31	81	1.91	8	0.19
2,500–4,999	448	10.56	396	9.33	10	0.24	63	1.48	4	0.09
Under 2,500	210	4.95	235	5.54	5	0.12	28	0.66	1	0.02
Geographic region										
New England	64	1.51	134	3.16	1	0.02	233	5.49	23	0.54
Mid-Atlantic	242	5.70	206	4.85	19	0.45	1	0.02	4	0.09
East North-Central	425	10.01	397	9.35	8	0.19	2	0.05	8	0.19
West North-Central	292	6.88	218	5.14	14	0.33	0	0.00	0	0.00
South Atlantic	131	3.09	427	10.06	5	0.12	0	0.00	0	0.00
East South-Central	148	3.49	48	1.13	2	0.05	0	0.00	0	0.00
West South-Central	119	2.80	266	6.27	0	0.00	0	0.00	2	0.05
Mountain	106	2.50	169	3.98	0	0.00	0	0.00	1	0.02
Pacific Coast	84	1.98	383	9.02	0	0.00	0	0.00	0	0.00

*based on a survey of 7,867 municipalities with a 53.9% response rate.

Source: *Adapted from K. L. Nelson,* Structure of American Municipal Government: Special Data Issue, *Washington, D.C.: International City/County Management Association, 2002, Table 2, p. 2.*

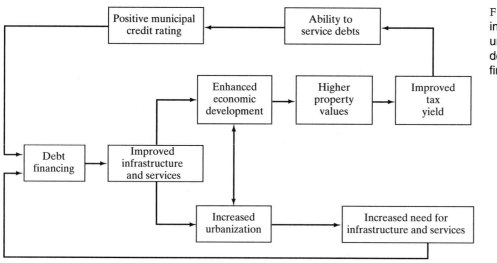

FIGURE 16.1   The interdependence between urbanization, economic development, and debt financing.

In the early nineteenth century local government was quietly dominated by a small pool of "natural" leaders drawn from "established" local families, few of whom regarded their role as extending beyond the provision of a reasonably sanitary and lawful environment. (In 1798, for example, New York City spent 60 percent of its budget on the night watch, poorhouse, and prison; about 10 percent on street lighting; and the remainder on miscellaneous services and expenses.[4]) Local governments covered their expenditures, such as they were, directly from revenues from property taxes, fines, and fees. The budget was balanced at the end of each year, and it was unusual for overspending to occur. The main disadvantage of this system was that capital expenditures for costly undertakings such as roads, sewers, and bridges were limited to the amount that could be covered by annual tax levies. It was a disadvantage that became acutely apparent as the pace of urbanization quickened. It was solved by **incorporation**, an act that changed urban government from a passive, regulatory activity to an active agent in economic development.

The origins of this change can be found in the first part of the nineteenth century, rooted in state constitutions that ensured that any group of people that petitioned could gain articles of incorporation as a village, town, or city. What was so important about incorporation was that it conferred limited liability and, therefore, the opportunity for **debt financing** through issuing bonds. Bond repayments for incorporated places were relatively secure: They could reasonably be expected to be met from property taxes, revenues from which could be expected to increase as a result of the growth stimulated by the new infrastructure and improved services financed by bond issues (Figure 16.1). The mercantile cities of the period were all incorporated in order to allow them to compete with one another as central places. As a result, American urban governance and politics came to be founded on a fragile web of debt financing, economic growth, and increasing property values.

MUNICIPAL SOCIALISM AND THE RISE OF MACHINE POLITICS (1840–1875)

The second phase in the evolution of urban governance and urban politics, roughly coincident with the era of the early industrial city (1840 to 1875), saw the erosion of laissez-faire ideology, an increase in the range and strength of local government powers, and an increase in expenditures (paid for through an expansion of debt financing), all in response to the continuing threats posed by an "unholy trinity" of urban problems: fires, disease, and the mob. This was also the age when **machine politics** emerged, with charismatic leaders controlling hierarchical (and often corrupt) political organizations that drew on working class support for more paternalistic urban governance.

The social and economic upheavals associated with the incipient industrialization and rapid urbanization of this period brought acute problems that could be tackled effectively only by increasing the scope and the scale of government. Crowding and uncontrolled development had resulted in an intolerable vulnerability to fires and epidemics. The free market, meanwhile, had created an impoverished workforce whose living conditions were so bad that rebellion and disorder emerged as a constant threat and an occasional reality. The combination was sufficient to convince a majority of people of the need for **municipal socialism**—local government intervention in the marketplace in order to impose standards and to ensure the provision of key services and basic amenities (what is now more often referred to as **collective consumption**). Governance thus became involved with efforts (1) to moralize the lives of citizens through regulating the physical and social environment, and (2) to develop a framework of welfare provision that would systematically change the conditions that caused individual impoverishment and fostered social disorder.

This period of radical change attracted a variety of new, self-made entrepreneurs—merchants, manufacturers, building contractors, and real estate speculators—to vie for political control in order to steer the expanded range of public policies in directions suited to their own interests. In this volatile and rather confused phase of urban politics, it was increasingly apparent that the key to success was *organization*. Political parties and trade unions became increasingly important, and they in turn fostered the emergence of new kinds of political leaders whose careers were based on close attention to the concerns of their constituents. In larger cities with sizable working-class populations, organization began to take the form of "machines" that delivered votes in return for the patronage jobs and benefits that could be granted by successful politicians.[5] These machines brought a distinctive dynamic to big-city politics:

> Machines provided jobs and favors to their constituency and offered utility franchises, contracts, tax adjustments, and favorable legislation for selected members of the business community. Machine politicians used these resources to build a personal political following. By resourcefully distributing patronage and material benefits, machine politicians also built centralized party organizations that overcame the fragmentation of authority characteristic of the formal governmental structure. The machines acted as brokers among contending political interests; they supplied material and symbolic rewards to immigrant and working-class residents who otherwise lacked resources; and by centralizing political authority, they brought some semblance of social and political order to the cities.[6]

This dynamic helped to ensure that patronage and corruption became firmly embedded in urban governance and politics. Paradoxically, the corruption of the machine system helped in some ways to accelerate the modernization of cities. Machine bosses were astute enough to appreciate the general economic benefits of infrastructure improvements, as well as the personal kudos that could be derived from being associated with Great Works. In more pragmatic terms, the more housing codes, building regulations, and municipal ordinances that were passed to regulate the emerging infrastructure, the more opportunities there were to solicit bribes in return for not enforcing them.

Machines took over most large cities for a time, though in many they managed to retain power for only two or three elections.[7] The excesses of machine organization, meanwhile, began to become issues in themselves, with some machine bosses who milked city treasuries to the tune of hundreds of millions of dollars being convicted on numerous counts of fraud and bribery.

It should be noted that Southern cities were for the most part an exception to this trend toward organized machine politics. Because they lagged in terms of industrialization, did not attract large numbers of immigrants, and did not allow African Americans to vote, their working-class vote could not be delivered in the same way as in the industrial centers of the Northeast. Throughout this phase, and well into the next, therefore, a continuation of the commercial elite that had risen to prominence in the early nineteenth century dominated their politics. It follows that the expansion of the scale and scope of government was less exaggerated in the South than elsewhere, with moralization and welfare provision also emerging much later.

BOOSTERISM AND THE POLITICS OF REFORM (1875–1920)

The era of the industrial city (1875–1920) saw a struggle between the interests of ethnic and working-class groups and those of the new urban elite. Downtown business and professional interests organized themselves to restore urban rule to the "the better elements" and to shift the emphasis of urban government to activities that might attract investors—**boosterism**. The expanding middle class, meanwhile, made its presence felt in the latter part of this period through the politics of the Progressive Era (1895–1920), lending its support to the reform movement that was aimed at reducing corruption, ridding urban governance of machine politics, and establishing the basis for a more progressive approach to urban development. Boosterism and the suburbanization of the middle classes also contributed to the spatial expansion of urban jurisdictions through municipal annexation.

Meanwhile, growing numbers of the business and professional classes were becoming concerned about investors' reactions to the political power of the working classes that was inherent to machine politics. *Corruption became the symbolic issue that this new urban elite chose in seeking to undermine the legitimacy of the machines and to replace them with a political leadership more in tune with the needs of industry.* It was no coincidence that this period also saw the **social construction** of immigrant neighborhoods (the power base of the machines) as breeding grounds for social problems (p. 408). Editors, novelists, and reformers competed with each other to draw attention to the corrupting moral influence of the "Great Unwashed" and to emphasize the theme of Anglo-Saxon superiority.

Meanwhile, the economic recessions of the 1870s and 1890s changed the context of urban politics decisively, stirring the new urban elite to form alliances that would eventually replace the corrupt machines and install representatives who would improve the atmosphere of the city for business and investment. Boosterism, which had always been a characteristic element of American urban development, became rampant. The leadership of every town and city, big and small, soon came to realize that, under the dramatic spatial reorganization involved in forging a modern industrial economy, their economic future depended on outdoing their competitors in attracting investors in manufacturing and business. The result was that boosterism escalated beyond promotional hype to become an important element in local politics, particularly in smaller and medium-sized cities that had not yet acquired sound economic foundations. Rather than directly exploiting potentially explosive divisions between classes or ethnic groups, politicians tended to avoid them, realizing that a facade of unity was essential for economic development. Class conflict and ethnic conflict were endemic but were (for the most part) kept muted in political terms for the sake of the city's economic prospects. What is particularly significant about this period for the subsequent evolution of American cities is that the public interest came to be defined more in terms of aggregate economic benefits than in terms of equity, social justice, civil rights, democracy, or community well-being.

Against this background, the common theme of reform emerged in the 1870s and 1880s. Religious and moral reform focused on temperance as the way to combat gambling, prostitution, and social malaise. Educational reform focused on school attendance, administrative reorganization, and revised curricula as the way to imbue the urban masses with standards of citizenship, democracy, and diligence. Housing and sanitary reformers saw improved physical conditions as a prerequisite for moral and social well-being and as a necessary step toward improved efficiency and productivity in the workplace.

The City Beautiful movement, as we have seen (p. 250), embodied this impulse to reform urban life, promising as it did to deliver moral and social order (through the noble idealism of Beaux Arts neoclassical urban design) within an efficient framework for economic development (through formal, rationalized spatial organization). The more elitist municipal reformers wanted to restrict voting rights to property owners. Fortunately, democratic principles and the political power of the working classes were too deeply entrenched. Consequently, municipal reformers turned to strategies that would weaken political machines and lessen the impact of immigrant and working-class voters without interfering with voting rights: tighter controls over both elections and the use of campaign funds, and more open nomination procedures.

In a few cities leaders who campaigned for such reforms, appealing to both immigrants and the middle classes, built political coalitions. The best example was in Detroit, where between 1889 and 1897 Mayor Hazen Pingree stamped out the legacy of corruption in public affairs that years of machine politics had installed. He went on to reconstruct the city's sewerage system, improve the streets from the worst to the best in the country, expand the park system, establish public control over utilities, build a public bath, reform the school board and expand the public school system, end street tolls, and force down the price of gas, telephone, ferry, and streetcar services.

THE PROGRESSIVE ERA

A few inspiring examples such as this were sufficient to encourage reformers all over the country. From the mid-1890s to 1920 the reform movement dominated local politics nationwide. This was the Progressive Era. In Philadelphia in 1894 the First Annual Conference for Good City Government organized the National Municipal League as a forum and coordinating body with the objective of formulating a plan that could be supported nationally. Nearly 200 affiliated societies joined it at once. Together, they developed a theory of good government that rested on four elements:[8]

1. The existence of a public interest that can be defined objectively, to the benefit of all citizens. Central to the public interest, in the reformers' view, was efficiency in the delivery of public services.

2. Strict separation of the day-to-day administration of public services from politics, elections, and representation.

3. A permanent civil service made up of experts with specialized training to run the city's business.

4. The application of principles of scientific management (**Taylorism**: see p. 11) to run government like a business.

This theory was to be implemented through a recommended package of reforms and new forms of government organization. The most important elements in the reform package were[9]

- A civil service based on a merit system that used written examinations to determine suitability for employment. This system, it was hoped, would eradicate political influence and ensure better-educated employees who would be more able to resist the temptations of corruption and to implement scientific methods.

- Detailed, standardized accounting procedures and mandatory competitive bidding for purchasing and construction contracts.

- The constitution of independently elected boards or commissions, such as school boards, to insulate key areas from city hall politics.

- At-large elections, with every city councilor being selected by citywide votes. These elections would replace ward-based systems in which each district had its own councilor. This method of electing city leaders, it was argued, would reduce both opportunities for patronage and the number of representatives who depended on the vote of specific ethnic groups.

The preferred new form of government organization involved the abolition of the elected chief executive post and its replacement with an appointed professional, a "city manager" trained and experienced in business principles and administration.

The Progressive movement was reinforced by several concurrent trends. The need for expertise in municipal government was increasingly apparent with every technological innovation and every increment in the size and scope of public activities. Cities had to regulate the laying of pipes and cables and the setting up of streetcar systems. They needed inspectors to regulate and inspect new building methods, elevators, and gas and electricity supplies. They needed accountants who could handle increasingly complex city budgets and administrative officers who could coordinate and professionalize civic boosterism, and so on. These growing occupations sought to establish and legitimize themselves with professional associations and trade magazines, and they realized from the start that it was in their own best interests to support the ideals of the Progressive movement. Their voices were joined by those from another emergent group: social scientists and their professional organizations, who saw in the Progressive movement a ready audience for the careful documentation and analysis of every facet of urban life. At the same time, falling paper prices and technical advances in printing allowed a great expansion in the circulation of newspapers and popular magazines. Although the ideals of the Progressive movement did not often make for good copy, scandals, exposés, and shocking revelations about slum life did. The consequent crusades had the incidental effect of publicizing the objectives of Progressive reform.

By the early years of the twentieth century, the Progressive movement had achieved some tangible success. Galveston, Tex., became the first city to be run on the business model when an emergency situation after a hurricane prompted the state legislature to replace the city council and mayor with a commission of business leaders, each of whom was responsible for administering a different branch of municipal affairs. This form of government was subsequently adopted on a permanent basis. By 1915 more than 450 towns and cities had adopted the commission format of urban governance (with the mayor usually being the commissioner who is responsible for public safety). Nonpartisan, citywide balloting was meanwhile instituted in a number of cities—Akron, Boston, Detroit, Los Angeles, and Nashville being among the first.

In 1908 the city manager form of government was introduced in Staunton, Va. After Dayton, Ohio, adopted it in 1914, the city manager form of government spread rapidly. The elected council and the chief elected official (e.g., mayor) are responsible for making policy, and the city manager, who is appointed by the council, has full responsibility for the day-to-day operations of the government. About half of all cities in the United States now operate with a city manager (Table 16.1). A growing number of cities continue to adopt this form of government, at the expense of the commission and the mayor-council (with an elected council and a chief elected official [e.g., mayor] who is generally elected separately) forms of government (Table 16.2). The mayor-council form of government—the oldest and most prevalent form until the early twentieth century and the Progressive movement reforms—remains prevalent in the largest cities in the United States. New York, Los Angeles, Chicago, Houston, Philadelphia, and Detroit all have the mayor-council form of government. There is usually either a strong mayor who

TABLE 16.2

The Forms of Local Government in Municipalities with Populations of 2,500 or Greater in the United States: 1984–2004

Form of government	1984	1988	1992	1996	2000	2004
Mayor-Council	3,686 (56%)	3,686	3,635	3,319	2,988	3,089 (44%)
City Manager-Council	2,290 (35%)	2,356	2,441	2,760	3,302	3,453 (49%)
Commission	176 (3%)	173	168	154	143	145 (2%)
Town Meeting	370 (6%)	369	363	365	334	338 (5%)
Rep. Town Meeting	81 (1%)	82	79	70	65	63 (1%)

Source: *International City/County Management Association,* Municipal Year Book. *Washington, D.C.: ICCM (various years).*

holds budgetary, appointment, and veto powers or a weak mayor who is subject to council control of many of the city's administrative functions.

The most widespread reform between the mid-1890s and 1920, however, was the introduction of the civil service model, which embodied the principles of specialization and scientific management that were central to the ethos of the industrial era. By 1920 some form of civil service was in operation in nearly all the larger and more industrialized cities. The chief hold-outs and exceptions, as in so many other aspects of urban governance and politics, were Southern cities.

Annexation

In keeping with the spirit of boosterism and efficiency that pervaded the period, the geography of cities was transformed by an acceleration of **annexation**—the addition of unincorporated land. Boosterism was one important motive for this annexation. With the advent of streetcar systems and new developments in home construction techniques, city populations had begun to decentralize, and city pride dictated that these numbers be recaptured.

Progressive ideals also played a part. Annexation, it was argued, created the framework for **economies of scale** in government—economies that would benefit both the suburbs and the central city as the costs of infrastructure provision and bureaucratic expertise were spread over a larger population. It also helped the Progressive cause that the addition of middle-class voters from outlying districts would weaken the potential strength of machine-style politics. Even more compelling (for the annexing governments, at least) was the importance of capturing the tax base represented by the growing suburbs. Annexation enabled cities to maintain, perhaps even strengthen, the delicate web of economic growth, increasing property values, and debt financing (Figure 16.1). Finally, real estate interests and development-related entrepreneurs (e.g., builders

and their suppliers) were strong advocates of annexation because it enhanced opportunities for speculative development. Annexation of undeveloped land was particularly attractive to them, since it carried the implicit promise that the suburb would eventually be equipped with all the utilities and amenities of the city.

Annexation occurred throughout the nineteenth century, but the scale and pace of annexation at the turn of the century and the fact that it was so closely related to the realities of sociospatial change made it especially significant in the Industrial Era. Chicago, for example, annexed 133 square miles of territory in 1889. The annexed land (now the city's far South Side) already contained about 225,000 people living in suburban and **exurban developments** such as Hyde Park, Kenwood, Pullman, and Woodlawn and in industrial satellites such as Grand Crossing. But it also contained large stretches of undeveloped land that were later able to accommodate a substantial part of the city's phenomenal growth. By 1920 more than a million people were living in the territory that had been annexed in 1889.

The growth of Chicago in fact helped to spur the most dramatic annexation of all. New Yorkers' fears that Chicago would become the nation's largest city contributed to the creation of a Greater New York City in 1898, when Brooklyn (itself the fourth largest city in the United States at the time) joined with Manhattan, most of Queens, Staten Island, and part of Westchester County (the Bronx). For the most part, however, annexation involved much more modest acquisitions, with central cities experiencing a series of small additions.

Overall, the annexation movement must be regarded as positive and successful. For pioneer suburbanites the services gained as a result of being annexed represented a significant positive change in their quality of life. These suburbanites had access to better quality urban services (city water, sewer, police and fire protection, etc.) than they could afford to provide themselves.[10] For the central cities, meanwhile, annexation provided important economies of scale and, above all, prevented the loss of their middle classes from political life and from the tax rolls.

Suburbanites, of course, recognized that they had to pay higher taxes once annexed. They also had to share schools and municipal services with city residents, which sometimes led to conflict (Figure 16.2). As suburban speculative development increased in sophistication (with improved packages of amenities) and suburban self-consciousness became stronger, opposition to annexation increased. Citizen participation in such decisions was still some way off, however, and state legislatures generally took the view that no small territory should be allowed to impede metropolitan development. It would not be long, however, before state legislatures began to retreat from forced annexation in the face of suburbia's mounting electoral power and political influence.

FIGURE 16.2   Parents and children protesting against proposed changes in school catchment areas in St. Louis, 1933.

METROPOLITAN FRAGMENTATION AND THE FORMATION OF PROGROWTH COALITIONS (1920–1945)

Between 1920 and 1945 the suburbanization associated with the mass production of automobiles gave rise to the incorporation of suburban jurisdictions that quickly developed a politics of their own: exclusionary and competitive. An important corollary of this shift was that central cities lost not only a large section of better-educated middle-class voters and taxpayers but also a significant component of the pool of potential political leaders. In the early 1930s the Depression helped to generate a climate of opinion that was much more favorable to governmental provision of goods and services, resulting in an extension in the roles and legal powers of both federal and local governments and in a shift in the nature of urban politics.

Beginning in the 1920s, the pace of suburbanization quickened. As we saw in Chapter 6, the combination of cheap, mass-produced automobiles and federal mortgage insurance policies resulted in a major shift in metropolitan spatial dynamics. For the first time, population growth in the suburbs outpaced that of the central cities in both relative and absolute terms. Middle-class suburbanites sought not only to escape from proximity to the slums and the **ghettos** of central cities but also to escape from the tax burdens of central cities and to establish a distinctive setting for governance and politics in which middle-class values and preferences might flourish.

Rather than waiting to fight annexation, many suburban communities took the preemptive strategy of petitioning for **suburban incorporation** for themselves (not necessarily as cities, since incorporation as a village was sufficient to protect their reputation, status, and independence). At the same time, state legislatures, recognizing the changing spatial distribution of voters, became increasingly reluctant to alienate suburbanites by approving annexation. Gradually, the prevailing legal view came to be that annexation should be a voluntary affair, with the clear support of the residents of the affected area. The result was that central cities came to be encircled by a ring of independent and often hostile governments (Figure 16.3). Geopolitically, America's metropolitan areas were fragmented, and that *metropolitan fragmentation undercut the ability of central cities to deal with urban change and development even as it intensified sociospatial segregation.*

THE CONSEQUENCES OF METROPOLITAN FRAGMENTATION

There were three main consequences of metropolitan fragmentation: (1) central cities were hampered in their ability to deal with urban change and development, (2) metropolitan sociospatial segregation was intensified, and (3) metropolitan political life became fragmented.

Metropolitan fragmentation *diminished the ability of central cities to deal with urban change and development in several ways:*

- The suburbanization of the middle classes reduced the pool of experienced and educated political leaders and reformers. Suburbanites with an impulse to become involved in civic affairs now did so in suburban jurisdictions, channeling their efforts into establishing and preserving a rather narrow and sectionalized version of the public interest.

- The loss of a large percentage of the middle classes, meanwhile, not only directly weakened the central city's tax base but also confined it by putting future property taxes out of reach. This loss, of course, made debt financing more difficult, which impaired the ability of central cities to provide safe and efficient settings for economic development. Without room for new development, central cities were left with a steadily aging environment requiring increasing levels of maintenance and expensive fire protection.

- At the same time, the loss of middle-class population and the continuing influx of low-income migrants and immigrants forced central city governments to face even greater burdens in terms of the provision of welfare-related services and social control functions.

- Central city governments continued to be responsible for providing metropolitan-wide amenities such as galleries and museums and for expenditures on roads, parking space, utilities, and policing incurred by suburban commuters and shoppers.

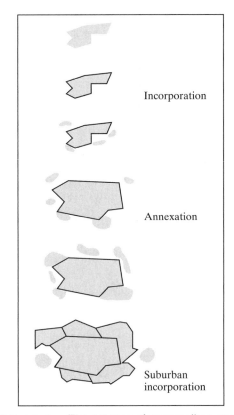

FIGURE 16.3   The process of metropolitan growth and political fragmentation.

All these changes and trends put central cities in an unavoidable **fiscal squeeze** (Figure 16.4) as revenue capacity fell while expenditure demands increased. The issues and conflicts associated with this squeeze have shaped the politics and geography of central cities ever since.

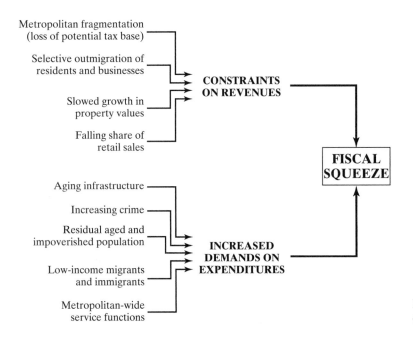

FIGURE 16.4   The fiscal squeeze on central city governments.

The second consequence of metropolitan fragmentation, *intensified sociospatial segregation*, was a result of the political independence of the middle classes, which enabled them to erect barriers to keep out lower-income groups. The reasons for wanting to exclude lower-income households had to do with the perceived threat to property values and to the lifestyles and the kind of moral order the middle-class suburbanites sought to achieve in their schools and communities.

The chief instrument at their disposal was **exclusionary zoning**. By carefully framing their **land use zoning** ordinances, suburban jurisdictions were able to make it difficult or impossible for "undesirables" to move in. The zoning tactics invented in San Francisco to discriminate against the Chinese and refined in New York in 1916 to discriminate against undesirable factories (see p. 125) were soon deployed by suburban communities to exclude undesirable social groups and unwanted land uses. By the spring of 1918, New York had become a place of pilgrimage for citizens and officials wanting to find out how zoning worked.[11] In 1924 the U.S. Department of Commerce drafted a model zoning law as a basic instrument that could be used across the country, and in 1926 the Supreme Court gave its approval to zoning in the landmark case of *Euclid v. Ambler*. The case was actually brought because the Ambler Real Estate Company objected to the decision by the government of Euclid, Ohio, to zone some of the company's land as residential—the company had anticipated substantial profits from industrial development.

The Supreme Court's decision to uphold Euclid's zoning ordinance was significant because it recognized land use zoning as a proper use of the **police powers** of local governments. But the Court actually went further in its judgment, explicitly ruling that it was legitimate for local governments to protect the character of neighborhoods of single-family dwellings against the threat of lower property values and negative externalities associated with apartment housing, commercial activities, and industrial land use. Reformers and planners endorsed this view because they felt that zoning provided a means of controlling urban growth and fostering efficient patterns of land use. But the chief motivation for introducing zoning ordinances in suburban jurisdictions was to keep out undesirable social groups by artificially raising the cost of housing. Normal practice quickly came to involve limits on the amount of new construction, minimum lot sizes, maximum densities, and bans on apartment houses. *Such barriers have underpinned metropolitan sociospatial fragmentation ever since.*

The political independence of middle-class suburbs also contributed to the third consequence of metropolitan fragmentation, the *fragmentation of metropolitan political life*. Politics in suburban jurisdictions tended to be relatively straightforward and free of controversy. The chief concerns were to maintain vigilance over zoning, to maintain law and order and protect property, and to develop educational settings appropriate to middle-class aspirations. Since the electorate tended (for a long time, at least) to be relatively uniform in its views on these matters, there was little scope for conflict. Looked at in broader perspective, metropolitan fragmentation can be interpreted as having defused political conflict in general, though it is a moot point as to whether quashing political conflict is a good thing for democracy.[12] The spatial fragmentation of the metropolis into separate central city and suburban communities also makes it difficult to make, or even to think about, area-wide decisions for area-wide problems such as transportation and affordable housing. Small issues rule the day for want of a political structure that can handle anything larger.

Metropolitan political fragmentation has imposed a legacy of competition rather than cooperation between neighboring governments. Not only do suburban jurisdictions compete with central cities for land users with a high tax yield and low demands for public services, they also compete with one another. In addition to land use zoning, local governments seek to gain a competitive edge through their tax packages and the "bundles" of services that they offer. The result has been termed **fiscal mercantilism**, an allusion to the fierce and unrelenting competition between trading nations in the sixteenth and seventeenth centuries.

THE URBAN LEGACIES OF THE NEW DEAL

The economic distress of the Depression in the early 1930s brought about new federal roles that not only restructured the relationship between the federal government and the country's cities but also brought about a realignment of politics at the metropolitan level. Both have left legacies that have been important in shaping the dynamics of contemporary metropolitan areas.

There had in fact been long-standing precedents of federal support for urban development. Throughout the nineteenth century the federal government had, for example, assisted many cities in dredging harbors and rivers and subsidizing canal and railroad development. What was so significant about the changes ushered in by the New Deal was their *scale*. After the New Deal, cities found themselves the beneficiaries of heavy federal expenditure on a series of programs, and this fiscal relationship, in turn, brought about a close interdependence between federal and local politics that lasted for almost 50 years.

These developments were attributable in part to the geography of distress during the Depression years.

As we have seen (p. 75), cities were certainly hit hard and the more specialized industrial cities experienced acute economic and social distress. Many cities attempted individually to provide relief, expand public employment, and initiate public works programs. But economic recession had caused a slump in property values, which triggered a precipitous decline in city revenues from property taxes. To make matters worse, unemployment and failing businesses led to property tax delinquencies. In 1930 tax delinquency in large and medium-sized cities averaged 11 percent. Worse still, many cities had floated large bond issues in the 1920s that were coming due in the 1930s.

Faced with the prospect of municipal bankruptcy, cities desperately petitioned state legislatures for help. Rural representatives, however, still controlled most state legislatures. In the summer of 1932, Detroit mayor Frank Murphy convened a meeting of representatives from 29 big cities in order to organize themselves to bypass their state governments and appeal directly to the federal government for assistance. "The mayors' demands for federal assistance marked a historic turning point for cities. *Historically, cities had been mere creatures of the states, with no direct ties to Washington.* Only a few months before, most of the mayors had declined to attend a mayors' conference suggested by the mayor of Milwaukee. Many officials felt it was illegitimate to ask the federal government for help, and others feared any aid, lest it lead to a loss of local autonomy. But desperation finally overcame tradition."[13]

To a degree, then, federal initiatives can be seen as a response to urban needs. But rural deprivation and distress were also acute. There were two fundamental reasons for the particularly close relationship between cities and the federal government during the Depression:

1. The federal government's acceptance of responsibility for national economic growth, the maintenance of aggregate demand, and the minimization of unemployment made cities important targets for **Keynesian** strategies of economic management because of the potential **multiplier effects** of public expenditures in metropolitan settings.

2. Urbanization had made metropolitan votes much more important to the composition of the Electoral College that determines presidential elections. The 1932 presidential election signaled this effect in clear terms. The coalition that swept Roosevelt to power was based on an alliance between big-city voters and traditional southern Democrats. As a result, both presidential politics and Congressional negotiations with incumbent presidents have since paid more explicit attention to the needs of metropolitan voters.

The specific policies and programs introduced under the New Deal are discussed in greater detail in Chapter 17. Several dimensions were of particular importance to the political economy of urbanization. One was federal spending on public works and infrastructure, part of a broader package of measures to stimulate economic growth. A second was federal expenditure to provide relief from poverty and unemployment, bolster personal incomes, and defuse urban social unrest. A third was federal expenditure aimed at clearing slums and resettling slum dwellers. A fourth was federal support for home mortgage insurance, another part of the strategy to stimulate economic growth. Together, these policies and programs not only had a significant impact on patterns of urban development, but they also changed the pattern of intergovernmental relations and fostered new political alignments within cities; both of these shifts have left their mark on the trajectory and outcomes of American urbanization.

The New Deal saw the "coming of age" of cities in American national politics. New Deal programs made it legitimate for local officials to solicit federal assistance directly, and in so doing they began a dialogue and established a working relationship that gave cities a voice in Washington. The federal government, newly sensitive to the significance of urban voters, published a report entitled *Our Cities: Their Role in the National Economy*,[14] which argued for a more extensive federal role in urban affairs, including increased responsibility for housing provision, more planning, urban renewal, and local government reform. The cooperation among cities that began with the meeting convened by Detroit's mayor Murphy was formalized in 1932 by the creation of the United States Conference of Mayors, which has since financed a permanent office in Washington to lobby for urban programs. Organizations such as the National Municipal League, the American Municipal League, and the International City-County Management Association have extended this cooperation and ensured that cities now constitute a significant lobbying force in Washington.

Meanwhile, the New Deal also fostered a new coalition *within* cities: that between liberal reformers and the blue-collar constituencies that had previously supported the political machines. Political scientist John Mollenkopf has argued persuasively that this alliance represented a **progrowth coalition** that came to be established across most of the country and that formed the basis of modern party politics:

> Domestic urban development programs . . . became the principal means through which the modern Democratic party was created. Progrowth coalition building thus became a central feature of national as well as local politics. National politicians and the federal government

became important actors in local politics, and this involvement, in turn, integrated local politicians, program administrators, and program beneficiaries into a new national political framework.[15]

The significance of the New Deal can be summarized in terms of three interrelated and mutually reinforcing trends:[16]

- It brought previously unrepresented constituencies to bear on national government.
- It reconciled these constituencies with more traditional participants to form a powerful new coalition of forces that spanned urban working-class and middle-class interests.
- It cemented this coalition by shifting the initiative in economic development away from the business community and toward the public sector.

After World War II these trends were to form the basis for a new political economy in which public involvement in housing provision, urban planning, urban renewal, and local government reform became far more prominent than the authors of *Our Cities: Their Role in the National Economy* could ever have thought possible.

CITIES AS GROWTH MACHINES AND SERVICE PROVIDERS (1945–1973)

The period of strong (Type A) economic growth after World War II (see p. 13) saw a consolidation of the welfare and service provision functions of urban governments, an expansion of their budgets, and an extension of their activities to include large-scale planning and management. Cities also became the beneficiaries of unprecedented amounts of federal aid designed to address the "crisis" of central city decay and social disorder. Economic decline and the depopulation of central cities resulted in the simultaneous emergence of (1) a distinctive ethnic politics and (2) a resurgence of political activity by downtown business interests in support of urban renewal. In the dynamics of urban politics, downtown business interests came to play important roles in the progrowth coalitions that came to prominence in many cities. Meanwhile, grassroots reactions to the outcomes of urban renewal, together with protests by African Americans against discrimination and disadvantage, helped to shape a new wave of federal initiatives and to heighten concern about certain aspects of the spatial organization of metropolitan government.

The relative affluence of the period after World War II, combined with changes in people's expectations about government, led to a steady rise in municipal expenditures. This rise was much faster than the growth of city populations. Expenditures on every kind of service were doubled (even allowing for inflation), then doubled again, then again. The fastest growth was in spending on education, sanitation, recreational services, and (at first) highways.

To a large extent, the growth of these municipal services was a consequence of rapid economic development. Spending on education was driven by the need for a more technically sophisticated workforce, by the increased numbers of schoolchildren (the first Baby Boomers reached kindergarten age in 1951), and by the increased amount of federal funds available for education (the embarrassment of the Soviet Union having been first to launch a space satellite successfully—Sputnik, in 1957—was an important factor here). Spending on sanitation was driven more simply by increasing affluence: the more households consumed, the more waste they generated, and the more expensive its collection and disposal became. Affluence and lifestyle changes also drove spending on recreational services as more and more people demanded access to golf courses, tennis courts, and softball diamonds. Spending on highways, meanwhile, was largely a function of increased levels of automobile ownership.

Expenditure on some other services, meanwhile, increased because the affluence of the period allowed an increased sensitivity to the plight of the needy. As a result, many new municipal services emerged to cater to welfare needs and to the needs of special groups such as the visually impaired and those facing learning challenges. Between 1960 and 1978 alone, city public welfare expenditures rose by nearly 450 percent.

GROWTH MACHINES AND URBAN RENEWAL

One of the immediate carryovers of New Deal politics was a slum clearance and public housing program that had been initiated under the 1937 Housing Act (see p. 508). The Housing Act of 1949 provided assistance to cities toward the costs of clearing "blighted" areas and assembling land for redevelopment. It also authorized the production of more than 800,000 public housing units. (For more details of federal urban policy and an interpretation in relation to urbanization and theories of urban change, see Chapter 17.) As we saw in Chapter 13, the public housing program faced overwhelming opposition from real estate, banking, and labor interests.

The provisions for **urban renewal**, in contrast, provided an ideal platform for the development of progrowth coalitions. During the 1950s downtown revitalization became the rallying cry for civic leaders around the country. The coalitions varied in character from city to city, but they typically included business elements (developers, bankers, and financiers), blue-collar interests (labor unions), liberal interests (planners and welfare agencies), and representatives of both local government (city managers and political leaders) and the federal government (urban renewal executives). It soon became clear that the central actors were to be the investors who had the economic power to organize the coalitions into would-be **growth machines**[17] that might restore prosperity to the central cities.

Although we might expect that the investors and allied business interests would naturally have preferred to operate in a free enterprise environment, they were instead quite happy to work with city and federal government representatives—in order to exploit the opportunities under the provisions of the 1949 Housing Act to close the inner-city **rent gap** (see p. 372). Through the machinery of urban governance, federal funds could be secured, the mechanics of slum clearance and land assembly could be efficiently organized, and discipline and cooperation could be instilled into the whole coalition. Where the New Deal alliance between liberal reformers and blue-collar communities was insufficient to sustain political support for this role, the specter of competition from other cities—for investment dollars as well as federal dollars—was usually enough to rally support. In addition, some support sprang from less noble motives: For some politicians, involvement in

the operation of a growth machine offered almost as much opportunity for graft as the ward-based machines of a century before.

Growth machines could not flourish simply on the basis of an alliance between business interests and local politicians, however. The urban renewal projects that were the preoccupation of urban governance and politics during the 1950s were carried forward within a web of interdependent interests (Figure 16.5). These interests and their motivations included:[18]

- *Investors.* The speculators and developers whose drive for profit were at the heart of the growth machine.

- *Downtown business interests.* Large retailers, banks, and other financial institutions and the local Chamber of Commerce, all of which sought to recapture the business that had been decentralizing to the suburbs.

- *Exchange professionals and self-employed professionals.* Real estate and other professionals involved directly in the property market and other self-employed professionals, such as doctors and lawyers, who supported growth both ideologically and because their high salaries enabled them to participate in property syndicates.

- *Political leaders* who sought to recapture jobs and tax dollars by remaking central cities into places that were better settings for business. The authority of city hall was critical in designating "blighted" areas. The legitimacy of city hall was important in establishing renewal projects as being in the public interest.

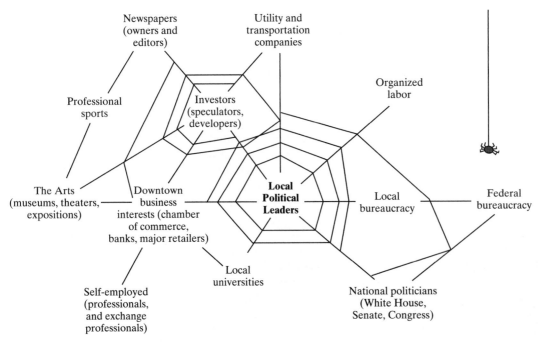

FIGURE 16.5
The web of growth machine politics.

- *Organized labor* was supportive because of the prospect of increased employment.
- *Utility and transit companies* supported revitalization because of their huge fixed investments in central cities.
- *Local and federal bureaucrats* saw the urban renewal schemes as a means of establishing or enhancing professional turf. For planners, in particular, renewal struck a chord with professional outlooks that were strongly influenced by design determinism and Modernist ideas of efficiency and progress (see p. 513).
- *National politicians* backed the urban renewal programs as a means of gaining the support of a wide spectrum of urban voters.
- *Universities* were often important landholders and sometimes active developers, and, in the case of land-grant institutions, had a duty to support growth and development. Universities with campuses or facilities in inner-city locations provided the growth machine with a substantial middle-class element that often provided the first wave of gentrifiers, while the physical expansion of universities themselves was sometimes an important first step in rejuvenating a deteriorated inner-city area.
- *Newspaper owners and editors* generally had close links with local business interests and could bolster popular support for renewal as a means of economic growth and as a matter of civic pride.
- *Professional sports.* Because of their ability to attract visitors and to put a locality on the map as a "big league" city, new stadiums were commonly subsidized as an anchor for some renewal projects.
- *The arts.* Museums, theaters, galleries, and concert halls were often located at the edge of the CBD, uncomfortably near decaying neighborhoods for potential visitors and customers. As strong upscale cultural attractions, they were often included in redevelopment schemes based on the assumption that they would prosper from a symbiotic relationship with hotels, offices, restaurants, and upscale residential developments.

The nature of the alliances between these different interests dictated that most renewal schemes were located in or very close to CBDs. Examples can be found in almost every large city.

- In San Francisco the growth machine was institutionalized in the form of the San Francisco Planning and Urban Renewal Association (SPUR), which undertook the redevelopment of large areas around the CBD, including the city's produce market, its Japanese neighborhood, and the deteriorating neighborhood of the Western Addition.
- In Minneapolis, Minn., a growth machine led by the Minneapolis Housing and Redevelopment Authority, that was governed by a board of community leaders, undertook the redevelopment of a huge area that involved demolishing about 200 buildings across 25 blocks, representing about 40 percent of the entire downtown area at the time (Figure 16.6).
- In Pittsburgh a growth machine led by the wealthy industrialist R. K. Mellon organized itself

FIGURE 16.6   Downtown Minneapolis in 1962. Urban renewal in this city during the 1950s and 1960s involved the demolition of about 200 buildings across 25 blocks, representing about 40 percent of the downtown area. This photo shows the Guaranty Loan Building being razed and what were intended to be temporary surface parking lots on the sites of buildings that had already been demolished.

into the Allegheny Conference on Community Development, with the aim of developing a new CBD (which became known as the Golden Triangle).

- In Boston a group called "The Vault" (because it met in the boardroom of the Boston Safe Deposit and Trust Company) sponsored the creation of the Boston Redevelopment Authority (BRA), which explicitly worked toward "a reversal of the present trend toward increasing proportions of low-income groups and non-whites in the core city."[19] The centerpiece of this strategy was the South End Urban Renewal Plan, which called for the demolition of more than 5,000 dwellings in a mixed ethnic neighborhood covering more than 600 acres and containing over 30,000 people.

BACKLASH: GRASSROOTS ACTIVISM AND PROTEST

By the early 1960s it had become increasingly apparent that growth machines such as these were producing results that were not entirely beneficial. Although most members of the growth machines were happy, an increasing number of citizens began to feel that the benefits of growth were being channeled unevenly and that their cities were being stripped of their identity as blank Modernist offices, apartment blocks, and mixed-use developments came to take the place of distinctive old settings.

The biggest losers were the residents of neighborhoods that were officially designated as blighted. Because of successful opposition to the public housing provision of the 1949 Housing Act, the net effect was almost always that more housing units were destroyed than were built in renewal projects. By the spring of 1963 nearly 610,000 people had been forced to relocate, only half of them receiving any compensation for property loss or moving expenses.[20] In particular, the lack of *affordable* units in renewal projects effectively dismantled whole communities; their members were scattered across the city to make room for upscale housing and commercial land uses. By removing the structure of social and emotional support provided by the neighborhood and by forcing people to rebuild their lives separately among strangers elsewhere, renewal imposed serious psychological costs on low-income households. At the same time, displaced families typically faced a steep increase in rents because of their forced move upmarket. In an early study of Boston's West End, the median rent of displaced families rose by over 70 percent. In addition, more than 25 percent of these displaced families ended up in substandard housing.[21]

Another cause of unfavorable reactions to renewal schemes was that they too often seemed to result in the downgrading of adjacent neighborhoods. Yet renewal officials seemed often to be, at best, indifferent to the harmful side effects of their programs. The public interest had come to be defined rather narrowly in terms of business and efficiency. Political scientist John Mollenkopf illustrated these problems in a study in which he described the renewal program in San Francisco's Western Addition. After World War II the Western Addition had a thriving black commercial life. As renewal began, according to some local businesspeople, the SFRA (San Francisco Redevelopment Agency) held off purchasing their successful establishments, instead driving them out of business by displacing their customers through demolishing much of the surrounding housing. By driving down commercial property values in this way, the SFRA reduced acquisition costs and destroyed the commercial center that had given the Western Addition its identity. In addition, many owners abandoned maintenance on their homes, with the result that many more housing units were demolished than the SFRA had anticipated. Real estate investors rapidly acquired the attractive Victorian housing stock and converted it into single-family homes or rental apartments. Tenants, often elderly and uninformed about their rights, were evicted without the benefit of SFRA relocation services or payments.[22]

Community disruption such as this eventually provoked a new kind of militant activism strong enough to change the dynamics of urban politics. Although renewal agencies had been careful to enlist support from established neighborhood organizations—indeed, they had to in order to qualify for federal assistance—they found increasing opposition from *new* organizations and coalitions of grassroots protest groups. Remember too that by the 1960s the Baby Boom generation, with its rebellious counterculture of iconoclastic politics and a collectivist approach to the public interest, was coming into young adulthood (see p. 100). In Boston's South End white activist ministers, poverty program employees, African-American social workers, Puerto Rican community organizers, and ideologically motivated young white professional and college students joined together in opposition to the Boston Redevelopment Authority. A similar mix of militant activists emerged in San Francisco and in nearly every city that had an urban renewal program.

Finding themselves outside the framework of both formal politics and the web of interdependent interests that constituted growth machines, these activists and their supporters came to represent the vanguard of **urban social movements** that epitomized the "people power" that permeated the urban politics of the 1960s. Their methods were chiefly those of door-to-door campaigns, community-wide meetings, and militant but (mostly) peaceful demonstrations. They were eventually successful in turning the tide of public opinion against renewal programs. Bolstered by books such as Jane Jacobs's *The Death and Life of Great American Cities*,[23] antirenewal, slow-growth, community-oriented, and environmentally sensitive perspectives increasingly came to be incorporated into urban politics.

Community groups gained representation on renewal agencies; accommodations began to be made; and a few outright victories were achieved, such as the abandonment of construction on San Francisco's Embarcadero freeway (Figure 16.7), which threatened to cut through several established neighborhoods and retailing districts. **NIMBYism** (Not In My Backyard-ism) was born (though, subsequently, it was in middle-class suburbs rather than central city neighborhoods that NIMBYism and the more extreme BANANAism [Build Absolutely Nothing Anywhere Near Anything] came of age). Equally important, the interaction between community organizations and city officials that took place over redevelopment issues resulted in much stronger local leadership networks. Over time, these leadership networks became less preoccupied with fending off unwanted change and increasingly focused on achieving positive change in terms of service delivery and collective consumption. In addition, this new community leadership came to represent role models for community youth (although many were accused of "selling out" and becoming "poverty pimps").

FIGURE 16.7   A monument to the power of citizen protest: the unfinished Embarcadero Freeway in San Francisco.

Black Power and Black Politics

Because of the social geography of cities, black communities and black businesses were typically the chief casualties of renewal programs. Indeed, the removal of black communities from the immediate environs of CBDs was very often an implicit goal of growth machines (and occasionally explicit, as in Boston), since the key participants were acutely aware of the need to protect property values and to maintain an attractive environment for upscale retailing and leisure aimed at a predominately white clientele. But African-American community activism over renewal projects has to be seen against the context of wider challenges to discrimination and segregation.

Urbanization had exposed African Americans to the promises and frustrations of mass consumer society and to the American Dream of opportunity and social mobility. Urbanization had also brought enough African Americans to the cities for collective action to be effective. In the early 1960s this action took the form of nonviolent protest. The model had been the boycott of the bus company in Montgomery, Ala., inspired by the personal protest of Rosa Parks in 1955 and sustained for a year by Dr. Martin Luther King, Jr. The tactics that led the Supreme Court to eventually overturn Alabama's segregation laws inspired boycotts, sit-ins, and demonstrations elsewhere. In 1964, after John Kennedy's assassination, President Lyndon Johnson was finally successful in getting Congress to pass the Civil Rights Act, which outlawed racial discrimination. The following year, Congress authorized federal agents to register voters who had been illegally kept off the electoral rolls in parts of the South.

As we have seen (p. 417), 1964 was also the year that marked the beginning of ghetto riots. The stress of ghetto life, exacerbated by rapidly rising but unfulfilled expectations in African-American communities and by antiblack activism (provoked by white fears about losing hegemony over schools, housing, and jobs), was sparked into violence and disorder by incidents of (white) police aggressiveness. In 1965, Malcolm X, leader of the militant and separatist Black Muslims, was assassinated. In 1968, Dr. King was assassinated, as was Robert Kennedy, a presidential candidate who as Attorney General in his brother's administration had spearheaded the White House drive for civil rights legislation.

The effect on black politics was decisive. Integration through nonviolent protest was displaced by self-advancement in the context of collective "black power." Racial pride, inspired by the emergence of independent black countries in Africa, became the basis for political, social, and cultural solidarity. African-American leaders such as the Reverend Jesse Jackson organized aggressive boycotts to persuade white-owned businesses to

Box 16.2   Milwaukee Demolishes the "Freeway to Nowhere"[24]

"Taking down a freeway without help from an earthquake is remarkable," said one of the judges from the Congress for the New Urbanism who awarded the City of Milwaukee's Park East Redevelopment Plan a Charter Award in 2003. What was commonly referred to as the "Park East Freeway," however, was not really a freeway at all. It was the remnant of an abandoned plan to encircle Milwaukee's downtown with freeways. About half of the Park East Freeway had been built when citizen protest in the 1970s halted construction of the connecting highway segments,

including one that was planned to run along the lakefront. This left the Park East Freeway as a spur—a half-mile of elevated roadway—that actually limited access to the downtown. The underutilization of the freeway and the potential redevelopment opportunities of the underutilized land beneath and surrounding this freeway spur prompted the city of Milwaukee to demolish it and adopt a tax increment finance district (see p. 477) to provide financing for the infrastructure needed to reconnect the land with the local street grid (Figure 16.8).

FIGURE 16.8   The Park East "freeway to nowhere"—the remnant of an abandoned plan to encircle Milwaukee's downtown with freeways in the 1970s—in the process of being removed in 2003 (at right); the 22-acre former Pabst Brewing Co. headquarters (behind) awaits planned redevelopment as a $300 million residential, retail, and entertainment complex.

hire African Americans and to do business with black-owned businesses. At the same time, they organized overt registration drives to maximize black power at the ballot box. Black culture, black history, and black issues demanded—and received—respect and attention as black voters and community leaders made their presence felt in urban politics.

Meanwhile, as more and more white households moved out to suburban jurisdictions, the power of the nonwhite electorate increased appreciably in those central cities. By 1992, with increased African-American voter participation, there were 338 black mayors in the United States, though only 30 of them represented cities of 50,000 or more. Under the leadership of black mayors such as Harold Washington, mayor of Chicago from 1983 to 1987, the community development activism that had started as a backlash to renewal programs matured into powerful and effective coalitions

geared to more progressive forms of urban revitalization. Chicago was one of the few cities where an old-style political machine had survived into the 1960s and 1970s. Harold Washington put together a multiracial coalition that drove out the machine of the late Richard J. Daley, replaced the ideology of growth through real estate development with one of economic development based on "good" jobs, and replaced the rigid hierarchy and patronage of city politics with a commitment to broad-based community participation and redistributive city programs.[25]

REFORM: THE STRUGGLE FOR SOCIAL JUSTICE AND SPATIAL EQUITY

Black politics was not the only source of change and reform, however. Even as the riots and civil disorder of the 1960s were forging the bases of a new black politics, the New Deal coalition between urban blue-collar classes and liberal reformers was reaching maturity. John F. Kennedy had concluded that the presidential election of 1960 would be decided by the votes delivered in key cities in a few industrial states and that the problems facing cities were "the great unspoken issue" in the election. He was, moreover, ideologically committed to social reform and urban regeneration. Once in office, Kennedy spelled out an ambitious program of domestic reforms that encompassed schools, housing, medical care, urban transportation, pollution control, and planning: a "New Frontier" for national development.

After Kennedy's assassination in 1963, the program was carried forward by President Johnson, who initiated a "War on Poverty" in an attempt to build the "Great Society" implicit in the New Deal vision. In 1966, Johnson created the U.S. Department of Housing and Urban Development (HUD), which was a landmark in urban affairs because of the subsequent proliferation of national urban policies and the expansion of federal aid to cities (see Chapter 17). By 1969 over 500 federal grant programs were aimed at urban America, with total annual appropriations amounting to $14 billion. At the beginning of the decade, there had been only 44 such programs, with appropriations amounting to less than $4 billion.[26]

The political mood fostered by these initiatives had the effect of intensifying certain long-standing issues related to metropolitan governance. One of these issues was the difficulty of matching the boundaries of electoral units to the shifting population distributions inherent in metropolitan growth—the **malapportionment** problem. Another was related to the conflicts and contradictions inherent to metropolitan areas with spatially fragmented systems of government. A third was related to the difficulty of ensuring equity in the delivery of public services. We will discuss each of these in turn.

Reapportionment

Unless there is an at-large voting system, variations in the number of voters between electoral districts represent an erosion of democracy. To take a simple example: The influence of a single voter in a constituency of 2,000 voters is proportionally much greater than that of another voter who happens to live in a constituency of 3,500 voters. This situation is an example of malapportionment. It arises almost inevitably as urbanization redistributes people across the map of electoral districts from year to year; it can also arise quite deliberately if the legislative bodies that control electoral boundaries are able to create larger-than-average constituencies in areas where opposing political groups are known to have the support of a majority of voters.

The most flagrant cases of deliberate malapportionment were in fact at the level of congressional districts and state senatorial and assembly districts, where the bias was decidedly in favor of rural districts and against metropolitan areas. In a series of decisions between 1962 and 1965 the Supreme Court ruled against the practice of malapportionment, beginning a "reapportionment revolution" that was based on the strict criterion that the number of voters in each constituency should vary by no more than one-half of 1 percent. In addition to representing an improvement in the natural justice of the situation, these rulings had the effect of adding decisively to the importance of the urban voter in national and statewide politics.

The Supreme Court's ruling did not apply to local jurisdictions, however, with the result that large variations (as much as 30 percent in Atlanta, Chicago, Philadelphia, and St. Louis) persisted in the size of the electorate from one city ward to another. This variation effectively disenfranchises large numbers of registered voters. If, as is often the case, the malapportioned group involves the inner-city poor, the problem assumes more serious proportions. Policies affecting rent control and waste collection and questions such as the location of undesirable land uses and facilities or the imposition of a commuter tax will be decided in favor of suburban communities.[27]

A related geopolitical issue is that of **gerrymandering**. This occurs where a specific group or political party deliberately manipulates the spatial configuration of election boundaries in relation to known concentrations of supporters or opponents. It is quite possible to satisfy fairly strict apportionment criteria but still bias the outcome of election results by carefully drawing election boundaries in relation to the social ecology of the city. This maneuver can take several forms:

- Creating "stacked" districts, which usually end up with a grotesque shape that encompasses pockets of support for the gerrymandering party.
- Creating "packed" districts in which the opposition party's supporters are concentrated into a few sacrificial seats.
- Creating "cracked" districts, in which opposition voting strength is fragmented, leaving the voters neutralized as a minority in a large number of seats.
- Operating a "silent" gerrymander, which involves the preservation of electoral boundaries over a long period, thereby creating malapportionment as the spatial distribution of population changes.

The problem is that it is very difficult to establish conclusive proof of gerrymandering. There is strong circumstantial evidence, however, of gerrymandering against

African-American communities. Research suggests that in cities with relatively small African-American neighborhoods the most common type of gerrymandering has involved cracked districts. In contrast, in cities with African-American neighborhoods that would be extensive enough to warrant two or more elected representatives it has been common for the electoral map to minimize the number of African-American majority districts, with the rest of the African-American electorate scattered across predominately white districts; the U.S. Congressional districts of New York and Chicago, the state senatorial districts of Milwaukee and Philadelphia, and the city council districts of Atlanta have been cited as examples of these in the past.[28]

Fragmentation Compounded: Special Districts

The community activism of the 1960s spilled over into a more generalized sentiment. "Power to the People," the catchphrase of inner-city black communities fighting renewal schemes, was appropriated by middle-income suburban white communities who wanted greater local autonomy in order to protect their communities from unwanted change. The sentiment was one that threatened to exacerbate the spatial fragmentation of metropolitan areas that had begun with the suburban incorporation of the 1920s. A series of cases illustrates the trend. In 1967 the borough president of Staten Island established a commission to study the possibility of secession from New York City; in 1968 the United Black Front of Roxbury demanded independence from Boston; and in 1969 the middle-class property owners of Glen Park sought secession from Gary, Ind., in order to prevent falling property values.[29]

Meanwhile, the pace of metropolitan change resulted in an increase in the number of **special districts**—single-purpose jurisdictions such as port authorities. Such jurisdictions were not new. The need for independent school boards was recognized early in the nineteenth century, and examples of special sanitary districts, police boards, and so on, grew steadily in response to the evolving needs of metropolitan areas. With the acceleration of population decentralization at the end of the nineteenth century, special districts became a popular means for suburban communities to secure high-quality sewerage, water, educational, and law-enforcement services without having to become involved with a more elaborate apparatus of government. The spirit of local autonomy that characterized the 1960s accelerated the creation of special districts, resulting in a confusing overlay of jurisdictions.

Special districts were seen as an attractive solution to a wide range of problems because they were not subject to the statutory limitations on financial or legal powers that apply to municipalities. In particular, a community can increase its debt or tax revenue by creating an additional layer of government for a specific purpose. Special districts also have the potential advantage of being customized to correspond closely to specific functional areas and, therefore, of being finely tuned to local social organization and participation. Between 1942 and 1972 the number of nonschool special districts in the United States increased from 6,299 to 23,885. By 1977 the largest 35 metropolitan areas in the United States each had an average of 293 separate jurisdictions of one kind or another (many of them with very narrow functions, such as the one created to provide for the St. Louis Zoo).[30]

The increasing partitioning and layering of metropolitan space created more problems than it solved, however. Taking a metropolitan-wide view, it was increasingly apparent that local autonomy and administrative convenience were more than offset by fiscal imbalances, inefficient and inequitable distributions of public services, the creation of complex, competing, and expensive bureaucracies, and the juxtaposition of conflicting policies. Opportunities for economies of scale and coordination in areas such as transportation, planning, water supply, housing, and public health were forfeited. The sheer proliferation of governmental jurisdictions made it difficult for citizens to keep up with the politics and policies that affected them, leaving the control of many special districts to a self-selected knot of individuals and special interest groups.

Together with the continuing fragmentation of metropolitan governance associated with the incorporation of suburban communities, these problems make it increasingly difficult to sustain the notion of *metropolitan* governance and management. There are several ways in which reforms might recapture the fit between governmental structures and socioeconomic reality.[31] These reforms include transferring key functions to a higher tier of government, instituting intergovernmental agreements where a number of jurisdictions jointly provide certain services, and consolidating two or more jurisdictions into a unitary, multifunction authority (see Chapter 17, pp. 520–522). In general such reforms have been tentative and exceptional: There are simply too many vested interests in the status quo.

Social Justice and Spatial Equity in Municipal Service Delivery

Although municipal expenditures have risen dramatically since World War II, the benefits arising from them have not always been distributed evenly across metropolitan areas. Unevenness in municipal service delivery occurs because:

- Many public services have to be tied to specific locations, so that **distance-decay effects** confer differential benefits. The amenities of parks, libraries, fire stations, and so on are greatest for those living nearby and tend to diminish with distance: the so-called **tapering effect**. On the other hand, the location of certain services—landfills, garbage incinerators, crematoria—brings a local disamenity.

- Municipalities struggle to reconcile spatial equity with economic efficiency. Because larger, centralized facilities are often more cost effective, and because municipalities have to be fiscally responsible, spatial equity was often traded off for net gains in cost efficiency.

- The fragmentation of metropolitan governance resulted in suburban jurisdictions being, quite simply, better able to pay for public services, while central cities, with much higher and more varied needs for public services, were less able to do so.

- Deliberate bias in the delivery of services resulted in white, middle-class, or activist neighborhoods getting a bigger share than African-American, Hispanic, impoverished, and politically passive communities. In Shaw, Miss., in the 1960s, 98 percent of the unpaved roads served African-American households and 19 percent of African-American homes were not served by sanitary sewer lines, compared with 1 percent of white homes. This disparity led to a landmark court case, *Hopkins v. Town of Shaw*, in which the U.S. Fifth Circuit Court of Appeals decided in 1971 that the town of Shaw had engaged in a racially motivated denial of equal rights in its provision of street, water, and sewer services.

Studies of the geography of public service provision have provided ample evidence of unevenness. The *pattern* of unevenness is both clear and pronounced at the metropolitan scale, where interjurisdictional differences are reflected in very different "packages" of services. Within the boundaries of individual municipalities there also tend to be significant disparities, but these disparities tend to be "unpatterned" in that they do not correspond systematically with neighborhood variation in race or income.[32]

Nevertheless, even unpatterned inequity may be enough to cause concern. It also raises some important issues with regard to concepts of economic efficiency, spatial equity, and social justice. So, for example, equity in expenditure may not result in equity in outcomes due to spatial variations in the intensity of needs and in the quality of services that can be provided for the same amount of money in different settings. How, then, is social justice to be served? Equal shares of public outlays seem fair at face value but only preserve the status quo of urban social geography; only through redistributional policies can differences in needs be addressed, but the more affluent neighborhoods that would pay more and receive less may not regard such policies as "fair."[33]

Nor can efficiency in public service delivery be traded off in a straightforward way against equity, however defined. Efficiency itself can be measured not only in terms of cost effectiveness but also in terms of delivery criteria (e.g., waiting time, physical accessibility) or in terms of meeting needs (the degree to which services reach those most in need [**horizontal efficiency**] or the proportion of services allocated to people or communities in need [**vertical efficiency**]). Before either theoreticians or practitioners could resolve these niceties, however, the political economy of American cities was disrupted by a new conservatism that made their resolution moot. City services were to be cut back substantially, and the ideas of social justice and spatial equity were to be the first casualties.

FISCAL CRISIS AND ENTREPRENEURIAL POLITICS (1973–PRESENT)

In the most recent phase in the evolution of urban governance and politics (1973–present), economic restructuring and spatial reorganization have prompted major changes in the nature of urban governance and politics. Aggravated by the **stagflation** of the 1970s, central city decline reached the point where governments faced mounting expenditures on everything from infrastructure provision and repair to welfare services—while at the same time they were losing the capacity to raise money through taxes and bond issues. The result was an extension of the "fiscal crisis" that first became apparent in New York in 1975. This fiscal crisis, in turn, served to reinforce increasingly widespread feelings that government at all levels had grown too big and too expensive, so that a **new conservatism** and **neoliberal ideologies** began to permeate urban politics. Neoliberal ideologies are predicated on a minimalist role for government, assuming the desirability of free markets as the ideal condition not only for economic organization, but also for political and social life (see Chapter 17, pp. 514–516).

One outcome of these ideological shifts was a retrenchment in service provision. Another was the emergence of a more "entrepreneurial" approach to urban governance, with a specific emphasis on joint public-private initiatives and a wholesale abandonment of the paternalistic stewardship of the public interest that had been such a strong element during the previous 100 years. Meanwhile, the **globalization** of the economy in most cities fostered a much greater dependency on out-of-town decision making, thus removing from the local arena some of the influence of the traditional business elite. City governments were forced to be highly competitive with one another—and with cities in other countries—in their search for investment dollars from nationally and internationally "mobile" companies.

The 1970s saw the disintegration of traditional progrowth politics. When Richard Nixon took office as president in 1969 he began to undermine the base of the old alliance by reducing the dependency of the cities on programmatic federal support. Under this "New Federalism" the role of federal agencies in distributing federal funding was cut back, with a revenue-sharing system set up in order to augment directly the budgets of local governments. This system was based on a strict formula that undercut the political influence of urban voters and city politicians. It also eliminated the major source of funding for most of the programs that involved neighborhood participation in policy-making, thus removing an important source of organization and leadership at the grassroots level. At the same time, Nixon sought to consolidate support for his own Republican coalition by cultivating Southern conservatives. He did this by discouraging civil rights reform and by attacking the welfare-dependent poor. The working-class support that had been one of the cornerstones of progrowth politics since the New Deal was thus split, with nonwhites set against whites and the working poor set against the unemployed.

As these changes began to weaken the old order, other changes were under way that were soon to change urban governance and politics decisively. Among these were the effects of **deindustrialization**, economic recession, corporate restructuring, the geographic decentralization of manufacturing, the rebuilding of local economies, the militarization of the national economy, social and economic polarization, changing patterns of immigration, and the aging of the population—all of which, as we have seen in earlier chapters, contributed to radical changes in the geography of the **urban system**, in metropolitan form, and in the economic, social, and cultural geography of urban America. In this context it is not surprising that urban governance and politics also changed. In 1973 the OPEC oil embargo triggered a recession that brought to a head a series of longer-term economic problems. Within two years, New York City found itself teetering on the brink of bankruptcy, the harbinger of acute fiscal problems that came to afflict many of the country's central cities and to move them to reprogram their governance and politics.

FISCAL CRISIS

As we have seen, metropolitan fragmentation left many cities with a legacy of chronic fiscal problems as a result of the fiscal squeeze—between falling revenues and increasing demands on municipal expenditures (Figure 16.4). After World War II this pressure intensified, particularly in the older industrial cities of the Northeast. Freeways and metropolitan sprawl drew out even more of their middle classes and employers, and with them went their tax dollars. Only a relatively small portion of the properties they left behind was replaced in renewal programs by newer, bigger, or more-luxurious structures. The remainder continued to deteriorate, the worst falling into a **spiral of decay** and abandonment (Figure 15.5). The aggregate yields of property taxes fell accordingly. Sales taxes also fell, the suburbanization of the middle-income groups having drawn retailers along in their wake. During the 1960s retail sales in larger central cities, when adjusted for inflation, fell by between 15 and 30 percent. Meanwhile, the demand for central city services increased at an unprecedented pace. In addition to the overall trend toward providing an increased range of services that was inherent to the progrowth coalition's welfare state, central cities faced increasing expenditures for several reasons:

- Their original infrastructure of roads, sewers, water mains, bridges, and so on had reached the end of its designed life of 75 to 100 years (see Chapter 15); repairs and replacements began to impose an increasing burden on city expenditures.

- Rising crime rates induced corresponding increases in spending on policing, incarceration, and the courts.

- Rising proportions of immigrants, the lone elderly, single-parent families, and the unemployed in older inner-city neighborhoods required increased expenditure on specialized public services and amenities.

- The costs of federal welfare programs began to be passed on to local governments. Federal Medicaid, for example, provided free medical care for welfare-dependent people, but state governments were required to contribute 50 percent of the costs. It did not take long for states to begin to pass along part of this burden to their cities.

The fiscal problems of American cities in the 1970s were rooted, then, in a long-standing and intensifying fiscal squeeze. But the fiscal squeeze on its own did not add up to crisis. What pushed the fiscal squeeze to the edge of crisis was the transition that had taken place in the restructuring of metropolitan economic geography and, with this restructuring, a shift in the political economy of central cities:

> On the one hand, new economic growth in the central cities did not provide sufficient employment and income benefits to the central city's residents. Industrial jobs were taken by suburbanized union workers.

BOX 16.3 TAX INCREMENT FINANCING (TIF)[34]

Tax increment financing, or TIF, is a mechanism used by cities to finance redevelopment efforts that is directly tied to the success of those efforts. If an area in a city can be made more attractive to private developers and new development occurs, the tax revenue collected from that area would be expected to rise. Tax increment financing taps into any increase in tax revenues by using the tax "increment" (the difference between the taxes after redevelopment and the expected taxes without redevelopment) to finance the improvements and other activities that stimulated the redevelopment to occur in the first place.

In U.S. states with statutes authorizing the use of TIF in areas that are "blighted" or that will generate significant economic benefits, the first step for a city is to ascertain the property tax revenue that is being collected in a particular area before redevelopment. Following feasibility studies and cost-benefit analyses, the city establishes a tax increment district (TID) for a specified period (often 20 years), formulates a development plan, enters into development agreements with private developers, and borrows money (through loans or by selling bonds) to use in various ways to improve the development prospects of that district: property acquisition, site preparation, loans to developers and new businesses, capital improvements such as new roads and street lights, and new services such as improved street cleaning and security patrols. As redevelopment occurs, tax revenues increase, and the tax above the preredevelopment property tax revenue is used to pay off the city's loans or bonds.

Although tax increment financing sounds very attractive in theory—the local government does not lose any revenue because the tax increment would not have occurred without the redevelopment efforts financed by that increment—it is not without potential drawbacks. Obviously, cities can find themselves in trouble if the redevelopment does not produce the estimated increment (that has been earmarked to pay off the loans or bonds). In situations where the TIF involves a general cap at pre-TIF levels on property valuations or tax assessments, other public entities that normally receive property tax revenues—school districts, special districts, the county—do not receive their part of the tax increment for the life of the TID or until the loans or bonds are paid off, even in situations where the redevelopment creates additional demands for their services. If TIF is used unnecessarily—in areas where redevelopment would have occurred in the absence of a TID—some or all of the tax increment represents revenues that the local government would have collected anyway and that now instead cuts into general revenue because it has to go to pay off the loans or bonds. There may also be a lack of "transparency" for ordinary citizens—because the TIF bonds are not "general obligation" bonds, voter approval is not required. This has led to concerns about high levels of TIF debt due to an overdependence on TIF by cities.

Construction work was dominated by restrictive craft unions. And the new office economy was drawing on the better educated, better heeled suburban workforce. Industrial investments were now part of vast multilocational networks of plants, thus weakening the local multipliers from local plant investments. This export of the income benefits of local economic growth meant a continuous reservoir of poor, structurally unemployed people who turned to city governments for jobs and services.

On the other hand, the rising office economy of the central city required a restructuring of urban space to move people and information most efficiently. This required a massive investment in public capital for mass transit, parking, and urban renewal, [as well as] the more traditional forms of infrastructure.[35]

These infrastructure investments were often effectively insulated from conflict and political debate by the new forms of administration and financing that had evolved with the rise of growth machines: autonomous special districts, banker committees, and new forms of revenue and **tax increment finance** bond issues. As a result, two spheres of city expenditures emerged: one oriented to constructing the new infrastructure necessary for profitable private development, the other to providing services and public employment for the city's residents. These two worlds—of social wage and social capital—were structurally separated, with the first governed by electoral politics and the excesses of patronage and the second housed in bureaucratic agencies dominated by public administrators who survived by their efficiency.[36]

Although the fiscal squeeze and a transformed political economy served as the preconditions for crisis, the immediate causes can be found in a combination of trends (Figure 16.9):

- Fearful of increasing racial violence, local officials in many cities expanded the welfare rolls and increased expenditures on social programs for inner-city blacks in the mid-1960s.

Fiscal squeeze + {
Costs of growth machine restructuring of urban space

Cutbacks in federal grants

Inflationary pay raises

Welfare spending driven up by racial tensions

Rollover of operating budget deficits
} = Downgraded credit ratings, banks sell off city bonds = **Fiscal crisis**

FIGURE 16.9   The origins of fiscal crisis.

- The overall economic prosperity of the 1960s encouraged city workers (among others) to seek big pay increases that often exceeded increases in productivity.

- Federal grants to cities were cut back at the beginning of the 1970s as revenue-sharing reforms and the New Federalism of the Nixon administration came into effect.

- Operating deficits caused by increased costs were "rolled over" from year to year, with short-term bonds (which carry higher interest rates) used to cover debt servicing on bonds coming due. In addition, creative budgeting (e.g., advancing the last payday of the year into the next fiscal year) was used to minimize the apparent deficits. This evidence of cash-flow problems led to downgraded credit ratings, which made it difficult (and more expensive) for cities to issue long-term bonds to finance urban services and infrastructure. With creditors increasingly concerned about cities' ability to pay off debts, banks began to sell off bonds, setting off a panic that effectively prevented fiscally distressed cities from further borrowing.

FISCAL RETRENCHMENT

The fiscal well-being of central cities was not the only aspect of urban governance and politics to be affected by the economic and metropolitan restructuring of the mid-1970s. The political economy as a whole was altered beyond recognition within just a few years. Partly in response to longer-term trends in the economy and society and partly in response to the jarring economic and fiscal experiences of the mid-1970s, a new conservatism emerged in local politics. This was a politics that was based on freeing up the economy for recovery by reducing government regulation and control and lowering taxes by cutting back on expenditures for public services. This conservatism was widely supported by middle-class voters who were seeing their incomes stagnating after years of steady growth and who had come to believe in the argument that all they had been given in return for their tax

dollars was a bloated group of meddlesome and inefficient bureaucrats and an army of welfare "freeloaders." The new conservatism was also, of course, supported by the entire spectrum of the business community, which naturally saw lower taxes and less regulation as catalysts for renewed profitability.

So a new alliance was born, between private business interests and the middle socioeconomic groups. It was an alliance that met relatively little opposition. The circumstances of its creation had damaged its would-be opponents. Working-class solidarity, and organized labor in particular, had been seriously weakened by deindustrialization and the associated changes in economic geography. Corporate restructuring in an ever more centralized and internationalized economy increasingly outmaneuvered workers and left them to operate in local arenas, while companies were able to switch production and investment between cities, regions, and countries.[37] The liberal and progressive element of the middle socioeconomic groups had been demoralized by the obvious failures of welfare provision and Modernist technocratic reform to combat crime, poverty, and environmental degradation. It had also been weakened by the defections of people who no longer felt as progressive or liberal once they personally began to feel the pinch of economic recession.

The first and most striking evidence of the new alliance was manifest in "taxpayer revolts." In California, Proposition 13 cut property taxes by 60 percent. In Massachusetts, Proposition $2\frac{1}{2}$ limited property taxes to 2.5 percent of assessed real estate value. Similar proposals were passed in Idaho and Nevada. In Arizona, Hawaii, Michigan, and Texas, voters approved bills that severely limited public spending, and everywhere there was increasing reluctance to sanction bond issues for municipal development. In 1980, Ronald Reagan, who had championed this new conservatism as governor of California, was elected president, marking the success of the new alliance at the national level. The Reagan administration revived the New Federalism that had been initiated under the Nixon administration, devolving responsibility (but not always the funds) for welfare and service provision to state and local governments.

BOX 16.4   AMERICA'S AILING CENTRAL CITIES

New York City was the first to reach the crisis point of fiscal distress. Between 1965 and 1970 the city's budget had doubled, as had the number of people on its welfare rolls. By 1975 its accumulated operating deficit was $2.5 billion and it faced an unfunded budget gap that could not be filled by borrowing. In the end, New York was bailed out not by the federal government but by the New York State Legislature, which turned over control of the city's finances to an Emergency Financial Control Board, headed by the governor and dominated by representatives of the major banks. It also created the Municipal Assistance Corporation (which inevitably came to be referred to as Big MAC) to lend New York City the funds it needed to service its debts. These institutions imposed a number of austerity measures that were to foreshadow the subsequent retrenchment of other local governments.

Between 1975 and 1978 New York laid off more than 60,000 public employees, initiated salary rollbacks and wage freezes, established user fees for some municipal services, reduced budget allocations, and consolidated or eliminated several municipal departments. Meanwhile, banking interests, appalled at the scale of the problem, persuaded a reluctant federal government to make loans and guarantees available. It did so, making it very clear that the sole reason was to avoid city bankruptcies that could trigger a national banking crisis. This development was important because it *refocused the public interest once more in terms of economics, leaving the interpretation of urban affairs to the representatives of banks, real estate investors, and corporate financiers.*

New York's experience was meanwhile repeated, albeit in somewhat less dramatic fashion, in a number of cities, including Baltimore, Boston, Cleveland, Detroit, Philadelphia, and St. Louis. A combination of bailouts, retrenchment, and a national economic recovery put an end to the crisis, but fiscal squeeze persisted, leaving central cities vulnerable. This vulnerability can be expressed quantitatively as an index of "standardized fiscal health" (the difference between a city's revenue-raising capacity and its need, measured by a standardized index, expressed

as a percentage of revenue-raising capacity). Using this index, a study of 71 central cities found that, overall, cities were in worse fiscal health in 1988 than they had been in 1972.[38]

Meanwhile, fiscal squeeze continues to leave central cities in a vulnerable position. In 1991, for example, Bridgeport, Conn., was threatened with bankruptcy, as was Chelsea (in the Boston metropolitan area). In Philadelphia the city treasury was so short of cash that weekly meetings were convened to decide which bills could be paid. Almost 20 years after its first crisis, New York was again facing a growing deficit and another crisis, even though its role as a provider of services had by now become secondary to its role as facilitator of economic development. This time, the city's enormous deficit—forecast to eventually be in the $3 to $4 billion range—is partly the result of simultaneous events outside the city's control—the September 11th attacks, the decline of the stock market, the national recession, and the problems and scandals of the financial industry that is so crucial to New York's economy. Nevertheless, the federal government is no more inclined to help (other than 9/11 reparations) than it was in 1975. The former chairman of Big MAC compared the city's plight in 2003 to that in 1975:

> The size of the City's budget gap is proportionally similar to 1975, . . . the recent growth of its budget has consistently exceeded inflation rates. According to the Citizens Budget Commission, from 1997 to 2001 City spending increased from $33 to $43 billion, an increase of 26%, or 13% above inflation. In that period, the number of City workers increased from 236,000 to 247,000. The City is burdened with a heavy debt load, about $35 billion, similar in relative size to 1975 . . . for the last 2 fiscal years the City has borrowed to pay for operating expenses, a practice avoided since the 1970s. In 2001–02 the City borrowed $600 million and in 2002–03 the City is borrowing over $2 billion . . . . Since 1995, similar to the early 1970s, the salaries of City employees, along with their fringe benefits, have consistently increased above the rate of inflation.[39]

The result was that cities were forced to reevaluate their roles, regardless of whether they were in fiscal trouble. The outcome of this reevaluation was a general **fiscal retrenchment** in service provision. In Oakland, for example, the city government responded to the new conservatism imposed by Proposition 13 by closing a fire station and four branch libraries; reducing

the police department's budget for criminal investigations, park maintenance, after-school recreation programs, library and museum hours, and street sweeping and maintenance; and eliminating over 100 administrative positions.[40] By the mid-1980s, many older central cities had cut the number of municipal employees by 10 to 15 percent. Boston and

Cincinnati had cut their workforce by more than 20 percent, while St. Louis had eliminated more than 40 percent of its employees.

Other trends accompanied this sort of retrenchment as it was replicated across the country. Cost efficiency in service delivery and public administration came to be emphasized at the expense of need, spatial equity, and social justice. Voluntarism came to be an important means of maintaining service provision, particularly in relation to neighborhood security, elementary education, and library services. But voluntarism is an uneven phenomenon, and for the most part it only served to underline the increasing metropolitan sociospatial polarization and fragmentation.[41] Other aspects of retrenchment included cost-recovery programs, coproduction arrangements (where the public sector works with nonprofit organizations), and "intensification" (increasing labor productivity through managerial and organizational changes). Perhaps the most important single trend in the United States, however, was that of the privatization of public services (as opposed to that privatization that involves selling off public assets, which is more common in other countries), because it reflected so clearly the new ethos of urban governance.

THE PRIVATIZED CITY[42]

In 1988 the Report of the President's Commission on Privatization concluded that **privatization** "may well be seen by future historians as one of the most important developments in American political and economic life in the late 20th century."[43] Since the commission was reporting to the Reagan administration, its conclusions are perhaps not surprising. There is, nevertheless, some substance to support the conclusion. The Council of State Governments has estimated that whereas under $30 billion of services were furnished annually by the private sector in the 1970s, the figure had risen to over $80 billion by the early 1980s and $150 billion by the 1990s, when the Local Government Center's database included over 35,000 specific instances of state or local privatization.

These data reflect a trend that has to be seen in terms of the perceived benefits for both the public and the private sector. The particular advantages of privatization as seen by city governments are:

- reducing direct municipal outlays.
- sharing financial risks with the private sector.
- accessing skilled staff not available in the public sector.
- reducing costs for taxpayers through private-sector efficiencies such as savings in construction

costs and time, operational productivity, and economies of scale.
- maintaining service levels with no increases in tax rates or user fees.
- improving service quality.
- increasing flexibility from the reduction in bureaucratic complexity and procedures.

For private companies, privatization has provided new markets, new outlets for investment capital, and new entrepreneurial opportunities. Timing is of critical importance here. It was no coincidence that private capital was available for public-private partnerships just as cities were facing retrenchment and fiscal stress: Both were the product of a phase of **overaccumulation**. *In short, changing circumstances in international, national, and real estate markets brought the private sector to the public sector as much as ideological and fiscal shifts brought the public sector to the private sector.*

In the United States privatization has been most pronounced in relation to infrastructure projects. As we saw in Chapter 15, both federal and local governments are increasingly aware of the extent of infrastructure decay and deficit and its implications for metropolitan economic development. Privatization has been seen as a means of funding projects that might otherwise have been left on the drawing board. Overall, the provision of basic public works facilities accounts for the greatest dollar volume of privatization activity. The most important projects have been large-scale schemes for wastewater and sewage handling, water provision, resource recovery, and waste-to-energy plants. During the 1970s, when the U.S. Environmental Protection Agency (EPA) increased water quality standards, many cities turned to the private sector to operate and maintain such plants. With the introduction in 1981 of tax incentives for businesses involved in privatization, the interest of major engineering companies intensified. By 1985, 15 municipalities had opted for privatized wastewater projects. By 1988 one-third of the local governments in a national survey had privatized some roads, bridges, or tunnels.[44] Almost as many had privatized water supplies, nearly one-quarter had privatized water treatment facilities, and almost one-fifth had privatized municipal buildings or garages.

The privatization of services has meanwhile accelerated to reach a significant level. Surveys of local governments suggest that more than 80 percent of U.S. cities now use some form of privatization in their service provision. The use of contracting by local governments varies significantly depending on the service, however. More than 50 percent of local governments contract out commercial waste collection, vehicle towing and storage, legal services, and the operation of hospitals, daycare facilities, and homeless shelters (Table 16.3).[45] Other services that are contracted privately on a significant scale include domestic waste and recycling collection, bus and ambulance operations and maintenance, street repair, street lighting, data processing, landscaping, and vehicle fleet repair. Concerns about quality and public accountability make contracting out other kinds of government services more difficult. Less than 5 percent of local governments contract out crime prevention, police and fire communications, fire prevention and suppression, traffic control and parking enforcement, sanitary inspection, or prison and jail services (Table 16.3).

Meanwhile, the retrenchment of public-sector activities and the unwillingness and inability of urban governments to take responsibility for new or expanded service needs caused another kind of privatization to emerge: the provision of services by the private sector *without contracting or formally cooperating with the public sector*. The most striking example of private-sector services is in the area of security: private security "officers" patrol shopping malls and office buildings, and private security forces protect upscale residential developments.

TABLE 16.3

Percentage of Local Governments (Cities and Counties) Involved in Privatization in the United States

	1988	1992	1997
Most commonly contracted-out services			
Vehicle towing and storage	80	86	82
Operation of daycare facilities	34	88	79
Operation/management of hospitals	24	61	71
Operation of homeless shelters	43	59	66
Commercial solid waste collection	38	55	60
Drug and alcohol treatment programs	34	54	56
Legal services	55	50	53
Least commonly contracted-out services			
Crime prevention	4	2	1
Police/fire communications	1	2	1
Traffic control/parking enforcement	1	2	2
Fire prevention/suppression	1	4	3
Prisons/jails	1	1	3

Source: *J. Greene,* Cities and Privatization: Prospects for the New Century, *Upper Saddle River, N.J.: Pearson Education Inc., 2002, Appendix B, pp. 159–62.*

Advantages and Disadvantages of Privatization

Some caution is needed in evaluating the effects of privatization. In the political climate of fiscal conservatism there has inevitably been a rush to report successes. It is true that there is empirical evidence to support the claim that privatization can reduce the *costs* of both capital projects and service provision. Studies by accounting companies have shown that financing infrastructure facilities through **public-private partnerships** is between 20 and 40 percent less expensive than through public financing alone.

The empirical evidence for cost savings through contracting out services is more limited and certainly more mixed. A 1993 survey of public agencies by the Reason Foundation found that the estimated cost savings—of less than 20 percent—for social services were lower than those for other services—estimated at between 20 and 40 percent. A 1996 U.S. General Accounting Office (GAO) report examined the cost effectiveness of privatized child support enforcement services at selected local sites in a small number of states. The cost-effectiveness results were mixed. For example, the privatized operations were 18 percent and 60 percent more cost effective at the sites in Arizona and Virginia respectively. In Tennessee one public operation was 52 percent more cost effective than its privatized counterpart, while at the other site, the cost effectiveness of the private and public operations was about the same. Some observers note that the cost estimates do not include the public transaction costs (such as the expense of effective contract design and monitoring) involved in the contracting process. In addition, the cost comparisons between the public and private sectors do not always control for crucial variables, such as the proportion of clients using social services who are "difficult to serve."[46]

Research on the *quality* of services is similarly limited, and, like that on cost savings, is mixed. The empirical evidence, though limited, suggests that the quality of privatized services may be the same or somewhat higher than that provided by the public sector. For example, the 1996 GAO report on the privatization of child support enforcement services concluded that the privatized operations achieved performance levels as good as or better than those of public child support enforcement programs in locating noncustodial parents, establishing paternity and support orders, and collecting support owed. Analysts point out, however, that the studies may be biased in favor of the private sector because privatization often occurs only when public services are particularly ineffective, providing a point of comparison that may not be typical of public-sector service provision.[47]

Consequently, some serious questions remain concerning the effectiveness and the desirability of privatization in the longer run. In particular, it remains to be seen whether private contractors will be both able and willing to keep costs down. The obvious danger, of course, is replacing a public monopoly with a private monopoly. The inefficiency of some Department of Defense contracting (the most privatized portion of any aspect of government activity in the United States) continues to provide horror stories that make many people skeptical about privatization. Parallel examples at the municipal level are not hard to find. A public employee union, the American Federation of State, County, and Municipal Employees (AFSCME), has drawn attention to six causes for concern about privatization:[48]

1. "Low-balling," in which contractors initially bid lower than their actual costs in order to secure a contract and then, after the city disbands its own delivery system, raise the price to recover their initial costs and establish a monopoly.

2. Lack of control and accountability arising from cities' inexperience in writing quantifiable performance measures.

3. Disruption of service resulting from bankruptcy or labor disputes. Clauses can be included in privatization contracts that hold the contractor liable for the remainder of the contract term, but the problem remains as to how cities might deal with disruption of service if they themselves have relinquished the capacity to provide the service.

4. The replacement of "primary" jobs in the public sector by "secondary" jobs in the private sector, with women and minorities in particular experiencing the effects of less security, lower pay, and fewer benefits. The effects of privatization on career civil servants could include career disruption and dislocation and reduced morale and productivity.

5. The risk of corruption associated with the interdependence of leading bureaucrats, politicians, and local contractors.

6. The emphasis on cost efficiency implicit in privatization; not a bad thing in itself, but cause for concern in that it reinforces the idea of a rather narrowly defined public interest.

CRISIS? WHAT CRISIS? CIVIC ENTREPRENEURIALISM AND THE POLITICS OF IMAGE

After a decade of economic and metropolitan restructuring, by the early 1980s the national economy had begun to recover from the stagflation crisis. With economic recovery, local property values went up, as did local tax receipts. Having trimmed municipal workforces, reduced services, and improved efficiency, many cities were in a position to expand their expenditures without risking fiscal crisis. Municipal expenditures rose sharply after 1982, continuing the trend that had begun after World War II. In many ways, the pattern of expenditure remained as before: Sanitation, recreation, education, highways, police, libraries, fire services, and so on remained the staples of municipal expenditure. Nevertheless, the period since 1982 has been distinctive because cities have added to their activities a number of high-profile and rather expensive undertakings designed to secure and enhance their share of national and international economic growth.

As we saw in Chapter 4, the economic recovery of the 1980s was based on a much more flexible economic geography. Civic leaders and local business interests were quick to appreciate this flexibility. They were also astute enough to appreciate that the globalization of the economy made it harder for the federal government to stage-manage patterns of investment. *It was thus left to local governments to negotiate with large corporate investors and to attempt to stimulate and attract private enterprise by creating the right conditions for profitable investment.*

Cities, in short, had to become entrepreneurial. The cornerstone of this new ethos of governance has been the public-private partnership (not to be confused with the privatization of projects and services).[49] Public-private partnerships can take various forms, but they are often managed by quasi-public development organizations known variously as Development Authorities, Economic Development Corporations, and Local Development Corporations:

> These institutions legitimate and organize an overt alliance between the local state and particular capital fractions under the rubric "partnership." They usually implement their decisions without referenda or legislative approval of specific projects. They have a great deal of discretion over the use of public funds and the granting of tax concessions.[50]

In seeking to create the preconditions for economic growth, public-private partnerships have subsidized private development through a variety of mechanisms: tax abatements, tax-exempt industrial revenue bonds, lease financing, sales tax exemptions, and tax "holidays" (whereby businesses are exempted from local taxes from a number of years as a reward for local investment). Partnerships may also involve using public capital as risk capital in true joint ventures, exercising public powers of **eminent domain** in assembling land, constructing public infrastructure customized to the needs of private development, rewriting city ordinances to accommodate private development, and accessing federal funds for urban development.

Ironically, the privatization of urban governance and politics was fostered in this final respect by the last fling of the Democratic progrowth coalition between liberal reformers and blue-collar interests. Under the Carter administration in 1978 the federal government introduced the Urban Development Action Grant (UDAG) program, mainly in the hope of easing fiscal stress by stimulating central city economic development. The UDAG program made federal funds available to distressed cities, allowing wide discretion over their use so long as they leveraged private investment that would create new jobs and taxes. With UDAG funds, cities could enter into partnerships with private developers and leverage millions of investment dollars. In spite of the new conservatism that characterized the subsequent Reagan and first Bush administrations, the UDAG program was maintained with significant levels of funding into the late 1980s. The Reagan and two subsequent Bush administrations also did what they could to facilitate the privatization of urban governance by reforming federal tax laws and relaxing various regulatory functions. Civil rights, labor laws, occupational health and safety, and environmental protection were all enforced less aggressively.

Strategies for Urban Economic Development

We can recognize four basic approaches to entrepreneurial governance in a global economy, each of which involves heavy reliance on public-private partnership and a high level of local boosterism and image-making.[51]

1. *Attempts to maintain the attractiveness of cities as settings for production and manufacturing.* This approach typically involves an extension of traditional investments in the kind of physical and social infrastructure that is important to industry: everything from roads and bridges to high-tech industrial parks and schools with specialized curricula. It also involves the provision of substantial packages of customized inducements such as tax breaks, subsidies, and purpose-built infrastructure. Such packages often pit one city against another in bidding wars for internationally "mobile" investment in major new developments by **transnational corporations**.

2. *Attempts to capitalize on federal government expenditures.* The federal government is a major employer and contractor. We saw in Chapter 4 that there are some Sunbelt cities whose economic base is dominated by federal outlays, particularly for the military and aerospace. With the "militarization" of the economy during the 1980s and the "war on terror" after September 11, 2001, the flow of federal outlays can represent a particularly important catalyst for urban development in conjunction with the localized multiplier effects from government contractors and subcontractors. Although the pattern of such flows is largely a function of existing geographies of defense and aerospace industries and the geography of congressional pork barreling, local boosters look to public-private partnerships to enhance the infrastructure in appropriate ways—research parks and research universities, for example.

3. *Attempts to capture or retain the key command and control functions in corporate management, government, and financial and business services.* This approach typically involves the provision of some rather expensive urban infrastructure: airports, international communications networks, convention centers, hotels, and so on. It also requires cities to enter into public-private partnerships to ensure an adequate supply of office space of appropriate quality, together with surrounding upscale amenities.

   Cities themselves have been chiefly responsible, however, for the proliferation of one of the key symbols of status as a **command and control center**. Convention and exhibition centers bring business and professional visitors who not only fill hotel rooms and generate trade (and taxes) in shops and restaurants but also are exposed to the city and its opportunities for business. Public-private partnerships all over the country—from Gary, Ind., to New York City—compete with one another with their large and sumptuous convention centers.

4. *Attempts to improve the attractiveness of cities as places of consumption.* This aspect of entrepreneurialism is also important in reinforcing the other three. In American society consumption and materialism add up to "quality of life," something that not only generates jobs, incomes, and tax revenues by itself but that also enhances the prospects of being able to secure investments in any activity that involves well-paid workers—whether in production activities, government research and contracting, or management and business services.

   It is this aspect of entrepreneurialism, too, that has been most obvious in the changing landscapes of cities. Public-private partnerships have fostered the entire range of settings that are now seen as essential for the proper transaction of the new materialism (see p. 100) of American society: cultural "anchors" such as renovated turn-of-the century theaters and

1920s movie palaces, for instance. Examples include the restoration of the defunct Woodward Avenue entertainment district in Detroit, which contains the restored 1928 Fox Theatre; Comerica Park, the new Detroit Tigers baseball stadium; and Ford Field, the new Detroit Lions football stadium. Another example is the creation of a cultural district featuring a performing arts center that houses the Pittsburgh Symphony Orchestra in the renovated shell of a 1927 movie palace in Pittsburgh's Golden Triangle.[52]

Another seemingly essential component for cities seeking to consolidate their status as top-level **central places** is the major-league sports franchise with top-of-the-line spectator facilities. Just as stadiums for professional sports had sometimes been key elements of renewal schemes, so entrepreneurial city governments saw them as magnets for tourists, regional visitors, and potential investors with a choosy workforce or clientele to please. Thus (to take just a few examples): Minneapolis, faced with the possible loss of the (baseball) Twins and (football) Vikings in the late 1970s, worked with the team owners, floating a special bond in order to fund a stadium in the downtown area. Baltimore, having lost the (football) Colts to Indianapolis, agreed to subsidize a new ballpark for the (baseball) Orioles (Figure 16.10) and relinquish the city's stake in incomes from ballpark concessions.

Large mall complexes, gallerias, and festival marketplaces are a third important component.

The archetypes of these developments were Boston's Quincy Market, opened in 1976, and Philadelphia's Gallery at Market East, opened in 1977. No city of any size could do without some similar form of development. Well-known examples include Baltimore's Harbor Place (Figure 16.11), Riverwalk in New Orleans, Pioneer Square in Seattle, and South Street Seaport in New York (Figure 16.12). Usually the product of some form of public-private partnership, these set-piece developments create focal settings for integrated packages of upscale offices, tourist shops, "impulse" retailing, restaurants, concert halls, and art galleries. They are important assets for entrepreneurial cities because of their scale and their consequent ability to stage—or merely to *be*—the spectacular. They are described by David Harvey as the "carnival mask" of contemporary urbanization, their spectacular spaces being a means to attract capital and people (of the right sort) while diverting attention from the continuing problems of urban decay and social deprivation in nearby (but unseen) neighborhoods.[53] They are settings for "events" (such as concerts, ethnic festivals, and outdoor exhibits) and carefully planned "animation" (based on farmers' markets, street entertainment, and the like), all subsidized by the publicity budget of the city or the developer, or both.

All four approaches have fostered a *dealmaking ethos* that is reflected in some radical changes in the nature of urban planning and policy outcomes. These

FIGURE 16.10   Oriole Park at Camden Yards in Baltimore. A ballpark built and operated with the help of city funds.

FIGURE 16.11
Harbor Place, Baltimore.
Part of the "carnival
mask" of urbanization
put in place through civic
entrepreneurialism and
public-private partnership.

FIGURE 16.12   South Street Seaport in New York City.

include the emergence of more flexible approaches to land use zoning and the encouragement of historic preservation (p. 267) and **gentrification** (pp. 370–374). Land use zoning will be examined in greater detail in Chapter 17, but it is important to note at this point that the roots of land use zoning, historic preservation, and gentrification are to be found in the political economy of entrepreneurial cities.

The Politics of Packaging

The urban politics associated with entrepreneurial governance has been characterized by the emergence of a distinctive kind of political leadership characterized as "Entrepreneurial Mayors." The best-known examples include Baltimore's Donald Schaefer, New York's Ed Koch, Pittsburgh's Richard Caliguiri, Buffalo's James Griffin, Detroit's Coleman Young, Boston's Kevin White, and St. Louis's Vincent Schoemehl. An important common denominator among these mayors was the importance they attached to image making and sophisticated hype. Encouraged by the staggering success of the **I ♥ NEW YORK** campaign, entrepreneurial leaders devoted a lot of attention to repairing tarnished images and "selling" their cities. It was, perhaps, no more than a sign of the times—the image and style consciousness of postmodernity (see p. 326) that lent itself perfectly to recycling all manner of themes from the urban past.

The message was that cities could be fun and profitable. "Even signs of blight and social disorder could be repackaged in positive language favorable to the

BOX 16.5   URBAN REGENERATION IN LONDON'S DOCKLANDS[54]

Between the late eighteenth and the mid-twentieth centuries, London's docklands became the trading heart of Britain's empire. Occupying the north bank of the River Thames for several miles downstream of Tower Bridge, the docksides and wharves developed into distinctive settings (Figure 16.13). Immediately downstream from Tower Bridge were elegant multistory Georgian warehouses for high-value goods linked to London's commodity trade as the imperial capital: ivory, teas, furs, tobacco, plant and flower oils, spices, and other exotic imports. Further downstream were facilities for handling and storing bulkier cargoes, such as tropical fruit and vegetables, coal, cattle feed, chemicals, cement, and paper. Still further downstream, newer, larger docks and huge, refrigerated warehouses were added later, providing a modern infrastructure for global trade. This commercial activity depended upon a huge labor force of dockworkers, who were crowded with their families into cheap and often substandard housing in the neighborhoods surrounding the docks.

Relatively little of this urban setting and community now remains. The docklands fell into a steep decline in the 1960s and 1970s as a result of labor problems; competition from Rotterdam and other European ports; and the construction of container port facilities further downstream. Employment fell from 30,000 in its heyday in the 1950s to 2,000 by 1980. The disused and derelict docks, so near the center of London, represented both an embarrassment to the national government and a huge potential property asset. So in 1979 the government,

taking up an idea by geographer Peter Hall, established an experimental **enterprise zone** in the Isle of Dogs, in the heart of the docklands, in an effort to attract new businesses to the area by suspending certain taxes and regulations. The following year, the government created the London Docklands Development Corporation (LDDC) and charged it with planning the economic regeneration of an eight-and-a-half square mile area of the docklands (Figure 16.13). This urban development corporation was given substantial resources and powers that the national government believed were necessary to regenerate such a large area in a relatively short time. The LDDC took over the planning powers for the area from the London boroughs of Tower Hamlets, Newham, and Southwark and was given the power to acquire land through eminent domain if necessary.

It was at this point that London's financial markets began to respond to globalization. Automation and information-based financial dealings required large, modern office units, flexible in plan, with deep floor plates to accommodate state-of-the-art technology in suspended ceilings and underfloor cabling. In the heart of the medieval city, London's central financial district had little such space and few opportunities to create it. Then in 1985 the world's largest property development company, Canadian-based Olympia and York, put together an ambitious redevelopment scheme for the docklands that would provide several million square feet of new office space. Although the redevelopment

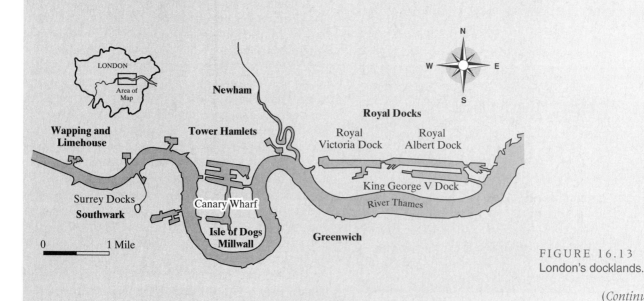

FIGURE 16.13
London's docklands.

(Continued)

BOX 16.5   URBAN REGENERATION IN LONDON'S DOCKLANDS   *(Continued)*

scheme ran into financial difficulties for a while (and Olympia and York fell into disastrous debt), regeneration efforts continued (Figure 16.14).

Between 1981 and 1998, when the government closed down the LDDC, national grants of £1.86 billion were spent on land acquisition and reclamation (including environmental cleanup) and the provision of utilities and transportation infrastructure (including new roads and a driverless light railway system). This public funding leveraged £7.7 billion in private investment so that today London's dockland area has a new image—created by the largest single urban redevelopment scheme in the world.

Its new image centers on the sleek office tower of Canary Wharf and its associated office complexes that are populated by day by the office workers in the international financial and publishing companies. The new development is surrounded by cleaned-up waterfronts and restored warehouses that have been converted into expensive condominiums and retail complexes (Figure 16.14). The new business and professional services jobs, high-end retail establishments, and upscale housing, however, do not address the needs of the remaining long-time residents—many of whom are unemployed dockworkers.

FIGURE 16.14
The new docklands in London. Aerial view of the Canary Wharf office complex under construction by Olympia and York in the late 1980s. The opening photograph in Chapter 11 shows the finished development.

city. The vandalism of New York City's subways by spray can-wielding youths was proclaimed 'graffiti art,' a blossoming of indigenous culture rather than a symptom of decay and lawlessness."[55] City life was in vogue again; industrial decay could be repackaged as romantic; downtown could be reinscribed with its former glamor; unexceptional neighborhoods could be "reclaimed" for the city's tax base by yuppies sold on the idea of a "frontier" urban experience.[56, 57] Projects like Detroit's Renaissance Center (Figure 16.15) were successfully portrayed as part of the regeneration of the economic foundations and social fabric of once-dead central cities. The power of their imagery, reinforced by themes in advertising for everyday products from jeans to automobiles, must be credited with a substantial (but unmeasurable) contribution to the recentralization of higher-order retail and service activity in central cities.

Architectural Convergence and Economic Volatility

One very striking outcome of this phase of entrepreneurial governance has been the *serial production* of new and renovated urban settings. The more cities attempted to develop an urbane stylishness, the more malls, festival marketplaces, world trade centers, and downtown galleries became commonplace. The more that cities attempted to exploit a historic distinctiveness, the more exposed brickwork, wrought ironwork, old-fashioned street furniture, and pedestrianized streets caused them to converge in appearance.

Equally important has been the increased instability within the urban system that is at least in part a result of intensified inter-urban competition (the other prime cause being the more-flexible economic

FIGURE 16.15   The Renaissance Center, Detroit.

geography associated with the "informational" mode of development: see p. 86). This volatility has been associated with **splintering urbanism** within the urban system as the very uneven success of the new projects has changed the face of urbanization. The combination of a construction boom in the 1980s and increasingly intense inter-urban competition created a surplus of office space, convention center seats, exhibition space, retail floor space, and cleaned-up harbor frontage. Only the biggest, glitziest projects in the most-favored locations have been able to compete successfully. Intensifying competition has driven up the stakes, pushing cities to make ever bigger deals with private-sector developers.

CONCEPTUAL AND THEORETICAL PERSPECTIVES ON GOVERNANCE, POLITICS, AND URBAN CHANGE

Having reviewed the evolving relationships between urban change, urban governance, and urban politics, we are in a position to summarize the most important conceptual and theoretical perspectives. We draw here, of course, on some of the substance of political science, though we will touch on only a small portion of that discipline's conceptual and theoretical content. The urban experience that we have reviewed raises three broad questions. One concerns *power*: How is power structured within the urban arena and how does its structure change with the changing dynamics of urbanization? The second concerns *governance*: How can we interpret the role of local government—the "local state" in theoretical parlance—in relation to urban development? The third concerns the *issues* that drive local politics: How can we interpret local conflict in relation to patterns and processes of urban change?

THE STRUCTURE OF LOCAL POWER

We should recognize at the outset that in practice different kinds of power structures exist in different cities and that these variations have a great deal to do with variations in cities' economic base and demographic composition. The focus here, however, is on generalized patterns and structures of power. At this level of abstraction, one of the classic models is the **elitist model**.[58] In elitist cities a handful of people who stand at the top of a stable pyramid of power make nearly all the important decisions. These individuals, drawn mainly from business and industrial circles, constitute a strongly entrenched and select group: Without their support or consent, little of significance is ever accomplished. Elected officials are not so much the puppets of this leadership as the "understructure" of power, dependent on the patronage and tolerance of powerful individuals, many of whom keep a low public profile. The elitist model assumes that the legitimacy of this leadership is ceded from below rather than imposed from above. That is, the passivity of the bulk of the population is a reflection of their voluntary acceptance of a relationship in which they indulge the leadership's domination in return for the latter's pursuit of the public interest, with the ballot box providing a register for any serious abuse of power.

A critical question raised by this perspective is whether such a consensus over the public interest is consciously and freely agreed to by the bulk of the population or whether it is the product of the systematic suffocation of opposition. The latter is the basis of the **neo-elitist model**, which suggests that there are three main ways in which potential opposition to the preferences of powerful elites are defused:[59]

1. The failure of disgruntled groups to press their demands, either because they anticipate a retaliatory response from the powerful elite or because they believe it would be futile.

2. The refusal of those in powerful positions to respond meaningfully to the political demands of less powerful groups. Response can effectively be denied, for example, by establishing committees or inquiries that take a long time to reach a conclusion.

3. The "mobilization of bias." This involves the manipulation by the elite of the values, beliefs, and opinions of the general public. So, for example, demands for change may be denied legitimacy by being branded a threat to the freedom of the individual or to economic development. In this context, influence over the mass media is particularly important. Given such influence, the elite group is able to restrict public consideration to those matters considered "safe" and, as a result, the demands or grievances of some sections of the community may remain muted.

An entirely different model of local power structure is the **pluralist model**, which posits that power is dispersed, with different kinds of interests dominant at different times over different issues.[60] In this model the business elites that have been found to be in control in some cities are only one among several "power clusters" comprising diverse, autonomous, and nonhierarchical groups. The structure of power is essentially competitive, drawing on a wide range of participants and ensuring a fundamental element of democracy through the need for elites to acquire mass loyalty. This competition keeps society in rough equilibrium, with labor acting as a counterbalance to business, consumers offsetting the power of retailers, tenants constraining the power of landlords, and so on. These groups, additionally, have overlapping membership, which promotes intergroup contact and makes for tolerance and moderation in local politics.

This pluralist model can be used to describe a *generalized political life cycle* of American cities. In this life cycle there are four successive "dynasties." The first, roughly coincident with the era of the mercantile city, was dominated by an oligarchy of "Patricians." Then, during the era of the early industrial city, the traditional elite of landowners and commercial leaders was pushed aside by the entrepreneurial leaders of immigrant communities—the bosses and their machines. The third "dynasty" flourished in the Reform era at the beginning of the twentieth century, when new middle-class and business interests captured the power base. The fourth "dynasty" saw the rise of the professional politician, the "Ex-Plebe," as a coalition builder whose job it is to take advantage of the natural pluralism of American urban politics.

The **corporatist model** of local power structures elevates government itself to the role of a key player in the power game.[61] In this model the basic framework of power rests on a symbiotic relationship between private organizations (labor organizations, community groups, business leadership clubs) and various arms of local government. Key organizations become incorporated into the formal decision-making process (via representation on committees, appointments to boards, licensing, subsidies, franchises, etc.), and city governments delegate a certain amount of authority in return for cooperation and support (moderating members' demands, contributing to public campaigns). The corporatist model is one that sees society in terms of segmented socioeconomic organizations that are taken under the wing of professional politicians and technocrats who are then able to expand and consolidate the scope of their power and authority.

A final model derives from the socioeconomic fragmentation of contemporary metropolitan America. It follows from the sociocultural fragmentation and metropolitan spatial restructuring since the 1980s, which sometimes made the pluralism that had formed the basis of progrowth coalitions vulnerable to unstructured, multilateral conflicts in which many different groups fought continuously with one another over a broad range of issues. The rupturing of working-class interests along cleavages of race and ethnicity; the division of middle-class interests between growth, no-growth, and slow-growth perspectives; and the split between downtown business interests and suburban developers, between big business and small businesses, and so on, have resulted in a **hyperpluralistic** situation in which unstable power relations are reflected in unstructured and multilateral conflict. Power is exercised over narrow areas and for limited time periods by a variety of special interest groups who go their own way in seeking narrow gains and who are less restrained in their conduct and less likely to accommodate one another than under the **pluralist model**. Coalitions are essential to the hyperpluralist model, but they are short-lived and ad hoc.

These models should be seen for what they are: crude portrayals of real situations in which power relations are complex and fluid. They are constructs, or "ideal types," that help us conceptualize certain aspects of urban dynamics. In this context, another useful concept is that of **urban regimes**—the idea of power structures that rest on slowly changing coalitions of dominant groups and interests that are represented by city officials (both elected and appointed) who sustain both their own power and that of the coalition by ensuring a variety of particular benefits, policy outcomes, or "side benefits" to the key groups involved.[62] Using this perspective, we can use two variables—the strength of political leadership and the cohesion of business elites—to categorize the models of power and chart the shifts that occur in the regimes in different cities.

Figure 16.16 maps the urban regimes of 13 of the largest cities in the United States. It shows, first of all, the variety of urban regimes; all the models of power structure are represented.

Figure 16.16 also shows a good deal of change. Mayors come and go, as do business executives, and conditions change. Cities may be in transition between different types of regimes. But the tendency in these 13 cities has been toward a pluralist or hyperpluralist model due to the gradual reduction in the political base of urban government (decline of political parties, competition from suburban communities) and in business concentration (diversified economies, "mobile" capital).[63] This trend has placed a much greater responsibility upon city mayors for coalition building within urban regimes—thus helping to explain the emergence of "Entrepreneurial Mayors" (see p. 485).

THE ROLE OF THE LOCAL STATE

Each of the models of power structures that we have described has an implicit interpretation of the role of the local state in relation to urban development.

From an **elitist** perspective the state is seen as an institution that can be manipulated, influenced, diverted, or somehow orchestrated to conform to the

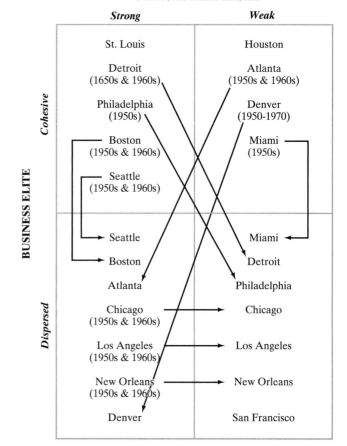

FIGURE 16.16   Political regimes in 13 U.S. cities.

central interests of those in the top levels of economic and social life. For those at the lower levels, therefore, the state can be an oppressor, the instrument of business interests.

From a **pluralist** or **hyperpluralist** perspective, the local state is seen as a passive institution that registers and responds to diverse demands. It cannot be captured or manipulated for long; it is a broker and a mediator rather than an oppressor.

From a **corporatist** perspective, the state is seen as a much more important institution: the only means of sustaining linkages between the divisions among the various levels within society. It is a common means to a multiplicity of ends. In this view, then, the local state can be characterized as an orchestrator and director.

Another school of thought emphasizes the autonomy of the technocrats and bureaucrats in the local civil service. This view originated with sociologist Max Weber's observations about the "dictatorship of the official" that accompanied nineteenth-century modernization. In this view urbanization had become so complex that elected officials had to rely increasingly on the expertise of civil servants. Taken a step further, the local state is seen to be controlled, on a day-to-day basis, by civil servants whose professional ideology and departmental allegiances are crucial in determining a good deal of the shape of local government activity. This interpretation can be characterized as a **managerialist model**, in which civil servants effectively become social gatekeepers by mediating policy implementation.

These interpretations are all somewhat narrow, however, in that they do not take into account the relationships between the local state and the national state, nor do they address the broader issues of the role of the local state in relation to the overall trajectory of economic and social change. For such perspectives we have to turn to a *structuralist* perspective developed by neo-Marxian scholars (see p. 5).[64] From this perspective the local state is seen as a relatively autonomous adjunct of the national state, with both acting in response to the prevailing balance of class interests in society. In broad terms, the local state safeguards the interests of big business while "buying off" the working classes through reformist strategies. In practice, the activities of the local state are interpreted as serving one of three key roles in relation to the economy and society:

1. Enhancing and sustaining private production and capital accumulation through
   a. The provision of infrastructural elements.
   b. Easing the spatial aspects of the reorganization of production—through the planning process, for example, or through the initiation of renewal projects.
   c. Investment in human capital—through public schools and technical training programs.
   d. "Demand orchestration" which, through public works contracts, for instance, brings stability and security to markets that might otherwise be volatile and unpredictable.
2. Reproducing labor power through collective consumption by
   a. Ameliorating the living conditions of the poor (e.g., subsidized public housing and transportation).
   b. Enhancing social and cultural amenities (e.g., public libraries, parks, art galleries).
3. Maintaining order and social cohesion through
   a. Police services.
   b. Welfare programs and social services.
   c. "Agencies of legitimation" such as local schools and citizen participation schemes.

PATTERNS OF LOCAL CONFLICT

However the role of the local state is interpreted, a certain amount of local conflict is inevitable—not just among the principal protagonists in the overall struggle for power but also among individuals, social groups, and communities as they react to the changes imposed by urbanization as mediated through the institutions of government. It is this latter kind of conflict that is in question here, the stuff of local politics—both formal and informal—from day to day. At this level most conflict centers on the public regulation of privately initiated change and in the manner and quality of provision of various services and amenities. It will come as no surprise to find that those with the greatest stake in a particular activity or "turf" are the ones who become most involved. Although no formal theories deal with the patterns of such conflict, some useful generalizations can be made by way of a model of locational conflict (Figure 16.17).

The model assumes a classic **concentric zone** pattern of land use and **social ecology**. Conflict is generated by change: the changing dynamics of the urban economy, changes associated with the aging of people and places, and changes in people's values and expectations. More specifically, three kinds of sociospatial change generate locally focused conflicts:

- Metropolitan decentralization (new suburban and exurban development).
- Physical restructuring (land use change and redevelopment in inner-city areas).
- Changing neighborhood composition (through residential mobility, aging, etc.).

Given this scenario, we can see that different settings are associated with different kinds of conflicts. **CBDs**, for example, are the location of conflict involving redevelopment, historic preservation, and transportation; established residential areas are striking for their relative absence of conflict, except over schooling; and suburban and exurban residential areas provide the settings for every kind of conflict except those involving historic preservation and redevelopment. We can also see how specific types of conflict vary in intensity across the metropolitan area. Conflict over transportation, for example, tends to be localized in two settings: in areas of suburban and exurban expansion and in the neighborhoods adjacent to the reorganizing city center. Conflict over "cultural" issues (e.g., the encroachment of one ethnic group on the territory of another), in contrast, tends to decline steadily toward the suburbs from a peak in the **zone in transition**.

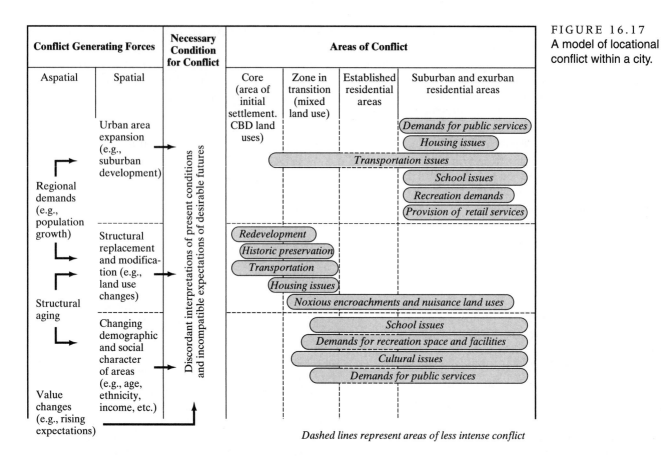

FIGURE 16.17
A model of locational conflict within a city.

Dashed lines represent areas of less intense conflict

FOLLOW UP

1. Be sure that you understand the following key terms:

    annexation
    boosterism
    civic entrepreneurialism
    corporatist model
    debt financing
    elitist model
    exclusionary zoning
    fiscal mercantilism
    fiscal retrenchment
    gerrymandering
    growth machines
    hyperpluralism
    incorporation
    machine politics
    malapportionment
    managerialist model
    metropolitan fragmentation
    municipal socialism
    neo-elitist model
    NIMBYism
    pluralist model

    privatization
    Progressive Era
    progrowth coalitions
    public-private partnerships
    reapportionment
    revenue sharing
    special districts
    urban regimes

2. Be sure that you understand the significance of the following in relation to American cities:

    Assassination of Dr. Martin Luther King (1968)
    Hopkins v. Town of Shaw (1971)
    Nixon's New Federalism
    Civil Rights Act of 1964
    New York's fiscal crisis (1975)
    Proposition 13 (1978)
    War on Poverty

3. Consider the way in which the idea of the "public interest" changed as cities themselves changed. How would you yourself define the public interest?

4. On p. 463 it is noted that "Central cities came to be encircled by a ring of independent and often hostile governments." Can you find evidence of this

sort of situation today? If so, what sorts of conflicts are involved between the different jurisdictions?

5. Can you identify any examples of entrepreneurial governance in your city or town? Your local government may have an economic development department with a website that publicizes its efforts. The business section of local newspapers also reports on entrepreneurial governance efforts. What kinds of projects (including high-profile ones such as a convention center or sports stadium) have been developed, and what can you find out about the initiators and supporters of these projects?

6. Add to your *portfolio*. The material covered in this chapter can be illustrated best with firsthand accounts from city newspapers. If you have access to old newspapers, find examples of political cartoons that illustrate topics such as machine politics, annexation, or reform. In current newspapers the business section contains news stories that cover city council meetings, routine announcements about zoning and highway repairs, and perpetual discussions about local economic development. What examples can you find, and how are they related to the context and concepts discussed in this chapter?

KEY SOURCES AND SUGGESTED READING

Anderson, M. 1964. *The Federal Bulldozer: A Critical Analysis of Urban Renewal, 1949–1962*. Cambridge, Mass.: MIT Press.

Brownell, B. A. and W. E. Stickle, eds. 1973, *Bosses and Reformers: Urban Politics in America, 1880–1920*. Boston: Houghton Mifflin.

Cisneros, H. 1993. *Interwoven Destinies: Cities and the Nation*. New York: Norton.

Ebner, M. and E. Tobin, eds. 1977. *The Age of Urban Reform: New Perspectives on the Progressive Era*. Port Washington, N.Y.: Kennikat Press.

Fainstein, S. 1996. Social Justice and the Creation of Urban Space. pp. 18–44 in *The Urbanization of Injustice*, eds. A. Merrifield and E. Swyngedouw. London: Lawrence & Wishart.

Greene. J. D. 2002. *Cities and Privatization: Prospects for the New Century*. Upper Saddle River, N.J.: Pearson Education.

Harvey, D. W. 1989. From Managerialism to Entrepreneurialism: Transformation in Urban Governance in Late Capitalism. *Geografiska Annaler* 71B: 3–17.

Jonas, A. E. G., and D. Wilson, eds. 1999. *The Urban Growth Machine: Critical Perspectives, Two Decades Later*. Albany, N.Y.: State University of New York Press.

Judd, D., and T. Swanstrom. 1994. *City Politics: Private Power and Public Policy*. New York: Harper-Collins.

Lauria, M. ed. 1997. *Reconstructing Urban Regime Theory: Regulating Urban Politics in a Global Economy*. Thousand Oaks, Calif.: Sage.

Peterson, G. E. 1994. *Big-City Politics, Governance, and Fiscal Constraints*. Washington, D.C.: The Urban Institute Press.

Schiesl, M. J. 1977. *The Politics of Municipal Reform: Municipal Administration and Reform in America, 1880–1920*. Berkeley, Calif.: University of California Press.

Squires, G. 1991. Partnership and the Pursuit of the Private City. pp. 196–221 in *Urban Life in Transition*, eds. M. Gottdiener and C. G. Pickvance. Newbury Park, Calif.: Sage.

Stave, B. M., and S. A. Stave. 1984. *Urban Bosses, Machines, and Progressive Reformers*, 2nd ed. Malabar, Fla.: Krieger.

Teaford, J. 1975. *The Municipal Revolution in America: Origins of Modern Urban Government, 1650–1825*. Chicago, Ill.: University of Chicago Press.

Weiss, M. 1980. The Origins and Legacy of Urban Renewal. pp. 53–80 in *Urban and Regional Planning in an Age of Austerity*, eds. P. Clavel, J. Forester, and W. Goldsmith. New York: Pergamon.

RELATED WEBSITES

U.S. Department of State, International Information Programs:
http://usinfo.state.gov/products/pubs/outusgov/ch7.htm
The website of the U.S. Department of State's International Information Programs contains information on the different forms of government found in the United States. See in particular the *Outline of U.S. Government* for information on local government structure.

National League of Cities: *http://www.nlc.org/*
The National League of Cities is the oldest and largest national organization representing municipal governments in the United States. Its website contains information and publications concerning policies that affect cities, including urban governance and fiscal health.

Privatization Link: *http://www.privatizationlink.com/*
This World Bank agency—which encourages

privatization—offers information on its website about the privatization activities in different countries. It also includes links to publications on privatization on the World Bank's main website (*http://www.worldbank.org/*) and from other sources such as the Organisation of Economic Cooperation and Development and its 2002 *Overview of Privatization Developments in OECD Member Countries During 2001* that is available online (*http://www.oecd.org/*).

The LDDC History Pages: *http://www.lddc-history.org.uk/*
The London Docklands Development Corporation (LDDC) was established to regenerate London's docklands between 1981 and 1998. Although the LDDC is no longer in operation, this website contains the LDDC's story of the docklands' growth, decline, and revitalization from its own property-led regeneration strategy perspective.

17

URBAN POLICY
AND PLANNING

The geography of contemporary cities is imprinted with layer after layer of the outcomes—both intentional and unintentional—of policies and plans. It took a long time for the idea of managing urban change and planning city development to become accepted, but in the process there emerged a series of ideas and precedents that left their mark on cities. Each successive phase of economic development and urban change brought new challenges—along with new interpretations of old problems—that brought innovations in policymaking and planning. These innovations are significant because of their legacy for today's policymakers and planners. The lessons derived from past responses to urban challenges condition our whole approach to urban policy and planning. Even the institutional apparatus that we deploy in urban policymaking and planning is part of this legacy. Like all other aspects of urbanization, it must be understood in relation to continuous, interdependent processes of sociospatial change.

CHAPTER PREVIEW

This chapter retraces the evolution of metropolitan regions one last time. The objective is to highlight urban policy and planning as both products of urban change and modifiers of urban dynamics. Having reviewed the changing problems attached to urbanization (Chapter 15) and the interactions between urban change, governance, and politics (Chapter 16), we can now appreciate the main issues and turning points. The beginnings of urban policy and planning were rooted in a long, slow struggle with the central paradox of urbanization: the unwanted physical and social side effects of cities that were the crucibles of free-enterprise competitive capitalism. The end of the beginning phase was marked by the severe crisis of the Great Depression and the consequent moderation of

free-enterprise principles through government intervention. A relatively brief "golden era" for urban policy and planning followed World War II. Subsequently, another **overaccumulation crisis** challenged the idea that cities, along with the economy and society as a whole, could be successfully planned and managed. The pressures of economic restructuring across fragmented metropolitan space marked the beginning of the end of the golden era of urban policy and planning. This is not to deny the very considerable importance of urban policymaking and planning in contemporary metropolitan regions. Rather, as we shall see, they have, for the time being at least, become fragmented and divorced from any broad sense of the public interest.

THE ROOTS OF URBAN POLICY AND PLANNING

The roots of modern Western urban policy and planning can be traced to the Renaissance and Baroque periods in Europe (between the fifteenth and seventeenth centuries), when artists and intellectuals dreamed of ideal cities and rich and powerful regimes used urban design to produce extravagant symbols of wealth, power, and destiny. Inspired by the classical art forms of ancient Greece and Rome, Renaissance urban planning sought to recast cities in a deliberate attempt to show off the power and the glory of the state and the church. Spreading slowly from its origins in Italy at the beginning of the fifteenth century, Renaissance urban planning had diffused to most of the larger cities of Europe by the end of the eighteenth century.

Dramatic advances in military ordnance (cannon and artillery) during the Renaissance brought a surge of planned redevelopment that featured impressive fortifications, geometric-shaped redoubts, or strongholds, and an extensive *glacis militaire*—a sloping, clear zone of fire. Inside new walls, cities were recast according to a new aesthetic of grand design—geometrical plans, streetscapes, fancy palaces, and gardens that emphasized views of dramatic perspectives. These developments were often of such a scale that they effectively

fixed the layout of cities well into the eighteenth and even into the nineteenth century, when walls or glacis, or both, eventually made way for urban redevelopment in the form of parks, railway lines, or beltways.

As societies and economies became more complex with the transition to **competitive capitalism**, national rulers and city leaders looked to urban planning and policy to impose order, safety, and efficiency, as well as to symbolize the new seats of power and authority. One of the most important early precedents was the comprehensive program of urban redevelopment that Baron Georges Haussmann carried out in Paris between 1853 and 1870 (see p. 124).

Haussmann's ideas were widely influential and extensively copied until the emergence of the Modern movement early in the 20th century (see p. 255), with its idea that buildings and cities should be designed and run like machines and that urban design, planning, and policy should not simply reflect dominant social and cultural values but, rather, help to create a new moral and social order.

The history of urban planning and policy since the inception of the Modern movement is best read as a story of successive crises and responses (Figure 17.1)

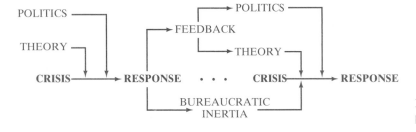

FIGURE 17.1   A crisis-response model of policy-making and planning.

rather than as the rise and fall of a movement or as a succession of ideas and experiments. Each perceived crisis—economic, social, political, environmental—has elicited a response that has modified both the theory and practice of urban policy and planning. Successive responses have been shaped not only by the nature of the immediate crisis but also by the experience of previous reactions to earlier crises—the *feedback loop.*

THEMES AND PERSPECTIVES

Running through this story of crisis and response is a continuing struggle not only with the central paradox of urbanization (cities as economic necessities but potential social and environmental evils) but also with the tensions and contradictions inherent in a property-owning democracy. (And in the United States the primacy accorded under law to private

---

BOX 17.1   THE VISIBLE LEGACY OF URBAN POLICY AND PLANNING IN EUROPEAN CITIES[1]

Although no two cities are identical, the strong legacy of urban policy and planning is visible as a set of common characteristics that make European cities distinctive.

High Density and Compact Form: The constraints of the city walls kept the density of development high during the Middle Ages. A number of factors perpetuated this compact, densely built-up form that is now characteristic of large cities in Western Europe. A long tradition of planning that restricted low-density urban sprawl dates back to the application of strict city building regulations in the earliest suburbs. The compact urban form also reflects the relatively late introduction and widespread use of the automobile in Europe.

Complex Street Pattern: The unplanned narrow streets and alleys of the medieval core in many cities evolved in the preautomobile era. Outside the walled town during the medieval period, suburban areas grew around the long-distance roads that radiated out from the city gates. Influenced by Haussmann's ideas in the nineteenth century, cities like Munich, Marseille, and Madrid made radial or tangential boulevards the axes of their planned suburbs.

Town Squares: The town square, the heart of the Greek, Roman, and medieval towns, has often survived as an important open space. Some medieval town squares in Western Europe boast a continuous tradition of regular open-air markets. The large open square, typical of socialist cities in Eastern Europe, was used for public gatherings and political rallies and symbolized the open, communal heart of the city. Today, many central squares and their historic buildings have been adapted to economic and social change and contain modern commercial functions such as tourist offices and restaurants.

Major Landmarks: The historic landmarks in West European city centers have traditionally been symbols of religious, political, military, educational, and cultural interests. Many cathedrals, churches, and statues continue to serve their original purpose. Some town halls, royal palaces, and artisan guildhalls have been converted for use as libraries, art galleries, and museums. Medieval castles and city walls have become tourist attractions. Today, of course, the major landmarks are expressions of economic power—no longer city halls, but the offices of **transnational corporations**; not cathedrals, but sports stadiums.

In Eastern Europe the hallmarks of socialist cities were the massive buildings in "wedding cake" style, red stars, and "heroic" statues. Since the late 1980s the red stars that had adorned most public buildings have been removed and the many statues of socialist leaders, especially Lenin, were either destroyed in protest or taken away for storage. Street names, such as "Lenin Boulevard," have been changed to their former designation, to a completely new name, or to honor individuals or events associated with the fall of Communism.

Low-Rise Skylines: For North American visitors, the most striking aspect of the older parts of West European cities is the general absence of skyscraper offices and high-rise apartments (Figure 17.2). The urban core was developed long before reinforced steel construction and the elevator made high-rises feasible. Master plans and building codes designed to minimize the spread of fire maintained building heights between three and five stories during the industrial period. Paris fixed the building height at 65 feet in 1795, while other large cities introduced height restrictions in the nineteenth century. Still regulated today, high-rises are found only in redevelopment areas or on land at

(Continued)

Box 17.1   The Visible Legacy of Urban Policy and Planning in European Cities   *(Continued)*

FIGURE 17.2   Amsterdam's low-rise skyline. A distinctive aspect of the older parts of West European cities is the general absence of skyscraper offices and high-rise apartments.

cities were usually international hotels, massive Houses of the People, or TV towers.

Bustling Downtowns: The high density and compact nature of European cities create downtowns that bustle with activity. The vitality of the city center is reinforced by the widespread use of public transportation systems (buses, trains, subways) that converge on the core.

In larger cities distinct functions dominate particular districts. Institutional districts house government offices and universities. Financial and office districts contain banks and insurance companies. A pedestrianized retail zone leads to the railroad station. Cultural districts offer museums and art galleries. Entertainment areas include theater and red-light districts. Many downtown buildings have multiple uses. Apartments are found above shops, offices, and restaurants. Large department stores, such as Harrods and Printemps of central London and Paris, are prominent features in most European downtowns. Modern downtown malls include Les Halles in Paris and Eldon Square in Newcastle in the northeast of England.

Suburban malls are becoming prevalent. Many coastal or riverine cities have also refurbished old port and industrial buildings to house mixed-use waterfront redevelopments like the one at Liverpool's Albert Dock. Other cities have renovated obsolete historic structures, such as London's Covent Garden, as festival marketplaces with specialized shops, restaurants, and street performers. In keeping with trends in the West, department, fashion, fast food, and even video rental stores began to appear in the downtowns of East European cities during the 1980s, becoming more widespread since the fall of Communism in the late 1980s. Likewise, shopping malls have appeared, mainly at the intersections of major transportation lines. Pedestrian shopping streets, containing the most exclusive shops, have become an important part of the retail structure of East European cities.

the periphery of the city. Skyscrapers have also been built in the central commercial and financial districts of some of the very largest cities, including London and Paris.

In socialist Eastern Europe there was no private ownership of land and so no urban land market. With few transnational corporations doing business in these countries, socialist cities were usually devoid of tall commercial buildings that mark the **CBD**. The tallest buildings in most East European

property ownership makes urban policy and planning much more difficult than in many other developed countries.) We saw in Chapter 5 that, although the sanctity of private property rights rigorously protects certain rights of certain individuals (i.e., owners of real estate), cities are less orderly than they might be for many businesses, less convenient than

they could be for most residents, and less healthy (in the widest possible sense) than they should be for everyone.

A second source of tension and contradiction for city planners and policymakers centers on the need, on the one hand, to exert some kind of public control over urban space (in order to make cities more efficient, convenient, and healthy) and, on the other, to avoid stifling capitalist enterprise with too much regulation.

The history of urban policy and planning begins with the increasing expression of *rationalism* in its various manifestations: concerns with efficiency, order, goals and objectives, cost effectiveness, and, above all, the idea of "progress." But as the story gains momentum and complexity, we shall describe the interactions between rationalism and several other -isms: the **environmental determinism** of most early social science, the antiurbanism of high-brow culture, the elitism of pioneer urban designers, the managerialism of technocrats, the pragmatism and entrepreneurialism of local politicians, and so on. The result is a complex legacy of thought and practice that continues to shape and modify urban space.

Finally, it is important not to stereotype urban policy and planning as antibusiness or anticapital. True, their history shows that they developed initially in response to criticisms of the laissez-faire approach to economic and urban development. But as planning historian Richard Fogelsong has suggested, it should not be inferred from this that urban planning was anticapitalist either in its origins or in its effects. Although urban planning and policy emerged in response to forces that were endogenous to capitalism, their interventions served to stabilize the process of urban development and to mitigate the effects of market forces in ways that contributed to the overall maintenance of the capitalist system.[2] It was the middle classes and business interests that led governments to acquire the power to protect public health, safety, and welfare through planning regulations, public works, and policies.

THE BEGINNING: PHILANTHROPY AND REFORM

There is no need here to reiterate in detail the miserable conditions of nineteenth-century urbanization. In Europe the lives of city dwellers were famously described by Charles Dickens and catalogued by Charles Booth and Friedrich Engels.[3] Physical conditions in American cities were equally miserable:

> Pigs still served as garbage scavengers. There were miles of crude paving, usually of cobble, granite or wood. Sewers were infrequent and flush toilets rare. Wooden construction, still prevailing in the centers of most cities, made tenements a firetrap. In consequence, waterworks were designed more for the needs of fire-fighting than from any concept of pure water or public health needs.[4]

It was abundantly clear that these conditions not only affected the inhabitants of crowded slums and industrial districts but also were a serious threat to everyone else. Fire, disease, and mob violence spilled all too frequently and easily out of the slums for anyone to feel secure. Employers were unhappy that their employees were sickly (and therefore less productive than they might be) and fearful of mob protest turning into mob rule (as revolutionary events in Bohemia, France, Germany, Hungary, and Switzerland in the 1840s showed all too clearly). Property owners were unhappy at being perpetually vulnerable to fires. The respectable middle classes were worried about the breakdown of moral order and the contaminating effects of crime and drunkenness. Everyone was unhappy at being constantly exposed to life-threatening diseases such as influenza, tuberculosis, whooping cough, and scarlet fever and to epidemics of smallpox, cholera, typhoid, and even bubonic plague.

EARLY EUROPEAN TRADITIONS

These concerns were the preconditions for a sense of crisis that was mobilized in Europe early in the nineteenth century. In Britain the first real focus of concern came from the efforts of one individual: Edwin Chadwick, first secretary of the Poor Law Board. His campaigning led to the Royal Commission on the Health of Towns. The publication of the commission's report in 1842 prompted progressive-minded liberals to form themselves into voluntary associations such as the Health of Towns Association, the Association for Promoting Cleanliness Amongst the Poor, the Society for Improving the Condition of the Labouring Classes, and the Metropolitan Association for Improving the Dwellings of the Industrious Classes. At first they limited themselves to discussion and passing resolutions, but soon they began to lead by example in constructing model housing: demonstration projects that sought to show how decent housing could be built for working-class households with affordable rents that still yielded a profit.

The problem was that the profit margins were far too modest to attract much serious attention from investors or developers, and the urban crisis intensified. By the 1860s the liberal response was dominated by wealthy philanthropists such as George Peabody and the Guinness family, who were willing and able to build worker housing that yielded profits that were, for the period, strikingly low: typically around 5 per cent. On this basis the Peabody Trust alone built more than 20,000 dwellings in London between 1860 and 1890, some of which are still in use (Figure 17.3).

FIGURE 17.3   Peabody Trust housing, near Covent Garden, London.

Such philanthropy did not affect the poorest of the poor (who could not afford the rents), nor did it make significant inroads into the ever-worsening conditions of industrial cities; but it did help to bring some attention in polite society to the issue of how to address the conundrum of industrial cities: sustaining environments in which investors could make a profit in both the labor market and the housing market without leaving large numbers of households so poor that their slum neighborhoods threatened the health and security of everyone else.

Ebenezer Howard and the Garden City Concept

The broader Victorian cultural response to industrialization and its urban problems was a Romanticism that dominated literature, poetry, art, and architecture. Central to this Romanticism was an idealized view of the countryside and rural life. When this Romanticism combined with the strong sense of paternalism and noblesse oblige that the Victorian bourgeoisie had inherited from the European aristocracy, it led to the idea of planned communities that might combine industrial production with the virtues of arcadian environments. Early in the nineteenth century socialist industrialist Robert Owen had built a model industrial community—New Lanark, in Scotland—with a strict paternalistic regime. By the mid-nineteenth century, paternalism, tempered by philanthropy, produced a flurry of model communities. Among these were the company towns of Saltaire, England, built by Titus Salt, and Margarethenhöhe in Germany, built by steel magnate Alfred Krupp, and the larger and more ambitious company suburbs of Bourneville, England, built by chocolate manufacturer George Cadbury, and Port

Sunlight, England, built by W. H. Lever, whose products included Sunlight Soap.

In 1902, Ebenezer Howard published *Garden Cities of Tomorrow,* drawing on the experience and the idealism of these and other projects and adding a rationale that appealed strongly to the Romantic ideals of the time: Garden Cities could be planned and designed so as to combine the best of the countryside (nature, beauty, tranquility) with the best of the city (jobs, amenities, society) while avoiding the worst of both (Figure 17.4). Howard's goal was harmonious, self-governing communities that would grow into cities of manageable size.

These new self-contained towns would be located outside the commuting range of existing cities. They would be big enough to provide residents with jobs so that they did not have to commute to other urban centers, but not so big as to have the social and environmental problems of the large industrial cities of the time. Although not achieved, Howard's idea was that a number of Garden Cities would form a polycentric **urban system** that would be the functional equivalent of the existing congested metropolis.

Howard's ideal plan limited each Garden City to 30,000 people on 6,000 acres (about 9 square miles). The built-up area was to be about 1,000 acres ($1\frac{1}{2}$ miles in diameter), at the center of which were public gardens surrounded by civic buildings—the town hall, courthouse, library, museums, and hospital—easily accessible by radial boulevards (Figure 17.5). This walking city would contain a concentric zone of housing and also a separate but accessible zone of factories and warehouses connected by a circumferential rail line. The built-up area was to be surrounded by a permanent green belt that restricted urban sprawl and offered recreational opportunities while at the same time protecting agriculture.

Howard invested his own money in the concept, cofounding the Garden City Pioneer Co., Ltd., which developed Letchworth (north of London), the first full Garden City. Using Howard's schematic plans, the company laid out roads, parks, and factory sites and invited private developers to build housing (within carefully regulated standards) on prepared sites. The scheme was supported by liberal reformers because it involved utopian ideals that invoked pastoralism and social order, by practical reformers because it involved land use control and centralized direction, and by conservatives because it gave private business more scope to develop real estate. Other Garden Cities were built, including Hellerau near Dresden, Germany; Floreal near Brussels, Belgium; and Tiepolo, in Italy. Although Howard's logic was based on stand-alone settlements, the appeal of his urban design principles also gave rise to Garden Suburbs. Examples include

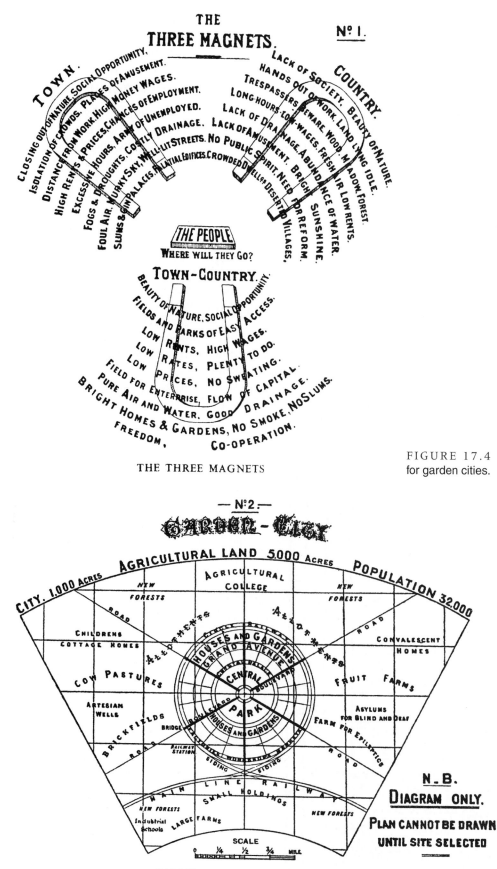

THE THREE MAGNETS

FIGURE 17.4   Ebenezer Howard's argument for garden cities.

GARDEN CITY AND RURAL BELT

FIGURE 17.5
Ebenezer Howard's plan for a garden city.

Wythenshaw, a suburb of Manchester, England; Hampstead Garden Suburb in northwest London; Romerstadt near Frankfurt, Germany; Hirzbrunnen in Basle, Switzerland; and Milanino outside Milan, Italy.

The popularity of the Garden City concept led to its inclusion as an important element in the formative stages of American city planning. Sunnyside Gardens and Radburn, the planned communities sponsored by the Regional Planning Association of America (see p. 145), were among the first examples, and they were highly influential. It was an American sociologist-planner, however, who developed Howard's principles into the concept of the **neighborhood unit**—a concept that was to become one of the most widely adopted city planning devices throughout the world. Clarence Perry worked for the New York-based Russell Sage Foundation and lived in one of the foundation's model railroad suburbs, Forest Hills Gardens, which had been built in 1911. Perry was strongly influenced by the ideas of Jane Addams and the settlement house movement, seeing the neighborhood as the key setting in which to foster moral order and community spirit. Living in Forest Hills Gardens underlined for Perry the importance of physical design in neighborhood life. The result was a blueprint for urban design (Figure 17.6), a neighborhood unit bounded by roads carrying heavier flows of automobile traffic and centered on a local elementary school, local shops, and community institutions.

Patrick Geddes and Scientific Planning

In the early years of the twentieth century, when science was setting discovery after discovery before an increasingly appreciative world, the social sciences in general and the study of urbanization in particular were still in their infancy. In this context the voice of Patrick Geddes, professor of biology at the University of Dundee, Scotland, was persuasive. As a scientist, he commanded considerable respect, and he was able to speak out on social issues without being pigeonholed as a liberal do-gooder. He was an active campaigner for housing reform and had good connections in all the relevant societies. He was fascinated by cities and appalled by what he saw. He likened them to grease stains and to coral reefs, growing organically with little order or purpose. In this respect he was a product of his time, intuitively antiurban. More than anything, he believed that cities needed to be managed, just as a farmer might manage fields of crops or herds of animals. The bad had to be eradicated in order for the good to prosper. There had to be a plan

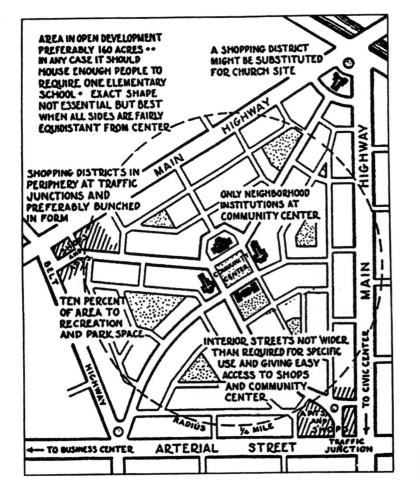

FIGURE 17.6   Clarence Perry's neighborhood unit concept.

for growth and development, which in turn implied an inventory of present resources.

It was this idea of an inventory, or survey, that was Geddes's first major contribution to his amateur interest. He had his own model for undertaking such surveys in his Outlook Tower in Edinburgh, which contained a camera obscura and a collection of photographs of urban life. Influenced by the writings of French sociologist Frederic Le Play, Geddes believed that the information gathered from urban surveys should clarify the availability of resources, the nature of human responses to the physical environment, and the nature of the local cultural matrix.

Following the writings of French geographer Vidal de la Blache, Geddes believed that this inventory should be undertaken in the context of a city's regional framework. This idea was his second major contribution to planning. The region, he argued in his 1915 book *Cities in Evolution*, had to be the basis for the reconstruction of economic, social, and political life. Cities and regions needed each other; they had to be planned and managed together, particularly, he noted, in view of the decentralizing forces of the new "neotechnic" technologies of electrical power and the internal combustion engine.

NORTH AMERICA: JACOB RIIS AND THE TENEMENT COMMISSIONS

In the United States the dangers of unplanned, uncontrolled urbanization were first highlighted in a book published by Jacob Riis in 1890. In his classic *How the Other Half Lives*, Riis graphically described the squalor and depravity of New York's tenement slums—*terra incognita* to many "respectable" families—while adding fuel to popular prejudices against cities in general and immigrants in particular: "beaten men from beaten races."[5] Though by no means the first example of a sense of crisis over the quality of life in American cities, it provides a convenient starting point—earlier crises had failed to elicit any decisive response. Throughout the nineteenth century there had been bursts of hand-wringing and pamphleteering by liberal reformers and do-gooders, but with relatively little effect either on mass opinion or on governmental practice. By the time Riis published his book, informed opinion had reached a critical point. It was clear that urban housing markets, left to themselves, were not going to achieve any kind of natural balance; nor did city governments seem to be able to address the problem of the slums, even where city bosses and political machines owed their position to the support of immigrants and low-income voters.

In response to the attention focused by Riis on crowded tenement life, two successive Tenement House Commissions were set up, the first in 1894 and the second in 1900. Both confirmed what people already knew: The slums were haunts of vice, the root of urban degeneracy, and a threat to moral order. The first commission's recommendations—for the introduction of by-laws that would prevent overbuilding—were quickly derailed by speculative developers.

The response of the second commission was a defining moment for American city policymaking and planning, even though it is not usually celebrated as such in the history of planning. In contrast to parallel commissions in Europe, the 1900 commission came down decisively against direct public intervention in the housing market, arguing that it would bring about a ponderous bureaucracy, intensified political patronage, and the discouragement of private capital, all merely to "better the living conditions of a favored few."[6] The best answer, the commission asserted, was tighter physical regulation of private developers through codified *building standards*.

But although building standards might prevent the embarrassment of "instant" slums, it still left the poor without access to decent, sanitary housing at an affordable rate. It was thus left to volunteers, philanthropists, and visionaries to address the fundamental problems of poverty, ignorance, and neighborhood decline and to propose the nature of change. As a result, the imprint of the upper middle classes came to be firmly stamped on the reforms of the Progressive Era—reforms that were to be woven into the fabric of twentieth-century urban policymaking and planning. In the first instance, this imprint could be seen in three interrelated movements: settlement houses, urban parks, and the City Beautiful movement.

PROGRESSIVE ERA REFORMS

The middle-class response to urban malaise was initially expressed through voluntarism and private philanthropy (Figure 17.7). By the end of the century, New York City alone boasted over 1,300 organized forms of philanthropic aid.[7] Some of these charitable institutions were based on ethnic lines and therefore exhibited a certain degree of spatial and functional coherence. But in general charity was fragmented, uncoordinated, and directed almost entirely toward individuals and families. At the same time, it was increasingly clear to those who actually worked among the poor that the slums were a collective problem, involving issues that extended to the sociogeographic contexts of neighborhood, school, and factory. This growing realization led to an influential movement—the settlement house movement—based on concepts of neighborhood and community as the focus for charity.

FIGURE 17.7   Urban policy as charity—for a long time the only option, and then only for the "deserving" poor.

Settlement Houses

The settlement house movement's exponents believed that by living among the people they hoped to serve they could develop a fellowship, learn about people's needs, and more effectively socialize the poor into middle-class standards of family life and educational achievement. The movement was modeled on a residence called Toynbee Hall, which has been set up in the slums of London, England, in 1884. One of the first settlement houses in the United States was the Neighborhood Guild (later renamed the University Settlement), established in a New York slum tenement by Stanton Coit in 1886; another well-known early example was Hull House (Figure 17.8), established by Jane Addams in Chicago.

The idea proved to be just what the eager volunteers—mainly women—were looking for. They organized continuing education for school dropouts, summer camps and neighborhood playgrounds to expose children to the virtues of nature, day nurseries for working women, educational programs to save "fallen" women, and clubs for the elderly. Volunteers also undertook social surveys in order to buttress their arguments for reform, and they campaigned for the prohibition of alcohol. In 1891 there were 6 settlement houses in the United States; by the turn of the century there were more than 100; and by 1910 there were over 400.

It was no coincidence that the social theorists of the time emphasized the dangers of moral disorder and **social disorganization**. Robert Park (the leader of the Chicago School of Human Ecology; see Chapter 12) endorsed the idea of neighborhood centers that would help to mold people's character and instill discipline and control in community life. "The purpose of social settlements," he pointed out, was nothing less than "to reconstruct city life."[8]

FIGURE 17.8   Hull House, a settlement house in Chicago opened by Jane Addams (1860–1935), a leader of the settlement house movement and co-winner of the 1931 Nobel Peace Prize.

Unfortunately, the settlement house volunteers' sense of love, sacrifice, and service was too often vulnerable to the corruption of condescension, paternalism, and evangelical zealotry. Novelist Sinclair Lewis saw them as "upholding a standard of tight-smiling prissiness."[9] Other charity workers accused them of gush and sentimentality and became worried that inexperienced charity workers were themselves becoming a threat to the emergence of an appropriate moral order.

In response, societies were set up in many cities to organize and coordinate charitable activities and to train relief workers to conform to strict sets of rules and record-keeping practices that were designed to make caregiving more effective. This training involved, among other things, learning to spot and weed out the "deserving" from the "undeserving" poor. This distinction, a legacy that has survived to the present day, permeated not only the ideology of professional planners and policymakers but also conventional social wisdom.

The Park Movement

Another Progressive Era legacy, much more tangible than our attitudes to charity and welfare, is the network of parks and public open spaces in cities. As we saw in Chapter 10, the park movement was rooted in the Arcadian Classicism inspired by pastoral ideals and the concept of nature as a spiritual wellspring. Early manifestations of these ideals were in the "rural" cemeteries of the early nineteenth century and in the mid-century creation of New York's Central Park under the direction of Frederick Law Olmsted, Sr.

The reform movement of the late nineteenth century drew on these same ideals, adopting the design motifs and intellectual underpinnings of Arcadian Classicism (see p. 248) in order to cultivate feelings of honesty, beauty, wholesomeness, cleanliness, and natural order among the laboring classes. Parks and public playgrounds (Figure 17.9) were one early step in this endeavor. Jane Addams opened one of the first real children's playgrounds in 1893 at her Chicago settlement house, and the example turned out to be the beginning of a trend. For the next 20 years, American cities were inscribed with all kinds of parks, playgrounds, and public open spaces, not just for children, but for everyone.

During this period, the rationale for park building quickly moved away from the relief they provided from the smells and diseases of the slums to the advantages they afforded in terms of recreation (for the masses), aesthetic pleasure (for the middle classes), and enhanced property values (for the upper-middle classes).

But for the reformers the decisive attribute of parks and public open spaces was their potential for spreading civilized values and social order within urban society (Figure 17.10). By bringing the laboring classes into contact not only with the spiritual energy of nature but also with the enlightened manners and comportment of other classes, parks could become a kind of universal moral force, a source of democratic and fraternal feelings: "It was believed that parks would breed a desire for beauty and order, spreading a benign, tranquilizing influence over their surroundings."[10]

Endorsed by leading intellectuals like Andrew Jackson Downing, Walt Whitman, and Herman Melville, the park movement rapidly gained momentum. By the turn of the century large parks had been established in Baltimore, Boston, Chicago, Cleveland, New York, Philadelphia, St. Louis, San Francisco, and Washington, D.C. New York had 11 parks of 100 acres

FIGURE 17.9
A settlement house playground in the early 1900s.

FIGURE 17.10   Public parks such as this one (Lincoln Park) in Los Angeles were a deliberate attempt to promote "civilized values" and social order.

FIGURE 17.11   Gramercy Park, New York, an exclusive vest-pocket park.

or more. Boston had initiated the first *planned* metropolitan system of parks: an "emerald necklace" of parks and parkway links. Smaller parks proliferated in cities everywhere: ornamental parks, zoological parks, parks for strolling, boating, lunching, skating and team sports, waterfront parks, downtown vest-pocket parks (Figure 17.11), and neighborhood parks.

The overall effect was significant for several reasons. The success of the park movement consolidated the ethos of *paternalism* in urban policymaking and planning. It established environmental determinism as a strong element in city planning. It brought about real change in the shape and pattern of daily life. And, last but not least, it established the infant art of city planning as a semiautonomous activity: Park boards and commissions were typically given responsibility for supervising the design of parks, financing them, building and maintaining them, and enacting regulations governing their use. These boards and commissions were the forerunners of city planning commissions that were appointed early in the twentieth century and of the **special districts** that compounded metropolitan fragmentation.

The City Beautiful Movement

The same logic that carried the park movement forward also nourished the City Beautiful movement that emerged at the end of the nineteenth century. The Beaux Arts style exploited by Daniel Burnham in the Chicago Exposition of 1893 (see p. 250) was right in line with reformers' desire to extend the civilizing influence of the park movement to the broader frame of reference provided by the city's built environment. The neoclassical vocabulary preferred by Burnham and other City Beautiful designers brought a strong element of *conservatism* to the whole enterprise. The symbols and motifs of Beaux Arts architecture and monuments not only suggested a link with the great European cities of the past but also helped to legitimize America's Anglo-Saxon ruling classes and institutions at a time of massive immigration and profound socioeconomic change.

At the same time, the broad boulevards and malls and radiating road networks that framed City Beautiful projects were welcomed by civic boosters as providing an orderly physical framework for economic development and by landowners whose property values escalated in anticipation of the implied redevelopment of large tracts of central city land.

The City Beautiful movement also spurred the development of *city planning commissions* and created an awareness of the need for technical experts to prepare plans. For the most part this task fell to architects, who brought with them the egotistical grandiosity of their profession. "Make no little plans," advised Daniel Burnham. His own plan for Chicago, published in 1909, was a magnificently illustrated volume that was truly metropolitan in scope, with proposals for a system of regional ring roads connected to downtown Chicago by a series of radial highways and **parkways**, for over 60,000 acres of monumental parkland, and for a series of parks, marinas, and developments along the lakeshore. It was a scheme for expansion, a framework for speculation, a design to embellish the city, to flatter its leaders, and to impress its inhabitants. *It did nothing, however, to address the fundamental problems of slum housing and social malaise.*

THE CITY PRACTICAL

It was at this point, early in the twentieth century, that liberal intellectuals in the United States began to take a broader view of urban problems, influenced in part by Patrick Geddes's ideas on scientific approaches to city planning. The first National Conference on City Planning was held in Washington, D.C., in 1909. Most of those attending the conference were government officials or representatives of charitable agencies; a few were architects, landscape architects, or engineers. What they had in common was a conviction that the process of city building was occurring piecemeal through speculative real estate deals without the guidance or benefit of any kind of public policy to protect or enhance the long-term interests of the community.

There was strong agreement that some form of government policymaking concerning urban development was necessary—preferably the sort that was proactive rather than reactive. There was, however, no consensus on the shape of such policies. As Richard Fogelsong has observed, "These planners were searching for a directive system, a method of decision making—a system they called 'planning' without knowing precisely what they meant by the term."[11]

Having slowly inched away from laissez-faire urbanization, struggled with housing reform, rushed ahead with parks, and flirted with monumentalism and neoclassical aesthetics, the urban policy and planning community now found itself a clear objective: the city as an efficient and disciplined spatial framework for both economic and social life, the "City Practical." This perspective required a shift of focus, away from the unwanted consequences of urbanization (poverty, slums, disorder) and toward the role of cities as generators of prosperity and enlightenment:

> A change in orientation began to occur that eventually would bring a reevaluation of the potency of environmental reform and the beginning of an idea that the

American City might be disciplined by the progressive development of human knowledge, state regulatory mechanisms, and public welfare provisions.[12]

Achieving those goals required, first of all, working toward a more economical and efficient system of land use and transportation and a carefully coordinated system of infrastructure provision. Though costly, this aim found widespread support among the business community. It was no coincidence that the notion of the City Practical coincided with the introduction of **Fordism** and **Taylorism**, and local business leaders were quick to appreciate the indirect benefits that this kind of planning would have on profitability and capital accumulation.

While city planning commissions busied themselves with implementing the City Practical, intellectuals with an interest in urban policy and planning began to address the implications of the regional economic reorganization and metropolitan decentralization that were beginning to reshape both the American urban system and the form of individual cities. It was in this context that Lewis Mumford came to prominence, articulating very effectively the ideas of Geddes and Howard and helping to draw attention to the need to address urban policy and planning at a larger scale: the alternative being "more and more of worse and worse." As we saw in Chapter 6, Mumford was instrumental in the formation of the Regional Planning Association of America (RPAA) in 1923. Although the direct achievements of the RPAA in the planned communities of Sunnyside Gardens and Radburn (see p. 145) were relatively modest, the indirect impact of the RPAA was profound, for the ideas promulgated by its members in the 1920s were to be a strong influence on the New Deal that gave policymaking and planning a much more central role in the economic and social life of the nation.

THE NEW DEAL

The New Deal, President Roosevelt's response to the **stagflation** crisis of the early 1930s, gave urban policy and planning its modern shape. The New Deal did *not* represent a coherent or integrated policy or planning framework. In the scramble for national economic recovery, programs were often ad hoc and sometimes contradictory. Nevertheless, the New Deal moved the nation toward a social democracy that sought to manage the national economy, defuse urban social unrest, and create an environment for coordinating production and consumption.

The New Deal era was a time of unprecedented experimentation in policymaking and planning. From the tentative conceptual, technical, and practical achievements of a few reformers and pioneering professionals, policymaking and planning became a central feature of the political economy of urbanization.[13] Indeed, policymaking and planning became integral to the entire political economy. The National Planning Board, established in 1933 (and renamed the National Resources Committee in 1935), brought the rationality of scientific, survey-based planning to everything from public works improvement to education, unemployment, health, old-age insurance, regional development, and technological assessment.

Urban policy and planning were specifically shaped by several elements of the New Deal package:

- The Civil Works Administration (CWA) and Public Works Administration (PWA) made grants to local jurisdictions for public works projects such as building highways, bridges, airports, and public buildings (Figure 17.12) and sewer and water projects that, it was hoped, would counter the Depression directly by creating jobs and indirectly by improving the infrastructure for economic development.

- The Works Progress Administration (WPA) directly employed workers through a network of local offices in order to achieve a more immediate effect on unemployment rates. Its efforts went mainly into constructing highways and streets, water and sewer systems, and recreational facilities.

- The Federal Emergency Relief Administration (FERA) made direct grants to states for unemployment relief and matching loans in order to stimulate public works projects, the majority of which were urban-related.

- The **Federal Housing Administration**, created under the National Housing Act of 1934,

succeeded the Home Owners Loan Corporation (see p. 146) as a means of providing federal mortgage insurance in order to prevent mortgage defaults and creating jobs by encouraging private residential construction.

- The Emergency Housing Division of the PWA, which became the U.S. Housing Authority through the U.S. Housing Act of 1937, undertook slum clearance and public housing construction in order to relieve the acute social distress of big-city slum neighborhoods.

- The **Resettlement Administration**, in its brief lifetime, was the most powerful and revolutionary planning authority ever created in the United States. Under the leadership of Rexford Guy Tugwell, RPAA ideas were vigorously pursued through a program of **greenbelt cities** that, it was hoped, would attract enough city dwellers to enable inner-city slums to be turned into parks. As we saw in Chapter 6, however, political opposition limited the scale of the program and brought a premature end to the Resettlement Administration itself.

Although the New Deal established strong governmental intervention as accepted rather than exceptional practice, public intervention was really rather piecemeal. The imperatives of economic management and social relief increasingly blurred the vision of the City Practical. The New Deal resulted in fragmentation

FIGURE 17.12
A road construction project in Manhattan funded by the Works Progress Administration (WPA).

and differentiation rather than a legacy of comprehensive planning. Planning and policy administration became separated from implementation, spatial policies were increasingly separated from social policies, and short-term policies were pursued without reference to long-term goals. It was by default, therefore, that policies initially drafted in relation to economic management and employment relief were to contribute to a Golden Age of urban policymaking and planning after World War II.

FORDIST ERA POLICY AND PLANNING

For roughly three decades after World War II, urban policymaking and planning were central elements of Western political economies. It was an era of technocratic policymaking and planning, whose practitioners enjoyed more public confidence than ever before or since in their ability to re-create cities as equitable and efficient settings for people, commerce, and industry. In Europe the emphasis was on postwar reconstruction and economic renewal. In the United States the emphasis was on economic growth. In both, urban policymaking and planning, urban governance, and urban development all presupposed one another in the rationalization of space required by Fordist systems of production.

EUROPE: PLANNING FOR RENEWAL

The context for professionalized planning in post-World War II Europe was heavily influenced by three sociopolitical movements with little in common except for a belief in the need for planning. One of these was dominated by paternalistic and idealistic liberal urban reformers who carried the heritage of Geddes and Howard. Another was dominated by conservative rural preservationists who were concerned about the encroachment of urbanization and industrialization on the limited amount of farmland and the associated heritage of picturesque landscapes. The third was dominated by the political representatives of industrial regions whose economic base had been severely damaged in the war and was now threatened by overseas competition.

One of the most effective advocates for professionalized planning was Patrick Abercrombie, who had affiliations with all three movements in the United Kingdom. Abercrombie's main concern was the unprecedented rate of urban sprawl. Because of the postwar economic recession, building materials and manual labor were cheap, and house prices were more affordable in relation to wages than at any time before or since. The consequent rate of suburban sprawl was unprecedented, with alarming losses of agricultural land, increased traffic congestion, and longer journeys to work. Highlighting these issues, Abercrombie advocated the idea of managing urban development. His strategy, embodied in his highly influential plan for London (Figure 17.13), involved three main elements:

severe restrictions on sprawl in a "Green Belt" surrounding the city, slum clearance and industrial regeneration in central areas, and the decentralization of industry and households to a series of New Towns located well beyond the Green Belt.

Abercrombie's plan for London involved decentralizing more than a million people to eight New Towns. It was implemented by the radical Labor government that came to power in 1947. Well suited to the technocratic Modernism of Le Corbusier (see Chapter 10), Abercrombie's ideas quickly became widely influential. In Scotland new towns were built at East Kilbride and Cumbernauld to accommodate decentralization from Glasgow. In France growth was redirected to five new towns near Paris and to Lille-East outside Lille. In the Netherlands, overspill population was channeled to new towns such as Zoetermeer near The Hague and Bijlmermeer outside Amsterdam. New towns were also built to decentralize population from Copenhagen in Denmark and Stockholm in Sweden.

In Germany overspill population was directed to new towns such as Märkische Viertel near Berlin, Perlach outside Munich, and Nordurestadt near Frankfurt. The new towns of Wulfen and Marl were built to accommodate planned population growth in an area of the northern Ruhr that was designated for mining industry expansion. In contrast, some new towns that were built to stimulate growth in declining or remote regions, such as Cwmbran in South Wales, Glenrothes in Scotland, and Aycliffe and Peterlee in northeast England, were not successful despite government subsidies for new residents and businesses.

THE UNITED STATES: PLANNING FOR GROWTH

After World War II there was a "closure" in the public policy dynamic in the United States that can best be understood as a product of a "class accord" between big business and organized labor, an accord that was regulated by the federal government.[14] It was in many ways the high point of Fordist capitalism. The foundation of the accord was the deal tacitly struck between employers and labor unions: higher pay and improved working conditions in return for management-driven improvements in productivity.

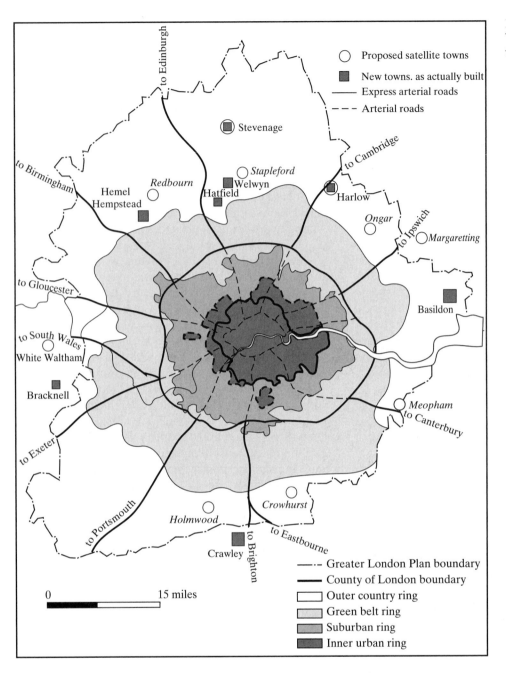

FIGURE 17.13
The Abercrombie Plan
for London.

This accord seemed quite natural in the context of the sustained economic boom of the postwar period. The policy objectives stemming from the accord were twofold: (1) to consolidate the economic growth on which the accord depended and (2) to redistribute the benefits of this growth at the margins, quelling (or, if possible, preempting) the social and spatial unevenness of the overall growth.

We have already encountered several of the principal outcomes of this approach. In Chapter 4 we noted the impact of the interstate highway system in facilitating the development of the urban system as the framework for an integrated national economy. In Chapter 6 we

saw how federal mortgage insurance carried over from the New Deal to underpin managed economic growth and development in the form of postwar suburban expansion. In Chapter 16 we saw how the **progrowth coalition** generated by New Deal politics carried over into the planned renewal of the fabric and economic base of central cities, how rapid economic development and increasing affluence induced massive increases in spending on **collective consumption** (education and other public services), and how increasing expectations and the backlash from persistent exclusion and inequality resulted in even greater increases in spending on welfare and income support.

In this section we round out the picture of Fordist urban policy and planning by summarizing the main policy initiatives of the period. First, however, we must acknowledge the role of the courts in establishing some of the principles on which the new policy dynamic was based.

THE COURTS AND URBAN POLICY IN THE UNITED STATES

The U.S. Supreme Court was a major factor in translating the social and political currents of the postwar period into landmark decisions that had important effects in framing urban policy and planning. These land-mark decisions were based largely on interpretations of the Fourteenth Amendment, which reads, in part:

> No state shall make or enforce any law which shall abridge the privileges and immunities of the Citizens of the United States; nor shall any State deprive any Person of life, liberty or property without due process of law; nor deny to any person within its jurisdiction the equal protection of the laws.

This constitutional provision directly affects cities, as legal agents of their state governments. The pressures of postwar economic, social, and metropolitan change, and in particular the racial conflicts of the 1960s, brought a stream of cases to the Court. Three issues were of particular importance in shaping the urban

---

BOX 17.2 PLANNING THE SOCIALIST CITY IN EASTERN EUROPE[15]

The cities in the countries of Eastern Europe that were part of the Soviet bloc following World War II—Poland, East Germany, Czechoslovakia, Hungary, Romania, Yugoslavia, Bulgaria, and Albania—developed differently from their West European counterparts partly because of historical legacy, but also because of their Communist totalitarian governments until around 1989.

These governments were guided by the basic tenets of Marxist-Leninist planning, namely eliminating excessive population concentration in large cities, achieving balanced urban and infrastructure systems, removing the "contradiction" between urban and rural standards of living, and creating a classless society. Matching these urban and social planning goals were economic directives that called for the rapid development of heavy industry and the collectivization of agriculture.

In order to reduce the size of large cities, governments in most East European countries placed restrictions on the growth of their capitals. In the majority of large cities, however, industrial expansion continued to draw workers from smaller towns or the countryside, thus swelling their populations.

With the aim of increasing overall industrial capacity, balancing the urban system, and providing urban functions for underserved areas, governments implemented a program of new town construction. These cities were developed around a large industrial facility, typically an iron and steel mill or chemical processing plant, were consciously sited away from existing towns, and were located either on a vacant piece of land or on the site of an existing village. Whereas residential capacity had been the most important element that influenced the planning and design of the new towns in Western Europe, accommodating industrial capacity and employment was the major issue for the authoritarian planners of the new socialist towns.

New towns in Eastern Europe included Eisenhüttenstadt in East Germany, Nova-Huta in Poland, and Dimitrovgrad in Bulgaria. These places grew rapidly between 1950 and 1970, eventually becoming important not only for the industrial products that they produced, but also for their role as urban centers in the **hinterlands** that they served and in creating a more balanced national urban system for their countries.

By the 1970s and 1980s planners in Eastern Europe had turned their attention from promoting large-scale industry and new towns to developing light industries and filling out the national urban systems. In many countries central place theory became either an explicit or implicit guide to development efforts as planners tried to create multitiered urban hierarchies that provided goods and services to particular regions according to their size and function. This kind of national planning and its outcome, however, varied among the different countries. Although most countries built new industrial towns, this form of planning was especially well developed in Hungary, where the National Settlement Development Strategy established the blueprint for developing the settlement network.

Since the early 1990s, East European cities have been affected by the national reform strategies that have involved the *privatization* of state-owned housing and industry and promoting new industries and companies. National policy continues to evolve to address urban problems such as the unemployment and homelessness caused by economic restructuring to capitalism.

policy dynamic of the period: school desegregation, open housing, and voting rights.

School Desegregation

In relation to school desegregation, the landmark decision was the 1954 case of *Brown v. Board of Education* in Topeka, Kans., in which the Supreme Court ruled segregated schools to be unconstitutional. This decision left a certain ambiguity, however, in that the ruling merely outlawed the use of government power or funds resulting in segregation; it did not require integration. This ambiguity was clarified in the 1968 case of *Green v. County School Board of New Kent County*, in which the Court ruled that school districts must achieve as much racial mixing as possible and that districts that had not been integrated because of past practices must take affirmative action to achieve balanced racial enrollments. This ruling, in turn, required judicial decision makers to confront the realities of American urban geography. The result was the adoption of school busing plans to secure balanced enrollments (*Swann v. Charlotte Mecklenberg Board of Education*, 1971). For the most part, however, the Court has shied away from decisions that would ensure metropolitan-wide desegregation.

Restrictive Covenants

The fundamental housing policy case was *Shelley v. Kraemer*, in which the Supreme Court ruled in 1948 that racially **restrictive covenants** represented a violation of the Fourteenth Amendment. The Court broadened this decision in 1968 in the case of *Jones v. Mayer* to ensure the freedom for people of any race "to buy whatever a white can buy, live wherever a white can live."

The Court also found itself confronting issues raised by public housing and **exclusionary zoning**, though its decisions in this area have been much less rigid in their effect on the practice of local policy and planning. In *NAACP v. Mt. Laurel* in 1972, the Court struck down an exclusionary zoning ordinance, arguing that every municipality has an obligation to provide a reasonable share of the areawide need for low-income housing. In the 1977 case of *Metropolitan Housing Development Corp. v. Arlington Heights* in Chicago, the Court decided that exclusionary zoning with discriminatory effects was not, *a priori*, illegal; Only where discriminatory intent could be shown were such ordinances to be outlawed. This decision, of course, left huge scope for interpretation and effectively allowed a continuation of the metropolitan social polarization discussed in Chapter 16.

Civil Rights

We have already noted the role of the Supreme Court in initiating the "reapportionment revolution" of the 1960s (p. 473). The landmark reapportionment case was *Baker v. Carr* in 1962. After the passage of civil rights legislation in the form of the 1965 Federal Voting Rights Act, the Court confirmed the principle of equality in the weight of each person's vote in balloting in the specific case of racial differences (*Allen v. State Board of Elections*, 1969). Another aspect of electoral geography was addressed in 1973, when in the case of *White v. Regester* the Court ruled that multimember districts (an issue that we encountered in Chapter 16) violated the Fourteenth Amendment. In 1986 the Court strengthened minority voting rights by interpreting vote dilution in a broader context than the mere geometry and arithmetic of electoral subdivisions, also taking into account the past record of local voting patterns and minority success or failure in gaining representation.

FEDERAL POLICY INITIATIVES

The **Fordist** era saw a sharp increase in government expenditures on social objectives, as embodied in the United States, for example, in the New Frontier of the Kennedy administration and the War on Poverty and Great Society of the Johnson administration (see Chapter 16). In practice, most of these federal expenditures took the form of grants-in-aid to state and local governments. During the 1960s and the 1970s, party political differences over domestic policy and planning were largely confined to disagreements over the distribution of these grants and the conditions attached to them: Democrats sought to perpetuate the progrowth accord between big business and organized labor by channeling grants-in-aid toward central cities and blue-collar communities, while Republicans sought to channel them more toward rural and suburban areas and to Sunbelt cities. Among the many grant programs and policy initiatives that addressed urban issues were the following:

- The Housing Act of 1949, which was the baseline legislation for assistance to cities in clearing blighted areas, assembling land for redevelopment, and building public housing.
- The Housing Act of 1959, which, among other things, provided for federal support for the preparation of comprehensive plans at the metropolitan level.
- The Federal Aid Highway Act of 1962, which mandated urban transportation planning as a condition for receiving federal funds in urban areas. As a result, it provided a major stimulus to local land use and transportation planning.
- The Economic Opportunity Act of 1964, which created the Office of Economic Opportunity (OEO) as a vehicle for neighborhood-based economic development in central cities. Only 25 percent of OEO funding went to public

agencies, the rest being channeled directly to neighborhood groups and private organizations (universities, churches, civil rights groups, settlement houses, family services agencies, United Way organizers) engaged in projects that were designed to assist the casualties of economic and metropolitan restructuring.

- Project Head Start, initiated in 1965 in an attempt to break the inner-city **cycle of poverty** (Figure 15.7) through funding educational programs designed to improve cognitive development and social, emotional, and physical well-being among younger low-income children.

- The Demonstration Cities and Metropolitan Development Act of 1966, an attempt to facilitate the concentration and coordination of federal, state, and local public and private efforts to improve the quality of urban life and combat deprivation and disadvantage. As such, it was very much a product of the intellectual climate of the time, an attack on "place poverty" rather than "people poverty" through "area-based positive discrimination." In this respect, the Model Cities program was strongly influenced by the concepts of spirals of *neighborhood decay* (Figure 15.5) and localized *cycles of poverty* (Figure 15.7).

- The National Environmental Policy Act of 1969 and the Environmental Quality Improvement Act of 1970 required federal agencies to adopt a systematic, interdisciplinary approach to planning and policymaking affecting the environment. It also required the preparation of an Environmental Impact Statement (EIS) containing information on the likely environment impacts of proposed projects. The Clean Air Act of 1970 created the Environmental Protection Agency (EPA) and empowered it to set ambient air quality standards.

- The Urban Mass Transportation Assistance Act of 1970 provided the first long-term federal commitment to financing urban mass transit, and the 1970 Federal Aid Highway Act significantly increased the influence of local jurisdictions in highway planning. The Urban Mass Transportation Assistance Act of 1974 authorized for the first time the use of federal funds for transit operating costs.

There is no doubt that these federal policies significantly affected the trajectory of urbanization. Although they did not always have the effects that were planned, their very existence inevitably changed both the pattern of development and the social, political, and economic context in which subsequent change took place. They were, in other words, yet another manifestation of the **sociospatial dialectic**.

This is not to say, however, that any dimension of Fordist-era urbanization could be seen as having been successfully brought under the spell of rational, planned change or as having been successfully managed because of the policymaking apparatus. Apart from the sheer complexity of the task, the structure of national and local governance works against the chances of success, especially in the United States, where the pork barrel effect tends to dilute and distort the intended effects of policy, and the grant-in-aid format makes innovative policies and plans unlikely to succeed because of the number of layers of government and the variety of outside participants involved.

EVANGELICAL BUREAUCRATS

But before these shortcomings came to be appreciated, there had been an unprecedented growth in the number of urban planners, in their collective power as a profession, and in their self-confidence concerning the possibility of delivering better, safer, nicer, and more-efficient cities. Fordist-era government programs for transportation planning, environmental quality, housing, **urban renewal**, **land use zoning**, and comprehensive planning projects all provided jobs for planners and enhanced the profession's visibility and growth. Planners became indispensable links between various kinds of projects and various layers of government.

Higher education shifted into gear: The number of planning programs in the United States increased from about 20 in the mid-1950s to around 90 by the end of the 1970s, with more than 20 programs offering doctoral degrees in planning. The output of formally qualified planners increased over the same period from about 100 per year to over 1,500 per year. At the top of the profession, a small but powerful and influential group of can-do, will-do planners and renewal executives emerged: Robert Moses, whose power base in New York had already been established by World War II; Edward J. Logue, who worked for Richard Lee's administration in New Haven, Conn.; Edmund N. Bacon (Figure 17.14) in Philadelphia; Dave Loeks in Minneapolis–St. Paul; and William Ryan Drew in Milwaukee. These were the men who seemed chosen to bring the bold and inspirational ideas of Howard, Geddes, Corbusier, Wright, and Stein to fruition in a golden age of technocratic planning on a truly heroic scale.

There was an evangelical spirit to the entire profession. Cities *should* be better places. They *could* be. Planners shared a professional orientation that was a product of both the long, slow beginnings and the heady, explosive growth of the postwar era. Thus they carried with them, through self-selection and through formal education, strong elements of liberal idealism, of utopianism, of environmental determinism and design determinism, a penchant for sweeping, futuristic solutions, and a concern for the aesthetics of the urban environment.

FIGURE 17.14   Edmund Bacon, Director of Planning for Philadelphia in the 1960s.

In addition, the realities of the world they trained for and in which they practiced forced them to be increasingly specialized and technically proficient. Different spheres of activity required specialized knowledge, skills, and techniques. Mirroring the latest developments in research and theory, planners came to acquire the languages and toolkits of regional economics, regional science, quantitative geography, systems analysis, and transportation modeling.

The combination produced a professional make-up that proved tragically unsuited to the ideals that they espoused. Even in their finest hour, they were forced to watch themselves fail. Their evangelism and environmental determinism led them to get bogged down in "bureaucratic offensives" like urban renewal and highway construction, to the point where the communities whose lives they had hoped to improve were angry and afraid. Before they knew it, their rationality and their predilection for efficient and tidy land use patterns had led them to become social gatekeepers. In their rush to grasp the opportunities of an expanding but highly specific labor market, they set aside their synthetic, progressive perspective in favor of a fragmented, formalistic, and reductionist view of urbanization drawn along functional and programmatic lines. There were many, of course, who prospered professionally under these circumstances, enjoying the unprecedented status and power with the complacent vanity of the evangelist. It was to prove short-lived.

NEO-FORDIST POLICY AND PLANNING

As we have argued in earlier chapters, the **Neo-Fordist** era was firmly established after 1973. By 1978 the Carter administration had abandoned the idea of an urban policy framework for the United States. The first *National Urban Policy Report*, published in 1978, promoted the disengagement of the federal government from urban affairs. President Carter himself, in his State of the Union address that year, reflected the new attitude very clearly. "Government," he said, "cannot solve our problems. It can't set our goals. It cannot define our vision. Government cannot eliminate poverty, or provide a bountiful economy, or reduce inflation; *or save our cities.*"[16]

It remained only for the apparatus of urban policy and planning to be dismantled. The Reagan administration enthusiastically undertook this job. The objective was now to free private enterprise as much as possible from the constraints of government at every level. Cities were no longer seen as the appropriate scale at which to focus initiatives to promote economic growth.

In the new, flexible geometry of an "informational" mode of development (see p. 86), the whole national space had to be deregulated and disengaged from government dependence. A major government report, *Urban America in the Eighties*, spelled out the logic. Free-enterprise markets are the best means of allocating urban land uses and making investment decisions that maximize the benefit for everyone. Government aid to distressed cities and neighborhoods, the report argued, only hampered the long-term efficiency of free-enterprise markets.[17]

NEOLIBERAL APPROACHES TO POLICY AND PLANNING

The neoliberal ideologies that have dominated the political economy of Western countries since the mid-1970s are predicated on a minimalist role for the state, assuming the desirability of free markets as the ideal condition not only for economic organization, but also for political and social life. Neoliberal ideology holds that free trade and investment across global markets allow more and more people to share in the prosperity of a growing world economy. Economic and political interdependence among cities, regions, and nations, it is argued, creates shared interests that help prevent conflict and foster support for common values. Democracy and human rights, it is asserted, will spread to billions of people in the wake of neoliberal policies that promote open markets and free trade.

Neoliberalism views policies designed to redistribute resources to disadvantaged areas or individuals as necessitating excessive taxation of the wealthy, thereby reducing investment capital and undermining productivity. If necessary, national social goals and standards have to be sacrificed to ensure that business has the maximum latitude for profitability. The rising tide of economic development would then float all boats, urban and rural, central city and suburban. If the tide does not happen to rise high enough in any given location, people can "vote with their feet" and move to more prosperous cities or regions of the country.[18]

This kind of thinking has led to a dramatic cutback in government aid and a corresponding increase in deregulation and **privatization** (see Chapter 4 and Chapter 16). Reductions in federal outlays for urban and regional development programs between 1981 and 1988 amounted to nearly 60 percent. Studies of the urban impacts of these reductions have established, to nobody's surprise, that the outcome was an increase in social polarization and fiscal stress, with the most adverse effects being felt in larger, older central cities. Indeed, the increased inequality soon reached the point where it was described in terms of a "new class war."[19]

Meanwhile, federal deregulation, in combination with the burgeoning of **civic entrepreneurialism** (Chapter 16), led to a dramatic resurgence in the scale and intensity of *corruption* in urban affairs. Under the Reagan administration the U.S. Department of Housing and Urban Development is thought to have lost over $4 billion in waste, fraud, and influence peddling. The deregulation of the savings and loan industry, also under the Reagan administration, allowed new investors to use federally insured savings deposits to fund thousands of speculative projects, including office buildings, shopping malls, mixed-use developments, and resort complexes, as well as junk bonds and risky land deals.

The Property Rights Movement

The free-enterprise, neoliberal climate engendered by the Reagan administration encouraged landowners and developers to challenge the fundamental power of governments to protect public health, safety, and welfare through planning regulations and public works that are in the public interest. In the United States this power was always constrained by the Fifth and Fourteenth Amendments to the Constitution, which bar public authorities from "taking" private property for public use without just compensation.

Until the neo-Fordist era, the courts had tended to support the public interest across a broad range of community issues, especially when they concerned health, safety, and welfare issues and were based on technical analysis and a publicly adopted comprehensive plan. Beginning in the 1980s, these rulings have been challenged with increasing vigor by a property rights movement that has also pressed for legislation that would provide compensation for landowners for *any* loss of value resulting from *any* government actions that restrict or curtail the use of their property. Through the 1980s and 1990s the Supreme Court decided several cases in favor of property owners against state agencies (*First English Evangelical Lutheran Church v. County of Los Angeles*, 1987; *Nolan v. California Coastal Commission*, 1987; *Lucas v. South Carolina Coastal Council*, 1992; *Dolan v. City of Tigard*, 1994), although development moratoria were validated in the case of *Tahoe Sierra Preservation Council v. Tahoe Regional Planning Agency* in 2002. Meanwhile, the property rights movement continues to sponsor the introduction in state legislatures of "takings" bills that would, if passed, exert an enormously inhibiting effect upon regulatory agencies at all levels of government.

PLANNING AS DEALMAKING

The **neoliberal ideologies** that have flourished during the neo-Fordist era have undermined the professional identity and credibility of urban policymaking and planning. As a result, they have become fragmented, pragmatically tuned to economic and political constraints, and oriented to stability rather than being committed to change through comprehensive plans. Practice has become estranged from theory and divorced from any broad sense of the public interest. It has become increasingly geared to the needs of producers and the wants of consumers and less concerned with overarching notions of rationality or criteria of public good.

Unfortunately, this loss of faith has occurred at a time when urban restructuring portends a particularly urgent need for some form of overall policy framework. Cities, as we have seen in earlier chapters, are having to accommodate a wholly new mixture of industry and employment within their splintering form. Economic transformation and social polarization would seem to call for a comprehensive approach to managed change. Yet the principal aim of policy and planning is now job creation rather than producing the city practical, the city beautiful, or, indeed, any kind of *city* at all.

The city and the metropolis have faded from the discourse of policymaking and planning as practitioners have fought for professional survival in the hostile atmosphere of the **new conservatism**. Specifically, as city governance has become entrepreneurial, city policymakers and planners have been drawn increasingly into the business of **public-private partnerships** (see p. 284). Planning has become entrepreneurial and planners have become dealmakers rather than regulators.

One policy development that has carried this dealmaking ethos forward is the idea of **enterprise zones**. Based on the work of geographer Peter Hall and first advanced in Britain, this idea sought to designate special areas in cities that would combine as many concessions as possible—tax breaks, subsidized factory space, and so on—with a relaxation of government controls and regulations. Inspired by the laissez-faire success of Hong Kong in the postwar world economy, it was an attempt to foster small businesses in inner-city settings. There is, however, little real evidence that enterprise zones have been effective in terms of job creation and no conclusive evidence that they have been effective in terms of inner-city regeneration.[20]

Mixed-Use Developments and Cluster Zoning

Perhaps the single most important aspect of recent changes in planning practice has been the way that planning has accommodated developers' need for flexibility through new approaches to land use zoning. Once the cornerstone of city planning practice, single-purpose "Euclidian" zoning (based on the landmark case *Euclid v. Ambler*; see p. 143) is now seen as wasteful, monotonous, and overly rigid. Planners and planning committees now see mixed-use zoning, which is critical for developers wanting to produce large set-piece projects like the Westlake Center in Seattle (Figure 17.15), as a means of achieving everything from enhancing a city's tax base to initiating urban revitalization and increasing ridership on public transit systems. In downtown areas mixed-use zoning is being combined with incentive deals in which developers are granted additional height or density allowances in exchange for specified building features (e.g., elaborate facades or building tops) or facilities such as daycare centers, residential space, and space for services that might help restore variety and vitality to downtown districts.

In suburban jurisdictions the solution to the rigidity of Euclidian zoning has been cluster zoning, whereby regulations are applied to an entire parcel of land (a Planned Unit Development, or PUD) rather than to individual building lots. With a PUD, developers can calculate densities and profits on a projectwide basis, allowing the clustering of buildings to make room for open spaces (such as golf courses) or to preserve attractive site features (such as ponds or old barns). Cluster zoning facilitates a blend of residential and nonresidential elements and

FIGURE 17.15   The Westlake Center in Seattle, a mixed-use development (MXD) that incorporates retail, office, hotel, parking, and cultural amenities in a carefully planned package where each element is designed to complement the others.

a mixture of housing types that can be adjusted as sales dictate. For developers, PUDs offer **economies of scale** plus **economies of scope** for product diversity and flexibility, all within a predictable regulatory framework—the basis for the packaged landscapes that we described in Chapter 10. For planners, cluster zoning offers the prospect of high tax-yield development with services and amenities provided at no cost to the taxpayer.

SPLINTERING URBANISM AND THE ARTFUL FRAGMENT: POSTMODERN PLANNING

The rise of neoliberalism and the diminished standing of the policy and planning professions has meant that the **splintering urbanism** of the polycentric metropolis (see p. 158) has been accompanied by a disorganized approach that has led to a collage of highly differentiated urban spaces and settings.

BOX 17.3   COMPETITIVE REGIONALISM[21]

Civic entrepreneurialism is commonly found in cities in the United States. Individual localities compete for private investment for goals such as jobs and tax base. Yet this competition has drawbacks, including the enormous resources diverted into incentives; instances of fewer gains for an urban economy than projected; the inequitable distribution of the costs and benefits within cities; and poorer cities paying more, with the costs outweighing the benefits for some. Because only one city can win, this competition is a zero or negative sum at a regional or national scale if it merely relocates investment between places at public expense. If public funds are diverted from education, technology, and the transportation and communications infrastructures, cities become less competitive nationally and internationally.

Aware of these drawbacks, government officials agree that eliminating wasteful competition is rational. Yet, acting individually, each local government can feel forced to offer incentives to companies because there is no certainty that others will not. Theories of cooperation (e.g., iterated game theory and the prisoners' dilemma game) provide insights into this collective inaction problem. In a simple two-person game, each player chooses to cooperate or defect. The best outcome for both is if they both cooperate. The worst outcome is if they both defect. Yet each opts to defect because neither can guarantee that the other will cooperate.

Regional cooperation—competitive regionalism—may offer an alternative to this prisoner's dilemma at a metropolitan scale. The argument is that

> America's economy should now be seen as a *common market of metropolitan-based local economic regions*. These regions are indeed strongly interdependent, but they also compete with each other and with the rest of the world . . . the new leadership coalitions and networks recognize that the geographic focus of their efforts has to be the metropolis as a whole, not just the central city or suburbs independently.[22]

Competitive regionalism has the potential to better market an area for investment and reduce wasteful competition, help a metropolitan region to mobilize its strengths and address socioeconomic divisions that can weaken its chances of success, and find a profitable niche in the international economy.

> Around the country, there is a dawning awareness that regional approaches to development—and particularly, cooperation between central cities and their suburbs—may be vital to economic survival . . . .

> This is not . . . the same thing as the movement to regionalize service delivery, or yet another round of drum-beating for metropolitan government. Rather, cities and their suburbs are trying to sort out ways in which a variety of public and private entities can work on behalf of the economic development of an entire region, instead of in competition with one another.[23]

An example from the United States is the Metro Denver Network, a public-private partnership that promotes economic development while reducing wasteful competition for private-sector investment among the localities in the six-county Denver metropolitan area. An example from Europe is Communauté Urbaine de Lille (CUdL) in France—a network of 86 municipalities with joint regional decision making for metropolitan Lille. CUdL uses the funds from tax-base sharing to promote economic development and new investment in the region. Mechanisms to discourage wasteful competition include the "25% measure"—where a locality that independently attracts a company from elsewhere must transfer 25 percent of the taxes from the new development to the CUdL.

The advantages for individual localities include channeling the public funds not provided as incentives into more productive activities, such as improving labor force skills. While providing fewer unnecessary incentives benefits individual localities, reducing company relocations that occur purely as a result of public subsidies benefits the national economy by minimizing this unproductive use of public and company funds.

This solution, however, may merely shift the scale from a contest between localities to one between regions. Unless regional **competitive advantage** is promoted through enhancing conditions for business, and economic specialization occurs between regions, competing as a region for mobile investment eliminates competition only between the jurisdictions *within* particular regions—the battle for investment would continue *between* regions. As in the private marketplace, competition and cooperation are not mutually exclusive—cooperation through strategic alliance is still competition.[24] Jonas[25] found this in Southern California where the "beggar-thy-city-neighbor" competition of the 1980s became "beggar-thy-region-neighbor" in the 1990s. The drawbacks of competition at a local scale can be replicated at regional scales. Intensified competition between regions will not necessarily produce new investment or a more productive economy at a national scale.

Fragmented elements of the city are planned or redeveloped as autonomous elements, with little relationship to the whole and with direct concern only for adjacent elements. Dealmaking between planners and developers tends to pay attention to the artful fragment but pays scant attention to the city as a whole.

We have seen the imprint of these artful fragments in the form of upscale malls, galleries, festival settings, and mixed-use developments (see pp. 482–488). This phenomenon has been described as the "Rousification" of cities, after private-sector developer-planner James Rouse, whose successes in Baltimore's Inner Harbor, New York's South Street Seaport (Figure 16.12), Boston's Faneuil Hall Marketplace, New Orleans's Riverwalk, Jacksonville's Jacksonville Landing, and Miami's Bayside have served as the models for countless other ventures. Like architecture, policymaking and planning have thus experienced a rupture with their past. The rationalism, idealism, utopianism, functional separation, determinism, and technological experimentation of Modernity have been eclipsed by the populist, historicist, scenographic, and eclectic impulses of postmodernity.

Place Marketing

Globalization has prompted communities in many parts of the world to become much more conscious of the ways in which tourists, businesses, media firms, and consumers perceive them. As a result, places are increasingly being reinterpreted, reimagined, designed, packaged, and marketed. Sense of place can become a valuable commodity through place marketing. Seeking to be competitive in the global economy, many places have sponsored extensive makeovers of themselves, including the creation of pedestrian plazas, cosmopolitan cultural facilities, festivals, and sports and media events. An increasing number of places have also set up websites containing maps, information, photographs, guides, and virtual tours in order to promote themselves in the global marketplace for tourism and commerce.

Place marketing depends heavily on promoting traditions, lifestyles, and arts that are locally rooted. The re-creation and refurbishment of historic districts and settings are so widespread that they have become a mainstay of a "heritage industry." This industry, based on the commercial exploitation of the histories of peoples and places, is now worldwide, as evidenced by the involvement of the United Nations Economic, Social, and Cultural Organization (UNESCO) in identifying places for inclusion on World Heritage Lists. One important consequence of the heritage industry is that urban cultural environments are vulnerable to a debasing and trivializing Disneyfication process. As the United Nations Centre for Human Settlements (UNCHS) *Global Report on Human Settlements 2001* noted:

> The particular historic character of a city often gets submerged in the direct and overt quest for an international image and international business. . . . Local identity becomes an ornament, a public relations artifact designed to aid marketing. Authenticity is paid for, encapsulated, mummified, located and displayed to attract tourists rather than to shelter continuities of tradition or the lives of its historic creators.[26]

As a result, contemporary landscapes contain increasing numbers of inauthentic settings that are often influenced more by images and symbols derived from movies, advertising, and popular culture than by cultural heritage.

PLANNING FOR HEALTHY AND LIVABLE CITIES

Not surprisingly, there have been some important counter-movements to laissez-faire neoliberalism and the consequent attenuation of Modernist-Fordist planning. Broadly speaking, liberal ideals now find expression in a variety of movements and initiatives that are concerned with planning for healthy and livable cities. One example is the Healthy Cities Project, launched in Europe in 1987 by the European Office of the World Health Organization (WHO). The goal of the WHO European Healthy Cities network is to promote a good overall quality of life (in terms of people's social and economic conditions) as well as physical environments that promote good health. The core strategies of Healthy Cities are set out in Health21 and Local Agenda 21 (WHO's main policy frameworks), the Athens Declaration of Healthy Cities, and the Aalborg Charter.

These documents aim to engage local governments across Europe in the creation of healthy urban settings through a variety of processes of political commitment, institutional change, and innovative action. Their focus is on addressing environmental justice, the social dimensions of sustainability, community empowerment, and urban planning.

By 2002, 49 European cities were designated as WHO Healthy Cities, while more than 1,300 other cities were linked through national, regional, metropolitan, and thematic healthy cities networks. Cities participating in these networks have developed and implemented a wide range of initiatives, including urban environmental health profiles and city health plans, community development programs, and programs that address the needs of vulnerable groups.

SUSTAINABILITY AND GREEN URBANISM

The Healthy Cities network is part of a broader movement concerned with people's urban environments. Planning for **sustainable urban development** and **green urbanism** involves facilitating major lifestyle changes (such as walking, bicycling, and reductions in material consumption), the preservation and restoration of the natural environment, and the application of new technologies (such as public transit, district heating, and green building and design).

As with the healthy cities idea, most of the applications and innovative examples of sustainability and green urbanism are found in Europe, especially in northwestern Europe. The European Union has strongly endorsed sustainability through its legislation and directives—in particular its Fifth Environmental Action Program and the Treaty of Amsterdam (1997). Most western European countries have prepared national sustainability strategies, and many local governments have implemented a variety of environmentally sensitive policies and strategies. Among these are the pedestrianization of central city precincts; restrictions on automobile access; traffic "calming" (slowing) by modifying street patterns; land use planning for compact urban development; urban greenways and forests; bicycle paths and bicycle expressways with priority access to schools, shopping areas, and green space; ecovillages; cohousing (typically, housing clustered around a pedestrian-only common street or courtyard, with a common house where residents can meet, eat together, or organize collective activities such as child care); and integrated rail, tram, metro, and bus systems.

In Italy a *Slow City* (CittaSlow) movement has emerged, seeking to promote urban livability. It is a clear reaction to big business and globalization, though its driving motivation is not so much political as ecological and humanistic. Although in its early stages, the movement has a charter and a formal process for membership. The charter covers many aspects of urban design and planning. Candidate cities must be committed not only to supporting traditional local arts and crafts but also to supporting modern industries whose products lend distinctiveness and identity to the region. They must also be committed to the conservation of the distinctive character of their built environment and must pledge to plant trees, create more green space, increase cycle paths and pedestrianized streets, keep piazzas free of advertising billboards and neon, ban car alarms, reduce noise pollution, light pollution, and air pollution, foster the use of alternative sources of energy, improve public transport, promote eco-friendly architecture in any new developments, and introduce a range of measures from the promotion of organic agriculture to the creation of centers where visitors can sample local traditional food. They must also take steps to protect the sources and purity of the raw ingredients and to fend off the advance of fast food and cultural standardization.

Most American towns and cities have made no such ambitious attempts at sustainable planning or green urbanism. Portland, Ore., is a notable exception to the general trend of sprawling, auto-dependent urban development that far exceeds the rate of population growth. New Urbanism embodies sustainable principles of walkable and transit-oriented communities, but in practice has become a recipe for packaged, exclusionary fragments of suburbia (see Chapter 10). American towns and cities consequently have high carbon dioxide emissions, draw in excessively large amounts of energy and resources, produce large amounts of waste, destroy large amounts of sensitive habitat and productive farmland, generate high levels of traffic congestion, and impose high costs of infrastructure provision.

SMART GROWTH

In the United States livability issues have been central to an intense debate about the quality, pace, and shape of metropolitan growth. Among the issues raised by the emergence of polycentric metropolises (see Chapter 6) is the failure of urban transportation systems to serve **edge cities**. James Vance, Jr. has suggested that the consequent "pathological" gridlock of automobile traffic has fostered an accelerated flight of households to **asylum suburbs** in the countryside, exurbs, private master-planned communities, and small towns in the penumbra of metropolitan development. As more and more households have sought refuge from the congestion of the metropolis, "A new stage has emerged, an asylum pattern under which those who can move out of the metropolis completely have done so."[27]

Central to the debate is the idea of smart growth. Smart growth is a package of suburban land use planning principles designed to curb sprawl and the proliferation of asylum suburbs. Its advocates claim that growth restrictions raise the quality of life, increase the efficiency of urban infrastructure, and protect the environment. Among the key principles of smart growth are the following:

- Preserving large areas of open space and protecting the quality of the environment by setting aside large fringe areas where development is prohibited.
- Redeveloping inner suburbs and infill sites with new and renovated structures to make them more attractive to middle-income and upper-income households.

- Reducing dependency on private automotive vehicles—especially one-person cars—by requiring higher-density development, clustering high density around transit stops, raising gasoline taxes, and increasing public investment in light-rail transit systems.

- Encouraging innovative urban design and land use zoning regulations that create pedestrian-friendly communities, mixed land uses, and commercial centers located at transit stops.

- Creating a greater sense of community within individual localities and a greater recognition of regional interdependence and solidarity.

Smart growth advocates include anti-growth or slow-growth and environmental organizations, together with central city and inner suburban leaders and interest groups, such as mayors, downtown business groups, and community-based organizations. Opposed to smart growth principles are most developers, homebuilders, major landowners, and chambers of commerce. Their objections to smart growth are partly based on the contention that smart growth costs more, and partly on an ideological objection to planning controls that inhibit opportunities for profit through real estate development.

METROPOLITAN GOVERNANCE AND PLANNING

As we saw in Chapter 16, the political fragmentation of metropolitan areas that has been associated with **suburban incorporation** in the United States has

---

Box 17.4 Metropolitan Governance in the Twin Cities

In the mid-1960s the Twin Cities (Minneapolis–St. Paul) metropolitan area faced a host of serious urban challenges: inadequately treated sewage was being released into public waters and contaminating groundwater supplies; the privately owned regional bus system was rapidly deteriorating due to rising fares, declining ridership, and an aging fleet; urban sprawl was threatening recreational and agricultural activities; and growing fiscal disparities among communities was making it difficult for some localities to provide essential public services. With some 272 separate local governments—including seven counties and 188 cities and townships—the metropolitan area was ill equipped to deal with problems that transcended local boundaries.[28]

The Minnesota legislature responded in 1967 by creating the Twin Cities Metropolitan Council to plan and coordinate development in the seven-county metropolitan area. The legislature also created regional wastewater and transit systems and enacted tax-base sharing legislation to reduce fiscal disparities among communities.

The Metropolitan Council's *Regional Development Framework* is a *development management system* (as opposed to a *growth management system*, because it plans for both growth and decline). The goals of this framework include:

- Preserving the urban core.

- Phasing new development outside the urban core based on population, households, and employment forecasts, and consistent with the region's capacity to provide services.

- Expanding regional facilities—such as water treatment plants and airports—on a planned development schedule.

- Protecting some outlying land for agricultural, rural, and recreational uses.

The Metropolitan Council instituted a metropolitan urban service area (MUSA) within which development and redevelopment are encouraged (Figure 17.16). Improvements in the regional system of sewers, transportation, parks, and airports are made to meet the needs of people living within the MUSA according to the comprehensive metropolitan-wide plans in the *Regional Development Framework*. A low-density rural service area was established where the goal is to preserve agriculture. The framework includes a fully developed urban area, a developing urban area, freestanding growth centers, rural centers, a commercial agricultural area, and a general rural use area. Each local government is required to develop a detailed land-use plan that is consistent with this framework. Although the MUSA has been extended incrementally over time to accommodate growth and has helped restrict uncontrolled urban sprawl, some new development has leap-frogged into the adjacent counties—particularly along the interstate highways—outside the seven-county area.

The Fiscal Disparities Act of 1971 introduced metropolitan tax-base sharing. This diverts 40 percent of the increase in commercial and industrial property taxes into a central fund that is then redistributed based on each locality's population and the ratio between the locality's per capita valuation

(Continued)

BOX 17.4   METROPOLITAN GOVERNANCE IN THE TWIN CITIES   *(Continued)*

of property and that of the metropolitan area. This tax-base sharing has lowered fiscal disparities across the metropolitan area. Tax revenues generated by the expansion of commercial and industrial property (mostly in suburban communities) have gone disproportionately to localities (at the urban core) that have lower tax bases but a high demand for services.

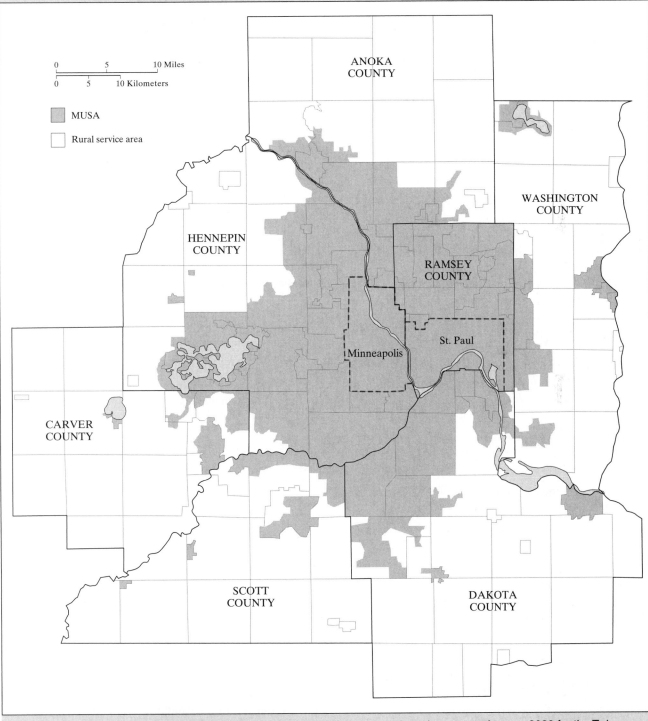

FIGURE 17.16   The metropolitan urban service area (MUSA) and rural service area to the year 2020 for the Twin Cities of Minneapolis–St. Paul metropolitan area.

made it difficult to sustain the notion of *metropolitan* governance and planning. The advantages of metropolitan-wide planning—including economies of scale in service provision and coordinated infrastructure planning—have led to some governance reforms.[29] Given the strong tradition of local government control over urban planning and the many vested interests in the status quo, however, these reforms have been limited and exceptional.

- Two or more jurisdictions have been consolidated into a unitary government, as occurred, for example, in the city-county consolidation in Indianapolis that created Unigov in 1969.
- Councils of Government, such as the Association of Bay Area Governments for the San Francisco metropolitan area, have been established. Local government membership is strictly voluntary, however, and each Council of Government acts in an advisory capacity only.
- Intergovernmental agreements have been instituted where one locality provides a service (such as water) to others in the metropolitan area or a group of jurisdictions jointly provides certain metropolitan-wide functions or services (such as fire and emergency response).
- Key functions have been transferred to a higher (metropolitan) level of government. One of the most ambitious and longer-established examples in the United States is in the three counties and 24 cities of the Portland, Ore., metropolitan area, where Metro, the metropolitan council, has responsibility for protecting open space and parks, planning for land use and transportation, and managing waste disposal and recycling.[30]

Metro manages an urban growth boundary that is designed to restrict uncontrolled urban sprawl. Another high-profile example of metropolitan government is in the Twin Cities (of Minneapolis and St. Paul) (see the Box entitled "Metropolitan Governance in the Twin Cities").

The current concern with planning for healthy and livable cities suggests that if we once expected too much of urban policy and planning, we now perhaps expect too little. Although we have learned that urbanization is a broad, deep, and complex collection of interrelationships that we cannot hope to "shape" or "control," we have also seen that reform is possible and that urban outcomes can be successfully modified. But what is clear from the previous chapter is that reform has to come from society at large rather than from the policy and design professions. Sociologist Janet Abu-Lughod put it this way:

If we are to get better cities, if we are to deserve better cities, then we cannot wait patiently for a technological or administrative "fix." We must find out what really creates our cities, both physically and socially. We must reach a deeper understanding of the forces and processes that have shaped them—that have caused them to become what they are now and that daily and continually reproduce and modify them. Such forces are well beyond the control of planners and administrators, although they are the collective outcomes of all our attitudes, institutions, and social structures. To get changed cities—to change our cities into a better image of society—we must change the society itself.

Knowing the causes of urban outcomes we do not like, and then working to change the conditions that lead to them—that is our task as social scientists and citizens.[31]

FOLLOW UP

1. Be sure that you understand the following key terms:

    building standards
    collective consumption
    enterprise zones
    garden cities
    green urbanism
    neighborhood unit concept
    scientific planning

2. Be sure that you understand the significance of the following people and events for urban change and development:

    Jane Addams
    Baker v. Carr (1962)
    Brown v. Board of Education (1954)

    Daniel Burnham
    Civil Works Administration
    Economic Opportunity Act (1964)
    Patrick Geddes
    Ebenezer Howard
    National Planning Board (1933)
    National Urban Policy Report (1978)
    Clarence Perry
    Jacob Riis
    Shelley v. Kraemer (1948)

3. Do you think that there should be a single planning authority that covers entire metropolitan areas? What sort of case might be made for the advantages of such jurisdictions? Would there be any damaging consequences of establishing metropolitan-wide jurisdictions?

4. Work on your *portfolio*. With the background you have now acquired in urban geography, you should be able to look at your own town or city and "read" it. What examples can you find of federal policies that have been "written into" the urban landscape?

Can you find examples of the nature or pattern of urban development having been influenced by local plans or policies? Are there aspects of your city's development that could or should have been planned in order to achieve a different outcome?

KEY SOURCES AND SUGGESTED READING

Beately, T. 2000. *Green Urbanism: Learning From European Cities*. Washington, D.C.: Island Press.

Boyer, M. C. 1983. *Dreaming The Rational City: The Myth of American City Planning*. Cambridge, Mass.: MIT Press.

Cullingworth, J. B. 1993. *The Political Culture of Planning: American Land Use Planning in Comparative Perspective*. New York: Routledge.

Fogelsong, R. E. 1986. *Planning the Capitalist City*. Princeton, N.J.: Princeton University Press.

Freestone, R., ed. 2000. *Urban Planning in a Changing World*. New York: Routledge.

Hall, P. 2002. *Cities of Tomorrow*, 3rd ed. New York: Basil Blackwell.

Hall, P. 2002. *Urban and Regional Planning*, 3rd ed. London: Routledge.

Jones, E. T. 2003. *The Metropolitan Chase: Politics and Policies in Urban America*. Upper Saddle River, N.J.: Prentice Hall.

Judd, D. and S. S. Fainstein, eds. 1999. *The Tourist City*. New Haven, Conn.: Yale University Press.

Kamal-Chaoui, L. 2003. *Metropolitan Governance in OECD Countries*. Paris: Organization for Economic Cooperation and Development. (*http://www.oecd.org/dataoecd/59/40/6100078.pdf*).

Katz, B. 2002. *Smart Growth: The Future of the American Metropolis?* Washington, D.C.: The Brookings Institution.

(*http://www.brook.edu/es/urban/publications/20021104katzlse2.htm*).

Krueckeberg, D. A., ed. 1983. *Introduction to Planning History in the United States*. New Brunswick, N.J.: Center for Urban Policy Research, Rutgers University.

Levy, J. 2002. *Contemporary Urban Planning*. 6th ed. Englewood Cliffs, N.J.: Prentice Hall.

Mondale, T. and W. Fulton. 2003. *Managing Metropolitan Growth: Reflections on the Twin Cities Experience*. Washington, D.C.: The Brookings Institution. (*http://www.brook.edu/es/urban/publications/200309_fulton.htm*).

Randolph, J. 2003. *Environmental Land Use Planning and Management*. Washington, D.C.: Island Press.

Schaffer, D., ed. 1988. *Two centuries of American Planning*. Baltimore, Md.: Johns Hopkins University Press.

Ward, S. V. 2002. *Planning the Twentieth Century City*. Chichester, UK: John Wiley.

Wolman, H., T. Swanstrom, M. Weir, and N. Lyon. 2004. *The Calculus of Coalitions: Cities and States and the Metropolitan Agenda*. Washington, D.C.: The Brookings Institution (*http://www.brook.edu/urban/publications/20040422_coalitions.htm*).

RELATED WEBSITES

Lowell National Historical Park (National Park Service): *http://www.nps.gov/lowe/*
Lowell, Mass., now a National Historic Park, began life as America's first large-scale planned industrial community in 1826. By the mid-nineteenth century, this textile mill town contained massive five-story and six-story brick mills that lined the Merrimack River for nearly a mile. Connecting the mills to the river was a system of nearly 6 miles of canals that was used for transporting goods and powering the machinery in the mills. The growth and decline of the city and its textile industry—from early industrialization to deindustrialization—are described and illustrated on this website.

Smart Growth Network: *http://www.smartgrowth.org/* and Smart Growth America: *http://www.smartgrowthamerica.com*
These are two of a number of websites that are good resources for more information and best practice for smart growth. The U.S. Environmental Protection Agency and several nonprofit and government agencies formed the Smart Growth Network in 1996. Smart Growth America is a coalition of nearly 100 advocacy

organizations supports citizen-driven planning that coordinates development, transportation, open space preservation, and the revitalization of urban cores.

The Metropolitan Council: *http://www.metrocouncil.org/* and Metro: *http://www.metro-region.org/*
The websites of the Metropolitan Council (of the Twin Cities [Minneapolis and St. Paul] metropolitan area) and Metro (for the Portland, Ore., metropolitan area) offer detailed information on the policy and practice of regional planning in these two metropolitan areas.

Urban and Regional Information Systems Association (URISA): *http://www.urisa.org/*
This website contains publications and resources generated by URISA, a nonprofit association of nearly 7,000 professionals worldwide who use Geographic Information Systems (GIS) and other information technologies to address problems in planning, public works, the environment, emergency services, and utilities throughout state, regional, and local governments. The website of the American Planning Association is another useful website, although access to some of its publications is restricted to members (*http://www.planning.org/*).

18

URBAN FUTURES

The subject of urban futures is often associated with speculative scenarios—imagined futures with fantastic technologies, astonishing engineering achievements, and incredibly innovative architecture. In one common scenario, a new and harmonious social order enables everyone to adapt to the new settings. In another, an Orwellian social order is maintained through technological innovation and exclusionary (but vaguely specified) social geographies. As we look toward the next quarter-century of urban development, such scenarios are not entirely unreasonable. A quarter of a century is a long time. There will be some technological developments that are scarcely credible today; there will be some impressive achievements in engineering; there will be innovations in architecture and urban design; and there will be social and geographical change. *But they will all be accommodated within a framework that already exists, alongside developments whose seeds have already taken root.*

CHAPTER PREVIEW

In this final chapter we explore the implications for urban change in the early twenty-first century of the intersection of current economic, demographic, social, cultural, and political trends in the context of prospective technological and environmental changes. The objective is to draw on the framework established in Chapter 1 to explore a few key aspects of urban change. The context for change is presented as a function of (1) increasing international economic interdependence, (2) the rhythm of economic long waves, and (3) the continuation of already established demographic, social, cultural, and political dynamics. Projecting this context a couple of decades into the future, we can anticipate greater volatility within urban systems, a continuation of both decentralizing and centralizing trends affecting metropolitan form and land use, and the persistence of problems of infrastructure provision, environmental degradation, fiscal stress, and slums.

A PERSPECTIVE ON FUTURE URBAN CHANGE

Cities of mile-high skyscrapers, spaceship-style pods, or bioecological harmony are for the coffee table books of dreamers. We now know that the triumphs of progress and the conquests of technology are uneven in their imprint on urbanization. The notion of a futuristic metropolis (of whatever sort: benign or malevolent, ecological or technological) is a chimera, an unrealizable dream/nightmare. Rather, bits and pieces of the metropolis and of urban life a couple of decades from now will bear the imprint of radical change. The basic fabric of the metropolis and urban life, however, will be a continuation and evolution of what is currently in place.

The future is already here, embedded in the physical, institutional, and social structures of the early twenty-first century city. It is a future that will grow from the various dimensions of urbanization that have been described in this book. Although we cannot map the future in detail, we can be fairly confident about the general shape of our urban future, what it will and will not involve. We know, for example, a good deal about the demographics of the next quarter century, given present populations, birth and death rates, and so on. Similarly, we know a great deal about the stock and life expectancy of the physical infrastructure of cities and their housing, about the basic parameters of labor markets, about people's occupational skills, about educational needs and resources, about environmental resources and constraints, and about the legal and political frameworks within which urbanization will likely take place. We know at least the immediate trajectory of economic, demographic, social, cultural, and political change, and we know something of prospective technological and environmental changes.

We know also that we are on the cusp of a particularly significant change in the political economy of urbanization and that some of the characteristic features of twentieth-century urbanization are now history. We have already entered a distinctive era—of informational cities, polycentric metropolises, and splintering urbanism—whose dynamics have yet to be fully inscribed on the face of cities. Although change has been a constant theme in the story of urbanization told in this book, the current pace of change is remarkable.

Since 1950 the world's urban population has increased threefold, reaching 2.9 billion in 2000. At current rates of urban growth, this figure is projected to rise to 5 billion by 2030, when more than 60 percent of the world's population will be urbanized. The bulk of this urban growth is expected to occur in the less developed countries. In fact, almost all of the population increase between now and 2030 will be absorbed by the less developed countries, whose urban population is expected to rise from almost 2 billion in 2000 to nearly 4 billion or almost 80 percent of the world's urban population by 2030.

As current events rapidly leave behind some of the old urban stereotypes, the classic textbook geometry of **sectors** and **concentric zones** is increasingly difficult to discern in the landscapes, **social ecologies**, and **bid-rent** patterns of cities in the developed countries. The classic mosaic of central city neighborhoods is increasingly blurred as new lifestyle and cultural preferences are superimposed upon the old dividing lines of socioeconomic background, ethnicity, and family status. In the United States long-established central city neighborhoods either are falling into decay and **social disorganization** or are being "reclaimed" by members of the professional middle-income groups. In the cities of many developed countries central business districts (CBDs) are experiencing a selective centralization of economic activity marked by a rash of speculative building and the re-creation of idealized versions of the urban past. Beyond the central city, suburban strips and subdivisions have been joined by **exurban** corridors, office parks, business campuses, privately planned residential communities, and outlying commercial centers large enough to be called **edge cities**—all on a scale big enough to justify the notion of a "galactic metropolis."

Although global influences are generating certain similarities in urban processes and structure worldwide, the patterns of land use and functional organization in cities in different regions of the world remain quite distinct, reflecting variations in factors such as historical legacies, levels of technology, and environmental and cultural influences. Many less developed countries are experiencing a process of **overurbanization** in which cities and their populations grow more rapidly than the jobs and housing that they can sustain. Unprecedented rates of urbanization have been associated with the growth of **megacities**. These megacities are usually characterized by **primacy** and a high degree of **centrality** within their national urban systems.

A variety of historic and contemporary processes—including colonialism, rural-to-urban migration, and overurbanization—have shaped the urban form and land use that we see today in the cities of less developed countries. Meanwhile, the local effects of splintering urbanism are transforming these historical patterns of urban land use and spatial organization.

Economic and cultural globalization is generating new urban landscapes of innovation, economic development, and cultural transformation. A major problem is that the uneven diffusion of the networked infrastructures of information and communications technologies has intensified economic and social inequalities between urban residents in and between cities around the world.

So just as it might have seemed that our theories and models had captured the essential truths of urban geography, the transformation of cities themselves is making some of the models obsolete, forcing a reevaluation of theory, and raising new issues. As we enter our urban future, therefore, we must be ready to modify or even abandon some of our current theories and to develop new ones. This being said, we can still be fairly certain that our basic conception of urbanization, as depicted in Figure 1.4, will continue to hold. It follows from this model that in surveying the likely outcomes of future urbanization we should look first to the implications of the broader sweep of economic change.

ECONOMIC CHANGE

THE GLOBALIZATION OF THE WORLD ECONOMY

The most important point about the broader sweep of economic change concerns the **globalization** of the world economy. The emergence of **transnational corporations**, the development of new and faster forms of transportation and telecommunication, and the geopolitical changes that have extended the realm of capitalist development are all conspiring to modify the role of cites as engines of economic growth. Economic roles that were based on specialization, comparative advantage, and the forces of **cumulative causation** within a particular *national* framework now have to be revised and conceptualized within an *international* economy. We saw in Chapter 4 how this change is already reflected in the U.S. and European urban systems in several ways: the centralization of key corporate control functions in a few **world cities**, for example, and the **deindustrialization** of the traditional manufacturing regions as transnational corporations have shifted productive capacity to cheaper locations in the less developed countries. As world economic geography adjusts to the scope of new technologies and new geopolitical realities, cities around the world are becoming increasingly interdependent.

This interdependence will have several important dimensions. At the top of the urban hierarchy, we will see an intensification of the interdependence of the world cities that act as basing points for capital and as corporate **control centers**. Table 18.1 lists 20 world cities that have been identified as constituting a *polycentered global urban network* on the basis of the locations of the headquarters of the largest transnational corporations. These cities are likely to become increasingly interconnected not only in terms of their economic functions but also in terms of the mutual concerns and interests of their civic leadership and their business and professional services workers. They are also likely to become increasingly alike in their socioeconomic structure and their political economy and increasingly different from other metropolitan areas.

Although world cities will be central to the "space of flows" (p. 88) of knowledge and information, other metropolitan areas will play their part in various ways. Informational functions will develop and grow in most metropolitan areas, with most of the growth at first being channeled into suburban and exurban settings—the **stealth cities** of **boomburbs** and edge cities in the United States, for example. Routine informational functions such as services delivery and data retrieval and distribution will be standardized and will suffuse throughout the urban systems at national and international scales, broadening the network of flows that lead to key **informational cities** and to world cities.

In parallel fashion, the time-space compression of long-distance transportation and communications technologies will allow for much greater flexibility in the location of many industries, both informational and manufacturing. This is not to say that there will be a homogenization of space. There will

TABLE 18.1

The Polycentered Global Urban Network

Region	20 top-ranked metropolitan areas	Number of headquarters of the 500 largest corporations*
North America	New York	56
	San Francisco	17
	Chicago	14
	Dallas	10
	Los Angeles	9
	Atlanta	9
	Philadelphia	8
	Minneapolis	7
	Washington, D.C.	7
	Pittsburgh	6
	St. Louis	6
	Boston	5
	Montreal	5
Asia	Tokyo	80
	Osaka	23
	Hong Kong	11
	Singapore	6
Europe	London	29
	Paris	17
	Munich	7

*Listed by Business Week *based on market value (which gives greater weight to financial institutions and real estate companies, as opposed to the greater weight given to manufacturing and other service firms when based on revenue).*

Source: *B. J. Godfrey and Y Zhou. Ranking world cities: Multinational corporations and the global urban hierarchy.* Urban Geography *20(3), 1999, Table 2, p. 272.*

still be unevenness in national and international urban systems, but the unevenness will involve different patterns, will be more volatile (because the increased flexibility will make locational switching of investments easier), and will therefore involve increasing interdependence between places as urban fortunes come to depend increasingly on what happens elsewhere. In addition, as in industrial development, there will be economic specialization in the informational mode of development. There are already some examples. At the general level in the United States, for example, we can contrast the specialization in financial and marketing services of New York and Chicago with the technology-oriented services of the San Francisco Bay area. More specifically, we can recognize the specialization in insurance services in Hartford, Conn., and, at the other end of the spectrum of information activities, home shopping services in Roanoke, Va.

The most critical of the specialized places (beyond the world cities themselves) in the coming informational economy will be **innovative milieux**—places with concentrations of research and development (R&D) activity and innovative venture capital. Such places "are the innovative powerhouses of the new economy, the sources of its inherent dynamism. They command a complex chain of interdependencies and, if anything, they become more exclusive, more secluded, and more central in the process."[1] In Japan a government-sponsored program—the Technopolis program—aims to create innovative milieux in "science cities" spread around the country. Given the recent history of planning and policy in the United States (Chapter 17), it is unlikely that any such program will be undertaken there. Rather, we can expect the established centers of innovation (Silicon Valley around San Jose in California, the Route 128/495 corridor around Boston, and the Research Triangle area between Raleigh, Durham, and Chapel Hill in North Carolina) to maintain their dominance while newer innovative milieux develop around some of the larger research universities located in smaller campus towns that prove themselves to be adaptable, physically and economically, to high-tech industries and services.

THE NEXT ECONOMIC SWINGS

Another important point about the broader sweep of economic change concerns the behavior of economic long waves, particularly the **Kuznets cycles** and **Kondratiev cycles** described in Chapter 1. We should be very cautious here, since our knowledge of how these long waves work is imperfect. The fact that the capitalist economy has experienced an impressively regular series of cycles over the past 200 years is not a sufficient basis for believing that the sequence will be repeated in the twenty-first century. The unprecedented interdependence of the world economy may itself affect the rhythm or even the nature of economic long waves. At the same time, there is no concrete reason to expect that the sequence will *not* continue.

Assuming for the moment that it will, we may experience a burst of economic growth around 2010 (Figure 18.1).[2] What might this cycle mean for the urban future? If events follow previous patterns, we should look to the deflationary growth cycle of the recent past for the emergence of the key clusters of innovations that should power the inflationary growth cycle that may begin around 2010. These innovations include, of course, current advanced telecommunications systems—satellite systems, fiber-optic systems, computer networks, and so on—that are already an integral part of urban change through the globalization of the economy. Together with innovations in the organization of industry, these telecommunications systems will form the basis of a new **technology system** that will exploit

further developments in electronics (semiconductors, superconductors, artificial intelligence), bioengineering, and materials technology that will determine the shape and character of the next epoch of urbanization.

This analysis reinforces the case for predicting the trends noted above in relation to the informational mode of development. At the same time, we should note that much of this restructuring is scheduled to take place before the next economic upturn, which has its own implications for urban life. It means, for example, that social polarization (already noted by urbanists such as Manuel Castells in relation to world cities and dominant informational cities) is likely to spread and intensify in and between cities around the world. It means that we should not expect to see any real increase in the resources available to governments to spend on welfare provision or ameliorative social policies. In the developed countries it means that the living standards of many people are unlikely to show any real improvement until the acceleration in economic growth after about 2010. For the poorest city dwellers in the less developed countries, it means that their future could be very bad indeed. It follows from this that housing availability and affordability is likely to loom larger and larger as a social and political issue. Just how the issue is resolved in different countries, in the marketplace, and through policy making will likely determine the character of a good deal of the change that we will see as cities develop.

It also follows that the political economy in many developed countries stands some chance of being weaned away from the addiction to growth that set in

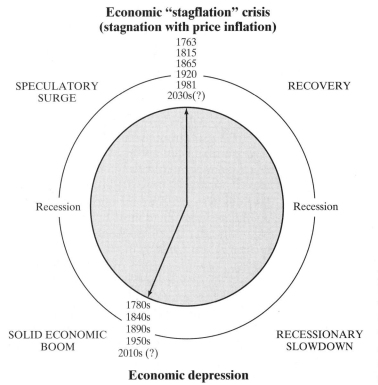

**Economic "stagflation" crisis
(stagnation with price inflation)**

1763
1815
1865
1920
1981
2030s(?)

SPECULATORY
SURGE

RECOVERY

Recession

Recession

1780s
1840s
1890s
1950s
2010s (?)

SOLID ECONOMIC
BOOM

RECESSIONARY
SLOWDOWN

Economic depression

FIGURE 18.1   The long-wave "clock." In the past, long-wave economic cycles have been fairly regular, with phases of depression occurring at 50- to 55-year intervals, interspersed with episodes of stagflation. At that rate, the next economic boom should begin around 2010, after which there should be a phase of inflationary speculation that leads to the next stagflation crisis, sometime in the 2030s.

after World War II, with attitudes shifting to a more inconspicuous consumption mentality in which materialism may become a less dominant sociocultural sensibility.[3] At the same time, consumption is rising among the new waged workforce in **newly industrializing countries** such as South Korea and Taiwan.

We should not expect to see any dramatic shift, however, away from the conservatism of the political economy that has dominated the urban governance, politics, planning, and policy for much of the past 30 years of countries such as the United States and international bodies such as the **World Bank** or the **International Monetary Fund**—unless social polarization becomes so acute or the recessionary slowdown so bad that a reformist reaction sets in. At this point we need to look again at the new technology system that will be maturing all the while, because it is quite plausible that another effect of the informational economy will be to soften the problems of speculative overshoot and undershoot that are endemic to the capitalist enterprise.

DEMOGRAPHIC CHANGE

Woven into this broader sweep of economic change in future cities will be patterns of demographic change whose basic design is already in place. In many less developed countries, the main feature of demographic change will be the high rates of **natural increase** of the population and the **rural-to-urban migration** that continue to fuel urban growth. In many developed countries the main feature of demographic change will be the aging of the massive Baby Boom generation. Currently in their 40s and 50s, this group is already established (for better or worse) in labor and housing markets, but it will continue to give shape to housing, consumer markets, and political values during the next few decades and beyond. This is the generation in the United States that was largely born and raised in suburban tract houses and that ultimately gave critical mass to suburbia; its future residential choices will define urban housing structure and geography.[4] In the immediate future the peaking incomes of the middle-aged Baby Boomers will maintain (for a while, at least) the momentum of materialism in American and European culture. Soon after, as the Baby Boomers reach old age, there will be unprecedented demand for specialized housing, recreational amenities, social services, and medical-care settings. Because not everyone will be able to afford them, this demand will probably set off jarring reverberations throughout urban political life.

The Post-Boom generation of young adults is also in place, and it too must age. The decline in entrants to the higher education market and to labor markets that bottomed out in the mid-1990s is being followed by a noticeable upswing in the first decade of the twenty-first century (the "baby-boomlet" resulting from the delayed childrearing of the "thirtysomething" Baby Boomers). For some time, therefore, there will be a relatively small cohort of young adult households. Consequently, there will be a fall in demand for entry-level housing—enough to significantly reduce the homebuilding industry's contribution to **gross domestic product** and enough to be reflected in a changing metropolitan form. Faced with a shrinking market, the homebuilding industry, along with its financial backers, may likely support reforms that will provide subsidies for additional low-income and moderate-income housing. Soon, as the Post-Boom cohort matures and the Baby Boom generation retires, there will be an acute shortage of labor that will be alleviated only through immigration. This will occur not only in countries that have traditionally been open to immigrant workers, such as the United States, but also across Western Europe, where resistance to immigration has been manifested quite starkly by the temporary restrictions on labor migration from countries that joined the European Union in 2004—including Poland, the Czech Republic, Slovakia, Slovenia, Hungary, Latvia, Lithuania, and Estonia.

That such immigration will occur seems almost inescapable—despite the impacts on international migration of the global "war on terrorism"—given U.S. and West European labor needs and the accelerating gap between rich and poor countries in terms of demographic growth and standards of living. For the United States we can expect the bulk of these immigrants to come from Central and South America and from South and Southeast Asia. Like all previous waves of immigrants, they will gravitate to older, cheaper housing. Unlike previous waves, however, they will find much of this housing outside the central city, in older suburban settings. In contrast to previous waves, they will enjoy relatively good facilities, and if they are able to take advantage of newer, inexpensive communications systems, they may be less obliged to become assimilated into the dominant culture. Immigration will be another source of strong political reverberations. Current estimates by the U.S. Bureau of the Census suggest that whites will account for well under 60 percent of the total population by 2030, when about 20 percent of the total will be Hispanic, 14 percent will be black, and 6 percent will be Asian. They will, of course, be spread unevenly across the **urban system** and across metropolitan space, so that ethnicity will become even more important as an element in the urban geography of the United States.

SOCIAL CHANGE

We should be able to anticipate a continuation and extension of two important social "revolutions" that began in the 1960s but that still have some way to go before they are completed in both the developed and less developed countries. The first is the change in the roles and opportunities for women in society. As women continue to gain power, respect, and educational status, the consequences will be increasingly apparent not only in labor markets and political life but also in domestic life, lifestyle patterns, and urban social geography. The second is the continuing struggle for civil rights. We can be less sanguine about the progression of this trend. Though society is clearly past the point of no return, incremental improvements to civil rights seem likely to be relatively slow and perhaps increasingly conflictual, not least because of the shifting ethnic balance noted above and the efforts of gay rights activists in countries such as the United States and Canada, but also because of the social transformations that have come about as many less developed countries have attempted to address the legacy of colonialism.

Meanwhile, we can expect a further fragmentation of lifestyle groupings at every socioeconomic level and in every major ethnic group in and between cities. In large measure this fragmentation will be a function of the **splintering urbanism** that is associated with the informational mode of development, which will open up all sorts of new opportunities for some city dwellers in terms of patterns of work, socializing, health, sport, arts, ideas, and popular culture, while large numbers of other people will continue to have nonexistent or rudimentary access to information and communications technologies. In addition, the economic situation before the upturn and burst of growth beginning about 2010 offers the possibility for the emergence of new lifestyles based on the commercial exploitation of attitudes associated with less conspicuous consumption in the developed countries in general and the United States in particular—"green" lifestyles, or lifestyles of "conspicuous frugality," perhaps, instead of conspicuous consumption.

CULTURAL CHANGE

The economic, demographic, and social changes affecting cities will be reflected increasingly in the culture of cities. In addition to the changes that are occurring as a direct result of the convergence of occupational structures and the mutual concerns and interests of the leadership in world cities, there is already evidence of an emerging *global culture* that is manifest not only in the larger metropolitan centers around the world but also at every social level and in smaller and more remote urban areas.[5] It is a "promotional culture"[6] that is increasingly evident in the urban built environment (e.g., "Disneyfication" and postmodern architecture that make for an increasing convergence in the appearance of cities in the developed countries). As a

result, some cities seem set to develop into "variations on a theme park."[7]

Not surprisingly, there is also evidence for persisting and even increasing local counter movements in the face of global trends toward homogenization and standardization, including the revival or creation of regional identities in Catalonia, Lombardy, Punjab, Quebec, Scotland, and Ukraine. Similarly, the Slow City (CittaSlow) movement that has emerged in Italy—involving a commitment to conserve the distinctive character of each city's built environment and to fend off the advance of fast food and cultural standardization—is a clear reaction to big business and globalization (Chapter 17).

POLITICAL CHANGE

Given the problems of political fragmentation in the context of splintering urbanism and the galactic metropolis, opportunities for metropolitan governance will continue to be considered. The upsurge in the number and depth of challenges that crosscut the boundaries of individual local governments has been exacerbated by the mismatch between the metropolitan scope of these challenges and the fragmented structure of local government powers, policy, and practice in metropolitan regions. Certain challenges can be addressed more effectively at a metropolitan

scale because individual local governments lack the capacity or resources to address some issues without the cooperation of neighboring jurisdictions. Urban economic development issues that will continue to present opportunities for metropolitan cooperation include strategic economic planning and promotion, education and work force preparedness, R&D, transportation and communications infrastructure, parks and recreation, urban growth management, and environmental protection, as well as social services, healthcare, and emergency response systems.

One of the most interesting changes as cities around the world have become increasingly interdependent is the development of **municipal foreign policy**: the direct involvement of local governments in international relations (usually with overseas localities) and international affairs (with national governments, home or abroad). It is a direct manifestation of the notion that we should "Think Globally, Act Locally." Much of the early activity in municipal foreign policy was related to arms control issues (e.g., the creation of nuclear-free zones, local initiatives and referenda on nuclear-free investing, opposition to military spending), human rights (e.g., local divestment from South Africa during apartheid, human rights resolutions concerning Central America), and the environment (e.g., local opposition to nuclear power, efforts to protect the ozone layer). Recently, however, increasing global economic interdependence has resulted in the expansion of economic aspects of municipal foreign policy. Sister-city arrangements and city networking are now oriented increasingly toward economic linkages,[8] and there has been a proliferation of municipally funded world trade centers and trade commissions, local development commissions with an international scope, local international affairs agencies, and local initiatives to promote international tourism. It seems likely that such aspects of municipal foreign policy will become increasingly important to the governance and politics of large metropolitan areas around the world—and, indeed, to our notion of citizenship and to our conception of the public interest in cities of the twenty-first century.[9]

FUTURE VISIONS

So: what will future cities be like? What will be the main changes to the outcomes of the urbanization process during the next few decades? We do not have 20/20 vision even when we look backward at history, so we must be circumspect here. We cannot reasonably expect to be able to picture the details of the built environment, neighborhood change, or urban politics. We can, however, anticipate certain changes to the geography of urban systems and to the nature of metropolitan form, and we can identify certain issues that are likely to persist in urban social and political life.

THE VOLATILITY OF URBAN SYSTEMS

It follows from the discussion above that a few cities will consolidate and maintain their dominant status in their national urban system as they are drawn increasingly into the polycentered networks of world cities and nationally important informational cities. Unless there is some as yet unforeseen development in telecommunications or corporate structure, they are likely to enjoy stable or improving levels of economic well-being. For most other cities, however, the future will bring the probability of a good deal of volatility as **flexible production systems**, global telecommunications, Internet shopping, and so on bring about a constantly shifting geometry of locational advantages and disadvantages for many aspects of production, distribution, and consumption. In addition, the immediate future is likely to involve an extra dimension of volatility as the world economy in general and the U.S. economy in particular continue to work through the restructurings and adjustments following the end of the Cold War and the initiation of the "war on terrorism." Since September 11, 2001, some cities that had developed an economic base dominated by military or defense-related activities during the Cold War are no longer poised to be cut back dramatically.

The outcome of this volatility is likely to depend a great deal not only on comparative advantages in terms of labor, markets, raw materials, and capital at any one time but also, as in previous epochs of urbanization, on the development of technological infrastructure. In practice, the best-placed metropolitan areas will be those that are first to develop fiber-optic networks for use with "smart" telecommunications systems and "smart" highways to accommodate the next generation of trucks and automobiles. The extension of high-speed railway systems with trains that run at 300 mph or more will be crucial to the economic prospects of regions of high-density metropolitan development across Europe and Asia in particular. In the United States they are not likely to extend to a national network because the distances are so great relative to the carrying capacity of rail systems. Where they do develop—most likely in California, Florida, and Texas (and in contrast to the Northeast Corridor with its Acela high-speed rail network that already links Boston, New York, Philadelphia, and Washington, D.C.), they are likely to link up with airports rather than downtowns, so that their effect on the urban system will be to push a few edge cities rapidly up the urban hierarchy.

Volatility will also continue to be a feature of the national urban systems in many less developed countries. The uneven impacts of economic globalization and the activities of transnational corporations will be a factor, especially in the newly industrializing countries. Although economic opportunity is improving for some workers, transnational corporations can always subcontract elsewhere should local labor conditions or environmental regulations prove unacceptable. Political and environmental circumstances will also be important in some countries. Wars can cause hundreds of thousands of refugees to flee to cities, while deforestation and overgrazing along with government inaction can force people to move to the cities as the expanding desert overtakes entire villages.

METROPOLITAN FORM

An appreciation of splintering urbanism and associated edge cities, stealth cities, **urban realms**, galactic metropolises, megacities, and **extended metropolitan regions** (pp. 158–168 and 214–215) provides some good clues about the future form of metropolitan areas around the world. We need not review this material again here, except to note that such trends will be heavily influenced by gasoline prices on the one hand and the development of inexpensive and more environmentally friendly new propulsion systems for private vehicles on the other. More generally, we must acknowledge that metropolitan form will be influenced by the resolution of two sets of forces: those that foster decentralization and those that foster concentration (centralization).

Several trends will tend to encourage *decentralization*:[10]

- New telecommunications systems.
- More-flexible production and distribution systems.
- Continued suburbanization of "back-office" jobs (e.g., the accounts departments of banks).
- Increasing social preferences for lower-density, park-like, or rural environments.
- Congestion and deteriorating infrastructure in central city areas.
- Corporate and worker concerns about terrorist targeting of high-profile buildings and **CBDs**.

This decentralization, however, will, as always, be *selective*. Thus the insurance industry is likely to continue to decentralize within metropolitan areas in the developed countries (and to selected cities elsewhere like Bangalore and Hyderabad in India) because daily decisions in the insurance business are very largely dependent on documents or electronically filed information. In contrast, only a certain proportion of banking jobs will decentralize because of the importance in banking of the kind of trust and quick responses that can be derived only from face-to-face interaction. Similarly, residential decentralization in countries such as the United States will involve only those who want to participate in the "packages" offered toward the edge of the polycentric metropolis and who are able to afford the costs of housing and transportation.

The factors that will be important in facilitating *centralization* include:

- Continuing needs for face-to-face interaction.
- The continued attraction of revitalized and gentrified inner-city settings to some members of the younger middle-income groups.
- Increasing energy costs or uncertain energy supplies.
- Increasingly protective yet flexible environmental cleanup and legal liability legislation combined with improved funding for **brownfield** redevelopment.
- The persistence of poverty and the continuation of large-scale immigration.
- The locational inertia of major institutions such as museums, hospitals, and universities.

The overall outcome will of course depend on the relative strength and balance of the two sets of forces. It will also depend on the degree to which the older and more central parts of metropolitan areas are redeveloped to allow new patterns of land use and new structures to accommodate centralizing jobs and residents. This redevelopment, as we have seen (p. 342), will depend on relative patterns of physical deterioration and structural and technological obsolescence. Given that much of the inner core of metropolitan areas is nearing the end of its physical life cycle, we can expect a period of unusual restlessness in the built environment of inner metropolitan rings. This restlessness could well reach the point in the United States where the sharp distinctions that have separated central cities from suburbs and CBDs from decaying **zones in transition** become blurred.

What is more certain is that as new outer rings of development are consolidating on the fringes of the future metropolis, earlier rings of development will be revised and remade. Geographer John Borchert addressed the nature of this change for the United States, concluding that we may well see a metropolis in which the newest structures and the lowest population densities are in the innermost ring and the oldest structures and highest population densities are in the automobile-era (1920–1970) suburbs.[11] More specifically, Borchert suggested that:

- The market position of traditional CBDs will be strengthened as the surrounding pre-automobile ring is redeveloped.

- Inner-urban areas (those that were built up before about 1920) will experience dramatic change, with many older structures being replaced. At first, much of this replacement will take place at the metropolitan fringe, leaving extensive tracts of land abandoned or unused. The great variety of available sites and their relative accessibility, however, will ensure that the next upswing of economic growth will see their redevelopment—for a mixture of residential, office, and institutional uses and for newly created public open space.

- The aging ring of suburbs and industrial districts built between 1920 and 1970 will become much more varied in its social and economic geography as existing cohorts of residents reach the end of their life cycle and housing filters down the socioeconomic ladder (see p. 364). The oldest of these districts could be experiencing significant levels of physical deterioration and abandonment by as early as the 2010s.

PERSISTENT PROBLEMS

This last point should remind us that the future will bring no escape from the problems of urbanization. As with the overall shape of the urban future, we can identify the main sources of these problems from the current trajectory of urban change. So, for example, it is almost inevitable that problems of traffic congestion will persist and probably intensify because polycentric development and central city redevelopment will make the spatial relationships between jobs and homes more complex, leaving little prospect of a net reduction in commuting distances. It is even more certain that problems of inadequate and decayed *infrastructure* will intensify: a large proportion of water lines, sewer lines, bridges, and so on are currently at or near the end of their expected life span in some developed countries; in many less developed countries there are limited municipal funds for providing adequate infrastructure to meet the demands of rapid urban growth.

Problems of *pollution and environmental degradation* are a little less certain, although these problems are escalating rapidly in those cities in the less developed countries that are unable to devote many resources to environmental issues due to pressing problems of poverty, poor housing, and inadequate service and transportation infrastructures. Political change could conceivably encompass **sustainable urban development** and a **green urbanism** movement that might significantly moderate some of the air and water quality problems through tough legislation in developed countries such as the United States. The aging Baby Boomers, whose political track record already includes the radical counterculture movements of the 1960s, represent a potent source for change in the developed countries in this respect. Now established with homes, families, and careers, they may feel able to afford to rekindle their radicalism around the cause of environmental quality. Figure 18.2 shows some of the changes that will need to be made to ensure the future sustainability of metropolitan areas. It has to be acknowledged, however, that there is no real evidence to suggest that a dramatic shift in attitudes among many people in countries where car ownership is entrenched is actually about to happen. A more sober prognosis, therefore, would be that problems of pollution and environmental degradation will spread and intensify along with the spread of urbanization in both developed and less developed countries.

Unless current patterns and structures of urban governance are radically reformed, the future will also see a continuation of problems of *fiscal stress* and, consequently, of problems associated with inadequate *municipal service provision* in both the developed and the less developed countries. In the United States central city governments will face increasing difficulty in raising the necessary resources to fund schools, police, and other important public services so long as their populations continue to be dominated by low-income migrants and immigrants and so long as they continue to lose semiskilled, mid-level jobs. Both are trends that appear set to continue through the near future, at least until inner-city areas begin to be redeveloped on a large scale, at which point low-income migrants and immigrants will be pushed into older suburban settings. This trend will not put an end to problems of fiscal stress—it will merely displace the problem geographically, so that in the future it is likely that some older suburban jurisdictions will experience the kind of fiscal stress that so far has been associated only with central cities.

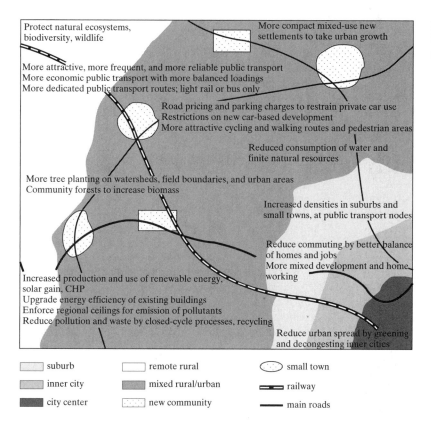

Protect natural ecosystems, biodiversity, wildlife

More attractive, more frequent, and more reliable public transport
More economic public transport with more balanced loadings
More dedicated public transport routes; light rail or bus only

More compact mixed-use new settlements to take urban growth

Road pricing and parking charges to restrain private car use
Restrictions on new car-based development
More attractive cycling and walking routes and pedestrian areas

Reduced consumption of water and finite natural resources

More tree planting on watersheds, field boundaries, and urban areas
Community forests to increase biomass

Increased densities in suburbs and small towns, at public transport nodes

Reduce commuting by better balance of homes and jobs
More mixed development and home-working

Increased production and use of renewable energy, solar gain, CHP
Upgrade energy efficiency of existing buildings
Enforce regional ceilings for emission of pollutants
Reduce pollution and waste by closed-cycle processes, recycling

Reduce urban spread by greening and decongesting inner cities

suburb   remote rural   small town
inner city   mixed rural/urban   railway
city center   new community   main roads

FIGURE 18.2   Some of the changes that will need to be made to ensure the sustainability of metropolitan areas.

Finally, we can be confident that problems of *poverty* will persist. Among them will be problems of **slums** and **squatter settlements** although, like fiscal stress, they may move around metropolitan areas in response to the broad changes that will occur in metropolitan form. Along with these fundamental and seemingly intractable problems of economic and social inequality, we can also expect certain other problems to persist: in many developed countries, the particular sociocultural problems experienced by the most disadvantaged members of some ethnic minority groups, the problems of containing criminal violence, and the problems of dealing with drug abuse; in the less developed countries, the growing number of people in cities subsisting below the international poverty line of US$1 a day, the large proportion of the population dependent on the **informal sector** of the economy, and the problems of dealing with urban health issues. Given the probable dynamics of the long-wave clock in the immediate future, it is reasonable to expect that poverty-related problems will not only persist but intensify. This prognosis holds the possibility that another crisis (or perhaps merely a continued sense of imminent crisis in wake of September 11, 2001) will trigger a new round of activity in terms of policy and planning. The critical question, of course, is: Will we be able to find more effective ways to plan and manage urban change?

KEY SOURCES AND SUGGESTED READING

Davis, M. 2002. *Dead Cities and Other Tales*. New York: The New Press.

Fishman, R. 2000. *Cities in the 21ˢᵗ Century*. Washington, D.C.: Urban Land Institute.

Graham, S., ed. 2003. *The Cybercities Reader*. New York: Routledge.

Harvey, D. 2000. *Spaces of Hope*. Edinburgh, UK: Edinburgh University Press.

Jacobs, J. 2004. *Dark Age Ahead*. New York: Random House.

Sassen, S., ed. 2002. *Global Networks: Linked Cities*. New York: Routledge.

Scott, A. J., ed. 2001. *Global City-Regions: Trends, Theory, Policy*. Oxford, UK: Oxford University Press.

Soja, E. W. 2000. *Postmetropolis: Critical Studies of Cities and Regions*. Oxford, UK: Blackwell.

Squires, G. D., ed. 2002. *Urban Sprawl: Causes, Consequences and Policy Responses*. Washington, D.C.: The Urban Institute Press.

RELATED WEBSITES

Sustainable Communities Network:

http://www.sustainable.org/

SCN is a forum for sharing information and best practice about sustainable development. The website includes publications and reports as well as case studies from around the United States. Topic areas include living sustainably, sustainable economies, and smart growth.

UN-Habitat (United Nations Human Settlements Program) Sustainable Cities Program: *http://www.unhabitat.org/ programmes/sustainablecities/default.asp*

The UN-Habitat Sustainable Cities Program (SCP) was established to build capacities in urban environmental planning and management as a basis for achieving sustainable urban development and growth around the world. The SCP promotes information sharing, community involvement, and inter-agency partnerships in dozens of cities in Latin America, Africa, and Asia, as well as in Eastern Europe and Russia.

Global Urban Research Unit (GURU):

http://www.ncl.ac.uk/guru/

The GURU website is an interdisciplinary center for urban research at Newcastle University in the United Kingdom. This site focuses on a number of themes, including cities and international development; neighborhood diversity and social cohesion; and urban technology, design, and development.

Center for Advanced Spatial Analysis (CASA) at University College London (UCL):

http://www.casa.ucl.ac.uk/news/index.htm

CASA is concerned with developing innovative computer technologies in several disciplines that deal with geography, space, location, and the built environment. Its website contains information on projects that include building computer models of the virtual city, urban remote sensing, urban dynamics involving sustainability and density, and the use of Geographic Information Systems (GIS) technology and spatial analysis in the study of the archaeology of ancient cities. This website also includes "The Geography of Cyberspace Directory" with its "Atlas of Cyberspaces," which contains maps and graphic representations of cities and the Internet, the World Wide Web, and other cyberspaces.

NOTES

CHAPTER 1

1. R. J. Johnston, "The World Is Our Oyster," *Transactions of the Institute of British Geographers* 9 (1984): 444.
2. For a more detailed review, see S. Aitkin, D. Mitchell, and L. Staeheli, "Urban Geography," in *Geography in America at the Dawn of the 21st Century*, eds. G. L. Gaile and C. J. Willmott (Oxford, UK: Oxford University Press, 2003), 237–63; D. Ley, *A Social Geography of the City* (New York: Harper & Row, 1983), Chapter 1; and R. J. Johnston, "City Structures," *Progress in Human Geography* 1 (1977): 118–23.
3. Ley, *A Social Geography*, 6.
4. For a more detailed explanation of structuration theory, see A. Giddens, *The Constitution of Society: Outline of the Theory of Structuration* (Cambridge, UK: Polity Press, 1984).
5. See U.S. Department of Commerce, *Census 2000 Basics* (MSO/02-C2KB, September 2002) (*http://www.census.gov/main/www/cen2000.html*).
6. See U.S. Census Bureau's website: *http://www.census.gov/population/www/estimates/metroarea.html*.
7. See R. E. Lang and D. Dhavale, "Micropolitan America: A Brand New Geography," *Census Note* 04:01 (March 2004), part of the Census Notes series of the Metropolitan Institute at Virginia Tech (*http://www.mi.vt.edu*).
8. Ibid.
9. Ibid.
10. B. J. L. Berry introduced the term "urban system" in "Cities as Systems Within Systems of Cities," *Proceedings of the Regional Science Association* 13 (1964): 147–63.
11. For a more detailed discussion of these changes, see P. Knox, J. Agnew, and L. McCarthy, *The Geography of the World Economy*, 4th ed. (London: Edward Arnold, 2003), Chapter 1.
12. E. Swyngedouw, "Neither Global nor Local: 'Glocalization' and the Politics of Scale," in *Spaces of Globalization: Reasserting the Power of the Local*, ed. K. Cox (New York: Guildford, 1997), 137–66.
13. G. Ritzer, *The McDonaldization of Society: An Investigation into the Changing Character of Contemporary Social Life*, rev. ed. (Thousand Oaks, Calif.: Pine Forge Press, 1996).
14. The term "disorganized capitalism" was introduced by S. Lash and J. Urry in *The End of Organized Capitalism* (Cambridge, UK: Polity Press, 1987). For an extended summary, see Knox, Agnew, and McCarthy, *The Geography of the World Economy*, Chapter 6.
15. For a more detailed discussion, see W. van Vliet in "Cities in a Globalizing World: From Engines of Growth to Agents of Change," *Environment & Urbanization* 14 (2002): 31–40; United Nations Centre for Human Settlements (UN-Habitat), *Cities in a Globalizing World: Global Report on Human Settlements 2001* (Nairobi, Kenya: UN-Habitat, 2001).
16. There has been a long history of research on this topic in geography and regional science. One important landmark was Walter Isard's paper, "A Neglected Cycle: The Transport-Building Cycle," *Review of Economic Statistics* 24 (1942): 149–58. Another was John R. Borchert's paper, "American Metropolitan Revolution," *Geographical Review* 57 (1961): 301–32. The summary presented here draws on the work of Brian J. L. Berry in *Long-Wave Rhythms in Economic Development and Political Behavior* (Baltimore, Md.: Johns Hopkins University Press, 1991).
17. Berry, *Long-Wave Rhythms*.
18. United Nations Human Settlements Programme (UN-Habitat), *The State of the World's Cities Report 2001* (Nairobi, Kenya: UN-Habitat, 2001).

CHAPTER 2

1. P. Wheatley, *The Pivot of the Four Quarters* (Edinburgh: Edinburgh University Press, 1971), xviii.
2. W. Sjoberg, "The Origin and Evolution of Cities," *Scientific American*, September (1965): 55–56.
3. V. G. Childe, "The Urban Revolution," *Town Planning Review* 21 (1950): 3–17.
4. H. Carter, *An Introduction to Urban Historical Geography* (London: Edward Arnold, 1983), 2.
5. Childe, "The Urban Revolution."
6. L. Woolley, "The Urbanisation of Society," in *History of Mankind*, eds. J. Hawkes and L. Woolley, *The Beginnings of Civilisation*, Vol. 1, Part 2 (Paris: UNESCO, 1963), 109–67.
7. K. A. Wittfogel, *Oriental Despotism: A Comparative Study of Total Power* (New Haven, Conn.: Yale University Press, 1957).
8. E. Boserup, *Population and Technology* (Oxford, UK: Blackwell, 1981).
9. J. Jacobs, *The Economy of Cities* (New York: Random House, 1969).
10. M. Weber, *The City.* Translated and edited by D. Martindale and G. Neuwirth (New York: The Free Press, 1958).
11. Wittfogel, *Oriental Despotism*.
12. Wheatley, *The Pivot of the Four Quarters*, 298–99.

13. W. Sjoberg, *The Pre-Industrial City, Past and Present* (Glencoe, Ill.: Free Press, 1960).
14. Wheatley, *The Pivot of the Four Quarters,* 298–99.
15. Ibid., 318.
16. A. E. J. Morris, *History of Urban Form: Before the Industrial Revolution,* 3rd ed. (London: Pearson Education, 1994).
17. Sjoberg, *The Pre-Industrial City, Past and Present.*
18. L. Woolley, *Ur of the Chaldees: A Record of Seven Years of Excavation* (New York: W. W. Norton & Company, 1965), 423–25.
19. Morris, *History of Urban Form.*
20. See Sjoberg's excellent account of urban expansion in *The Pre-Industrial City, Past and Present.*
21. C. Chant and D. Goodman, eds., *Pre-Industrial Cities & Technology* (London: Routledge, 1999), 56–58.
22. A. Kriesis, *Greek Town Building* (Athens: F. Constantinidis & C. Michalas, 1965), 126.
23. Morris, *History of Urban Form,* 56–61.
24. Ibid., 58.
25. Sjoberg, *The Pre-Industrial City, Past and Present,* 55.
26. Morris, *History of Urban Form,* 92.
27. Ibid., 98.
28. M. Bishop, *The Penguin Book of the Middle Ages* (Harmondsworth, UK: Penguin Books, 1971), 209.
29. F. Engels, *The Condition of the Working Class in England* (Translated and edited by W. O. Henderson and W. H. Chaloner, Stanford: Stanford University Press, 1968).
30. J. G. Kohl, *England and Wales* (London: Frank Cass, 1844. Reprinted in 1968 by Augustus M. Kelley, Publishers, New York), 146.
31. M. Pacione, *Glasgow: The Socio-Spatial Development of the City* (Chichester, UK: John Wiley & Sons, 1995), 84–86.

CHAPTER 3

1. D. Ward, *Poverty, Ethnicity, and the American City, 1840–1925: Changing Conceptions of the Slum and the Ghetto* (Cambridge, UK: Cambridge University Press, 1989).
2. G. K. Zipf, *Human Behavior and the Principle of Least Effort* (Reading, Mass.: Addison-Wesley, 1949).
3. M. Jefferson, "The Law of the Primate City," *Geographical Review* 29 (1939): 231.
4. W. Christaller, *Central Places in Southern Germany,* trans. C. W. Baskin (Englewood Cliffs, N.J.: Prentice Hall, 1966). Originally published by Gustav Fischer, Jena, 1933.
5. See A. Lösch, *The Economics of Location,* trans. W. H. Woglom and W. F. Stolper (New Haven, Conn.: Yale University Press, 1954). Originally published by Gustav Fischer, Stuttgart, 1940.

6. J. U. Marshall, *The Structure of Urban Systems* (Toronto: University of Toronto Press, 1989), 278.
7. J. Gottman, *Megalopolis* (New York: The Twentieth Century Fund, 1961).
8. N. Smith, *Uneven Development* (New York: Blackwell, 1984), 150.
9. See B. J. L. Berry, *Long-Wave Rhythms in Economic Development and Political Behavior* (Baltimore, Md.: Johns Hopkins University Press, 1991), 32–33.

CHAPTER 4

1. J. Borchert, "Major Control Points in American Economic Geography," *Annals of the Association of American Geographers* 68 (1978): 230.
2. The list of Fortune 500 companies in 2002 showed that New York City was home to the headquarters of 40 of "America's largest corporations," Chicago had 11, Houston 18, and Atlanta 12 (*Fortune*, April 14, 2003, F-1 to F-19). It should be noted, however, that Houston's city limits, in particular, are extensive enough to capture those companies located in its suburbs. In contrast, New York and Chicago are "underbounded," and, since the 1970s, have "lost" headquarter offices to nearby suburban and exurban locations outside their city limits. For example, Sears Roebuck relocated its headquarters within the Greater Chicago region in the late 1980s—from the Sears Tower in downtown Chicago to Hoffman Estates, 30 miles to the northwest.
3. In constant 1990 dollars. See U.S. Congress, House Committee on Ways and Means, *Green Book, 1991* (Washington, D.C.), Appendix F, Table 34.
4. M. Castells, *The Informational City* (Cambridge, Mass.: Basil Blackwell, 1989).
5. B. Ó hUallacháin and N. Reid, "The Location and Growth of Business and Professional Services in American Metropolitan Areas, 1976–1986," *Annals of the Association of American Geographers* 81 (1991): 254–70.
6. Ibid., 10.
7. Castells, *The Informational City.*
8. OECD, *Understanding the Digital Divide* (Paris: Organisation for Economic Cooperation and Development, 2001).
9. United States Department of Commerce, *A Nation Online: How Americans Are Expanding their Use of the Internet,* (Washington, D.C.: Economics and Statistics Administration; National Telecommunications and Information Administration, February 2002) (*http://www.ntia.doc.gov/ntiahome/dn/*).
10. M. Pastore, "West Coast Cities among Most Wired in U.S.," *CyberAtlas* (*http://cyberatlas.internet.com*).
11. U.S. Census Bureau, *Home Computers and Internet Use in the United States: August 2000,* by Eric C.

Newburger (Washington, D.C.: U.S. Department of Commerce, September 2001) (*http://www. census.gov/prod/2001pubs/p23-207.pdf*).

12. B. Warf, "Segueways into Cyberspace: Multiple Geographies of the Digital Divide," *Environment and Planning B: Planning and Design* 28 (2001): 3–19.

13. S. Graham, "Bridging Urban Digital Divides? Urban Polarization and Information and Communications Technologies (ICTs)," *Urban Studies* 39 (2002): 33–56.

14. See J. O. Wheeler and R. L. Mitchelson, "Information Flows among Major Metropolitan Areas in the United States," *Annals of the Association of American Geographers* 79 (1989): 523–43; R. L. Mitchelson and J. O. Wheeler, "The Flow of Information in a Global Economy: The Role of the American Urban System in 1990," *Annals of the Association of American* Geographers 84 (1994): 87–107; J. O. Wheeler, "Local Information Links to the National Metropolitan Hierarchy: The Southeastern United States," *Environment and Planning A* 31 (1999): 841–54.

15. UN-Habitat (United Nations Centre for Human Settlements), *Cities in a Globalizing World: Global Report on Human Settlements 2001* (London: Earthscan Publications, 2001), 6.

16. S. Graham and S. Marvin, *Splintering Urbanism* (New York: Routledge, 2001).

17. UN-Habitat, *Cities in a Globalizing World*.

18. T. Noyelle, "The Implications of Industry Restructuring for Spatial Organization in the United States," in *Regional Analysis and the New Division of Labor*, eds. F. Moulaert and P. W. Salinas (The Hague: Kluwer Nijhof, 1983), 126.

19. The classification used here draws heavily on the analysis of the employment structure and functional characteristics of the 140 largest U.S. cities by Noyelle and Stanback (1984) (see Key Sources and Suggested Reading).

20. D. Savageau with Ralph D'Agostino, *Places Rated Almanac*, millennium edition (Foster City, Calif.: IDG Books Worldwide Inc., 2000).

21. See L. McCarthy and D. Danta, "Cities of Europe," in *Cities of the World: World Regional Urban Development*, eds. S. Brunn, J. Williams, and D. Zeigler, 3rd ed. (Lanham, Md.: Rowman & Littlefield, 2003), 168–221.

22. J. K. Galbraith, *The Affluent Society*, 4th ed. (Boston: Houghton Mifflin, 1984).

23. Between 1970 and 1990, the U.S. metropolitan population increased by 38 percent, while the number of households in metropolitan areas increased by 62 percent.

24. Between 1973 and 2001, the real median income of all households headed by someone under 35 increased only by just under 10 percent.

25. W. H. Frey, "Boomer Havens and Young Adult Magnets," *American Demographics* (September 2001).

26. Ibid.

27. Ibid.

28. W. H. Frey, "Seniors in Suburbia," *American Demographics* (November 2001), 18.

29. B. J. L. Berry, ed., *Urbanization and Counterurbanization* (Beverly Hills, Calif.: Sage, 1976); B. J. L. Berry, *The Changing Shape of Metropolitan America: Commuting Patterns, Urban Fields, and Decentralization Processes* (Cambridge, Mass.: Ballinger, 1977).

30. U.S. Census Bureau, *Demographic Trends in the 20th Century*, by Frank Hobbs and Nicole Stoops (Washington, D.C.: U.S. Department of Commerce, November 2002) (*http://www.census. gov/prod/2002pubs/censr-4.pdf*).

CHAPTER 5

1. W. Sjoberg, *The Pre-Industrial City, Past and Present* (Glencoe, Ill.: Free Press, 1960).

2. J. E. Vance, Jr., "Land Assignment in Pre-Capitalist, Capitalist and Post-Capitalist Cities," *Economic Geography* 47 (1971): 101–120.

3. J. P. Radford, "Testing the Model of the Pre-Industrial City: The Case of Ante-Bellum Charleston, South Carolina," *Transactions of the Institute of British Geographers* 4 (1979): 392–410.

4. J. S. Adams, "Residential Structure of Midwestern Cities," *Annals of the Association of American Geographers* 60 (1970): 37–62.

5. S. B. Warner, Jr., *The Urban Wilderness* (New York: Harper & Row, 1972), 19.

6. Ibid.

7. See, for example, the analysis of locational conflict within Worcester, Mass., during the nineteenth century in W. B. Meyer and M. Brown, "Locational Conflict in a Nineteenth Century City," *Political Geography Quarterly* 8 (1989): 107–22.

8. L. Mumford, *The City in History* (London: Secker and Warburg, 1961), 461.

9. C. G. Kennedy, "Commuter Services in the Boston Area 1835–1860," *Business History Review* 26 (1962): 277–87.

10. J. E. Vance, Jr., "Housing the Worker: The Employment Linkage as a Force in Urban Structure," *Economic Geography* 42 (1966): 294–325.

11. Mumford, *The City in History*, 460.

12. R. McKenzie, *The Metropolitan Community* (New York: McGraw-Hill, 1933).

13. R. Hurd, *Principles of City Land Values* (New York: The Record and Guide, 1903), 13.

14. H. Hoyt, *The Structure and Growth of Residential Neighborhoods in American Cities* (Washington, D.C.: Federal Housing Administration, 1939).

15. W. Alonso, "A Theory of the Urban Land Market," *Papers and Proceedings of the Regional Science Association* 6 (1960): 149–58.

CHAPTER 6

1. P. J. Hugill, "Good Roads and the Automobile in the United States, 1880–1929," *Geographical Review* 72 (1982): 327–49.
2. H. P. Chudacoff and J. E. Smith, *The Evolution of American Urban Society*, 5th ed. (Upper Saddle River, N.J.: Prentice Hall, 2000), 219.
3. H. Reasoner, CBS News, *60 Minutes*, cited in *Building American Cities*, eds. J. Feagin and R. Parker, 2nd ed. (Englewood Cliffs, N.J.: Prentice Hall, 1991), 156.
4. Ibid., 157.
5. D. Schaffer, *Garden Cities for America: The Radburn Experience* (Philadelphia: Temple University Press, 1982), 173–74, 177.
6. Jackson, *Crabgrass Frontier*, 193.
7. W. Leach, "Transformations in a Culture of Consumption," *Journal of American History* 71 (1984): 326, cited in Chudacoff and Smith, *The Evolution of American Urban Society*, 221.
8. C. D. Harris and E. L. Ullman, "The Nature of Cities," *Annals, American Academy of Political and Social Science* CCXLII (1945): 7–17.
9. P. Hall, *Cities of Tomorrow: An Intellectual History of Urban Planning and Design in the Twentieth Century*, 3rd ed. (New York: Basil Blackwell, 2002), 316–19.
10. D. W. Meinig, "Symbolic Landscapes," in *The Interpretation of Ordinary Landscapes*, eds. D. W. Meinig et al. (New York: Oxford University Press, 1979), 169, 171.
11. A. Scott, *Metropolis: From the Division of Labor to Urban Form* (Berkeley, Calif.: University of California Press, 1989).
12. J. A. Casazza et al., *Shopping Center Development Handbook* (Washington, D.C.: Urban Land Institute, 1985), 16.
13. S. Graham and S. Marvin, *Splintering Urbanism* (New York: Routledge, 2001).
14. P. Lewis, "The Galactic Metropolis," in *Beyond the Urban Fringe*, eds. R. Platt and G. Macuriko (Minneapolis: University of Minnesota Press, 1983), 23–49.
15. P. Hall, "Global City-Regions in the Twenty-First Century," in *Global City-Regions: Trends, Theory, Policy*, ed. A. J. Scott (New York: Oxford University Press, 2001), 59–77.
16. R. Fishman, *Bourgeois Utopias* (New York: Basic Books, 1987).
17. The discussion of Japanese cities draws heavily on J. F. Williams and K. W. Chan, "Cities of East Asia," in *Cities of the World: World Regional Urban Development*, eds. S. D. Brunn, J. F. Williams, and D. J. Zeigler, 3rd ed. (Lanham, Md.: Rowman & Littlefield, 2003), 412–55.
18. R. J. Stimson and S. Baum, "Cities of Australia and the Pacific Islands," in *Cities of the World: World Regional Urban Development*, eds. S. D. Brunn, J. F. Williams, and D. J. Zeigler, 3rd ed. (Lanham, Md.: Rowman & Littlefield, 2003), 456–88.
19. This discussion of Australian cities draws heavily on B. Badcock, *Making Sense of Cities: A Geographical Survey* (London: Arnold, 2002).
20. B. Badcock, "The Imprint of the Post-Fordist Transition on Australian Cities," in *Globalizing Cities: A New Spatial Order?* eds. P. Marcuse and R. van Kempen (Oxford: Blackwell, 2000), 225.

CHAPTER 7

1. See Box: "Core, Semi-Periphery, and Periphery in the World-System."
2. United Nations Centre for Human Settlements (UN-Habitat), *Cities in a Globalizing World: Global Report on Human Settlements 2001* (London: Earthscan Publications Ltd., 2001), Table A.2.
3. Ibid.
4. See J. Friedmann, "The End of the Third World," *Third World Planning Review* 14 (1992): iv–viii.
5. I. Wallerstein, *The Politics of the World-Economy* (Cambridge, UK: Cambridge University Press, 1984).
6. W. W. Rostow, *The Stages of Economic Growth: A Non-Communist Manifesto* (Cambridge, UK: Cambridge University Press, 1960).
7. G. Myrdal, *Economic Theory and Underdeveloped Regions* (London: Duckworth, 1957).
8. A. O. Hirschman, *The Strategy of Economic Development* (New Haven, Conn.: Yale University Press, 1958).
9. F. Perroux, "Note Sur la Notion de Pole de Croissance," (1955), in *Development Economics and Policy: Selected Readings*, ed. I. Livingstone, 1979 (London: Allen and Unwin, 1979), 182–87.
10. J. Friedmann, *Regional Development Policy* (Cambridge, Mass.: MIT Press, 1966).
11. M. Lipton, *Why Poor People Stay Poor: Urban Bias in World Development* (Cambridge, Mass.: Harvard University Press, 1977).
12. A. G. Frank, *Capitalism and Underdevelopment in Latin America: Historical Studies in Chile and Brazil* (New York: Monthly Review Press, 1967), 146–47.
13. Wallerstein, *The Politics of the World-Economy*.
14. See Chapter 1.
15. A. Gilbert and J. Gugler, *Cities, Poverty and Development: Urbanization in the Third World*, 2nd ed. (Oxford, UK: Oxford University Press, 1992), 44.

16. D. A. Smith, *Third World Cities in Global Perspective: The Political Economy of Uneven Urbanization* (Boulder, Colo.: Westview Press, 1996), 8.

17. A. Gilbert and J. Gugler, *Cities, Poverty and Development: Urbanization in the Third World*, 2nd ed. (Oxford, UK: Oxford University Press, 1992), 44.

18. Drakakis-Smith's account forms the basis of this discussion. See D. Drakakis-Smith, *Third World Cities*, 2nd ed. (London: Routledge, 2000), 46–49.

19. A. D. King, *Urbanism, Colonialism, and the World Economy: Cultural and Spatial Foundations of the World Urban System* (London: Routledge, 1990).

20. J. Dickenson, B. Gould, C. Clarke, S. Mather, M. Prothero, D. Siddle, C. Smith, and E. Thomas-Hope, *A Geography of the Third World*, 2nd ed. (London: Routledge, 1996).

21. D. Simon, *Cities, Capital and Development: African Cities in the World Economy* (London: Belhaven Press, 1992), 25–33

22. Drakakis-Smith, *Third World Cities*.

23. Myrdal, *Economic Theory and Underdeveloped Regions*.

24. Drakakis-Smith's excellent account of the sequence of phases of colonial urbanization forms the basis for this discussion. See Drakakis-Smith, *Third World Cities*.

25. E. J. Taaffe, R. L. Morrill, and P. R. Gould, "Transport Expansion in Underdeveloped Countries: A Comparative Analysis," *Geographical Review* 53 (1963), 503–29.

26. Rostow, *The Stages of Economic Growth*.

27. S. Graham and S. Marvin, *Splintering Urbanism* (New York: Routledge, 2001).

CHAPTER 8

1. C. S. Sargent, "The Latin American City," in *Latin America and the Caribbean: A Systematic and Regional Survey*, eds. B. W. Blouet and O. M. Blouet (New York: John Wiley & Sons, Inc., 2002), 167–68.

2. See L. R. Ford, "A New and Improved Model of Latin American City Structure," *The Geographical Review* 86 (1996): 437–40. This model is a "new and improved" version of the earlier one in E. Griffin and L. Ford, "A Model of Latin American City Structure," *Geographical Review* 70 (1980): 397–422.

3. W. K. Crowley, "Modeling the Latin American City," *Geographical Review* 88 (1998): 127–30; W. K. Crowley, "Order and Disorder—A Model of Latin American Urban Land Use," in *Yearbook of the Association of Pacific Coast Geographers* 57, ed. D. E. Turbeville III (Corvallis, Ore.: Oregon State University Press, 1995), 9–31.

4. D. D. Arreola and J. R. Curtis, *The Mexican Border Cities: Landscape Anatomy and Place Personality* (Tucson, Ariz.: The University of Arizona Press, 1993), 69.

5. M. Hays-Mitchell and B. J. Godfrey, "Cities of South America," in *Cities of the World: World Regional Urban Development*, eds. S. D. Brunn, J. F. Williams, and D. J. Zeigler, 3rd ed. (Lanham, Md.: Rowman & Littlefield, 2003), 122–67.

6. C. C. Diniz, "Polygonized Development in Brazil: Neither Decentralization nor Continued Polarization," *International Journal of Urban and Regional Research* 18 (1994): 293–314.

7. United Nations, *Urban Land Policies and Land-Use Control Measures: Volume 1. Africa* (New York: UN Department of Economic and Social Affairs, 1973), 10.

8. A. O'Connor, *The African City* (London: Hutchinson & Co., 1983).

9. R. J. Davis, "The Spatial Formation of the South African City," *GeoJournal* Supplementary Issue 2 (1981): 59–72.

10. T. M. Wills, "Pietermaritzburg," in *Homes Apart: South Africa's Segregated Cities*, ed. A. Lemon (London: Paul Chapman Publishing Ltd., 1991), 91–103.

11. A. Lemon, ed., *Homes Apart: South Africa's Segregated Cities* (London: Paul Chapman Publishing Ltd., 1991).

12. S. Aryeetey-Attoh, "Urban Geography of Sub-Saharan Africa," in *Geography of Sub-Saharan Africa*, ed. S. Aryeetey-Attoh, 2nd ed. (Upper Saddle River, N.J.: Pearson Education, Inc., 2003), Chapter 10, 254–97.

13. A. Mehretu and C. Mutambirwa, "Cities of Sub-Saharan Africa," in *Cities of the World: World Regional Urban Development*, eds. S. D. Brunn, J. F. Williams, and D. J. Zeigler, 3rd ed. (Lanham, Md.: Rowman & Littlefield, 2003), 290–328.

14. D. J. Zeigler, "Cities of the Greater Middle East," in *Cities of the World: World Regional Urban Development*, eds. S. D. Brunn, J. F. Williams, and D. J. Zeigler, 3rd ed. (Lanham, Md.: Rowman & Littlefield, 2003), 254–91.

15. S. Lowder, *The Geography of Third World Cities* (Totowa, N.J.: Barnes & Noble Books, 1986), 29–31.

16. D. J. Zeigler, *Cities of the World*, 269–71.

17. Ibid., 274–76; Zeigler's description of these cities forms the basis for this discussion.

18. Ibid., 275.

19. A. K. Dutt and G. M. Pomeroy, "Cities of South Asia," in *Cities of the World: World Regional Urban Development*, eds. S. D. Brunn, J. F. Williams, and D. J. Zeigler, 3rd ed. (Lanham, Md.: Rowman & Littlefield, 2003), 330–71.

20. J. Goss, "Urbanization," in *Southeast Asia: Diversity and Development*, eds. T. R. Leinbach and R. Ulack (Upper Saddle River, N.J.: Prentice Hall, 2000), 110–132.

21. T. G. McGee, *The Southeast Asian City: A Social Geography of the Primate Cities of South East Asia* (New York: Frederick A. Praeger, Publishers, 1967).

22. J. Goss, "Urbanization," in *Southeast Asia: Diversity and Development*.

23. T. G. McGee, *The Southeast Asian City*, 128.

24. L. R. Ford, "A Model of Indonesian City Structure," *The Geographical Review* 83 (1993): 382.

25. T. G. McGee and I. M. Robinson, *The Mega-Urban Regions of Southeast Asia* (Vancouver: UBC Press, 1995), ix–x.

26. T. G. McGee, "The Emergence of Desakota Regions in Asia: Expanding a Hypothesis," in *The Extended Metropolis: Settlement Transition in Asia*, eds. N. Ginsburg, B. Koppel, and T. G. McGee (Honolulu: University of Hawaii Press, 1991), 3–25.

27. See N. Ginsburg, B. Koppel, and T. G. McGee, *The Extended Metropolis*.

28. Ibid.

29. J. F. Williams and K. W. Chan, "Cities of East Asia," in *Cities of the World: World Regional Urban Development*, eds. S. D. Brunn, J. F. Williams, and D. J. Zeigler, 3rd ed. (Lanham, Md.: Rowman & Littlefield, 2003), 412–55.

30. C. P. Lo, "Shaping Socialist Chinese Cities: A Model of Form and Land Use," in *China: Urbanization and National Development*, Department of Geography Research Paper No. 196, eds. C-K. Leung and N. Ginsburg (Chicago: University of Chicago, 1980), 130–55.

31. A. K. Dutt, Y. Xie, F. J. Costa, and Z. Yang, "City Forms of China and India in Global Perspective," in *The Asian City: Processes of Development, Characteristics and Planning* (Dordrecht, The Netherlands: Kluwer Academic Publishers, 1994), 41.

32. C. P. Lo, "Shaping Socialist Chinese Cities," in *China: Urbanization and National Development*, eds. C-K. Leung and N. Ginsburg, 154.

33. J. F. Williams and K. W. Chan, "Cities of East Asia," in *Cities of the World: World Regional Urban Development*, eds. S. D. Brunn, J. F. Williams, and D. J. Zeigler, 3rd ed. (Lanham, Md.: Rowman & Littlefield, 2003), 412–55.

34. Ibid.

35. A. K. Dutt, Y. Xie, F. J. Costa, and Z. Yang, "City Forms of China and India in Global Perspective," in *The Asian City: Processes of Development, Characteristics and Planning*, 43.

36. Q. Luo, "Shanghai: The 'Dragon Head' of China's Economy," *Issues & Studies* 33 (1997): 17–32.

37. Ibid.

38. C. Olds, *Globalization and Urban Change: Capital, Culture, and Pacific Rim Mega-Projects* (Oxford, UK: Oxford University Press, 2001).

39. Q. Luo, "Shanghai: The 'Dragon Head' of China's Economy," *Issues & Studies* 33 (1997): 17–32.

40. V. F. S. Sit and C. Yang, "Foreign-Investment-Induced Exo-Urbanisation in the Pearl River Delta, China," *Urban Studies* 34 (1997): 647–77.

41. Ibid.

42. See C. C. Diniz, "Polygonized Development in Brazil: Neither Decentralization nor Continued Polarization," *International Journal of Urban and Regional Research* 18 (1994): 293–314.

CHAPTER 9

1. United Nations Centre for Human Settlements (UN-Habitat), *Cities in a Globalizing World: Global Report on Human Settlements 2001* (Nairobi, Kenya: UN-Habitat, 2001).

2. D. Drakakis-Smith, *Third World Cities*, 2nd ed. (London: Routledge, 2000).

3. R. B. Potter and S. Lloyd-Evans, *The City in the Developing World* (Harlow, UK: Longman, 1998).

4. Ibid.

5. D. Drakakis-Smith, *Third World Cities*. 2nd ed. (London: Routledge, 2000).

6. United Nations Centre for Human Settlements (UN-Habitat), *Cities in a Globalizing World*.

7. J. Stewart and P. Balchin, "Community Self-Help and the Homeless Poor in Latin America," *The Journal of the Royal Society for the Promotion of Health* 122 (2002): 99–107.

8. Drakakis-Smith, *Third World Cities*.

9. Ibid.

10. United Nations Centre for Human Settlements (UN-Habitat), *Cities in a Globalizing World*.

11. See the United Nations website: *http://www.unaids.org*.

12. ILO (International Labor Office), *HIV/AIDS: A Threat to Decent Work, Productivity and Development* (Geneva: ILO, 2000).

13. United Nations Population Fund, *The State of the World Population 1996: Changing Places: Population, Development and the Urban Poor* (New York: United Nations, 1996).

14. E. Msiyaphazi Zulu, F. Nii-Amoo Dodoo, and A. Chika-Ezeh, "Sexual Risk-Taking in the Slums of Nairobi, Kenya, 1993–98," *Population Studies* 56 (2002): 311–23.

15. P. Newman, J. Kenworthy, and P. Vintila, *Housing, Transport and Urban Form* (Canberra, Australia: Institute for Science and Technology Policy, Murdoch University, 1992).

16. See G. S. Eskeland and T. Feyzioglu, "Rationing Can Backfire: The 'Day without a Car' in Mexico City," *The World Bank Economic Review* 11 (1997): 383–408.

17. Ibid.

18. S. W. Williams, "'The Brown Agenda:' Urban Environmental Problems and Policies in the Developing World," *Geography* 82 (1997): 17–26; World Bank, *Housing: Enabling Markets to Work*, A World Bank Policy Paper (Washington, D.C.: World Bank, 1993).

19. D. Drakakis-Smith, "Third World Cities: Sustainable Urban Development, 1," *Urban Studies* 32 (1995): 659–77.

20. C. Bartone, J. Bernstein, J. Leitmann, and J. Eigen, *Toward Environmental Strategies for Cities: Policy Considerations for Urban Environmental Management in Developing Countries* (Washington, D.C.: UNDP/UNCHS/World Bank, Urban Management Programme, 1994).

21. United Nations Centre for Human Settlements (UN-Habitat), *Cities in a Globalizing World*, 58.

22. Ibid.

23. United Nations Centre for Human Settlements (UN-Habitat), *An Urbanizing World: Global Report on Human Settlements, 1996* (Oxford, UK: Oxford University Press for UN-Habitat, 1996), 331, Box 9.25.

24. United Nations Centre for Human Settlements (UN-Habitat), *Cities in a Globalizing World*, 125, Box 10.4.

25. United Nations Centre for Human Settlements (UN-Habitat), *An Urbanizing World*, 313, Box 9.18.

26. United Nations Centre for Human Settlements (UN-Habitat), *An Urbanizing World: Global Report on Human Settlements, 1996* (Oxford, UK: Oxford University Press for UN-Habitat, 1996), 323, Table 9.3.

27. F. Schuurman and T. van Naerssen, *Urban Social Movements in the Third World* (London: Routledge, 1989).

28. M. Castells, *The City and the Grassroots: A Cross-Cultural Theory of Urban Social Movements* (London: Arnold, 1983).

29. Schuurman and Naerssen, *Urban Social Movements in the Third World*.

30. Y. Corcoran-Nantes, "Women and Popular Urban Social Movements in São Paulo, Brazil," *Bulletin of Latin American Research* 9 (1990): 249–64.

31. United Nations Centre for Human Settlements (UN-Habitat), *Cities in a Globalizing World*, 167, Box 14.4.

32. United Nations Centre for Human Settlements (UN-Habitat), *An Urbanizing World*, 422.

CHAPTER 10

1. L. Mumford, *The Culture of Cities* (New York: Harcourt, Brace and World, 1938), 403.

2. R. Glass, "Urban Sociology in Great Britain," in *Readings in Urban Sociology*, ed. R. E. Pahl (Oxford, UK: Pergamon, 1968), 21–46.

3. R. Darke and J. Darke, "Towards a Sociology of the Built Environment," *Architectural Psychology Newsletter* 11 (1981): 12.

4. H. Lasswell, *The Signature of Power* (New York: Transaction Books, 1979).

5. D. Appleyard, "The Environment as a Social Symbol," *Ekistics* 278 (1979): 272.

6. The argument is made in Emerson's *Nature* (Boston: James Munroe & Co., 1836). See also J. Woods, "Build, Therefore, Your Own World": The New England Village as Settlement Ideal," *Annals of the Association of American Geographers* 81 (1991): 32–50.

7. H. D. Thoreau, *Walden* (New York: Holt, Rinehart and Winston, 1961). First published in 1854.

8. F. J. Turner, *The Frontier in American History* (New York: Henry Holt. 1920).

9. L. Marx, *The Machine in the Garden* (New York: Oxford University Press, 1964), 12.

10. Ibid.

11. B. Rubin, "Aesthetic Ideology and Urban Design," *Annals of the Association of American Geographers* 69 (1979): 342.

12. For a detailed review of the movement and its legacy, see W. H. Wilson, *The City Beautiful Movement* (Baltimore: Johns Hopkins University Press, 1989).

13. M. Tafuri and F. D. Co, *Modern Architecture: 1* (New York: Rizzoli, 1986), 40.

14. M. Domosh, "The Symbolism of the Skyscraper," *Journal of Urban History* 14 (1988): 321–45.

15. R. Hughes, *The Shock of the New* (New York: Knopf, 1980).

16. Tafuri and Co, *Modern Architecture*, 116.

17. E. Relph, *The Modern Urban Landscape* (London: Croom Helm, 1987), 118.

18. P. Hall, *Cities of Tomorrow: An Intellectual History of Urban Planning and Design in the Twentieth Century*, 3rd ed. (New York: Basil Blackwell, 2002), 219.

19. Relph, *The Modern Urban Landscape*, 198.

20. F. L. Wright, *When Democracy Builds* (Chicago: University of Chicago Press, 1945).

21. Hall, *Cities of Tomorrow*, 314–15.

22. J. Jacobs, *The Death and Life of Great American Cities* (New York: Vintage Books, 1961).

23. O. Newman, *Defensible Space* (New York: Macmillan, 1972).

24. R. Venturi, *Complexity and Contradiction in Architecture* (New York: Museum of Modern Art, 1966).

25. T. Wolfe, *From Bauhaus to Our House* (New York: Farrar, Straus, Giroux, 1981), 109–10.

26. D. Harvey, *The Condition of Postmodernity* (Oxford, UK: Blackwell, 1989).

27. V. Scully, *American Architecture and Urbanism* (New York: Henry Holt, 1988), 287–88. Historic preservation involves the identification, evaluation, and protection of historic and archaeological resources.

28. Aside from recognition of significance, the benefits of National Register listing include eligibility for federal investment tax credits for renovating

income-producing properties and consideration in projects involving federal funding. Although National Register listing does not protect against alteration or demolition by private property owners using their own funds, listing by local historic district commissions does.

29. D. Listokin, "Landmark Designation: An Emerging Form of Land-Use Control." Paper presented to the annual meeting of the Association of Collegiate Schools of Planning, Portland, Oregon, 1989; National Park Service, *National Register of Historic Places* (*http://www.cr.nps.gov/nr/about.htm*).

30. M. Davis, *City of Quartz: Excavating the Future of Los Angeles* (New York: Verso, 1990), 226.

31. Ibid., 238.

CHAPTER 11

1. Harvey first outlined these ideas in a 1975 essay, "The Political Economy of Urbanization in Advanced Capitalist Societies: The Case of the United States," in *The Social Economy of Cities*, eds. G. Gappert and H. Rose (Beverly Hills, Calif.: Sage Publications), 119–163. He has subsequently developed them in a series of books and articles. See "The Urban Process Under Capitalism," *International Journal of Urban and Regional Research* 2 (1978): 101–31; *The Limits to Capital* (Oxford, UK: Basil Blackwell, 1982); and *The Urbanization of Capital* (Oxford, UK: Basil Blackwell, 1985).

2. R. J. King, "Capital Switching and the Role of Ground Rent: 1, Theoretical Problems," *Environment & Planning A* 21 (1989): 450.

3. *Property* here means land or buildings, or both.

4. M. Edel, "Marx's Theory of Rent: Urban Applications," *Kapitalistate* 4–5 (1976): 100–124. Summarized in King, "Capital Switching," 445–62.

5. A. Haila, "Four Types of Investment in Land and Property," *International Journal of Urban and Regional Research* 15 (1991): 348.

6. Ibid., 349–51.

7. Harvey, *The Limits to Capital*, 347.

8. Ibid., 368.

9. See A. Haila, "Land as a Financial Asset: The Theory of Urban Rent as a Mirror of Economic Transformation," *Antipode* 20 (1988): 79–101.

10. E. J. Kaiser and S. Weiss, "Public Policy and the Residential Development Process," *Journal of the American Institute of Planners* 36 (1970): 30–37.

11. The following description is based on sections of T. Baerwald, "The Site Selection Process of Suburban Residential Builders," *Urban Geography* 2 (1981): 339–57; and on P. Knox and S. Pinch, *Urban Social Geography*, 4th ed. (Harlow, UK: Pearson Education, 2000), 180–200.

12. J. Logan and H. Molotch, *Urban Fortunes: The Political Economy of Place* (Berkeley: University of California Press, 1987), 29–31.

13. Ibid., 30.

14. Figure 11.5 and the accompanying discussion are based on the outline of the development process given by Harvey Rabinowitz in *Urban Planning*, eds. A. J. Catanese and J. C. Snyder, 2nd ed. (New York: McGraw-Hill, 1988), 245–49.

15. Baerwald, "The Site Selection Process," 351.

16. The Southern Pacific Company lost money on operations consistently in the 1980s and 1990s and was acquired by the Union Pacific and merged into its operations in 1995.

17. M. Davis, *City of Quartz* (New York: Verso, 1990), 131–32.

18. In 1996 the Rouse Company acquired the Hughes Corporation.

19. J. R. Feagin and R. Parker, *Building American Cities. The Urban Real Estate Game* (Englewood Cliffs, N.J.: Prentice Hall, 1991), 21. Emphasis added.

20. The following account draws on D. Mangan, "Consolidation of the Shopping Center Industry," *Urban Land* 49, no. 6 (1990): 30–31.

21. The Canadian company, Trizec Corporation, purchased Hahn in 1980 and operated as TrizecHahn Corporation until 2002, when it was reorganized as Trizec Properties (a publicly traded U.S. real estate investment trust [REIT]) and Trizec Canada (this ownership structure enables non-U.S. investors to invest in Trizec Properties through Trizec Canada). The Rouse Company, another REIT, is one of the U.S.'s largest owners of commercial real estate, with some 50 shopping centers (including Baltimore's The Gallery at Harborplace, Boston's Faneuil Hall Marketplace, and New York's South Street Seaport) and approximately 125 office, mixed-use, and industrial properties; Rouse acquired the Hughes Corporation in 1996. Melvin Simon & Associates (which completed the Mall of America in 1992) went public in 1993 and folded most of its properties into the Simon Property Group. Then, following a $3-billion merger with DeBartolo Realty Corp. (founded by Edward DeBartolo, Sr.) in 1996 it became Simon DeBartolo Group but has since reverted to the Simon Property Group (now the largest publicly traded REIT and shopping center owner in the United States) after it split with DeBartolo in 1998 following Edward DeBartolo, Jr.'s resignation from the board amid reports that he was being investigated by a federal grand jury looking into whether he committed fraud in connection with a Louisiana casino license he won. Meanwhile, DeBartolo Jr. established the

commercial real estate company, the DeBartolo Property Group (DPG) in 2001.

22. General Growth Properties, a REIT, purchased Homart Development Co.'s portfolio of regional malls and community shopping centers from Sears for $1.85 billion in 1995; the Simon Property Group purchased May Centers in 1993.

23. "Securitization" is the conversion of assets into a financial obligation that has readily identifiable characteristics and can be accordingly rated to risk in the international capital markets. Such rating can be effected through the same commercial rating services that rate corporate bonds or through credit enhancement by a third party. In the past Japanese banks have been very competitive in this process because of their own high credit ratings, low funding costs, and the low fees they have been willing to charge in order to promote their global business expansion. See J. Logan, "Cycles and Trends in the Globalization of Real Estate," in *The Restless Urban Landscape*, ed. P. L. Knox (Englewood Cliffs, N.J.: Prentice Hall, 1993), 35–54.

24. L. McCarthy, "The Brownfield Dual Land-Use Policy Challenge: Reducing Barriers to Private Development while Connecting Reuse to Broader Community Goals," *Land Use Policy* 19 (2002): 287–96.

25. J. J. Ori, "Real Estate Strategies for Profitability in the 1990s," *Real Estate Review* 20 (1990): 94–96; J. J. Ori, "New Strategies for Real Estate Industry Success," *Real Estate Review* 28 (1998): 34–37.

26. J. Hammer, "In Practice: The Entrepreneurial Mind," *Urban Land*, May (2002) (*http://planet.uli.org/Content/University/Univ_GradCtr.htm*). The limitations and dangers of continuing consolidation within the real estate industry are discussed, for example, in articles in the journal *Real Estate Review*.

27. D. R. Suchman, "Housing and Community Development," in *Development Trends 1989*, ed. D. Schwanke (Washington, D.C.: Urban Land Institute, 1990), 35.

28. Ibid., 23.

29. *Architectural Record* (June 1937), 9. Quoted in S. Zukin, "The Postmodern Debate over Urban Form," *Theory, Culture and Society* 5 (1988): 437–38.

CHAPTER 12

1. Melvin Webber first advanced this argument in the early 1960s. See his essay, "The Urban Place and the Nonplace Urban Realm," in *Explorations into Urban Structure*, M. M. Webber *et al.* (Philadelphia, Pa.: University of Pennsylvania Press, 1964), 79–153.

2. S. Keller, *The Urban Neighborhood: A Sociological Perspective* (New York: Random House, 1968).

3. R. J. Johnston, *The American Urban System: A Geographical Perspective* (New York: St. Martin's Press, 1982).

4. G. Suttles, *The Social Construction of Communities* (Chicago, Ill.: University of Chicago Press, 1972).

5. For a discussion of human territoriality from a geographer's perspective, see R. Sack, "Human Territoriality: A Theory," *Annals of the Association of American Geographers* 73 (1983): 55–74.

6. L. H. Lofland, *A World of Strangers* (New York: Basic Books, 1973).

7. For a detailed discussion of class, see A. Giddens, *The Class Structure of the Advanced Societies*, 2nd ed. (New York: Harper & Row, 1981).

8. R. Walker, "Class, Division of Labor, and Employment in Space," in *Social Relations and Spatial Structures*, eds. D. Gregory and J. Urry (London: Macmillan, 1985), 187. Emphasis added.

9. D. W. Harvey, "Class Structure in a Capitalist Society and the Theory of Residential Differentiation," in *Process in Physical and Human Geography*, eds. R. Peel, M. Chisholm, and P. Haggett (London: Heinemann, 1975).

10. See R. L. Morrill and F. R. Pitts, "Marriage, Migration, and the Mean Information Field," *Annals of the Association of American Geographers* 57 (1967): 401–22.

11. M. Farnsworth Riche, *The Implications of Changing U.S. Demographics for Housing Choice and Location in Cities* (Washington, D.C.: The Brookings Institution, 2001).

12. This discussion is based on P. Knox and S. Pinch, *Urban Social Geography*, 4th ed. (Harlow, UK: Pearson Education, 2000), 228–29.

13. See F. W. Boal, "Ethnic Residential Segregation," in *Social Areas in Cities*, vol. 1, eds. D. T. Herbert and R. J. Johnston (New York: Wiley, 1976).

14. Ibid., 109–110.

15. W. Bell, "Social Choice, Lifestyles, and a Theory of Social Choice," in *The Suburban Community*, ed. W. Dobriner (New York: Putnam, 1958).

16. See D. Timms, "Quantitative Techniques in Urban Social Geography," in *Frontiers in Geographical Teaching*, eds. R. Chorley and P. Haggett (London: Methuen, 1965), 239–265.

17. J. Iceland and D. H. Weinberg, *Racial and Ethnic Residential Segregation in the United States: 1980–2000* (Washington, D.C.: U.S. Department of Commerce, 2002).

18. E. L. Glaeser and J. L. Vigdor, *Racial Segregation in the 2000 Census: Promising News* (Washington, D.C.: The Brookings Institution, 2001).

19. U.S. Department of Housing and Urban Development (HUD), *Discrimination in Metropolitan Housing*

Markets: National Results from Phase 1 of the Housing Discrimination Study (HDS) (Washington, D.C.: HUD, 2002) (*http:www.huduser.org/publications/hsgfin/phase1.html*). This study used "paired testing," in which 4,600 sets of paired individuals—one minority and the other white—posed as otherwise identical homeseekers and visited real estate or rental agents to inquire about the availability of advertised housing units across 23 U.S. metropolitan areas. This methodology is designed to provide direct evidence of differences in the treatment experienced by minorities and whites in their search for housing.

20. R. E. Park, "The City: Suggestions for the Investigation of Human Behavior in an Urban Environment," *American Journal of Sociology* 20 (1916): 608. The following description of human ecology draws on my summary in *Urban Social Geography*, 2nd ed. (London: Longman, 1987), 59–63. For more-detailed discussions, see A. H. Hawley, *Human Ecology* (New York: Ronald Press, 1950); and J. N. Entrikin. "Robert Park's Human Ecology and Human Geography," *Annals of the Association of American Geographers* 70 (1980): 43–58.

21. H. Zorbaugh, *The Gold Coast and The Slum* (Chicago: Chicago University Press, 1929).

22. E. W. Burgess, "The Growth of the City: An Introduction to a Research Project," in *The City*, eds. R. E. Park *et al.* (Chicago, Ill.: University of Chicago Press, 1925).

23. W. Firey, "Sentiment and Symbolism as Ecological Variables," *American Sociological Review* 10 (1945): 140–48.

24. A. Hawley, *Human Ecology: A Theory of Community Structure* (New York: Ronald Press, 1950); L. F. Schnore, *The Urban Scene* (New York: Free Press, 1965).

25. See P. Knox and S. Pinch, *Urban Social Geography*, 4th ed. (Harlow, UK: Pearson Education, 2000), 105–19. For a more detailed discussion of the methodology, see W. K. D. Davies, *Factorial Ecology* (Aldershot, UK: Gower Press, 1984).

26. R. A. Murdie, *Factorial Ecology of Metropolitan Toronto, 1950–1960*, Research Paper no. 116 (Chicago: University of Chicago Department of Geography, 1969), 168.

27. M. White, *American Neighborhoods and Residential Differentiation* (New York: Russell Sage Foundation, 1987).

28. W. K. D. Davies and R. A. Murdie, "Consistency and Differential Impact in Urban Social Dimensionality: Intra-Urban Variation in the 24 Metropolitan Areas of Canada," *Urban Geography* 12 (1991): 55–79.

29. Davies, *Factorial Ecology*.

30. This discussion is based on L. McCarthy and D. Danta, "Cities of Europe," in *Cities of the World: World Regional Urban Development*, eds. S. Brunn, J. Williams, and D. Zeigler, 3rd ed. (Lanham, Md.: Rowman & Littlefield, 2003), 196–202.

31. Davies, *Factorial Ecology*.

32. An Urban Land Institute conference in Dallas in 2003 discussed the trend toward "hiving" and away from "cocooning." See the Urban Land Institute's website: *http://www.experts.uli.org/Content/PressRoom/press_releases/2003/PR_038.htm*.

33. W. H. Frey, *Melting Pot Suburbs: A Census 2000 Study of Suburban Diversity* (Washington, D.C.: The Brookings Institution, 2001).

34. S. Christopherson, "Labor Flexibility in the U.S. Service Economy," *Transactions of the Institute of British Geographers* 14 (1989): 131–143.

35. W. H. Frey, *Melting Pot Suburbs: A Census 2000 Study of Suburban Diversity* (Washington, D.C.: The Brookings Institution, 2001).

36. Ibid.

37. J. R. Logan, *From Many Shores: Asians in Census 2000* (Albany, N.Y.: University of Albany, Lewis Mumford Center for Comparative Urban and Regional Research, 2001) (*http://mumford1.dyndns.org/cen2000/AsianPop/AsianReport/AsianDownload.pdf*).

38. J. R. Logan, *The New Ethnic Enclaves in America's Suburbs* (Albany, N.Y.: University of Albany, Lewis Mumford Center for Comparative Urban and Regional Research, 2001) (*http://mumford1.dyndns. org/cen2000/suburban/SuburbanReport/SubReport.pdf*).

39. W. Li, "Anatomy of a New Ethnic Settlement: The Chinese *Ethnoburb* in Los Angeles," *Urban Studies* 35 (1998): 479–501.

40. W. Li, "Los Angeles's Chinese *Ethnoburb*: From Ethnic Service Center to Global Economy Outpost," *Urban Geography* 19 (1998): 502–17.

41. J. Fields and L. M. Casper, *America's Families and Living Arrangements* (Washington, D.C.: U.S. Department of Commerce, U.S. Census Bureau, 2001).

42. This section draws on P. L. Knox, ed., *The Restless Urban Landscape* (Englewood Cliffs, N.J.: Prentice Hall, 1993), 1–34.

43. A. Heller, "Existentialism, Alienation, Postmodernism: Cultural Movements as Vehicles of Change in Patterns of Everyday Life," in *Postmodern Conditions*, eds. A. Milner *et al.* (New York: Berg, 1990), 1–13.

44. D. Ley, "Modernism, Postmodernism, and the Struggle for Place." Paper presented to the seminar series *The Power of Place* (Department of Geography, Syracuse University, 1986), 28.

45. G. Debord, *The Society of the Spectacle* (New York: Zone Books, 1994), para. 167.

46. K. Butler, "Paté Poverty," *Utne Reader* (Sept.–Oct., 1989), 77.

47. Ibid., 74, 77.

48. B. Ehrenreich, *Fear of Falling* (New York: Pantheon, 1989), 229.

49. Ibid., 228.

50. B. J. L. Berry, *The Human Consequences of Urbanization* (New York: St. Martin's Press, 1973).

51. M. J. Weiss, *The Clustered World: How We Live, What We Buy, and What It All Means About Who We Are* (Boston: Little, Brown and Company, 2000).

52. M. J. Weiss, "Inconspicuous Consumption," *American Demographics*, 24 (2002): 31–32.

53. Ibid., 32.

54. Ibid., 37.

55. Experiencing two or more of the following: poverty-level incomes, unemployment, no telephone, incomplete kitchen or plumbing facilities, and crowded living conditions. See P. Knox and R. Rohr-Zanker, *Economic Change, Demographic Change, and the Composition and Distribution of Vulnerable and Disadvantaged Households in the United States* (Washington, D.C.: U.S. Department of Commerce, Economic Development Administration, 1989).

56. M. Castells, *The Informational City* (Cambridge, Mass.: Blackwell, 1990).

57. W. J. Wilson, *The Truly Disadvantaged* (Chicago, Ill.: University of Chicago Press, 1987).

58. M. Dear and J. Wolch, *Landscapes of Despair* (Princeton, N.J.: Princeton University Press, 1987).

59. S. Zukin, "Urban Lifestyles: Diversity and Standardization in Spaces of Consumption," *Urban Studies* 35 (1998): 835–37.

60. Weiss, *The Clustered World*.

61. The Claritas Inc. PRIZM "segmentation" system webpage (see *http://www.clusterbigip1.claritas.com*) invites U.S. users to enter the 5-digit ZIP code for their own neighborhood and examine the results for themselves.

62. Weiss, *The Clustered World*, 32–33.

63. See Claritas Inc. webpage: *http://www.clusterbigip1.claritas.com*.

64. Weiss, *The Clustered World*, 180 and 300.

65. L. M. Quinn and J. Pawasarat, *Confronting Anti-Urban Marketing Stereotypes: A Milwaukee Economic Development Challenge* (Milwaukee, Wis.: University of Wisconsin–Milwaukee, 2001) (*http://www.uwm.edu/Dept/ETI/purchasing/markets.htm*).

66. J. Pawasarat and L. M. Quinn, *Exposing Urban Legends: The Real Purchasing Power of Central City Neighborhoods* (Washington, D.C.: The Brookings Institution) (*http://www.brook.edu/dybdocroot/es/urban/pawasarat.pdf*), vi.

67. This discussion is based on an industry interview, "Location Analysis Tools Help Starbucks Brew Up New Ideas," *Business Geographics* 8 (October 2000): 32–34; and P. L. Knox and S. A. Marston, *Places and Regions in Global Context: Human Geography*, 3rd ed. (Upper Saddle River, N.J.: Pearson Education, Inc., 2004), 94–95.

CHAPTER 13

1. L. Bourne, *The Geography of Housing* (London: Edward Arnold, 1981), 23.

2. Sweat equity is a term generally used to describe an increase in the value of residential property that results from householders' unpaid labor in remodeling and refurbishing.

3. D. W. Harvey, *Society, the City and the Space-Economy of Urbanism* (Washington, D.C.: Association of American Geographers, College Resource Paper no. 18, 1972), 16.

4. Bourne, *The Geography of Housing*, 14–15.

5. R. J. Johnston, *City and Society: An Outline for Urban Geography* (Harmondsworth, UK: Penguin, 1980), 186.

6. Ibid.

7. J. Vance, Jr., *The Continuing City: Urban Morphology in Western Civilization* (Baltimore, Md.: Johns Hopkins University Press, 1990).

8. U.S. National Committee on Urban Problems, *Building the American City* (Washington, D.C.: U.S. Government Printing Office, 1968), 401.

9. J. Goodman, *Housing Affordability in the United States: Trends, Interpretations, and Outlook* (Washington, D.C.: U.S. Congress, Bipartisan Millennial Housing Commission, 2001).

10. Freddie Mac, *Automated Underwriting: Making Mortgage Lending Simpler and Fairer for America's Families* (McLean, Va.: Freddie Mac, 1996) (*http://freddiemac.com/corporate/reports/Moseley/mosehome.htm*).

11. Goodman, *Housing Affordability in the United States*.

12. Joint Center for Housing Studies of Harvard University. *State of Our Nation's Housing: 2000* (Cambridge, Mass.: Harvard University, 2000).

13. Ibid.

14. Bipartisan Millennial Housing Commission, *Meeting Our Nation's Housing Challenges* (Washington, D.C.: U.S. Congress, 2002), 2.

15. J. S. Fuerst, ed., *Public Housing in Europe and America* (London: Croom Helm, 1974).

16. Bourne, *The Geography of Housing*, 91.

17. This discussion is based on L. McCarthy and D. Danta, "Cities of Europe," in *Cities of the World: World Regional Urban Development*, eds. S. Brunn,

J. Williams, and D. Zeigler, 3ʳᵈ ed. (Lanham, Md.: Rowman & Littlefield, 2003), 189–96.

18. J. S. Adams, "Housing Submarkets in an American Metropolis," in *Our Changing Cities*, ed. J. F. Hart (Baltimore, Md.: Johns Hopkins University Press, 1991), 108–26.

19. D. W. Harvey and L. Chatterjee, "Absolute Rent and the Structuring of Space by Governmental and Financial Institutions," *Antipode* 6 (1974): 22–36.

20. This discussion is based on L. McCarthy and D. Danta, "Cities of Europe," in *Cities of the World: World Regional Urban Development*, eds. S. Brunn, J. Williams, and D. Zeigler, 3ʳᵈ ed. (Lanham, Md.: Rowman & Littlefield, 2003), 189–96.

21. D. Ley, *A Social Geography of the City* (New York: Harper & Row, 1983), 238–39. Emphases added.

22. M. T. Cadwallader, "A Unified Model of Urban Housing Patterns, Social Patterns and Residential Mobility," *Urban Geography* 2 (1981): 115–30.

23. P. H. Rossi, *Why Families Move: A Study in the Social Psychology of Urban Residential Mobility* (New York: Free Press, 1955).

24. L. A. Brown and E. G. Moore, "The Intra-Urban Migration Process: A Perspective," *Geografiska Annaler* 52B (1970): 1–13.

25. E. W. Butler *et al. Moving Behavior and Residential Choice: A National Survey.* National Cooperative Highway Research Program Report no. 81 (Washington, D.C.: Highway Research Board, 1969).

26. See, for example, J. O. Huff, "Geographic Regularities in Residential Search Behavior," *Annals of the Association of American Geographers* 76 (1986): 208–27.

27. See, for example, W. A. V. Clark, "Residential Segregation in American Cities: A Review and Interpretation," *Population Research and Policy Review* 5 (1986): 95–127; G. Galster, "Residential Segregation in American Cities: A Further Response to Clark," *Population Research and Policy Review* 8 (1989): 181–92; and W. A. V. Clark, "Residential Segregation in American Cities: Common Ground and Differences in Interpretation," *Population Research and Policy Review* 8 (1989): 193–97.

28. Economist Albert Hirschman first formulated this framework as a model of consumer strategies when faced with a declining service or inferior product. See A. O. Hirschman, *Exit, Voice and Loyalty* (Cambridge, Mass.: Harvard University Press, 1970).

29. Political scientists John Orbell and Toro Uno adapted Hirschman's "loyalty, voice, and exit" framework to the context of neighborhood change. See J. M. Orbell and T. Uno, "A Theory of Neighborhood Problem Solving: Political Action Versus Residential Mobility," *American Political Science Review* 66 (1972): 471–89.

30. See, for example, S. Kim and G. D. Squires, "The Color of Money and the People Who Lend It," *Journal of Housing Research* 9 (1998): 271–84, which found that the likelihood of home loan approval for black or Hispanic applicants increases as the proportion of black or Hispanic employees, and particularly administrative and professional workers, increases at a lending institution. This finding is consistent with the "cultural affinity" hypothesis: When more employees at a lending institution are from a particular community, they may be better placed to obtain information from marginal applicants and approve applications that do not clearly meet all the objective indicators. At the same time, other reasons for the relationship could be the minority loan officer's knowledge of, or sympathy with, fair lending concerns, the influence of minority coworkers on white loan officers, and self-selection.

31. R. E. Pahl, *Whose City?* 2nd ed. (Harmondsworth, UK: Penguin, 1975).

32. R. Palmer, *Realtors as Social Gatekeepers*. Ph.D. diss., Yale University, 1955.

33. Ibid., 77.

34. P. M. Downing and L. Gladstone, *Segregation and Discrimination in Housing: A review of Selected Studies and Legislation*, Congressional Research Report 89-317 (Washington, D.C.: The Library of Congress, 1989), 25–26.

35. The 2000 study results are available in U.S. Department of Housing and Urban Development (HUD), *Discrimination in Metropolitan Housing Markets: National Results from Phase 1 of the Housing Discrimination Study*, Washington, D.C.: HUD, 2002 (*http://www.huduser.org/publications/hsgfin/phase1.html*). The comparability of the 2000 data with the 1989 *Housing Discrimination Study* results is discussed in Annex 5 of the 2002 study.

36. G. Galster, F. Freiberg, and D. Houk, "Racial Differentials in Real Estate Advertising Practices: An Exploratory Case Study," *Journal of Urban Affairs* 9 (1987): 199–215.

37. M. E. Stone, "Housing, Mortgage Lending, and the Contradictions of Capitalism," in *Marxism and the Metropolis*, eds. W. K. Tabb and L. Sawers (New York: Oxford University Press, 1978), 190.

38. J. Darden, "Lending Practices and Policies Affecting the American Metropolitan System," in *The American Metropolitan System: Present and Future*, eds. S. D. Brunn and J. O. Wheeler (New York: Winston, 1980), 93–110.

39. G. B. Canner and D. S. Smith, "Home Mortgage Disclosure Act: Expanded Data on Residential Lending," *Federal Reserve Bulletin* 77 (1991): 859–81. This study was based on Home Mortgage Disclosure Act (HMDA) data for 5.3 million mortgage applications.

40. Association of Community Organizations for Reform Now (ACORN), *The Great Divide: Home Purchase Mortgage Lending Nationally and in 68 Metropolitan Areas* (Washington, D.C.: ACORN, 2002) (*http://www.acorn.org/reporter_pub/index.php*). This study was based on HMDA data released by the Federal Financial Institutions Examination Council (FFIEC). ACORN is the largest U.S. community organization of low-income and moderate-income families (more than 120,000) organized into 600 neighborhood chapters in 45 cities.

41. Canner and Smith, "Home Mortgage Disclosure Act."

42. ACORN, *The Great Divide.*

43. For a review of racial discrimination in underwriting in U.S. cities, see G. D. Squires, "Racial Profiling, Insurance Style: Insurance Redlining and the Uneven Development of Metropolitan Areas," *Journal of Urban Affairs* 25 (2003): 391–410.

44. Insurance Research Council, *Homeowners Loss Patterns in Eight Cities* (Wheaton, Ill.: IRC, 1997).

45. R. W. Klein, "Availability and Affordability Problems in Urban Homeowners Insurance Markets," in *Insurance Redlining: Disinvestment, Reinvestment, and the Evolving Role of Financial Institutions*, ed. G. D. Squires (Washington, D.C.: The Urban Institute Press, 1997), 43–82.

46. S. L. Smith and C. Cloud, "Documenting Discrimination by Homeowners Insurance Companies through Testing," in *Insurance Redlining: Disinvestment, Reinvestment, and the Evolving Role of Financial Institutions*, ed. G. D. Squires (Washington, D.C.: The Urban Institute Press, 1997), 97–117.

47. T. C. Pittman, "Rejoinder to Racial Profiling, Insurance Style: A Spirited Defense of the Insurance Industry," *Journal of Urban Affairs* 25 (2003): 411–22.

48. C. Hamnett, "Gentrification and Residential Location Theory: A Review and Assessment," in *Geography and the Urban Environment*, eds. D. T. Herbert and R. J. Johnston (Chichester, UK: Wiley, 1984), 284.

49. N. Smith, "New Globalism, New Urbanism: Gentrification as Global Urban Strategy," *Antipode* 34 (2002): 427–50.

50. For reviews of this research, see N. Smith and P. Williams, eds., *Gentrification of the City* (Boston: Allen and Unwin, 1986); and the special issue of *Urban Studies* 12 (2003) devoted to gentrification.

51. R. T. LeGates and C. Hartman, "The Anatomy of Displacement in the United States," in *Gentrification of the City*, eds. N. Smith and P. Williams (Boston: Allen and Unwin, 1986), 197.

52. D. Ley, *The New Middle Class and the Remaking of the Central City* (Oxford, UK: Oxford University Press, 1996).

53. See N. Smith, *The New Urban Frontier: Gentrification and the Revanchist City* (London: Routledge, 1996); and N. Smith, "New Globalism, New Urbanism: Gentrification as Global Urban Strategy," *Antipode* 34 (2002): 427–50.

54. C. Hamnett, "The Blind Men and the Elephant: The Explanation of Gentrification," *Transactions of the Institute of British Geographers* 16 (1991): 173–89.

55. Ibid., 186.

56. Ibid., 187.

57. L. Bondi, "Gender Divisions and Gentrification: A Critique," *Transactions of the Institute of British Geographers* 16 (1991): 190–98; A. Warde, "Gentrification as Consumption: Issues of Class and Gender," *Society and Space* 9 (1991): 223–32.

58. See L. Knopp, "Sexuality and the Spatial Dynamics of Capitalism," *Environment and Planning D: Society and Space* 10 (1992): 651–69; M. Lauria and L. Knopp, "Towards an Analysis of the Role of Gay Communities in the Urban Renaissance," *Urban Geography* 6 (1985): 152–69.

59. S. Zukin, "The Postmodern Debate over Urban Form," *Theory, Culture and Society* 5 (1988): 431–46; R. Deutsche, "Uneven Development: Public Art in New York City," *October* 47 (1988): 3–53.

60. N. Smith, "*New Globalism, New Urbanism,*" 440.

CHAPTER 14

1. See, for example, Y-F. Tuan, "The City: Its Distance from Nature," *Geographical Review* 68 (1978): 1–12; M. C. Jaye and A. C. Watts, eds., *Literature and the Urban Experience: Essays on the City and Literature* (New Brunswick, N.J.: Rutgers University Press, 1981).

2. E. Soja, "The Socio-Spatial Dialectic," *Annals of the Association of American Geographers* 70 (1980): 207–25.

3. See D. Ley, "Social Geography and the Taken-for-Granted World," *Transactions of the Institute of British Geographers* 2 (1977): 498–512; and A. Buttimer, "Grasping the Dynamism of the Lifeworld," *Annals of the Association of American Geographers* 66 (1976): 277–92.

4. Tönnies's work was first published in 1887 and was translated into English by C. Loomis in 1957 as *Community and Society* (East Lansing, Mich.: Michigan State University Press).

5. Durkheim's work was first published in 1893 and translated by G. Simpson in 1933 as *The Division of Labor in Society*. It was published by the Free Press, New York, in 1964.

6. This discussion is based on Knox and Pinch, *Urban Social Geography*, 314–15.

7. J. Perkin, *Victorian Women* (London: John Murray, 1993).

8. L. Bondi, "Sexing the City," in *Cities of Difference*, eds. R. Fincher and J. M. Jacobs (New York: Guildford Press, 1998), 177–200.

9. N. Duncan, "Renegotiating Gender and Sexuality in Public and Private Spaces," in *Bodyspace*, ed. N. Duncan (London: Routledge, 1996), 127–145.

10. E. Durkheim, *Suicide* (New York: Free Press, 1897).

11. G. Simmel, "The Metropolis and Mental Life," in *Classic Essays on the Culture of Cities*, ed. P. Sennett (New York: Appleton-Century-Crofts, 1961), 47–60.

12. See, for example, C. Beaudelaire, *The Painter of Modern Life and Other Essays* (Oxford, UK: Phaidon Press, 1964); W. Benjamin, *Das Passagen-Werk*, 2 vols., ed. R. Tiedermann (Frankfurt: Suhrkamp, 1982).

13. M. Foucault, "What is Enlightenment?" in *The Foucault Reader*, ed. P. Rabinow (Harmondsworth, UK: Penguin, 1984), 41–42.

14. L. Wirth, "Urbanism as a Way of Life," *American Journal of Sociology* 44 (1938): 1–24.

15. A. Toffler, *Future Shock* (New York: Bantam, 1970).

16. S. Milgram, "The Experience of Living in Cities," *Science* 167 (1970): 1461–68.

17. For a general review of this work see P. Knox and S. Pinch, *Urban Social Geography*, 4th ed. (Harlow, UK: Pearson Education, 2000), Chapter 10; for a more-detailed review, see C. S. Fischer, "The Public and Private Worlds of City Life," *American Sociological Review* 46 (1981): 306–16; C. S. Fischer, *The Urban Experience* (New York: Harcourt Brace Jovanovich, 1976).

18. L. Lofland, *A World of Strangers* (New York: Basic Books, 1973), 178. Emphases added.

19. Fischer, "Public and Private Worlds," 306.

20. This discussion is based on Knox and Pinch, *Urban Social Geography*, 316–21.

21. B. Weightman, "Commentary: Towards a Geography of the Gay Community," *Journal of Cultural Geography* 1 (1981): 106–12.

22. L. Knopp, "Some Theoretical Implications of Gay Involvement in an Urban Land Market," *Political Geography Quarterly* 9 (1990): 337–52; "Exploiting the Rent-Gap: The Theoretical Significance of Using Illegal Appraisal Schemes to Encourage Gentrification in New Orleans" *Urban Geography* 11 (1990): 48–64.

23. J. Egerton, "Out but Not Down: Lesbians' Experience of Housing," *Feminist Review* 36 (1990): 75–88; H. P. M. Winchester and P. White, "The Location of Marginalized Groups in the Inner City," *Environment and Planning D: Society and Space* 6 (1988): 37–54.

24. G. Valentine, "Out and About: Geographies of Lesbian Landscapes," *International Journal of Urban and Regional Research* 19 (1995): 96–111.

25. S. Kirby and I. Hay, "(Hetero)sexing Space: Gay Men and 'Straight' Space in Adelaide, South Australia," *Professional Geographer* 49 (1997): 295–305.

26. L. Johnston and G. Valentine, "Wherever I Lay my Girlfriend, That's My Home: The Performance and Surveillance of Lesbian Identities in Domestic Environments," in *Mapping Desire: Geographies of Sexuality*, eds. D. Bell and G. Valentine (London: Routledge, 1995), 99–113; L. McDowell "Body Work: Heterosexual Gender Performances in City Workplaces," in *Mapping Desire: Geographies of Sexuality*, eds. D. Bell and G. Valentine (London: Routledge, 1995), 75–95; L. McDowell, "Spatializing Feminism: Geographic Perspectives," in *Bodyspace: Destabilizing Geographies of Gender and Sexuality*, ed. N. Duncan (London: Routledge, 1996), 28–44.

27. H. J. Gans, *The Urban Villagers* (New York: Free Press, 1962).

28. In controlled laboratory experiments crowding has been shown to cause rat populations to exhibit aggression, listlessness, promiscuity, homosexuality, and the rodent equivalent of juvenile delinquency. Projecting these ideas directly to human behavior leads to the notion of crowded urbanites as "killer apes." But people are not rats and in any case never approach the degree of crowding experienced by laboratory animals. Evidence from studies of environmental psychology is mixed, and the whole debate continues to attract controversy in all of the social and environmental disciplines. For a brief summary, see Knox and Pinch, *Urban Social Geography*, 281–83.

29. Gans, *The Urban Villagers*, 249.

30. C. S. Fischer, "Toward a Subcultural Theory of Urbanism," *American Journal of Sociology* 80 (1975): 1319–41.

31. See, for example, C. R. Shaw and H. D. McKay, *Juvenile Delinquency and Urban Areas* (Chicago, Ill.: University of Chicago Press, 1942).

32. L. Mumford, *The Culture of Cities* (London: Secker and Warburg, 1940), 215.

33. R. S. Lynd and H. M. Lynd, *Middletown* (New York: Harcourt, Brace and World, 1956); W. L. Warner and P. S. Lunt, *The Social Life of a Modern Community* (New Haven, Conn.: Yale University Press, 1941).

34. H. J. Gans, *The Levittowners* (London: Allen Lane, 1967).

35. P. Bourdieu, *Distinction: A Social Critique of the Judgement of Taste* (London: Routledge & Kegan Paul, 1984).

36. M. Maffesoli, "Affectual Postmodernism and the Metropolis," *Threshold* 4 (1988): 1; M. Maffesoli, "Jeux de Masques: Postmoderne Tribalisme," *Design Issues* 4 (1988): 1–2.

37. U.S. National Commission on Neighborhoods, *People, Building Neighborhoods*. Final Report to the President and Congress of the United States (Washington, D.C.: U.S. Government Printing Office, 1979), 7.

38. K. Lynch, *The Image of the City* (Cambridge, Mass.: MIT Press, 1960), 47–48. Emphases added. See also K. Lynch, "Reconsidering 'The Image of the City,'" in *Cities of the Mind*, eds. R. M. Hollister and L. Rodwin (New York: Plenum, 1984), 151–62.

39. R. J. Johnston, "Spatial Patterns in Suburban Evaluations," *Environment and Planning A* 5 (1973): 385–95; F. M. Carp *et al.*, "Dimensions of Urban Environmental Quality," *Environment and Behavior* 8 (1976): 239–64.

40. T. Lee, "Urban Neighborhood as a Socio-Spatial Schema," *Human Relations* 21 (1968): 241–68.

41. P. Jackson and S. J. Smith, *Exploring Social Geography* (Boston: Allen and Unwin, 1984), 9.

42. Y-F. Tuan, "Place: An Experiential Perspective," *Geographical Review* 65 (1975): 152.

43. G. Pratt and S. Hanson, "On Theoretical Subtlety, Gender, Class and Space," *Society and Space* 9 (1991): 241.

44. M. Marsh. *Suburban Lives* (New Brunswick, N.J.: Rutgers University Press, 1990).

45. F. M. L. Thompson, "The Rise of Suburbia," in *The Rise of Suburbia*, ed. F. M. L. Thompson (Leicester, UK: Leicester University Press, 1982), 13. Emphasis added.

46. This discussion is based on Knox and Pinch, *Urban Social Geography*, 322–26.

47. D. C. Park, J. P. Radford, and M. H. Vickers, "Disability Studies in Human Geography," *Progress in Human Geography* 22 (1998): 208–33.

48. See R. Butler and S. Bowlby, "Bodies and Spaces: An Exploration of Disabled People's Experience of Public Space," *Environment and Planning D: Society and Space* 15 (1997): 411–33; M. Dear, R. Wilton, S. L. Gaber, and L. Takahashi, "Seeing People Differently: The Sociospatial Construction of Disability," *Environment and Planning D: Society and Space* 15 (1997): 455–80.

49. M. Oliver, "Theories of Disability in Health Practice and Research," *British Medical Journal* 7170 (1998): 1446–49.

50. R. Imrie, *Disability and the City: International Perspectives* (London: Paul Chapman, 1996).

51. V. Chouinard and A. Grant, "On Being Not Even Anywhere Near 'The Project:' Ways of Putting Ourselves in the Picture," in *Bodyspace*, ed. N. Duncan (London: Routledge, 1996), 170–93.

See also K. England, "Disabilities, Gender and Employment: Social Exclusion, Employment Equity and Canadian Banking," *The Canadian Geographer* 47 (2003): 429–50.

52. See U.S. Census Bureau, *Disability Status: 2000*, Census 2000 Brief (Washington, D.C.: U.S. Department of Commerce) for details on how the Census Bureau counts people with disabilities (*http://www.census.gov/prod/2003pubs/c2kbr-17.pdf*).

53. B. Gleeson, "Justice and the Disabling City," in *Cities of Difference*, eds. R. Fincher and J. M. Jacobs (New York: Guildford Press, 1998), 89–119.

54. R. Imrie and P. E. Wells, "Disablism, Planning and the Built Environment," *Environment and Planning C: Government and Policy* 11 (1993): 213–31.

55. A. J. Downing, *The Architecture of Country Houses* First published in 1850; reprinted in 1969 by (Dover Publications, New York).

56. G. Wright, *Building the Dream: A Social History of Housing in America* (New York: Pantheon, 1981).

57. C. Rock, S. Torre, and G. Wright, "The Appropriation of the House: Changes in House Design and Concepts of Domesticity," in *New Space for Women*, ed. G. R. Wekerle (Boulder, Colo.: Westview Press, 1980), 84.

58. A. Giddens, *The Constitution of Society: Outline of the Theory of Structuration* (Cambridge, UK: Polity Press, 1984).

59. M. Dear and J. Wolch, "How Territory Shapes Social Life," in *The Power of Geography: How Territory Shapes Social Life*, eds. M. Dear and J. Wolch (Boston: Unwin Hyman, 1989), 6.

60. D. Ley, *A Social Geography of the City* (New York: Harper & Row, 1983), 135.

61. A. Schutz, "The Social World and the Theory of Social Action," *Social Research* 27 (1960): 205–21.

62. R. Williams, *Marxism and Literature* (Oxford, UK: Oxford University Press, 1977).

63. P. Jackson, *Maps of Meaning* (Boston: Unwin Hyman, 1989), 39.

64. D. Seamon, *A Geography of the Lifeworld* (London: Croom Helm, 1979).

65. For an introduction to time-geography, see T. Carlstein, *Time, Resources, Society and Ecology* (Boston: Allen and Unwin, 1982); and T. Carlstein, D. Parkes and N. Thrift, eds., *Human Activity and Time Geography* (London: Arnold, 1978).

66. R. Miller, "Selling Mrs. Consumer: Advertising and the Creation of Suburban Socio-Spatial Relations. 1910–1930," *Antipode* 23 (1991): 263–301.

67. Marsh, *Suburban Lives*, 184–85.

68. B. Friedan, *The Feminine Mystique* (New York: W.W. Norton, 1963).

69. R. M. Law and J. R. Wolch, "Social Reproduction in the City: Restructuring in Time and

Space," in *The Restless Urban Landscape*, ed. P. L. Knox (Englewood Cliffs, N.J.: Prentice Hall, 1993), 165–206.

70. Ibid., 194.

71. This discussion is based on Knox and Pinch, *Urban Social Geography*, 187–88.

72. E. Wilson, *The Sphinx in the City: Urban Life, the Control of Disorder, and Women* (Berkeley, Calif.: University of California Press, 1991), 94.

73. See D. Spain, *Gendered Spaces* (Chapel Hill, N.C.: University of North Carolina Press, 1992); L. K. Weisman, *Discrimination by Design* (Urbana, Ill.: University of Illinois Press, 1992).

74. L. Bondi, "Gender, Symbols and Urban Landscapes," *Progress in Human Geography* 16 (1992): 157–70; N. Gregson and M. Lowe, "Home-Making: On the Spatiality of Daily Social Reproduction in Contemporary Middle-Class Britain," *Transactions of the Institute of British Geographers* 20 (1995): 224–35.

75. R. Madigan, M. Munro, and S. Smith, "Gender and the Meaning of Home," *International Journal of Urban and Regional Research* 16 (1992): 625–27.

CHAPTER 15

1. This distinction was first made by D. T. Herbert and R. J. Johnston in *Social Areas in Cities: Spatial Processes and Form* (London: Wiley, 1976).

2. *Report of the Special Committee of the Board of Supervisors of San Francisco on the Conditions of the Chinese Quarter and the Chinese in San Francisco* (San Francisco, Calif.: San Francisco Board of Supervisors, July 1885).

3. J. B. Trauner, "The Chinese as Medical Scapegoats in San Francisco, 1870–1905," *California History* 57 (1978): 70–87.

4. A. F. Weber, *The Growth of Cities in the Nineteenth Century: A Study in Statistics* (Ithaca, N.Y.: Cornell University Press, 1963), 368 (first published in 1899 by the Macmillan Company, New York).

5. Ibid., 369.

6. Ibid., 368. The quote is from M. S. Nordau, *Degeneration* (New York: D. Appleton, 1895), 35.

7. S. Smith, *The City That Was* (New York: Allaben, 1911). Quoted in D. Ward, *Poverty, Ethnicity and the American City, 1840–1920* (New York: Cambridge University Press, 1989).

8. J. Kirkland, "Among the Poor of Chicago," in *The Poor in Great Cities*, eds. R. A. Woods *et al.* (New York: Arno Press, 1971), 198. Originally published in New York by Scribner's in 1895.

9. D. Gaines, *Teenage Wasteland: Suburbia's Dead End Kids* (New York: Pantheon, 1991).

10. M. J. Dear, "Abandoned Housing," in *Urban Policy-making and Metropolitan Dynamics*, ed. J. S. Adams (Cambridge, Mass.: Ballinger, 1976), 59–99.

11. Sociologist Oscar Lewis first outlined the concept in relation to village life in Mexico, but it has since been deployed in relation to the slums of American cities. See O. Lewis, "The Culture of Poverty," *Scientific American* 215 (1966): 19–25.

12. Despite criticisms, the federal poverty standard continues to be used both as an administrative measure to determine government program eligibility and in academic research. See M. Orshansky, "Counting the Poor: Another Look at the Poverty Profile," *Social Security Bulletin* 28 (1965): 3–29; G. M. Fisher, "The Development and History of the Poverty Thresholds," *Social Security Bulletin* 55 (1992): 3–14; and the U.S. Department of Health and Human Services website (*http://aspe.hhs.gov/poverty/poverty.shtml*).

13. For brief summaries, see K. Fox, Chapter 5 in *Metropolitan America: Urban Life and Urban Policy in the United States, 1940–1980* (New Brunswick, N.J.: Rutgers University Press, 1985); and J. C. Teaford, Chapter 6 in *The Twentieth Century American City* (Baltimore, Md.: Johns Hopkins University Press, 1986). For a more-detailed review, see J. R. Feagin and H. Hahn, *Ghetto Revolts* (New York: Macmillan, 1973).

14. G. T. Kingsley and K. L. S. Pettit, *Concentrated Poverty: A Change in Course* (Washington, D.C.: Urban Institute, 2003) (*http://www.urban.org/UploadedPDF/310790_NCUA2.pdf*).

15. P. A. Jargowsky, *Stunning Progress, Hidden Problems: The Dramatic Decline of Concentrated Poverty in the 1990s* (Washington, D.C.: The Brookings Institution) (*http://www.brookings.edu/es/urban/publications/jargowskypoverty.pdf*).

16. Kingsley and Pettit, *Concentrated Poverty*.

17. Ibid.

18. Jargowsky, *Stunning Progress*.

19. Ibid.

20. Ibid.

21. Kingsley and Pettit, *Concentrated Poverty*.

22. Jargowsky, *Stunning Progress*.

23. W. J. Wilson, *The Truly Disadvantaged: The Inner City, The Underclass, and Public Policy* (Chicago, Ill.: University of Chicago Press, 1990), 8.

24. M. Van Reitsma, "A Conceptual Definition of the Underclass," *Focus* 12 (1989): 28.

25. J. Kasarda, "Jobs, Migration, and Emerging Urban Mismatches," in *Urban Change and Poverty*, eds. M. G. H. McGeary and L. E. Lynn. Jr. (Washington, D.C.: National Academy Press, 1988), 148–98; see also C. Jencks and S. E. Mayer, "Residential Segregation, Job Proximity, and Black Job Opportunities," in *Inner-City Poverty in the United States*, eds. L. E. Lynn, Jr. and M. G. H. McGeary (Washington, D.C.: National Academy Press, 1990), 187–222.

26. See S. Raphael and M. A. Stoll, *Modest Progress: The Narrowing Spatial Mismatch between Blacks and Jobs in the 1990s* (Washington, D.C.: The Brookings Institution, 2002) (*http://www.brook.edu/es/urban/ publications/ raphaelstoll_spatialmismatch.htm*).

27. M. Castells, *The Informational City* (Cambridge, Mass.: Blackwell, 1989), 227.

28. See U.S. Bureau of Justice Statistics website (*http://www.ojp.usdoj.gov/bjs/*).

29. R. B. Taylor, "Urban Communities and Crime," in *Urban Life in Transition*, eds. M. Gottdiener and C. G. Pickvance (Newbury Park, Calif.: Sage, 1991), 106–34.

30. C. F. Schmid, "Urban Crime Areas," *American Sociological Review* 25 (1960): 678.

31. S. Smith, *Crime, Space and Society* (Cambridge, UK: Cambridge University Press, 1986).

32. O. Newman, *Defensible Space* (New York: Macmillan, 1972).

33. Drug Policy Information Clearinghouse, *Washington, D.C.: Profile of Drug Indicators* (Washington, D.C.: White House Office of National Drug Policy, 2003) (*http://www.whitehousedrugpolicy.gov/statelocal/dc/ dc.pdf*).

34. D. Davis, Jr., *Violent Crimes in the Capital City.* Paper presented at the annual meeting of the American Association for the Advancement of Science (Washington, D.C., February 14–19, 1991).

35. P. L. Knox, "The Postmodern Urban Matrix," in *The Restless Urban Landscape*, ed. P. L. Knox (Englewood Cliffs, N.J.: Prentice Hall, 1993).

36. See the U.S. Department of Health and Human Services, National Institutes of Health (NIH), National Institute on Drug Abuse (NIDA) website (*http:// www.nida.nih.gov/Infofax/treatmeth.html*).

37. This discussion is based on S. A. Marston, P. L. Knox, and D. M. Liverman, *World Regions in Global Context: Peoples, Places, and Environments* (Upper Saddle River, N.J.: Prentice Hall, 2002), 181.

38. Knox, "The Postmodern Urban Matrix."

39. See Taylor, "Urban Communities and Crime," 127–28.

40. Knox, "The Postmodern Urban Matrix."

41. L. Duke and D. M. Price, "A Microcosm of Despair in Washington, D.C.," *Washington Post* (April 2, 1989): A7.

42. This discussion is based on H. V. Savitch and G. Ardashev, "Does Terror Have an Urban Future?" *Urban Studies* 38 (2001): 2515–33; P. L. Knox and S. A. Marston, *Places and Regions in Global Context: Human Geography*, 2nd ed. (Upper Saddle River, N.J.: Pearson Education, Inc., 2004), 406.

43. J. Coaffee, "Rings of Steel, Rings of Concrete and Rings of Confidence: Designing out Terrorism in Central London pre and post September 11th," *International Journal of Urban and Regional Research* 28 (2004): 201–11.

44. J. Harrigan and P. Martin, "Terrorism and the Resilience of Cities," *Economic Policy Review* 8 (2002): 97–116.

45. Coaffee, "Rings of Steel."

46. Ibid.

47. P. Rossi, *Down and Out in America: The Origins of Homelessness* (Chicago, Ill.: University of Chicago Press, 1989).

48. The Bring LA Home, The Partnership to End Homelessness, Economic Roundtable estimated in 2004 that on a typical night in Los Angeles County, 10 percent of homeless residents were doubled up with friends and relatives; 11 percent were in rehabilitation facilities, jail, or hospital; 24 percent were in emergency shelters and transitional housing; and the sleeping arrangements of the remaining 55 percent were sidewalks, cars, public transit, empty buildings, roadways, or parks (see *http:// bringlahome.org/*).

49. M. Burt and L. Aron, *Homelessness: Programs and the People they Serve* (Washington, D.C.: The Urban Institute, 1999) (*http://www.urban.org/*).

50. A *Los Angeles Times* article from June 28, 2001, reported that Census Bureau officials did not want a repeat of the 1990 count, in which homeless advocates immediately blasted the bureau's estimate of the homeless at 750,000, calling it drastically low.

51. M. Rosler, "Fragments of a Metropolitan Viewpoint," in *If You Lived Here: The City in Art, Theory, and Social Activism*, ed. B. Wallis (Seattle, Wash.: Bay Press, 1991), 20.

52. Rossi, *Down and Out in America*, Chapter 2; Burt and Aron, *Homelessness*.

53. Ibid.

54. U.S. Conference of Mayors, *Hunger and Homelessness Survey: A Status Report on Hunger and Homelessness in America's Cities: A 25-City Survey* (Washington, D.C.: U.S. Conference of Mayors, 2003) (*http:// www.usmayors.org/uscm/hungersurvey/2003/ onlinereport/HungerAndHomelessnessReport2003.pdf*).

55. Ibid.

56. See M. J. Dear and J. R. Wolch, *Landscapes of Despair* (Princeton, N.J.: Princeton University Press, 1987).

57. C. Dolbeare, "Housing Policy: A General Consideration," in *Homelessness in America*, National Coalition for the Homeless (Washington, D.C.: Oryx Press, 1996).

58. U.S. Conference of Mayors, *Hunger and Homelessness Survey*; P. Koegel *et al.*, "The Causes of Homelessness," in *Homelessness in America*, National Coalition for the Homeless (Washington, D.C.: Oryx Press, 1996); G. Laws and S. Lord, "The

Politics of Homelessness," in *Geographic Dimensions of United States Social Policy*, eds. J. Kodras and J. P. Jones III (London: Edward Arnold, 1990), 59–85; Rossi, *Down and Out in America*, 182.

59. See U.S. Conference of Mayors, *Hunger and Homelessness Survey*.

60. U.S. Environmental Protection Agency, *TRI (Toxics Release Inventory) Explorer: Releases: Chemical Report* (*http://www.epa.gov/triexplorer/*). The Toxics Release Inventory contains information on releases of nearly 650 chemicals and chemical categories from industries including manufacturing, metal and coal mining, electric utilities, and commercial hazardous waste treatment.

61. I. Douglas, "The Rain on the Roof: A Geography of the Urban Environment," in *Horizons in Human Geography*, eds. D. Gregory and R. Walford (Totowa, N.J.: Barnes and Noble, 1989), 217–38.

62. Ibid., 220.

63. R. J. Naiman, J. J. Magnuson, and P. L. Firth, "Integrating Cultural, Economic, and Environmental Requirements for Fresh Water," *Ecological Applications* 8 (1998), 1338–39; U.S. Geological Survey, *Estimated Use of Water in the United States in 2000*, USGS Circular 1268 (Washington, D.C.: USGS, 2004) (*http://water.usgs.gov/pubs/circ/2004/circ1268*); C. Keenan, R. Lee, and M. Wu, *Water Reuse and Conservation* (Berkeley, Calif.: University of California at Berkeley, 2003).

64. United Nations Environmental Programme, *GEO-2000: Global Environment Outlook* (*http://www.unep.org/geo2000/English/0099.htm*).

65. J. Bebow, "Aging Water Pipes Leak $23 million a Year," *The Detroit News* (July 22, 2002) (*http://www.detnews.com/2002/specialreport/0209/04/a01-543158.htm*).

66. Congressional Budget Office, *Future Investment in Drinking Water and Wastewater Infrastructure* (Washington, D.C.: U.S. Congress, 2002). The CBO's estimates of water investment averaged between two different scenarios: a $11.6 billion low-cost case and a $20.1 billion high-cost case.

67. Quoted in B. Howard, "Scientists and Citizens Are Stymied by Water Crisis," *E/The Environmental Magazine* (January 8, 2004) (*http://www.enn.com/news/2004-01-08/s_9210.asp*).

68. M. Lavelle and J. Kurlantzick, "The Coming Water Crisis," *U.S. News and World Report* (August 12, 2002) (*http://www.bcwaternews.com/*); P. L. Knox and S. A. Marston, *Places and Regions in Global Context: Human Geography*, 2nd ed. (Upper Saddle River, N.J.: Pearson Education, Inc., 2004), 431.

69. Congressional Budget Office, *Future Investment in Drinking Water and Wastewater Infrastructure* (Washington, D.C.: CBO, 2002).

70. Ibid. The CBO's estimates of wastewater investment averaged between two different scenarios: a $13.0 billion low-cost case and a $20.9 billion high-cost case.

71. U.S. Environmental Protection Agency, *Latest Findings on National Air Quality: 2002 Status and Trends* (Washington, D.C.: Office of Air Quality Planning and Standards, 2003) (*http://www.epa.gov/airtrends/2002_airtrends_final.pdf*).

72. This discussion is based on P. L. Knox and S. A. Marston, *Places and Regions in Global Context: Human Geography*, 3rd ed. (Upper Saddle River, N.J.: Pearson Education, Inc., 2004), 131–32.

73. R. J. Earickson and I. H. Billick, "The Areal Association of Urban Air Pollutants and Residential Characteristics: Louisville and Detroit," *Applied Geography* 8 (1988): 5–23; U.S. Environmental Protection Agency, *Latest Findings on National Air Quality*.

74. P. Choate and S. Walter, *American in Ruins: The Decaying Infrastructure* (Durham, N.C.: Duke University Press, 1983).

75. M. Kaplan *et al.*, *Hard Choices: A Report on the Increasing Gap Between America's Infrastructure Needs and Our Ability to Pay for Them* (Washington, D.C.: US Government Printing Office, 1984).

76. This discussion is based on S. A. Marston, P. L. Knox, and D. M. Liverman, *World Regions in Global Context: Peoples, Places, and Environments* (Upper Saddle River, N.J.: Prentice Hall, 2002), 128; and PLANCO Consulting GmbH, *Ten—Invest: Final Report* (Brussels: European Commission, 2003) (*http://europa.eu.int/comm/ten/transport/documentation/index_en.htm*); Union of European Railway Industries, *High Speed Trains in Europe* brochure (Brussels: UNIFE, 2003) (*http://www.uic.asso.fr/d_gv/publications/ brochure/brochure-europe_en.pdf*).

77. American Society of Civil Engineers, *Renewing America's Infrastructure, 2001; 2003 Progress Report for America's Infrastructure* (*http://www.asce.org/reportcard/*).

78. A. G. Hevesi, *Dilemma in the Millennium: Capital Needs of the World's Capital City* (New York: Office of the Comptroller, City of New York, August 1998) (*http://www.comptroller.nyc.gov/press/archive_releases/98-08-056.shtm*); J. Obser, "The Road Ahead for New York City," *Future.Newsday.Com* (May 16, 1999) (*http://future.newsday.com/5/fqtp0516.htm*).

79. A. C. Revkin, "Federal Study Calls Spending on Water Systems Perilously Inadequate," *The New York Times* (April 10, 2002).

80. With an average speed of less than 35 mph.

81. Texas Transportation Institute, *2003 Urban Mobility Study* (College Station, Tex.: Texas A&M University System) (*http://www.mobility.tamu.edu/ums/report/*).

82. United States General Accounting Office, *Highway Infrastructure*, Report to the Chairman, Committee

on Transportation and Infrastructure, House of Representatives (Washington, D.C.: GAO, 2002) (*http://www.gao.gov/new.items/d02571.pdf*).

83. This discussion is based on "Congestion Charging: Ken Livingstone's Gamble," *The Economist* (February 13, 2003) (*http://www.economist.com/World/Europe/displayStory.cfm?story_id=1578107*); BBC News, "Congestion Charging: In London" (*http://news.bbc.uk/1/shared/spl/hi/uk/03/congestion_charge/exemptions_guide/html/what.stm*).

CHAPTER 16

1. M. Edelman, *The Symbolic Uses of Politics* (Urbana, Ill.: University of Illinois Press, 1985).
2. H. Lasswell, *Politics: Who Gets What, When, How* (New York: McGraw-Hill, 1936).
3. United Nations Human Settlements Programme, *The Global Campaign on Urban Governance: Principles of Good Urban Governance* (New York: UN-Habitat, 2002) (*http://www.unhabitat.org/campaigns/governance/Principles.asp*).
4. P. Studenski and H. E. Kroos, *Financial History of the United States*, 2nd ed. (New York: McGraw-Hill, 1963), 59-60, cited in E. Monkkonen, *America Becomes Urban* (Berkeley, Calif.: University of California Press, 1988).
5. This evolution drew on a long history of patronage and corruption in American urban politics. The organizational influences of early trade union and socialist politics were also important. See, for example, M. Shefter, "The Emergence of the Political Machine," in *Theoretical Perspectives on Urban Politics*, eds. W. D. Hawley *et al.* (Englewood Cliffs, N.J.: Prentice Hall, 1976).
6. H. P. Chudacoff and J. E. Smith, *The Evolution of American Urban Society*, 5th ed. (Upper Saddle River, N.J.: Prentice Hall, 2000), 159.
7. M. C. Brown and C. Halaby, "Machine Politics in America, 1870–1945," *Journal of Interdisciplinary History* 17 (1987): 587–612.
8. D. Judd, *The Politics of American Cities*, 3rd ed. (Glenview, Ill.: Scott. Foresman, 1988), 97.
9. L. J. R. Herson and J. M. Bolland, *The Urban Web: Politics, Policy and Theory* (Chicago, Ill.: Nelson Hall, 1990), 64.
10. Ibid., 146–47.
11. S. I. Toll, *Zoned America* (New York: Grossman, 1969).
12. K. Newton, "Conflict Avoidance and Conflict Suppression: The Case of Urban Politics in the United States," in *Urbanization and Conflict in Market Societies*, ed. K. R. Cox (London: Methuen, 1978), 84.
13. Judd, *The Politics of American Cities*, 136. Emphasis added.

14. C. Dykstra, *Our Cities: Their Role in the National Economy* (Washington, D.C.: Committee on Urbanism, National Resources Committee, U.S. Department of the Interior, 1937).
15. J. H. Mollenkopf, *The Contested City* (Princeton, N.J.: Princeton University Press, 1983), 3.
16. Ibid., 55.
17. Ibid., 55.
18. See J. R. Logan and H. L. Molotch, *Urban Fortunes. The Political Economy of Place* (Berkeley, Calif: University of California Press, 1987), 62–84; M. Lauria, ed., *Reconstructing Urban Regime Theory: Regulating Urban Politics in a Global Economy* (Thousand Oaks, Calif.: Sage, 1997).
19. Boston Redevelopment Authority, *1965 Plan*, cited in Mollenkopf, *The Contested City*, 170.
20. M. Anderson, *The Federal Bulldozer* (Cambridge, Mass.: MIT Press, 1964), 8.
21. C. Hartman, "The Housing of Relocated Families," *Journal of the American Institute of Planners* 30 (1964): 266–86.
22. Mollenkopf, *The Contested City*, 182–83.
23. J. Jacobs, *The Death and Life of Great American Cities* (New York: Random House, 1961).
24. This discussion is based on information in the City of Milwaukee's Department of City Development website (*http://www.mkedcd.org/parkeast/*).
25. P. Clavel and W. Wiewel, eds., *Harold Washington and the Neighborhoods* (New Brunswick, N.J.: Rutgers University Press, 1991).
26. H. P. Chudacoff and J. E. Smith, *The Evolution of American Urban Society*, 5th ed. (Upper Saddle River, N.J.: Prentice Hall, 2000), 286.
27. J. O'Loughlin, "Malapportionment and Gerrymandering in the Ghetto," in *Urban Policymaking and Metropolitan Dynamics*, ed. J. S. Adams (Cambridge, Mass.: Ballinger, 1976), 539–65.
28. Ibid., 540.
29. K. Jackson, *Crabgrass Frontier* (New York: Oxford University Press, 1985), 154.
30. J. C. Bollens and H. J. Schmandt, *The Metropolis*, 4th ed. (New York: Harper & Row, 1982).
31. R. D. Honey, "Metropolitan Governance," in *Urban Policymaking and Metropolitan Dynamics*, ed. J. S. Adams (Cambridge, Mass: Ballinger, 1976), 425–62.
32. See, for example, R. L. Lineberry, *Equality and Urban Policy* (Beverly Hills, Calif.: Sage, 1977); B. D. Jones, S. Greenberg, and J. Drew, *Service Delivery in the City* (New York: Longman, 1980); and *The Politics of Urban Public Services*, ed. R. C. Rich (Lexington, Mass.: Lexington Books, 1982).
33. See D. W. Harvey, *Social Justice and the City* (London: Arnold, 1973).
34. See American Planning Association, *Growing Smart: Legislative Guidebook* (Washington, D.C.: APA,

2002); and the State and Local Government Law website of Washington University Law School for American Planning Association *Growing Smart Project, Model Tax Increment Financing Statute: Commentary: Tax Increment Financing* (*http://ls.wustl.edu/statelocal/tifstat.htm*).

35. R. Friedland, "Central City Fiscal Stress: The Public Costs of Private Growth," *International Journal of Urban and Regional Research* 5 (1981): 370–71. Emphasis added.

36. Ibid., 371.

37. F. F. Piven and R. A. Cloward, *Regulating the Poor: The Functions of Public Welfare*, updated ed. (New York: Vintage Books, 1993).

38. H. F. Ladd and J. Yinger, *America's Ailing Cities: Fiscal Health and the Design of Urban Policy* (Baltimore, Md.: Johns Hopkins University Press, 1991).

39. F. G. Rohatyn, *New York's Fiscal Crises: 1975–2003.* M. Moran Weston II Distinguished Lecture in Urban and Public Policy, School of International and Public Affairs, Columbia University, New York, February 26, 2003 (*http://www.columbia.edu/cu/sipa/NEWS/Rohatyn%20speech.pdf*).

40. C. H. Levine *et al.*, *The Politics of Retrenchment: How Local Governments Manage Fiscal Stress* (Beverly Hills, Calif.: Sage, 1981).

41. For a detailed examination of voluntarism, see J. Wolch, *The Shadow State: Government and the Voluntary Sector in Transition* (New York: The Foundation Center, 1990).

42. This section is based on P. L. Knox, "Public-Private Cooperation: A Review of the Experience in the U.S.," *Cities* 5 (1988): 340–46.

43. President's Commission on Privatization, *Privatization: Toward More Effective Government* (Washington, D.C.: U.S. Government Printing Office, 1988), 251.

44. R. W. Poole, Jr. and P. E. Fixler, Jr., "Privatization of Public-Sector Services in Practice: Experience and Potential," *Journal of Policy Analysis and Management* 6 (1987): 612–25.

45. E. R. Gerber, C. K. Hall, and J. R. Hines, Jr., *Privatization: Issues in Local and State Service Provision* (Ann Arbor, Mich.: Center for Local, State, and Urban Policy, University of Michigan, 2004) (*http://www.closup.umich.edu/research/reports/pr-1-privatization.pdf*).

46. P. Winston, A. Burwick, Sheena McConnell, and R. Roper, *Privatization of Welfare Services: A Review of the Literature.* Report prepared by Mathematica Policy Research, Inc. for the U.S. Department of Health and Human Services, Washington, D.C., 2002 (*http://aspe.hhs.gov/hsp/privatization02/*).

47. Ibid.

48. J. Hanrahan, *Passing the Bucks: The Contracting Out of Public Services* (Washington, D.C.: American Federation of State, Federal, and Municipal Employees, 1983).

49. D. W. Harvey, "From Managerialism to Entrepreneurialism: The Transformation of Urban Governance in Late Capitalism," *Geografiska Annaler* 71B (1989): 3–17.

50. P. F. Smith, *City, State and Market: The Political Economy of Urban Society* (New York: Blackwell, 1988), 210.

51. Harvey, "From Managerialism to Entrepreneurialism."

52. These examples, together with several of those that follow, are detailed in J. Teaford, *The Rough Road to Urban Renaissance* (Baltimore, Md.: Johns Hopkins University Press, 1990).

53. D. W. Harvey, *The Condition of Postmodernity* (Oxford, UK: Blackwell, 1989).

54. This discussion is based on S. A. Marston, P. L. Knox, and D. M. Liverman, *World Regions in Global Context: Peoples, Places, and Environments* (Upper Saddle River, N.J.: Prentice Hall, 2002), 136–37.

55. Teaford, *The Rough Road*, 254.

56. See N. Smith, "New City, New Frontier: The Lower East Side as Wild, Wild, West"; and M. C. Boyer, "Cities for Sale: Merchandising History at South Street Seaport," in *Variations on a Theme Park: The New American City and the End of Public Space*, ed. M. Sorkin (New York: Hill and Wang, 1992), 61–93 and 181–204, respectively.

57. Harvey, "From Managerialism to Entrepreneurialism," 13.

58. F. Hunter, *Community Power Structures: A Study of Decision Makers* (Chapel Hill, N.C.: University of North Carolina Press, 1953).

59. P. Bachrach and M. Baratz, *Power and Poverty* (New York: Oxford University Press, 1970).

60. R. Dahl, *Who Governs?* (New Haven, Conn.: Yale University Press, 1961).

61. See P. Schmitter, *Comparative Political Studies* (Beverly Hills, Calif.: Sage, 1977); and H. V. Savitch, *Post-Industrial Cities: Politics and Planning in New York, Paris and London* (Princeton, N.J.: Princeton University Press, 1988).

62. C. N. Stone, *Regime Politics: Governing Atlanta, 1946–1988* (Lawrence, Kans.: University Press of Kansas, 1989); A. E. G. Jonas and D. Wilson, eds., *The Urban Growth Machine: Critical Perspectives, Two Decades Later* (Albany, N.Y.: State University of New York Press, 1999).

63. H. V. Savitch and J. C. Thomas, "Conclusion: The End of Millennium Big City Politics," in *Big City Politics in Transition*, eds. H. V. Savich and J. C. Thomas (Newbury Park, Calif.: Sage, 1991), 248.

64. See P. Saunders, *Urban Politics: A Sociological Interpretation* (London: Hutchinson, 1979); and J. O'Connor, *The Fiscal Crisis of the State* (New York: St. Martin's Press, 1973).

CHAPTER 17

1. This discussion is based on L. McCarthy and D. Danta, "Cities of Europe," in *Cities of the World: World Regional Urban Development*, eds. S. Brunn, J. Williams, and D. Zeigler, 3rd ed. (Lanham, Md.: Rowman & Littlefield, 2003), 189–96.

2. R. Fogelsong, *Planning the Capitalist City* (Princeton, N.J.: Princeton University Press, 1986).

3. Charles Dickens found the subject matter for novels like *Oliver Twist* in the teeming slums of London (see the U.N.'s description of the appalling living conditions in London and New York in the 1800s in "A Tale of Dickens's London" in *The State of the World's Cities Report, 2001* (Nairobi: UN-Habitat) (*http://www.unchs.org/Istanbul+5/20-27.pdf*)); Charles Booth, a philanthropic Liverpool shipowner, produced the prominent 17-volume *Life and Labour of the People in London (1886–1903)* to highlight the plight of the poor (for more information about Booth, see the Charles Booth Online Archive at the London School of Economics Library (*http://booth.lse.ac.uk*); Fredrich Engels, shocked by the living conditions of the poor, wrote in 1844 the influential *Condition of the Working Classes in England*.

4. M. C. Boyer, *Dreaming the Rational City* (Cambridge, Mass.: MIT Press, 1983), 18.

5. J. A. Riis, *How the Other Half Lives: Studies among the Tenements of New York* (New York: Scribner's, 1890).

6. R. W. DeForest and L. Veiller, eds., *The Tenement House Problem. Including the Report of the New York State Tenement House Commission* (New York: Macmillan, 1903). Quoted in P. Hall, *Cities of Tomorrow: An Intellectual History of Urban Planning and Design in the Twentieth Century*, 3rd ed. (New York: Basil Blackwell, 2002), 39.

7. Boyer, *Dreaming the Rational City*, 27.

8. R. E. Park, "The City: Suggestions for the Investigation of Human Behavior in the City Environment," *American Journal of Sociology* 20 (1915): 580.

9. Quoted in Hall, *Cities of Tomorrow*, 42.

10. Boyer, *Dreaming the Rational City*, 39.

11. Fogelsong, *Planning the Capitalist City*, 4.

12. Boyer, *Dreaming the Rational City*, 60.

13. J. Hancock, "The New Deal and American Planning: The 1930s," in *Two Centuries of American Planning*, ed. D. Schaffer (Baltimore, Md.: Johns Hopkins University Press, 1988), 197–230.

14. R. Florida and A. Jonas, "U.S. Urban Policy: The Postwar State and Capitalist Regulation," *Antipode* 23 (1991): 349–84.

15. This discussion is based on L. McCarthy and D. Danta, "Cities of Europe," in *Cities of the World: World Regional Urban Development*, eds. S. Brunn, J. Williams, and D. Zeigler, 3rd ed. (Lanham, Md.: Rowman & Littlefield, 2003), 219–20.

16. Quoted in W. K. Tabb, "The Failure of National Urban Policy," in *Marxism and the Metropolis*, eds. W. K. Tabb and L. Sawers (New York: Oxford University Press, 1984), 257. Emphasis added.

17. President's Commission for a National Agenda for the Eighties, Panel on Policies and Prospects, *Urban America in the Eighties: Perspectives and Prospects* (Washington, D.C.: U.S. Government Printing Office, 1980).

18. See H. Wolman, "The Reagan Urban Policy and Its Impacts," *Urban Affairs Quarterly* 21 (1986): 311–35; and *Reagan and the Cities*, eds. G. E. Peterson and C. W. Lewis (Washington, D.C.: The Urban Institute Press, 1986).

19. F. F. Piven and R. A. Cloward, "The New Class War in the United States," in *Cities in Recession*, ed. I. Szelenyi (Beverly Hills, Calif.: Sage, 1984), 26–45.

20. R. E. Green, ed., *Enterprise Zones* (Newbury Park, Calif.: Sage, 1991).

21. This discussion is based on L. McCarthy, "The Good of the Many Outweighs the Good of the One: Regional Cooperation Instead of Individual Competition in the USA and Western Europe?" *Journal of Planning Education and Research* 23 (2003): 140–52.

22. H. G. Cisneros, *Urban Entrepreneurialism and National Economic Growth* (Washington, D.C.: United States Department of Housing and Urban Development, 1995), 3, 19.

23. R. Gurwitt, R. "The Painful Truth about Cities and Suburbs: They Need Each Other," *Governing* 5 (1992): 56.

24. E. Goetz and T. Kayser, "Competition and Cooperation in Economic Development: A Study of the Twin Cities Metropolitan Area," *Economic Development Quarterly* 7 (1993): 63–78.

25. A. E. G. Jonas, "Regulating Suburban Politics: 'Suburban-Defense Transition,' Institutional Capacities, and Territorial Reorganization in Southern California," in *Reconstructing Urban Regime Theory: Regulating Urban Politics in a Global Economy*, ed. M. Lauria (Thousand Oaks, Calif.: Sage, 1997), 222.

26. United Nations Centre for Human Settlements, *Cities in a Globalizing World* (London: Earthscan, 2001), 38.

27. J. E. Vance, Jr., *The Continuing City* (Baltimore, Md.: Johns Hopkins University Press, 1990), 83.

28. This discussion is based on information in the Metropolitan Council's website (*http://www.metrocouncil.org/*).

29. See, for example, M. Orfield, *Metropolitics: A Regional Agenda for Community and Stability* (Washington, D.C.: The Brookings Institution Press,

1997); and D. Rusk, *Cities without Suburbs* (Washington, D.C.: Woodrow Wilson Press, 1993).

30. This discussion is based on information in Metro's website (*http://www.metro-region.org/*).

31. J. Abu-Lughod, *Changing Cities* (New York: Harper Collins, 1991), 379.

CHAPTER 18

1. P. Hall, "Three Systems, Three Separate Paths," *Journal of the American Planning Association* 57 (1991): 18.

2. B. J. L. Berry, *Long-Wave Rhythms in Economic Development and Political Behavior* (Baltimore, Md.: Johns Hopkins University Press, 1991), 125–27.

3. G. Gappert, "Future Urban America: Post-Affluent or Advanced Industrial Society," in *Cities in the 21st Century*, eds. G. Gappert and R. Knight (Beverly Hills, Calif.: Sage, 1982), 99–134.

4. G. Sternlieb and J. Hughes, "Rediscovering America," in *America's New Market Geography*, eds. G. Sternlieb and J. Hughes (New Brunswick, N.J.: Center for Urban Policy Research, 1988), 5.

5. M. Featherstone, ed., *Global Culture* (Newbury Park, Calif.: Sage, 1991).

6. A. Wernick, *Promotional Culture* (Newbury Park, Calif.: Sage, 1992).

7. M. Sorkin, ed., *Variations on a Theme Park* (New York: Hill and Wang, 1992).

8. W. Zelinsky, "The Twinning of the World: Sister Cities in Geographic and Historic Perspective," *Annals of the Association of American Geographers* 81 (1991): 1–31.

9. S. Marston and A. Kirby, "The Impact of the Municipal Foreign Policy Movement on the Discourse of National Peace and Security." Unpublished Project Description, Department of Geography, University of Arizona, 1992.

10. N. Pressman, "Forces for Spatial Change," in *The Future of Urban Form*, eds. J. Brotchie *et al.* (London: Croom Helm, 1985), 349–61.

11. J. Borchert, "Futures of American Cities," in *Our Changing Cities*, ed. J. F. Hart (Baltimore, Md.: Johns Hopkins University Press, 1991), 218–50.

Glossary

The glossary words and terms defined below are highlighted in bold text throughout the book as well as the index. All words in italics below are defined.

A

acid rain Precipitation that has mixed with airborne pollutants (particularly sulfur dioxide and nitrogen oxides from burning fossil fuels) to produce rain that contains levels of acidity—often in the form of sulfuric acid—that are harmful not only to vegetation and aquatic life but also to some building surfaces and statues in cities.

advanced capitalism The term for the most recent or "disorganized" phase of capitalism in which the relationships between business, labor, and government are more flexible, largely because a great deal of corporate activity has escaped the framework of national states and their institutions that still constrain organized labor and most government functions. See *flexible production systems*.

agglomeration diseconomies The negative economic effects of urbanization and the local concentration of industry, including higher rents, traffic congestion, and air pollution.

agglomeration economies Economic advantages that individual companies enjoy because of their location clustered together among functionally related activities, thereby sharing the use of specialized servicing industries, financial and other services, and public utilities. See also *localization economies* and *urbanization economies*.

American Renaissance The literary culture of the American Renaissance—rooted in the works of Ralph Waldo Emerson and Henry David Thoreau—that strongly influenced the American design professionals of the mid-nineteenth century. The American Renaissance took nature as a fundamental spiritual wellspring, defining the ideal as a setting in which mankind and nature had achieved a state of balance. This attitude led to a vision of ideal urban landscapes that combined the morality attributed to nature with the enriching and refining influences of cultural, political, and social institutions.

annexation The addition of unincorporated land into a municipality, especially during the nineteenth century in the United States. Pioneer suburbanites gained better services from being annexed by the municipality; the municipality achieved important *economies of scale* and prevented the loss of their middle income groups from the tax rolls and political life.

anomie Literally "normlessness"—often associated with the isolation of urban life, but it is more accurate to think of those social conditions in which the norms of personal and social behavior are so weak or muddled that some people become socially isolated, confused, or uncertain about how to behave, while others are easily able to challenge or ignore social conventions. One of the consequences of such circumstances is an increase in *deviant behavior*. See also *Gesellschaft*.

Art Nouveau A movement in the 1890s and early 1900s in France (called *Jugendstil* in Central Europe) that emphasized sensuous shapes and organic themes as an alternative to Victorian extravagance. The Art Nouveau style became widely used in graphic art, but in architecture it was limited to the embellishment of buildings rather than being reflected in their form.

Arts and Crafts Movement Led by William Morris in England, this movement sought to reestablish the importance of craftsmanship during the increasing mechanization and mass production of the second half of the nineteenth century. Morris wanted to bring artists and craftspeople together in order to synthesize honest, simple, and popular forms. In architecture, this meant drawing heavily on vernacular themes, something that proved convenient and attractive to the planners of the first Garden Cities.

asylum suburb The term coined by geographer James Vance, Jr., to capture the nature of those suburbs in the U.S. countryside in which higher income households seek refuge and complete escape from the problems and congestion of a growing polycentric metropolis.

B

backward linkage One of the interrelated effects associated with the self-propelling growth process of urban-industrial development. Given a certain amount of original employment growth in a particular industry, backward linkages occur as new companies arrive in order to provide the growing industry with components, supplies, specialized services, or facilities. See *forward linkage*.

backwash effects In Gunnar Myrdal's model, the negative spillover effects on an urban area or region—including outmigration and the loss of capital—caused by the economic growth of some other urban area or region. Also known as polarization effects. See *spread effects* and *cumulative causation*.

balloon-frame construction Balloon framing uses standardized, machine-cut 2" x 4" studs nailed together by inexpensive, machine-cut nails. By spreading the stress over a large number of light boards, the balloon frame has strength and stability far beyond its insubstantial appearance. Using standardized components and requiring only semi-skilled labor, balloon framing can be around 40 percent less expensive than traditional methods that used heavy beams and corner posts held together by mortise and tenon joints.

Bauhaus School Founded by Walter Gropius in 1919, the Bauhaus School did more than any other Modernist movement to promote the ideal of architecture and design as agents of social redemption. Through industrialized production, modern materials, and functional design, architecture could be produced inexpensively, become available to all, and so improve the physical, social, moral, and aesthetic condition of cities. The unifying themes were simplicity of line, plain surfaces, and suitability for mass production.

behavioral assimilation The process in which members of a minority group adopt the culture—language, norms, and values—of the wider society (or charter group), and so become acculturated to the mainstream life of the city. Contrast with *structural assimilation*.

bid-rent theory William Alonso's neoclassical economic model of urban land use based on the idea that companies and households compete for spaces in a way that maximizes their utility. The bid-rent curve on a graph plots the relationship between how much rent companies and households are prepared to pay for land relative to accessibility to a specific point—usually the center of a city. Commercial establishments are expected to be able and willing to outbid households for central sites because the extra income accruing to a central location through increased trade is likely to outweigh the savings in commuting costs obtained at the same site by a household. The bid-rent curve slopes downward with decreasing accessibility, reflecting decreasing rents with distance from the center of a monocentric (single-centered) city. See *concentric zone model*.

bird-dogging A tactic used by larger commercial and industrial companies with a compelling need to acquire land at a rapid rate in order to keep their organization fully employed. It involves searching out and bidding for suitable land before it has been put on the market (and before any thought has been given to project conceptualization). Compare with *land banks*.

block busting The practice by some real estate agents in which prices are deliberately driven down, temporarily, allowing them or their associates to buy up as many properties as possible before restoring equanimity to the market and then selling to a new group of purchasers. In U.S. cities, for example, some real estate agents have introduced black purchasers into predominantly white areas in the hope that the latter will move out and sell to the realtors or their agents at deflated prices, and have then resold the properties to new black families at higher prices. See *steering*.

boomburb The fastest-growing suburban jurisdictions, typically located along the interstate beltways that ring large metropolitan areas primarily in the western United States. Defined by Robert Lang of Virginia Tech's Metropolitan Institute as: places with more than 100,000 residents; not the largest city in their metropolitan areas; maintaining double-digit rates of population growth in recent decades. Although possessing elements found in most cities (housing, retailing, entertainment, and offices), they are not typically patterned in a traditional urban form (e.g., they almost always lack a dense business core). Distinct from traditional cities not so much in function but in their low density and loosely configured spatial structure. Contrast with *edge cities*.

boosterism Attempts by local governments to develop their local economies by attracting companies and investment into the city from elsewhere and through partnerships with private-sector sources of capital. See also *civic entrepreneurialism*, *growth machines*, *progrowth coalitions*, *public-private partnerships*.

branch plant Subsidiary factories that are responsible for standardized, routine production or assembly; located to take advantage of factors of production including labor costs and skills, and the availability and cost of factory space. See *transnational corporation*.

Broadacre City Frank Lloyd Wright's response to the challenge of machine age, automobile-based urbanization. Based on the idea that, through planned metropolitan decentralization, cities could be spared from the sprawl and congestion threatened by the automobile, allowing people to live instead in affordable "Prairie Style" homes designed to be in harmony with their low-density semirural surroundings. His ideal city—Broadacre City—was to be built on the basis of two new technologies: the automobile and mass-production building technology using high-pressure concrete, plywood, and plastic.

brownfields Abandoned or underutilized industrial and commercial facilities where expansion or redevelopment is complicated by real or perceived contamination. Many brownfields are found in central cities and industrial suburbs with a history of traditional manufacturing. These sites can be abandoned industrial and railroad facilities or manufacturing plants that are operating but show signs of pollution. Brownfields can also be small commercial or even residential lots with only suspected contamination. Contrast with *greenfield site*.

business services See *producer services*.

C

CBD The central business district of an urban area. This central nucleus of commercial land uses has a concentration of office, government, retail, and cultural activities. As the most accessible part of a monocentric (single centered) urban area, land values and building densities are high.

central business district See *CBD*.

central places Urban centers—hamlets, villages, towns, and cities—that provide goods and services to their surrounding *hinterlands*. A central place system comprises a hierarchy of central places—ranging from a small number of very large central places (cities) offering higher-order goods (expensive and infrequently purchased items, such as designer furniture and jewelry) to a large number of small central places (hamlets) offering low-order goods (inexpensive, frequently purchased, everyday necessities, such as newspapers and milk). Walter Christaller sought to explain this size and spacing of towns and cities in his central place theory.

centrality The functional dominance of cities within an *urban system*. Cities that account for a disproportionately high share of economic, political, and cultural activity have a high degree of centrality.

centrifugal movement Outward movement within an urban area as a result of forces—high land costs, congestion, and so on—that displace economic and other activities toward the periphery. Contrast with *centripetal movement*.

centripetal movement Inward movement within an urban area as a result of forces—accessibility to suppliers and customers, specialized services, workers—that attract economic and other activities toward the center. Contrast with *centrifugal movement*.

chain migration A migration process in which migrants who were encouraged and assisted in their move by friends and relatives from their place of origin encourage other friends or relatives to join them, helping them to find accommodation and jobs when they arrive.

circuits of capital investment Geographer David Harvey's notion of circuits of capital investment within economies between which capital is switched depending on market and other economic conditions—for example, from the *primary* circuit of industrial production to the *secondary* circuit of fixed capital assets (machinery and equipment, buildings) or to the *tertiary* circuit of education or technology research.

city-state An independent political unit comprising a city that controls a surrounding *hinterland*. Early examples included the Sumerian and Greek city-states and medieval city-states such as Florence, Genoa, and Venice in Italy and Bremen, Hamburg, and Lübeck in Germany. A contemporary example of a city-state is Vatican City.

civic entrepreneurialism This term captures how local governments have become more proactive in promoting economic development since the early 1970s. They now negotiate with large corporate investors and attempt to stimulate and attract private enterprise with fiscal and other incentives and by attempting to create attractive conditions for profitable investment. See *boosterism, growth machines, progrowth coalitions, public-private partnerships*.

class formation A process that results in conscious collectivities of people who experience class as it is built into their lives in particular ways. They come to realize the force of class in and through the immediate circumstances that they can experience and understand directly. See *class fractions, class structuration, class structure*.

class fractions Narrower categories ("the professions," for example) within the overall concept of *class structure*. See *class formation, class structuration, class structure*.

class structuration Dimensions, in addition to *class structure* and *class fractions*, that David Harvey has identified as contributing to people's experiences that constitute the process of *class formation*. These include the division of labor that determines the formal class structure, institutional barriers to social mobility, the system of authority, and the dominant consumption patterns of a particular time and place. See *class formation, class fractions, class structure*.

class structure The formal categorization of class positions in a society at any given time. It is based on the positions people hold within the *division of labor* and the framework of economic organization. The broad categories of class structure are cast in terms of fairly heterogeneous groups: the "middle class," for example, consisting of a great variety of occupations. See *class formation, class fractions, class structuration*.

cognitive distance A measure of the perceived (rather than just the physical) distance that is generated from the brain's perception of the distance between visible objects in an urban area, a perception that is influenced by land use patterns, the distinctiveness of objects, and the impact of symbolic representations of the environment such as maps and road signs. An important concept in understanding mental maps.

collective consumption Usually refers to goods and services provided by the public sector. Less often refers to services that, literally, have been consumed by a group of people in a collective manner (such as a lecture). The term originated in a neo-Marxist (or Marxian) theory formulated by Manual Castells that argues that there are certain services (schools, hospitals) that are crucial for the maintenance of capitalism but that can be too expensive for individual capitalist enterprises to provide and so require provision through non-market means by the public sector.

colonial cities   Cities that were deliberately established or developed as administrative or commercial centers by colonial or imperial powers.

colonies (ethnic)   A type of ethnic minority residential cluster that results from situations in which a particular area of a city serves as a port-of-entry for an immigrant (or migrant) ethnic group. It is a temporary phenomenon, a base from which ethnic group members are culturally assimilated and spatially dispersed. Contrast with *enclave, ghetto*.

comparative advantage   A principle to explain patterns of trade and specialization in which places and regions specialize in activities for which they have the greatest advantage in production relative to other regions—or for which they have the least disadvantage. Contrast with *competitive advantage*.

competitive advantage   The advantage acquired in economic competition by some locations because of the benefits that accrue from an early start in production of a particular good and the continuing defense of that historic base through superior organization and adaptability. Other locations can suffer from a competitive disadvantage due to the absence of an *initial advantage* that would have allowed them to develop and maintain an ongoing competitive advantage. See *initial advantage*. Contrast with *comparative advantage*.

competitive capitalism   The earliest phase of industrial capitalism—from the late eighteenth to the end of the nineteenth century—the heyday of free enterprise, with markets characterized by comparatively high levels of free-market competition between small-scale businesses, consumers, and workers who acted almost completely independently, and with few constraints or controls imposed by government. See *organized capitalism*.

concentric zone model   The idealized model of city structure devised by Ernest Burgess based on Chicago in the 1920s in which socioeconomic status increases in a series of concentric zones moving outward from the *CBD*. Compare with the *sector model* and *multiple-nuclei model*.

congregation   The residential clustering of an ethnic minority through choice, rather than the involuntary segregation brought about by structural constraints and discrimination. See *enclave*.

consumer services   Personal services, including retailing, medical care, personal grooming, leisure and recreation, eating out, and entertainment. Contrast with *producer services*.

contagion effect   The negative contagious process that contributes to the *spiral of decay* in neighborhoods in which abandonment and long-term vacancy have a depressing effect on the value and desirability of adjacent properties. Nearby property owners seek to sell or move out and remaining property owners are discouraged from investing in the maintenance and upkeep of their depreciating properties—that can lead to further abandonment. See *spiral of decay*.

containerization   The process in which automated cranes efficiently load and unload massive standardized containers filled with large amounts of cargo between the ships and the flatbeds of nearby trains and trucks—improving the operation of cargo shipping and handling by reducing the turnaround time at ports. Previously, ships, trains, and trucks sat in port while major items of cargo were loaded and unloaded individually by a sizeable workforce.

control centers   Cities with a high proportion of company headquarters where a large amount of corporate decision-making is done and from where national and transnational business operations are controlled. See *world cities*.

core countries   Developed countries within the core regions of the world, such as the United States and Germany, that dominate trade, control the most advanced technologies, and have high levels of productivity within diversified economies. Contrast with *peripheral countries* and *semiperipheral countries*.

core-periphery model   A model that describes the unequal relations—economic, political, technological, and so on—between a dominant core and a dependent periphery. At a global scale, a strong core is centered on North America and Western Europe. Core-periphery patterns can also exist at the scale of continental areas, for example, in Europe or Japan, or in individual countries such as the United Kingdom and Italy. Immanuel Wallerstein and John Friedmann made important contributions to shaping ideas about core-periphery models. See *initial advantage*.

corporatist model of urban governance   A model of local power structure that rests on a symbiotic relationship between the various arms of local government and private organizations (labor organizations, community groups, etc.). Key organizations become incorporated into the formal decision-making process and city governments delegate a certain amount of authority in return for cooperation and support. This model sees society in terms of segmented socioeconomic organizations that are taken under the wing of professional politicians and technocrats who are then able to expand and consolidate the scope of their power and authority. Contrast with *elitist model, managerialist model, neoelitist model, pluralist model, hyperpluralist model*.

counterurbanization   The slowdown in population growth, and in some cases the decline, of larger metropolitan areas due to the loss of people to smaller towns and rural areas. Many people still commute into metropolitan areas for work while others telework from home. Compare with *deurbanization*.

creative destruction   The process inherent to capitalism involving the withdrawal of investment from activities (and places) that yield low rates of profit, in order to reinvest in new activities (and places) due to, for example, shifts in the profitability of "old" industries in established cities relative to new industries in rapidly growing cities. As soon as the differential is large enough, some disinvestment will take place within established cities, leading to *deindustrialization*; meanwhile, the capital is invested in the new ventures located elsewhere. Eventually, if deindustrialization of the established cities is severe enough, the relative cost of their land, labor, and infrastructure may decline to the point where substantial flows of new investment are attracted back. See *deindustrialization, uneven development*.

cultural transmission   The process in which distinctive values and norms of behavior, sometimes ones that would be seen as different by society at large, are passed from one generation to another within local cultures within local environments. See *culture of poverty, cycle of poverty, neighborhood effects*.

culture of poverty   The argument that emphasizes the role of values and attitudes in perpetuating the *cycle of poverty* so that a culture of poverty eventually emerges. A vicious cycle links lack of opportunity and lack of aspiration. The poor see the improbability of achieving any kind of material or social success and so adapt their expectations and behavior, becoming so accustomed to deprivation, so defeated, and so withdrawn that they become unable and unwilling to seize opportunities for educational or occupational advancement. Indolence, distrust, suspicion, and introversion become a way of life, defining a distinctive culture of poverty. See *cultural transmission, cycle of poverty, neighborhood effects*.

cumulative causation   In Gunnar Myrdal's model, the self-propelling spiral of growth that occurs in specific settings like cities as a result of the build-up of advantages from the development of *economies of scale, agglomeration economies*, and *localization economies*. Cumulative causation results in some other cities and regions experiencing negative consequences called *backwash effects*. See *spread effects*.

cycle of poverty   The transmission of poverty from one generation to another through a cycle that begins with unhealthy conditions: low incomes, poor housing, and overcrowding. Overcrowding contributes to susceptibility to physical ill-health, which is compounded by poor diets that are also a result of low incomes. Ill-health causes absenteeism from work, which results in decreased income. Absenteeism from school through illness may also contribute to the cycle of poverty by constraining educational achievement and limiting occupational skills, leading to low wages. See *contagion effect, neighborhood effects, spiral of decay*.

D

Dark Ages   The period of stagnation and decline in economic and city life in Western Europe from the collapse of the Roman Empire in the fifth century until the eleventh century.

debt financing   The issuing of bonds by municipalities to finance new infrastructure and improved services to support more urban growth. Bond repayments are met from property taxes, the revenues from which are expected to increase as a result of the growth stimulated by the spending financed by the bond issues.

defensible space   Physical spaces and settings that people can identify with or exert control over. Related to architect Oscar Newman's attack on Modern architecture in which he argued that much of the vandalism, burglary, and criminal violence of inner-city neighborhoods was related to the "designing out" of territorial definition and delineation in Modernist apartment blocks: Once the space immediately outside a dwelling becomes public, nobody feels obliged to supervise it or defend it against intruders.

deindustrialization   A relative decline in industrial employment associated with a shift to services in a country or region where industry had traditionally been a significant component of the economy. It may result from secular shifts in an economy that are related to technological change and/or the globalization of the economy in which some mass production and assembly line manufacturing are moved to lower cost locations in less developed countries. Such trends may involve not only a relative decline but also an absolute one and may involve declining industrial output as well as employment. See also *Fordism, neo-Fordism, informational city*.

Demographic Transition   A trend in birth rates and death rates—from high to low levels—over time. This model suggests that improved diets, public health, and scientific medicine cause a steady decline in death rates with increasing levels of economic development. Birth rates decline later, and more slowly, as sociocultural practices take time to adjust to these new circumstances. The result is a sharp increase in population growth until birth rates fall to relatively low levels. Based on the experience of developed countries like the United Kingdom, not all less developed countries should be expected to follow this demographic path.

depreciation curve   A decline in the value of housing as it ages over time within each new subdivision or tract. Such curves are averages that mask local unevenness in physical deterioration. Housing of the same age and initial quality wears unevenly because of variations in maintenance and improvements and the localized effects of road works, abandonment, and so on. See *spiral of decay*.

deurbanization   A scenario in which the traditional sequence of urbanization—with high growth in both population and households in metropolitan areas, followed first by moderate population growth and continuing high growth in the number of households and then by only moderate growth in both population and households—is followed by three stages of deurbanization: (1) population decline but moderate growth in households, (2) declining numbers of both population and households, and (3) "land decline"—the physical shrinkage of metropolitan areas due to reduced numbers of people and households. Compare with *counterurbanization*.

Deutscher Werkbund   A future-oriented school of thought, led by Peter Behrens (architect and chief designer for the German electrical manufacturer AEG) founded in 1906 in response to the challenge of the Machine Age. This loose coalition of architects, artists, and craft firms wanted to reform the relationship between artists and industry, their guiding principle being that quantity and quality should complement one another. Behrens saw industrialization as the manifest destiny of the German nation, and his factory designs were muscular temples to technological power. Other members of the Werkbund, such as Walter Gropius, emphasized the need to overcome the alienating aspects of traditional society by leaving behind all the architectural refinements and symbolism of the old order and replacing them with a no-nonsense style. See also *Secessionists*.

deviant behavior   Social behavior, such as criminal activity, that deviates from social norms.

diagonal economic integration   A form of business organization in which a company diversifies its interests through corporate mergers or acquisitions of companies that are engaged in separate and distinct enterprises, producing different goods or services for different markets. An automobile manufacturer, for example, may buy into energy, advertising, or entertainment companies.

Digital Divide   The gap in opportunities, among different segments of the population and different places, to access advanced information and communication technologies, particularly the Internet, for a variety of activities.

disorganized capitalism   See *advanced capitalism*.

distance-decay effect   The rate at which a particular process or activity diminishes with increasing distance, such as the tendency for those who live furthest away from the sources of goods and services to consume them less often because of the increased travel costs or the increased time involved in visiting the source of supply. See also *tapering effect*.

divisions of labor   The specialization of different workers, countries, or regions in particular kinds of tasks and economic activities. Social divisions of labor reflect the social characteristics of the people who undertake different types of work (e.g., related to age, ethnicity, gender). Spatial divisions of labor involve regional economic specialization, based on the distribution of resources and markets and on the exploitation of *economies of scale*, *agglomeration economies*, and *localization economies*.

Dual City   A large metropolitan area characterized by disparities in wealth and status or a trend toward increasing social inequality, or both. Manuel Castells argued that the economic, social, and metropolitan restructuring since the 1960s has resulted in both qualitative and quantitative changes in American slum and poverty-area settings that have produced new patterns of vulnerability, fragmentation, and disadvantage while intensifying the occupational and spatial polarization of cities. See *world cities*.

dust dome   See *urban dust dome*.

E

economies of scale   Cost advantages for companies from large-scale production; equivalent to increasing returns to scale in which an increase in inputs (raw materials, labor, etc.) by x percent results in an increase in output of more than x per cent. A crucial part of mass production in *Fordism*.

economies of scope   Cost advantages from large-scale flexible organization of a range of commodities; created by the capacity of companies to provide entirely new products or services, or both, through the flexible use of the same production or service network. A crucial part of *neo-Fordism*.

edge city   A term coined by journalist Joel Garreau to describe recent decentralized urban development—that sometimes overshadows the old downtown—comprising nodal concentrations of shopping and office space situated on the outer fringes of large metropolitan areas, typically near major highway intersections. Contrast with *boomburbs*.

elitist model of urban governance   A model that sees most important decisions made by a handful of powerful business people with elected officials as the "understructure" of power, dependent on the patronage and tolerance of these powerful individuals. The legitimacy of this leadership is ceded from below rather than imposed from above; the passivity of most people reflects their voluntary acceptance of the leadership's domination in return for its pursuit of the public interest, with the ballot box providing a register for any serious abuse of power. Compare with *neoelitist model* and contrast with *corporatist model, managerialist model, pluralist model, hyperpluralist model*.

eminent domain The right of government to take private property under due process of law for legitimate public purpose with just compensation. See *police powers*.

EMR See *extended metropolitan region*.

enclaves (ethnic) A type of ethnic minority residential cluster—generally not as segregated as a *ghetto*—that stays in existence over several generations mainly because the inhabitants choose to congregate for functional reasons. Contrast with *colony* and *ghetto*.

enterprise zone (EZ) An officially designated economically distressed area where public fiscal and other incentives—tax exemptions or reduced planning regulations—are available to promote economic development by inducing companies to locate and create jobs there.

entrepôt A port that specializes in the trade of goods for reexport. They operate primarily as intermediary trading centers—receiving goods from foreign countries for transshipment to other countries. Hong Kong, Singapore, and Rotterdam are the world's top three entrepôts.

environmental determinism The theory that human behavior is determined by physical environment (especially climate). In an urban context, an approach that draws on behaviorist notions to argue that city living affects behavior. See *Gesellschaft*.

EPZ See *export processing zone*.

ethnoburb Suburban ethnic clusters of residential areas and business districts characterized by vibrant ethnic economies that depend on large numbers of local ethnic minority consumers; they also have strong ties to the global economy that reflects their role as outposts in the international economic system through business transactions, capital circulation, and flows of entrepreneurs and other workers. They are multiethnic communities in which one ethnic minority group has a significant concentration but does not necessarily comprise a majority. The clear ethnic imprint on the suburban landscape differentiates the ethnoburb from the typical U.S. suburb in which members of ethnic minorities are more dispersed among the white charter group. Contrast with *colonies, enclaves, ghettos*.

exchange value The value of a commodity, such as housing, as measured by the amount that it can command on the market when it is sold or exchanged. Compare with *use value*.

exclusionary zoning Zoning tactics invented in San Francisco to discriminate against the Chinese and refined in New York in 1916 to discriminate against undesirable factories that were soon deployed by suburban communities to exclude undesirable social groups and unwanted land uses. By carefully framing their *land use zoning* ordinances, suburban jurisdictions can restrict certain types of activity and people from moving into a locality. See *restrictive covenant*.

exo-urbanization Urbanization that is exo(genous)—that is, foreign-investment-induced urbanization characterized by export oriented, trade-creating foreign investment, and labor-intensive assembly manufacturing. See *branch plant*.

export processing zone (EPZ) A small, closely defined area in which government creates especially favorable investment and trading conditions to attract export-oriented, usually foreign-owned, industries. These conditions include the availability of factory space and warehousing at subsidized rents, low tax rates, and exemption from export duties. See *free trade zone*.

Expressionism A style of art and literature that, instead of depicting objective reality, depicts the subjective emotions aroused by objects or events. Accomplished by distorting and transforming nature, rather than imitating it.

extended metropolitan region (EMR) The recent extension of *megacities* in parts of Asia and Latin America beyond their city and metropolitan boundaries to form extended metropolitan regions. This results in an urban form without set boundaries that can stretch for 50 miles from the urban core. These have emerged as single, economically integrated regions comprising the central city, developments within transportation corridors, satellite towns, and other peripheral developments, but no single government body is responsible for overall land use planning.

external economies of scale Benefits that translate into cost savings that accrue to producers from associating with similar producers in places that offer the services that they need, such as specialist suppliers.

exurban development Residential and other development that occurs outside a city and beyond its suburbs. See *centrifugal movement, counterurbanization, edge cities*.

EZ See *enterprise zone*.

F

FDI   See *foreign direct investment.*

Federal Home Loan Bank Act   Legislation passed in 1932 by which the Hoover administration created a reserve for mortgage lenders in an effort to respond to declining residential construction activity with the onset of the Great Depression and the large number of households losing their properties through foreclosure.

Federal Housing Administration (FHA)   An agency created by the Roosevelt administration in 1934 in an effort to reduce unemployment resulting from the Great Depression by stimulating construction jobs in the private sector. It stabilized the mortgage market and facilitated sound home financing on reasonable terms. Instead of lending money, it insured mortgage home loans made by private institutions. The insurance gave banks and savings and loan associations the confidence to disburse more mortgages and to charge one or two percentage points less than before, stimulating demand for home ownership. FHA guarantees also stimulated demand as a result of terms that required smaller down payments and extended repayment periods.

feminization of poverty   The process in which women are more likely to be poor, malnourished, and otherwise disadvantaged due to gender inequalities within households, communities, and countries.

Fertile Crescent   A region of early agriculture and early urbanization in a crescent-shaped area of fertile land stretching from Mesopotamia (the land between the Tigris and Euphrates rivers) in modern Iraq to the lower Nile valley in Egypt.

FHA   See *Federal Housing Administration.*

filtering   Homer Hoyt's thesis that the primary motor behind residential mobility is the chain of moves initiated by the construction of new dwellings for the wealthy and involving the filtering of households up the housing scale and the consequent filtering of houses down the social scale. The construction of new housing for the affluent leads to their out-migration from older properties and the subsequent occupation of this housing by persons of lower socioeconomic background. See *sector model* and *vacancy chain.* Contrast with *gentrification.*

fiscal mercantilism   Efforts by local governments to increase local revenues by attracting—through the use of fiscal and other incentives—companies that create lucrative taxable land uses. Similar to *boosterism* and *civic entrepreneurialism.*

fiscal retrenchment   Cutting back by governments on expenditures for public welfare and services as a result of trends such as economic downturns or voter aversion to higher taxes. See *new conservatism* and *privatization.*

flexible labor strategies   A set of policies designed to increase the capacity of companies to adjust their outputs to variations in market demand and/or to reduce the costs of production through industrial organization involving labor practices involving part-time temporary workers, flexible working hours, and so on. See *flexible production systems* and *neo-Fordism.*

flexible production systems   Practices that allow flexibility in manufacturing in terms of what, when, how, and where production occurs. These practices include exploiting various enabling technologies, subcontracting arrangements, different labor markets, different market niches for products, and the development of new labor processes using flexible working hours, part-time workers, etc. See *flexible labor strategies* and *neo-Fordism.*

Fordism   A method of organizing economic production, income distribution, consumption, and public goods and services that centers on the mutual reinforcement of mass production and mass consumption. Named after Henry Ford because of his innovations and philosophy concerning automobile manufacture, it features a highly specialized and differentiated division of labor based on scientific management and assembly-line production geared to the provision of standardized, affordable goods for mass markets. Associated with *Keynesianism* and *Taylorism.*

foreign direct investment (FDI)   The total overseas investment made by private companies involving direct investment in a company or companies in one or a number of foreign countries (e.g. takeovers, new subsidiaries, etc. rather than portfolio investment) in order to achieve managerial and production control. See *transnational corporations.*

forward linkage   One of the interrelated effects associated with the self-propelling growth process of urban-industrial development. Given a certain amount of original employment growth in a particular industry, forward linkages can occur as new companies arrive that take the finished products of the growing industry and use them as inputs in their own processing or assembly, finishing, packaging, or distribution. Together with the initial growth, the growth in these linked industries helps to create a threshold of activity large enough to attract ancillary industries and activities such as maintenance, repair, and security. See *backward linkage.*

free trade zone   An area delimited by a government unit in an effort to promote jobs and investment in which goods may be manufactured or traded without customs duties. See *export processing zone.*

Futurists   A subset of the Modern Movement in architecture that developed in the 1920s that wanted nothing to do with the past. Led by Filippo Marinetti and represented in the field of architecture and urban design by Antonio Sant'Elia, the Futurists sought to provoke social and institutional change using cities as the stages for permanent revolt, with huge and spectacular edifices that were monuments to both the masses and technology.

G

gateway cities Cities that because of their location serve as links between one country or region and others.

GDP See *Gross Domestic Product*.

Gemeinschaft Roughly translated as "community," tight-knit social relationships that are based around family and that are of the kind that Tönnies argued were evident in traditional agrarian environments. Contrast with *Gesellschaft*.

generalized housing market The housing market—identified by historical geographer James Vance, Jr.—fueled by the ownership and control of rental property by landlords comprising entrepreneurial shopkeepers, small merchants, and people in professional occupations.

gentrification A process involving the renovation of housing in older, centrally located lower-income neighborhoods through an influx of more affluent people (e.g., middle-income professionals) seeking the character and convenience of less expensive and well-located residences. It displaces poorer people through eviction, rising real estate values and rents, and increasing property taxes. Contrast with *filtering*.

gerrymandering The practice in redistricting of manipulating the boundaries of electoral subdivisions to gain political advantage for a particular party or candidate.

Gesellschaft Roughly translated as "society," loose-knit social relationships that are of the kind that Tönnies argued were evident in urbanized environments. Contrast with *Gemeinschaft*.

ghetto A type of ethnic minority residential cluster—generally more segregated than an *enclave*—that stays in existence over several generations mainly because of the constraints of charter group attitudes and discrimination, which are often institutionalized through the operation of housing markets. Usually applied to ethnic minorities but may also refer to other groups such as older people or gays and lesbians. Historically, the section of a European city to which Jews were restricted. Contrast with *enclave* and *colony*. See also *service-dependent ghetto*.

GI Bill See *Servicemen's Readjustment Act*.

globalization The increasing interconnectedness of different people and places around the world through common processes of economic, political, and cultural change. Also, the tendency for the emergence of a global culture (that is, a universal trend that is sweeping all countries). See *world cities*.

green urbanism A process that involves planning for cities in a way that facilitates major lifestyle changes (such as walking, bicycling, and reductions in material consumption), the preservation and restoration of the natural environment, and the application of new or collective technologies (such as public transit, district heating, and green building and design).

greenbelt cities Cities containing government-sponsored low-cost housing that were intended to allow the U.S. government to plan suburban development by drawing people from central cities to make room for slum clearance and redevelopment. Three were built in the United States in the 1930s and sold off to nonprofit corporations after World War II, to be swallowed up in the sprawl of automobile suburbs. See *Resettlement Administration*.

greenfield site Building sites—former farms, golf courses, and so on—at the edge of a city's built-up area that are ripe for development. Contrast with *brownfields*.

greenhouse effect The process in which the earth's atmosphere is warmed due to the trapping of solar radiation by water vapor and so-called greenhouse gases, such as carbon dioxide. The resulting "global warming" may have dire consequences due to the melting of the polar ice caps, rising sea levels, and climatic changes.

gridiron street pattern A street pattern in which the streets are laid out at right angles to one another. Contrast with *organic growth*.

gross domestic product (GDP) An estimate of the total value of all materials, foodstuffs, goods, and services produced by a country in a particular year. It does not include the value of profits from overseas investments and profits accruing to foreign investors.

ground rent Also known as economic rent or location rent, the surplus paid (to land owners, for example) above the minimum amount that would be necessary to use the land at all. Often defined in practice as the total revenue that can be generated by a particular activity on a particular parcel of land, less the total production and transportation costs associated with that same parcel of land.

growth machine Partnerships of private-sector and public-sector interests that implement strategies to enhance the economic development of cities and regions, largely through attracting inward investment, mostly from the private sector but also from public funds. See *boosterism, civic entrepreneurialism, progrowth coalitions, public-private partnerships*.

growth pole Introduced as a concept by the French regional economist François Perroux in the 1950s, a growth pole is usually planned around one or more highly integrated high-growth industries that are organized around a propulsive leading sector, benefits from *agglomeration economies*, and can spread prosperity to nearby regions through *spread effects*. For example, eight *metropoles d'équilibre* (balancing metropolises) were planned in France (including Lyon, Marseille, and Bordeaux) to stimulate regional development while redirecting some economic activity away from Paris to reduce its *primacy*.

H

habitus   A term coined by French sociologist Pierre Bourdieu to describe the culture associated with people's life-world that involves both material and discursive elements. A social group has a habitus if it has a distinctive set of values, ideas, and practices: a collective perceptual and evaluative schema that derives from its members' everyday experience that operates at a subconscious level, through commonplace daily practices, dress codes, use of language, and patterns of consumption.

heat island   The microclimate of a city typically is slightly warmer than the temperature of the surrounding countryside. The built environment has a lower albedo than natural earth surfaces, is less able to reflect incoming solar radiation, and so absorbs more heat. The atmosphere in cities is also warmed by the release of heat from fossil fuels and, in summer, from air conditioning units. The aggregate effect is for cities to become heat islands, with average temperatures one or two degrees Fahrenheit above those of the surrounding countryside. See also *urban dust dome*.

hinterland   A market area—the sphere of economic influence of an urban area. The urban area serves its hinterland with goods and services and its hinterland in turns supplies the urban area with products for processing or for export.

horizontal economic integration   A form of business organization in which a company tries to capture the market for a single stage of production, a single good or service or an entire industry, and achieve *economies of scale*, by using corporate mergers or acquisitions of companies that formerly competed in the same market(s) with similar goods or services. A successful automobile manufacturer, for example, might buy out other automobile manufacturers.

horizontal efficiency   In the provision of public services, the degree to which these services reach those most in need. Contrast with *vertical efficiency*.

horsecar suburb   Suburbs that sprang up in the nineteenth century at the edge of the walking city around the terminals of radial horsecar and omnibus routes.

household lifecycle   A lifecycle that reflects the relationship between household types and residential segregation. For middle-income households, for example, there are recognizable stages, each with a distinctive household composition that is associated with particular space needs, specific preferences in terms of accessibility, type of housing tenure, and locational setting, and different propensities to move.

housing submarkets   Distinctive types of housing in localized areas of cities that, through various institutional mechanisms, tend to be inhabited by people of a particular type (in terms of socioeconomic background, age, ethnicity, etc.). See *culture of poverty*.

hyperpluralist model of urban governance   The sociocultural fragmentation and metropolitan spatial restructuring since the 1980s can turn the pluralism of *progrowth coalitions* into a hyperpluralistic situation in which unstable power relations are reflected in unstructured and multilateral conflict. Power is exercised over narrow areas and for limited periods by a variety of special interest groups who go their own way in seeking narrow gains and who are less restrained in their conduct and less likely to accommodate one another than under the *pluralist model*. Coalitions are essential to the hyperpluralist model, but they are short-lived and ad hoc. Compare with the *pluralist model* and contrast with the *corporatist model*, *elitist model*, *managerialist model*, *neoelitist model*.

I

IDL   See *international division of labor*.

IMF   See *International Monetary Fund*.

import substitution   The development of domestic companies capable of producing goods or services formerly provided by foreign companies.

incorporation   The right of a group of people to petition their state for articles of incorporation as a village, town, or city in the United States. This confers limited liability and, therefore, the opportunity for *debt financing* through issuing bonds to pay for new urban infrastructure and services to support further urban growth. See *suburban incorporation* and *annexation*.

industrial capitalism   The phase of capitalism, beginning with the Industrial Revolution and lasting until the early 1970s, that comprised (1) an early phase, *competitive capitalism*, characterized by comparatively high levels of free-market competition, with many small-scale producers, consumers, and workers who acted almost completely independently with relatively little government intervention, and (2) a later phase, *organized capitalism*, characterized by comparatively highly structured relationships between labor, government, and corporate enterprise.

informal sector The sector of an economy that involves a wide variety of activities that are not subject to formalized systems of regulation or remuneration. In addition to domestic labor, these activities include strictly illegal activities such as drug peddling and prostitution as well as a wide variety of legal activities such as casual labor in construction crews, domestic piece work, street trading, and providing personal services such as shoe-shining.

informational city Also referred to as "post-industrial," Manuel Castells' term for a city that acts as a focus for information, and whose productivity and competitiveness depends on high-technology networks. A large percentage of the workers are engaged in high-tech communications or services jobs that generate new knowledge or access and process a steady flow of information (scientific, technical, and fashion-related).

initial advantage The critical advantage acquired in economic competition by some locations because of the benefits that accrue from an early start in economic development.

innovative milieux Places that are inventive powerhouses and an important source of the economy's dynamism because they have concentrations of research and development (R&D) activity and innovative venture capital.

international division of labor (IDL) The idea of the organization of spatial *divisions of labor*, principally at the national scale until the late twentieth century, in which each country specialized in certain sectors of the economy, such as industry in Western Europe or agriculture and raw materials in many African countries.

International Monetary Fund (IMF) A United Nations affiliate established in 1945 to help encourage international monetary cooperation, ensure international currency exchange stability, promote economic and employment growth, and, while not a development bank, provide temporary economic assistance to countries experiencing balance of payment problems. In 2004 the IMF had about US$107 billion in credit and loans to 87 of its 184 member countries. See *World Bank*.

inverse concentric zone pattern The land use pattern found in some less developed countries, such as in Latin America, that contrasts with the concentric zone pattern identified by Burgess for cities in the United States in the 1920s. In the inverse pattern residential quality decreases with distance from the *CBD*. Contrast with the *concentric zone model*.

K

Keynesian suburb Suburbs that were developed by the U.S. government in the period of Keynesian macroeconomic management (during and after the Great Depression) in an effort to stimulate the economy by promoting housing construction by the private sector and to provide affordable housing to U.S. households. See *Keynesianism*.

Keynesianism A doctrine of macroeconomic management closely associated with the British economist John Maynard Keynes, who advocated the use of fiscal policy (e.g., budget deficits) and economic *multiplier effects* to achieve and maintain full employment. This set of policies underpinned the welfare state in developed countries such as the United States in the 1950s and 1960s. The objective was to manage economies by countering the lack of demand during recessions through government spending—hence the term "demand management." This approach was undermined by inflation and high unemployment in the 1970s. A key element of *Fordism*.

Kondratiev cycles Identified by the Russian economist Nikolai Kondratiev in the 1920s, these cyclical long-waves of 50–55 years in duration have characterized the rate of change in price inflation (a decline in the value of money because prices keep rising) within the capitalist world economy for the past 250 years. Their origins and significance remain controversial, but in recent years they have been widely recognized to be closely tied in to distinctive phases of political-economic development. Contrast with *Kuznets cycles*.

Kuznets cycles Business cycles of approximately 25 years in duration that have characterized the pattern of acceleration and deceleration in economic growth. Named after Ukrainian-born economist Simon Kuznets, who established their existence in the 1920s, they are cycles of activity in investment and building. Contrast with *Kondratiev cycles*.

L

labeling A situation in which all residents of a neighborhood are characterized based on the poor image of their surroundings. This can have negative consequences, such as restricting their employment opportunities. See *cycle of poverty*.

land banks Land bought and held by companies mainly to ensure a supply of developable land. Many of the parking lots on the edge of downtown areas, for example, are held primarily for their speculative value rather than for their earning capacity as parking lots. Compare with *bird-dogging*.

land use zoning The public regulation of land and building use to control the character of a place.

localization economies Cost savings that accrue to companies as the output of their particular industries increase as a result of clustering together at a specific location. See also *agglomeration economies* and *urbanization economies*.

M

machine politics   A form of paternalistic urban governance that emerged with incipient industrialization and rapid urbanization in the late 1800s in the United States that was characterized by charismatic leaders controlling hierarchical political organizations that drew on working class support. See *municipal socialism*.

malapportionment   The creation of electoral subdivisions of unequal size. See *gerrymandering*.

managerialist model of urban governance   A model that sees civil servants effectively as social gatekeepers and "street-level" bureaucrats who mediate policy implementation. It emphasizes the autonomy of the technocrats and bureaucrats in the local civil service and the fact that urbanization had become so complex that elected officials had to rely increasingly on the expertise of civil servants; as a result, the local state is controlled by civil servants whose professional ideology and departmental allegiances are crucial in determining a good deal of the shape of local government activity. Contrast with *corporatist model, elitist model, neoelitist model, pluralist model, hyperpluralist model*.

maquiladora   Literally, "mill" in Spanish. These factories take advantage of Mexico's low-cost labor and lax environmental regulations and locate in *export processing zones* (*EPZs*) in Mexico, mainly along the U.S. border. They are often owned or built with foreign capital, and assemble U.S. components for reexport to the United States as finished products free from customs duties.

mechanical solidarity   One of the two forms of social "solidarity" identified by Emile Durkheim, it is based on similarities between people. Contrast with *organic solidarity*.

megacity   A very large city—with a population of 10 million or more—characterized by both primacy and centrality within its national economy. Manuel Castells used this term for large cities in which some people are connected to global information flows while others are disconnected and "information poor." See *Dual City* and *world city*.

megalopolis   French geographer Jean Gottman's term for multicity, multicentered, urban regions. The name was first given to the dominant urban corridor in the United States that extends along the eastern seaboard from Boston to Washington, D.C.

merchant capitalism   The initial phase of capitalism. As the feudal system disintegrated, it was replaced by an economy that was dominated by market exchange, in which communities came to specialize in the production of the goods and commodities that they could produce most efficiently in comparison with other communities. The key actors were the merchants who supplied the capital required to initiate the flow of trade—hence the label merchant capitalism. See *comparative advantage*.

metropolitan consolidation   The process in which key economic functions (e.g., headquarter offices and research and development [R&D] laboratories) tend to become increasingly localized in larger metropolitan areas in response to corporate reorganization following mergers and acquisitions and the differential growth and changing patterns of economic specialization within urban systems. Contrast with *regional decentralization*.

multiple-nuclei model   Harris and Ullman's model of urban city structure characterized by decentralization and nodes of different economic and residential areas. Contrast with the *concentric zone model* and the *sector model*.

multiplier effects   The extra industries, companies, incomes, and employment in various sectors of the economy generated by new economic activity in one sector of the economy can be said to result from that activity's multiplier effects. Primary multiplier effects comprise four *localization economies*: *backward linkages*, *forward linkages*, the pool of labor in the existing interrelated industries that has skills and experience attractive to additional firms, and the likelihood of new innovations that provide further stimulus to growth as a result of the linkages between existing industries that promote interactions among professional and technical personnel and sustains R&D facilities, research institutes, and so on. Secondary multiplier effects are the result of the spiral of growth resulting from the initial spiral of growth caused by the primary multiplier effects.

municipal foreign policy   The direct involvement of local governments in international relations (usually with overseas localities) and international affairs (with national governments, home or abroad). Much of the early activity was related to arms control issues, human rights, and the environment. More recently, increasing global economic interdependence has resulted in the expansion of economic aspects of municipal foreign policy, such as sister-city arrangements and city networking, municipally funded world trade centers and trade commissions, local development commissions with an international scope, and local initiatives to promote international tourism.

municipal socialism   A form of urban governance that emerged with incipient industrialization and rapid urbanization in the late 1800s in the United States and Europe that was characterized by local government intervention in the marketplace in order to impose standards and to ensure the provision of key services and basic amenities. See *collective consumption* and *machine politics*.

N

natural increase The surplus of births relative to deaths in a country or place (calculated as the difference between the crude birth rate and crude death rate). It does not take into account migration.

neighborhood effects The concept that residential environments both influence and reflect local cultures within cities: People tend to conform to what they perceive as local norms in order to gain or maintain the respect of their local peer group. See *cultural transmission*.

neighborhood unit American sociologist-planner Clarence Perry's concept that became one of the most widely adopted city planning building units for planned new towns throughout the world. The neighborhood unit was centered on a local elementary school, shops, and community institutions within walking distance of homes and bounded by roads carrying heavier flows of automobile traffic in order to provide a safer, traffic-free environment at its center.

neoclassical economics A conceptualization of how economic activity operates in capitalist society. The economy comprises many small producers and consumers who act rationally, although none is large enough to affect significantly the operation of the market. Companies are seen as atomistic agents with full information in a world of pure markets (with no entry barriers) and all have exactly the same resources, technological capability, and market power with deviations regarded as market "imperfections." Companies utilize factors of production (land, labor, and capital) to maximize their profits; consumers sell theirs (especially labor) to purchase goods and services that maximize their individual preferences. The market is an abstract space in which companies and consumers set the prices and the forces of supply and demand cause economic resources to be used in the most efficient way possible. Involves normative model-building—constructing simplified versions of how the real world ought to operate.

neoelitist model of urban governance This model accepts the premise of the *elitist model* that in some cities nearly all important decisions are made by a handful of powerful business people but questions the assumption that a consensus on the public interest is consciously and freely agreed to by the bulk of the population or whether rather it is the product of the systematic suffocation of opposition. Compare with the *elitist model* and contrast with *corporatist model, managerialist model, pluralist model, hyperpluralist model*.

Neo-Fordism Sometimes referred to as "Post-Fordism" or "flexible accumulation"; a term that identifies a way of organizing economic production, income distribution, consumption, and public goods and services that is intended to overcome the problems inherent to *Fordism*. Rather than being predicated on the mutual reinforcement of mass production and mass consumption, neo-Fordism depends on *flexible production systems* to exploit specific market segments or niches, or both.

neoliberal ideologies A particular conceptualization that economies can be made more competitive through the implementation of various types of New Right strategies based on the desirability of free markets as the ideal condition not only for economic organization but also for political and social life.

neoliberal policies Policies designed to make economies more competitive through the implementation of various types of New Right strategies based on the desirability of free markets as the ideal condition not only for economic organization but also for political and social life. These policies involve reducing the role and budget of government, including removing subsidies, deregulation, and the *privatization* of formerly publicly owned and operated operations, such as utilities.

new conservatism More authoritarian than neoliberalism (New Right) ideologies, new conservatism involves a commitment to high military spending and the global assertion of national values (especially in the United States by, for example, the Reagan and both Bush presidencies). See also *neoliberal ideologies* and *neoliberal policies*.

new international division of labor (NIDL) The idea of the reorganization of the *international division of labor*, formerly principally at the national scale, to a global scale based on international production and marketing systems. Involves the decentralization of traditional manufacturing production from core regions to some semiperipheral and peripheral countries. Contrast with *international division of labor*.

newly industrializing countries (NICs) Less developed countries, formerly peripheral within the world-system, that have acquired a significant industrial sector, usually through *foreign direct investment* (*FDI*) by *transnational corporations*. Examples are Mexico, Brazil, and Taiwan. See also *semiperipheral countries*.

NICs See *newly industrializing countries*.

NIDL See *new international division of labor*.

NIMBYism Not In My Backyard-ism. Characteristic of middle-income suburbs that are averse to the location of locally unwanted land uses (LULUs) such as refuse burners within their jurisdictions. The more extreme version of NIMBYism is BANANAism (Build Absolutely Nothing Anywhere Near Anything).

nodal center Urban centers that are important nodes—in terms of their concentrations of businesses and economic activities—for a region or urban system.

O

organic growth Urban growth that evolves in an unplanned manner—as when homes are built along preexisting winding pedestrian paths—rather than in a predetermined way based on some planned approach. Contrast with *gridiron street pattern*.

organic solidarity One of the two forms of social "solidarity" identified by Emile Durkheim, it is based on the existence of differences stemming from specialized economic roles—the *division of labor*. This term was based on a biotic analogy, the idea that in a complex organism (modern, urban society), every single "organ" (social group) is mutually dependent on the rest. Contrast with *mechanical solidarity*.

organized capitalism The later phase of industrial capitalism that began in the late nineteenth century and was characterized by comparatively highly structured relationships between labor, government, and corporate enterprise. These relationships were mediated through legal and legislative instruments, formal agreements, and public institutions. See *competitive capitalism*.

overaccumulation crises Distinctive crisis phases in the long-term dynamics of capitalist economies, characterized by unused or underutilized capital and labor. An inevitable outcome of the difficulty of matching supply to demand under changing conditions, these crises represent critical moments for the political economy of capitalism. They can be recognized by the appearance of idle productive capacity, excess inventories, gluts of commodities, surplus money capital, and high levels of unemployment.

overurbanization A condition experienced in many contemporary less developed countries in which cities grow more rapidly than the jobs and housing they can sustain. See *megacities*.

P

parkway A limited-access recreational highway developed by master planner Robert Moses in the United States.

peripheral countries Less developed countries within the peripheral regions of the world characterized by dependent and disadvantageous trading relationships, inadequate or obsolescent technologies, and undeveloped or narrowly specialized economies with low levels of productivity. Contrast with *core countries* and *semiperipheral countries*.

pluralist model of urban governance This model suggests that power is dispersed, with different kinds of interests dominant at different times over different issues. The structure of power is essentially competitive, drawing on a wide range of participants and ensuring a fundamental element of democracy through the need for elites to acquire mass loyalty. Society is kept in rough equilibrium by this competition, with labor acting as a counterbalance to business, consumers offsetting the power of retailers, tenants constraining the power of landlords, etc. These groups also have overlapping membership, which promotes intergroup contact and makes for tolerance and moderation in local politics. Compare with the *hyperpluralist model* and contrast with the *corporatist model, elitist model, managerialist model, neoelitist model*.

police powers The powers of government to take actions (including exerting control over the development of privately owned property and land by local governments) in order to protect the health, safety, and public welfare. See *eminent domain*.

postmodern generation The generation of Baby Boomers that came of age in the 1970s and 1980s and were characterized by materialism and a pluralism of taste.

poverty area Areas within metropolitan areas defined by the U.S. Census Bureau as contiguous census tracts in which at least 20 percent of households have an income below the official poverty level. "Extreme poverty areas" are census tracts with 40 percent or more below the poverty level.

primacy A condition in which the population of the largest city in an urban system is disproportionately large in relation to the second-largest and third-largest cities in that system. See also *primate city*.

primary products Products derived from natural resources, as in agriculture, mining, forestry, and fishing. These products are often important as raw materials in manufacturing.

primate city A country's leading city as evidenced by measures of its *primacy* including its significantly larger population compared to other cities and by other characteristics reflecting the primate city's national importance and influence, such as economic and political activity and power. See also *primacy* and *megacities*.

privatization A diverse set of policies designed to introduce private ownership and/or private market allocation mechanisms to goods and services previously allocated and owned by the public sector. Involves the sale of government assets (such as key industries) to private owners and the contracting out of services formerly provided by government to private companies in an effort to increase government efficiency and save public money.

producer services Also known as business services, these services enhance the productivity or efficiency of other companies' activities or enable them to maintain their specialized roles. Examples include advertising, personnel training, recruitment, finance, insurance, and marketing. Contrast with *consumer services*.

product-cycle model   A model that describes how the locational requirements for production change as products move from being novel and expensive to being standardized and cheaper. In particular, labor costs can become more important than adjacency to markets.

progrowth coalition   Another name for *growth machine*. See *boosterism, civic entrepreneurialism, public-private partnerships*.

public-private partnership   Coalitions of private-sector businesses and/or business leaders and public-sector officials and agencies, with others such as unions and chambers of commerce, that seek to promote economic growth and the prosperity of their urban area. See *boosterism, civic entrepreneurialism, growth machines, progrowth coalitions*.

pull factors   In migration studies, forces of attraction that induce migrants to move *to* a particular location. Usually operate in combination with *push factors*.

push factors   In migration studies, negative events and conditions that impel an individual to move *from* a location. Usually operate in combination with *pull factors*.

R

rank-size rule   A statistical regularity in city-size distribution, set out by Zipf in 1949, such that the population of a particular city is equal to the population of the largest city divided by the rank of the particular city. So if the largest city in an urban system has a population of 1 million, the fifth-largest city should have a population of 200,000, the hundredth-ranked city should have a population of 10,000, etc.

redlining   The practice by commercial lenders (e.g., banks) of withholding loans for properties in areas of cities that are perceived as high risk. The practice involves literally drawing a red line around high-risk neighborhoods on a city map and then using the map as the basis for determining loans. This results in a bias against minorities, female-headed households, and other vulnerable groups because they tend to be localized in high-risk neighborhoods. Redlining can also become a self-fulfilling prophecy, as neighborhoods starved of mortgage funds become progressively more run-down and increasingly high-risk for loans. Compare with *steering*.

regional decentralization   The process in which improved transportation and communications infrastructures, coupled with the attraction of cheaper land, lower taxes, lower energy costs, *boosterism*, and cheaper and less-militant labor, allow cities located in traditionally underdeveloped regions of a country to grow rapidly as a result of their growing relative attractiveness to private investment compared to the cities in established industrial regions. Contrast with *metropolitan consolidation*.

relative location   The location of an urban area, although absolute (as measured, for example, by latitude and longitude), can also be relative, fixed in terms of its "situation." Situation refers to the location of a city relative to other places; its accessibility to routeways, for example, or to other cities.

rent gap   The disparity between the rents actually charged for run-down inner-city areas and their potential market rents (as predicted by *bid-rent theory*) following renovation. If large, the rent gap can lead to urban redevelopment and *gentrification*.

Resettlement Administration   An agency created in 1935 by the Roosevelt administration to help plan suburban development and draw people from central cities in order to make room for slum clearance and redevelopment. See *greenbelt cities*.

restrictive covenant   A contractual provision in the sale of a property that limits the nature of subsequent development. Such covenants have usually been used to discriminate against low-income groups and locally unwanted land uses (LULUs). See *exclusionary zoning*.

rural-to-urban migration   The movement of rural residents to larger towns and cities in search of a better life, as during the Industrial Revolution in England. In many less developed countries today, rural residents migrate to urban areas because of the desire for employment and the prospect of access to public facilities and services that are often unavailable in rural regions. See *overurbanization*.

S

Secessionists   A future-oriented school of thought that emerged in the early 1900s in response to the challenge of the machine age. Headed by Austrian architect Adolf Loos, the principal motivation of the Secessionists was the elimination of all "useless" ornamentation from architecture. See also *Deutscher Werkbund*.

sector model   The model of urban residential land use structure advocated by Homer Hoyt that suggests that class differences in residential areas are arranged in wedges (sectors) based on transportation routes. See *filtering*. Contrast with the *concentric zone model* and *multiple-nuclei model*.

semiperipheral countries   Less developed countries (including *newly industrializing countries* [*NICs*]) within semiperipheral regions that are able to exploit *peripheral countries* but are themselves exploited and dominated by *core countries*. Contrast with *core countries* and *peripheral countries*.

service-dependent ghetto   A concentration of dependent groups such as ex-psychiatric patients in inner-city areas close to community-based services. Also known as "asylum-without-walls."

Servicemen's Readjustment Act (GI Bill)   Legislation passed in 1944 to facilitate the readjustment of U.S. veterans returning from World War II. Created the Veterans Administration, one of the major goals of which was to facilitate homeownership for returning veterans.

setbacks   *Land use zoning* ordinances, beginning with New York in 1916, that required the upper stories of skyscrapers to be stepped back beyond a certain height in order to allow light and air to reach the street. This restriction resulted in the characteristic wedding-cake profile of interwar skyscrapers. See *land use zoning*.

shock city   A city that is seen as the embodiment of surprising and disturbing changes in economic, social, and cultural life—such as Manchester, England, during the Industrial Revolution.

single-room occupancy (SRO) hotel   Hotels in the run-down parts of cities that rent out units for occupancy by one person. These units may contain food preparation or bathroom facilities, or both.

social construction   The idea that most differences between people are the result not of their inherent characteristics but of the way they are treated by others in society. Race, for example, is a social construction that artificially categorizes people into separate groups based on characteristics that include physical appearance (particularly skin color), ancestral heritage, and cultural history.

social disorganization   A situation in which some communities find themselves weakly equipped to maintain social order because of high levels of poverty or inequality, high rates of population turnover, or high levels of diversity in social, demographic, cultural, and ethnic attributes that can inhibit the formation of local ties and weaken the effectiveness of local social institutions.

social distance   The difference between people based on factors such as socioeconomic background and power leading to separation in social life. May be the result of mutual desire or predominantly the wishes of the powerful. Often expressed in terms of physical distance and residential differentiation.

social ecology   The social and demographic composition of neighborhoods.

social reproduction   The various elements that are necessary to reproduce the workforce and the consumers needed to keep a capitalist society functioning (e.g., family, schools, health services, welfare state).

sociospatial dialectic   Ed Soja's term for the mutually interacting process in which people shape the structure of cities and at the same time are themselves affected by the structure of those cities.

spatial mismatch   The mismatch in the location of low income lower-skilled people and jobs that has developed as many of the low-skilled jobs traditionally found in inner-city areas where many low income people are concentrated have been relocated to suburban areas, to be replaced mainly by jobs requiring higher skills.

special district   A single-purpose jurisdiction, such as a special sanitary district, established as an additional layer of government for a specific purpose. Special districts are not subject to the statutory limitations on financial or legal powers that apply to municipalities. They also have the potential advantage of being customized to correspond closely to specific functional areas.

spiral of decay   The process in which deferred or makeshift maintenance and repair, in conjunction with an accumulation of garbage and graffiti, can set a low-income neighborhood containing substandard housing on a downward spiral of decay that results in it becoming a *slum*. See *contagion effect*.

splintering urbanism   The term devised by geographers Stephen Graham and Simon Marvin to describe the fragmentation of the economic, social, and material fabric of cities as a result of the selective impact of new technologies and networked information and communications infrastructures.

spread effects   In Gunnar Myrdal's model, the positive spillover effects on an urban area or region—including capital investment and the inmigration of skilled workers—from the economic growth of some other urban area or region. Also known as *trickle-down effects*. See *backwash effects* and *cumulative causation*.

squatter settlement   Residential development on land that is neither owned nor rented by its occupants. Often found in the cities of less developed countries, squatting also occurs in developed countries, including in some European cities.

SRO hotel   See *single-room occupancy hotel*.

stagflation   An episode of economic recession accompanied by comparatively high rates of price inflation (a decline in the value of money because prices keep rising).

stealth cities   Urban developments at the edge of metropolitan areas, including some *edge cities*, that have grown so quickly that they have not been *incorporated* as separate jurisdictions and do not have official names that appear on street maps or government statistical tables. They are called stealth cities because they are invisible administratively and politically, without their own chambers of commerce, libraries, town halls, or courthouses. But they are very real, showing up clearly both on the skyline and on plots of land values.

steering   The practice by realtors of deterring households from moving into neighborhoods occupied by households of a different socioeconomic background or ethnicity in order not to jeopardize local prices. Until 1968, when it was made illegal in the United States, it resulted in "slammed door" discrimination, mainly against African Americans. Compare with *redlining*.

streetcar suburb   A suburb that sprang up at the edge of the city around the terminal of a radial streetcar line. By making it feasible to travel up to 10 miles from the *CBD* in 30 minutes or so, the streetcar greatly increased the territory available for residential development and opened the way for the growth of streetcar suburbs.

structural assimilation   A process involving the diffusion of members of a minority ethnic group through the social and occupational structure of the wider society (or charter group). Contrast with *behavioral assimilation*.

subsistence activities   Activities that comprise an economic system, usually farming, in which the producers (farmers), and their families, consume most of what is produced, leaving little surplus for trade or sale.

suburban incorporation   The petitioning for articles of incorporation by suburban communities to protect their reputation, status, and independence—often a preemptive strategy to avoid annexation of their community by a central city. As the pace of suburbanization quickened from the 1920s in the United States, middle-income suburbanites sought to escape not only proximity to central city *slums* and *ghettos* but also central city tax burdens and to establish a distinctive setting for governance and politics in which middle-income values and preferences might flourish. See *incorporation* and *annexation*.

sustainable urban development   A vision of urban development and resource use that seeks a balance among economic growth, environmental impacts, and social equity in a manner that can be sustained in the long run for future generations.

T

tapering effect   The effect associated with the unevenness of municipal service delivery such that the amenities of parks, libraries, fire stations, and so on are greatest for those living nearby and tend to diminish with distance because many public services have to be tied to specific locations, so that *distance-decay effects* confer differential benefits.

tax increment financing (TIF)   A mechanism used by cities in the United States to finance redevelopment efforts that is directly tied to the success of those efforts. If an area within a city can be made more attractive to

private developers and new development occurs, the tax revenue collected from that area would be expected to rise. Tax increment financing taps into any increase in tax revenues by using the tax "increment" (the difference between the taxes after redevelopment and the expected taxes without redevelopment) to finance the improvements and other activities that stimulated the redevelopment in the first place.

Taylorism   The name (after analyst F. W. Taylor) for a form of labor organization in manufacturing industries in which the planning and control of work are given over entirely to management, leaving production workers to be allocated specialized tasks that are subject to careful analysis—"scientific management" using techniques such as time-and-motion studies. Adopted by Henry Ford in the early twentieth century to mass produce automobiles in Detroit. See *Fordism*.

technology systems   Distinctive "packages" of technologies, energy sources, and political-economic structures that represent the most efficient means of organizing production at any given phase of economic development. Based on key sets of interdependent technologies, they represent the underpinnings of successive modes of organizing such things as economic production, income distribution, consumption, and public goods and services over several decades at a time.

territoriality   The tendency for particular groups within a society to attempt to establish some form of control, dominance, or exclusivity within a localized area. Group territoriality depends primarily on the logic of using space as a focus and symbol for group membership and identity as a means of regulating social interaction.

TIF   See *tax increment financing*.

TNC   See *transnational corporation*.

transnational corporation (TNC)   Similar to a multinational corporation (MNC), a company that engages in production, distribution, and marketing that span international boundaries, with subsidiary companies, factories, offices, or facilities in a number of countries. TNCs have the power to coordinate and control operations in several countries; control may be exercised without the necessity for legal ownership, such as through subcontracting arrangements. Production is carried out on a global scale in such a way as to maximize profits. For example, as part of a global assembly system, a low-skill labor-intensive stage of the production process may be located in a less developed country where wages and unionization levels are low. Many headquarters of the largest TNCs are concentrated in the *world cities* of London, New York, and Tokyo. See *globalization*.

trickle-down effects   The economic growth that spreads—trickles down—to more remotely located cities and regions that is induced by high levels of demand in more centrally located economically vibrant cities and regions. Also known as *spread effects*.

U

underemployment   A situation in which people work less than full time even though they would prefer to work more.

uneven development   The spatial outcome, within and between countries, of the continuous seesawing of capital from one set of opportunities to another on the basis of particular local mixes of skills and resources. Capital is invested unevenly over time and across space because, whenever possible, development will occur wherever businesses judge that their investment will yield the highest return. When businesses try to exploit differences between places, they create a continuously variable geometry of labor, capital, production, markets, and management.

"Untouchables"   At the bottom of all caste systems of kinship groupings are the group whose members deal with human waste and dead animals. Mohandas Gandhi, the inspirational leader of India before independence, crusaded to dissociate this group from this demeaning term. He called them Harijans, meaning "children of God," but today most of these people prefer to be referred to as Dalits, meaning "the oppressed"; the Indian government refers to them as "Scheduled Castes." Traditionally, the Dalits were forced to live outside the main community because they were deemed capable of contaminating food and water by their touch. They were denied access to water wells used by other groups, refused education, banned from temples, and subject to violence and abuse. Although these practices were outlawed by India's constitution in 1950, discrimination and violence against Dalits is still routine in many rural areas.

urban dust dome   The *heat island* effect leads to a distinctive pattern of local air circulation on days when there is no regional air movement. Light surface winds are drawn toward the city center, then rise and descend slowly at the edge of the built-up area. This pattern is associated with the development of an urban dust dome in which fumes, soot, and chemicals are trapped in the air over the city. This phenomenon becomes particularly pronounced when a temperature inversion develops and a lid of relatively warm air effectively flattens the dust dome, keeping pollution low over the city. When exposed to strong sunshine, such concentrations of pollution can be transformed through photochemical reactions into *smog*. See also *heat island*.

urban heat island   See *heat island*.

urban realms   The term coined by geographer James Vance, Jr., to capture the fact that many of the world's largest metropolitan regions consist of a loose coalition of urban realms or economic subregions bound together through urban freeways that tend to function semi-independently, and contain a broad mix of land uses (retail, commercial, and residential). As a result, the central urban core loses its economic dominance and becomes just another realm.

urban regime   The term coined by Clarence Stone to capture the fact that local *public-private partnerships* involve power structures that rest on slowly changing coalitions of dominant groups and interests that are represented by city officials who sustain both their own power and that of the coalition by ensuring a variety of benefits and policy outcomes to the key groups involved.

urban renewal   The revitalization of run-down sections of central cities. In the United States the federal Urban Renewal Program was a slum clearance and public housing program initiated under the 1937 Housing Act. The 1949 Housing Act subsequently provided assistance to central cities toward the costs of clearing "blighted" areas and assembling land for redevelopment. Due to problems including the negative reactions of residents of neighborhoods that were officially designated as blighted, it was terminated by Congress in 1973. In 1974 the Housing and Community Development Act introduced the block grant as the principal form of federal aid for local community development.

urban social movements   Pressure groups and organizations with varying degrees of public support that petition for change, often outside conventional political channels. Sometimes termed new social movements. These form an important part of the theory of *collective consumption*.

urban system   The complete interdependent set of urban settlements of different sizes that exists within a particular territory, such as a region or a country.

urbanism   The forms of social interaction, patterns of behavior, attitudes, values, and ways of life that develop in urban settings.

urbanization economies   The economic benefits that companies enjoy because of the package of infrastructure, ancillary activities, labor, and markets typically associated with urban settings. See also *agglomeration economies* and *localization economies*.

use value   The value of a commodity, such as housing, as measured by its owners when they consume it. Compare with *exchange value*.

Usonia   Architect Frank Lloyd Wright's physical framework for a new American way of life in which the working class would be emancipated from the trap of congested but expensive cities, living instead in a semirural setting spread out at low densities in differentiated and individualized houses designed to be in harmony with their natural surroundings.

V

vacancy chain The chain of movement resulting from properties becoming available through factors such as new building, the subdivision of properties, and the death or out-migration of existing occupants. See *filtering*.

vertical economic integration A form of business organization in which a single large company tries to control all aspects of the same industry and capture a greater proportion of the final selling price by using corporate mergers or acquisitions of companies that formerly were engaged in different stages of the same industry (from research and design through production to sale). An automobile manufacturer, for example, might take over companies that make specialized components like engines or car navigation systems or that distribute or sell automobiles. See also *Fordism*.

vertical efficiency In the provision of public services, the proportion of these services allocated to people or communities in need. Contrast with *horizontal efficiency*.

W

White City The temporary plaster city built for the World's Columbian Exposition in Chicago in 1893, designed by a team lead by architect Daniel Burnham, that showcased the Beaux Arts style of architecture and later City Beautiful movement and belief in the role of the built environment as an uplifting and civilizing influence. The White City had uniform building heights and imposing avenues with dramatic perspectives.

World Bank Along with its main component, the International Bank for Reconstruction and Development, a United Nations affiliate established in 1948 to finance productive projects that further the economic development of its 184 member countries from Afghanistan to Australia and from the United States and the United Kingdom to Ukraine, the United Arab Emirates, Uganda, and Uruguay. In 2003, for example, the World Bank loaned about US$11 billion to less developed countries. See *International Monetary Fund*.

world cities A term coined by Patrick Geddes to describe those cities in which a disproportionate share of the world's most important business—economic, political, and cultural—is conducted and that serve as headquarters to *transnational corporations*. Typically, London, New York, and Tokyo are identified as the leading tier of world cities, although other cities such as Paris, Frankfurt, Chicago, Los Angeles, and Zurich also have important global roles. Characterized by social polarization. See *control centers*.

world-system A term coined by historian Immanuel Wallerstein to describe any spatially extensive interdependent economic system comprising a number of countries that have a single *division of labor* but multiple cultural systems.

Z

zone in transition The name given by Ernest Burgess and the Chicago School of Human Ecology to the concentric ring of land uses lying between the *CBD* and the inner ring of working class residential areas. Characterized by a mixture of industry, low-status commercial activity, and high-density poor-quality rental accommodation, often inhabited by recent immigrants. The *concentric zone model* identifies it as a zone of disinvestment despite its central location, and consequently the likely location of a *rent gap*.

zoning See *land use zoning*.

CREDITS

9.11   Corbis/Bettmann photo by Kevin R. Morris
9.13   Corbis/Bettmann photo by Caroline Penn
10.0   Corbis/Bettmann photo by Wes Thompson
10.1   Corbis/Bettmann photo by GE Kidder Smith
10.2   Library of Congress
10.3   Corbis/Bettmann
10.4   Visuals Unlimited photo by John Sohlden
10.5   Corbis/Bettmann photo by Alexander Alland
10.6   National Archives and Records Administration
10.7   Corbis/Bettmann photo by Ferdinand S. Hirsh
10.8   Library of Congress
10.10   Corbis/Bettmann
10.11   Corbis/Bettmann photo by G. E. Kidder Smith
10.12   Corbis/Bettmann photo by Paul Almasy
10.15   Corbis/Bettmann photo by Thomas A. Heinz
10.16   Chicago Historical Society photo by Bill Hedrich, Hedrich-Blessing
10.17   Corbis/Bettmann photo by Angelo Hornak
10.18   Corbis/Bettmann photo by GE Kidder Smith
10.19   Getty Images Inc. - Hulton Archive Photos
10.20   Photolibrary.Com photo by CLIFF PHILIPIAH
10.21   Corbis/Bettmann photo by Thomas A. Heinz
10.22   Corbis/Bettmann photo by Grant Smith
10.23   Saint Louis Post-Dispatch
10.24   Photolibrary.Com photo by CLAVER CARROLL
10.25   Corbis/Bettmann photo by Harald A. Jahn
10.26   Margherita Spiluttini photo by Margherita Spiluttini
10.27a   Corbis/Bettmann photo by Alan Schein Photography
10.27b   Peter Arnold, Inc. photo by Manfred Vollmer/Das Fotoarchiv
10.28   Corbis/Bettmann photo by Richard Bickel
10.29   Corbis/Bettmann photo by Lee Snider
10.31   PhotoEdit photo by A. Ramey
10.32   Corbis/Bettmann photo by Roger Ressmeyer
11.0   Corbis/Bettmann photo by Jason Hawkes
11.1   PhotoEdit photo by Tony Freeman
11.6   Corbis/Bettmann
11.7   Corbis/Bettmann photo by Luiz C. Ribeiro
11.8   Paul L. Knox
11.9   Corbis/Bettmann photo by Grant Smith
11.10   Corbis/Bettmann photo by Morton Beebe
11.11   Linda McCarthy
11.12   Corbis/Bettmann photo by Bob Rowan
11.13   Corbis/Bettmann photo by Patrick Ward
11.14   Corbis/Bettmann photo by Craig Tuttle
12.0   Corbis Royalty Free
12.3   Corbis/Bettmann photo by Tim Zinnemann
12.4   Omni-Photo Communications, Inc. photo by Grace Davies
12.5   Paul L. Knox photo by Paul Knox
12.7   Corbis/Bettmann photo by Michael S. Yamashita
12.8   Corbis/Bettmann
12.9   Corbis/Bettmann photo by Dave G. Houser
12.11   Corbis/Bettmann photo by Gail Mooney
12.12   Corbis/Bettmann photo by Ed Kashi
12.13   AP Wide World Photos
12.15   Library of Congress photo by Lee Russell
12.32   Alamy Images Royalty Free

13.0   Corbis/Bettmann photo by David Turnley
13.5   Getty Images Inc. - Hulton Archive Photos
13.7   Getty Images - Photodisc photo by C Borland
13.10   Corbis/Bettmann
13.11   Library of Congress
13.12   Aurora & Quanta Productions Inc photo by Nina Berman
13.13   Polaris Images photo by Cale Merege
13.15   Paul L. Knox
13.19   AP Wide World Photos photo by Charles Rex Arbogast
13.21   Corbis/Bettmann photo by Phil Schermeister
14.0   PhotoEdit photo by Rudi Von Briel
14.1   Corbis/Bettmann
14.2   PhotoEdit photo by Spencer Grant
14.3   AP Wide World Photos photo by Wally Santana, Stringer
14.5   Pearson Education/PH College photo by Laimute Druskis
14.8   Corbis/Bettmann photo by Dave G. Houser
14.14   PhotoEdit photo by Rudi Von Briel
14.19   Corbis/Bettmann photo by H. Armstrong Roberts
14.20   Library of Congress
15.0   Corbis/Bettmann
15.1   Museum of the City of New York photo by Jacob A. Riis
15.2   Corbis/Bettmann photo by Jacob August Riis
15.3   Library of Congress photo by Arnold Genthe
15.4   AP Wide World Photos
15.6   Corbis/Bettmann photo by Joseph Sohm
15.8   Corbis/Bettmann photo by Joseph Sohm
15.9   Corbis/Bettmann photo by Ted Soqui
15.18   Getty Images Inc. - Hulton Archive Photos
15.22   Chris Brown
15.24   The Image Works photo by Bob Daemmrich
15.25a   Corbis/Bettmann photo by Steve McDonough
15.25b   Al Dodge photo by Al Dodge
15.28   AP Wide World Photos
15.29   Corbis/Bettmann photo by Joel Stettenheim
15.34   Corbis/Bettmann photo by Joseph Sohm
15.35   Corbis/Bettmann photo by Nik Wheeler
15.36   FEMA photo by Andrea Booher
15.40   Corbis/Bettmann photo by Bill Varie
15.44   Corbis/Bettmann photo by Russell Boyce
16.0   Getty Images, Inc. - Taxi photo by Peter Gridley
16.6   Minneapolis Public Library
16.7   Corbis/Bettmann photo by Lloyd Cluff
16.8   Linda McCarthy
16.11   Rob Crandall photo by Rob Crandall
16.12   Corbis/Bettmann photo by Gail Mooney
16.14   Robert Harding World Imagery photo by Financial Times
16.15   SuperStock, Inc.
17.0   Corbis Royalty Free
17.2   Anthony Bayford
17.3   Paul L. Knox
17.7   Library of Congress photo by Charles F. Bretzman
17.8   Corbis/Bettmann
17.9   Library of Congress

17.10 Corbis/Bettmann
17.11 Albert H. Teich photo by Al Teich
17.12 Corbis/Bettmann
17.14 Getty Images/Time Life Pictures
17.15 Corbis/Bettmann photo by Morton Beebe
18.0 Dorling Kindersley Media Library photo by Peter Wilson

LINE ART CREDITS

1.6 B. J. L. Berry, Long-Wave Rhythms in Economic Development and Political Behavior, Baltimore, Md.: Johns Hopkins University Press, 1991, Fig. 51, p. 100
2.3 L. Woolley, "The Urbanisation of Society," in J. Hawkes and L. Woolley (eds), *History of Mankind*, Vol. 1, Part 2 *The Beginnings of Civilisation*, Paris: UNESCO, 1963, Fig. 61, p. 424
2.4 L. Woolley, "The Urbanisation of Society" in J. Hawkes and L. Woolley (eds), *History of Mankind*, Vol. 1, Part 2 *The Beginnings of Civilisation*, Paris: UNESCO, 1963, Fig. 73, p. 456 (after Mackay)
2.7 Based on A. E. J. Morris, *History of Urban Form: Before the Industrial Revolution*. Harlow, UK: Pearson Education, 3rd ed., 1994, Fig. 2.11, p. 44 and Fig. 2.12, p. 45
2.9 N. J. G. Pounds, *An Historical Geography of Europe*, Cambridge, UK: Cambridge University Press, 1990, Fig. 2.11, p. 56
2.10 H. Carter, *An Introduction to Urban Geography*, London: Arnold, 1983, Fig. 2.15, p. 33
2.14 A. E. J. Morris, H*istory of Urban Form: Before the Industrial Revolution*, Harlow, UK: Pearson Education, 3rd ed., 1994, Fig. 4.11, p. 104
2.15 A. E. J. Morris, *History of Urban Form: Before the Industrial Revolution*, Harlow, UK: Pearson Education, 3rd ed., 1994, Fig. 4.13, p. 106
2.16 A. E. J. Morris, *History of Urban Form: Before the Industrial Revolution*, Harlow, UK: Pearson Education, 3rd ed., 1994, Fig. 4.17, p. 108
2.20 P. Knox, J. Agnew, and L. McCarthy, *The Geography of the World Economy*, London: HodderArnold, 4th ed., 2003, Fig. 4.3, p. 66 (based on P. Hugill, *World Trade Since 1431*. Baltimore, Md.: Johns Hopkins University Press, 1993, Fig. 2.5, p. 50)
2.25 P. L. Knox and S. A. Marston, *Places and Regions in Global Context: Human Geography*, Upper Saddle River, N.J.: Pearson Education Inc., 3rd ed., 2004, Fig. 10.12, p. 400
3.9 J. E. Vance, Jr., The Merchant's World: The Geography of Wholsaling, Englewood Cliffs, N.J.: Prentice Hall, 1970, p. 151
3.14 Association for American Railroads
3.19 P. Knox *et al., The Geography of the World Economy.* 4th ed., London: Arnold, 2003, Fig. 5.3, p. 156
3.20 From Megapolis, by Jean Gottman © 1960 by the Twentieth Century Fund Press, New York
4.13 J. Wheeler and R. L. Mitchelson, "Information Flows among Major Metropolitan Areas in the U.S., *Annals, Association of American Geographers* 79 (1989): Fig. 7, p. 534
4.16 Updated from P. Knox, J. Agnew, and L. McCarthy, *The Geography of the World-Economy*, London: Edward Arnold, 4th ed., 2003, Fig. 7.4, p. 227

5.1 J. P. Radford, Testing the model of the pre-industrial city, *Transactions, Institute of British Geographers,* 4, 1979, Fig. 1, p. 394
5.18 E. M. Horwood and R. R. Boyce, *Studies of the Central Business District and Urban Freeway Development*, Seattle, Wash.: University of Washington Press, 1959, p. 21
5.21 Adapted from Adapted from F. P. Stutz and A. R. de Souza, *The World Economy*, Upper Saddle River, NJ: Prentice Hall, 1998, Fig. 6.5(a), p. 285
5.22 Adapted from F. P. Stutz and A. R. de Souza, *The World Economy*, Upper Saddle River, NJ: Prentice Hall, 1998, Fig. 6.5(b), p. 285
5.23 C. D. Harris and E. L. Ullman, The Nature of Cities, *Annals, American Academy of Political and Social Science,* CCXLII, 1945, Fig. 5
6.4 After F. P. Stutz and A. R. de Souza, The World Economy, Upper Saddle River, N.J.: Prentice Hall, 1998, Fig. 4.28, p. 187
6.11 C. D. Harris and E. L. Ullman, The Nature of Cities, *Annals, American Academy of Political and Social Science,* CCSLII, 1945, Fig. 5
6.23 Adapted from E. Soja, Postmetropolis, Oxford, UK: Blackwell, 2000, Fig. 4.1, p. 113
6.28 J. E. Vance, Jr., The Continuing City, Baltimore, Md.: Johns Hopkins University Press, 1990, p. 504
6.29 P. O. Muller, Contemporary Suburban America, Englewood Cliff, N.J.: Prentice Hall, 1981, Fig. 1.3, p. 10
6.31 P. L. Knox and S. A. Marston, *Human Geography: Places and Regions in Global Context,* 3rd ed., Upper Saddle River, N.J.: Prentice Hall, 2002, Fig. 11.C, p. 461
7.6 P. L. Knox and S. A. Marston, *Human Geography: Places and Regions in Global Context,* 3rd ed., Upper Saddle River, N.J.: Prentice Hall, 2002, Fig. 7.12, p. 268
7.7 P. Knox, J. Agnew, and L. McCarthy, *The Geography of the World Economy,* London: HodderArnold, 4th ed., 2003, Fig. 7.12, p. 243
7.8 J. Dickenson, B. Gould, C. Clarke, S. Mather, M. Prothero, D. Siddle, C. Smith, and E. Thomas-Hope, *A Geography of the Third World,* London: Routledge, 2nd ed., 1996, Fig. 1.16, p. 21
7.9 J. Dickenson, B. Gould, C. Clarke, S. Mather, M. Prothero, D. Siddle, C. Smith, and E. Thomas-Hope, *A Geography of the Third World,* London: Routledge, 2nd ed., 1996, Fig. 1.17, p. 23
7.10 D. Drakakis-Smith, *Third World Cities,* London: Routledge, 2nd ed., 2000, Fig. D.1, p.46
7.11 D. Drakakis-Smith, *Third World Cities,* London: Routledge, 2nd ed., 2000, Fig. D.2, p.48
7.12 A. D. King, *Urbanism, Colonialism, and the World Economy: Cultural and Spatial Foundations of the World Urban System,* London: Routledge, 1990, Fig. 2.1, p. 24
7.13 Based on S. Lowder, *The Geography of Third World Cities,* Totowa, N.J.: Barnes & Noble, 1986, Fig. 3.1, p. 53; D. Drakakis-Smith, *Third World Cities,* London: Routledge, 2nd ed., 2000, Table 2.1, p. 32
7.17 Taaffe, E. J., Morrill, R. L., and Gould, P. R., "Transport Expansion in Underdeveloped Countries: A Comparative Analysis," in *Geographical Review* 53, 1963, Fig. 1, p, 504
8.1 C. S. Sargent, "The Latin American City" in B. W. Blouet and O. M. Blouet, eds., *Latin America and the*

Caribbean: A Systematic and Regional Survey, New York: John Wiley & Sons, Inc., 2002, Fig. 6.4, p. 167

8.2 L. F. Ford, "A New and Improved Model of Latin American City Structure," *The Geographical Review* 86(2), 1996, Fig. 1, p. 439

8.3 Crowley, W. K. 1995. Order and Disorder—A Model of Latin American Urban Land Use. In: D. E. Turbeville III (ed.), *Yearbook of the Association of Pacific Coast Geographers* 57, Corvallis, Ore.: Oregon State University Press, Fig. 8, p. 28

8.4 Arreola, D. D. and Curtis, J. R., *The Mexican Border Cities: Landscape Anatomy and Place Personality*, Tucson, Ariz.: The University of Arizona Press, 1993, Fig. 3.9, p. 69

8.5 C. C. Diniz, "Polygonized Development in Brazil: Neither Decentralization or Continued Polarization," *International Journal of Urban and Regional Research* 18, 1994, Fig. 2, p. 295

8.7 R. J. Davis, 1981. The Spatial Formation of the South African City. *GeoJournal* Supplementary Issue 2, Fig. 2, p. 64, and Fig. 8, p. 69

8.8 S. Aryeetey-Attoh (ed.), *Geography of Sub-Saharan Africa*, 2nd ed., Upper Saddle River, N.J.: Pearson Education, Inc., 2003, Fig. 10.6, p. 267

8.9 D. J. Zeigler, "Cities of the Greater Middle East," in: S. D. Brunn, J. F. Williams, and D. J. Zeigler (eds.) *Cities of the World: World Regional Urban Development*, 3rd ed., Lanham, Md.: Rowman & Littlefield, 2003, Fig. 7.2, p. 257

8.10 D. J. Zeigler, "Cities of the Greater Middle East," in: S. D. Brunn, J. F. Williams, and D. J. Zeigler (eds.), *Cities of the World: World Regional Urban Development*, 3rd ed., Lanham, Md.: Rowman & Littlefield, 2003, Fig. 7.7, p. 269

8.11 A. K. Dutt and G. M. Pomeroy, "Cities of South Asia," in: S. D. Brunn, J. F. Williams, and D. J. Zeigler (eds.), *Cities of the World: World Regional Urban Development*, 3rd ed. Lanham, Md.: Rowman & Littlefield, 2003, Fig. 9.5, p. 341

8.12 A. K. Dutt and G. M. Pomeroy, "Cities of South Asia," in: S. D. Brunn, J. F. Williams, and D. J. Zeigler (eds.), *Cities of the World: World Regional Urban Development*, 3rd ed. Lanham, Md.: Rowman & Littlefield, 2003, Fig. 9.8, p. 345

8.16 T. G. McGee, *The Southeast Asian City: A Social Geography of the Primate Cities of South East Asia*, New York: Frederick A. Praeger, Publishers, 1967, Fig. 25, p. 128

8.17 L. R. Ford, "A Model of Indonesian City Structure," *The Geographical Review 83(2)*, 1993, Fig. 3, p. 382

8.18 G. Hugo, "Indonesia," in: T. R. Leinbach and R. Ulack (eds.), *Southeast Asia: Diversity and Development*, Upper Saddle River, N.J.: Prentice Hall, 2000, Map 12.8, p. 326

8.19 T. G. McGee, "The Emergence of Desakota Regions in Asia: Expanding a Hypothesis," in *The Extended Metropolis: Settlement Transition in Asia*, (eds.) N. Ginsburg, B. Koppel, and T. G. McGee, Honolulu, Hawaii: University of Hawaii Press, 1991, Fig. 1.1, p. 6

8.20 A. K. Dutt, Y. Xie, F. J. Costa, and Z. Yang, "City Forms of China and India in Global Perspective," in: *The Asian City: Processes of Development, Characteristics and Planning*, Dordretcht, The Netherlands: Kluwer Academic Publishers, 1994, Fig. 3:7, p. 41

8.21 C. P. Lo, "Shaping Socialist Chinese Cities: A Model of Form and Land Use," in: C-K. Leung and N. Ginsburg (eds.) *China: Urbanization and National Development*, Department of Geography Research Paper No. 196, Chicago, Ill.: University of Chicago, 1980, Fig. 14, p. 154

8.22 A. K. Dutt, Y. Xie, F. J. Costa, and Z. Yang, "City Forms of China and India in Global Perspective," in: *The Asian City: Processes of Development, Characteristics and Planning*, Dordretcht, The Netherlands: Kluwer Academic Publishers, 1994, Fig. 3:8, p. 43

8.25 After A. Bertaud, *Spatial Structure of the Pearl River Delta Metropolitan Region: A Study Proposal*, unpublished manuscript, 2001, Fig. 1, p. 2

8.26 V. F. S. Sit and C. Yang, "Foreign-Investment-Induced Exo-Urbanisation in the Pearl River Delta, China," *Urban Studies* 34, 1997, Fig. 3, p. 657

9.5 Adapted from D. Drakakis-Smith, London: Routledge, 2000, 2nd ed,, Figure 6.4, p. 160

9.9 P. Newman, J. Kenworthy, and P. Vintila, *Housing, Transport and Urban Form*, Australia: Institute for Science and Technology Policy, Murdoch University, 1992, Fig. 1.1, p. 3; Fig. 1.2, p. 5; Fig. 1.3, p. 6

9.12 D. Drakakis-Smith. 1995. "Third World Cities: Sustainable Urban Development," 1. *Urban Studies* 32(4-5), Fig. 2, p. 665

11.2 A. Haila, "Four types of investment in land and property," *International Journal of Urban and Regional Research* 15, 1991, Fig. 1, p. 349

12.6 Adapted from W. A. V. Clark, *Human Migration.* Scientific Geography Series, No. 7, Beverly Hills: Sage Publications, 1987, Fig. 2.6, p. 39

12.16 R. J. Johnston, *Urban Residential Patterns.* London: Bell, 1971, Fig. 6.3, p. 253

12.17 R. E. Park, E. W. Burgess, and R. D. McKenzie, *The City.* Chicago, Ill.: University of Chicago Press, 1925, p. 53

12.18 R. A. Murdie, *Factorial Ecology of Metropolitan Toronto, 1951–1961*, Research Paper no. 116. Department of Geography, University of Chicago, 1969, p. 8

12.19 L.. McCarthy and D.Danta, "Cities of Europe," in *Cities of the World: World Regional Urban Development*, eds. S. D. Brunn, J. F. Williams, and D. J. Zeigler. Lanham, Md.: Rowman & Littlefield, 2003, 3rd ed., Fig 5.23, p. 198

12.23 W. Li, "Anatomy of a New Ethnic Settlement: The Chinese *Ethnoburb* in Los Angeles," *Urban Studies* 35 (1998): Fig. 2, p. 481

12.25 C. Hamnett, "Gentrification and the Middle-class Remaking of Inner London, 1961–2001," *Urban Studies* 40 (2003): Fig. 1, p. 2407

12.30 M. A. Hughes, "Formation of the Impacted Ghetto. Evidence from Large Metropolitan Areas, 1970–1980," *Urban Geography* 11 (1990): Figs. 3, 4, pp. 279-80

12.33 P. L. Knox and S. A. Marston, places and Regions in Global Context: Human Geography, Upper Saddle River, N. J.: Pearson Education, Inc., 3rd ed., 2004, Fig. 3.B, p. 95

13.3 E. G. Moore, Residential Mobility in the City, Washington, D.C.: Association of American Geographers, Commission on College Geography, Resource Paper no. 13, 1972, p. 33

13.14 J. Adams, "Housing Submarkets in an American Metropolis," in *Our Changing Cities*, ed. J. F. Hart, Baltimore, Md.: Johns Hopkins University Press, 1991, Fig. 7.2, p. 113

13.17 Adapted from W. A. V. Clark and J. Onaka, "Life Cycle and Housing Adjustment as Explanation of Residential Mobility," *Urban Studies* 20 (1983): Fig. 2, p. 50

13.18   Adapted from B. Robson, Urban Social Areas, Oxford, UK: Oxford University Press, 1975, p. 33

13.20   Developed from a diagram by L. Bourne, *The Geography of Housing*, London: Edward Arnold, 1981, p. 85

14.12   D. Ley, The Black Inner-City as Frontier Outpost, Washington, D.C.: Association of American Geographers, Monograph Series no. 7, Fig. 36, p.22

14.13   After *New York Magazine*.

14.15   After D. Gregory, in *Horizons in Geography*, ed. D. Gregory and R. Walford, Totowa, N.J.: Barnes & Noble, 1989, Fig. 1.4.4, p. 82

14.16   A. Pred and R. Palm, "The Status of American Women: A Time-Geographic View," in An Invitation to Geography, ed. D. A. Lanegran and R. Palm, New York: McGraw-Hill, 1978, p. 103

14.17   D. Gregory, in Horizons in Geography, ed. D. Gregory and R. Walford, Totowa, N.J.: Barnes & Noble, 1989, Fig. 1.4.3, p.81

15.11   P. L. Knox, "The Vulnerable, the Disadvantaged, and the Victimized: Who They Are and Where They Live," in Social Problems and the City, eds. D. T. Herbert and D. Smith, Oxford, UK: Oxford University Press, 1989, Fig. 2.1, p. 44

15.12   P. A. Jargowsky, Stunning Progress, Hidden Problems: The Dramatic Decline of Concentrated Poverty in the 1990s, Washington, D.C.: The Brookings Institution, 2003, Fig. 1, p. 4

15.15   P. A. Jargowsky, Stunning Progress, Hidden Problems: The Dramatic Decline of Concentrated Poverty in the 1990s, Washington, D.C.: The Brookings Institution, 2003, Fig. 5, p. 7

15.16   P. A. Jargowsky, Stunning Progress, Hidden Problems: The Dramatic Decline of Concentrated Poverty in the 1990s, Washington, D.C.: The Brookings Institution, 2003, Fig. 6, p. 7

15.17   P. A. Jargowsky, Stunning Progress, Hidden Problems: The Dramatic Decline of Concentrated Poverty in the 1990s, Washington, D.C.: The Brookings Institution, 2003, Fig. 7, p. 8

15.19   S. Raphael and M. A. Stoll, *Modest Progress: The Narrowing Spatial Mismatch between Blacks and Jobs in the 1990s*, Washington, D.C.: The Brookings Institution, 2002, Fig. 1, p. 3

15.20   S. Raphael and M. A. Stoll, Modest Progress: *The Narrowing Spatial Mismatch between Blacks and Jobs in the 1990s*, Washington, D.C.: The Brookings Institution, 2002, Fig. 3, p. 6

15.27   J. Coaffee, "Rings of Steel, Rings of Concrete and Rings of Cinfidence: Designing out Terrorism in Central London pre and post September 11th," *International Journal of Urban and Regional Research* 28 (2004), Fig. 2, p. 204

15.30   Developed from a diagram by G. Laws and S. Lord, in Geographic Dimensions of United States Social Policy, eds. J. Kodras and J. P. Jones III, London: Edward Arnold, 1990, Fig. 3.1

15.43   The Economist, "Congestion Changing: Ken Livingstoneís Gamble," *The Economist,* Fig. 1 (February 13, 2003)

16.13   S. A. Marston, P. L. Knox, and D. M. Liverman, World Regions in Global Context: Peoples, Places, and Environment, Upper Saddle River, N.J.: Prentice Hall, 2002, Fig. 1, p. 136

16.16   Redrawn from H. V. Savitch and J. C. Thomas, Big City Politics in Transition, Newbury Park, Calif.: Sage, 1991, Fig. 15.1, p. 249

16.17   Adapted from D. G. Janelle and H. A. Millward, "Locational Conflict Patterns and Urban Ecological Structure," *Tijdschrift voor Economische en Sociale Geografie* 67 (1976): 10

17.13   P. Hall, Urban & Regional Planning, London: Routledge, 4th ed. , 2002, Fig. 4.1, p. 65

18.1   Adapted from B. J. L. Berry, Long-Wave Rhythms in Economic Development and Political Behavior, Baltimore, Md.: Johns Hopkins University Press, 1991, Fig. 70, p. 126

18.2   A. Blowers and K. Pain, "The Unsustainable City?" in Unruly Cities? Order/Disorder, eds. S. Tile, C. Brook, and G. Mooney, London: Routledge, 1999, Fig. 6.12, p. 282

INDEX